Characterizing Biotherapeutics

Characterizing Biotherapeutics

Analytical Methods for Diverse Modalities

Edited by

Jennie R. Lill
Proteomic and Genomic Technologies
Genentech Inc
South San Francisco, CA

Wendy Sandoval
Botanica
Sebastopol, CA, United States

WILEY

Published by John Wiley & Sons, Inc., Hoboken, New Jersey.
Published simultaneously in Canada.

For general information on our other products and services or for technical support, please contact our Customer Care Department within the United States at (800) 762-2974, outside the United States at (317) 572-3993 or fax (317) 572-4002.

Wiley also publishes its books in a variety of electronic formats. Some content that appears in print may not be available in electronic formats. For more information about Wiley products, visit our web site at www.wiley.com.

Library of Congress Cataloging-in-Publication Data applied for:

Hardback ISBN: 9781394236114

Cover Design: Wiley
Cover Images: © DrAfter123/Getty Images, Olga Pankova/Getty Images

Set in 9.5/12.5pt STIXTwoText by Straive, Chennai, India

Contents

About the Editors

Jennie R. Lill has been a leader in the biophysical space, enabling Drug Discovery and Development for over 30 years. She currently leads various teams in deciphering new drug targets and characterizing a broad range of therapeutic modalities. Jennie has been at Genentech for the past 21 years, leading their Genomic and Proteomic initiatives. Jennie has two sons, Joe and Charlie, who inspire her to do her best every day!

Wendy Sandoval worked for over 30 years in Research at Genentech where she led the Translational Mass Spectrometry team and focused on native mass spectrometric methods to characterize complex biotherapeutics. She has coauthored over 100 manuscripts and 2 books. She currently owns and runs Botanica (https://botanicanurseryandgardens.com/) in Sebastopol, CA, a public botanical garden, and takes her meetings under a massive Dawn Redwood canopy. Her mother, Ame, and daughters, Nicolina and Olivia, assist in managing the 7.5 acre oasis.

List of Contributors

Corey E. Bakalarski
Proteomic and Genomic Technologies
Genentech Inc.
South San Francisco, CA
USA

Mercedesz Balasz
Biochemical and Cellular Pharmacology
Genentech
South San Francisco
CA, USA

Marcus Bantscheff
Target Discovery, GSK
Heidelberg
Germany

and

Research and Early Development
Roche Pharma
Basel
Switzerland

Hayley M. Bennett
Department of Proteomic and Genomic Technologies
Genentech Research and Early Development
South San Francisco, CA
USA

Maureen Beresini
Department of Biochemical and Cellular Pharmacology
Genentech Inc.
South San Francisco, CA
USA

Tess Branon
Proteomic & Genomic Technologies Department
Genentech Inc.
South San Francisco, CA
USA

Ashley Byrne
Department of Proteomic and Genomic Technologies
Genentech Research and Early Development
South San Francisco, CA
USA

Julien Camperi
Cell Therapy Engineering and Development
Genentech
South San Francisco, CA
USA

Shengya Cao
Development
Inceptive
Palo Alto, CA
USA

Naincy R. Chandan
Discovery Technology
Nurix Therapeutics, Inc.
San Francisco, CA
USA

Jieming Chen
gRED Computational Sciences
Genentech
South San Francisco, CA
USA

Sivan Cohen
Bioanalytical Sciences
Genentech, Inc.
South San Francisco, CA
USA

Phillip Chu
Genentech, Inc.
South San Francisco, CA
USA

Michael Dillon
Department of Antibody Engineering
Genentech Research and Early Development
South San Francisco, CA
USA

H. Christian Eberl
Target Discovery, GSK
Heidelberg
Germany

Alberto Estevez
Thomson Instrument Company
Carlsbad, CA
USA

Amin Famili
Department of Synthetic Molecule Analytical Chemistry
Genentech
South San Francisco, CA
USA

Abel Ferrel
Independent Researcher
San Francisco, CA
USA

Emily Freund
Department of Molecular Biology
Genentech
South San Francisco, CA
USA

Manasi Gaikwad
Center for Mass Spectrometry and Optical Spectroscopy (CeMOS)
Mannheim University of Applied Sciences
Mannheim
Germany

Carolina Galan
Department of Molecular Biology
Genentech
South San Francisco, CA
USA

Amy Gilbert
Adaptive Biotechnologies
Seattle, WA
USA

Alissa D. Guarnaccia
Department of Proteomic and Genomic Technologies
Genentech
South San Francisco, CA
USA

and

Proteomic and Genomic Technologies
Genentech Inc.
South San Francisco, CA
USA

Axel Guilbaud
Protein Analytical Chemistry
Genentech
South San Francisco, CA
USA

Albert J.R. Heck
Biomolecular Mass Spectrometry and Proteomics
Bijvoet Center for Biomolecular Research and Utrecht Institute for Pharmaceutical Sciences
University of Utrecht
Utrecht
The Netherlands

and

Netherlands Proteomics Center
Utrecht
The Netherlands

Suk-Joon Hyung
Genentech, Inc.
South San Francisco, CA
USA

Saeed Izadi
Pharmaceutical Development
Genentech Inc.
South San Francisco, CA
USA

Eric Janezic
Biochemical and Cellular Pharmacology
Genentech Inc.
South San Francisco, CA
USA

Christine C. Jao
Department of Structural Biology
Genentech Inc.
South San Francisco, CA
USA

Hiruni S. Jayasekera
Department of Chemistry and Biochemistry
The University of Arizona
Tucson, AZ
USA

Robert S. Jones
Drug Metabolism and Pharmokinetics
Genentech, Inc.
South San Francisco, CA
USA

S. Cyrus Khojasteh
Genentech, Inc.
South San Francisco, CA
USA

Sascha Knecht
Target Discovery, GSK
Heidelberg
Germany

Felix Kuhne
Analytical Characterization, Pharma Technical Development
Roche Diagnostics GmbH
Penzberg
Germany

M. Violet Lee
Bioanalytical Sciences
Genentech
South San Francisco, CA
USA

Paola Di Lello
Department of Structural Biology
Genentech, Inc.
South San Francisco, CA
USA

Dennis H. Leung
Genentech, Inc.
South San Francisco, CA
USA

Amritha Lewis
Adaptive Biotechnologies
Seattle, WA
USA

Yinyin Li
Biochemical and Cellular Pharmacology
Genentech
South San Francisco, CA
USA

Jennie R. Lill
Department of Proteomic and Genomic Technologies
Genentech
South San Francisco, CA
USA

and

Proteomic and Genomic Technologies
Genentech Inc.
South San Francisco, CA
USA

and

Proteomic and Genomic Technologies
Genentech
South San Francisco, CA
USA

Rachel P. Liu
Protein Analytical Chemistry
Genentech Inc.
South San Francisco, CA
USA

T. Noelle Lombana
Department of Antibody Engineering
Genentech Research and Early Development
South San Francisco, CA
USA

and

ReCode Therapeutics
Menlo Park, CA
USA

Kelly Loyet
Department of Biochemical and Cellular Pharmacology
Genentech Inc.
South San Francisco, CA
USA

Patrick Lupardus
Department of Structural Biology
Genentech, Inc.
South San Francisco, CA
USA

Bin Ma
Genentech, Inc.
South San Francisco, CA
USA

Michael T. Marty
Department of Chemistry and Biochemistry
The University of Arizona
Tucson, AZ
USA

Rebecca Melmon
Adaptive Biotechnologies
Seattle, WA
USA

Madyson Migliozzi
Department of Pharmaceutical Development
Genentech
South San Francisco, CA
USA

Zora Modrusan
Department of Proteomic and Genomic Technologies
Genentech Research and Early Development
South San Francisco, CA
USA

Farhana Afrin Mohona
Department of Chemistry and Biochemistry
The University of Arizona
Tucson, AZ
USA

Huy Nguyen
Genentech, Inc.
South San Francisco, CA
USA

Emma Pelegri-O'Day
Amgen Research, Amgen Inc.
South San Francisco, CA
USA

Qui T. Phung
Proteomic and Genomic Technologies
Genentech
South San Francisco, CA
USA

Emile Plise
Genentech, Inc.
South San Francisco, CA
USA

Markus A. Queisser
Modality Platform Technologies, GSK
Medicines Research Centre
Stevenage
UK

Roshan M. Regy
Pharmaceutical Development
Genentech Inc.
South San Francisco, CA
USA

Amber D. Rolland
Biomolecular Mass Spectrometry and Proteomics
Bijvoet Center for Biomolecular Research and Utrecht
Institute for Pharmaceutical Sciences
University of Utrecht
Utrecht
The Netherlands

and

Netherlands Proteomics Center
Utrecht
The Netherlands

Ola M. Saad
Genentech, Inc.
South San Francisco, CA
USA

and

Bioanalytical Sciences
Genentech
South San Francisco, CA
USA

Sudeshna Sadhu
Adaptive Biotechnologies
Seattle, WA
USA

Wendy Sandoval
Department of Proteomic and Genomic Technologies
Genentech
South San Francisco, CA
USA

and

Proteomic and Genomic Technologies
Genentech Inc.
South San Francisco, CA
USA

Rahamthulla Shaik
Adaptive Biotechnologies
Seattle, WA
USA

Whitney Shatz-Binder
Department of Pharmaceutical Development
Genentech
South San Francisco, CA
USA

Christoph Spiess
Department of Antibody Engineering
Genentech Research and Early Development
South San Francisco, CA
USA

Alavattam Sreedhara
Department of Pharmaceutical Development
Genentech
South San Francisco, CA
USA

Nithya Srinivasan
Amgen Research
Amgen Inc.
Thousand Oaks, CA
USA

William Stephenson
Department of Proteomic and Genomic Technologies
Genentech Research and Early Development
South San Francisco, CA
USA

Yu Tong Tam
Department of Pharmaceutical Development
Genentech
South San Francisco, CA
USA

Devin B. Tesar
Department of Pharmaceutical Development
Genentech
South San Francisco, CA
USA

John C. Tran
Genentech, Inc.
South San Francisco, CA
USA

Egemen Tutuncuoglu
Adaptive Biotechnologies
Seattle, WA
USA

Lingfei Wang
Protein Analytical Chemistry
Genentech Inc.
South San Francisco, CA
USA

Xiangdan Wang
Bioanalytical Sciences
Genentech Inc.
South San Francisco, CA
USA

Aaron T. Wecksler
Protein Analytical Chemistry
Genentech Inc.
South San Francisco, CA
USA

and

Department of Protein Analytical Chemistry
Genentech
South San Francisco, CA
USA

Sara Wichner
Department of Protein Chemistry
Genentech
South San Francisco, CA
USA

Zhaojun Yin
Bioanalytical Sciences
Genentech, Inc.
South San Francisco, CA
USA

Qinying Yu
Genentech, Inc.
South San Francisco, CA
USA

Xing Zhang
Genentech, Inc.
South San Francisco, CA
USA

Zhenru Zhou
Proteomic and Genomic Technologies
Genentech
South San Francisco, CA
USA

Preface

The focus of this book is to introduce and describe analytical approaches for biologic drug modalities. The intervention of diseases once deemed an unsurmountable task is now becoming a reality due to recent breakthroughs in our understanding of human biology, therapeutic intervention approaches, and drug delivery mechanisms. Such transformative medicines require sophisticated analytical tools to identify new pathways for intervention, profile the therapeutic moieties, discover biomarkers, and monitor pharmacokinetic and dynamic profiles after patient administration. Developing these analytical approaches and often the software or algorithms associated with them requires the heroic collaboration of diverse disciplines, working together to develop and refine these custom tools that can transform medicine. Biologics, or biotherapeutics, are medicines made up of biological material and are often large and complicated molecules, sometimes heterogeneous mixtures. Although sub-microscopic, biologics are much larger than typical chemical drug molecules, presenting particular challenges for characterization, production, and stability. Biologic drug modalities encompass a wide swath of molecules, such as bispecific antibodies, IgM-based antibodies, peptides, RNA, and cell and gene therapies, each with unique challenges that require carefully designed analytical strategies. In this book, we discuss a variety of biologic drugs, their unique analytical challenges, and the methods that enable their analysis. By delving into these advancements, we aim to shed light on how innovative bioanalytical tools are pivotal in transforming scientific discoveries into effective, safe, and accessible treatments for patients worldwide.

May 2025

Jennie R. Lill
Wendy Sandoval

1

Introduction

Alissa D. Guarnaccia, Wendy Sandoval, and Jennie R. Lill

Department of Proteomic and Genomic Technologies, Genentech, South San Francisco, CA, USA

Exploring a Diversity of Biotherapeutic Approaches

The focus of this book is to introduce and describe analytical approaches for biologic drug modalities. Biologics, or biotherapeutics, are the types of medicines that are made up of biological material. Biologics are often large and complicated molecules and sometimes can be heterogeneous mixtures. Although still on the submicroscopic scale, biologics are much larger than typical chemical drug molecules, and the larger size of biologics presents a particular challenge for characterization, production, and stability. Biologic drug modalities encompass a wide swath of molecules; bispecific antibodies, IgM-based antibodies, peptides, RNA, and cell and gene therapies all have unique challenges that require carefully designed analytical strategies to bring them from the bench to the patient. In this book, we discuss a variety of biologic drugs, their unique analytical challenges, and the methods that enable their analysis.

Antibody-Based Therapies

Antibody-based therapies, such as bispecific antibodies and IgM-based therapies, are powerful, specific, and often durable treatments. Bispecific antibodies (BsAbs) bind to two different targets simultaneously, enabling profound specificity in therapeutic targeting. BsAbs are a therapeutic game-changer in cancer, immune diseases, ocular diseases, and neurological disorders [1]. Analytically, BsAbs pose special challenges due to their multiple polypeptide chains and their dual-target specificity. In Chapter 2, we explore how advances in mass spectrometry have significantly enhanced our ability to characterize BsAbs. Next, we move to the realm of even larger biological therapeutics and focus on IgM-based therapies which involve complexes of six antibodies (Chapter 3). There are two key advantages to hexameric IgM-based therapies. First, IgM complexes tend to have higher avidity compared to IgG antibodies, enabling better disease targeting. Second, IgM complexes can induce cell agglutination and complement activation, which is crucial for destroying diseased cells [2]. Analytically, IgM modalities are quite challenging due to their size, polymeric nature, extensive glycosylation, and intricate disulfide bond networks. But advances in MS instrumentation, data analysis software, and sample preparation techniques are enabling us to overcome such challenges and advance antibody-based therapies as powerful medicines (Chapters 4 and 5).

Peptide-Based Therapies

In Chapter 6, we cover peptide therapies, both linear and cyclic. Peptide modalities offer unique advantages for specific target engagement due to their flexible structures their stabile pharmacokinetic properties due to modifications with a wider range of amino acids, i.e. non-natural amino acids [3, 4]. Advances in peptide technology have enabled peptide therapies for a variety of diseases in the clinic, but their development is not trivial. Peptide therapeutics need to have optimal solubility and lipophilicity and require various *in vitro* bioanalytical assays to refine the characteristics that are suitable for a therapeutic. In Chapter 6, the authors discuss formulation development, optimization, and examples of approved peptide therapeutics.

RNA-Based Therapies

RNA-based therapies are rapidly advancing [5] and can be engineered in remarkable ways. RNA in the forms of antisense oligonucleotides (ASOs) and siRNA can be used to degrade transcripts or stabilize transcripts. RNA in the

Characterizing Biotherapeutics: Analytical Methods for Diverse Modalities, First Edition. Edited by Jennie R. Lill and Wendy Sandoval.

form of lipid-packaged mRNAs can be used as vaccines. A remarkable example of RNA-based therapy advances is the technologies that enabled the COVID-19 mRNA vaccines [6] and garnered the 2023 Nobel Prize in Physiology or Medicine [7, 8]. Currently, the transformative potential of mRNA vaccines is expanding beyond infectious diseases to treat cancer [9]. RNA-based therapeutics come with their own unique set of technical, analytical, and delivery challenges. Analytical methods like UV spectroscopy, Reverse Transcriptase-Polymerase Chain Reaction (RT-PCR), capillary electrophoresis, High performance liquid chromatography (HPLC), and mass spectrometry are used to characterize RNA-based therapeutics. As RNA technology evolves rapidly, so too are the analytical tools needed to ensure high-quality production. We discuss this and more in Chapters 13 and 15.

Cells as Therapies

Cell-based therapies have shown enormous potential as medicines [10] and are being implemented in oncology and autoimmunity. For example, T-cell-based therapies have transformed treatment for various malignancies and immune disorders, offering new hope for patients. Cell therapies are arguably the most complex type of therapy we discuss, and in Chapter 14, we describe the invaluable analytical methods that are used to characterize these modalities, from preclinical development to post-treatment monitoring in patients.

Small Molecules as Therapies

Although this book focuses on biotherapeutics, we would be remiss to exclude small molecules. Small molecules, typically chemical compounds of <900 Da, are an essential modality in targeting disease. Recently, the area of small molecules in drug discovery and development has been expanding. Bifunctional molecules in particular are opening up new therapeutic opportunities because these enable specific targeting of a protein to a particular location including to the proteasome, to the lysosome, to an RNA, or to a particular cellular compartment [11, 12]. Currently, the most clinically advanced form of bifunctional molecules are protein degraders, such as Proteolysis-targeting chimeras (PROTACs), that chemically tether a target protein to the proteasome to induce its degradation [13, 14]. Bifunctional molecules are enabling scientists to address challenging targets that may have previously been considered "undruggable." In terms of biotherapeutics, combining small molecules with antibodies has been transformative for drug development. Engineered antibody–drug conjugates (ADCs) enable more targeted, less toxic drug delivery to specific tissues [15, 16]. Analytically, the complexity of ADCs requires diverse analytical approaches to investigate their synthesis, stability, and pharmacological profiles, which is covered in depth in Chapter 19. Altogether, many modalities on their own can be transformative, but combining different modalities can also lead to innovation in medicine.

Delivering Diverse Biotherapeutics to Expand Patient Treatment

Alongside the development of biotherapeutic drug modalities, there has also been significant advancement in drug delivery strategies. Many therapeutic approaches once deemed failures are now being revived by optimized delivery systems. Oral delivery often does not work well for larger, complex modalities such as antibodies or cells, which instead require intravenous, subcutaneous, or intramuscular routes of delivery. Additionally, innovative delivery mechanisms such as nanoparticles [17] and ocular implants [18] are enabling certain therapies to have longer and localized activity that requires fewer dosages and can improve patient access and quality of life. In Chapter 20, we explore how innovations in delivery strategies are enhancing efficacy while minimizing side effects.

Unlocking Innovation: The Role of Biophysical Techniques in Drug Discovery and Development

To unlock the new modes of drug development, we must first understand the molecular details of how cells and proteins function. Various methods enable this, including biophysical methods that discern not only the shape of proteins but also how proteins interact with one another. In Chapters 7 and 8, we explore how structural techniques – such as nuclear magnetic resonance (NMR) spectroscopy, X-ray crystallography, and cryo-electron microscopy (Cryo-EM) – provide high-resolution insights that can inform and drive drug discovery. We also discuss how mass-spectrometry-based approaches (Chapter 16) and non-mass-spectrometry approaches (Chapter 17) enable molecular characterizations of cellular interactions and cell–cell communication processes.

Mass spectrometry is a workhorse technology in biotherapeutic development. Because of its importance – and applicability to so many therapeutic modalities – we delve deep into how mass spectrometry advances are powering biotherapeutic development (Chapters 4 and 5). In order to characterize complex biotherapeutic molecules, mass spectrometry often requires other methods in tandem, such as separation techniques and ionization techniques. Hyper-specialized approaches are often needed to successfully analyze a particular protein class. For example, membrane proteins – with strong hydrophobicity and high levels of post-translational modifications – pose unique challenges, and we dedicate Chapter 18 to discussing these specialized analytical methods. With high sensitivity and versatility in its application, mass spectrometry has proven invaluable in analyzing biotherapeutics and continues to advance in sensitivity and capability.

Increasing Throughput: The Role of High-Quality Analytical Assays in Characterizing Biotherapeutics

In Vitro Assays

Getting the right assay is crucial for knowing if a potential therapeutic is safe and on-target. *In vitro* assays that measure binding or function are critical for the targeted analysis of biotherapeutics. Tailored methods enable end-to-end validation throughout the drug development process, from initial validation of a target, to optimization of the best candidate molecule to pursue, to quality control later in the development process. These *in vitro* assays are simplified systems with minimal components (*in vitro* means in a test tube) that enable quantitative measurements. Tailored assays can be used to identify biophysical properties such as binding kinetics (Chapter 9), biological activity (Chapter 10), and a variety of other custom characteristics that inform therapeutic potential (Chapter 12). We discuss each of these in detail in the relevant chapters. Ultimately, we want to highlight that the iterative process of assay development, assay implementation, and assay validation is essential for advancing preclinical and clinical candidates.

In Silico Assays

Although many assays are physically performed, sometimes assays can be orchestrated computationally with simulations, or *in silico*. *In silico* assays can sometimes be a critical aid in the development of biotherapeutics, particularly when there is a large amount of data or complicated patterns involved. In Chapter 11, we discuss how adverse immune reactions related to biotherapeutic administration can be monitored in patients throughout the drug development process using an *in silico* approach. The ability of a substance to elicit an immune response is called immunogenicity, and when immunogenicity occurs toward a therapeutic, it can be detrimental – causing more harm than good. When a patient's immune system recognizes a biological drug as a foreign entity, the immune system generates antibodies against the drug, inducing an anti-drug antibody (ADA) response. Adverse immune responses like ADA responses are becoming an increasing concern as more highly engineered biotherapeutics and delivery devices enter the clinic. *In silico* algorithms can be used to recognize immunogenic patterns and pinpoint particularly immunogenic regions in a protein sequence. Predicting immunogenicity is a key example of how *in silico* assays can enable therapeutic molecules to be optimized as safe and effective medicines.

Sequencing-Based Assays

Sequencing-based assays are often crucial in drug development. DNA sequencing technologies have revolutionized the development and characterization of therapeutics, and sequencing is also essential to analyze and inform the quality control of therapies that rely on DNA or RNA sequences. In Chapter 15, we discuss ASOs, small interfering RNAs (siRNAs), DNA and mRNA vaccines, and gene therapies. All of these require sequencing-based assays to ensure the purity, integrity, and efficacy of these therapeutics. Sequencing enables the detection of synthesis-related errors, off-target effects, and structural variations. And sequencing is also important for understanding therapeutic potential of certain drugs for specific patients. Genomic sequencing of patients can help identify biomarkers that can inform patient response to a particular biotherapeutic. Thus, sequencing-based assays are enabling more precise applications of medicine.

The Next Frontier: Additional Bioanalytical Tools and Targets on the Horizon

In the drug discovery and development industry, the mantra "drug the undruggable" has become a guiding principle. But what exactly is this "undruggable" space? New technical capabilities and biological insights have turned targets once thought of as undruggable – KRAS (named as it was found in Kirsten rat sarcoma virus) for example [19] – into breakthrough therapies. The idea of "undruggable" does not need to be a limitation in the way we think about exploring disease-causing targets.

One approach to expand the view of therapeutic targets is by exploring the dark genome, encoding the dark proteome [20]. Dark genomic sequences encode potential regulatory molecules that do not conform to conventional rules of transcription or translation and have therefore largely been overlooked. Today, we have tools and understandings that enable us to probe these types of regions. For example, ribosome profiling (Ribo-seq) enables unbiased translational profiling of all sequences engaged by a ribosome [21]. And mass spectrometry-based proteomics enables us to detect proteins that are not part of the canonical reference genome. For mass spectrometry-based methods, there are unique analytical challenges, such as a larger search space for peptide sequences and enzymatic digestion limitations. But technical advances and computational algorithms are overcoming these limitations [22]. Deeper biological insights, such as those obtained by interrogating the dark genome, could lead to future breakthrough therapies.

A second example of promising therapeutic targets on the horizon is protein isoforms. Protein isoforms are the different forms of a particular protein that are the result of alternative splicing, germline mutations, or posttranslational modification. Many modern drug targets are a mixture of protein isoforms, such as Tau in neurodegeneration or various glycosylated moieties on cell surface proteins such as Her2. Sometimes, only one of the protein isoforms will be pathogenic, but teasing out these variations and how they impact disease is challenging. Interrogating protein isoforms is a worthwhile effort because targeting one protein isoform specifically can be the key to developing an effective medicine. Omics methods are promising analytical tools for elucidating protein isoforms. For example, long-read next-generation sequencing technologies (Chapter 15) can delineate alternative splicing mechanisms [23], and top-down proteomics (Chapter 4) can determine the details of post-translational modifications [24]. Additionally, emerging technologies such as single-molecule sequencing enable the investigation of individual protein copies from a minimal number of cells [25]. Importantly, it is not only the development of new tools that will enable the next wave of therapeutic innovation but also the combinations of existing tools that will enhance our understanding of the molecular underpinnings of disease and inform the development of exquisitely targeted therapeutics.

Artificial Intelligence and Machine Learning: A Transformative Shift for Computational Tools

Alongside the therapeutic advancements we have discussed is a data revolution that promises to redefine personalized medicine. Computational systems are gaining new capabilities in analyzing, integrating, and handling data. Machine learning (ML) and artificial intelligence (AI) are enabling computer capabilities that are changing many aspects of human life, including healthcare and the drug development process. AI is being applied to protein structure [26, 27], to predicting protein interaction sites [28] and to interrogating drug–drug interactions in patients [29]. Notably, the 2024 Nobel Prize in Chemistry was awarded for the development of AlphaFold that has enabled amazing advances in protein structure prediction [30].

Data is also now being collected and analyzed at speeds and quantities never seen before. The prevalence of wearable monitoring devices and the significant increase in data collected from patients with specific diagnoses [31, 32] are enabling treatments to be tailored to the specific genetic, environmental, and lifestyle factors of individual patients. Algorithms are now being integrated into drug discovery with the promise of streamlining identification of drug candidates and optimal patient populations [33–35]. However, challenges such as insufficient training data, transparency in AI decision-making, and the integration of heterogeneous data into decision-making models must be carefully addressed. Despite these hurdles, new data technologies are here to stay, not as a replacement for human expertise, but as a powerful tool that aids researchers and healthcare professionals in developing superior medicines. Although the book does not have a specific chapter on this topic, many chapters touch upon AI and ML in drug discovery, and integrating AI and ML into data-driven decision-making will undoubtedly transform the ways we discover, analyze, and develop medicines.

A Bioanalytical Call to Action!

The promise of more personalized and effective treatments is becoming a reality, enabling us to imagine a future where healthcare is not only more advanced but also tailored specifically to the needs of each individual patient. One of the ways this can be achieved is with better analytical technologies and efficient integration of biological data. The bioanalytical mission is to provide the most efficient, accurate, and cost-effective analytical tools to characterize therapeutics at every step of the way, from their inception to their arrival on the market. In this book, we take a tour of different kinds of bioanalytical methods that enable innovative therapeutics, and we discuss the advantages and limitations of each of them. Notably, the analytical tools often evolve in parallel with the therapy, and new modalities prompt the development of analytical innovations. Ultimately, the persevering bioanalytical efforts are rewarded by enabling and accelerating the development of new medicines that transform the lives of patients.

Biologic drug modalities are a major driver of revenue growth in the biopharmaceutical industry. In 2021, the most prominent biotechnology products based on sales and FDA approvals were biotherapeutics [36]. Consistently, in 2023, four of the six top-selling biopharma products were peptide- or antibody-based therapeutics (pembrolizumab, adalimumab, semaglutide, and dupilumab). Clearly, therapeutic options are evolving toward biologic therapies. However, these advances are not without their challenges. As drugs are becoming more complicated, they are also becoming more expensive and more difficult to manufacture [37]. Once the biologic drugs are made, sophisticated storage infrastructure is often required to keep these therapies stable before they are administered to patients. Using biotherapeutics, we have realized the potential to revolutionize patient care, and now patient accessibility is a crucial part of this revolution that needs to be carefully considered. Investments from pharmaceutical companies and innovative research technologies will help to develop groundbreaking biotherapeutic drugs and get them to the people who need them. Bioanalytical scientists bear a piece of this responsibility to aid in reducing the costs associated with delivering therapies to patients. One approach to achieve this is by developing efficient analytical methods to identify better drugs on a faster timescale. Our hope is that the information in this book helps equip the next generation of scientists to develop excellent and innovative therapies that transform the treatment of disease and bring better access to patients.

References

1 Spiess, C., Zhai, Q., and Carter, P.J. (2015). Alternative molecular formats and therapeutic applications for bispecific antibodies. *Mol. Immunol.* 67 (2): 95–106.

2 Keyt, B.A., Baliga, R., Sinclair, A.M. et al. (2020). Structure, function, and therapeutic use of IgM antibodies. *Antibodies* 9 (4): 53.

3 Sharma, K., Sharma, K.K., Sharma, A., and Jain, R. (2023). Peptide-based drug discovery: current status and recent advances. *Drug Discov. Today* 28 (2): 103464.

4 Zaman, R., Islam, R.A., Ibnat, N. et al. (2019). Current strategies in extending half-lives of therapeutic proteins. *J. Control. Release* 301: 176–189.

5 Li, Y., Wang, M., Peng, X. et al. (2023). mRNA vaccine in cancer therapy: current advance and future outlook. *Clin. Transl. Med.* 13 (8): e1384.

6 Polack, F.P., Thomas, S.J., Kitchin, N. et al. (2020). Safety and efficacy of the BNT162b2 mRNA Covid-19 vaccine. *N. Engl. J. Med.* 383 (27): 2603–2615.

7 Krammer, F. (2020). SARS-CoV-2 vaccines in development. *Nature* 586 (7830): 516–527.

8 Rappuoli, R., Alter, G., and Pulendran, B. (2024). Transforming vaccinology. *Cell* 187 (19): 5171–5194.

9 Rojas, L.A., Sethna, Z., Soares, K.C. et al. (2023). Personalized RNA neoantigen vaccines stimulate T cells in pancreatic cancer. *Nature* 618 (7963): 144–150.

10 Chancellor, D., Barrett, D., Nguyen-Jatkoe, L. et al. (2023). The state of cell and gene therapy in 2023. *Mol. Ther.* 31 (12): 3376–3388.

11 Modell, A.E., Lai, S., Nguyen, T.M., and Choudhary, A. (2021). Bifunctional modalities for repurposing protein function. *Cell Chem. Biol.* 28 (7): 1081–1089.

12 Schreiber, S.L. (2024). Molecular glues and bifunctional compounds: therapeutic modalities based on induced proximity. *Cell Chem. Biol.* 31 (6): 1050–1063.

13 Nalawansha, D.A. and Crews, C.M. (2020). PROTACs: an emerging therapeutic modality in precision medicine. *Cell Chem. Biol.* 27 (8): 998–1014.

14 Guedeney, N., Cornu, M., Schwalen, F. et al. (2023). PROTAC technology: a new drug design for chemical biology with many challenges in drug discovery. *Drug Discov. Today* 28 (1): 103395.

15 Tsuchikama, K., Anami, Y., Ha, S.Y.Y., and Yamazaki, C.M. (2024). Exploring the next generation of antibody–drug conjugates. *Nat. Rev. Clin. Oncol.* 21 (3): 203–223.

16 Dumontet, C., Reichert, J.M., Senter, P.D. et al. (2023). Antibody–drug conjugates come of age in oncology. *Nat. Rev. Drug Discov.* 22 (8): 641–661.

17 Paunovska, K., Loughrey, D., and Dahlman, J.E. (2022). Drug delivery systems for RNA therapeutics. *Nat. Rev. Genet.* 23 (5): 265–280.

18 Laramy, M.N.O. and Nagapudi, K. (2022). Long-acting ocular drug delivery technologies with clinical precedent. *Expert Opin. Drug Deliv.* 19 (10): 1285–1301.

19 Huang, L., Guo, Z., Wang, F., and Fu, L. (2021). KRAS mutation: from undruggable to druggable in cancer. *Signal Transduct. Target. Ther.* 6 (1): 386.

20 Wright, B.W., Yi, Z., Weissman, J.S., and Chen, J. (2022). The dark proteome: translation from noncanonical open reading frames. *Trends Cell Biol.* 32 (3): 243–258.

21 Limbu, M.S., Xiong, T., and Wang, S. (2024). A review of ribosome profiling and tools used in ribo-seq data analysis. *Comput. Struct. Biotechnol. J.* 23: 1912–1918.

22 Kong, Y., Rose, C.M., Cass, A.A. et al. (2019). Transposable element expression in tumors is associated with immune infiltration and increased antigenicity. *Nat. Commun.* 10 (1): 5228.

23 Warburton, P.E. and Sebra, R.P. (2023). Long-read DNA sequencing: recent advances and remaining challenges. *Annu. Rev. Genomics Hum. Genet.* 24 (1): 109–132.

24 Roberts, D.S., Loo, J.A., Tsybin, Y.O. et al. (2024). Top-down proteomics. *Nat. Rev. Methods Prim.* 4 (1): 38.

25 Marchal, I. (2024). A strategy to load, rethread and read protein sequences through a nanopore. *Nat. Biotechnol.* 42 (10): 1499–1499.

26 Thompson, B. and Howe, N.P. (2024). Alphafold 3.0: the AI protein predictor gets an upgrade. *Nature*.

27 Fudge, J.B. (2023). Diffusion model expands RoseTTAFold's power. *Nat. Biotechnol.* 41 (8): 1072–1072.

28 Yang, A., Jude, K.M., Lai, B. et al. (2023). Deploying synthetic coevolution and machine learning to engineer protein-protein interactions. *Science* 381 (6656): eadh1720.

29 Demirsoy, I. and Karaibrahimoglu, A. (2023). Identifying drug interactions using machine learning. *Adv. Clin. Exp. Med.* 32 (8): 829–838.

30 Callaway, E. (2024). Chemistry nobel goes to developers of AlphaFold AI that predicts protein structures. *Nature* 634 (8034): 525–526.

31 Subramanian, M., Wojtusciszyn, A., Favre, L. et al. (2020). Precision medicine in the era of artificial intelligence: implications in chronic disease management. *J. Transl. Med.* 18 (1): 472.

32 Lawlor, A., Beyer, K., Russell, B. et al. (2024). PIONEER big data platform for prostate cancer: lessons for advancing future real-world evidence research. *Nat. Rev. Urol.* 1–9.

33 Gupta, R., Srivastava, D., Sahu, M. et al. (2021). Artificial intelligence to deep learning: machine intelligence approach for drug discovery. *Mol. Divers.* 25 (3): 1315–1360.

34 Vamathevan, J., Clark, D., Czodrowski, P. et al. (2019). Applications of machine learning in drug discovery and development. *Nat. Rev. Drug Discov.* 18 (6): 463–477.

35 Sarkar, C., Das, B., Rawat, V.S. et al. (2023). Artificial intelligence and machine learning technology driven modern drug discovery and development. *Int. J. Mol. Sci.* 24 (3): 2026.

36 Walsh, G. and Walsh, E. (2022). Biopharmaceutical benchmarks 2022. *Nat. Biotechnol.* 40 (12): 1722–1760.

37 Schuhmacher, A., Hinder, M., von Stegmann Und Stein, A. et al. (2023). Analysis of pharma R&D productivity – a new perspective needed. *Drug Discov. Today* 28 (10): 103726.

2

Analytical Characterization of Bispecific Antibodies and Bispecific Molecules

Michael Dillon[1], *T. Noelle Lombana*[1,2], *and Christoph Spiess*[1]

[1] *Department of Antibody Engineering, Genentech Research and Early Development, South San Francisco, CA, USA*
[2] *ReCode Therapeutics, Menlo Park, CA, USA*

Introduction

Bispecific antibodies (BsAbs) and bispecific molecules have specificity for two distinct protein antigens compared to a monospecific antibody (MAb) which binds only a single antigen. The idea of using a BsAb as a therapeutic has been around for almost as long as its monospecific counterpart, however, during this time, over 100 MAbs were clinically approved compared to only 15 BsAbs. One of the main reasons why BsAbs have lagged behind is that they are far more complex to produce. Over the last decade, significant advances in the production and characterization of BsAbs contributed to their approval and commercialization, and over 500 BsAbs have been under clinical investigation covering a range of therapeutic indications. The majority of the BsAbs target the oncology disease area [1, 2], while immunology, ocular, and neurobiology indications are also being explored.

Bispecific Antibody Applications

There are many therapeutic reasons to choose a bispecific despite the increased complexity of production and characterization versus the more straightforward development of a monospecific. Adding an additional layer of specificity has the advantage of bestowing antibodies with new modes of action, increasing potency or efficacy, or even combatting biological redundancy [3, 4]. BsAbs function in several ways, namely by bringing two cells together (retargeting effector cells to tumor cells), binding molecules on the same cell (to block, activate, or induce proximity), neutralizing two soluble receptor ligands, or delivering antibodies to the brain (Figure 2.1). Additionally, BsAbs can be utilized to improve cancer diagnostics and imaging.

The classical application of BsAbs is the retargeting of effector cells to kill tumor cells, which is achieved by binding a receptor on each cell type. In their original application, these molecules retargeted natural killer (NK) cells and macrophages via CD16 to eradicate target cells [5], but now mainly focus on redirecting cytotoxic T-cells [6]. More than half of the approved bispecifics are in this category and have anti-CD3 (T-cell receptor subunit)/anti-tumor specific antigen activity. Beyond T-cells and NK cells, neutrophils and macrophages have also been effectively retargeted to kill tumor cells [7–9] or phagocytose Epstein–Barr virus [10]. Another application of effector cell retargeting BsAbs, a fusion of a T-cell receptor and anti-CD3 scFv named ImmTAC, expands their capabilities to include binding intracellular proteins as peptides displayed on the major histocompatibility complex of cells [11].

Most of the remaining BsAbs in clinical development block or inactivate two receptors or ligands. Many of these include HER family member receptors such as EGFR, HER2, and HER3, checkpoint inhibitors such as PD-1 and CTLA-4, or proinflammatory cytokine ligands including TNF-alpha, IL-1, and IL-13 [2, 12]. There are also preclinical and clinical stage biepitopic BsAbs that bind to two different epitopes on the same target, as this approach can have better antagonistic [13–16] or agonistic properties [14]. In addition, in an effort to achieve even deeper pathway inhibition, proteolysis-targeting antibodies (PROTABs) were recently described that tether cell-surface E3 ligases and transmembrane receptors to promote efficient degradation and superior pathway inhibition over conventional antagonists [17]. Alternatively to inactivating signaling pathways, there is one BsAb approved (Hemlibra®; emicizumab)t that acts as a Factor VIII mimetic to dimerize Factor IXa and Factor X to induce

Characterizing Biotherapeutics: Analytical Methods for Diverse Modalities, First Edition. Edited by Jennie R. Lill and Wendy Sandoval.

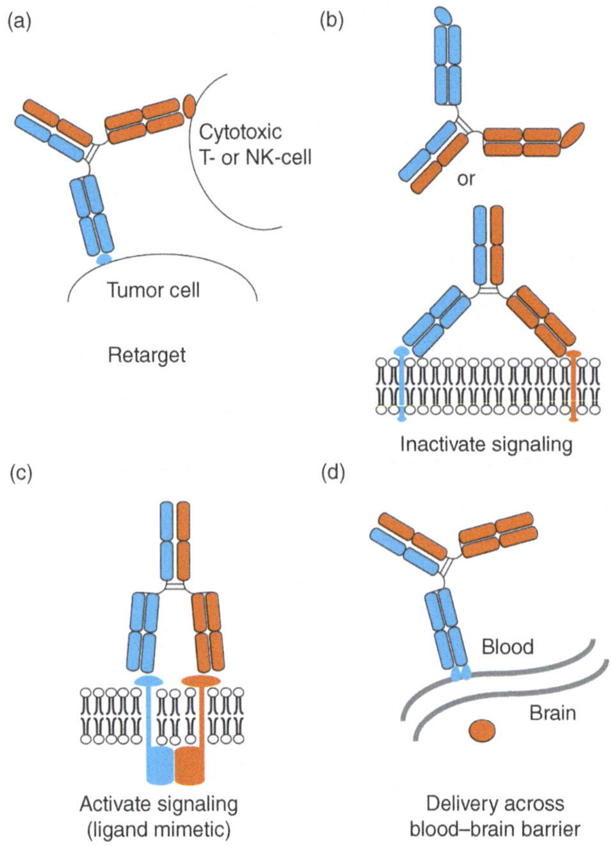

Figure 2.1 Bispecific antibody modes of action. (a) Retarget cytotoxic T- or NK-cells to kill tumor cells, (b) inactivate signaling of two soluble ligands or receptors, (c) activate signaling by a ligand mimetic, or (d) delivery across the blood–brain barrier.

signaling for the clotting disorder Hemophilia A [18]. Selective agonist activity is another potential application. A BsAb that binds a receptor complex, fibroblast growth factor receptor 1 and β-Klotho, demonstrated selective activation of the pathway and a promising therapeutic route for the treatment of type 2 diabetes and other obesity-linked disorders [19, 20]. Bispecific molecules are also a promising modality for the targeted delivery of other agonist biomolecules such as cytokines, for example, IL-2, IL-12, or IL-15. The antibody moiety facilitates targeting a desired cell subset that expresses the cytokine receptor and a tumor or cell-type specific marker on the same or proximal cell, thus reducing systemic toxicity or unwanted cell proliferation [21].

BsAbs can also function as delivery vehicles to target previously inaccessible antigens in the brain. To combat the difficulties of getting therapeutic antibodies across the tight epithelial cell junctions of the blood–brain barrier, a receptor-mediated transport arm can be combined with a therapeutic arm. A BsAb that binds the transferrin receptor and either the enzyme β-secretase [22] or amyloid beta directly [23, 24] can "hitchhike" on the transferrin-mediated endocytosis mechanism to increase brain uptake and decrease amyloid beta levels. A similar strategy was also employed to transmigrate a single-domain antibody conjugated to blood–brain barrier impermeable neuropeptides for inflammatory pain [25].

In addition to disease applications, BsAbs can be used as highly specific tools for infectious disease and cancer diagnostics as well as imaging. In a diagnostic assay, the first arm of the BsAb recognizes the pathogen or tumor cell, while the second arm is specific for a detection enzyme, such as horseradish peroxidase [26]. This approach allows one-step detection with no need to label the antibody. BsAbs can also improve cancer cell imaging by pre-targeting the BsAb to the cancer cell antigen and then binding a radioactive hapten-peptide, for example, which is added at a later time. This imaging technology shows higher sensitivity and specificity with lower background binding compared to radiolabeled antibodies [27].

Overview of the Various Bispecific Formats

There are more than 100 recombinant bispecific formats that have been created due to the modular nature of the immunoglobulin domain [2, 28]. Each format offers a distinct set of properties, such as size, antigen-binding valency, geometry of antigen-binding regions, flexibility, pharmacokinetic half-life, effector function, and biodistribution. The bispecific format and its properties should be carefully considered depending on the requirements of the therapeutic or diagnostic application. Another important consideration is the complexity of the bispecific molecule, as increasingly complex molecules complicate the mode of production and often require more extensive characterization.

A BsAb can be made in a single cell by co-expression of the two different heavy chains and two different light chains; however, there are a few challenges to overcome. Due to the proposed random assembly of the four individual protein chains [29], there can be up to nine unwanted species with only one correct species containing the properly assembled heavy chain heterodimer (heavy chain-pairing problem) and the correct pairing of each light chain with its cognate heavy chain (light chain-pairing problem) [30, 31] (Figure 2.2). Although a BsAb molecule can be produced this way, the desired bispecific is found in low yield and it is difficult to purify away the unwanted species [32]. To overcome these challenges, several solutions/formats were developed. Several BsAb formats are introduced below by the mode of production, as this has profound consequences on downstream analytical

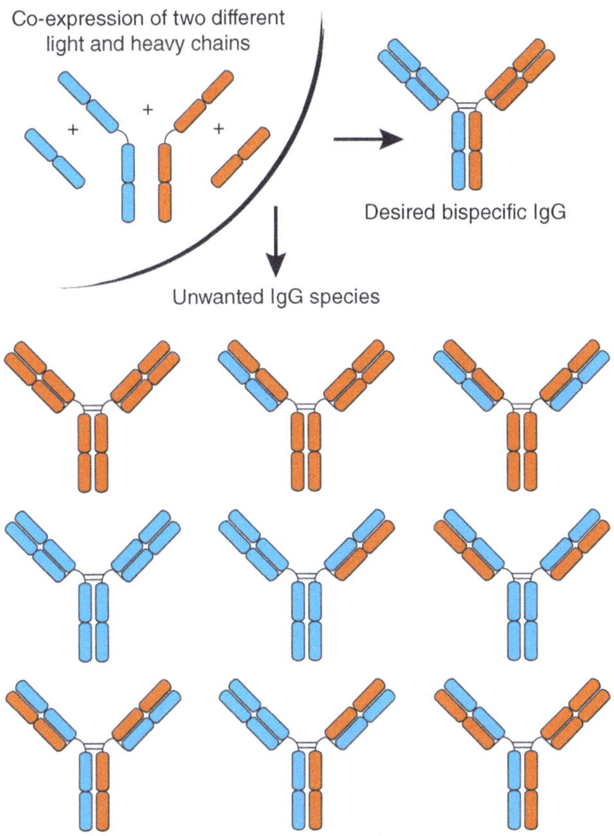

Figure 2.2 The bispecific antibody light- and heavy-chain pairing problems. Co-expression of two light and two heavy chains can result in nine unwanted species besides the desired bispecific antibody. An efficient bispecific platform requires correct light chain pairing with its cognate heavy chain, as well as heterodimeric heavy chain pairing.

characterization (see Figure 2.3). Here, they are classified into five groups: (i) purification from mixtures, (ii) tethered bispecific molecules and alternative scaffolds, (iii) untethered bispecific molecules with engineered amino acid substitutions, (iv) native bispecific IgG with dual binding behavior, and (v) BsAb conjugates. Their advantages and limitations are summarized in Figure 2.4.

Purification from Mixtures

A number of BsAbs have been purified from complex mixtures and gone into clinical development. Catumaxomab (Removab®), an anti-CD3/anti-epithelial cell adhesion molecule (EpCAM) that retargets T-cells to treat malignant ascites in EpCAM-positive tumors was the first approved BsAb in 2009 (now withdrawn from marketing). Catumaxomab was made by co-expression of a mouse and rat antibody (Triomab®) in a single cell (quadroma technology) [33]. Due to the preferential species-specific pairing of the light and heavy chains, the light chain-pairing problem is overcome [34]. To facilitate purification of the desired BsAb, protein A is used with a sequential pH elution to separate the BsAb from the mouse IgG, as only mouse antibodies can bind.

A common light chain approach combined with engineering distinct resin-binding specificities into each of the two heavy chains can be used to purify unwanted species away from the desired bispecific. In this approach, one light chain is engineered to pair with two different heavy chains with each pair having a different antigen binding specificity. Using this approach, the light chain-pairing problem is avoided and the heavy chain-pairing problem is solved by differentially tagging the Fc region for purification with protein A [35].

An alternative way to isolate a BsAb from unwanted species is to use a common heavy chain, which eliminates the heavy chain-pairing problem. The κλ-body format has a common heavy chain with kappa and lambda light chains where each light chain-heavy chain pair is engineered to bind a distinct antigen [36]. Two sequential affinity purification steps that bind to either the kappa- or lambda-specific light chain produce the desired BsAb. This two-step process disfavors binding to antibodies that only contain a single type of light chain, elegantly solving a complex purification process.

Isolating the preferred BsAb from other unwanted species in a mixture comes with the advantage that non-natural amino acid substitutions can be reduced or avoided but does come with purification challenges and compromises. It is a process with low efficiency due to the production of at least two other species besides the desired bispecific. Engineered preferential binding can streamline the purification of the wanted BsAb [35, 36], however, there is significant upfront engineering to use a common light or heavy chain. Additionally, this method may not be amenable for certain antigen targets and may limit epitope diversity and attainable antibody affinity. Therefore, alternative modes of production may be preferred.

Bispecific Antibodies and Alternative Scaffolds with Tethered Domains

One way to eliminate the pairing problem and simplify the characterization of the desired BsAb format is to fuse an additional antibody-variable region to the N- or C-terminus of the light chain or heavy chain. These molecules are full-length antibodies with additional domains added. One such molecule is the dual-variable-domain immunoglobulin (DVD-Ig™). This molecule has V_L and V_H domains having specificity to a second antigen added N-terminally by a short peptide linker onto the V_L and V_H domains,

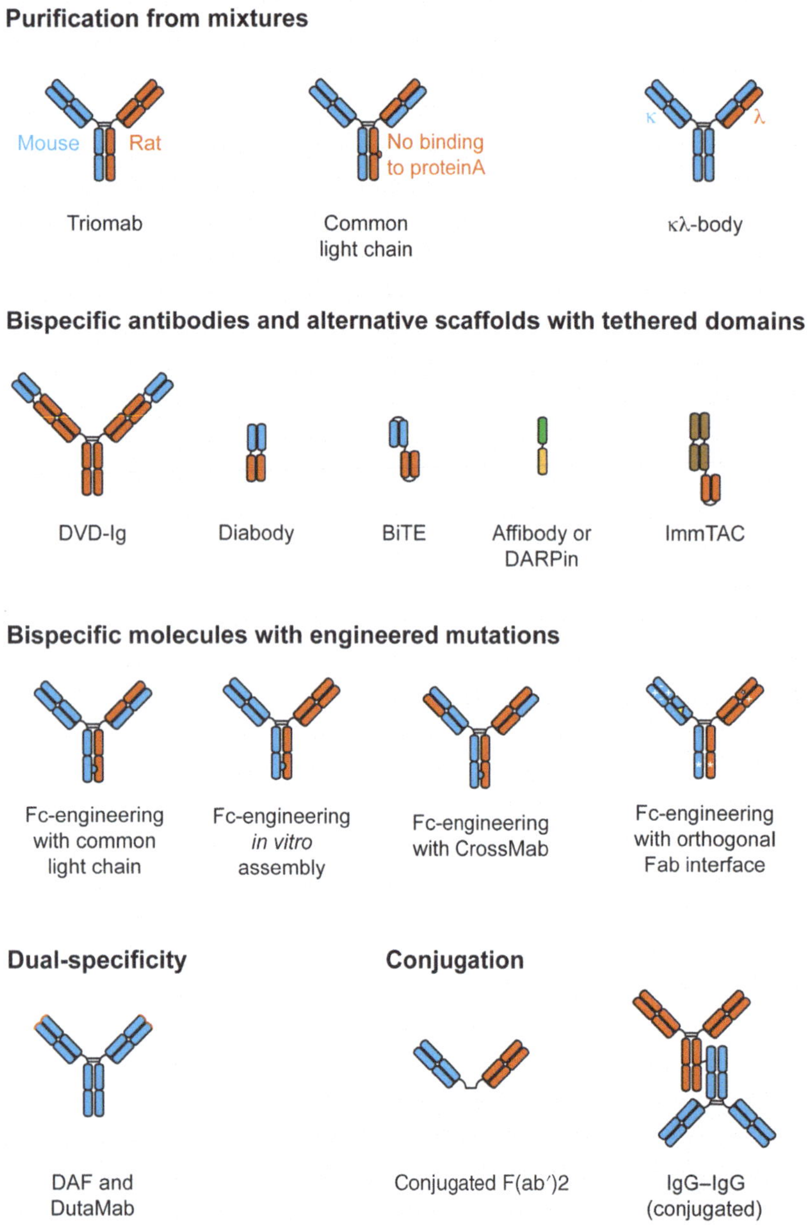

Figure 2.3 Examples of bispecific antibodies are classified by their mode of production. Antibody domains are colored red and blue, alternative scaffold domains in green and orange, and the T-cell receptor domains in brown.

respectively, of the full-length antibody molecule. In this format, the DVD-Ig is bivalent for each antigen, giving it double the valency than would be found in a traditional BsAb. The advantage of this format is that it only has one heavy and light chain, eliminating the production of unwanted species found with BsAbs having three or four different chains. The additional antibody domains can also be fused to the C-terminus of the antibody Fc region. Here, either antibody fragments, such as scFv [37], or Fab domains [24] have been successfully utilized.

The molecules discussed so far contain Fc regions, which are useful for purification with protein A, extending half-life, or having effector function. Antibody fragments have also been explored as a bispecific format to simplify production and remove effector function. These formats typically include only the antigen-binding Fv domains that are tethered to each other via short linkers. One of the earlier formats is the diabody that has two chains with linkages V_H(a)–V_H(b) and V_L(a)–V_L(b) [38]. The first approved BsAb fragment, the bispecific T-cell Engager (BiTE™), blinatumomab (Blincyto®), retargets T-cells to kill tumor cells. Blinatumomab is a tandem scFv fragment binding both CD3 and CD19. A second example with clinical impact is Kimmtrak, a fusion between a T-cell receptor (binding HLA-A*02:01/gp100 complex) and an anti-CD3 scFv.

In addition, single-domain antibodies (sdAb; also known as variable heavy antibody fragment [VHH] or Nanobody®)

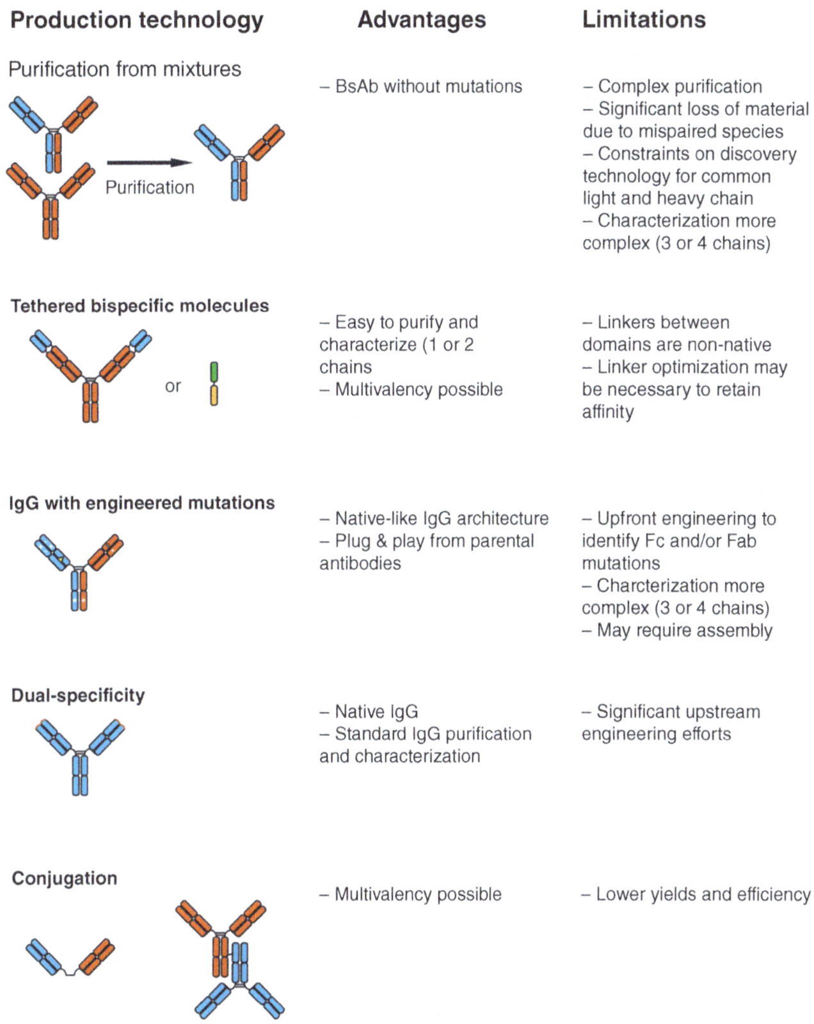

Figure 2.4 Production technologies of bispecific antibody formats and their advantages and limitations.

have recently gained popularity as BsAb building blocks that obviate the chain-pairing problem and are also easily strung together to generate bi- or multispecific fusion proteins. VHH domains can, for example, be identified from camelid animal immune libraries [39–41], naive libraries derived from animals [42], or synthetic libraries [43]. However, an isolated VHH domain has a short serum half-life due to its small size and lack of binding to the neonatal receptor FcRn. To prolong serum half-life, different solutions exist. To maintain the small size of the BsAb, a fusion to a VHH with human serum albumin (HSA) binding properties can be used. One example is the approved trivalent Nanobody Nanozora® (ozoralizumab) consisting of two different anti-TNF-alpha and one anti-HSA VHH [44]. Alternatively, fusion to the immunoglobulin Fc is frequently pursued. VHH is also readily amenable to fuse to an IgG to create multivalent bispecific molecules [45].

A hydrophobic region between the former variable/constant domain is exposed in the VHH domains and pre-existing antibodies in humans toward this region have been described. The contribution of pre-existing antibodies to the immunogenicity of VHH BsAbs is not fully understood, but it seems prudent to eliminate these liabilities upfront by protein design. Multiple solutions to reduce the binding of these pre-existing antibodies by introducing amino acid substitutions into the VHH have been described [46].

Alternative scaffolds have also been the focus of much attention in the bispecific molecule space. These scaffolds are generally well-behaved proteins that are engineered at multiple positions to recruit binding to an antigen. Most of these scaffolds also have the advantage of being modular in their architecture, allowing multiple single units to be linked together for the desired antigen-binding valency or extending binding specificity beyond two antigens. Due to these advantages, many different combinations can be easily screened for the best potency. Two examples of these scaffolds are the Affibody molecule [47], engineered from the Z-domain of protein A, and the DARPins [48], designed from the ankyrin repeat proteins.

With all the advantages that come with producing tethered bispecific molecules, there are some potential downsides to consider. For many of these formats, peptide linker optimization can be necessary to obtain the optimal antigen binding affinity [49] and reduced aggregation properties [50]. Additionally, when engineered linkers are used between antibody binding domains or small antigen binding domains are linked together in non-natural configurations, there is the possibility of increased immunogenicity and rapid proteolysis *in vivo* [50].

Bispecific Molecules with Engineered Amino Acid Substitutions

The untethered human BsAb format is typically the most utilized therapeutically due to the native-like antibody architecture, however, it is the most complex to design and characterize. These molecules have engineered, ideally non-solvent exposed amino acid substitutions that drive preferential pairing between heavy chains, while different solutions are implemented to solve the light chain-pairing problem. The earliest designed heavy chain heterodimerization solution was the "knobs-into-holes" technology [51, 52], which revolutionized the generation of BsAb molecules. This platform has differential engineering in each C_H3 domain, where one side has a large amino acid such as tryptophan to disfavor homodimerization (knob), while the other side has smaller compensating substitutions (hole), which also disfavor homodimerization. These asymmetric heavy chain arms are robust in their pairing with various Fab arms. Alternative strategies to favor heavy chain heterodimerization have also been developed that use different steric substitutions, electrostatic engineering, or exchanging structurally similar C_H3 strands of IgA and IgG to preferentially bind to each other but not to themselves [53–58]. While these bispecific engineering solutions were developed using the IgG1 isotype, some are also compatible with the IgG2 and IgG4 isotypes [56, 59]. This is not only useful for modulating the Fc-gamma receptor-mediated effector-function, but can also be leveraged to tune the activity of agonist antibodies [14].

When using substitutions that drive heavy chain heterodimerization, the simplest solution to avoid the light chain pairing problem is to use the same light chain (common light chain) [18, 60]. A common light chain can be identified by phage display or animals transgenic for a single light chain [35, 61–63]. Due to the difficult engineering efforts that go into finding a common light chain, the most common strategy to produce a BsAb is to express the two-component antibodies in separate cells, followed by *in vitro* assembly into the intact BsAb [56, 64–66]. However, strategies to express the BsAb in a single cell do exist. One solution is using domain swaps of the antibody constant or variable domains [67], and multiple molecules with this CrossMab format have reached clinical development [68]. More recently, orthogonal Fab interfaces have been engineered [15, 69–71] with substitutions in both the V_H/V_L and C_H1/C_L interfaces. When these substitutions are combined with a heavy chain heterodimerization strategy, they seem broadly applicable to allow efficient BsAb production from a single host cell [15, 69, 71, 72].

Native Bispecific IgG with Dual Binding Behavior

One way to obviate the heavy or light chain-pairing problems altogether is to engineer the Fab domain to bind to two distinct antigens via two different binding sites on the antibody (paratopes). This approach produces a molecule that is a regular antibody with only a single heavy and light chain, however, the antibody can now bind to two different antigens. There are two BsAb formats in this category, the Dual Action Fab (DAF) [73, 74] and the DutaMab [75]. The DAF paratopes use the light and heavy chain complementarity determining regions (CDRs) and somewhat overlap, whereas the DutaMab uses diversity in only three CDRs for each antigen, creating two separate paratopes. For certain applications, a dual-binding antibody could be very desirable as it allows for different types of antigen-binding valency, but when monospecificity is required, this approach will not work. The benefit of these formats is producing and characterizing a molecule with the same properties as a monospecific antibody. However, there is a large amount of upfront engineering that is required to recruit a second antigen binding to a Fab domain, and not all molecules may be amenable to this process.

Bispecific Antibody Conjugates and Protein Ligation

In the early 1960s before recombinant protein expression, antibodies or antibody fragments were chemically conjugated to create a BsAb. The first BsAbs produced were bispecific $F(ab')_2$ molecules that were monovalent for each antigen [76, 77]. These were made by digesting two different antibodies with pepsin followed by a reduction/oxidation procedure to conjugate into the bispecific $F(ab')_2$. However, these initial conjugation reactions had poor yields, with many undesirable side products. Thus, a more advanced conjugation strategy based upon cysteine-reactive crosslinking agents linking recombinantly-produced full-length IgG or Fab′ was used to improve efficiency [78–80]. These bispecific molecules can be bivalent or monovalent for each antigen, respectively. Chemical conjugation to amine groups has also been used to cross-link two different IgG into a bispecific format with bivalency to each antigen [81]. This strategy has been applied to a number of different tumor targets [82–84] as well as cytomegalovirus infection [85]. Although site-specific conjugation does increase BsAb yield, this process is still not very efficient with much optimization required to decrease unwanted side products [86]. Conjugation with orthogonal clickable functional groups has the potential to overcome these limitations and provides the modularity to rapidly assemble BsAb [87].

Protein ligation is an alternative strategy to chemical conjugation and has served as a facile route to generate larger panels of BsAbs during the preclinical screening of antibody combinations.

Targeted Cytokines and Other Hybrid Molecules

The specificity and pharmacologic properties of antibodies have been leveraged to target other protein therapeutics and cargoes such as cytokines, enzymes, toxins, or neuroprotective proteins to cells of interest to avoid widespread systemic activation and toxicity [88–90]. This creates a hybrid molecule between an antibody or antibody fragment and the functional cargo protein, where a variety of valencies and geometries are possible. For example, the cargo can be fused to either the antibody heavy or light chain creating a molecule that is bivalent for both modalities. Alternatively, Fc heterodimerization technologies (see Section "Bispecific Molecules with Engineered Amino Acid Substitutions") can be employed to create molecules that are monovalent for the antibody, the cargo, or both. In many cases, the cargo has a strong affinity to its cognate target (receptor), limiting the molecule to preferential distribution that is insufficient to overcome systemic toxicity. This property has constrained the initial utility for targeting cytokines. To increase the therapeutic index further, second-generation targeted cytokines with masked cytokines or attenuated receptor binding have been designed [91]. By introducing amino acid substitutions into the cytokine, the affinity to the cognate receptor is detuned to an extent that productive cytokine/receptor interaction only occurs with concurrent binding of the antibody on the target cell. In general, similar analytical and developability considerations apply to these hybrid molecules as described for BsAbs. The lower thermal stability of many cargos compared to antibodies, as well as their abundant N-linked glycosylation, can introduce additional challenges such as heterogeneity, limited formulation stability, and less ideal pharmacologic properties. While additional protein engineering by amino acid substitutions can address these issues, it will inherently increase the foreignness and potential risk of immunogenicity. Anti-drug antibodies (ADAs) toward a natural cargo can potentially cross-react with the endogenous protein. Neutralization of the natural ligand can be a consequence with severe safety implications, warranting careful consideration between hybrid molecules with a substantially engineered natural protein and a BsAb ligand mimetic.

Identifying the Best Parental Antibody Pairs to Combine in a BsAb

While high-throughput (HTP) antibody discovery platforms can routinely produce hundreds or thousands of MAb clones for screening [92], a challenge remains in how to best identify which formats and binding domains to combine for the desired activity. Many applications for BsAbs depend on the optimization of the epitope, affinity, valency, and geometry of all binding domains in combination to achieve the desired biological activity. Parental MAb properties alone are often not predictive of bispecific activity, necessitating the production and screening of clones in a relevant BsAb format to identify lead molecules [93, 94]. However, permuting hundreds of MAb clones across one or more bispecific formats requires the production of thousands of complex BsAb molecules in sufficient yield and purity for activity assays. While production and screening of a matrix of 40,000 BsAbs through traditional chain co-expression methods has been demonstrated [95], this requires considerable time and resources that limit the broad feasibility of this approach. Several combinatorial high throughput methods have been developed to address this bottleneck by leveraging modular component expression and subsequent matrixed assembly to produce BsAb candidates in a relevant format for activity screening.

Demonstrated techniques for efficient modular assembly of individual binding domains include chemical coupling using click chemistry [87, 96], non-covalent scFv-peptide interaction of tagged Fabs [97], Sortase-A-mediated domain coupling [98], SpyTag/SpyCatcher-mediated protein ligation [99–101], split intein protein splicing [102], and format chain exchange (FORCE) [94]. Due to the dependence of BsAb activity on factors such as valency and geometry, the chosen screening technology should ideally match as closely as possible to the properties of the intended final production format. Some of the above techniques, including protein ligation and FORCE, have additionally demonstrated utility in screening for optimal format valency and geometry along with preferred binding domain pairings [94, 99]. These techniques can be utilized to produce BsAbs that are representative of final production formats in high enough quality for screening using HTP, automation-friendly, miniaturized production techniques to maximize the number of combinations that can be evaluated while minimizing resource requirements.

Alternatives to Bispecific Antibodies: Antibody Mixtures

Depending on the requirements of the biological system of interest, both antigen-binding activities of a BsAb molecule may be "obligatory" or "optional" to reside in the same molecule. Examples of "obligatory" BsAbs are the retargeting molecules that simultaneously engage cytotoxic T-cells and tumor cells, targeting molecules that use a transcytosis receptor to transport an antibody to the brain, and ligand mimetics that dimerize cellular receptors. "Optional" BsAbs include those that block or neutralize independent signaling pathways or those that can have multiple molecules bound to an antigen to "decorate" it. Only "optional" but not "obligatory" BsAbs can be formulated as mixtures, where two or more antibodies are combined, or co-administered [103]. Even when not strictly "obligatory," it is possible a BsAb can further potentiate the agonist activity beyond the mixture of the individual component antibodies [14].

There are some advantages that mixtures or coadministration have over a BsAb. Besides both antibodies being bivalent and the benefits of avidity for antigen binding, the antibody ratios can be tuned. Additionally, coadministration allows each antibody to be dosed separately providing more flexible timing in each dosing schedule. Antibody mixtures can also mimic the natural polyclonal response to decorate the target antigen, and this may provide a solution to prevent drug resistance in oncology or infectious diseases. Moreover, using mixtures to test different antibody combinations for additive and synergistic effects is preferred in a research setting due to the simpler production and characterization of MAbs compared to BsAbs. However, there are also some disadvantages to using antibody mixtures, such as choosing the proper formulation and analyzing multiple antibodies in the same mixture, which introduces complexity to the manufacturing process. Some examples of antibody mixtures are the three antibody cocktail to the Ebola virus called ZMAb [104, 105], the mixture of two anti-EGFR antibodies binding to distinct epitopes which are superior at inhibiting EGFR-dependent cancer cells in preclinical [106] and early clinical studies [107], and antibody mixtures that target viruses or botulinum toxin [108–110]. Additionally, REGN-COV (Ronapreve), a combination of the two monoclonal antibodies casirivimab and imdevimab, reached approval for the treatment and prevention of COVID-19 [111].

Characterization of the Bispecific Molecule

Access to and establishing quantitative analytics is crucial during the engineering of the initial bispecific format and to ensure high product quality of the produced BsAb for research, development, and clinical use. For example, when engineering substitutions that steer heterodimerization of the two heavy chains or orthogonal pairing of the light chains, quantitative assays are paramount to determine success. Understandably, once a bispecific format is selected, careful characterization of the expressed or assembled bispecific product is of utmost importance. While mispaired light and heavy chains are commonly binding-inert, side-products such as homodimers can pose unwanted modes of action and safety liabilities [112, 113]. Also, unwanted partially assembled side products can contribute to lower activity levels, while "obligatory" BsAbs that rely on both arms of the BsAb for activity require high purity for correct interpretation of research data. Thus, BsAbs should always be characterized by chromatographic and mass spectrometric methods to ensure product quality in addition to performing binding and functional *in vitro* and *in vivo* assays to confirm that the desired properties of the parental antibodies were successfully transferred to the BsAb. In the following sections, the most common analytical techniques to characterize BsAbs are described and summarized in Table 2.1.

Characterization by Bioanalytical Methods

To assess the quality of the purified antibodies, denaturing capillary electrophoresis-sodium dodecyl sulfate (CE-SDS) or standard denaturing SDS-polyacrylamide gel electrophoresis (SDS-PAGE) and size-exclusion chromatography (SEC) analysis are commonly performed. Non-reducing electrophoresis can give information on the integrity of the antibody interchain and intrachain disulfides, whereas reducing conditions can provide an estimate of the relative chain distribution and purity of the material [59, 114]. SEC analysis can detect fragments and aggregates and in combination with a high-performance liquid chromatography system (SEC-HPLC), improved resolution can be achieved [115]. To increase resolution further, reverse phase-HPLC (RP-HPLC) can be used to evaluate assembly efficiency as it allows the separation of homodimers from the assembled BsAb [115]. These techniques are often used in combination with the additional analytical tools discussed below.

For purification and characterization of BsAbs both cation-exchange chromatography (CEX) and hydrophobic interaction chromatography (HIC) have proven

Table 2.1 Techniques to purify and characterize bispecific antibodies.

Analytical technique	Purpose	Advantages	Limitations
Gel or capillary electrophoresis (SDS-PAGE or CE-SDS)	• Verify integrity of interchain disulfide bonds • Evaluate purity	• High throughput (CE-SDS)	• Limited mass resolution
Size exclusion chromatography (SEC)	• Preparative: remove fragments and aggregates • Analytical: determine amount of fragments and aggregates	• Universal buffer conditions • Analytical: high throughput • Higher resolution with SEC-HPLC	• Not scalable to manufacturing
Reverse phase-high performance liquid chromatography (RP-HPLC)	• Determine mispaired species, fragments, and aggregates	• High throughput and high resolution	• Peak identity needs to be confirmed by other method • Only analytical
Hydrophobic interaction chromatography (HIC)	• Preparative: remove mispaired species, fragments, and aggregates • Analytical: determine amount of mispaired species, fragments, and aggregates	• Universal pH and salt gradient conditions	• Elution times need to be determined empirically • Species with similar hydrophobicity cannot be separated
Cation exchange chromatography (CEX)	• Preparative: remove mispaired species • Analytical: determine amount of mispaired species	• Elution times can be estimated from theoretical pI	• Species with similar pI cannot be separated
Differential scanning calorimetry/fluorimetry (DSC or DSF)	• Thermal stability	• Rapid formulation screening • High throughput (DSF)	• Individual domains, such as Fab or Fc, may be required for straightforward data interpretation
HPLC/UHPLC-ESI-Q-TOF	• Confirm expected mass of molecule • Detect chemical stability under accelerated stress conditions	• Increased resolution over HPLC separation (UHPLC)	
Orbitrap	• MS analysis of complex mixtures	• Increased resolution over ESI-Q-TOF (baseline separation of species close in mass)	
BV-ELISA, Protein Panel Profiling, Polyspecificity reagents	• Risk-mitigation to identify off-target binding and potential fast clearance	• High throughput	• Target-mediated clearance not predicted

particularly useful and versatile [36, 56, 64, 65]. CEX has the advantage that elution of the bispecific, as well as the mispaired species, can be estimated based on the theoretical pI of the component molecules. In contrast, HIC has the advantage of using universal pH and salt gradient conditions but elution times cannot be predicted. Most laboratories characterize BsAbs by CEX and/or HIC in combination with MS methods, which are described next.

Characterization by Mass Spectrometry Methods

General Considerations

To evaluate the molecular integrity of bispecific composition, the use of mass spectrometry (MS) for characterization is essential. However, the large number of off-pathway species creates large analytical complexity that requires high-resolution techniques to separate correctly paired chains from mispaired. There are a number of strategies that can be employed to reduce BsAb heterogeneity making the sample more amenable to MS. Removing glycosylation prior to analysis, such as deglycosylation of N-linked glycans with PNGase-F, reduces the heterogeneity of the BsAb [116]. For many basic engineering approaches, the presence of the N-linked glycan in the antibody C_H2-domain is unimportant, thus mutations that abrogate glycosylation are a simple engineering strategy to also improve product homogeneity [117, 118]. Furthermore, incomplete processing of the C-terminal lysine residue in the antibody heavy chain is another source of product heterogeneity [119–121]. To simplify analytical characterization by MS, deletion of the terminal lysine residue in the protein construct can also be considered [117].

The difficulty in MS characterization of BsAbs scales with the number of chains. The simplest BsAb to characterize are the ones containing only one chain, for example, the BiTE, as no pairing is required. BsAbs containing two chains that must pair, such as the DAF and DVD-IgG™, can be characterized like a regular monoclonal antibody. However, to create a BsAb with native-like architecture that has three or four different chains, it is important to note that mispairing can occur during expression or *in vitro* assembly. The assembly of a BsAb from multiple chains requires that two heavy-light chain pairs with similar masses be distinguished during the analytical process. To facilitate an easier analytical process, substitution steering chain pairing should be placed strategically onto each antibody chain to achieve the maximum mass differences between all possible species. However, certain mispaired species will have an isobaric mass with the correctly assembled bispecific and mass spectrometric methods unable to distinguish between them. One example is the BsIgG (cognate heavy chain (HC) and light chain (LC) pairing; HC1-LC1/HC2-LC2) and the light chain-scrambled (HC1-LC2/HC2-LC1) species, where the best way to approximate the amount of mispaired product is to estimate the purity based on the abundance of other mispaired species [117].

Purity Analysis of the Final Bispecific Antibody

In addition to using MS analysis for the characterization of the produced BsAb, it is essential to validate the lack of side products during the BsAb engineering process [15, 56, 122]. Woods et al. demonstrated the use of a high-performance liquid chromatography (HPLC)-MS-based assay using an advanced electrospray ionization and quadrupole time-of-flight (ESI-Q-TOF) mass spectrometer to characterize the purity of a heterodimeric antibody with a common light chain [123]. Relative quantification of homodimer and half-antibody in the samples was achieved by first preparing pure samples of the heterodimer and the separately expressed homodimers, then performing subsequent spiking experiments. One complication is that side products such as half-antibodies can have greater ionization than the intact antibody, which can lead to over-representation of the half-antibody and under-representation of the intact BsAb. Nevertheless, this heterodimer characterization assay allows purity assessment and can detect low levels (2%) of homodimers. Recently, multiple groups have also used HPLC-ESI-Q-TOF MS to validate the success of engineered amino acid substitutions in the Fab region that steer cognate heavy-light chain pairing in a single cell host [15, 69, 124].

Several bispecific strategies rely on the assembly of the intact BsAb from two separately expressed and purified parental homodimers [56, 65] or half-antibodies [64]. A redox reaction facilitates the efficient formation of the covalent heterodimers. Under these assembly conditions, the hinge-disulfides of the homodimer and the cyclized hinge-disulfides of half-antibodies are reduced, however, the disulfide between the antibody light and heavy chain is also reduced, resulting in potential light-chain exchange. Intact MS of the antibody cannot distinguish between the correctly paired and double-mispaired light chain species, therefore manipulation of the BsAb prior to MS analysis must be performed. One characterization option is using proteolytic digestion by LysC to release the individual Fab fragments for separate analysis [64]. This strategy has been applied to analyze assembled BsAbs, but could also be applied to unambiguously identify the success of BsAbs expressed in a single cell host.

MS has also been utilized to inform cell line development of BsAbs. During cell line development, high-throughput and rapid feedback is necessary to optimize transfection strategies and clone selection. In addition, this process should be coupled with purification screening to identify lead clones that balance the absolute purity of the product and the ability to remove individual side products during downstream purification. To analyze a BsAb with a common light chain, cell culture samples can be subjected to a protein A capture step linked to inline MS [125]. By estimating the ratio of each species from the deconvoluted spectra, the relative amounts of homodimer, heterodimer, as well as partially assembled species can also be estimated. Moreover, increased throughput can be achieved using a RapidFire mass spectrometry workflow [126]. As mass spectrometric analysis provides more detail of the side-products than other commonly used assays such as enzyme-linked immunosorbent assay (ELISA), it may enable a targeted approach in optimizing transfection or clone selection.

Antibody Mixtures

In-depth characterization is necessary to gain approval of composite antibody mixtures for clinical development. Traditionally, CEX and ELISA have characterized antibody mixtures, but the high resolution of native MS predisposes its use for this application. It has been demonstrated that native MS using an ESI-TOF platform is able to qualitatively and semi-quantitatively analyze a complex mixture of up to 10 antibodies [116]. The relative abundance of each antibody could be quantified with an accuracy of 3% and 9 out of 10 antibodies in the mixture were reliably identified with superior resolving power compared to CEX.

The automated peak fitting of the native MS technology combined with minimal sample amounts and handling requirements make it a fast and robust tool. In addition to the characterization of antibody integrity, native MS was used to interrogate the dynamics of half-antibody formation for IgG4 isotype antibodies [127] by providing kinetic details about natural bispecific molecules formed by Fab arm exchange [128, 129]. Furthermore, native MS using ESI-TOF has been used to determine the stoichiometry of antigen–antibody complexes [130, 131]. While these studies have been limited to monospecific antibodies, they suggest that large antigen sizes can limit occupancy. As many BsAbs have unique Fab geometry and valency, it seems intriguing to also apply this technology to bispecific molecules in order to elucidate the antigen occupancy.

Increasing Resolution

Often the masses of correctly paired and mispaired BsAb species are close and demand high-resolution analytics to reliably resolve differences. To increase resolution and separation compared to traditional reverse phase (RP)-HPLC columns used in HPLC-MS applications, RP-ultra high-performance chromatography (UHPLC) was newly established and optimized for analyzing variants of intact antibodies [132]. While this technology was applied to the characterization of a free thiol variant, the increased resolution has promise to streamline variant analysis of BsAbs as well. To also satisfy these high-resolution requirements, native Orbitrap MS has recently provided a powerful approach to the qualitative and semi-quantitative characterization of intact IgG [133, 134] and composite mixtures [135]. The high mass-resolving power allows baseline separation of antibodies close in mass. For example, Thompson et al. [135] could differentiate between two antibodies that differed by only 42.27 Da. This resolution enables not only superior identification of all products in a mixture but also the accurate relative quantification of desired and off-pathway species. In addition, the high resolution of the Orbitrap instrument allows the differentiation of antibodies without the need for enzymatic deglycosylation, which is particularly useful for BsAbs that are glycosylated or glycoengineered to enable or enhance antibody effector function, respectively. High sample throughput without sacrificing resolution can be achieved by coupling an HPLC instrument with a supermacroporous reverse-phase column to an exactive plus extended mass range (EMR) orbitrap mass spectrometer [117] or a RapidFire mass spectrometry workflow using electrospray ionization [126]. These improved workflows provide valuable insights during the research stages when engineering the bispecific antibody and testing many permutations but also in the downstream development workflows during cell line development to identify clones with the best product quality [126].

Characterization of Bispecific Antibodies by Binding Assays

For larger panels of BsAbs, it is easiest to screen binding using a bridging ELISA assay. Here, the first antigen is coated on the plate and after incubation with the BsAb, the second antigen is added and detected. To obtain binding kinetics of the BsAb, surface plasmon resonance (SPR) techniques can be used with a setup analogous to the bridging ELISA assay [15, 56, 67, 136]. For an "optional" BsAb, however, it is often sufficient to verify the affinity to each antigen independently by SPR [59]. If fully binding-competent BsAb reference material is available, both ELISA- and SPR-based analysis can determine the relative binding competency of the sample BsAb [70, 137]. This is particularly useful in engineering situations for ranking the success of efficient BsAb production. After confirming the binding properties by *in vitro* techniques, cell-based assays to study antibody potency should be performed.

When characterizing the binding and potency of a BsAb, there are a number of considerations. For BsAb formats that require monovalent target engagement, suitable parental antibodies should be selected based on their monovalent affinity as there is often a disconnect between monovalent and bivalent binding affinity. On the other hand, the decrease in affinity due to the monovalent nature of some BsAb formats can also be desirable. Trivalent BsAb molecules, for example, can allow avidity to one target (with bivalent binding), while preventing receptor dimerization or avidity to the second target (with monovalent binding). This was leveraged in the case of an anti-Abeta/Transferrin receptor BsAbs to modulate cellular trafficking by facilitating transcytosis instead of lysosome sorting [23]. Two other features that can impact target engagement are the order of binding domains and the length of the linker between them. This has been studied in detail with several DVD-Ig molecules [49] and single-chain diabodies [138] and was shown to be antigen or format dependent [139].

Developability Assessment of the Bispecific Antibody

The complex composition of BsAbs from two parental antibodies raises the bar for developability. A number of assays to select favorable properties of parental antibodies have been established and reviewed in detail elsewhere [140, 141]. In summary, to enable the developability of an antibody as a therapeutic, criteria including expression, favorable physicochemical properties such as thermal stability, aggregation, solubility, and viscosity, as well as chemical modifications, off-target binding and *in vivo* stability need to be considered. In general, both parental antibodies need to have favorable properties from a developability perspective as the combination into the BsAb rarely improves liabilities. Thus, during the discovery process and the selection of the two bispecific arms, it is necessary to first characterize both parental antibodies independently and then confirm developability with the intact bispecific molecule.

Expression

BsAbs have been successfully expressed in prokaryotic, yeast, mammalian, and cell-free systems [60, 66, 124, 142]. Ideally, multiple expression platforms should be compared to achieve optimal product quality and yield as each BsAb format has its own unique properties.

Physicochemical Properties

Since many BsAbs divert from their natural protein architecture or use amino acid substitutions to achieve the correct chain pairing, it is recommended to determine the stability of the BsAb [56, 58, 143]. Many proteins with long serum half-life are characterized by high thermostability and decreased thermal stability of proteins can impact serum half-life [144], expression [145], immunogenicity [146], and propensity to aggregate [147]. Commonly used techniques to determine protein stability are differential scanning calorimetry (DSC) and differential scanning fluorimetry (DSF) [147–149]. Historically, DSC is used to determine the thermostability of proteins. For antibody characterization, DSC and DSF lead to comparable melting temperatures [150] since DSF is a high throughput technique with minimal protein requirement, it is particularly attractive to guide the development and formulation of the BsAb [150, 151].

Subcutaneous delivery of biotherapeutics is becoming increasingly common due to its convenience for patients and physicians. Its advantages over intravenous delivery include reduced cost, reduced administration time, and enabled self-administration at home, which is particularly attractive for autoimmune and inflammation indications where multiple doses and long-term administration are required [152]. A successful drug product that uses subcutaneous injection requires low viscosity and high solubility. Viscosity is most accurately determined with a rheometer to ensure that sample processing during purification and delivery by syringe is possible [153, 154]. Commonly used techniques to determine solubility include filtration and precipitation, as well as dialysis, evaporation, and chromatographic methods [155].

For the selection and engineering of highly soluble antibodies, cross-interaction chromatography can also be applied [156]. In this technique, a purified monoclonal or polyclonal IgG is chemically coupled to a chromatography resin. The sample antibody is then passed over the column and its retention time is inversely correlated to its solubility. Thus, antibodies prone to precipitation show delayed retention times and broader peaks. As many BsAb formats deviate from the traditional architecture of a monoclonal antibody or contain linkers, they may be faced with solubility challenges. Viscosity and solubility are properties unique to the BsAb format and cannot be predicted; however, upfront evaluation of different formats can help to mitigate developability issues [157].

Chemical Modifications

Chemical modifications can introduce product variability during storage and impact *in vitro* and *in vivo* binding behavior. These modifications include isomerization, deamidation, succinimide formation of aspartate and asparagine residues, and oxidation of methionine and tryptophan residues. Chemical modifications can lead to reduced binding affinity or lead to a complete loss of binding. "Optional" BsAbs can still have partial function and therapeutic efficacy with these modifications, although this is undesired. However, "obligatory" BsAbs that rely on simultaneously engaging two cells or receptors require both binding partners to be fully functional.

Chemical modifications in antibodies are commonly analyzed after accelerated stress by MS identification and quantification of proteolytic fragments [140, 158]. This makes the chemical modifications time and labor intensive to identify and commonly, analysis is limited to only the antibody CDRs. Most BsAbs have 12 unique CDRs compared to only six in a monoclonal antibody and this introduces additional complexity to the analysis. As chemical modifications tend to translate from the parental CDRs into the BsAb, it is often easier to initially evaluate the parental antibodies. Recently, *in silico* methods have been described that aid the analysis of a larger number of antibodies. Using predicted solvent accessibility and conformational flexibility, the degradation propensity of asparagine and aspartate residues, as well as oxidation of methionine and tryptophan residues in the CDRs can be calculated [159, 160].

To study the impact of the identified chemical modification on the binding behavior, SPR- or cell-based assays can be performed. As demonstrated with an anti-Ang2/anti-VEGF BsAb, SPR-based binding assays could determine the fraction of fully binding competent BsAb [137]. As the signal depends on the simultaneous binding of both antigens, it can reliably determine the active concentration of BsAb after accelerated stress conditions. However, such studies are limited to BsAbs with at least one high-affinity Fab, as this ensures good association of the BsAb to the immobilized antigen and accurate kinetic assessment of the second antigen binding.

Characterization of In Vivo Properties

In addition to having the desired *in vitro* characteristics, the BsAb needs to have suitable *in vivo* properties. Before evaluating the BsAb in pharmacokinetic assays to determine the *in vivo* half-life, it is often helpful to first determine its serum stability at physiological temperatures [136, 161]. This is particularly valuable when evaluating a new bispecific format as it is a fast and easy assay. After the molecule demonstrates suitable serum stability, pharmacokinetic studies are typically performed in rodent and cynomolgus monkeys to determine the *in vivo* half-life.

The antibody Fc-region mediates the long pharmacokinetic half-life of many BsAbs. The immunoglobulin Fc binds to the neonatal Fc receptor, FcRn, providing for extended serum half-life [162]. Hence, bispecifics lacking the Fc region have a short serum half-life as they are not rescued after receptor-mediated endocytosis or pinocytosis. In addition, molecules of small size (below 70 kDa) are subject to rapid renal clearance [163, 164]. One possibility to overcome these limitations is to deliver the therapeutic by continued intravenous infusion. This has been successfully done with the tandem scFv blinatumumab (Blincyto) that has a serum half-life of ~two hours [165]. Alternatively, the half-life can be extended by strategies used for other protein therapeutics [166]. Commonly exploited routes include PEGylation [166] and fusion to human serum albumin or albumin-binding peptides [167–169]. Binding to human IgG can also extend pharmacokinetic half-life as shown for a diabody molecule that is an anti-carcinoembryonic antigen/anti-human IgG bispecific [170].

Similar to monospecific antibodies, off-target binding can account for reduced serum half-life of bispecific molecules. Several *in vitro* assays have been established for risk mitigation. Nonspecific binding to baculovirus particles (baculovirus ELISA [BV-ELISA]) can identify antibodies with an increased risk for fast clearance [171]. This correlation allowed the implementation of a simple, but high-throughput ELISA assay to rapidly evaluate a large number of candidates. Rapid clearance can also be predicted by *in silico* analysis based on the charge profile and hydrophobicity [160]. Protein panel profiling (*3P*), a fully automated, high-throughput immunoassay, test binding to a comprehensive set of proteins [172]. Alternatively, counter-selection against a polyspecificity reagent can be applied during the screening process [173]. These risk-mitigation assays should be performed during lead generation and optimization to identify parental antibodies with minimal off-target binding risk and confirmed with the final BsAb.

A long serum half-life is not always advantageous. For example, in ocular indications, it is desirable to minimize systemic exposure. Intravitreally injected antibody-based therapeutics have been used for the treatment of neovascular (wet) age-related macular degeneration and other neovascular diseases of the retina [174]. These antibodies either contain amino acid substitutions in the FcRn binding site that reduce the half-life [175] or have a format that lacks the Fc altogether, such as a Fab or $F(ab')_2$ [79, 176]. In addition, many alternative scaffolds display a short systemic half-life due to their small size, so they are particularly suitable for these indications as well [177].

Immunogenicity Considerations and Derisking of Bispecific Antibodies

Over the past few years, an increasing number of bispecific molecules have transitioned from the preclinical space into clinical development. Bispecific antibodies typically deviate from a traditional monoclonal antibody, with a more complex structure and the composition of multiple variable domains. This raises the question if BsAbs will have an increased immunogenicity risk compared to a conventional monoclonal antibody, leading to the formation of ADAs. To date, a spectrum of immunogenicity has been observed in clinical studies with examples of high [178–181] and low clinical immunogenicity [182]. Even when an ADA response occurs, the consequences can range from mild to severe. ADAs can be transient or persistent over the treatment and impact exposure, compromise efficacy, and in severe cases lead to anaphylactic reactions [179]. Many factors, such as sequence [183], product quality [184], and therapeutic indications, can influence the immunogenicity of a biologic and they have been extensively reviewed [185]. While these factors also apply to the bispecific antibody, there are also unique considerations to the bispecific molecule: (i) typically an increased foreign sequence content of the CDRs, (ii) non-natural design features of the bispecific format (e.g. linker sequences, substitutions to ensure chain pairing), (iii) biology of targets and mode of action (e.g. T-cell engagers), and (iv) target engagement stoichiometry and immune complex formation between the bispecific antibody and the targets.

Bispecific or multispecific antibodies are typically composed of two or multiple different Fab molecules. Immunogenicity is primed by an initial T-cell epitope found in the presentation of non-self peptides on MHC-class II receptors. The probability of finding such epitopes should be correlated with the amount of foreign sequence content, which is in part contributed by engineered or nonhuman antibody CDRs. Indeed, a positive correlation between the clinical ADA rate and mutations in the antibody sequence was observed [183]. Possibilities to reduce foreign sequence content are the use of a common light chain, dual binding Fabs, or the use of a VHH. To reduce the risk, the variable domain amino acid sequences can be initially assessed by *in silico* methods for potential T-cell epitopes prior to more labor-intensive *in vitro* assays [185]. T-cell proliferation assays [186], dendritic cell uptake assays, and MHC class II-associated peptide proteomics (MAPPS) are commonly used *in vitro* assays to risk-rank molecules or bispecific designs. Importantly, the signal in these *in vitro* assays seems comparable between research grade material and clinical material, allowing representative data to be generated during early discovery stages [187]. Unfortunately, the mode of action of many T-cell-engaging bispecific antibodies can interfere with the assay set-up as the bispecific intrinsically activates T-cells. One solution to this problem is to perform DC:T-cell assays thereby uncoupling the uptake and presentation by dendritic cells and the subsequent stimulation of T-cells. Another limitation of these assays is that they are typically performed outside of the context of antigen. Many bispecific antibodies engage their targets monovalently, however, if the target is a dimeric or multimeric protein, larger immune complexes can form potentially leading to increased uptake and immune response [185]. The broadly used knobs-into-holes Fc heterodimerization substitutions don't seem to increase the overall immunogenicity risk in a suite of *in vitro* assays suggesting that the engineering elements needed to create bispecific antibodies likely do not inherently increase the immunogenicity risk [187]. Non-natural junctions and linker sequences may also contribute to immunogenicity risk and should be carefully selected. For example, the identified linker sequence of a DVD-Ig molecule was represented in the pool of peptides identified in a MAPPS assessment [188]. As immunogenicity is a result of a complex interplay of many factors that are drug intrinsic, as well as patient, indication, dose, and route of administration related, an accurate prediction of clinical immunogenicity is currently not possible. Nonetheless, it seems prudent to minimize risk by applying preclinical risk mitigation strategies.

Conclusions

BsAbs are finally coming of age with 15 approvals and hundreds more in clinical development. A broad number of bispecific formats have evolved accordingly, each with varying challenges in production and characterization. In addition, therapeutic modes of action for BsAbs often require highly pure material at an early research stage, and careful characterization is necessary to interpret experimental results. Historically BsAbs were only characterized by bioanalytical methods, however, mass spectrometry is now playing a leading role in ensuring product quality and purity due to its increased resolution. This has enabled the generation of high-quality BsAbs and thus, reliable readouts from *in vitro*, pharmacokinetic, and pharmacodynamic experiments. With many BsAbs being explored in exciting new applications, the development and refinement of analytical characterization tools will continue to rapidly coevolve.

Abbreviations

BsAbs	bispecific antibodies
MAb	monospecific antibody
NK	natural killer
EpCAM	anti-epithelial cell adhesion molecule
DVD-Ig	dual variable domain immunoglobulin
BiTE	Bispecific T-cell Engager
DAF	Dual Action Fab
CDRs	complementarity determining regions
sdAb	single-domain antibody
VHH	variable heavy antibody fragment
Fc	fragment crystallizable
HTP	high throughput
CE-SDS	capillary electrophoresis-sodium dodecyl sulfate
SDS-PAGE	sodium dodecyl sulfate-polyacrylamide gel electrophoresis
SEC	size-exclusion chromatography
HC	heavy chain
LC	light chain
SEC-HPLC	size-exclusion high-performance liquid chromatography
RP-HPLC	reverse phase high-performance liquid chromatography
CEX	cation-exchange chromatography
HIC	hydrophobic interaction chromatography
MS	mass spectrometry
HPLC	high-performance liquid chromatography
ESI-Q-TOF	electrospray ionization and quadrupole time-of-flight
ELISA	enzyme-linked immunosorbent assay
UHPLC	ultra high-performance chromatography
SPR	surface plasmon resonance
DSC	differential scanning calorimetry
DSF	differential scanning fluorimetry
ADA	anti-drug antibody
MAPPS	MHC class II-associated peptide proteomics
BV-ELISA	baculovirus ELISA

References

1 Labrijn, A.F., Janmaat, M.L., Reichert, J.M., and Parren, P.W.H.I. (2019). Bispecific antibodies: a mechanistic review of the pipeline. *Nat. Rev. Drug Discov.* 18 (8): 585–608.

2 Spiess, C., Zhai, Q., and Carter, P.J. (2015). Alternative molecular formats and therapeutic applications for bispecific antibodies. *Mol. Immunol.* 67 (2): 95–106.

3 Brinkmann, U. and Kontermann, R.E. (2021). Bispecific antibodies. *Science* 372 (6545): 916–917.

4 Marvin, J.S. and Zhu, Z. (2005). Recombinant approaches to IgG-like bispecific antibodies. *Acta Pharmacol. Sin.* 26 (6): 649–658.

5 Weiner, L.M., Clark, J.I., Davey, M. et al. (1995). Phase I trial of 2B1, a bispecific monoclonal antibody targeting c-erbB-2 and Fc gamma RIII. *Cancer Res.* 55 (20): 4586–4593.

6 Satta, A., Mezzanzanica, D., Turatti, F. et al. (2013). Redirection of T-cell effector functions for cancer therapy: bispecific antibodies and chimeric antigen receptors. *Future Oncol.* 9 (4): 527–539.

7 Stockmeyer, B., Dechant, M., van Egmond, M. et al. (2000). Triggering FCα-receptor I (CD89) recruits neutrophils as effector cells for CD20-directed antibody therapy. *J. Immunol.* 165 (10): 5954–5961.

8 Li, B., Xu, L., Pi, C. et al. (2017). CD89-mediated recruitment of macrophages via a bispecific antibody enhances anti-tumor efficacy. *Onco. Targets Ther.* 7 (1): e1380142.

9 Heemskerk, N., Gruijs, M., Temming, A.R. et al. (2021). Augmented antibody-based anti-cancer therapeutics boost neutrophil cytotoxicity. *J. Clin. Invest.* 131 (6).

10 He, H., Lei, F., Huang, L. et al. (2023). Immunotherapy of Epstein-Barr virus (EBV) infection and EBV-associated hematological diseases with gp350/CD89-targeted bispecific antibody. *Biomed. Pharmacother.* 163: 114797.

11 Oates, B.K.J.J. (2013). ImmTACs: novel bi-specific agents for targeted cancer therapy. *Onco. Targets Ther.* 2 (2).

12 Kontermann, R.E. and Brinkmann, U. (2015). Bispecific antibodies. *Drug Discov. Today* 20 (7): 838–847.

13 Weisser, N.E., Sanches, M., Escobar-Cabrera, E. et al. (2023). An anti-HER2 biparatopic antibody that induces unique HER2 clustering and complement-dependent cytotoxicity. *Nat. Commun.* 14 (1): 1394.

14 Kim, H.S., Dunshee, D.R., Yee, A. et al. (2017). Tethered-variable CL bispecific IgG: an antibody platform for rapid bispecific antibody screening. *Protein Eng. Des. Sel.* 30 (9): 627–637.

15 Lewis, S.M., Wu, X., Pustilnik, A. et al. (2014). Generation of bispecific IgG antibodies by structure-based design of an orthogonal Fab interface. *Nat. Biotechnol.* 32 (2): 191–198.

16 Boersma, Y.L., Chao, G., Steiner, D. et al. (2011). Bispecific designed ankyrin repeat proteins (DARPins) targeting epidermal growth factor receptor inhibit A431 cell proliferation and receptor recycling. *J. Biol. Chem.* 286 (48): 41273–41285.

17 Marei, H., Tsai, W.-T.K., Kee, Y.-S. et al. (2022). Antibody targeting of E3 ubiquitin ligases for receptor degradation. *Nature* 610 (7930): 182–189.

18 Kitazawa, T., Igawa, T., Sampei, Z. et al. (2012). A bispecific antibody to factors IXa and X restores factor VIII hemostatic activity in a hemophilia A model. *Nat. Med.* 18 (10): 1570–1574.

19 Kolumam, G., Chen, M.Z., Tong, R. et al. (2015). Sustained brown fat stimulation and insulin sensitization by a humanized bispecific antibody agonist for fibroblast growth factor receptor 1/βKlotho complex. *EBioMedicine* 2 (7): 730–743.

20 Baruch, A., Wong, C., Chinn, L.W. et al. (2020). Antibody-mediated activation of the FGFR1/Klothoβ complex corrects metabolic dysfunction and alters food preference in obese humans. *Proc. Natl. Acad. Sci.* 117 (46): 28992–29000.

21 Rybchenko, V.S., Aliev, T.K., Panina, A.A. et al. (2023). Targeted cytokine delivery for cancer treatment: engineering and biological effects. *Pharmaceutics* 15 (2): 336.

22 Yu, Y.J., Zhang, Y., Kenrick, M. et al. (2011). Boosting brain uptake of a therapeutic antibody by reducing its affinity for a transcytosis target. *Sci. Transl. Med.* 3 (84): 84ra44.

23 Niewoehner, J., Bohrmann, B., Collin, L. et al. (2014). Increased brain penetration and potency of a therapeutic antibody using a monovalent molecular shuttle. *Neuron* 81 (1): 49–60.

24 Grimm, H.P., Schumacher, V., Schäfer, M. et al. (2023). Delivery of the Brainshuttle™ amyloid-beta antibody fusion trontinemab to non-human primate brain and projected efficacious dose regimens in humans. *MAbs* 15 (1): 2261509.

25 Farrington, G.K., Caram-Salas, N., Haqqani, A.S. et al. (2014). A novel platform for engineering blood-brain barrier-crossing bispecific biologics. *FASEB J*: official publication of the Federation of American Societies for Experimental Biology 28 (11): 4764–4778.

26 Bhatnagar, P.K., Das, D., and Suresh, M.R. (2008). Sequential affinity purification of peroxidase tagged

bispecific anti-SARS-CoV antibodies on phenylboronic acid agarose. *J. Chromatogr. B Analyt. Technol. Biomed. Life Sci.* 863 (2): 235–241.
27 Sharkey, R.M., Cardillo, T.M., Rossi, E.A. et al. (2005). Signal amplification in molecular imaging by pretargeting a multivalent, bispecific antibody. *Nat. Med.* 11 (11): 1250–1255.
28 Brinkmann, U. and Kontermann, R.E. (2017). The making of bispecific antibodies. *MAbs* 9 (2): 182–212.
29 Kaushik, A., Schulze, D.H., Bonilla, F.A. et al. (1990). Stochastic pairing of heavy-chain and kappa light-chain variable gene families occurs in polyclonally activated B cells. *Proc. Natl. Acad. Sci. U S A* 87 (13): 4932–4936.
30 Suresh, M.R., Cuello, A.C., and Milstein, C. (1986). Bispecific monoclonal antibodies from hybrid hybridomas. *Methods Enzymol.* 121: 210–228.
31 Schaefer, W., Völger, H.R., Lorenz, S. et al. (2016). Heavy and light chain pairing of bivalent quadroma and knobs-into-holes antibodies analyzed by UHR-ESI-QTOF mass spectrometry. *MAbs* 8 (1): 49–55.
32 Suresh, M.R., Cuello, A.C., and Milstein, C. (1986). Advantages of bispecific hybridomas in one-step immunocytochemistry and immunoassays. *Proc. Natl. Acad. Sci. USA* 83 (20): 7989–7993.
33 Lindhofer, H., Mocikat, R., Steipe, B., and Thierfelder, S. (1995). Preferential species-restricted heavy/light chain pairing in rat/mouse quadromas. Implications for a single-step purification of bispecific antibodies. *J. Immunol. (Baltimore, Md: 1950)* 155 (1): 219–225.
34 Chelius, D., Ruf, P., Gruber, P. et al. (2010). Structural and functional characterization of the trifunctional antibody catumaxomab. *MAbs* 2 (3): 309–319.
35 Smith, E.J., Olson, K., Haber, L.J. et al. (2015). A novel, native-format bispecific antibody triggering T-cell killing of B-cells is robustly active in mouse tumor models and cynomolgus monkeys. *Sci. Rep.* 5: 17943.
36 Fischer, N., Elson, G., Magistrelli, G. et al. (2015). Exploiting light chains for the scalable generation and platform purification of native human bispecific IgG. *Nat. Commun.* 6: 6113.
37 Kakiuchi-Kiyota, S., Ross, T., Wallweber, H.A. et al. (2022). A BCMA/CD16A bispecific innate cell engager for the treatment of multiple myeloma. *Leukemia* 36 (4): 1006–1014.
38 Holliger, P., Prospero, T., and Winter, G. (1993). "Diabodies": small bivalent and bispecific antibody fragments. *P. Natl. Acad. Sci. USA* 90 (14): 6444–6448.
39 Frenken, L.G.J., van der Linden, R.H.J., Hermans, P.W.J.J. et al. (2000). Isolation of antigen specific Llama VHH antibody fragments and their high level secretion by Saccharomyces cerevisiae. *J. Biotechnol.* 78 (1): 11–21.
40 Sriraman, S.K., Davies, C.W., Gill, H. et al. (2023). Development of an 18F-labeled anti-human CD8 VHH for same-day immunoPET imaging. *Eur. J. Nucl. Med. Mol. Imaging* 50 (3): 679–691.
41 Eyssen, L.E.-A., Ramadurai, S., Abdelkarim, S. et al. (2024). From Llama to nanobody: a streamlined workflow for the generation of functionalised VHHs. *Bio-Protoc.* 14 (6): e4962.
42 Wang, M., Wei, L., Xiang, H. et al. (2022). A megadiverse naïve library derived from numerous camelids for efficient and rapid development of VHH antibodies. *Anal. Biochem.* 657: 114871.
43 Sevy, A.M., Chen, M.-T., Castor, M. et al. (2020). *Structure-* and sequence-based design of synthetic single-domain antibody libraries. *Protein Eng. Des. Sel.* 33: gzaa028.
44 Keam, S.J. (2023). Ozoralizumab: first approval. *Drugs* 83 (1): 87–92.
45 Madsen, A.V., Kristensen, P., and Goletz, S. (2024). IgG-VHH bispecific fusion antibodies: challenges and opportunities as therapeutic agents. *Expert. Opin. Biol. Ther.* 24 (4): 203–206.
46 Johansson, M.U., Weinert, C., Reichardt, D.A. et al. (2023). Design of antibody variable fragments with reduced reactivity to preexisting anti-drug antibodies. *MAbs* 15 (1): 2215887.
47 Löfblom, J., Feldwisch, J., Tolmachev, V. et al. (2010). Affibody molecules: engineered proteins for therapeutic, diagnostic and biotechnological applications. *FEBS Lett.* 584 (12): 2670–2680.
48 Stumpp, M.T. and Amstutz, P. (2007). DARPins: a true alternative to antibodies. *Curr. Opin. Drug Discov. Devel.* 10 (2): 153–159.
49 Jakob, C.G., Edalji, R., Judge, R.A. et al. (2013). Structure reveals function of the dual variable domain immunoglobulin (DVD-IgTM) molecule. *mAbs* 5 (3): 358–363.
50 Whitlow, M., Bell, B.A., Feng, S.L. et al. (1993). An improved linker for single-chain Fv with reduced aggregation and enhanced proteolytic stability. *Protein Eng.* 6 (8): 989–995.
51 Ridgway, J.B.B., Presta, L.G., and Carter, P. (1996). 'Knobs-into-holes' engineering of antibody CH3 domains for heavy chain heterodimerization. *Protein Eng. Des. Sel.* 9 (7): 617–621.
52 Atwell, S., Ridgway, J.B.B., Wells, J.A., and Carter, P. (1997). Stable heterodimers from remodeling the domain interface of a homodimer using a phage

display library11Edited by P.E. Wright. *J. Mol. Biol.* 270 (1): 26–35.

53 Nardis, C.D., Hendriks, L.J.A., Poirier, E. et al. (2017). A new approach for generating bispecific antibodies based on a common light chain format and the stable architecture of human immunoglobulin G1. *J. Biol. Chem.* 292 (35): 14706–14717.

54 Davis, J.H., Aperlo, C., Li, Y. et al. (2010). SEEDbodies: fusion proteins based on strand-exchange engineered domain (SEED) CH3 heterodimers in an Fc analogue platform for asymmetric binders or immunofusions and bispecific antibodies. *Protein Eng. Des Sel.*: PEDS 23 (4): 195–202.

55 Gunasekaran, K., Pentony, M., Shen, M. et al. (2010). Enhancing antibody Fc heterodimer formation through electrostatic steering effects: applications to bispecific molecules and monovalent IgG. *J. Biol. Chem.* 285 (25): 19637–19646.

56 Strop, P., Ho, W.-H., Boustany, L.M. et al. (2012). Generating Bispecific Human IgG1 and IgG2 Antibodies from Any Antibody Pair. *J. Mol. Biol.* 420 (3): 204–219.

57 Labrijn, A.F., Meesters, J.I., de Goeij, B.E.C.G. et al. (2013). Efficient generation of stable bispecific IgG1 by controlled Fab-arm exchange. *Proc. Natl. Acad. Sci. USA* 110 (13): 5145–5150.

58 Kreudenstein, T.S.V., Escobar-Carbrera, E., Lario, P.I. et al. (2013). Improving biophysical properties of a bispecific antibody scaffold to aid developability: quality by molecular design. *MAbs* 5 (5): 646–654.

59 Spiess, C., Bevers, J., Jackman, J. et al. (2013). Development of a human IgG4 bispecific antibody for dual targeting of interleukin-4 (IL-4) and interleukin-13 (IL-13) cytokines*. *J. Biol. Chem.* 288 (37): 26583–26593.

60 Merchant, A.M., Zhu, Z., Yuan, J.Q. et al. (1998). An efficient route to human bispecific IgG. *Nat. Biotechnol.* 16 (7): 677–681.

61 Dhimolea, E. and Reichert, J.M. (2012). World Bispecific Antibody Summit, September 27–28, 2011, Boston, MA. *mAbs* 4 (1): 4–13.

62 de Kruif, J., Kramer, A., Visser, T. et al. (2009). Human immunoglobulin repertoires against tetanus toxoid contain a large and diverse fraction of high-affinity promiscuous VH genes. *J. Mol. Biol.* 387 (3): 548–558.

63 Murphy, A.J., Macdonald, L.E., Stevens, S. et al. (2014). Mice with megabase humanization of their immunoglobulin genes generate antibodies as efficiently as normal mice. *Proc. Natl. Acad. Sci.* 111 (14): 5153–5158.

64 Spiess, C., Merchant, M., Huang, A. et al. (2013). Bispecific antibodies with natural architecture produced by co-culture of bacteria expressing two distinct half-antibodies. *Nat. Biotechnol.* 31 (8): 753–758.

65 Labrijn, A.F., Meesters, J.I., Priem, P. et al. (2014). Controlled Fab-arm exchange for the generation of stable bispecific IgG1. *Nat. Protoc.* 9 (10): 2450–2463.

66 Jackman, J., Chen, Y., Huang, A. et al. (2010). Development of a two-part strategy to identify a therapeutic human bispecific antibody that inhibits IgE receptor signaling. *J. Biol. Chem.* 285 (27): 20850–20859.

67 Schaefer, W., Regula, J.T., Bahner, M. et al. (2011). Immunoglobulin domain crossover as a generic approach for the production of bispecific IgG antibodies. *Proc. National Acad. Sci.* 108 (27): 11187–11192.

68 Surowka, M., Schaefer, W., and Klein, C. (2021). Ten years in the making: application of CrossMab technology for the development of therapeutic bispecific antibodies and antibody fusion proteins. *MAbs* 13 (1): 1967714.

69 Liu, Z., Leng, E.C., Gunasekaran, K. et al. (2015). A novel antibody engineering strategy for making monovalent bispecific heterodimeric IgG antibodies by electrostatic steering mechanism. *J. Biol. Chem.* 290 (12): 7535–7562.

70 Mazor, Y., Oganesyan, V., Yang, C. et al. (2015). Improving target cell specificity using a novel monovalent bispecific IgG design. *MAbs* 7 (2): 377–389.

71 Dillon, M., Yin, Y., Zhou, J. et al. (2017). Efficient production of bispecific IgG of different isotypes and species of origin in single mammalian cells. *MAbs* 9 (2): 213–230.

72 Bagert, J.D., Oganesyan, V., Chiang, C.-I. et al. (2023). Robust production of monovalent bispecific IgG antibodies through novel electrostatic steering mutations at the CH1-Cλ interface. *MAbs* 15 (1): 2273449.

73 Bostrom, J., Yu, S.-F., Kan, D. et al. (2009). Variants of the antibody herceptin that interact with HER2 and VEGF at the antigen binding site. *Science* 323 (5921): 1610–1614.

74 Schaefer, G., Haber, L., Crocker, L.M. et al. (2011). A two-in-one antibody against HER3 and EGFR has superior inhibitory activity compared with monospecific antibodies. *Cancer Cell* 20 (4): 472–486.

75 Beckmann, R., Jensen, K., Fenn, S. et al. (2021). DutaFabs are engineered therapeutic Fab fragments that can bind two targets simultaneously. *Nat. Commun.* 12 (1): 708.

76 Nisonoff, A. and Rivers, M.M. (1961). Recombination of a mixture of univalent antibody fragments of different specificity. *Arch. Biochem. Biophys.* 93: 460–462.

77 Mandy, W.J., Rivers, M.M., and Nisonoff, A. (1961). Recombination of univalent subunits derived from rabbit antibody. *J. Biol. Chem.* 236: 3221–3226.

78 Cao, Y., Axup, J.Y., Ma, J.S.Y. et al. (2015). Multiformat T-cell-engaging bispecific antibodies targeting human breast cancers. *Angew. Chem. Int. Ed. Engl.* 54 (24): 7022–7027.

79 Gadkar, K., Pastuskovas, C.V., Couter, J.E.L. et al. (2015). Design and pharmacokinetic characterization of novel antibody formats for ocular therapeutics. *Invest. Opthalmol. Vis. Sci.* 56 (9): 5390.

80 Albrecht, H., Denardo, G.L., and Denardo, S.J. (2006). Monospecific bivalent scFv-SH: effects of linker length and location of an engineered cysteine on production, antigen binding activity and free SH accessibility. *J. Immunol. Methods* 310 (1–2): 100–116.

81 Sen, M., Wankowski, D.M., Garlie, N.K. et al. (2001). Use of anti-CD3 x anti-HER2/neu bispecific antibody for redirecting cytotoxicity of activated T cells toward HER2/neu+ tumors. *J. Hematother. Stem Cell Res.* 10 (2): 247–260.

82 Davol, P.A., Smith, J.A., Kouttab, N. et al. (2004). Anti-CD3 x anti-HER2 bispecific antibody effectively redirects armed T cells to inhibit tumor development and growth in hormone-refractory prostate cancer-bearing severe combined immunodeficient beige mice. *Clin. Prostate Cancer* 3 (2): 112–121.

83 Reusch, U., Sundaram, M., Davol, P.A. et al. (2006). Anti-CD3 × anti-epidermal growth factor receptor (EGFR) bispecific antibody Redirects T-Cell Cytolytic Activity to EGFR-Positive Cancers In vitro and in an Animal Model. *Clin. Cancer Res.* 12 (1): 183–190.

84 Yankelevich, M., Kondadasula, S.V., Thakur, A. et al. (2012). Anti-CD3 × anti-GD2 bispecific antibody redirects T-cell cytolytic activity to neuroblastoma targets. *Pediatr. Blood Cancer* 59 (7): 1198–1205.

85 Lum, L.G., Ramesh, M., Thakur, A. et al. (2012). Targeting cytomegalovirus-infected cells using T cells armed with anti-CD3 × anti-CMV bispecific antibody. *Biol. Blood Marrow Transplant.* 18 (7): 1012–1022.

86 Ellerman, D., and Scheer, J.M. (2011). Generation of Bispecific Antibodies by Chemical Conjugation. Springer Berlin Heidelberg. 47–63. doi:10.1007/978-3-642-20910-9_3.

87 Maruani, A., Szijj, P.A., Bahou, C. et al. (2020). A plug-and-play approach for the De Novo generation of dually functionalized bispecifics. *Bioconjug. Chem.* 31 (3): 520–529.

88 Silver, A.B., Leonard, E.K., Gould, J.R., and Spangler, J.B. (2021). Engineered antibody fusion proteins for targeted disease therapy. *Trends Pharmacol. Sci.* 42 (12): 1064–1081.

89 Holder, P.G., Lim, S.A., Huang, C.S. et al. (2022). Engineering interferons and interleukins for cancer immunotherapy. *Adv. Drug Deliv. Rev.* 182: 114112.

90 Fu, Y., Tang, R., and Zhao, X. (2023). Engineering cytokines for cancer immunotherapy: a systematic review. *Front. Immunol.* 14: 1218082.

91 Garcin, G., Paul, F., Staufenbiel, M. et al. (2014). High efficiency cell-specific targeting of cytokine activity. *Nat. Commun.* 5 (1): 3016.

92 Bos, A.B., Luan, P., Duque, J.N. et al. (2015). Optimization and automation of an end-to-end high throughput microscale transient protein production process. *Biotechnol. Bioeng.* 112 (9): 1832–1842.

93 Dickopf, S., Georges, G.J., and Brinkmann, U. (2020). Format and geometries matter: structure-based design defines the functionality of bispecific antibodies. *Comput. Struct. Biotechnol. J.* 18: 1221–1227.

94 Dengl, S., Mayer, K., Bormann, F. et al. (2020). Format chain exchange (FORCE) for high-throughput generation of bispecific antibodies in combinatorial binder-format matrices. *Nat. Commun.* 11 (1): 4974.

95 Sampei, Z., Igawa, T., Soeda, T. et al. (2013). Identification and multidimensional optimization of an asymmetric bispecific IgG antibody mimicking the function of factor VIII cofactor activity. *PLoS One* 8 (2): e57479.

96 Voigt, J., Meyer, C., and Bordusa, F. (2022). Synthesis of multiple bispecific antibody formats with only one single enzyme based on enhanced trypsiligase †. *Int. J. Mol. Sci.* 23 (6): 3144.

97 Bhatta, P., Whale, K.D., Sawtell, A.K. et al. (2021). Bispecific antibody target pair discovery by high-throughput phenotypic screening using in vitro combinatorial Fab libraries. *MAbs* 13 (1): 1859049.

98 Andres, F., Schwill, M., Boersma, Y.L., and Plückthun, A. (2020). High-throughput generation of bispecific binding proteins by sortase A–mediated coupling for direct functional screening in cell culture. *Mol. Cancer Ther.* 19 (4): 1080–1088.

99 Driscoll, C.L., Keeble, A.H., and Howarth, M.R. (2024). SpyMask enables combinatorial assembly of bispecific binders. *Nat. Commun.* 15 (1): 2403.

100 Sutherland, A.R., Alam, M.K., and Geyer, C.R. (2019). Post-translational assembly of protein parts into complex devices by using SpyTag/SpyCatcher protein ligase. *Chembiochem* 20 (3): 319–328.

101 Hentrich, C., Putyrski, M., Hanuschka, H. et al. (2024). Engineered reversible inhibition of SpyCatcher reactivity enables rapid generation of bispecific antibodies. *Nat. Commun.* 15 (1): 5939.

102 Hofmann, T., Schmidt, J., Ciesielski, E. et al. (2020). Intein mediated high throughput screening for bispecific antibodies. *MAbs* 12 (1): 1731938.

103 Robak, T. (2013). The emerging therapeutic role of antibody mixtures. *Expert Opin. Biol. Th.* 13 (7): 953–958.

104 Qiu, X., Wong, G., Audet, J. et al. (2014). Reversion of advanced Ebola virus disease in nonhuman primates with ZMapp. *Nature* 514 (7520): 47–53.

105 Qiu, X., Audet, J., Wong, G. et al. (2013). Sustained protection against Ebola virus infection following treatment of infected nonhuman primates with ZMAb. *Sci. Rep.* 3: 3365.

106 Koefoed, K., Steinaa, L., Søderberg, J.N. et al. (2011). Rational identification of an optimal antibody mixture for targeting the epidermal growth factor receptor. *MAbs* 3 (6): 584–595.

107 Dienstmann, R., Patnaik, A., Garcia-Carbonero, R. et al. (2015). Safety and activity of the first-in-class Sym004 anti-EGFR antibody mixture in patients with refractory colorectal cancer. *Cancer Discov.* 5 (6): 598–609.

108 Tolstrup, A.B., Frandsen, T.P., and Bregenholt, S. (2006). Development of recombinant human polyclonal antibodies for the treatment of complex human diseases. *Expert. Opin. Biol. Ther.* 6 (9): 905–912.

109 Nowakowski, A., Wang, C., Powers, D.B. et al. (2002). Potent neutralization of botulinum neurotoxin by recombinant oligoclonal antibody. *Proc. Natl. Acad. Sci.* 99 (17): 11346–11350.

110 Nayak, S.U., Griffiss, J.M., McKenzie, R. et al. (2014). Safety and pharmacokinetics of XOMA 3AB, a novel mixture of three monoclonal antibodies against botulinum toxin A. *Antimicrob. Agents Chemother.* 58 (9): 5047–5053.

111 Weinreich, D.M., Sivapalasingam, S., Norton, T. et al. (2021). REGEN-COV antibody combination and outcomes in outpatients with Covid-19. *N. Engl. J. Med.* 385 (23): e81.

112 Riethmüller, G. (2012). Symmetry breaking: bispecific antibodies, the beginnings, and 50 years on. *Cancer Immun.* 12: 12.

113 Couch, J.A., Yu, Y.J., Zhang, Y. et al. (2013). Addressing safety liabilities of TfR bispecific antibodies that cross the blood-brain barrier. *Sci. Transl. Med.* 5 (183): 183ra57-1-12.

114 Wu, C., Ying, H., Bose, S. et al. (2009). Molecular construction and optimization of anti-human IL-1alpha/beta dual variable domain immunoglobulin (DVD-Ig) molecules. *MAbs* 1 (4): 339–347.

115 Williams, A.J., Giese, G., and Persson, J. (2015). Improved assembly of bispecific antibodies from knob and hole half-antibodies. *Biotechnol. Prog.* 31 (5): 1315–1322.

116 Rosati, S., Thompson, N.J., Barendregt, A. et al. (2012). Qualitative and semiquantitative analysis of composite mixtures of antibodies by native mass spectrometry. *Anal. Chem.* 84 (16): 7227–7232.

117 Yin, Y., Han, G., Zhou, J. et al. (2016). Precise quantification of mixtures of bispecific IgG produced in single host cells by liquid chromatography-Orbitrap high-resolution mass spectrometry. *MAbs* 8 (8).

118 Sazinsky, S.L., Ott, R.G., Silver, N.W. et al. (2008). Aglycosylated immunoglobulin G1 variants productively engage activating Fc receptors. *Proc. Natl. Acad. Sci. USA* 105 (51): 20167–20172.

119 Liu, H., Gaza-Bulseco, G., Faldu, D. et al. (2008). Heterogeneity of monoclonal antibodies. *J. Pharm. Sci.* 97 (7): 2426–2447.

120 Dick, L.W., Qiu, D., Mahon, D. et al. (2008). C-terminal lysine variants in fully human monoclonal antibodies: investigation of test methods and possible causes. *Biotechnol. Bioeng.* 100 (6): 1132–1143.

121 Harris, R.J. (1995). Processing of C-terminal lysine and arginine residues of proteins isolated from mammalian cell culture. *J. Chromatogr. A* 705 (1): 129–134.

122 Wranik, B.J., Christensen, E.L., Schaefer, G. et al. (2012). LUZ-Y, a novel platform for the mammalian cell production of full-length IgG-bispecific antibodies. *J. Biol. Chem.* 287 (52): 43331–43339.

123 Woods, R.J., Xie, M., Kreudenstein, T.V. et al. (2013). LC-MS characterization and purity assessment of a prototype bispecific antibody. *mAbs* 5 (5): 711–722.

124 Xu, Y., Lee, J., Tran, C. et al. (2015). Production of bispecific antibodies in "knobs-into-holes" using a cell-free expression system. *MAbs* 7 (1): 231–242.

125 Prentice, K.M., Wallace, A., and Eakin, C.M. (2015). Inline protein A mass spectrometry for characterization of monoclonal antibodies. *Anal. Chem.* 87 (4): 2023–2028.

126 Sawyer, W.S., Srikumar, N., Carver, J. et al. (2020). High-throughput antibody screening from complex matrices using intact protein electrospray mass spectrometry. *Proc. National Acad. Sci.* 117 (18): 9851–9856.

127 Rose, R.J., Labrijn, A.F., van den Bremer, E.T.J. et al. (2011). Quantitative analysis of the interaction strength and dynamics of human IgG4 half molecules by native mass spectrometry. *Structure* 19 (9): 1274–1282.

128 van der Kolfschoten, M.N., Schuurman, J., Losen, M. et al. (2007). Anti-inflammatory activity of human IgG4 antibodies by dynamic Fab arm exchange. *Science* 317 (5844): 1554–1557.

129 Labrijn, A.F., Buijsse, A.O., van den Bremer, E.T.J. et al. (2009). Therapeutic IgG4 antibodies engage in

Fab-arm exchange with endogenous human IgG4 in vivo. *Nat. Biotechnol.* 27 (8): 767–771.

130 Oda, M., Uchiyama, S., Noda, M. et al. (2009). Effects of antibody affinity and antigen valence on molecular forms of immune complexes. *Mol. Immunol.* 47 (2–3): 357–364.

131 Oda, M., Uchiyama, S., Robinson, C.V. et al. (2006). Regional and segmental flexibility of antibodies in interaction with antigens of different size. *FEBS J.* 273 (7): 1476–1487.

132 Liu, H., Jeong, J., Kao, Y.-H., and Zhang, Y.T. (2015). Characterization of free thiol variants of an IgG1 by reversed phase ultra high pressure liquid chromatography coupled with mass spectrometry. *J. Pharm. Biomed. Anal.* 109: 142–149.

133 Rosati, S., Rose, R.J., Thompson, N.J. et al. (2012). Exploring an orbitrap analyzer for the characterization of intact antibodies by native mass spectrometry. *Angew. Chem. Int. Ed.* 51 (52): 12992–12996.

134 Rose, R.J., Damoc, E., Denisov, E. et al. (2012). High-sensitivity orbitrap mass analysis of intact macromolecular assemblies. *Nat. Methods* 9 (11): 1084–1086.

135 Thompson, N.J., Hendriks, L.J., de Kruif, J. et al. (2014). Complex mixtures of antibodies generated from a single production qualitatively and quantitatively evaluated by native orbitrap mass spectrometry. *MAbs* 6 (1): 197–203.

136 Johnson, S., Burke, S., Huang, L. et al. (2010). Effector cell recruitment with novel Fv-based dual-affinity re-targeting protein leads to potent tumor cytolysis and in vivo B-cell depletion. *J. Mol. Biol.* 399 (3): 436–449.

137 Gassner, C., Lipsmeier, F., Metzger, P. et al. (2015). Development and validation of a novel SPR-based assay principle for bispecific molecules. *J. Pharm. Biomed. Anal.* 102: 144–149.

138 Gall, F.L., Reusch, U., Little, M., and Kipriyanov, S.M. (2004). Effect of linker sequences between the antibody variable domains on the formation, stability and biological activity of a bispecific tandem diabody. *Protein Eng. Des. Sel.*: PEDS 17 (4): 357–366.

139 Schanzer, J.M., Wartha, K., Croasdale, R. et al. (2014). A novel glycoengineered bispecific antibody format for targeted inhibition of epidermal growth factor receptor (EGFR) and insulin-like growth factor receptor type I (IGF-1R) demonstrating unique molecular properties. *J. Biol. Chem.* 289 (27): 18693–18706.

140 Beck, A., Wagner-Rousset, E., Ayoub, D. et al. (2013). Characterization of therapeutic antibodies and related products. *Anal. Chem.* 85 (2): 715–736.

141 Jarasch, A., Koll, H., Regula, J.T. et al. (2015). Developability assessment during the selection of novel therapeutic antibodies. *J. Pharm. Sci.* 104 (6): 1885–1898.

142 Braren, I., Greunke, K., Umland, O. et al. (2007). Comparative expression of different antibody formats in mammalian cells and Pichia pastoris. *Biotechnol. Appl. Biochem.* 47 (4): 205.

143 Moore, G.L., Bautista, C., Pong, E. et al. (2011). A novel bispecific antibody format enables simultaneous bivalent and monovalent co-engagement of distinct target antigens. *MAbs* 3 (6): 546–557.

144 Sharma, B. (2007). Immunogenicity of therapeutic proteins. Part 1: impact of product handling. *Biotechnol. Adv.* 25 (3): 310–317.

145 Schaefer, J.V. and Plückthun, A. (2012). Transfer of engineered biophysical properties between different antibody formats and expression systems. *Protein Eng. Des. Sel.* 25 (10): 485–506.

146 Ohkuri, T., Nagatomo, S., Oda, K. et al. (2010). A protein's conformational stability is an immunologically dominant factor: evidence that free-energy barriers for protein unfolding limit the immunogenicity of foreign proteins. *J. Immunol.* 185 (7): 4199–4205.

147 Cheng, W., Joshi, S.B., He, F. et al. (2012). Comparison of high-throughput biophysical methods to identify stabilizing excipients for a model IgG2 monoclonal antibody: conformational stability and kinetic aggregation measurements. *J. Pharm. Sci.* 101 (5): 1701–1720.

148 Niesen, F.H., Berglund, H., and Vedadi, M. (2007). The use of differential scanning fluorimetry to detect ligand interactions that promote protein stability. *Nat. Protoc.* 2 (9): 2212–2221.

149 Chaudhuri, R., Cheng, Y., Middaugh, C.R., and Volkin, D.B. (2014). High-throughput biophysical analysis of protein therapeutics to examine interrelationships between aggregate formation and conformational stability. *AAPS J.* 16 (1): 48–64.

150 Goldberg, D.S., Bishop, S.M., Shah, A.U., and Sathish, H.A. (2010). Formulation development of therapeutic monoclonal antibodies using high-throughput fluorescence and static light scattering techniques: role of conformational and colloidal stability. *J. Pharm. Sci.* 100 (4): 1306–1315.

151 Menzen, T. and Friess, W. (2013). High-throughput melting-temperature analysis of a monoclonal antibody by differential scanning fluorimetry in the presence of surfactants. *J. Pharm. Sci.* 102 (2): 415–428.

152 Turner, M.R. and Balu-Iyer, S.V. (2018). Challenges and opportunities for the subcutaneous delivery of therapeutic proteins. *J. Pharm. Sci.* 107 (5): 1247–1260.

153 Connolly, B.D., Petry, C., Yadav, S. et al. (2012). Weak interactions govern the viscosity of concentrated antibody solutions: high-throughput analysis using the diffusion interaction parameter. *Biophys. J.* 103 (1): 69–78.

154 Jezek, J., Darton, N.J., Derham, B.K. et al. (2013). Biopharmaceutical formulations for pre-filled delivery devices. *Expert Opin. Drug Deliv.* 10 (6): 811–828.

155 Shire, S.J., Shahrokh, Z., and Liu, J. (2004). Challenges in the development of high protein concentration formulations. *J. Pharm. Sci.* 93 (6): 1390–1402.

156 Jacobs, S.A., Wu, S.-J., Feng, Y. et al. (2010). Cross-interaction chromatography: a rapid method to identify highly soluble monoclonal antibody candidates. *Pharm. Res.* 27 (1): 65–71.

157 Durben, M., Schmiedel, D., Hofmann, M. et al. (2015). Characterization of a bispecific FLT3 X CD3 antibody in an improved, recombinant format for the treatment of leukemia. *Mol. Ther.* 23 (4): 648–655.

158 Haberger, M., Bomans, K., Diepold, K. et al. (2014). Assessment of chemical modifications of sites in the CDRs of recombinant antibodies: susceptibility versus functionality of critical quality attributes. *MAbs* 6 (2): 327–339.

159 Sydow, J.F., Lipsmeier, F., Larraillet, V. et al. (2014). Structure-based prediction of asparagine and aspartate degradation sites in antibody variable regions. *PLoS One* 9 (6): e100736.

160 Sharma, V.K., Patapoff, T.W., Kabakoff, B. et al. (2014). In silico selection of therapeutic antibodies for development: viscosity, clearance, and chemical stability. *Proc. Natl. Acad. Sci.* 111 (52): 18601–18606.

161 Silverman, J., Liu, Q., Lu, Q. et al. (2005). Multivalent avimer proteins evolved by exon shuffling of a family of human receptor domains. *Nat. Biotechnol.* 23 (12): 1556–1561.

162 Roopenian, D.C. and Akilesh, S. (2007). FcRn: the neonatal Fc receptor comes of age. *Nat. Rev. Immunol.* 7 (9): 715–725.

163 Knauf, M.J., Bell, D.P., Hirtzer, P. et al. (1988). Relationship of effective molecular size to systemic clearance in rats of recombinant interleukin-2 chemically modified with water-soluble polymers. *J. Biol. Chem.* 263 (29): 15064–15070.

164 Tryggvason, K. and Wartiovaara, J. (2005). How does the kidney filter plasma? *Physiology* (Bethesda, Md.) 20: 96–101.

165 Oak, E. and Bartlett, N.L. (2015). Blinatumomab for the treatment of B-cell lymphoma. *Expert Opin. Investig. Drugs* 24 (5).

166 Kontermann, R.E. (2011). Strategies for extended serum half-life of protein therapeutics. *Curr. Opin. Biotechnol.* 22 (6): 868–876.

167 Stork, R., Müller, D., and Kontermann, R.E. (2007). A novel tri-functional antibody fusion protein with improved pharmacokinetic properties generated by fusing a bispecific single-chain diabody with an albumin-binding domain from streptococcal protein G. *Protein Eng Des. Sel.*: PEDS 20 (11): 569–576.

168 Müller, D., Karle, A., Meissburger, B. et al. (2007). Improved pharmacokinetics of recombinant bispecific antibody molecules by fusion to human serum albumin. *J. Biol. Chem.* 282 (17): 12650–12660.

169 Tijink, B.M., Laeremans, T., Budde, M. et al. (2008). Improved tumor targeting of anti-epidermal growth factor receptor nanobodies through albumin binding: taking advantage of modular nanobody technology. *Mol. Cancer Ther.* 7 (8): 2288–2297.

170 Holliger, P., Wing, M., Pound, J.D. et al. (1997). Retargeting serum immunoglobulin with bispecific diabodies. *Nat. Biotechnol.* 15 (7): 632–636.

171 Hötzel, I., Theil, F.-P., Bernstein, L.J. et al. (2012). A strategy for risk mitigation of antibodies with fast clearance. *MAbs* 4 (6): 753–760.

172 Frese, K., Eisenmann, M., Ostendorp, R. et al. (2013). An automated immunoassay for early specificity profiling of antibodies. *MAbs* 5 (2): 279–287.

173 Xu, Y., Roach, W., Sun, T. et al. (2013). Addressing polyspecificity of antibodies selected from an in vitro yeast presentation system: a FACS-based, high-throughput selection and analytical tool. *Protein Eng. Des. Sel.* 26 (10): 663–670.

174 Volz, C. and Pauly, D. (2015). Antibody therapies and their challenges in the treatment of age-related macular degeneration. *Eur. J. Pharm. Biopharm.* 95 (Pt B): 158–172.

175 Olafsen, T. (2012). Fc engineering: serum half-life modulation through FcRn binding. *Methods Mol. Biol.* (Clifton, NJ) 907: 537–556.

176 Avery, R.L., Castellarin, A.A., Steinle, N.C. et al. (2014). Systemic pharmacokinetics following intravitreal injections of ranibizumab, bevacizumab or aflibercept in patients with neovascular AMD. *Br. J. Ophthalmol.* 98 (12): 1636–1641.

177 Souied, E.H., Devin, F., Mauget-Faÿsse, M. et al. (2014). Treatment of exudative age-related macular degeneration with a designed ankyrin repeat protein that binds vascular endothelial growth factor: a phase I/II study. *Am. J. Ophthalmol.* 158 (4): 724–732.e2.

178 Staton, T.L., Peng, K., Owen, R. et al. (2019). A phase I, randomized, observer-blinded, single and multiple ascending-dose study to investigate the safety, pharmacokinetics, and immunogenicity of BITS7201A, a bispecific antibody targeting IL-13 and IL-17, in healthy volunteers. *BMC Pulm. Med.* 19 (1): 5.

179 Hellmann, M.D., Bivi, N., Calderon, B. et al. (2021). Safety and immunogenicity of LY3415244, a bispecific antibody against TIM-3 and PD-L1, in patients with advanced solid tumors. *Clin. Cancer Res.* 27 (10): 2773–2781.

180 Akpalu, D.E., Frederick, B., Nnane, I.P. et al. (2019). Pharmacokinetics, pharmacodynamics, immunogenicity, safety, and tolerability of JNJ-61178104, a novel tumor necrosis factor-alpha and interleukin-17A bispecific antibody, in healthy subjects. *J. Clin. Pharmacol.* 59 (7): 968–978.

181 Wong, C., Dash, A., Fredrickson, J. et al. (2022). FGFR1/Klothoβ agonist BFKB8488A improves lipids and liver health markers in patients with diabetes or NAFLD: a phase 1b randomized trial. *Hepatology* 78 (3): 847–862.

182 Schmitt, C., Emrich, T., Chebon, S. et al. (2021). Low immunogenicity of emicizumab in persons with haemophilia A. *Haemophilia* 27 (6): 984–992.

183 Hu, Z., Cohen, S., and Swanson, S.J. (2023). The immunogenicity of human-origin therapeutic antibodies are associated with V gene usage. *Front. Immunol.* 14: 1237754.

184 Alt, N., Zhang, T.Y., Motchnik, P. et al. (2016). Determination of critical quality attributes for monoclonal antibodies using quality by design principles. *Biologicals* 44 (5): 291–305.

185 Zhou, Y., Penny, H.L., Kroenke, M.A. et al. (2022). Immunogenicity assessment of bispecific antibody-based immunotherapy in oncology. *J. Immunother. Cancer* 10 (4): e004225.

186 Cohen, S., Myneni, S., Batt, A. et al. (2021). Immunogenicity risk assessment for biotherapeutics through in vitro detection of CD134 and CD137 on T helper cells. *MAbs* 13 (1): 1898831.

187 Tsai, W.-T.K., Li, Y., Yin, Z. et al. (2024). Nonclinical immunogenicity risk assessment for knobs-into-holes bispecific IgG1 antibodies. *MAbs* 16 (1): 2362789.

188 Deora, A., Hegde, S., Lee, J. et al. (2017). Transmembrane TNF-dependent uptake of anti-TNF antibodies. *MAbs* 9 (4): 680–695.

3

Analytical Challenges in Analyzing IgM Biotherapeutics

Amber D. Rolland[1,2] *and Albert J.R. Heck*[1,2]

[1] *Biomolecular Mass Spectrometry and Proteomics, Bijvoet Center for Biomolecular Research and Utrecht Institute for Pharmaceutical Sciences, University of Utrecht, Utrecht, The Netherlands*
[2] *Netherlands Proteomics Center, Utrecht, The Netherlands*

Introduction

IgM: The Most Unique Human Antibody

In the human body, all antibodies start out as IgM due to the gene encoding the μ heavy chain being first on the chromosome, making it expressed early in B cell differentiation well before class switching to other isotypes can occur. IgM is a major part of both innate and adaptive immunity, present in "natural" and antigen-induced (or adaptive) forms. Encoded by unmutated variable region gene segments and in the absence of antigen exposure, natural IgM antibodies conduct immunosurveillance against foreign pathogens and tumors [1], targeting both altered self- and nonself antigens present on dying cells for removal. In addition to maintaining homeostasis, natural IgM may help facilitate normal B cell development and prevent autoimmunity, and this form accounts for the bulk of IgM circulating in healthy individuals [2, 3]. Adaptive IgM, on the other hand, is the first antibody secreted following exposure to foreign antigens and thus serves as the body's first responder to defend against pathogens. These antigen-induced IgM antibodies can also provide long-term protection, as evidenced by the identification of long-lived IgM-expressing memory B cells [4].

Unlike most other isotypes, which are monomeric, IgM is secreted by B cells in healthy individuals as pentamers associated with an additional protein called the J chain [5], while hexamers lacking the J chain are the primary form produced under some specific pathological conditions [6–8]. This not only makes it our largest naturally occurring antibody but also confers uniquely high avidity due to the presence of 10 (or 12) antigen-binding sites in a single IgM assembly. Because natural IgM is polyreactive and adaptive IgM is produced before extensive affinity maturation can occur, the polyvalency of IgM is critical in compensating for the often low affinity and specificity with which it binds antigens. Capable of binding multiple identical epitopes at once, antibodies of this class are thus particularly well-suited to target otherwise "difficult" antigens, such as those present in low amounts and/or those with repeating motifs. Besides tumor cells, microbial surfaces and cell walls, in particular, feature such antigens [9] (e.g. carbohydrates, phospholipids, and lipopolysaccharides), and IgM is capable of clearing circulating antigens by both neutralization and opsonization, in addition to antibody-dependent cytotoxicity mechanisms through interactions with its Fc receptors, FcμR, and Fcα/μR [2, 3]. However, IgM is first and foremost recognized for its role as a robust activator of the complement system *via* the classical pathway to clear pathogens [10], with its polymeric structure allowing efficient engagement of C1q while bound to multiple cell surface targets.

IgM fulfills these vital roles in not only systemic but also mucosal immunity [11]. In serum, IgM has a half-life of ~5–10 days and is found at concentrations in the range of ~0.2–2.8 mg/mL with an average of ~1.5 mg/mL [12]. The presence of the J chain in IgM complexes enables transport across the epithelium into the mucosa, including the lining of the respiratory and gastrointestinal tracts. Here, secretory IgM provides the very first line of defense against foreign pathogens by preventing them from gaining entry into the circulatory system [10]. Although IgM is typically present in secretions at much lower levels than in serum (e.g. 0–18 μg/mL in tears, 0–64 μg/mL in saliva, and 50–610 μg/mL in milk and colostrum), this isotype has been shown to be a major factor in protecting against SARS-CoV-2, HIV, and other viruses that invade through the mucosa [4, 13–21].

All of these features – polyvalency, high avidity binding, targeting antigens with poor IgG immunogenicity, dual

Characterizing Biotherapeutics: Analytical Methods for Diverse Modalities, First Edition. Edited by Jennie R. Lill and Wendy Sandoval.

roles in systemic and mucosal immunity, and efficient activation of the complement system – make IgM an attractive target for development for therapeutic use [9, 10, 13, 17, 19, 20, 22–29]. Indeed, some of the first monoclonal antibodies (mAbs) to be tested clinically in the 1980s were of the IgM isotype. However, there are still no IgM mAbs approved for therapeutic use in humans today, as issues with demonstrating sufficient efficacy have proved a consistent problem. As reviewed by Keyt et al. [28], this can largely be attributed to the reliance on natural IgM as the source of most if not all of the ~20 or so IgM mAbs tested in humans so far, many of which were developed to treat sepsis. Fortunately, the development of IgM mAbs with high binding specificity is nowadays possible, thanks to advances in engineering mAbs to improve functionality, as well as efforts to identify and sequence antigen-specific adaptive antibodies [4, 30].

Until recently therapeutic IgM development had also been hampered by issues with production, limiting the available moieties to be clinically tested. IgM's large size, polymeric nature, and extensive glycosylation result in low expression levels using methods that are suitable for recombinant production of IgG-based mAbs [31]. Preservation of human-like glycosylation patterns is especially important for IgM due to the role of glycosylation in proper assembly and secretion, requiring the use of mammalian cell lines [32, 33]. Production of J chain-incorporated IgM mAbs also requires optimization of the ratios of the plasmids encoding the heavy chain, light chain, and J chain to avoid size heterogeneity [34]. However, breakthroughs have been made with the reported high expression of IgM in PER.C6 human cell lines [35] and glycoengineering in plants to produce functional IgM with human-like glycosylation [20, 36]. Several groups now have also shown the production of monodisperse populations of J chain-incorporated IgM pentamers from the co-expression of IgM and J chain in reported optimized ratios [29, 37, 38].

Low expression levels had also been further compounded by low recovery of recombinantly-expressed IgM [39]. Early IgM purification strategies relied on separation based on physicochemical characteristics shared by antibodies in general, such as their relatively basic isoelectric points and thiophilicity. Thus, isoelectric precipitation was a common first step, requiring an additional orthogonal chromatography step such as gel filtration to achieve complete isolation of IgM. Notably, widely used and commercially available options like Protein A and Protein G do not work for IgM purification, while immobilized mannan-binding lectin, which recognizes specific glycan structures, results only in partial purification of some IgM glycoforms [40]. More efficient chromatography materials are fortunately now available for the purification of IgM [32], with the development of mixed-mode resins, workflows utilizing ceramic hydroxyapatite chromatography followed by ion exchange, and peptoid-based affinity materials for purification of recombinant IgM. Most recently, CaptureSelect technology was used to develop the first and so far only IgM-specific affinity resin suitable for purification from serum, featuring immobilized recombinant single-domain antibody fragments derived from camelids that bind human, rat, and mouse IgM via the Fc domain.

With the production bottleneck overcome thanks to these improvements, and with methods in place for enhancing antigen specificity, the interest in IgM biotherapeutics has begun to bear fruit. More recent efforts have focused on intravenous delivery of IgM-enriched immunoglobulin preparations [18], engineering based on IgM's polymericity (e.g. HexaBody technology [41, 42] and ACE2 decoys [15, 43]), development of IgM antibody-drug conjugates (ADCs) [44, 45], and, of course, IgM mAbs, which have been developed not just to protect against bacterial infections but also to treat myriad diseases, including multiple types of cancer, lipid metabolism disorders, lupus, rheumatoid arthritis, and COVID-19. Current biotherapeutic IgM mAbs undergoing early phases of clinical trials have been designed to act as receptor cross-linking agonists to induce programmed cell death of cancer cells [46, 47], for targeted cytokine delivery, and as potent bispecifics capable of inducing T cell-dependent cytotoxicity of cancer cells. There have also been many exciting preclinical reports of IgM mAbs with enhanced antigen binding and neutralization capabilities that can be delivered by various administration routes [4, 14–16, 48].

With these advancements resulting in an increased number of IgM biotherapeutics with better functionality in the pipeline, now the field can move forward to face the next obstacle – the analytical characterization of IgM mAbs. This is required for regulatory approval, but the tools and techniques developed so far have chiefly been for the characterization of IgG (and even mostly just IgG1), as almost all of the ~140 mAbs approved for biotherapeutic use in humans in the EU and/or US are of this isotype. In this review, we describe the several unique features of IgM that make it a much more challenging molecule to analyze compared to IgG1, namely its larger size and polymericity, incorporation of additional protein subunits, extensive glycosylation, and complex disulfide bond network. Throughout we include a discussion of potential solutions for overcoming these challenges and implications for IgM mAb development, highlighting recent exciting developments. Taking advantage of the full therapeutic potential of IgM not only relies on the development of functional, specific mAbs but also on their critical evaluation, requiring adjusted workflows, re-optimization of the current mass spectrometry (MS)-based methods, or even new analytical tools for quality control.

Characterization of Biotherapeutic mAbs

Obtaining and retaining regulatory approval of mAbs for biotherapeutic applications requires careful characterization of their structural, physicochemical, immunological, and biological properties, in addition to quantification, purity, and stability. This includes identification not just of impurities and contaminants but also of product-related variants differing in, for example, size, charge, glycosylation, and/or amino acid sequence, that can arise at any stage from manufacture to storage to post-administration [49, 50]. Methods used in recombinant production of mAbs can induce several post-translational modifications (PTMs), and even minor modifications to a single amino acid residue can result in charge and size variants due to alterations to the antibody's surface charge distribution and interactions. For example, common PTMs such as deamidation, glycation, and cysteinylation are associated with the formation of acidic charge variants, while N-terminal pyroglutamate formation, C-terminal lysine processing, oxidation, and isomerization tend to yield basic variants. Further, glycosylation and disulfide bond formation are major contributors to proper mAb assembly and function, with disulfides linked to aggregate formation and the influence of the cell line used and culture conditions known to influence glycosylation patterns [33, 51, 52].

Characterization of this intrinsic heterogeneity is critically important because of the potentially serious health consequences resulting from changes to mAb structure, function, stability, biological activity, and pharmacokinetics [49]. While many techniques are used to assess these potentially altered structural and functional features, MS has for decades been the primary method for characterizing the underlying PTMs, including identification, quantitation, and localization within the mAb amino acid sequence [50, 53–58]. This is a result of the development of soft ionization techniques (MALDI and ESI) and continual improvements to sensitivity and resolution, with further advantages provided by fragmentation/dissociation capabilities and compatibility with online separation methods.

MS-based characterization of mAbs primarily utilizes a bottom-up proteomics approach, in which smaller mAb fragments (peptides) are generated through proteolytic digestion with general proteases such as trypsin, pepsin, GluC, and/or LysN and then subjected to RPLC-MS/MS for online separation, fragmentation, and detection. Most modern MS instruments are capable of performing tandem MS experiments through fragmentation and dissociation techniques such as CID/HCD, ETD/ECD, and/or UVPD to sequence and identify the peptides and the modifications they harbor. Separation of all peptides before MS analysis is usually performed on the basis of size, charge, or hydrophobicity/hydrophilicity and can be done either online or offline using chromatography (RPLC, SEC, HILIC, IEX, HIC, and PGC) or electrophoresis (CE, CZE, and SDS-PAGE). In bottom-up approaches, sample preparation often also involves reduction, denaturation, deglycosylation, labeling, and/or enrichment steps to generate peptides of suitable size, charge, and/or abundance for LC-MS analysis.

Alternative, more protein-centric MS-based strategies are also nowadays more frequently employed in the analytical characterization of mAbs, including top-down and middle-down proteomics and intact protein analysis [59, 60]. In such approaches, subclass-specific enzymes are often useful to generate intact Fab or $F(ab')_2$ and Fc fragments [54, 61]. While there are many commercially available IgG-specific proteases for digestion either above or below the hinge to release either Fab or $F(ab')_2$ molecules, IgM fragmentation was limited to the use of reduction or immobilized trypsin/pepsin with carefully optimized conditions until the first IgM-specific enzyme entered the market in 2023. IgMBRAZOR cleaves at one specific site at the N-terminus of the Cμ3 domain in human IgM, releasing $F(ab')_2$ fragments (~125 kDa) while leaving the pentameric Fc core (~327 kDa) intact [62]. However, the size of these fragments may still be prohibitively large for analysis with certain approaches, and generation of Fab molecules (~50 kDa for both IgG1 and IgM) with a specific protease is not yet possible for IgM. Size and heterogeneity are also both limiting factors in the analysis of intact mAbs, even for IgG, which is relatively small compared to IgM, as we detail in the following section.

IgM Sequence and Structure

All antibodies are glycoproteins composed of two heavy chains paired with two light chains, held together not only by non-covalent interactions but also featuring several disulfide linkages [63], as illustrated for IgM monomer subunits in Figure 3.1. For all human antibodies, the light chains can be of either isotype (i.e. κ or λ), but the identity of the heavy chain determines the antibody class. There are five classes of immunoglobulins in humans (IgM, IgD, IgG, IgA, and IgE, encoded by μ, δ, γ, α, and ε heavy chain genes, respectively), with multiple subclasses of IgG (IgG1, IgG2, IgG3, IgG4) and IgA (IgA1, IgA2). The amino acid sequences encoded by these heavy chain genes not only

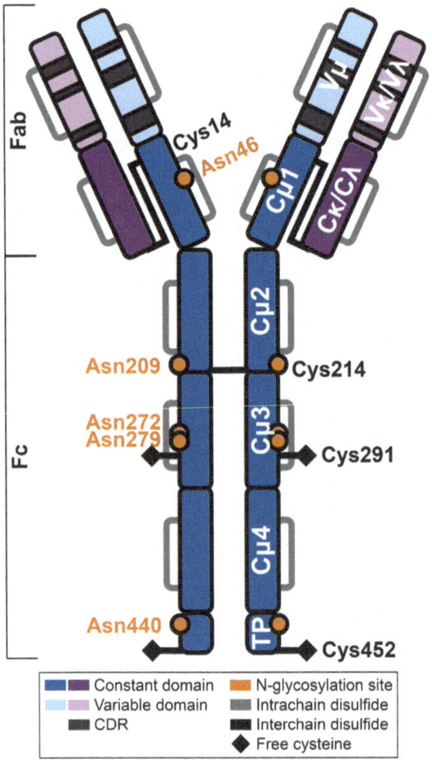

Figure 3.1 Schematic diagram of IgM monomer with domains, disulfide bonds, and N-glycosylation sites highlighted. IgM μ heavy chains are shown in blue and light chains (κ or λ) in purple, with darker shading for constant regions and lighter shading for variable regions which each further include three CDRs (dark grey). Disulfide bonds are indicated with light grey (intrachain) or black (interchain) lines. Cysteines that form interchain disulfides between subunits in polymeric IgM assemblies but remain free in IgM monomers are annotated with a black diamond. N-glycosylation sites are represented by orange circles. All indicated residues are for the IgM heavy chain, with the amino acid positions numbered to match the UniProt entry for the secreted form of IGHM (accession number P01871-1). Excluding the tailpiece (TP) region, the lengths of each domain are directly proportional to amino acid sequence length, with the locations of all highlighted features directly related to their relative amino acid positions within each domain/region.

contribute to differences in effector functions but also differing numbers of constant domains/features, disulfide bond networks, and glycosylation patterns (Figure 3.2). Light chains contain one variable (Vκ/λ) and one constant domain (Cκ/λ), while heavy chains contain one variable domain (VH) but multiple constant domains (CH), the number of which varies between antibody classes. Each of these domains (~125 amino acid residues) adopts the classic immunoglobulin fold (Ig-fold) of a compact two-layer β-sandwich with Greek key topology composed from seven to nine antiparallel β-strands. Antigen recognition and binding occur in the N-terminal Fab arms *via* the loops formed by the three complementarity-determining regions (CDRs) present in the variable region of each chain. The remainder of the heavy chain constant region sequence makes up the Fc domain and is responsible for effector functions and receptor binding.

IgM heavy chains are encoded by the μ gene and possess additional domains/regions that make it the largest human antibody. There are four constant domains in each IgM heavy chain, a feature shared only with IgE while all other heavy chains have only three, as shown in Figure 3.2. IgM (and IgA) heavy chains also feature an additional 18-residue stretch at the C-terminus called the TP responsible for polymerization via interaction with the J chain [5, 64–68]. While monomers are a major form of IgA, IgM is exclusively polymeric and secreted from B cells as J chain-incorporated pentamers (Figure 3.3), although other assemblies (predominantly hexamers) can be produced in the absence of the J chain or under adverse pathological conditions [69]. Thus while the monomer mass of IgM is already elevated compared to IgG1 due to additional domains, fully-formed polymeric IgM (pIgM) assemblies are nearly 10 times larger with a total mass of ~1 MDa (Table 3.1). Of course, the larger number of amino acids in IgM heavy chains (453 residues versus 330 in IgG1) inherently increases the potential for PTMs and thus also the potential formation of charge/size variants and multiple proteoforms in general. For example, IgM heavy chains are relatively rich in aspartic acid, methionine, and asparagine residues, all of which are sites of common modifications (isomerization, oxidation, and deamidation, respectively).

Besides increasing the mass and complexity of IgM molecules, the additional domains have structural and functional consequences. As shown in the heavy chain sequence alignment in Figure 3.2, the Cμ1, Cμ3, and Cμ4 domains of IgM correspond to Cγ1, Cγ2, and Cγ3 of IgG. The "extra" constant domain Cμ2 takes the place of a true hinge, a proline- and cysteine-rich region that serves to physically separate the antigen-binding Fab arms from the Fc region of all other subclasses except IgE. This provides flexibility and allows freedom of movement both from the Fc domain as well as from each other, allowing IgG and IgA Fab arms to bind antigens independently. Instead, within a monomer subunit of IgM, the two Cμ2 domains are disulfide-linked to each other, functioning more like a single, rigid homodimer unit [70]. While the Cμ2 dimer unit does separate the Fab arms from the IgM Fc core (encompassing the Cμ3, Cμ4, and TP regions) like a true hinge, the range of Fab motion is quite limited both with respect to the Fc core and to one another, with ~50° between each pair of Fab arms [71]. This makes the functional antigen-binding unit of IgM the V-shaped $(C\mu2)_2$-$F(ab')_2$ unit (hereafter referred to as the $F(ab')_2$). As a consequence of this, rather

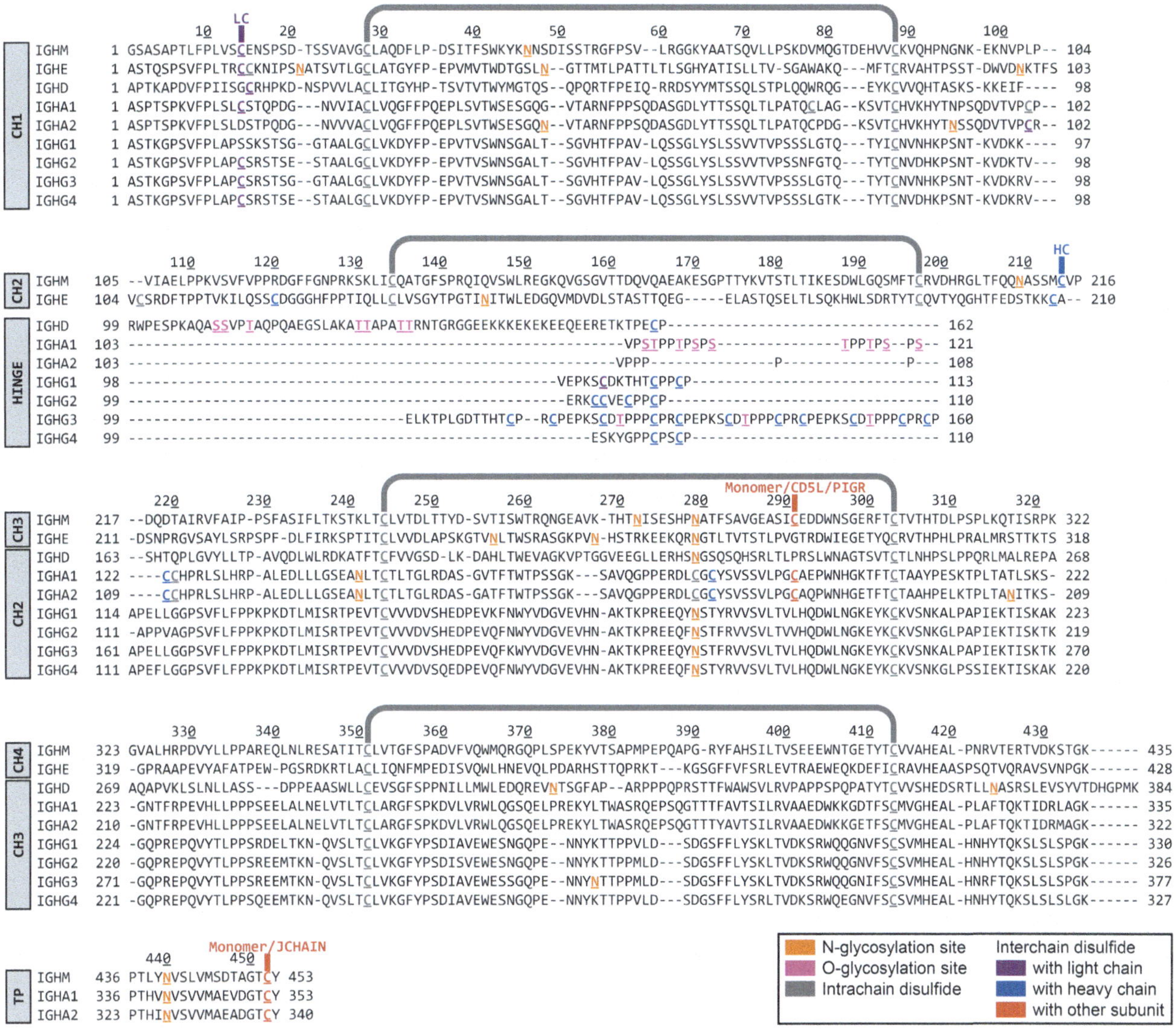

Figure 3.2 Sequence alignment of the heavy chain constant regions of human immunoglobulin isotypes with glycosylation sites and disulfide bonds highlighted. Intrachain disulfide bonds are shown in light grey, and cysteine residues involved in interchain disulfide bonds with the light chain, heavy chain, or another subunit are shown in purple, blue, and red, respectively. Glycosylation sites are highlighted in orange (N-linked) or pink (O-linked). The position of the first and last amino acids in each domain/region is indicated for each isotype. Periodic sequence numbering and disulfide bond annotations shown above each section are for IGHM. Immunoglobulin heavy chain sequences are numbered to match the corresponding UniProt entries for the secreted form (IGHM: P01871-1, IGHE: P01854-1, IGHD: P01880-1, IGHA1: P01876-1, IGHA2: P01877-1, IGHG1: P01857-1, IGHG2: P01859-1, IGHG3: P01860-1, and IGHG4: P01861-1). Sequences for each region were aligned with Clustal Omega and reordered as shown (note: IGHM and IGHE CH2 sequences were aligned separately from the hinge regions of IGHD, IGHA, and IGHG).

than utilizing just one Fab arm to target antigen, IgM exhibits pairwise surface binding with both arms of each subunit engaged simultaneously as revealed by cryo-ET and other imaging techniques [72].

The motion of the F(ab′)$_2$ units together with the large size of IgM overall limits the level of structural detail that can be determined for fully-assembled IgM. Thus the prevailing model of J chain-incorporated IgM pentamer structure – featuring the five subunits arranged symmetrically in a pentagon – was based initially on electron micrographs and lower-resolution data and remained relatively unchanged for roughly 50 years until 2018 when single-particle negative-stain EM analysis of J chain-coupled IgM Fc core pentameric assemblies provided conclusive evidence of a quasi-hexagonal subunit arrangement [73]. Here, Hiramoto et al. [73] revealed IgM to be an asymmetric pentagon, with a ~ 50° gap between subunits 1 and 5 (see Figure 3.3) occupied by the J chain

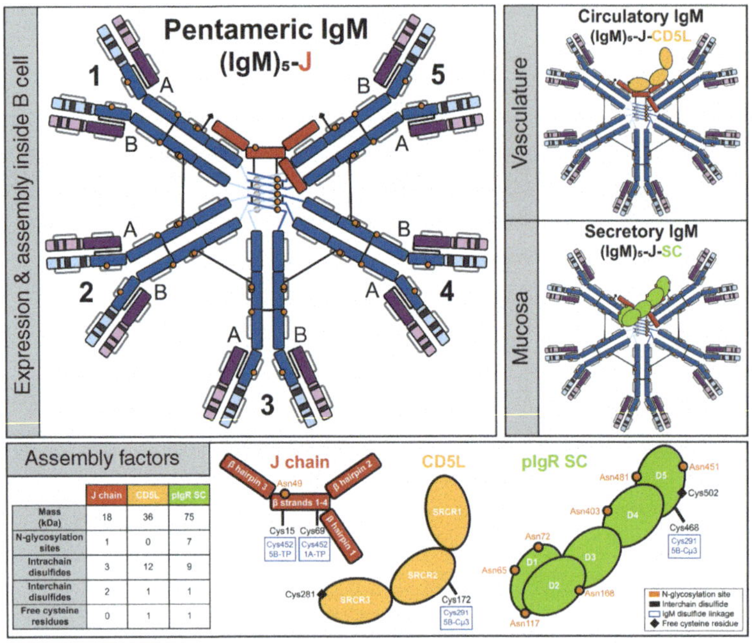

Figure 3.3 Schematic models of IgM pentameric assemblies formed endogenously (top) through association with one or more assembly factors (bottom). In healthy individuals, IgM is secreted by B cells as J chain-incorporated pentamers (top left panel) that can subsequently associate extracellularly with CD5L to form circulatory IgM present in the vasculature or with pIgR to form secretory IgM present in mucosal secretions (top right panels). All IgM features are represented in Figure 3.1, except for the tailpieces, which are elongated here to give a more structurally accurate depiction of the central two-layer β-sandwich core and positions of tailpiece glycans and disulfides as revealed in cryo-EM structures. IgM monomer subunits are numbered counterclockwise 1–5 and positioned to reflect the inherent asymmetry of pentameric IgM with a 50° gap between subunits 1 and 5. Each pair of heavy chains is further labeled as A or B to provide clarity to discussions within the main text. For each assembly factor (J chain, red; CD5L, gold; pIgR SC, green), the major domains, glycosylation sites, free cysteines, and interchain disulfide bonds with IgM are indicated in the top panel with more detailed information highlighted in the bottom panel. Assembly factor intrachain disulfides are omitted for clarity, and all other features are positioned to reflect their relative locations when bound to IgM pentamers.

Table 3.1 Comparison of IgG1 mAbs with various IgM assemblies to highlight differences in size, molecular complexity, glycosylation patterns, and disulfide bond networks.

Antibody assembly	IgG1	IgM monomer subunit	IgM pentamer	IgM hexamer	Circulatory IgM	Secretory IgM
Subunit composition	$H(\gamma)_2$-$L(\kappa/\lambda)_2$	$H(\mu)_2$-$L(\kappa/\lambda)_2$	$(IgM)_5$-J	$(IgM)_6$	$(IgM)_5$-J-CD5L	$(IgM)_5$-J-SC
Mass (kDa)	150	185	945	1,110	985	1,020
Peptide chains	4	4	21	24	22	22
N-glycosylation sites	2	10	51	60	51	58
Total disulfide bonds	16	17	98	114	111	108
Intrachain disulfide bonds	12	14	73	84	85	82
Interchain disulfide bonds	4	3	25	30	26	26
Free cysteine residues	0	4	2	0	2	2

in place of a sixth subunit (and thus also explaining why the J chain is absent from IgM hexamers). This paved the way for subsequent detailed structural analyses of J chain-coupled pentameric Fc core complexes produced recombinantly rather than full-length IgM [71, 74–80], and in 2020, Li et al. reported at long last the first 3D structure of IgM with cryo-EM of J chain-coupled pentameric IgM Fc cores [74]. This revealed additional asymmetry because of the way the J chain binds with one side or "face" featuring its N-terminal β hairpin 1 (which interacts primarily with subunit 5; +β face) and the other characterized by the C-terminal β hairpin 3 (which interacts with subunit 1; −β face) [40]. As illustrated in Figure 3.3, neighboring heavy chains from different IgM protomers are disulfide-linked to each other via their Cμ3 domains and TPs, with the J chain forming two disulfides with the TP cysteines of the two outermost heavy chains (1A and 5B) that line the gap. Together, the J chain and TPs form a central two-layer β-sandwich core held together not only by these disulfide linkages but also by substantial non-covalent interactions both between adjacent strands in each layer and between the two layers. These robust interactions are a major contributor to enhancing the stability of IgM assemblies and to the rigid, planar nature of the Fc core.

Results from both cryo-EM and cryo-ET indicate that free IgM is planar but adopts a staple- or dome-like configuration when bound to surface antigen with all $F(ab')_2$ units turned significantly downward (~120° for the innermost Fab arm) from the Cμ2-Cμ3 junction [72]. With the C1q binding site in the topmost Cμ3 domain of each heavy chain thus exposed and positioned quasi-hexagonally, IgM is able to efficiently engage five of the six globular head domains of C1q with simultaneous full utilization of its 10 antigen binding sites, ultimately giving rise to its primary role as a robust activator of the classical pathway to induce complement-dependent cytotoxicity (CDC). More recently, it has been proposed that J chain-incorporated IgM pentamers bind C1q and surface antigens each with a preferred face (−β and +β, respectively). As reported by Chen et al., cryo-EM density maps of full-length IgM revealed that the $F(ab')_2$ of one subunit (subunit 1 in Figure 3.3) exhibited uniquely rigid motion with a substantially biased orientation toward the −β side compared to those of subunits 2–5 [71]. The authors attributed this to extensive interaction of the J chain C-terminal β hairpin 3 with the Cμ3 domain of heavy chain 1A [62], altering the position of this domain such that it would be more energetically favorable to bind antigens on the cell surface, in turn predisposing the +β side to engage C1q. Notably, the J chain is absent from IgM hexameric assemblies, which are symmetric and exhibit superior CDC activity compared to J chain pentamers [7, 9, 29, 81]. Taken together, this highlights potential functional consequences of the compositional heterogeneity that results from association with the J chain, which is just one of IgM's canonical assembly factors.

IgM Assembly Factors

As illustrated in Figure 3.3, the major forms of healthy human IgM are each associated with a minimum of one other protein. The J chain facilitates the polymerization and assembly of IgM pentamers inside the B cell [5], and these complexes can subsequently associate extracellularly with a second protein that differs between serum (CD5L) and secretions (pIgR SC). Further challenges are introduced by the presence of these additional subunits in endogenous pIgM assemblies, as these assembly factors necessarily increase the size and complexity of features that must be characterized. These could be excluded from the production of IgM biotherapeutics but not without the structure, function, and/or stability of IgM being affected to some degree, in addition to potential interactions that would be expected to form *in vivo* following administration of such therapeutics. However, while these assembly factors make analysis more challenging and can even complicate the production of IgM, they also offer advantages to IgM biotherapeutics, not only through the unique functions and features they confer to IgM endogenously but also through the potential to further enhance activity.

The J Chain Facilitates Assembly of Pentameric IgM Inside B Cells

The immunoglobulin joining chain, or J chain, is a polypeptide containing 137 amino acid residues in its mature form with a backbone mass of ~15.6 kDa (UniProt accession number P01591, residues 23–159 renumbered here as 1–137). Despite its name, the J chain bears no significant homology in either structure or sequence to any other proteins and possesses no Ig-fold domains. Instead, the J chain is highly conserved across species from the animal kingdom, where its sole function appears to be facilitating the polymerization of TP-containing Igs [5]. J chain expression and incorporation into pIgs occurs inside the B cell before secretion of the fully-assembled J chain-containing polymeric complex, and free J chain is not released [82]. Recombinant production of J chain-incorporated IgM pentamers requires co-expression of the heavy chain, light chain, and J chain in carefully optimized ratios and with proper glycosylation patterns to avoid size heterogeneity, as IgM can also polymerize on its own [8]. The J chain is also critical for many functions and features that have prompted efforts to develop IgM biotherapeutics in the first place, including polyvalency and access to mucosal sites following intravenous delivery. The J chain is required for IgM binding CD5L [38, 73, 79], pIgR [68, 74, 75], and Fcα/μR [83], which all share similar binding sites on IgM such that the presence of one of these assembly factors/receptors on pentameric IgM-J chain complexes will reduce or even prevent binding of the other two [2, 38, 79, 83]. Both J chain-containing and J chain-deficient IgM assemblies and ICs can bind FcμR, possibly through distinct mechanisms [77, 78, 84–86]. The CDC activity of J chain-incorporated pentamers is reduced relative to IgM hexamers [81], although still very potent when compared to other isotypes, including IgG1 [29].

Incorporation of the J chain also increases the molecular complexity of IgM, making analytical characterization more challenging (Table 3.1). There is a single N-linked glycosylation site at Asn49 that is consistently occupied by complex-type N-glycans when bound to either pIgM or pIgA. The mass of glycosylated J chain is closer to ~18 kDa, with ~10% of the total mass contributed by the single attached glycan. The J chain also features eight cysteine residues, with six of these involved in its three intrachain disulfides. The other two, Cys15 and Cys69, form disulfide bonds with the IgM TP Cys452 residues on heavy chains 5B and 1A, respectively [87] (see Figure 3.3). Thus the J chain contributes an additional 5 disulfide bonds and 1 glycosylation site to IgM pentamers, bringing the total for fully-assembled J chain-incorporated IgM pentamers to 98 disulfide bonds and 51 N-linked glycosylation sites with two free sulfhydryl groups at Cys291 in heavy chains 1A and 5B (Table 3.1).

While the J chain introduces new challenges and difficulties to IgM biotherapeutic production and analysis, the highly stable binding of this extra protein subunit in the gap where steric hindrance is limited is ideal for engineering IgM mAbs with enhanced activity [24]. Indeed, the J chain has already been exploited to generate bispecific IgM biotherapeutics currently in clinical trials. One example is IGM-2323, an anti-CD20 IgM with a CD3 binding domain fused to the J chain, able to bind cancer antigens with potent T cell-dependent cytotoxic activity due to its "10 × 1" bispecificity.

CD5L Associates with IgM Assemblies in Serum

CD5 antigen-like, or CD5L (also referred to as apoptosis inhibitor expressed by macrophages, AIM; or soluble protein alpha, Spα), is secreted mainly by tissue macrophages as a ~36 kDa protein with 328 residues in the mature polypeptide sequence in humans (UniProt accession number O43866, residues 20–347 renumbered here as 1–328) [88, 89]. As illustrated in Figure 3.3, CD5L consists of 3 scavenger receptor cysteine-rich (SRCR) domains connected by short, flexible linker regions, with 8 cysteines in the first domain (SRCR1) and 9 in each of the other two. While the primary and best-defined function is inhibiting apoptosis of macrophages to support their survival, CD5L can also serve as a pattern recognition receptor for microbial pathogens, most likely through the presence of multiple binding sites in its three highly homologous SRCR domains [88–92]. CD5L can bind both Gram-negative and Gram-positive bacteria, as well as fungi, by targeting common epitopes (including lipopolysaccharide, lipoteichoic acid, peptidoglycan, mannan, β-D-glucan, and zymosan) and further increase clearance by inducing their aggregation and/or promoting phagocytotic uptake by macrophages.

Although CD5L was first reported to be associated with IgM in human serum more than 20 years ago [93], the strength and prevalence of this interaction were only recently discovered and mark exciting new developments in IgM research [38, 73, 79]. The 2018 negative-stain EM analysis by Hiramoto et al. not only revealed the asymmetry of IgM pentamers but also pinpointed the binding location of CD5L in the gap [73], with a pair of 2024 reports by Wang et al. and Chen et al. reporting cryo-EM structures of CD5L bound to J chain-coupled IgM-Fc pentamers and identifying significant interactions between CD5L with the J chain [79, 94]. Also in 2023, Oskam et al. reported results from bottom-up proteomics and anti-CD5L ELISA of healthy human donor samples, revealing that instead of just being a minor or transient interaction, virtually all CD5L molecules circulating in serum are in complex with IgM (circulatory IgM, see Figure 3.3) [38]. This finding was specific to serum, as CD5L was not detected in secretory IgM (sIgM) complexes from either saliva or milk, likely indicating that the binding sites of CD5L and pIgR overlap. The current view in the literature is that IgM acts as a transporter for inactive CD5L in circulation, protecting it from renal excretion until certain conditions trigger dissociation to allow CD5L to perform its function(s) [92]. The association of CD5L with IgM has been reported to occur in a Ca^{2+}-dependent manner by Wang et al. [79], and dissociation has been proposed to occur at sites of inflammation where released glutathione or thioredoxin could disrupt the disulfide bond formed between CD5L and IgM [38]. However, the exact details of the conditions under which CD5L associates and dissociates from pentameric IgM-J chain complexes in circulation remain elusive [92, 94].

Together, these two facts – extracellular association and lack of details surrounding the interaction – present additional considerations or challenges in the production of recombinant IgM mAb biotherapeutics. CD5L could be used to enhance the function of IgM biotherapeutics, either through its natural roles in inflammation or by fusing a binding domain to its N-terminal SRCR1 domain, where issues with steric hindrance are least likely to result due to it not being involved in the interaction with IgM, unlike SRCR2 and SRCR3, which are both critical (as illustrated in Figure 3.3) [73]. Production of CD5L-containing IgM biotherapeutics would require coculturing two different cell lines, one expressing the IgM-J complex and the other expressing CD5L [38]. On the other hand, because of the strong association observed natively in human serum that will likely occur post-administration *in vivo*, IgM biotherapeutics could be developed to either target CD5L or prevent binding completely. CD5L is implicated in many diseases and pathologies for which IgM mAbs have been developed, including cancer [95], sepsis [96], systemic lupus erythematosus [97], and other autoimmune conditions [98]. Recent work from two groups has indicated that while CD5L association does not affect IgM's ability to activate complement or interact with its specific Fc receptor FcμR, binding to both Fcα/μR and pIgR is reduced [38, 79]. As Fcα/μR mediates internalization of IgM [99], the strong association of CD5L with IgM in serum could not only protect CD5L from renal clearance but also protect IgM immune complexes (ICs) from internalization, thus promoting antigen presentation to B cells [92, 98].

Regardless of whether IgM biotherapeutics are produced with CD5L incorporated, the potential impact on function and stability of interaction with CD5L must be assessed, and thus analytical characterization is made more challenging with the added complexity of CD5L's features. Although human CD5L contains one potential O-linked glycosylation site in the SRCR1-SRCR2 linker region and a possible N-X-C N-linked glycosylation motif in SRCR2, no glycans have been detected following extensive deglycosylation treatment [100]. However, recombinant expression of human CD5L can result in multiple glycosylated isoforms, as shown for two isoforms with varying sialylation produced by HEK293-EBNA cells [89]. This represents an important point of consideration in the production of IgM biotherapeutics, as the introduction of an N-glycosylation site by mutagenesis in either SRCR2 or SRCR3 significantly disrupted CD5L's ability to associate with J chain-incorporated pentameric IgM Fc core complexes [79].

The lack of glycosylation in endogenous CD5L is more than made up for by the considerable complexity of its network of disulfide bonds. Although an experimentally-determined structure of human CD5L had not yet been reported at the time of this writing (with the cryo-EM structures of CD5L bound to IgM-Fc pentamers by Wang et al. and Chen et al. only recently published [79, 94]), some structural evidence has been provided by mutagenesis and other studies, and other features can reasonably be inferred due to SRCR domains being highly-conserved. Almost all of CD5L's 26 cysteines are predicted to form disulfides with each other in the AlphaFold structural model, which features 4 intrachain bonds localized within each SRCR domain, leaving one free cysteine each in the SRCR2 and SRCR3 domains to form an interchain disulfide with IgM. Of these two, experimental results from multiple independent studies indicate a single disulfide bond is formed between SRCR2 Cys172 and the free Cμ3 Cys291 residue on heavy chain 5B in J chain-incorporated IgM pentamers (Figure 3.3) [38, 79]. However, the relative importance of this disulfide bond is still under debate, and it is not clear whether it is essential to the association of CD5L with IgM or rather enhances the stability of circulatory IgM following their assembly. Regardless, the association of CD5L dramatically increases the number of disulfide bonds, with circulatory IgM assemblies most likely having a total of 111 disulfides (26 interchain, 85 intrachain) and 2 free cysteines, Cμ3 Cys291 on IgM heavy chain 1A and Cys281 in the SRCR3 domain of CD5L (see Table 3.1).

pIgR Transports IgM Assemblies into the Mucosa

The polymeric immunoglobulin receptor, or pIgR, contains 746 residues in its full-length mature form, corresponding to a backbone mass of ~81 kDa (UniProt accession number P01833, residues 19–764, renumbered here as 1–746). Expressed on the basolateral surfaces of mucosal epithelial cells, pIgR is a transmembrane protein with an extracellular N-terminal region (featuring five heavily-glycosylated immunoglobulin-like domains, D1–D5) that is connected by a disordered linker to the transmembrane region and an intracellular C-terminal cytoplasmic domain. The presence of the J chain in polymeric IgM and IgA and their ICs enables binding to the N-terminal domain of pIgR and subsequent transcytosis to the apical mucosal epithelial cell surface, preventing IC accumulation on the basolateral surface and helping to eliminate or neutralize circulating antigen [5, 101]. Upon reaching the apical membrane, pIgR is proteolytically cleaved between the most C-terminal extracellular domain (D5) and the transmembrane region, releasing the free or antigen-bound pIg into the mucosa with its N-terminal region still attached [74, 75]. This cleavage product that remains bound is referred to as the secretory component (SC), and together the resulting assembly consisting of the J chain, polymeric IgM, and SC is referred to as secretory IgM (sIgM, see Figure 3.3) [10, 11]. SC reportedly contributes a protective effect against cleavage and degradation, including against gastric acids and digestive enzymes present in the gastrointestinal tract, where pIgR is primarily found, and perhaps this enhanced stability explains why nearly all structures of J chain-coupled pentameric IgM Fc cores determined with atomic resolution (including the first 3D structure reported by Li et al. in 2020) are associated with SC and thus represent subassemblies of sIgM [74, 75, 77].

As illustrated in Figure 3.3, the SC consists of the first 585 amino acids of the mature pIgR polypeptide chain, encompassing all 5 Ig-like domains and the linker region, with a backbone mass of ~64.3 kDa. However, SC is heavily glycosylated with 7 N-linked glycosylation sites distributed among 4 of its 5 domains (indicated in Figure 3.3), which feature a particularly diverse array of glycans that can contribute up to 25% of the molecular mass of SC (~70–85 kDa with glycosylation) [102]. The pIgR amino acid sequence that corresponds to SC features 20 cysteine residues, with a few in each Ig-like domain (4 each in D1, D3, and D4, 2 in D2, and 6 in D5). All of these form disulfide bonds with other cysteines in the same domain for a total of 10 in free SC, but the intrachain disulfide bond between Cys468 and Cys502 in D5 can be disrupted in sIgM, with D5 Cys468 instead forming an interchain disulfide with Cys291 in the Cμ3 domain on IgM heavy chain 5B leaving a free sulfhydryl group on D5 Cys502 (Figure 3.3). Cryo-EM structures of sIgM only provide evidence of partial formation of this SC-IgM disulfide bond, in comparison to sIgA structures where a disulfide between SC Cys468 and the homologous cysteine in IgA is always identified, but the reason for this difference is unclear [74, 75, 101, 103]. Notably, 5B-Cμ3 Cys291 is also the same cysteine involved in the interchain disulfide bond with CD5L with its contributions to circulatory IgM assembly and/or stability under debate. Association with the SC of pIgR thus increases the complexity and heterogeneity of IgM assemblies with 58 N-glycosylation sites in addition to a total of 108 disulfide bonds and 2 free cysteines regardless of whether SC D5 Cys468 forms an interchain disulfide bond with IgM (26 interchain, 82 intrachain, with free sulfhydryls at IgM 1A-Cμ3 Cys291 and SC D5 Cys502 as indicated in Table 3.1 and Figure 3.3) or maintains its native intrachain interaction with Cys502 (25 interchain, 83 intrachain, with the Cμ3 Cys291 of heavy chains 1A and 5B remaining free). While IgM biotherapeutic preparations will likely not include SC as a cofactor, binding and transcytosis should still be assessed, as this is an important mechanism for eliminating circulating antigen, clearing IgM ICs, and enabling IgM to serve as the first line of defense against invading pathogens in the mucosa [5, 10, 11, 104]. The interaction with pIgR also enables J chain-containing IgM biotherapeutics administered intravenously to easily access mucosal sites, an important advantage over IgG mAbs in fighting viral infection [13, 18, 19, 21].

IgM Glycosylation Patterns

As for all other glycoproteins, antibody structure, function, and stability can be impacted by their glycosylation [27, 31, 50, 56, 105–111]. Notably, Fc receptor-mediated effector functions of IgGs are influenced by differences in glycosylation, even though each subclass (with the exception of IgG3) possesses just one site per heavy chain [112, 113] (Figure 3.2). A similar role has not yet been reported for IgM [114]. Instead, IgM glycosylation has been shown to impact various aspects of its assembly, structure, and function, although not to the same level of detail as for IgG1, with more general features of glycosylation correlated to outcomes rather than identification of specific glycoforms. Briefly, IgM glycosylation has been shown to be important for its secretion from B cells [107], its polymeric assembly with the J chain [64, 115], and internalization by T cells [1], as well as for increasing solubility and half-life *in vivo* [32]. IgM glycosylation has also been correlated to complement deposition and various pathologies, including COVID-19 and cancer [109, 116–118]. Glycoengineering [119] of mAbs including IgM has been done to modulate their immune roles [20, 36], effector functions [112], and half-life [120], as well as to generate site-specific ADCs [45]. More generally, altered glycosylation features (e.g. sialic acid, high-mannose oligosaccharides, and aglycosylation) can result in both charge and size variants and is a major feature expected for recombinantly-produced mAbs [49, 52, 119, 121–124]. Thus, it is critical to analyze IgM glycosylation, which can be studied at the molecular level of released glycans, glycopeptides, or intact glycoprotein using MS-based approaches to get varying degrees of glycan- and/or protein-rich information [56, 58, 111, 113, 121, 125–127]. The most in-depth characterization includes the determination of site-specific glycan composition and occupancy from glycopeptide analysis, although these features can also be analyzed in the context of bulk/total carbohydrate content or intact molecules [56, 128]. Features of glycans to characterize include the composition and order/sequence of monosaccharide subunits, branching patterns and type, configuration and position of glycosidic linkages, distinction between isomers, and the nature (N- or O-linked) and location on the protein.

As no O-linked glycosylation has been identified for IgM, this section will focus on the analysis of IgM N-linked glycosylation [121]. N-glycans share a common core consisting of two GlcNAc and three mannose residues and are synthesized following a strict regime. They are first produced in the endoplasmic reticulum as Man9 with a triglucose terminus and attached to protein sites with the consensus sequence motif N-X-S/T (where X can be any amino acid except proline, see Figure 3.2) before undergoing stepwise processing by enzymes in the Golgi apparatus. Thus, N-glycans can be divided into three types based on the extent of processing of their structures, from high-mannose (un- or minimally-processed) to hybrid (only one arm processed) to complex (both arms processed) [129]. This relatively simple structural diversity paired with knowledge of human glycobiology together makes the analysis of N-glycans an easier task than for O-glycans [111, 130].

However, while both IgG1 and IgM only feature N-glycans, there are significant differences in their glycosylation patterns. IgG1 heavy chains contain just one N-glycosylation site in the Cγ2 domain at Asn180, which is conserved in all human antibodies except IgA subclasses, as shown in Figure 3.2. By comparison, IgM is much more densely glycosylated with 5 N-glycosylation sites on each heavy chain (GS1–5): Asn46 in Cμ1, Asn209 in Cμ2, Asn272 and Asn279 both in Cμ3, and Asn440 in the C-terminal TP (see Figure 3.1). The J chain also has a single N-glycosylation site at Asn49, bringing the total to 51 N-glycosylation sites in J chain-incorporated pentamers (or even up to 58 with the addition of SC in sIgM) compared to a total of just 2 in IgG1 mAbs (see Table 3.1). With the potential for each site to be independently and differentially glycosylated, including not only whether the site is occupied but also the identity of the attached glycan, the intrinsic heterogeneity of protein glycosylation scales rapidly with the number of glycosites. Thus the dramatic difference in glycosylation makes characterization of IgM a significantly more challenging task compared to IgG mAbs due to increased size and complexity.

Site-specific characterization of IgG1 glycosylation is reasonably simple nowadays, with released glycan analysis being the gold standard [131]. This glycan-centric MS-based approach involves the identification and quantitation of glycans released from the glycoprotein, most commonly done enzymatically by using the enzyme PNGase F for N-glycans but can also be done chemically (which is more typical for O-glycans). Determination of the monosaccharide composition of each N-glycan is fairly straightforward and does not require extensive separation or fragmentation (typically CID), which is necessary for complete structural elucidation or distinguishing between isomers. Label-free quantitation is possible, but derivatization (most commonly via reductive amination with labeling reagent 2-AB) is often performed to improve stability, ionization efficiency, and accuracy. Details of glycan localization on the glycoprotein are lost with this method, but because IgG1 only has a single glycosylation site the analysis of released glycans inherently also provides site-specific information. Thus bulk glycan analysis methods, which are faster/higher throughput with less complex data analysis, cannot provide the same level of detail for IgM [40], requiring other approaches that involve more extensive time and effort at every step in the workflow.

Glycosylation can also be studied with glycoprotein-centric MS methods, which involve the analysis of intact glycoproteins [56, 57, 60, 106]. Conventional native MS is used for glycoform profiling, providing a picture of the total glycosylation on individual proteins [112] but also requiring extensive separation of high-purity samples, higher-energy fragmentation, and high resolution. More detailed localization information can be accessed with top-down or middle-down approaches with electron-based dissociation or alternative fragmentation methods (ETD, ECD, EThcD, and UVPD), but spectra generated in this approach can be incredibly complex even for smaller glycoproteins with only a few glycosylation sites and require specialized software for spectral deconvolution and data analysis. Still, intact glycoform profiling of IgG1 mAbs has been achieved [56, 132], but even with just one glycosylation site, recombinant IgG mAbs can exhibit more than 70 different glycoforms. The larger size and enhanced glycosylation complexity of IgM make it much more heterogeneous, which results in drastically broadened peaks in the mass spectrum that are far from baseline-resolved [133]. Identification of individual IgM glycoforms is effectively impossible as a consequence of this, and intact IgM glycoprotein analysis is typically limited to a determination of an average mass for the bulk population at best. Determination of the drug-to-antibody ratio in IgM-ADCs resulting from site-specific conjugation of azide groups to glycoengineered sialylated IgM has similarly been prohibited [45], despite intact or top/middle-down MS-based analysis being suitable for characterization of IgG-ADCs.

Thus, accessing site-specific information for IgM can only be done with glycoproteomics approaches, which offer a wealth of glycan- and protein-rich information, including site-specific glycan composition and occupancy [128]. Compared to site-specific IgG1 glycosylation analysis with glycan-centric approaches, IgM glycosylation characterization requires more extensive sample preparation, online separation, higher resolution and sensitivity, and data interpretation. Glycopeptide-centric MS analysis workflows begin with proteolytic digestion, after which glycopeptides are typically enriched to overcome their low abundance and poor ionization efficiency compared to non-glycosylated peptides and then separated and analyzed most commonly with RPLC-ESI-MS/MS. Typically glycopeptides are fragmented with CID/HCD with glycosidic bonds being more labile than those of the peptide backbone, resulting in b-ion oxonium glycan fragments and corresponding y-ions for the remaining glycopeptide fragments. Quantitation of glycopeptides can be done with or without labeling, but for mAbs, this is commonly achieved with isotopically labeled internal standards. Site-specific occupation is quantitatively determined by comparing the relative ion intensities of deglycosylated and non-glycosylated forms of the same peptide, which can be differentiated by mass. Deglycosylation with PNGase F converts Asn to Asp, inducing a mass shift of +1 Da, but because deamidation can also occur naturally, this is often performed in heavy water to prevent false positive glycopeptide identifications. An alternative method involves partial deglycosylation to leave only a single GlcNAc residue attached to the peptide with endoglycosidases or chemical methods.

Glycopeptide analysis and quantitation can be more challenging in the context of IgM mAbs even for the same site shared with IgG1. As shown by sequence alignment in Figure 3.2, IgG1's single glycosite Asn180 in the Cγ2 domain is homologous to IgM's Asn279 in the Cμ3 domain, but the type of glycans attached differs (complex-type in IgG1, high-mannose in IgM). This can be problematic because the intensity of b-ions resulting from the CID of oligomannose glycopeptides is often lower than for complex and hybrid N-glycans. The relatively higher degree of sialylation of IgM glycans can also complicate analysis due to signal suppression and the higher lability of negatively charged sialic acid groups. Enzymatic digestion is also highly important for accurate quantitation of glycopeptides, but the presence of glycans can reduce cleavage efficiency. With 5 sites on each heavy chain, IgM can reasonably be expected to be more challenging to fully digest and deglycosylate. Indeed, the two Cμ3 sites at Asn272 and Asn279 are so close together that they are localized to the same tryptic peptide (see Figure 3.2), which prevents site-specific analysis completely if alternative or additional enzymes are not used [134]. Trypsinization also notably results in two glycopeptides for the J chain, the correct tryptic peptide and one with a missed cleavage, which have been reported to carry different glycan compositions [135]. Differences in glycan and glycopeptide identification according to the protease used for digestion have been noted previously for IgM [134, 136], but preferential cleavage of certain glycoforms has even been documented for IgG and was not completely mitigated by denaturation [137].

Site-Specific Glycan Composition of Endogenous Human IgM

Despite these challenges, site-specific analysis of N-glycosylation has been performed on both endogenous IgM (from human serum [136, 138–140] and to a lesser extent milk [141–143]) and recombinant IgM (from human cell lines as well as from plants). Notably, the most comprehensive information comes from glycopeptide-centric studies utilizing multiple proteases to ensure comprehensive coverage of all glycosites, which has primarily been done for serum IgM, as well as in one in-depth analysis of recombinant pentamers and hexamers produced in human cell lines [134]. Results from analyses of IgM glycosylation in human milk have generally been consistent with those from serum described below, but these studies have lacked site-specific quantitation, typically due to only performing tryptic digestion, which prevents information about the Cμ3 sites [141–143]. J chain glycosylation appears rather consistent regardless of association with IgM or IgA or endogenous source. More in-depth site-specific analysis of milk IgM glycosylation should be performed, as glycan differences between milk and serum could potentially be related to increased clearance of ICs via transcytosis by pIgR, for example. SC glycosylation has also mostly been studied in the context of sIgA complexes from human milk and colostrum and not from sIgM assemblies directly, likely due to challenges related to the comparatively low abundance of endogenous sIgM and to the presence of numerous glycosites with impressive glycan heterogeneity on SC.

Besides having at least 49 additional glycosylation sites compared to IgG1 mAbs as detailed in Table 3.1, IgM glycosylation patterns are also more complex. In IgG1, glycans attached at the single glycosite are all complex-type, but the distribution of N-glycan types varies across IgM glycosylation sites. The three N-terminal sites (GS1–3) and the sole J chain glycosite feature predominantly complex-type glycans while glycans attached to the two C-terminal sites (GS4–5) are predominantly high-mannose (see Figure 3.4). This glycosylation pattern is consistent with the J chain and N-terminal regions of IgM being more accessible to the glycan-processing enzymes in the Golgi apparatus before secretion from the B cell, while the presence of predominantly unprocessed high-mannose glycans on the C-terminal Cμ3 site and TP may indicate steric hindrance preventing access, which could extend to enzymes used for proteolytic digestion [62]. Indeed, cryo-EM structures indicate that the 5 TP glycans on each side of the β-sandwich core are located directly next to each other, while the J chain glycan also in the core instead faces outward in the gap, as illustrated in Figure 3.3. Both monomeric and hexameric IgM glycans tend to be more processed than those attached to pentamers, which implicates the presence of the J chain in the reduced accessibility of GS4–5 (as well as emphasizes the importance of fully separating different IgM assemblies before analysis) [134].

Even glycosites that feature the same glycan type appear to be differentially processed, as indicated by variation in the degree of sialylation, fucosylation, bisection, and other features in glycans from GS1–3 and GS4–5. A representative example of this variation across IgM glycosites can be seen in Figure 3.4, which shows the total relative abundance of various features across all glycans identified at each IgM-J glycosite individually from the plasma of 19 healthy donors, from which IgM was purified and digested on-bead with trypsin and GluC to generate glycopeptides for LC-MS/MS analysis [144]. While complex N-glycans are still the predominant type, both hybrid and high-mannose glycan structures have been identified for GS1 and hybrid glycans have also been detected at low abundance for the GS2 [139], as illustrated by the results from Cheng et al. in Figure 3.4. By comparison, all site-specific analyses

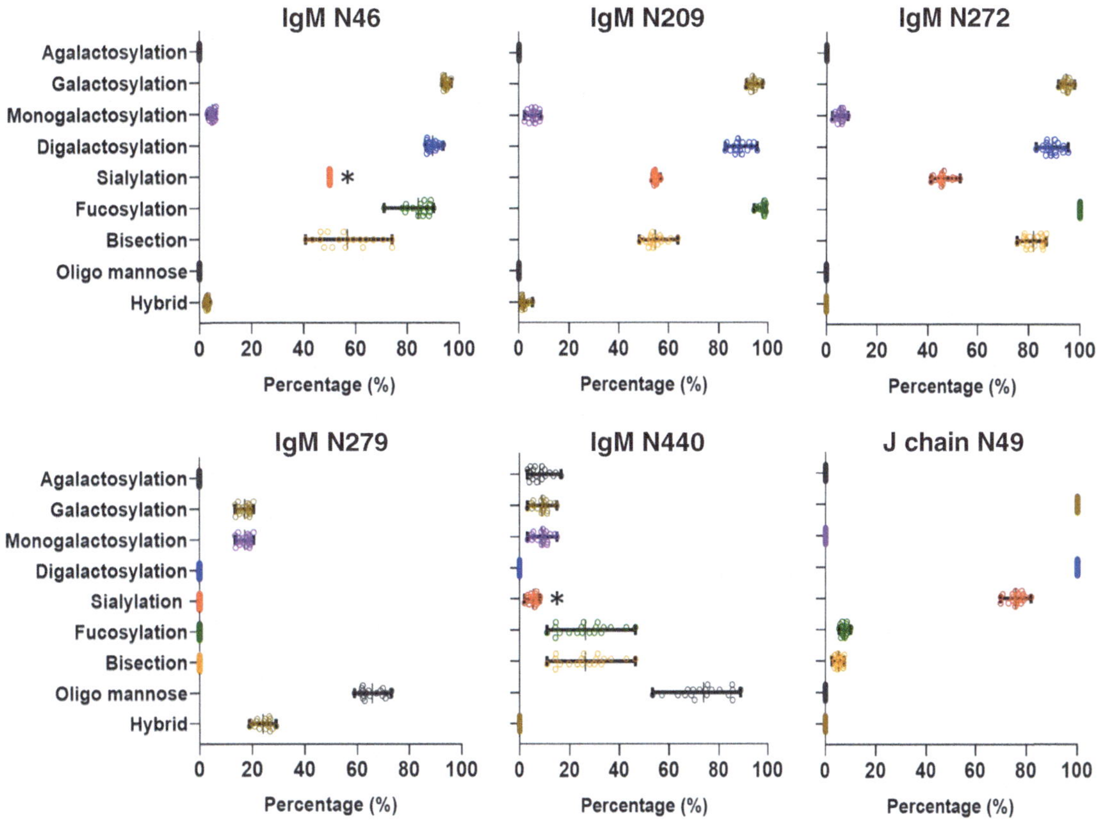

Figure 3.4 Relative abundances of glycan types and features determined from site-specific glycosylation analysis of endogenous IgM from the plasma of 19 healthy individuals. Values plotted as individual points correspond to the percentage of all glycans identified at each site for each donor that matches each type/feature category. Differences in the abundance of minor glycan types and features such as degree of fucosylation and bisection can be seen among IgM glycosites that predominantly feature the same major glycan type (complex-type, GS1–3 in the top row; high-mannose, GS4–5 in the first two panels of the bottom row). Source: Adapted from Cheng et al. [144].

have reported exclusively complex-type glycans attached to GS3, and the presence of bisecting GlcNAc structures is generally higher. The more extensive processing of the Cμ2 Asn209 and Cμ3 Asn272 glycans suggests that GS1 in the Cμ1 domain may also be less accessible to enzymes in the Golgi. The predominantly high-mannose GS4–5 also exhibit differences in their glycan types and features. TP GS5 glycans identified in site-specific analyses of endogenous IgM have been high-mannose, typically with Man6 as the most abundant. By comparison, the glycan composition of GS4 features Man5 as the most abundant glycan overall but always includes hybrid structures well above trace levels [139] (see Figure 3.4 for an example), and complex-type glycans have even been reported. Thus, glycans at GS4 have undergone more processing than those located on the TP at GS5. Interestingly, GS4 located at Asn279 in the Cμ3 domain does not appear to be accessible to all glycan-processing enzymes, as all hybrid and complex-type glycans reported have lacked sialic acid residues, including those analyzed by Cheng et al. in Figure 3.4.

The drastic difference in sialylation between the first three sites and the two C-terminal sites (exemplified in Figure 3.4) could indicate two different functional roles for glycosylation in IgM. Sialic acid, present on the predominantly complex-type GS1–3 in endogenous IgM, has been reported to affect C1q binding and internalization by T cells [1], in addition to being enriched in IgM isolated from the serum of cancer patients [117]. Of the other two sites, glycosylation at GS4 has been linked to complement activation, and TP GS5 glycosylation is critical for J chain incorporation and proper assembly formation. The natural glycosylation pattern of IgM could also facilitate the introduction of new functions in biotherapeutics, for example with the presence of sialic acids used to generate site-specific IgM-ADCs [45].

Site-Specific Glycan Composition of Recombinant IgM

The glycosylation patterns of recombinant mAbs are known to differ from endogenous antibodies, which

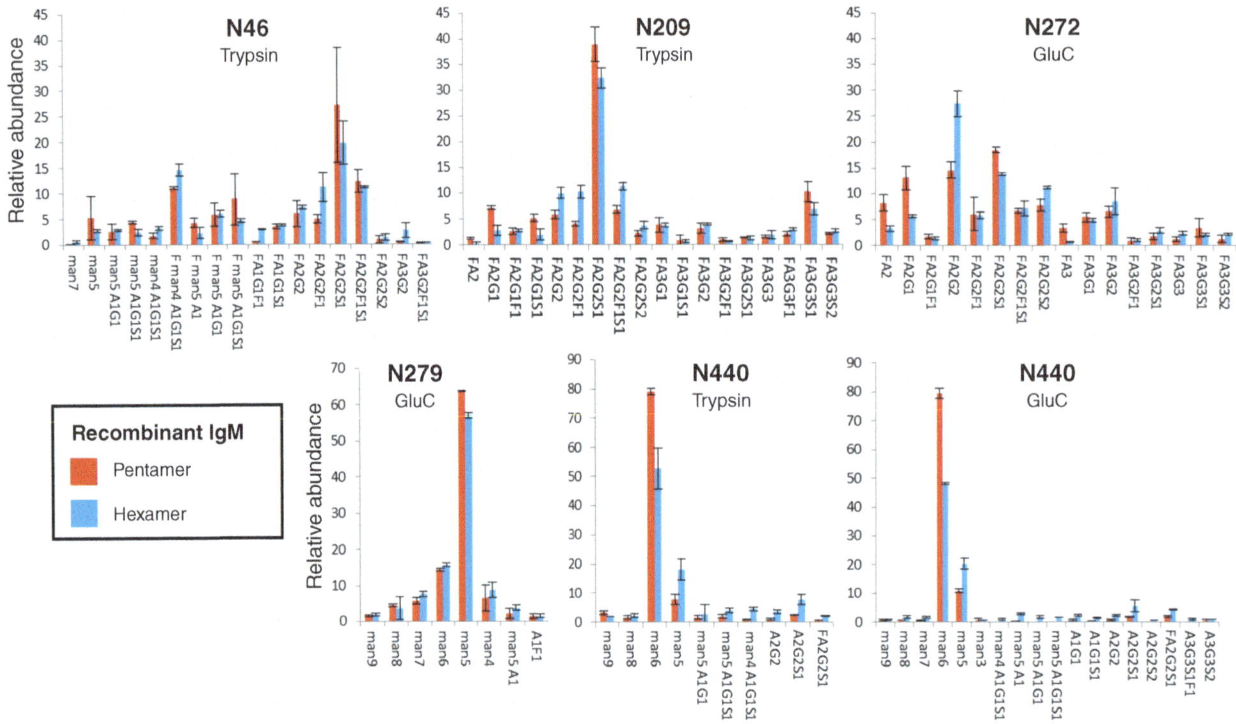

Figure 3.5 Site-specific glycan composition of recombinant IgM pentamers and hexamers. Individual glycan relative abundances are based on two technical replicates from analysis of PAT-SM6 human IgM J chain-incorporated pentameric and J chain-deficient hexameric clones expressed in PER.C6 human cell lines. Glycan nomenclature follows Oxford notation. Source: Adapted from Moh et al. [134].

can impact the function and stability of biotherapeutics which is especially true for IgM [31]. Several studies have compared the effects on glycosylation of recombinant production of human IgM with different mammalian cell lines and cell culture conditions (e.g. temperature and pH), but typically only report differences in glycan features or structures without identification and quantitation of individual glycans [32, 108, 110]. However, there has been at least one exception to this trend with an in-depth analysis of site-specific N-glycosylation of recombinant J chain-incorporated pentameric and J chain-deficient hexameric IgM reported by Moh et al. [134] (Figure 3.5). PAT-SM6 human IgM clones were expressed in the human cell line PER.C6 with the intent of better preserving native glycosylation patterns of serum IgM, and the purified proteins were subjected to reducing SDS-PAGE to separate the heavy chain and J chain of the pentamer. Glycopeptides from trypsin or GluC digests (thus enabling site-specific analysis for GS3–4 in the Cμ3, as indicated by the annotations in Figure 3.5) of the excised gel bands were enriched with ZIC-HILIC SPE and then analyzed by RPLC-MS/MS with either CID or ETD fragmentation, providing comprehensive data on all 5 IgM glycosites for both forms [134] and the sole J chain site in the pentameric mAbs [45].

As shown in Figure 3.5, the overall distribution of the most abundant glycan types at all 5 sites matched closely not just between the pentamer and hexamer but also with endogenous serum IgM (see Figure 3.4 for a comparison). That is, predominantly complex-type glycans with high levels of sialylation were identified at GS1–3 and predominantly high-mannose glycans at GS4–5. The relative abundances of the glycans identified from the single J chain glycosite in this study were later included in a separate report and also matched results from endogenous human IgM [45]. One novel finding, though, was the identification of hybrid, complex, and sialylated glycans at the TP site [134, 145] (see Figure 3.5) rather than exclusively high-mannose as observed in serum IgM studies, which could have functional consequences as indicated above. The individual glycan composition varied at each site for the pentamer and hexamer, but the only statistically significant differences observed between the two forms were found in the TP glycosite. Here, the relative abundance of bi-antennary complex glycans was higher for hexamers than pentamers, where instead high-mannose glycans were more typically found [134]. These results – indicating more extensive glycan processing at the TP site in general as well as in hexamers – could be related either to differences typical of recombinantly produced mAbs or to polymeric state (or both).

Another detailed glycosylation study performed by Loos et al. compared results for human serum IgM to this same PAT-SM6 IgM clone produced not only in the human PER.C6 cell line but also in *Nicotiana benthamiana* plants [36]. Similar to the majority of recombinant IgM glycosylation studies, the overall glycan compositions and relative abundances determined from glycopeptide analysis with LC-ESI-MS were reported for GS1–3 and GS4–5 as a whole rather than for each of the 5 sites individually. Still, comparison between IgM from different sources even without more detailed information is useful for future IgM biotherapeutic development. In particular, glycoengineering *in planta* allowed them to generate IgM decorated with glycans containing galactose and sialic acid residues found endogenously in humans and not in plants, where xylose and fucose are more typical. Importantly, while glycosylation of plant-derived IgM, which was noted to contain a higher proportion of hexamers, still did not exactly match that of serum-derived IgM, the authors reported similar antigen-binding properties [36]. More recently, this same *in planta* glycoengineering strategy was used to produce recombinant J chain-incorporated IgM pentamers with an anti-SARS-CoV-2 Fab isotype-switched from IgG1 that demonstrated markedly increased antigen binding and neutralization capabilities compared to the original IgG1 mAb [20].

Site-Specific Occupancy of the Tailpiece Glycosite

Site-specific occupancy is another important feature to characterize for mAbs and compare with endogenous antibodies. Based on the relative ion intensities of non-glycosylated and deglycosylated peptides following PNGase F digestion without enrichment, Moh et al. reported that all 5 sites in the recombinant human cell line-derived IgM were fully occupied with glycans except for Asn440 in the TP, which was ~60% glycosylated in both the pentamer and hexamer forms [134]. Other analyses of recombinant human IgM have found the TP glycosite to be glycosylated at levels comparable to Moh et al., including IgM produced in human–human–mouse hybridoma cell lines [145] (40–50% occupancy) and glycoengineered plant-derived IgM [20] (56% occupancy for J chain-incorporated pentamers, 31% for monomers). Site-specific analysis of serum IgM found ~36% of TPs to be occupied with glycans, averaged across four healthy human donors [138]. This contrasts several other endogenous IgM studies, which have found Asn440 to be fully occupied with N-glycans, as well as one study that determined aglycosylated TPs were present in just 0.1% of human serum IgM. Thus the occupancy of Asn440 is still somewhat a mystery, with TP aglycosylation reported at significant levels for both endogenous and recombinant IgM.

This becomes even more puzzling given that glycosylation of the TP is critical to the incorporation of the J chain to form pentameric IgM and that disruption of this glycosylation site results in increased production of hexamers lacking the J chain [64, 115]. However, the requirement of TP glycosylation for polymerization and J chain incorporation is still not fully characterized. In particular, the number of glycosylated TP sites required has not been clearly defined yet, and the interpretation of incomplete site occupancy in IgM is nuanced. Because variable occupancy of the TP site has been reported based on analysis of glycopeptides rather than intact glycoproteins or subunits, it is impossible to determine whether reported values describe IgM molecules individually or in bulk. For example, the 60% occupancy reported by Moh et al. could mean that 6/10 TPs in each IgM pentamer (and ~7/12 in each hexamer) are glycosylated, or that all 10 (or 12) TPs in 60% of IgM molecules are fully glycosylated [134]. The former seems most likely due to the role of glycosylation in J chain incorporation, but resolving this issue requires intact glycoprotein analysis or top-down approaches. This is currently intractable for IgM with conventional native MS due to the large size (~1 MDa) and extensive glycosylation (see Table 3.1). CDMS could be used to measure the masses of intact glycoforms, which could potentially confirm or rule out the possibility of some IgM molecules having all TPs fully glycosylated or not but would still be complicated by the presence of glycans on the 41 other sites.

Solving the mystery of IgM TP glycosylation occupancy – necessary to fully compare recombinantly-produced mAbs with endogenous IgM, as well as to complete our understanding of IgM biology and possible relevance in therapeutic antibody engineering – requires even newer tools or perhaps new combinations of existing tools, such as the methods developed by Yin et al. in a recent report. Proteolytic cleavage of recombinant human IgM with IgMBRAZOR, an IgM-specific protease cleaving below the Cμ2-Cμ2 disulfide bond, was monitored over time by online SEC hyphenated to CDMS [62]. Complete digestion resulted in the release of all $F(ab')_2$ subunits (which include GS1–2 for a total of 4 N-glycans), leaving the J chain-coupled pentameric Fc core intact with reduced glycosylation complexity compared to intact IgM assemblies (with all GS3–5 plus J chain Asn49 for a total of 31 out of 51 glycosites). This same approach could be applied to more conclusively determine the details of TP site occupancy, especially with comparison to fully deglycosylated IgM and/or mutants lacking the Cμ3 glycosites.

IgM Disulfide Bond Networks

Antibodies consist of multiple chains held together by multiple disulfide bridges within each heavy and light chain (intrachain), as well as between the chains (interchain), to enable efficient, stable Ab formation [63] (as illustrated for IgM monomers in Figure 3.1). Generally, each constant and variable domain of each heavy and light chain features one intrachain disulfide bond that is buried between the central β-sheets that typify Ig-fold domains (see Figure 3.2). An interchain disulfide links each pair of heavy and light chains via their constant domains in the Fab region, with a number of additional interchain bonds formed between the CH2 and/or hinge regions of the two heavy chains that vary depending on the subclass, as indicated in Figure 3.2. These disulfide bonds ensure proper protein folding, stability, and function, with potentially serious health consequences if proper disulfide formation is disrupted or altered [146–150]. Incomplete or mismatched disulfide bonding can result in size variants from large aggregates to half-bodies, disulfide scrambling and subunit exchange, or even prevent complex formation entirely [150–153]. Some disulfide bridges may be improperly formed without leading to size heterogeneity but instead, exhibit structural differences and reduced stability. Modifications such as cysteinylation, glutathionylation, and trisulfide bond formation can also occur, adding other possible consequences to function [149]. Besides being highly reactive with other sulfhydryl groups, free cysteines or selenocysteines resulting from sequence mutation can also be utilized to generate ADCs by site-specific conjugation [44, 154]. As such, it is critical to analyze correct disulfide connectivity in recombinant mAb preparations and determine whether any (undesirable) disulfide variants, including free cysteines, are present [149, 151, 154]. Specifically, the determination of which cysteine residues are linked to each other and identification of any connectivity variants (including free cysteines, partial native disulfide formation, and non-native disulfide bonds) and/or other modifications is essential for quality control.

While disulfide mapping and identification of the cysteine residues involved were originally achieved using Edman degradation methods, nowadays MS is the go-to method for protein disulfide bond characterization [148, 149, 152, 154]. However, proteins that are especially cysteine-rich and/or have complex connectivity can still be challenging to analyze by MS [61, 155, 156]. This is true for IgG antibodies but even more so for the larger and more complex network of disulfide bridges present in IgM (Table 3.1). With three constant heavy domains and two cysteines in its hinge, IgG1 features 12 intrachain and 4 interchain disulfide bonds (Table 3.1). As illustrated in Figure 3.1, the disulfide bond network of monomeric IgM is comparable with 14 intrachain (with one in each of its VL, CL, Vμ, and Cμ domains) and 3 interchain disulfides in total, which are formed between Cμ1 Cys14 and the light chain C-terminal cysteine in each Fab arm and between the two Cys214 residues to connect the heavy chains via their Cμ2 domains. However, IgM heavy chains also possess two additional cysteine residues (Cys291 in the Cμ3 domain and Cys452 in the TP, indicated with black diamonds in Figure 3.1) that form interchain disulfides between subunits only in polymeric IgM assemblies. This drastically increases not only the number of disulfide linkages but also the complexity of the IgM disulfide bond network because the connectivity of Cμ3 Cys291 and TP Cys452 varies not only between different IgM forms but also within J chain-incorporated pentamers. As shown in Figure 3.1, these cysteines have a free sulfhydryl group in IgM monomers. In IgM hexamers, each Cys291 and Cys452 forms a disulfide with the same residue on the heavy chain of a neighboring subunit. These disulfides are also formed between the eight innermost heavy chains (1B-5A) in J chain-incorporated pentamers illustrated in Figure 3.3. However, the other two TP Cys452 residues present on IgM heavy chains 1A and 5B form disulfides with the J chain at Cys69 and Cys15, respectively [87], while the Cys291 residues in the Cμ3 domains on each of these heavy chains remain free. Incorporation of an additional assembly factor alters the disulfide bond network further, with Cys291 on IgM heavy chain 5B instead forming an interchain disulfide with CD5L (via Cys172 in the SRCR2 domain) or the SC of pIgR (via Cys468 in D5), which each also possess a free cysteine (see Figure 3.3). With intrachain disulfides within each subunit as well, pIgM assemblies possess a grand total of no less than 98 disulfide bonds (J chain-incorporated pentamers) and up to as many as 114 (hexamers) as indicated in Table 3.1. Compared to IgG1, IgM has not only more residues and bonds to investigate but also more potential modifications, mismatches, and resulting effects on complex formation, structure, stability, and function.

The most common approach to characterize disulfide bonds in proteins including IgG mAbs involves proteolytic digestion under nonreducing conditions, with the resulting disulfide-bridged peptides subjected to RPLC-MS/MS analysis and assignment of disulfide bond connectivity is done using one of many available software tools. Conditions must be carefully controlled throughout each step in sample preparation to avoid disulfide scrambling or other experimentally induced disulfide artifacts, which can result from reactions between free and/or disulfide-bonded cysteine residues, especially at alkaline pH and/or high temperatures. Low levels of free cysteine can still be present even for proteins in which all cysteine residues are expected to be disulfide-bonded, including IgG mAbs [157], necessitating the use of an alkylating reagent such as NEM to cap any free cysteines as well as denaturation to access cysteines involved in intrachain disulfides buried within each domain. This is especially important for IgM, which possesses multiple free cysteines in native assemblies.

Additional challenges in the analysis of IgM disulfide networks can even arise during sample preparation. Deglycosylation is necessary if cysteines and glycosylation sites will co-localize to the same peptide. Enzymatic digestion of the alkylated, deglycosylated protein represents the final step in sample preparation and should be performed using a carefully selected enzyme to give the "simplest" interchain disulfide-linked peptides, ideally containing two chains of ~4–15 amino acid residues linked by only one disulfide bond. A number of tools are thus available for *in silico* digestion to compare peptides resulting from the use of various enzymes. Trypsin, Lys-C, Asp-N, and pepsin are common choices, with the latter having high cleavage efficiency at very acidic pH where non-native disulfide formation is largely reduced. Alternatively, a combination of enzymes may also be used, either to yield more ideal peptides or to improve digestion efficiency. Compared to IgG1, the extensive glycosylation and disulfide networks of IgM assemblies require deglycosylation and the use of additional or different enzymes compared to IgG1, not just to combat possible reduced digest efficiency due to the large polymeric structure [62] but also to generate peptides that can be analyzed by LC-MS. As an example, trypsinization of J chain-incorporated pentameric IgM without deglycosylation or reduction would yield a variety of different disulfide-linked peptides with 1–4 N-glycans, as well as glycosylated peptides linked by two disulfides. Thus high-throughput methods developed for simultaneous monitoring of multiple features of IgG mAbs are not directly suitable for analysis of IgM [55].

Detection and quantitation of free cysteine and non-native disulfides is even more challenging by comparison due to their drastically low abundances compared to native disulfides [154]. Even with just 16 unique cysteine residues in IgG1 mAbs, there are 136 possible disulfide bond combinations, only 9 of which are native. While both IgG1 and IgM monomers form 9 different native disulfides, there are 19 unique cysteines in IgM heavy and light chains, translating to a total of 190 different disulfide permutations just in monomers. This number increases further to 378 and 1,431 in J chain-incorporated pentamers and CD5L-associated circulatory IgM, respectively, but only 13–14 of these correspond to native disulfide bonds. Efforts to improve analysis of low levels of non-native disulfides in IgG mAbs rely on post-column partial reduction of the disulfide-bonded peptides to ensure that both the remaining intact disulfide and newly-formed reduced peptides undergo MS analysis at exactly the same retention time [155, 158], thus simplifying data analysis, but there is still room for improvement.

The presence of cysteines that remain free and/or form canonical disulfide bonds with multiple partners, as described above, introduces further challenges for IgM, which can be aggravated further by the presence of multiple co-occurring assemblies during sample preparation [159]. Disulfide scrambling has also been reported for IgM, including the formation of intermediate non-native disulfides between the Cys452 residues in IgM monomers [65, 160, 161]. Indeed, to circumvent these issues in a recent report from Su et al. describing the heterogeneity in disulfide bond connectivity and free cysteines of teleost IgM (which, among other differences, does not possess an equivalent of human IgM Cμ3 Cys291 and assembles into various polymeric states), the authors first performed nonreducing SDS-PAGE of alkylated IgM to separate the different polymeric assemblies [162]. The bands corresponding to the various IgM forms were then excised and subjected to in-gel deglycosylation and digestion to generate assembly-specific peptides for LC-MS/MS analysis. As shown in Figure 3.6, HCD fragmentation confirmed that the TP cysteine residue was present in two different states – with a free sulfhydryl group or disulfide-bound to the same residue from another heavy chain – in monomeric and polymeric grass carp IgM [162]. A similar observation would result from the analysis of monomeric versus polymeric human IgM, but more complex due to there being three different disulfide bonds involving the TP Cys452 in J chain-incorporated pentamers (see Figure 3.3).

Role of Disulfide Bonds in Formation of IgM Assemblies

As exemplified by the use of SDS-PAGE to separate grass carp IgM assemblies above [162], the presence of multiple species varying in size or of one species differing in size from that expected for a fully-assembled complex can often be revealed with numerous methods requiring less extensive time and effort in sample preparation and analysis. This observation of size variants can, in turn, be used to more easily deduce the occurrence of improper disulfide formation than with MS analysis. However, proper assembly into IgM monomers or polymers can still proceed even with some degree of improper disulfide formation due to extensive non-covalent interactions and the somewhat redundant nature of the interchain disulfide bond network between chains or monomer subunits [37, 163–165].

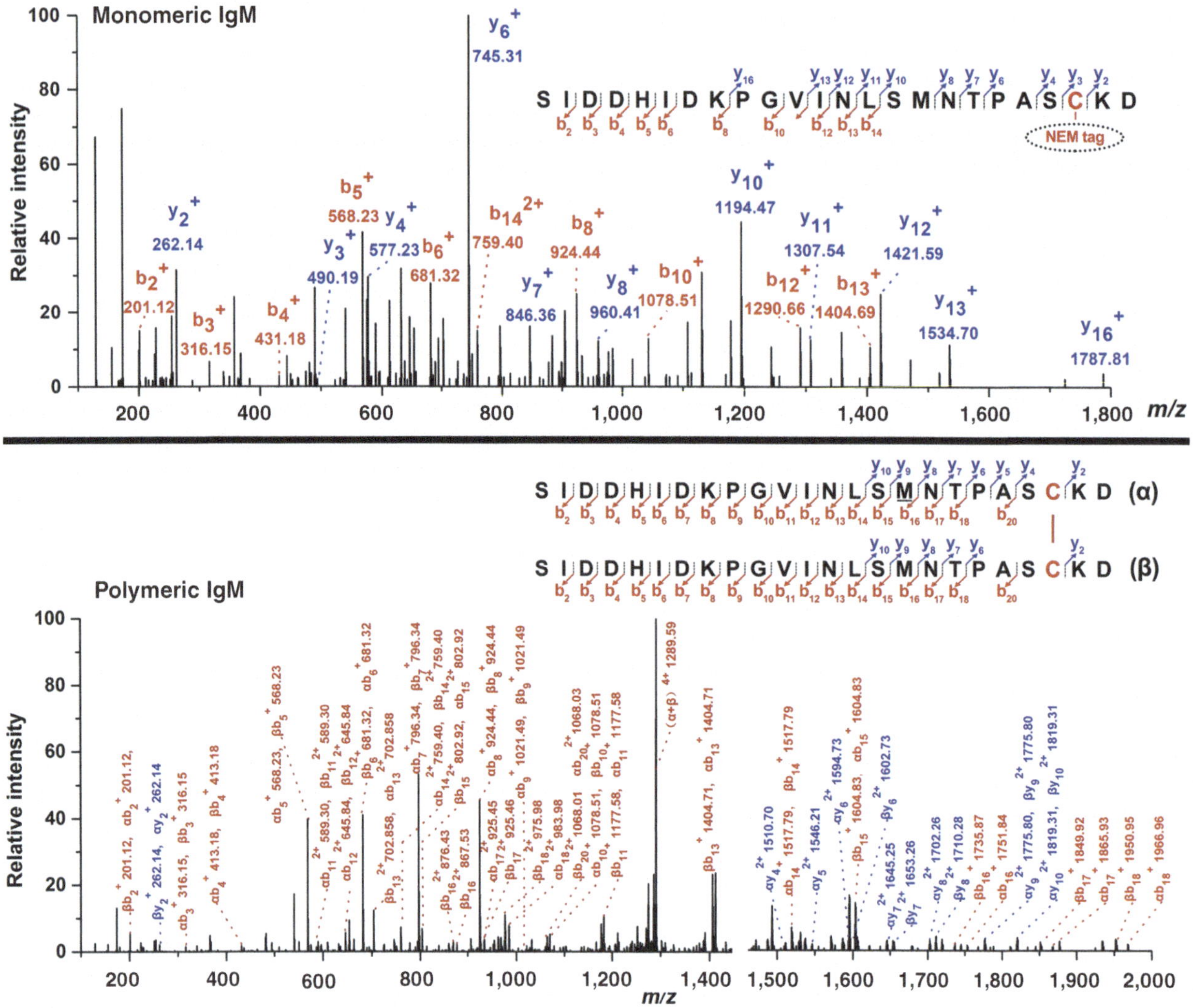

Figure 3.6 MS-based identification of the grass carp IgM tailpiece cysteine being present with a free sulfhydryl group in monomeric IgM (top) versus disulfide-bound in polymeric IgM (bottom). Cys574 in the tailpiece of grass carp IgM is similar to Cys452 in the tailpiece of human IgM. The different oligomeric isoforms of grass carp IgM were separated before the generation of tryptic peptides and subsequent LC-ESI-MS/MS analysis. Each panel shows a representative HCD-MS^2 spectrum for the corresponding assembly-specific Cys574-containing peptide precursor, with fragment b-ions annotated in red and y-ions in blue. Source: Adapted from Su et al. [162].

While this necessitates the use of MS-based methods to fully characterize IgM mAbs, it could also be exploited to develop biotherapeutics with more desirable structures and functional features while maintaining proper assembly into polymeric complexes by altering the cysteine and disulfide bond network. For example, mutation of the Cμ3 Cys291 residue which remains free on the heavy chains lining the gap in J chain-incorporated pentamers could be explored to generate IgM biotherapeutics with the advantages of polyvalency and high avidity binding but with reduced capacity to form certain interactions *in vivo* post-administration (e.g. association with CD5L, pIgR-mediated transport across the epithelium). However, almost all reports of the relative contributions of interchain disulfide bonds to the formation of IgM assemblies have been based on the mutation of mouse IgM with attempts to reconstitute polymeric IgM *in vitro* [81, 115]. More recent work provides much better insight directly applicable to engineering human IgM mAbs, as Hasegawa et al. importantly not only performed mutagenesis on human IgM produced in HEK293 cells with or without the J chain but also monitored the biosynthesis of IgM within these cells and characterized the secreted assemblies [37]. These results from recombinant human IgM suggest that the roles of the interchain disulfides formed between subunits via their Cμ3 domains and TPs are nonredundant and also differ between IgM hexamers and J-chain pentamers [161]. Perhaps this can be rationalized somewhat by a model proposed by Giannone et al., whereby hexamer assembly involves the formation of intermediate non-native disulfides between the TPs in IgM monomers that are subsequently attacked by a disulfide isomerase to promote successive addition of subunits covalently linked through their TPs [65, 160]. Because the heavy chains are also covalently linked near the C-terminus of their Cμ2 domains (see Figure 3.1), the formation of the intra-monomer TP disulfide contributes to a conformational change with outward bending at the Cμ3-Cμ4 junction [65]. Thus Cμ3 Cys291 residues likely play an outsized role in stabilizing interactions between subunits as new protomers are added sequentially during hexamer formation [163, 164].

Finally, several findings reported by Hasegawa et al. [37] also directly conflict with earlier studies with mice IgM, where, for example, the Cμ3 cysteine was found to have no influence on polymeric IgM assembly while the TP cysteine was reported to be essential [165]. While these discrepancies regarding the formation of mouse versus human IgM assemblies indicate that much more extensive characterization of human IgM disulfide/cysteine mutants is needed, the successful recombinant production of intact J chain-incorporated pentameric human IgM mAbs lacking one or both of the Cμ3 and TP cysteines holds promise both for developing disulfide/cysteine-engineered IgM biotherapeutics and for facilitating analytical characterization of their disulfide bond networks with reduced complexity.

Relative Susceptibility of IgM Disulfide Bonds to Reductive Cleavage

IgM disulfide bonds also exhibit differential susceptibility to reductive cleavage. Importantly, while this feature has also been reported for IgG mAbs, in IgM it can be exploited to facilitate analytical characterization not only by reducing sample size and complexity but also by enabling distinction between features of particular subunits and subassemblies with their structural integrity intact. This is because intrachain disulfide bonds, which stabilize the fold of each domain by covalently linking the two β-sheets, are more stable than interchain bonds [157]. Careful titration of a reducing reagent will thus first result in the dissociation of subunits or peptide chains from pIgM assemblies [161].

Results from multiple independent reports support a consensus ranking of the relative labilities of IgM's interchain disulfide bonds, broadly grouped into three categories: intersubunit, intrasubunit, and μ-J chain [161, 166, 167]. (Interchain disulfides formed between CD5L and pIgR SC with 5B-Cμ3 Cys291 have not been systematically investigated in this context, but the evidence so far suggests they are somewhat weaker but still comparable to the intersubunit disulfides. CD5L can be partially released from human serum IgM upon addition of just 0.125 mM DTT compared to a higher concentration of 1 mM being required for dissociation into mostly monomer subunits [38], and cryo-EM structures of sIgM only provide evidence for partial disulfide bond formation with the SC of pIgR to begin with.) Intersubunit disulfide bonds formed between IgM protomers via their Cμ3 and TP regions (i.e. Cys291–Cys291 and Cys452–Cys452 disulfides) are the least resistant to reductive cleavage, with J chain-incorporated pentamers dissociating first into monomers and J chain-coupled dimers [81, 161]. The exact ranking of the intrasubunit disulfides linking the heavy chains to each other (via Cμ2 Cys214) and to the light chains (via Cμ1 Cys14) varies somewhat between reports and may be influenced by light chain isotype [152], but the literature is in agreement that these interchain disulfides fall in the middle of the three types. Finally, disulfides formed between the J chain and the TPs of the two outermost heavy chains (1A and 5B as indicated in Figure 3.3) are the most resistant to reduction. Thus, limited reduction paired with a separation method like gel electrophoresis or gel filtration chromatography can be used to analyze the features of IgM monomer subunits and/or heavy chains that interact with the J chain separately from the rest.

Conclusions

Here we have described analytical challenges in characterizing IgM biotherapeutics, which represents the next major obstacle in the way of their widespread clinical use now that more suitable methods are in place to help overcome bottlenecks in the production of functional, specific IgM mAbs. Most advanced MS-based methods to analyze recombinant antibodies have been developed and optimized for IgG1, as this is the natural human template on which most therapeutic mAbs have been based. Many of these methods can also be used to analyze IgM. However, due to its increased molecular complexity, even common MS-based approaches, for example, bottom-up proteomics and glycoproteomics, are less straightforward when applied to the analysis of IgM. In particular, the increase in size, molecular weight, number of modifications, glycosylation sites, and disulfide bridges make IgM a tough nut to crack. Moreover, with its extended size and additional assembly factors (e.g. J chain, CD5L, and pIgR) the ability to form co-occurring assemblies and/or structural conformers grows exponentially. Nonetheless, improvements and new developments in instrumentation/hardware, data analysis software, and sample preparation workflows and tools have continually pushed the limits of the molecular heterogeneity that can be disentangled with MS-based approaches, consequently providing us with a wealth of structural and functional detail for both endogenous and recombinant IgM. As the development of IgM therapeutics continues to gain momentum, it becomes increasingly important to further advance the analytical toolbox, as quality control remains a key step before successful clinical use of this exciting antibody format.

Acknowledgments

A.J.R. Heck acknowledges support from the Netherlands Organization for Scientific Research (NWO) through the Spinoza Award (SPI.2017.028). A.D. Rolland is supported by the European Molecular Biology Organization (EMBO) through a long-term Postdoctoral Fellowship (ALTF 371-2022).

References

1 Colucci, M., Stöckmann, H., Butera, A. et al. (2015). Sialylation of N-linked glycans influences the immunomodulatory effects of IgM on T cells. *J. Immunol.* 194 (1): 151–157.

2 Klimovich, V.B. (2011). IgM and its receptors: structural and functional aspects. *Biochem. Mosc.* 76 (5): 534–549.

3 Liu, J., Wang, Y., Xiong, E. et al. (2019). Role of the IgM fc receptor in immunity and tolerance. *Front. Immunol.* 10: 529.

4 Hale, M., Netland, J., Chen, Y. et al. (2022). IgM antibodies derived from memory B cells are potent cross-variant neutralizers of SARS-CoV-2. *J. Exp. Med.* 219 (9): e20220849.

5 Matsumoto, M.L. (2022). Molecular mechanisms of Multimeric assembly of IgM and IgA. *Annu. Rev. Immunol.* 40: 221–247.

6 Petrušić, V., Stojanović, M., Živković, I. et al. (2010). Changes in composition of IgM polymers in patients suffering from recurrent urinary bacterial infections after bacterial immunization treatment. *Immunol. Investig.* 39 (8): 781–795.

7 Hughey, C.T., Brewer, J.W., Colosia, A.D. et al. (1998). Production of IgM hexamers by Normal and autoimmune B cells: implications for the physiologic role of hexameric IgM. *J. Immunol.* 161 (8): 4091–4097.

8 Eskeland, T. and Christensen, T.B. (1975). IgM molecules with and without J chain in serum and after purification, studied by ultra-centrifugation, electrophoresis, and electron microscopy. *Scand. J. Immunol.* 4 (3): 217–228.

9 Azuma, Y., Ishikawa, Y., Kawai, S. et al. (2007). Recombinant human hexamer-dominant IgM monoclonal antibody to ganglioside GM3 for treatment of melanoma. *Clin. Cancer Res.* 13 (9): 2745–2750.

10 Michaelsen, T.E., Emilsen, S., Sandin, R.H. et al. (2017). Human secretory IgM antibodies activate human complement and offer protection at mucosal surface. *Scand. J. Immunol.* 85 (1): 43–50.

11 Michaud, E., Mastrandrea, C., Rochereau, N., and Paul, S. (2020). Human secretory IgM: an elusive player in mucosal immunity. *Trends Immunol.* 41 (2): 141–156.

12 Gonzalez-Quintela, A., Alende, R., Gude, F. et al. (2008). Serum levels of immunoglobulins (IgG, IgA, IgM) in a general adult population and their relationship with alcohol consumption, smoking and common metabolic abnormalities. *Clin. Exp. Immunol.* 151 (1): 42–50.

13 Gong, S. and Ruprecht, R.M. (1943). Immunoglobulin M: an ancient antiviral weapon – rediscovered. *Front. Immunol.* 2020: 11.

14 Ku, Z., Xie, X., Hinton, P.R. et al. (2021). Nasal delivery of an IgM offers broad protection from SARS-CoV-2 variants. *Nature* 595 (7869): 718–723.

15 Liu, J., Mao, F., Chen, J. et al. (2023). An IgM-like inhalable ACE2 fusion protein broadly neutralizes SARS-CoV-2 variants. *Nat. Commun.* 14 (1): 5191.

16 Piepenbrink, M.S., Park, J.-G., Oladunni, F.S. et al. (2021). Therapeutic activity of an inhaled potent SARS-CoV-2 neutralizing human monoclonal antibody in hamsters. *Cell Rep. Med.* 2 (3): 100218.

17 Pisil, Y., Yazici, Z., Shida, H., and Miura, T. (2021). Is SARS-CoV-2 neutralized more effectively by IgM and IgA than IgG having the same fab region? *Pathogens* 10 (6): 751.

18 Rahmel, T., Kraft, F., Haberl, H. et al. (2022). Intravenous IgM-enriched immunoglobulins in critical COVID-19: a multicentre propensity-weighted cohort study. *Crit. Care* 26 (1): 204.

19 Zhang, J., Zhang, H., and Sun, L. (2022). Therapeutic antibodies for COVID-19: is a new age of IgM, IgA and bispecific antibodies coming? *mAbs* 14 (1): 2031483.

20 Kallolimath, S., Palt, R., Föderl-Höbenreich, E. et al. (2023). Glyco engineered pentameric SARS-CoV-2 IgMs

show superior activities compared to IgG1 orthologues. *Front. Immunol.* 14: 1147960.

21 Wolbank, S., Kunert, R., Stiegler, G., and Katinger, H. (2003). Characterization of human class-switched polymeric (immunoglobulin M [IgM] and IgA) anti-human immunodeficiency virus type 1 antibodies 2F5 and 2G12. *J. Virol.* 77 (7): 4095–4103.

22 Oostindie, S.C., Lazar, G.A., Schuurman, J., and Parren, P.W.H.I. (2022). Avidity in antibody effector functions and biotherapeutic drug design. *Nat. Rev. Drug Discov.* 21 (10): 715–735.

23 de Jong, R.N., Beurskens, F.J., Verploegen, S. et al. (2016). A novel platform for the potentiation of therapeutic antibodies based on antigen-dependent formation of IgG hexamers at the cell surface. *PLoS Biol.* 14 (1): e1002344.

24 Samsudin, F., Yeo, J.Y., Gan, S.K.-E., and Bond, P.J. (2020). Not all therapeutic antibody isotypes are equal: the case of IgM versus IgG in pertuzumab and trastuzumab. *Chem. Sci.* 11 (10): 2843–2854.

25 Shinde, S.H., Sandeep, and Pande, A.H. (2024). Polyvalency: an emerging trend in the development of clinical antibodies. *Drug Discov. Today* 29 (1): 103846.

26 Chiu, M.L., Goulet, D.R., Teplyakov, A., and Gilliland, G.L. (2019). Antibody structure and function: the basis for engineering therapeutics. *Antibodies* 8 (4): 55.

27 Goulet, D.R. and Atkins, W.M. (2020). Considerations for the design of antibody-based therapeutics. *J. Pharm. Sci.* 109 (1): 74–103.

28 Keyt, B.A., Baliga, R., Sinclair, A.M. et al. (2020). Structure, function, and therapeutic use of IgM antibodies. *Antibodies* 9 (4): 53.

29 Oskam, N., Ooijevaar-de Heer, P., Derksen, N.I.L. et al. (2022). At critically low antigen densities, IgM hexamers outcompete both IgM pentamers and IgG1 for human complement deposition and complement-dependent cytotoxicity. *J. Immunol.* 209 (1): 16–25.

30 Melani, R.D., Des Soye, B.J., Kafader, J.O. et al. (2022). Next-generation serology by mass spectrometry: readout of the SARS-CoV-2 antibody repertoire. *J. Proteome Res.* 21 (1): 274–288.

31 Maiorella, B.L., Winkelhake, J., Young, J. et al. (1993). Effect of culture conditions on IgM antibody structure, pharmacokinetics and activity. *Nat. Biotechnol.* 11 (3): 387–392.

32 Hennicke, J., Lastin, A.M., Reinhart, D. et al. (2017). Glycan profile of CHO derived IgM purified by highly efficient single step affinity chromatography. *Anal. Biochem.* 539: 162–166.

33 She, Y.-M., Dai, S., and Tam, R.Y. (2022). Highly sensitive characterization of non-human glycan structures of monoclonal antibody drugs utilizing tandem mass spectrometry. *Sci. Rep.* 12 (1): 15109.

34 Vorauer-Uhl, K., Wallner, J., Lhota, G. et al. (2010). IgM characterization directly performed in crude culture supernatants by a new simple electrophoretic method. *J. Immunol. Methods* 359 (1): 21–27.

35 Tchoudakova, A., Hensel, F., Murillo, A. et al. (2009). High level expression of functional human IgMs in human PER.C6® cells. *MAbs* 1 (2): 163–171.

36 Loos, A., Gruber, C., Altmann, F. et al. (2014). Expression and glycoengineering of functionally active heteromultimeric IgM in plants. *Proc. Natl. Acad. Sci.* 111 (17): 6263–6268.

37 Hasegawa, H., Wang, S., Kast, E. et al. (2024). Understanding the biosynthesis of human IgM SAM-6 through a combinatorial expression of mutant subunits that affect product assembly and secretion. *PLoS One* 19 (6): e0291568.

38 Oskam, N., den Boer, M.A., Lukassen, M.V. et al. (2023). CD5L is a canonical component of circulatory IgM. *Proc. Natl. Acad. Sci.* 120 (50): e2311265120.

39 Chromikova, V., Mader, A., Steinfellner, W., and Kunert, R. (2015). Evaluating the bottlenecks of recombinant IgM production in mammalian cells. *Cytotechnology* 67 (2): 343–356.

40 Arnold, J.N., Wormald, M.R., Suter, D.M. et al. (2005). Human serum IgM glycosylation: identification of glycoforms that can bind to Mannan-binding lectin. *J. Biol. Chem.* 280 (32): 29080–29087.

41 Overdijk, M.B., Strumane, K., Beurskens, F.J. et al. (2020). Dual epitope targeting and enhanced hexamerization by DR5 antibodies as a novel approach to induce potent antitumor activity through DR5 agonism. *Mol. Cancer Ther.* 19 (10): 2126–2138.

42 Overdijk, M.B., Strumane, K., Buijsse, A.O. et al. (2019). Abstract 2391: DR5 agonist activity of HexaBody®-DR5/DR5 (GEN1029) is potentiated by C1q and independent of fc-gamma receptor binding in preclinical tumor models. *Cancer Res.* 79 (13): 2391–2391.

43 Verstraete, M.M., Heinkel, F., Li, J. et al. (2023). Multivalent IgM scaffold enhances the therapeutic potential of variant-agnostic ACE2 decoys against SARS-CoV-2. *MAbs* 15 (1): 2212415.

44 Vire, B., Skarzynski, M., Thomas, J.D. et al. (2014). Harnessing the Fcμ receptor for potent and selective cytotoxic therapy of chronic lymphocytic leukemia. *Cancer Res.* 74 (24): 7510–7520.

45 Moh, E.S.X., Sayyadi, N., and Packer, N.H. (2019). Chemoenzymatic glycan labelling as a platform for site-specific IgM-antibody drug conjugates. *Anal. Biochem.* 584: 113385.

46 Wang, B., Wang, L., Kothambawala, T. et al. (2020). IGM-8444 as a potent agonistic death receptor 5 (DR5) IgM antibody: induction of tumor cytotoxicity, combination with chemotherapy and in vitro safety profile. *J. Clin. Oncol.* 38 (15_Suppl): 3595–3595.

47 Wang, B.T., Kothambawala, T., Wang, L. et al. (2021). Multimeric anti-DR5 IgM agonist antibody IGM-8444 is a potent inducer of cancer cell apoptosis and synergizes with chemotherapy and BCL-2 inhibitor ABT-199. *Mol. Cancer Ther.* 20 (12): 2483–2494.

48 Chen, Y., Paluch, M., Zorn, J.A. et al. (2020). Targeted IgMs agonize ocular targets with extended vitreal exposure. *MAbs* 12 (1): 1818436.

49 Mimura, Y., Saldova, R., Mimura-Kimura, Y. et al. (2021). Micro-heterogeneity of antibody molecules. In: *Antibody Glycosylation* (ed. M. Pezer), 1–26. Cham: Springer International Publishing.

50 Upton, R., Migas, L.G., Pacholarz, K.J. et al. (2019). Hybrid mass spectrometry methods reveal lot-to-lot differences and delineate the effects of glycosylation on the tertiary structure of Herceptin®. *Chem. Sci.* 10 (9): 2811–2820.

51 Gawlitzek, M., Estacio, M., Fürch, T., and Kiss, R. (2009). Identification of cell culture conditions to control N-glycosylation site-occupancy of recombinant glycoproteins expressed in CHO cells. *Biotechnol. Bioeng.* 103 (6): 1164–1175.

52 Liu, H., Nowak, C., Shao, M. et al. (2016). Impact of cell culture on recombinant monoclonal antibody product heterogeneity. *Biotechnol. Prog.* 32 (5): 1103–1112.

53 Zhang, H., Cui, W., and Gross, M.L. (2014). Mass spectrometry for the biophysical characterization of therapeutic monoclonal antibodies. *FEBS Lett.* 588 (2): 308–317.

54 Cotham, V.C. and Brodbelt, J.S. (2016). Characterization of therapeutic monoclonal antibodies at the subunit-level using middle-down 193 nm ultraviolet photodissociation. *Anal. Chem.* 88 (7): 4004–4013.

55 Wang, Y., Li, X., Liu, Y.-H. et al. (2016). Simultaneous monitoring of oxidation, deamidation, isomerization, and glycosylation of monoclonal antibodies by liquid chromatography-mass spectrometry method with ultrafast tryptic digestion. *MAbs* 8 (8): 1477–1486.

56 Yang, Y., Wang, G., Song, T. et al. (2017). Resolving the micro-heterogeneity and structural integrity of monoclonal antibodies by hybrid mass spectrometric approaches. *MAbs* 9 (4): 638–645.

57 Wohlschlager, T., Scheffler, K., Forstenlehner, I.C. et al. (2018). Native mass spectrometry combined with enzymatic dissection unravels glycoform heterogeneity of biopharmaceuticals. *Nat. Commun.* 9 (1): 1713.

58 Thompson, N.J., Rosati, S., Rose, R.J., and Heck, A.J.R. (2013). The impact of mass spectrometry on the study of intact antibodies: from post-translational modifications to structural analysis. *Chem. Commun.* 49 (6): 538–548.

59 Kline, J.T., Melani, R.D., and Fornelli, L. (2023). Mass spectrometry characterization of antibodies at the intact and subunit levels: from targeted to large-scale analysis. *Int. J. Mass Spectrom.* 492: 117117.

60 Schachner, L.F., Mullen, C., Phung, W. et al. (2024). Exposing the molecular heterogeneity of glycosylated biotherapeutics. *Nat. Commun.* 15 (1): 3259.

61 McClain, A., Zhang, Y., Yin, Y. et al. (2022). Using digestion by IdeS protease to improve quantification of degradants in monoclonal antibodies by non-reducing capillary gel electrophoresis. *Anal. Chem.* 94 (50): 17388–17395.

62 Yin, V., Deslignière, E., Mokiem, N. et al. (2024). Not all arms of IgM are equal: following hinge-directed cleavage by online native SEC-Orbitrap-based CDMS. *J. Am. Soc. Mass Spectrom.* 35 (6): 1320–1329.

63 Schroeder, H.W. and Cavacini, L. (2010). Structure and function of immunoglobulins. *J. Allergy Clin. Immunol.* 125 (2, Suppl 2): S41–S52.

64 Lalla, C.D., Fagioli, C., Cessi, F.S. et al. (1998). Biogenesis and function of IgM: the role of the conserved μ-chain tailpiece glycans. *Mol. Immunol.* 35 (13): 837–845.

65 Pasalic, D., Weber, B., Giannone, C. et al. (2017). A peptide extension dictates IgM assembly. *Proc. Natl. Acad. Sci.* 114 (41): E8575–E8584.

66 Smith, R.I.F., Coloma, M.J., and Morrison, S.L. (1995). Addition of a mμ-tailpiece to IgG results in polymeric antibodies with enhanced effector functions including complement-mediated cytolysis by IgG4. *J. Immunol.* 154 (5): 2226–2236.

67 Sopp, J.M., Peters, S.J., Rowley, T.F. et al. (2021). On-target IgG hexamerisation driven by a C-terminal IgM tail-piece fusion variant confers augmented complement activation. *Commun. Biol.* 4 (1): 1031.

68 Braathen, R., Sørensen, V., Brandtzaeg, P. et al. (2002). The carboxyl-terminal domains of IgA and IgM direct isotype-specific polymerization and interaction with the polymeric immunoglobulin receptor. *J. Biol. Chem.* 277 (45): 42755–42762.

69 Petrušić, V., Živković, I., Stojanović, M. et al. (2011). Antigenic specificity and expression of a natural idiotope on human pentameric and hexameric IgM polymers. *Immunol. Res.* 51 (1): 97–107.

70 Seifert, O., Plappert, A., Heidel, N. et al. (2012). The IgM CH_2 domain as covalently linked homodimerization module for the generation of fusion proteins with dual specificity. *Protein Eng. Des. Sel.* 25 (10): 603–612.

71 Chen, Q., Menon, R., Calder, L.J. et al. (2022). Cryomicroscopy reveals the structural basis for a flexible hinge motion in the immunoglobulin M pentamer. *Nat. Commun.* 13 (1): 6314.

72 Sharp, T.H., Boyle, A.L., Diebolder, C.A. et al. (2019). Insights into IgM-mediated complement activation based on in situ structures of IgM-C1-C4b. *Proc. Natl. Acad. Sci.* 116 (24): 11900–11905.

73 Hiramoto, E., Tsutsumi, A., Suzuki, R. et al. (2018). The IgM pentamer is an asymmetric pentagon with an open groove that binds the AIM protein. *Sci. Adv.* 4 (10): eaau1199.

74 Li, Y., Wang, G., Li, N. et al. (2020). Structural insights into immunoglobulin M. *Science* 367 (6481): 1014–1017.

75 Kumar, N., Arthur, C.P., Ciferri, C., and Matsumoto, M.L. (2021). Structure of the human secretory immunoglobulin M core. *Structure* 29 (6): 564–571.e563.

76 Akhouri, R.R., Goel, S., and Skoglund, U. (2023). Cryo-electron microscopy of IgM-VAR2CSA complex reveals IgM inhibits binding of plasmodium falciparum to chondroitin Sulfate a. *Nat. Commun.* 14 (1): 6391.

77 Li, Y., Shen, H., Zhang, R. et al. (2023). Immunoglobulin M perception by FcμR. *Nature* 615 (7954): 907–912.

78 Chen, Q., Menon, R.P., Masino, L. et al. (2023). Structural basis for Fc receptor recognition of immunoglobulin M. *Nat. Struct. Mol. Biol.* 30 (7): 1033–1039.

79 Wang, Y., Su, C., Ji, C., and Xiao, J. (2024). CD5L associates with IgM via the J chain. *Nat. Commun.* 15: https://doi.org/10.1038/s41467-024-52175-y.

80 Ji, C., Shen, H., Su, C. et al. (2023). Plasmodium falciparum has evolved multiple mechanisms to hijack human immunoglobulin M. *Nat. Commun.* 14 (1): 2650.

81 Wiersma, E.J., Collins, C., Fazel, S., and Shulman, M.J. (1998). Structural and functional analysis of J chain-deficient IgM. *J. Immunol.* 160 (12): 5979–5989.

82 Randall, T.D., Brewer, J.W., and Corley, R.B. (1992). Direct evidence that J chain regulates the polymeric structure of IgM in antibody-secreting B cells. *J. Biol. Chem.* 267 (25): 18002–18007.

83 Ghumra, A., Shi, J., Mcintosh, R.S. et al. (2009). Structural requirements for the interaction of human IgM and IgA with the human Fcα/μ receptor. *Eur. J. Immunol.* 39 (4): 1147–1156.

84 Kubagawa, H., Oka, S., Kubagawa, Y. et al. (2009). Identity of the elusive IgM fc receptor (FcμR) in humans. *J. Exp. Med.* 206 (12): 2779–2793.

85 Kubagawa, H., Skopnik, C.M., Zimmermann, J. et al. (2017). Authentic IgM fc receptor (FcμR). In: *IgM and its Receptors and Binding Proteins* (ed. H. Kubagawa and P.D. Burrows), 25–45. Cham: Springer International Publishing.

86 Rochereau, N., Michaud, E., Waeckel, L. et al. (2021). Essential role of TOSO/FAIM3 in intestinal IgM reverse transcytosis. *Cell Rep.* 37 (7): 110006.

87 Frutiger, S., Hughes, G.J., Paquet, N. et al. (1992). Disulfide bond assignment in human J chain and its covalent pairing with immunoglobulin M. *Biochemist* 31 (50): 12643–12647.

88 Sanchez-Moral, L., Ràfols, N., Martori, C. et al. (2021). Multifaceted roles of CD5L in infectious and sterile inflammation. *Int. J. Mol. Sci.* 22 (8): 4076.

89 Sarrias, M.-R., Padilla, O., Monreal, Y. et al. (2004). Biochemical characterization of recombinant and circulating human Spα. *Tissue Antigens* 63 (4): 335–344.

90 Maehara, N., Taniguchi, K., Okuno, A. et al. (2021). AIM/CD5L attenuates DAMPs in the injured brain and thereby ameliorates ischemic stroke. *Cell Rep.* 36 (11): 109693.

91 Sarrias, M.-R., Roselló, S., Sánchez-Barbero, F. et al. (2005). A role for human SPα as a pattern recognition receptor. *J. Biol. Chem.* 280 (42): 35391–35398.

92 Miyazaki, T., Yamazaki, T., Sugisawa, R. et al. (2018). AIM associated with the IgM pentamer: attackers on stand-by at aircraft carrier. *Cell. Mol. Immunol.* 15 (6): 563–574.

93 Tissot, J.-D., Sanchez, J.-C., Vuadens, F. et al. (2002). IgM are associated to Spα (CD5 antigen-like). *Electrophoresis* 23 (7–8): 1203–1206.

94 Chen et al. (2024). Cryo-EM reveals structural basis for human AIM/CD5L recognition of polymeric immunoglobulin M. *Nat. Commun.* 15: 9387. https://doi.org/10.1038/s41467-024-53615-5.

95 Sanchez-Moral, L., Paul, T., Martori, C. et al. (2023). Macrophage CD5L is a target for cancer immunotherapy. *EBioMedicine* 91: 104555.

96 Oliveira, L., Silva, M.C., Gomes, A.P. et al. (2024). CD5L as a promising biological therapeutic for treating sepsis. *Nat. Commun.* 15 (1): 4119.

97 Lai, X., Xiang, Y., Zou, L. et al. (2018). Elevation of serum CD5L concentration is correlated with disease activity in patients with systemic lupus erythematosus. *Int. Immunopharmacol.* 63: 311–316.

98 Arai, S., Maehara, N., Iwamura, Y. et al. (2013). Obesity-associated autoantibody production requires

AIM to retain the immunoglobulin M immune complex on follicular dendritic cells. *Cell Rep.* 3 (4): 1187–1198.

99 Shibuya, A., Sakamoto, N., Shimizu, Y. et al. (2000). Fcα/μ receptor mediates endocytosis of IgM-coated microbes. *Nat. Immunol.* 1 (5): 441–446.

100 Mori, M., Kimura, H., Iwamura, Y. et al. (2012). Modification of N-glycosylation modulates the secretion and lipolytic function of apoptosis inhibitor of macrophage (AIM). *FEBS Lett.* 586 (20): 3569–3574.

101 Pan, S., Manabe, N., and Yamaguchi, Y. (2021). 3D structures of IgA, IgM, and components. *Int. J. Mol. Sci.* 22 (23): 12776.

102 Huang, J., Guerrero, A., Parker, E. et al. (2015). Site-specific glycosylation of secretory immunoglobulin a from human colostrum. *J. Proteome Res.* 14 (3): 1335–1349.

103 Kumar, N., Arthur, C.P., Ciferri, C., and Matsumoto, M.L. (2020). Structure of the secretory immunoglobulin a core. *Science* 367 (6481): 1008–1014.

104 Turula, H. and Wobus, C.E. (2018). The role of the polymeric immunoglobulin receptor and secretory immunoglobulins during mucosal infection and immunity. *Viruses* 10 (5): 237.

105 Vattepu, R., Sneed, S.L., and Anthony, R.M. (2022). Sialylation as an important regulator of antibody function. *Front. Immunol.* 13: 818736.

106 Struwe, W.B. and Robinson, C.V. (2019). Relating glycoprotein structural heterogeneity to function – insights from native mass spectrometry. *Curr. Opin. Struct. Biol.* 58: 241–248.

107 Sitia, R., Rubartelli, A., and Hämmerling, U. (1984). The role of glycosylation in secretion and membrane expression of immunoglobulins M and A. *Mol. Immunol.* 21 (8): 709–719.

108 Hennicke, J., Schwaigerlehner, L., Grünwald-Gruber, C. et al. (2020). Transient pentameric IgM fulfill biological function – effect of expression host and transfection on IgM properties. *PLoS One* 15 (3): e0229992.

109 Haslund-Gourley, B.S., Woloszczuk, K., Hou, J. et al. (2024). IgM N-glycosylation correlates with COVID-19 severity and rate of complement deposition. *Nat. Commun.* 15 (1): 404.

110 Hennicke, J., Reinhart, D., Altmann, F., and Kunert, R. (2019). Impact of temperature and pH on recombinant human IgM quality attributes and productivity. *New Biotechnol.* 50: 20–26.

111 de Haan, N., Falck, D., and Wuhrer, M. (2019). Monitoring of immunoglobulin N- and O-glycosylation in health and disease. *Glycobiology* 30 (4): 226–240.

112 Wada, R., Matsui, M., and Kawasaki, N. (2019). Influence of N-glycosylation on effector functions and thermal stability of glycoengineered IgG1 monoclonal antibody with homogeneous glycoforms. *MAbs* 11 (2): 350–372.

113 Huhn, C., Selman, M.H.J., Ruhaak, L.R. et al. (2009). IgG glycosylation analysis. *Proteomics* 9 (4): 882–913.

114 Lloyd, K.A., Wang, J., Urban, B.C. et al. (2017). Glycan-independent binding and internalization of human IgM to FCMR, its cognate cellular receptor. *Sci. Rep.* 7 (1): 42989.

115 Wiersma, E.J., Chen, F., Bazin, R. et al. (1997). Analysis of IgM structures involved in J chain incorporation. *J. Immunol.* 158 (4): 1719–1726.

116 Torii, T., Kanemitsu, K., and Hagiwara, A. (2017). Sialic acid level is significantly elevated in IgM enriched protein fraction in sera of cancer patients. *J. Immunoassay Immunochem.* 38 (2): 127–139.

117 Ruhaak, L.R., Kim, K., Stroble, C. et al. (2016). Protein-specific differential glycosylation of immunoglobulins in serum of ovarian cancer patients. *J. Proteome Res.* 15 (3): 1002–1010.

118 Breen, L.D., Pučić-Baković, M., Vučković, F. et al. (2016). IgG and IgM glycosylation patterns in patients undergoing image-guided tumor ablation. *Biochim. Biophys. Acta* 1860 (8): 1786–1794.

119 Zhang, P., Woen, S., Wang, T. et al. (2016). Challenges of glycosylation analysis and control: an integrated approach to producing optimal and consistent therapeutic drugs. *Drug Discov. Today* 21 (5): 740–765.

120 Chia, S., Tay, S.J., Song, Z. et al. (2023). Enhancing pharmacokinetic and pharmacodynamic properties of recombinant therapeutic proteins by manipulation of sialic acid content. *Biomed. Pharmacother.* 163: 114757.

121 Shrivastava, A., Joshi, S., Guttman, A., and Rathore, A.S. (2022). N-glycosylation of monoclonal antibody therapeutics: a comprehensive review on significance and characterization. *Anal. Chim. Acta* 1209: 339828.

122 Beck, A. and Liu, H. (2019). Macro- and micro-heterogeneity of natural and recombinant IgG antibodies. *Antibodies* 8 (1): 18.

123 Song, T., Ozcan, S., Becker, A., and Lebrilla, C.B. (2014). In-depth method for the characterization of glycosylation in manufactured recombinant monoclonal antibody drugs. *Anal. Chem.* 86 (12): 5661–5666.

124 Füssl, F., Cook, K., Scheffler, K. et al. (2018). Charge variant analysis of monoclonal antibodies using direct coupled pH gradient cation exchange chromatography to high-resolution native mass spectrometry. *Anal. Chem.* 90 (7): 4669–4676.

125 Zauner, G., Selman, M.H.J., Bondt, A. et al. (2013). Glycoproteomic analysis of antibodies. *Mol. Cell. Proteomics* 12 (4): 856–865.

126 Ruhaak, L.R., Xu, G., Li, Q. et al. (2018). Mass spectrometry approaches to glycomic and glycoproteomic analyses. *Chem. Rev.* 118 (17): 7886–7930.

127 Plomp, R., Bondt, A., de Haan, N. et al. (2016). Recent advances in clinical glycoproteomics of immunoglobulins (Igs). *Mol. Cell. Proteomics* 15 (7): 2217–2228.

128 Hong, Q., Lebrilla, C.B., Miyamoto, S., and Ruhaak, L.R. (2013). Absolute quantitation of immunoglobulin G and its glycoforms using multiple reaction monitoring. *Anal. Chem.* 85 (18): 8585–8593.

129 Clerc, F., Reiding, K.R., Jansen, B.C. et al. (2016). Human plasma protein N-glycosylation. *Glycoconj. J.* 33 (3): 309–343.

130 Aldredge, D., An, H.J., Tang, N. et al. (2012). Annotation of a serum N-glycan library for rapid identification of structures. *J. Proteome Res.* 11 (3): 1958–1968.

131 Marie, A.-L., Ray, S., and Ivanov, A.R. (2023). Highly-sensitive label-free deep profiling of N-glycans released from biomedically-relevant samples. *Nat. Commun.* 14 (1): 1618.

132 Olivova, P., Chen, W., Chakraborty, A.B., and Gebler, J.C. (2008). Determination of N-glycosylation sites and site heterogeneity in a monoclonal antibody by electrospray quadrupole ion-mobility time-of-flight mass spectrometry. *Rapid Commun. Mass Spectrom.* 22 (1): 29–40.

133 Greisch, J.-F., den Boer, M.A., Lai, S.-H. et al. (2021). Extending native top-down electron capture dissociation to MDa immunoglobulin complexes provides useful sequence tags covering their critical variable complementarity-determining regions. *Anal. Chem.* 93 (48): 16068–16075.

134 Moh, E.S.X., Lin, C.-H., Thaysen-Andersen, M., and Packer, N.H. (2016). Site-specific N-glycosylation of recombinant pentameric and hexameric human IgM. *J. Am. Soc. Mass Spectrom.* 27 (7): 1143–1155.

135 Deshpande, N., Jensen, P.H., Packer, N.H., and Kolarich, D. (2010). GlycoSpectrumScan: fishing glycopeptides from MS spectra of protease digests of human colostrum sIgA. *J. Proteome Res.* 9 (2): 1063–1075.

136 Hong, Q., Ruhaak, L.R., Stroble, C. et al. (2015). A method for comprehensive glycosite-mapping and direct quantitation of serum glycoproteins. *J. Proteome Res.* 14 (12): 5179–5192.

137 Falck, D., Jansen, B.C., Plomp, R. et al. (2015). Glycoforms of immunoglobulin G based biopharmaceuticals are differentially cleaved by trypsin due to the Glycoform influence on higher-order structure. *J. Proteome Res.* 14 (9): 4019–4028.

138 Chandler, K.B., Mehta, N., Leon, D.R. et al. (2019). Multi-isotype glycoproteomic characterization of serum antibody heavy chains reveals isotype- and subclass-specific *N*-glycosylation profiles. *Mol. Cell. Proteomics* 18 (4): 686–703.

139 Pabst, M., Küster, S.K., Wahl, F. et al. (2015). A microarray-matrix-assisted laser desorption/ionization-mass spectrometry approach for site-specific protein N-glycosylation analysis, as demonstrated for human serum immunoglobulin M (IgM). *Mol. Cell. Proteomics* 14 (6): 1645–1656.

140 Pan, K.-T., Chen, C.-C., Urlaub, H., and Khoo, K.-H. (2017). Adapting data-independent acquisition for mass spectrometry-based protein site-specific N-glycosylation analysis. *Anal. Chem.* 89 (8): 4532–4539.

141 Goonatilleke, E., Huang, J., Xu, G. et al. (2019). Human milk proteins and their glycosylation exhibit quantitative dynamic variations during lactation. *J. Nutr.* 149 (8): 1317–1325.

142 Huang, J., Kailemia, M.J., Goonatilleke, E. et al. (2017). Quantitation of human milk proteins and their glycoforms using multiple reaction monitoring (MRM). *Anal. Bioanal. Chem.* 409 (2): 589–606.

143 Zhu, J., Lin, Y.-H., Dingess, K.A. et al. (2020). Quantitative longitudinal inventory of the N-glycoproteome of human milk from a single donor reveals the highly variable repertoire and dynamic site-specific changes. *J. Proteome Res.* 19 (5): 1941–1952.

144 Cheng, Y.-H., Lee, C.-H., Wang, S.-Y. et al. (2024). Multiplexed antibody glycosylation profiling using dual enzyme digestion and liquid chromatography-triple quadrupole mass spectrometry method. *Mol. Cell. Proteomics* 23 (2): 100710.

145 Monica, T.J., Williams, S.B., Goochee, C.F., and Maiorella, B.L. (1995). Characterization of the glycosylation of a human IgM produced by a human-mouse hybridoma. *Glycobiology* 5 (2): 175–185.

146 Feige, M.J., Hendershot, L.M., and Buchner, J. (2010). How antibodies fold. *Trends Biochem. Sci.* 35 (4): 189–198.

147 Liu, H. and May, K. (2012). Disulfide bond structures of IgG molecules. *MAbs* 4 (1): 17–23.

148 Lakbub, J.C., Shipman, J.T., and Desaire, H. (2018). Recent mass spectrometry-based techniques and considerations for disulfide bond characterization in proteins. *Anal. Bioanal. Chem.* 410 (10): 2467–2484.

149 Moritz, B. and Stracke, J.O. (2017). Assessment of disulfide and hinge modifications in monoclonal antibodies. *Electrophoresis* 38 (6): 769–785.

150 Ren, T., Tan, Z., Ehamparanathan, V. et al. (2021). Antibody disulfide bond reduction and recovery during biopharmaceutical process development – a review. *Biotechnol. Bioeng.* 118 (8): 2829–2844.

151 Coghlan, J., Benet, A., Kumaran, P. et al. (2022). Streamlining the characterization of disulfide bond shuffling and protein degradation in IgG1 biopharmaceuticals under native and stressed conditions. *Front. Bioeng. Biotechnol.* 10: 862456.

152 Hutterer, K.M., Hong, R.W., Lull, J. et al. (2013). Monoclonal antibody disulfide reduction during manufacturing. *MAbs* 5 (4): 608–613.

153 Lu, C., Liu, D., Liu, H., and Motchnik, P. (2013). Characterization of monoclonal antibody size variants containing extra light chains. *MAbs* 5 (1): 102–113.

154 Metcalfe, C. (2022). A review of methodologies for the detection, quantitation, and localization of free cysteine in recombinant proteins: a focus on therapeutic monoclonal antibodies. *Front. Mol. Biosci.* 9: 886417.

155 Kleinberg, A., Joseph, R., Mao, Y., and Li, N. (2022). Ultrasensitive disulfide scrambling analysis of mAbs by LC-MS with post-column reduction and glycine signal enhancement. *Anal. Biochem.* 653: 114773.

156 Liu, H., Lei, Q.P., and Washabaugh, M. (2016). Characterization of IgG2 disulfide bonds with LC/MS/MS and postcolumn online reduction. *Anal. Chem.* 88 (10): 5080–5087.

157 Natesan, R., Dykstra, A.B., Banerjee, A., and Agrawal, N.J. (2023). Heterogeneity in disulfide bond reduction in IgG1 antibodies is governed by solvent accessibility of the cysteines. *Antibodies* 12 (4): 83.

158 Li, X., Wang, F., Xu, W. et al. (2013). Disulfide bond assignment of an IgG1 monoclonal antibody by LC–MS with post-column partial reduction. *Anal. Biochem.* 436 (2): 93–100.

159 Wang, Y., Lu, Q., Wu, S.-L. et al. (2011). Characterization and comparison of disulfide linkages and scrambling patterns in therapeutic monoclonal antibodies: using LC-MS with electron transfer dissociation. *Anal. Chem.* 83 (8): 3133–3140.

160 Giannone, C., Chelazzi, M.R., Orsi, A. et al. (2022). Biogenesis of secretory immunoglobulin M requires intermediate non-native disulfide bonds and engagement of the protein disulfide isomerase ERp44. *EMBO J.* 41 (3): e108518.

161 Chapuis, R.M. and Koshland, M.E. (1974). Mechanism of IgM polymerization. *Proc. Natl. Acad. Sci.* 71 (3): 657–661.

162 Su, Y., Wang, B., Zhang, Y. et al. (2019). Mass spectrometric determination of disulfide bonds and free cysteine in grass carp IgM isoforms. *Fish Shellfish Immunol.* 95: 287–296.

163 Davis, A.C., Roux, K.H., Pursey, J., and Shulman, M.J. (1989). Intermolecular disulfide bonding in IgM: effects of replacing cysteine residues in the mu heavy chain. *EMBO J.* 8 (9): 2519–2526.

164 Sørensen, V., Rasmussen, I.B., Sundvold, V. et al. (2000). Structural requirements for incorporation of J chain into human IgM and IgA. *Int. Immunol.* 12 (1): 19–27.

165 Wiersma, E.J. and Shulman, M.J. (1995). Assembly of IgM. Role of disulfide bonding and noncovalent interactions. *J. Immunol.* 154 (10): 5265–5272.

166 Kownatzki, E. (1973). Disulfide bonds of human IgM: differential sensitivity to reductive cleavage. *Scand. J. Immunol.* 2 (4): 433–438.

167 Mukkur, T.K.S. and Inman, F.P. (1970). Relative labilities of interchain disulfide bonds in human immunoglobulin M subunits. *Biochemist* 9 (4): 1031–1037.

4

Mass Spectrometric Characterization of Therapeutic Recombinant Proteins

Alissa D. Guarnaccia, Corey E. Bakalarski, Wendy Sandoval, and Jennie R. Lill

Proteomic and Genomic Technologies, Genentech Inc., South San Francisco, CA, USA

Introduction

A mass spectrometer, in its simplest form, is an instrument designed to determine the mass of individual atoms or molecules by measuring the mass-to-charge (m/z) ratio of ions [1]. This capability allows for the determination of the molecular composition of complex molecules, such as proteins, based on their m/z values. Antibodies and other recombinant proteins are prominent therapeutic agents due to their high specificity and reduced off-target side effects compared to traditional small-molecule drugs [2]. However, these biomolecules often exhibit heterogeneity due to various modifications, either synthetically induced or naturally occurring during the manufacturing process [3]. N-terminal heterogeneity, resulting from truncated or elongated termini due to aberrant proteolysis during secretion, is common in recombinant proteins and needs to be monitored as it can induce adverse effects therapeutically [4]. Additionally, co- or post-translational modifications, such as autophosphorylation in active kinases or glycosylation in IgG antibodies affecting Fc effector function, can indicate suboptimal manufacturing processes and impact the molecule's activity or stability [5]. Given these inherent heterogeneities, extensive analytical characterization is crucial to ensure the reproducible, reliable, and safe production of therapeutic proteins. Mass spectrometry, through intact molecular weight measurements or by bottom-up proteomic approaches, stands out as the fastest, easiest, most accurate, and sophisticated technology for assessing these modifications [6].

Instrumentation

Ionization

The initial step in the mass spectrometric analysis of proteins is ionization, a process that converts molecules from a liquid or solid phase into a gaseous state, making them detectable by the mass analyzer. Prior to ionization, samples undergo purification and dialysis into an MS-compatible buffer, typically a mix of methanol or acetonitrile with water and an ion-pairing reagent. Two prominent ionization techniques for large biomolecules are matrix-assisted laser desorption ionization (MALDI) and electrospray ionization (ESI). MALDI (Figure 4.1), first described by Hillenkamp and Karas in 1990 [7], involves co-crystallizing the sample with an organic matrix, which absorbs ultraviolet light from a laser and converts it to heat energy, vaporizing the matrix and sample together. This technique generally produces singly charged ions, making the resulting spectra easier to deconvolute and more tolerant to salts in sample buffers. MALDI is often coupled with time-of-flight (TOF) mass analyzers and is favored for its rapid analysis capabilities, providing spectra in seconds.

ESI (Figure 4.2), on the other hand, was first described by Fenn and colleagues in the late 1980s and earned Fenn a Nobel Prize in 2002 [8]. ESI generates gaseous ions from a liquid aerosol in the presence of a high voltage, typically producing multiply-charged ion species. Multiple charges allow for the analysis of higher mass species by generating lower m/z ratios for molecules of identical molecular weight. ESI is particularly well suited for coupling with separation devices such as capillary electrophoresis (CE) or high-performance liquid chromatography (HPLC), enabling online desolvation of ions. Variants of ESI, such as nanospray ionization, operate at lower flow rates and voltages, offering enhanced sensitivity for specific applications. Both MALDI and ESI are "soft" ionization techniques that preserve the integrity of biomolecules, retaining their molecular weight information during the ionization process, thus playing crucial roles in the mass spectrometric analysis of proteins.

Characterizing Biotherapeutics: Analytical Methods for Diverse Modalities, First Edition. Edited by Jennie R. Lill and Wendy Sandoval.

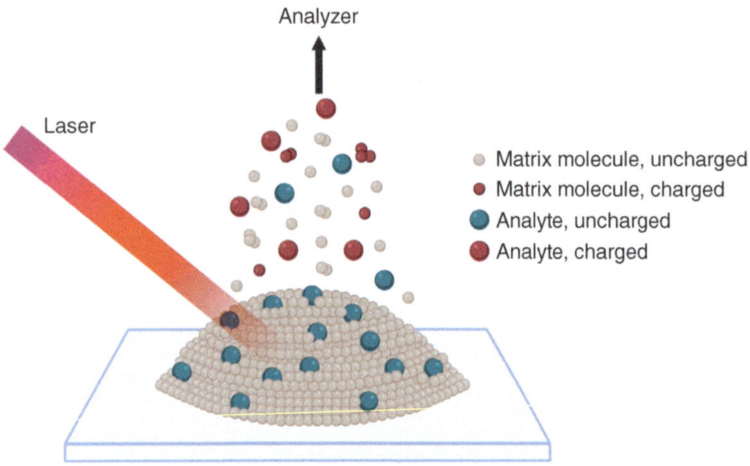

Figure 4.1 *MALDI ionization.* The uncharged analyte is mixed with an uncharged matrix molecule. Energy from a laser pulse charges the matrix, and the charged matrix donates a proton to the analyte via desorption. The charged analyte is then guided through the MALDI mass spectrometer for m/z analysis. Source: Created with BioRender.com.

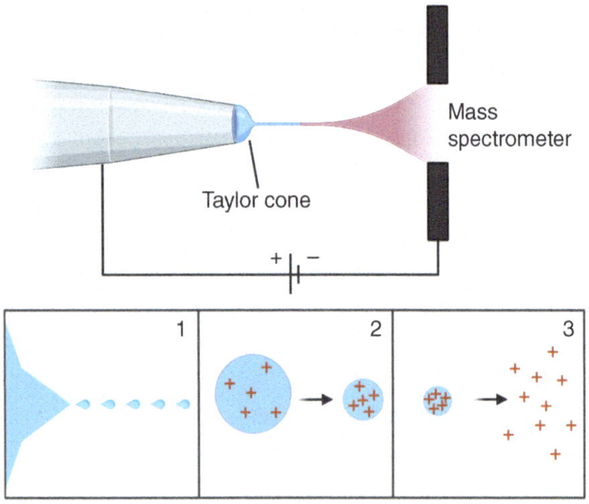

Figure 4.2 *Electrospray ionization.* Schematic of the path of the analyte from solvated through the charged outlet and droplet formation, to the mass spectrometer for analysis. (1) Under high voltage, the eluent from a syringe, tip, or HPLC creates a cone shape (a Taylor Cone) and represents the initiation of the transfer of ions from the liquid to the gas phase. (2) Evaporation of the droplets occurs leaving them increasingly highly charged. (3) When the charge exceeds the Rayleigh limit (the maximum amount of charge a liquid droplet could carry before throwing out fine jets of liquid), the droplet completely dissociates leaving a stream of charged gaseous ions that now enter the mass spectrometer for separation and detection. Source: Created with BioRender.com.

Mass Analyzers for Intact Molecular Weight Analysis of Biotherapeutics

A variety of mass spectrometers are available, with some specifically designed for high mass accuracy analysis of large biotherapeutic entities like antibodies and recombinant proteins. Choosing the right mass spectrometer depends on the type of analysis, required throughput, and system robustness. Below are two popular examples for intact molecular weight analysis of proteins, though many capable instruments exist on the market.

TOF mass spectrometers and their more sophisticated counterpart, the quadrupole time of flight (Q-TOF) instrument, have become the instruments of choice for determining intact molecular weight mass measurements for recombinant proteins [9]. TOF mass spectrometers (TOF-MS) are known for their simplicity, durability, rapid scan speeds, and virtually limitless mass range. These instruments determine the m/z ratio of ions by measuring the time it takes for them to travel from an ion source to a detector. Upon ionization, ions are accelerated through an electric field, imparting the same kinetic energy to each ion. They then traverse a drift tube before reaching the detector. The m/z ratio is calculated based on the "time of flight" for each ion. Essentially, ions with identical charges but different masses exhibit varying velocities: lighter ions travel faster and reach the detector sooner, while heavier ions move more slowly and arrive later.

To increase specificity, many TOF instruments are now coupled to a quadrupole (Q-TOF). Quadrupoles consist of four parallel metal rods with oscillating direct current (DC) and radio frequency (RF) voltages on each pair of rods. Ions accelerate through the quadrupole between the rods, and the trajectory of the ions is influenced by their m/z ratio. Figure 4.3 shows a schematic of a quadrupole time of flight mass spectrometer. Ions can therefore be selected based upon their m/z ratios, and the quadrupole can act either as a mass filter or, in broadband RF mode, increasing signal-to-noise by decreasing the transmission of superfluous chemical noise or neutral ions before they enter the TOF tube. Q-TOF mass spectrometers have high mass accuracy, within a few parts per million of the true value, and high resolution, up to ten times higher than a quadrupole instrument. Q-TOF mass spectrometers have a virtually unlimited acquisition mass range and fast acquisition speeds without sacrificing resolution.

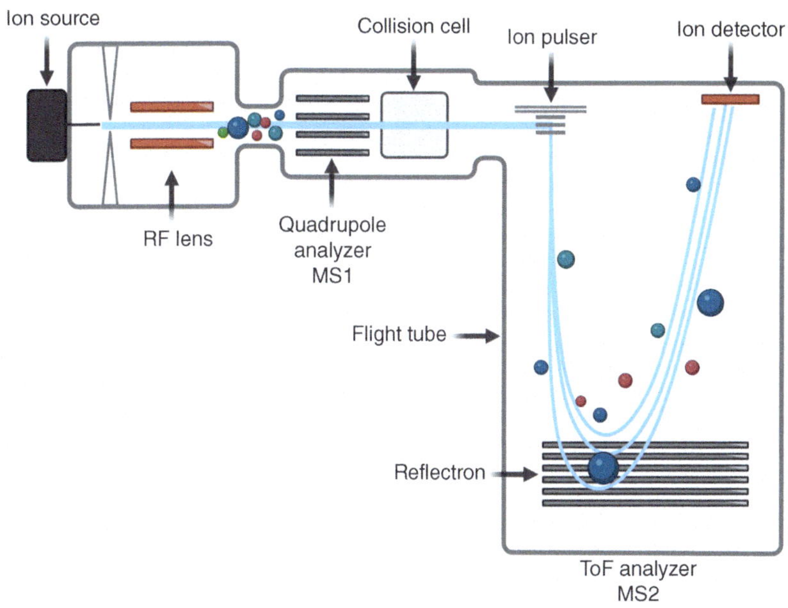

Figure 4.3 *Quadrupole time-of-flight (Q-TOF) mass spectrometer.* A Q-TOF is a hybrid instrument that combines the features of a quadrupole mass filter and a time-of-flight mass analyzer. The sample is first ionized (by ESI or MALDI). Ions are focused with an RF lens and then directed into a quadrupole mass filter, which uses alternating RF and DC voltages to selectively filter ions based on their m/z ratio. Optionally, ions may pass through a collision cell for collision-induced dissociation (CID) to obtain structural information. The ion pulser then accelerates the selected ions into the time-of-flight (TOF) analyzer where they travel a fixed distance and are reflected up toward the detector by a reflectron. The detector absorbs and records the signals of the ions, and the time of flight from the pulser to the detector is measured. Because the flight time for each mass is unique, the measured time corresponds to the m/z ratio that is recorded. The recorded data is processed to generate a mass spectrum, displaying the intensity of detected ions as a function of their m/z ratios, allowing for the identification and quantification of sample components. This combination of quadrupole filtering and high-resolution TOF analysis enables precise and accurate mass measurements, making Q-TOF instruments valuable in fields such as proteomics, metabolomics, and pharmaceutical analysis. Source: Created with BioRender.com.

Many analyses of biotherapeutic molecules are typically performed under denaturing conditions; however, a subset of proteins, such as bispecific antibodies, drug–protein conjugates, or non-covalent multimeric protein complexes, require the analysis of proteins in their native conformational state. This process is known as native mass spectrometry and was greatly accelerated in 2014 when Thermo Fisher Scientific introduced the Exactive Plus Extended Mass Range (EMR) Orbitrap mass spectrometer [10]. This instrument combines high resolution with accurate mass measurements over an EMR (up to 20,000 m/z), thereby allowing researchers to gain highly accurate m/z information for large biomolecules such as antibodies and their derivatized counterparts. In addition, the instrument is capable of interrogating non-covalent interactions between biomolecules and other molecular entities such as small molecules, metabolites, and lipids, as well as allowing characterization of the structure and topology of native-like tertiary and quaternary protein structures.

Typically, in native mass spectrometry, the original aqueous buffer solution of the biological analyte is exchanged with a more volatile solution. Ammonium acetate is often used because it mimics a native buffer that can maintain protein structure in solution but is amenable to rapid solvent evaporation during ESI. Ionized proteins under native conditions often have lower charge state distributions than peptides generated by ESI. Thus, ions generated by native mass spectrometry present at a higher m/z window than non-native counterparts. The EMR Orbitrap Plus from Thermo Fisher Scientific is capable of handling these challenging sample characteristics and generates data with high mass accuracy and high resolution. Remarkably, this instrument has proven to be a high-performing mass spectrometer for screening a variety of biomolecules such as antibody-drug conjugates, PEGylated species, and complex glycoforms of proteins [11–13].

An example of where native mass spectrometry on an EMR mass spectrometer has proved important is in the study of hexameric IgG complexes. IgG antibodies form ordered complexes of six (hexamers) and bind to antigens on cell surfaces. These cell-bound hexamers then engage complement C1 to initiate the process of complement-dependent target cell killing. There are various efforts underway to exploit such IgG antibody complexes as therapeutic moieties. For example, introducing different mutational variants can allow the antibodies to form hexamers without being on the cell surface, therefore adding more avidity to the antibody/antigen complex.

Given the large size of these hexameric structures (each monomer is ~150 kDa) and the non-covalent interactions between the monomers, these complexes are particularly challenging to study [14]. In order to investigate the suitability of the IgG variants for their potential development as hexameric biopharmaceuticals, a robust analysis method was needed that could detect solution-phase multimers. For this, we employed a triple-mutant antibody version of human IgG1 (E345R, E430G, S440Y) termed RGY [15]. These mutations enhance Fc domain interactions with one another, enabling RGY antibodies to readily form hexamers in solution, and increasing hexamerization on target cell surfaces [16]. In comparison with high-performance size-exclusion chromatography (HP-SEC) and multi-angle laser light scattering analysis of hollow fiber flow field-flow fractionation, native mass spectrometry proved to be the most successful analytical tool for detecting hexameric RGY complexes (Figure 4.4) [17].

Native mass spectrometry is employed in biotechnology and in drug discovery to analyze intact protein complexes and their interactions with ligands, such as drugs or lipids. Native mass spectrometry provides critical insights into protein structure, subunit stoichiometry, and the dynamics of protein-protein and protein-ligand interactions, enabling the analysis of the composition and stoichiometry of protein complexes, and providing insights into biological function and binding affinities of interactions [18]. Native mass spectrometry also can be used to identify proteoforms, such as post-translational modifications and splice variants, which can be crucial in the development of diagnostic biomarkers [19]. Additionally, native mass spectrometry can be used to monitor the purity and homogeneity of protein-based drugs, and for examining the structure and interactions of membrane proteins in their native lipid environment [20]. Finally, native mass spectrometry enables the analysis of post-translational modifications and provides structural information about protein complexes through m/z ratios and dissociation behavior [21]. Applications of native mass spectrometry enhance and enable drug development capabilities and have become an invaluable tool for advancing novel therapeutics.

Another tool that is commonly employed for the analysis of biotherapeutics is ion mobility spectrometry (IMS) which is a complementary analytical technique to mass spectrometry. IMS is used to separate and characterize ionized molecules based on their mobility through a carrier buffer gas. Gas-phase ion mobility provides a collisional cross section (CCS) for an ion as it drifts through an inert buffer gas under a low electrical field. A CCS (σ_c) is a quantitative measurement of the probability of a collision taking place between two or more particles. For two bodies in collision, the CCS represents the effective area of a particle as seen by another passing particle. During IMS, the ion collides with inert buffer gas molecules the time it takes for the ion to traverse the drift region is measured (Figure 4.5). Ions with larger cross sections drift more slowly through the inert gas compared to those with smaller cross sections. Thus, IMS can be employed to separate proteins and protein complexes based on their native conformational states. CCS can differentiate between the structures of two proteins sharing the same m/z but exhibiting different conformations in the gas phase. Various types of IMS include drift-tube ion mobility spectrometry (DT-IMS), traveling-wave ion mobility spectrometry (TW-IMS), and field asymmetric ion mobility spectrometry (FAIMS) which is alternatively known as differential mobility spectrometry (DMS).

Coupling ion mobility with mass spectrometry (IM-MS) can be used to analyze a variety of analytes. IM-MS is used in proteomics, lipidomics, and metabolomics, and it can be integrated with a variety of mass analyzers, chromatographic techniques, and fragmentation methods for the characterization of biotherapeutics [22]. For example, IMS-MS has been employed to characterize different disulfide variants in intact IgG2 monoclonal antibodies (mAbs) [23], to characterize the thermal stability of mAbs [24], and to measure drug-to-antibody ratios (DARs) in antibody-drug conjugates [25]. One notable example where IM-MS has proven itself is in the characterization of generic versions of biotherapeutics, termed biosimilars. IM-MS can distinguish analytical features that may not be evident by mass spectrometry alone, identifying modifications that could be crucial for altering the efficacy of a biosimilar. For a comprehensive review of recent developments in liquid-phase separations coupled with IM-MS and for the development and characterization of biotherapeutics, refer to Makey et al. [26].

Separation Devices for the Characterization of Biotherapeutics

High-performance liquid chromatography (HPLC) is an analytical technique used to separate and quantify components in a mixture [27]. HPLC operates by pumping a pressurized liquid containing the analytes through a column filled with a solid matrix, typically a silica-based stationary phase (e.g. a C18 matrix is often used for peptides and a C4 or C8 matrix is often used for proteins). Peptide and protein analytes in the sample are adsorbed onto the hydrophobic surface of the column. A gradient of an organic solvent (usually methanol or acetonitrile) is

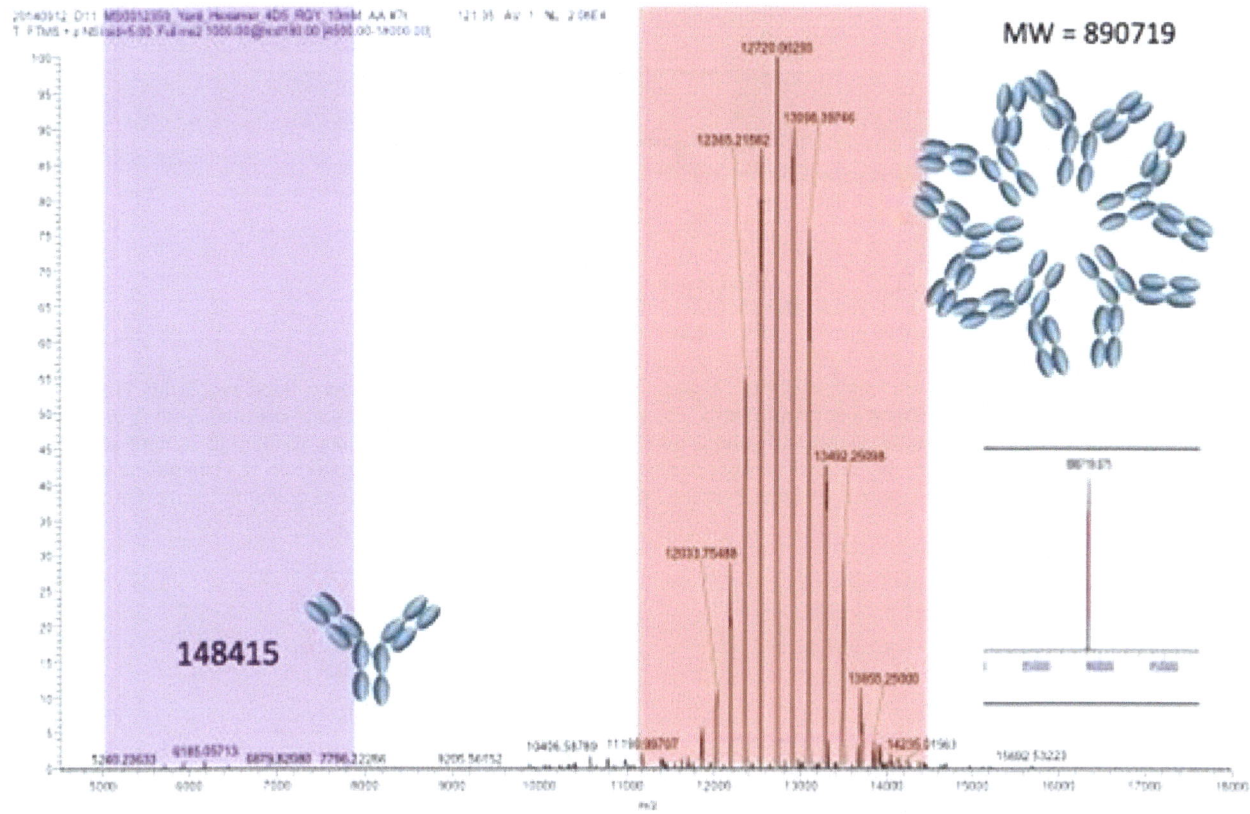

Figure 4.4 *Native MS analysis of an RGY-antibody hexamer*. Native mass spectrometry proves to be a robust method for analyzing the hexameric species of the RGY-IgG mutant. Analysis was performed under native conditions (10 mM ammonium acetate) on an Extended Mass Range Exactive Orbitrap, using high C-trap pressure. The charge envelope of the IgG monomer appears around 6,000 m/z and that of the IgG hexamer around 13,000 m/z.

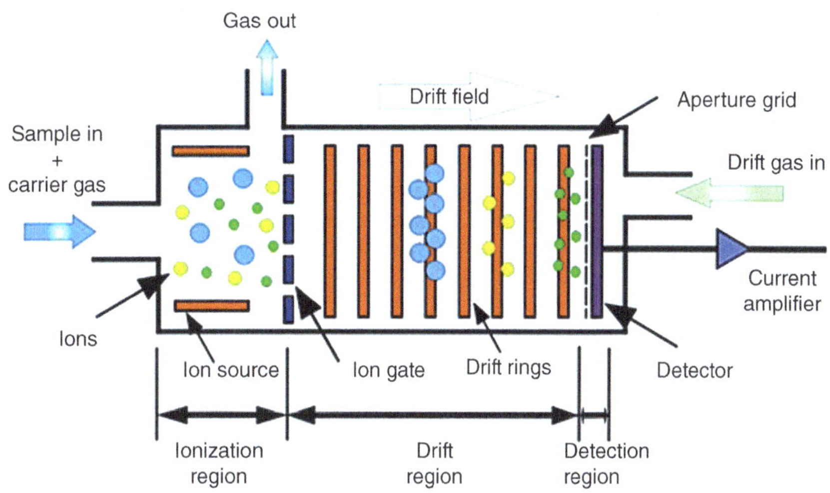

Figure 4.5 *Conventional drift-tube ion mobility spectrometry (DT-IMS)* system showing three ions of different sizes in the reaction region and then migrating at different velocities in the drift region.

then applied to the column, and different species elute at different points of the gradient, depending on their biochemical properties (Figure 4.6). More hydrophilic solutes will elute earlier in the gradient and more hydrophobic species will elute later. HPLC is often coupled to a mass spectrometer for direct analysis.

An alternative separation approach is capillary electrophoresis (CE). CE is an electrokinetic separation technique that uses a small-diameter capillary under an electric field to separate an analyte based on ion mobility. As an analyte migrates through the capillary, differentially charged species migrate at different rates based on their

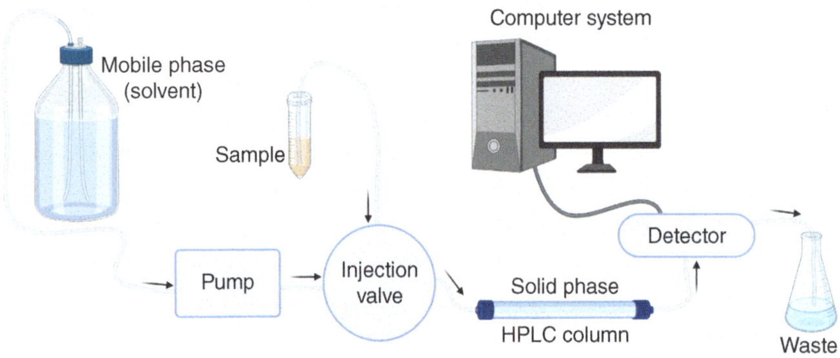

Figure 4.6 *High-performance liquid chromatography (HPLC).* Analytes are separated based on their hydrophobic properties. First, a sample is loaded onto a column containing an organic matrix (solid phase) which binds the proteins and peptides. Then, over the course of a gradual gradient, the sample is eluted with a liquid (mobile phase). Different species elute at different concentrations of the elution gradient, based on the hydrophobic property of peptide or protein. Chromatographic separation of a complex mixture can be analyzed by a detector (e.g. UV absorbance) that monitors the flow of the elution over the course of gradient. Source: Created with BioRender.com.

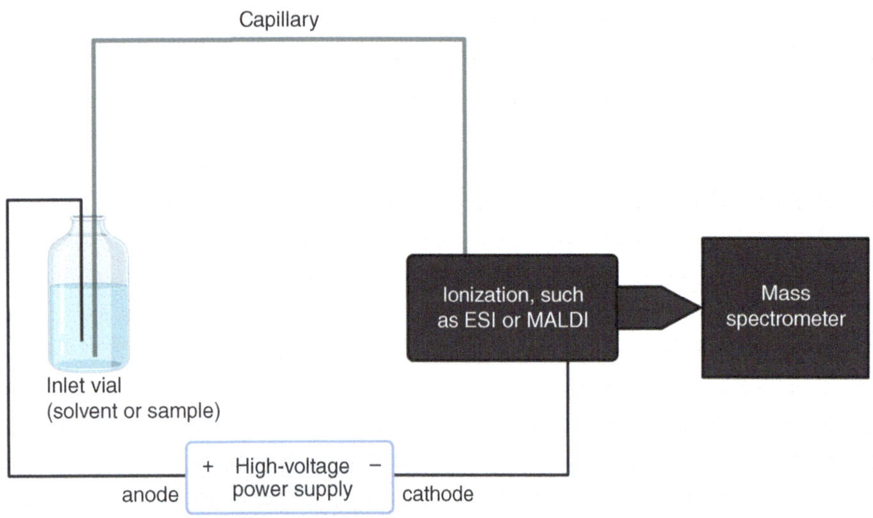

Figure 4.7 *Capillary electrophoretic system for mass spectrometry.* The analyte of interest is introduced into the CE system by placing the capillary inlet into the sample vial where it traverses through the capillary by capillary action, pressurization, siphoning, or through electrokinetic mechanisms. Migration of analytes occurs when an electric field (supplied by the electrodes) is applied between the source and destination, and in most CE systems, all ions, positive or negative, navigate through the capillary in the same direction by electroosmotic flow. Analytes are detected near the outlet end of the capillary, data is displayed as an electropherogram, and the separated sample can be directly analyzed by mass spectrometry. Source: Created with BioRender.com.

biophysical properties, enabling robust, and reproducible separation. A simple CE system includes a sample vial, source and destination vials, an electrophoretic capillary, electrodes, and a detector (Figure 4.7). CE was first coupled with mass spectrometry in 1990 [28] and proved to be an invaluable advance for characterizing biotherapeutic moieties due to its high sensitivity and minimal sample requirement. Various injection strategies, such as pressure or electrokinetic injection, concentrate analytes near the capillary inlet. CE instruments now exist in arrayed formats with 16 or 96 capillaries, allowing simultaneous analysis of multiple samples from 96-well plates, and enabling high throughput [29].

CE-MS has been extensively used for characterizing biotherapeutics. For example, CE has been used to monitor the deamidation and isomerization of biomolecules [30], dimerization of therapeutic antibodies [31], and the stability and pharmacokinetics of pegylated interferon-beta-1b [32]. The field continues to evolve with new multiplexed capillary formats and miniaturized systems [33].

Software for the Analysis of Intact Molecular Weight Measurements

As mass spectrometers measure not the overall mass of a molecule but rather its m/z ratio, the observed signal is dictated by a function of both mass and apparent charge state. Owing to an accumulation of chemical and electronic noise, this signal can be difficult to deconvolute from the observed spectrum. Several methods have been proposed for the deconvolution of the charge information from the uncharged protein mass. Many methods rely, at least in part, upon probabilistic approaches of maximum entropy [34] to find the most probable parent mass spectrum which would yield the observed, often convoluted, signal. Probabilistic approaches rely on Bayes' rule to yield the appropriate parent spectrum given the signal observed. While implementations of such probabilistic approaches varies in how they the observed noise is modelled, these approaches are commonplace in both individual tools [35–37] and integrated vendor software. Within the last decade, more sophisticated deconvolving methods have been developed that implement summing methods to increase signal-to-noise ratios, or that apply sliding windows across time to summarize results into more informative mass features and in an automated fashion [38–41].

Emerging Mass Spectrometric Approaches for Characterizing Molecular Heterogeneity in Biotherapeutics

Biomolecular therapeutic modalities are becoming more complicated in formulation, with conjugated molecules and fusion molecules entering the clinic and the market. With increased complexity comes increased heterogeneity during production. Many biotherapeutics, such as IgG-based and Fc-fusion biologics, are post-translationally glycosylated in ways that dramatically influence their therapeutic efficacy. Glycans, which modify over 50% of the eukaryotic proteome, play a crucial role in defining the pharmacological properties of biotherapeutics, including potency, stability, bioavailability, solubility, and immunogenicity. Glycosylation results from a complex network of metabolic inputs and enzymatic reactions that are difficult to predict and influenced by genetic, epigenetic, and environmental factors. But in order to develop safe and consistent biotherapeutics, in-depth characterization of these biomolecules must be performed. The promise of these therapeutics, together with formidable molecular complexity, presents an impressive analytical challenge for mass spectrometry.

One technique that addresses this bioanalytical challenge and is proton transfer charge reduction (PTCR). PTCR enables complicated spectra from a heterogeneous sample to be analyzed successfully. Here, a multiply charged cation interacts with a singly charged anion, resulting in a neutral molecule and a charge-reduced cation, effectively simplifying complex mass spectra by lowering the overall charge state of the analyzed molecule. PTCR is particularly useful for analyzing large biomolecules like proteins where high charge states can lead to spectral overlap [42]. This method decreases the charge of ions, which expands the m/z distribution and reduces spectral feature overlaps. With a wider m/z distribution, more analytical peaks can be resolved, enabling the observation of more of the charge states of the protein analyte. The observation of more charge states facilitates molecular weight measurement of more molecular forms of a biotherapeutic in a single intact MS experiment. PTCR on its own is powerful, but integrating PTCR with other techniques has shown further promise. By integrating PTCR with data-independent acquisition (DIA), custom scripts in UniDec [40], and glycoproteomics and glycomics analyses, researchers were able to obtain insights beyond molecular weight distributions, such as glycan composition of proteoforms. The DIA–PTCR method provides an unprecedented window into the molecular weights and overall distribution of proteoforms in heterogeneous biotherapeutic samples. These types of methods have crucial implications for enabling sensitive glycoform profiling and quality assessment of the therapeutic products that reach patients.

Tandem Mass Spectrometric Characterization of Biomolecules

The analyses described above focus on the intact molecular weight measurement of biotherapeutics. For more granular detail, tandem mass spectrometry (MS/MS) is employed. This technique involves two mass analyzing steps, first analyzing a parent ion in the mass spectrometer, and second, fragmenting the parent ion to analyze the product ions (daughter ions). Various MS/MS fragmentation methods include collision-induced dissociation (CID) [43], photo-induced dissociation (PID) [44], electron transfer dissociation (ETD) [45], and electron capture dissociation (ECD) [46].

In protein characterization, researchers distinguish between "bottom-up" and "top-down" analyses, the two most common methods of tandem mass spectrometry. There are pros and cons to each technique, which are particularly important to consider when characterizing biotherapeutics (Figure 4.8). Bottom-up analysis involves proteolytic cleavage of a protein into peptides, followed by MS/MS on these peptides to provide overall sequence coverage. Top-down analysis, on the other hand, performs MS analysis on the entire protein or large polypeptides directly. Bottom-up proteomics can lose contextual information, and some peptides may be too small or hydrophilic to be retained on the chromatography column or selected for

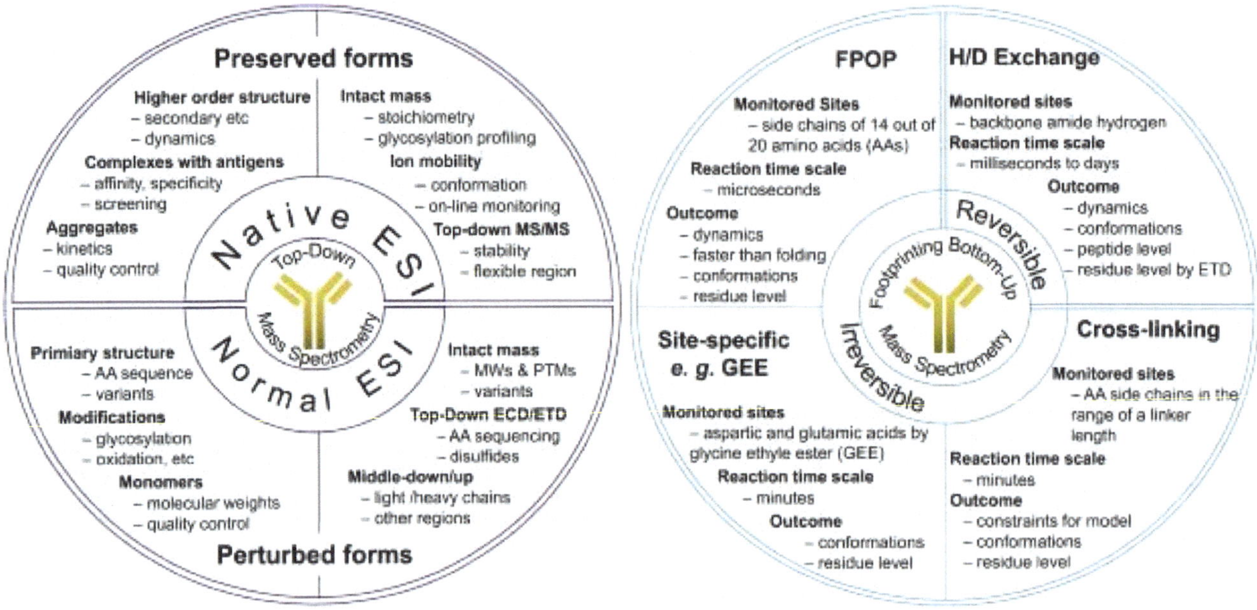

Figure 4.8 *Top-down and bottom-up mass spectrometry-based analysis of an antibody*. The left circle is the summary of top-down approaches. The right circle is the summary of bottom-up approaches. Source: Reproduced from Zhang et al. [47] with permission of Elsevier.

MS/MS fragmentation. Top-down analysis faces challenges such as dealing with protein isoforms, separating at the protein level, and obtaining spectra that can be adequately deconvoluted.

Bottom-up mass spectrometry involves MS/MS analysis at the peptide level. Proteins are digested with an enzyme, typically trypsin, which cleaves the peptide backbone after the charged amino acid residues lysine (K) or arginine (R) and results in peptides that are compatible with positive mode ionization. After digestion, peptides are separated (by methods like reverse-phase chromatography) and then ionized (by ESI or CID). Tandem mass spectrometry of a peptide generates a ladder of ions with charges retained at either the N- or C-terminus. Depending on the random fragmentation site along the peptide backbone, a series of N-terminally charged ions (*a*, *b*, and *c* ions) or C-terminally charged ions (*x*, *y*, and *z* ions) are formed. The types of fragment ions observed in an MS/MS spectrum depend on various factors, including the peptide's primary sequence, the amount of internal energy, the method of energy introduction for fragmentation, and the charge state. This nomenclature for fragment ions, initially proposed in 1984 by Roepstorff and Fohlman [48] and later refined by Johnson et al. [49], is widely recognized (Figure 4.9).

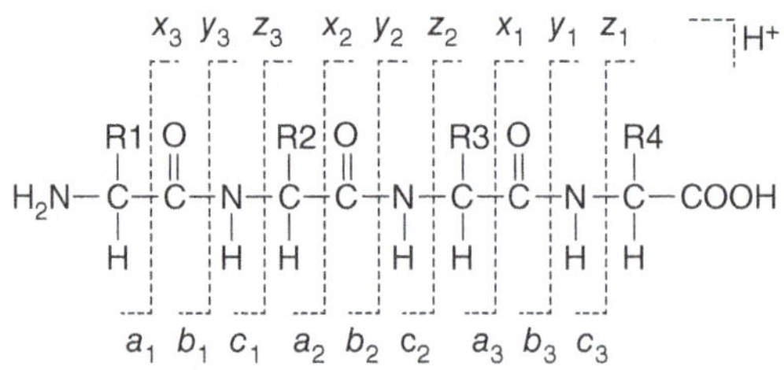

Figure 4.9 *Peptide fragment MS/MS ion series*. If the charge is retained on the N-terminal fragment, the ion is classed as *a*, *b*, or *c*. If the charge is retained on the C-terminal, the ion type is *x*, *y*, or *z*. A subscript indicates the number of residues in the fragment.

Typically, samples subjected to bottom-up mass spectrometry are separated online using reverse-phase chromatography followed by tandem mass spectrometry. For simple samples, such as a single recombinant protein, rapid chromatography can be completed in seconds to minutes, however, for more complex samples or protein mixtures, longer chromatography time is usually required to ensure comprehensive sequence coverage. Once separated, tandem mass spectrometry is performed by fragmenting peptides to enable sequence identification. CID is the predominant fragmentation method for peptide analysis in instruments like the Orbitrap and Q-TOF.

Top-down proteomics (TDP) is a method for analyzing intact proteins by mass spectrometry to identify and profile proteoforms in biological samples. TDP starts by determining the exact mass of the biotherapeutic protein after which the ion of interest is isolated and tandem mass spectrometry is performed. Traditionally, ECD or ETD was employed for protein fragmentation; currently, any MS/MS fragmentation method can be used. Despite technical challenges, particularly with heavily modified proteins, TDP is invaluable in the pharmaceutical industry for characterizing and ensuring the quality of large-molecule therapeutics like mAbs. Native top-down mass spectrometry (nTDMS) proteomics is an extension of TDP which preserves the protein's higher-order structure prior to fragmentation.

TDP and nTDMS complement bottom-up fragmentation methodologies by identifying and quantifying protein isoforms through the fragmentation of selected intact proteins. This enables the targeted characterization of sequence variants, post-translational modifications, and proteolysis products. Lastly, one additional mass spectrometric approach, termed middle-down analysis, works by generating larger polypeptides from a protein and then performing MS/MS. In order to generate such larger peptides, enzymes that are more specific than trypsin are typically used. The middle-down approach serves an intermediary method that is useful in validating therapeutic antibody sequences and in comparing biosimilars [50].

Data Acquisition Methods

There are various methods for collecting tandem mass spectrometric data, but the most dominant method is shotgun proteomics. Shotgun proteomics is a set of tools employed for analyzing and quantifying proteins in complex mixtures, such as those from tissues, cells, or body fluids. It involves digesting protein mixtures, separating peptides online using chromatography, and using tandem mass spectrometry for analysis. There are two main flavors of data acquisition: data-independent acquisition (DIA) and data-dependent acquisition (DDA) [51]. Both DDA and DIA are discovery-based approaches in bottom-up proteomics, where proteins are enzymatically digested into peptides, and the peptide data inform protein-level conclusions.

DIA [52] and DDA [51] differ primarily in the way that they collect tandem mass spectral data. DDA selectively fragments ions based on their intensity, which can lead to the omission of less abundant, but potentially important, peptides. In contrast, DIA fragments all ions within a predefined m/z range, ensuring comprehensive coverage and enabling the detection of low-abundance peptides. DIA either uses a spectral library to extract information from the rich data, or a more complex search algorithm. Reference to a spectral library or algorithm enables comprehensive peptide sampling, quantitative accuracy, and high reproducibility. In contrast, DDA selects and fragments ions based on intensity, not on a library. Thus, ions with greater intensity are more likely to be analyzed by DDA, making it less likely to detect low-abundance peptides. Separation techniques can surmount some of the coverage issues that are inherent to DDA, and this acquisition method is still valuable for generating libraries of spectra for databases. But because of its specificity and reproducibility, DIA is gaining traction and utility as an acquisition method [53].

Quantitative Mass Spectrometric Methods

There are various quantitative mass spectrometry methods, some of which rely on sophisticated labeling techniques. Quantitation is achieved in three main ways: first, by label-free quantitation that relies on comparing signals across different samples; second, by the incorporation of stable isotopes into the sample by metabolic labeling (e.g. stable isotope labeling by amino acids in cell culture (SILAC)); and third, labeling peptide samples with isobaric molecular tags (e.g. Isobaric Tags for Relative and Absolute Quantitation (iTRAQ) and Tandem Mass Tag (TMT)) [54].

Label-free quantification uses the intensity of detected peptide ions to estimate relative abundance [55]. This technique requires advanced data analysis to account for variations in ionization efficiency. In contrast to the label-based methods described below, label-free technologies do not require complicated sample preparation methods to enable quantification and can be easily paired with DIA data acquisition. Label-free methods are also more compatible with longitudinal studies where samples may be collected and analyzed over the course of weeks or months. Proper sample preparation and the use of internal standards are critical to ensure accurate quantification [56, 57]. The internal standards provide a benchmark to correct for variations in ionization efficiency and sample loading between different runs. Sophisticated data analysis software is required in order to reproducibly identify peptides and quantitate their relative abundances. However, because label-free methods require reproducible sample preparation and are quite sensitive to analytical variability between runs, these methods are generally not as accurate as the label-dependent methods described below [58].

In stable isotope labeling, molecules in a sample are metabolically labeled to enable direct, quantitative comparison between different samples. SILAC [59] is commonly used for cell culture-based quantitative proteomics. Cells are grown in media containing either heavy or light stable isotope-labeled amino acids. Because of the heavy and light labels the samples can be mixed together and analyzed in the same mass spectrometry runs. The signal intensity of peptides from the two different conditions can be directly measured together by comparing the signal intensities of the heavy- and the light-labeled peptides. Quantitative differences between samples are analyzed by comparing the intensities of peptide ions in the MS1 or survey scan, or can be coupled to multiple reaction monitoring (MRM). The quantitative capabilities of SILAC are robust; however, the number of samples that can be quantitatively analyzed in parallel is limited by the metabolic labels that can be incorporated, typically only two or three.

Chemical tags are the third quantitative method, and they overcome the multiplexing limitations of stable isotope labeling. Chemical tags now enable multiplexing dozens of samples for simultaneous analysis, resulting in reduced signal variability and more accurate quantitation. The two most common methodologies are iTRAQ labelling [60] and TMT labelling [61]. Both iTRAQ and TMT sample preparation methods involve chemically labeling peptides with isobaric tags. During mass spectrometry analysis, the isobaric tags are fragmented to produce distinct reporter ions to enable robust and reliable quantification. Today, TMT is typically favored over iTRAQ, mainly due to a higher multiplexing capability that quantification of more samples simultaneously. In biotherapeutic drug development, implementing isotope-labeled internal standards is a common practice in many pharmacokinetic analyses. Quantitative labeling methods such as SILAC, TMT, and iTRAQ are quite common in the discovery phase of drug discovery and development, such as in identifying new target pathways upon small molecule or genetic or disease perturbations, or in biomarkers discovery. For a comprehensive review of these and other quantitative techniques, refer to Bakalarksi et al. [54].

Computational Analysis of Tandem Mass Spectra

In all but the smallest of experiments, the assistance of computational tools and algorithms is necessary to match the acquired spectral information to the source biomolecules and to extract biological trends and insights. Tools for MS/MS analysis are specialized to the type of data acquired and quantitative method. Specialized algorithms are tailored to DDA or DIA analysis. Although it is possible to analyze tandem mass spectra *de novo* to generate likely sequence candidates for some data-dependent analysis spectra, incomplete fragment ion series and varying data quality make *de novo* analyses difficult and computationally intensive. Instead, a reverse approach is often employed, relying on a database of theoretical spectra generated from a set of known protein sequences. This database searching approach can be applied with more accuracy and speed to a dataset when the search space is well defined ahead of time.

Database searching tools have been employed for over 30 years, starting with the development of Sequest in 1994 [62]. Since then, dozens of search tools have been developed, and comprehensive reviews can be found in the literature. Here, we will highlight a few such tools which have proved reliable and found widespread adoption. Beyond Sequest and its reincarnation as Comet [63], tools such as Mascot [64], Andromeda [65], MS-GF+ [66], PEAKS [67], and more recently MSFragger [68], all operate in the same general manner: the input is a FASTA-formatted file of theoretical protein sequences and the collected spectral data, and the output is a list of peptide-spectral matches (PSMs). Each search tool differs mainly in its approach to matching potential PSMs by providing different scoring methods. After scoring, the raw results of a search tool are then often further analyzed by filtering to a desired protein or peptide false discovery rate (FDR), either through the use of internal methods or through the application of a standalone algorithm such as Percolator [69] or Philosopher [70]. The use of an FDR strikes a balance between sensitivity and specificity in the experiment. FDR calculation is often accomplished through the use of a target decoy database search strategy [71], where the typical database of known peptide sequences (the targets) is joined with a set of decoy sequences which are unlikely to occur naturally (e.g. reversed or randomized peptide sequences). The number of decoy hits assigned by the search algorithm then serves as a proxy for the number of false positives within the entire dataset. Typically reported FDR values are in the range of 1–5%, depending on the size of the dataset and the stringency desired for the experiment at hand. Similar methods can be applied to score the localization of post-translational modifications such as phosphorylation or ubiquitination [72–74].

Although database-matching algorithms are very successful in the matching of many spectra to known sequences, they do have their drawbacks. Limitations in search space and the requirement for *a priori* knowledge of the putative protein sequences make it impossible for these kinds of algorithms to identify novel proteins or protein variants. To tackle an unknown protein sequence, a larger search space and specialized tools must be employed, for example, Byonic [75] or PeaksAB [76].

While the ecosystem of tools for DDA data analysis is rich and well established, the landscape of tools for the interpretation of DIA data is much more nascent, due to the relative novelty of the field and the considerably tougher challenge of deconvoluting and identifying fragments from multiple spectra at once. While tools such as OpenSWATH [77] and DIA-Umpire [78] were the first to provide tractable solutions to DIA analysis, the field has enjoyed recent success with the advent of algorithms based at least in part on the use of deep neural networks. Two such popular approaches are DIA-nn [79] and Spectronaut [80]. Other end-to-end toolkits such as FragPipe-DIA [81] and Skyline [82] have incorporated similar elements or portions of these algorithms. Overall, these data analysis systems identify correlated fragments in DIA spectra, assemble them into groups, and then use that information to either match to a previously generated spectral library or to theoretically generated spectra from a source protein database. Results are then filtered through analogous methods to DDA to yield a final result at a desired FDR cutoff. The DIA analysis landscape is rapidly evolving and promises to provide similar results in terms of sensitivity and coverage to DDA approaches.

Regardless of the analysis method and how FDRs are used in PSM identification, for quantitative experiments, additional statistical methods are necessary to assess the quality of the quantification results. This can be done via Student's T-test, ANOVA, or through the use of linear models, which provide greater sensitivity and specificity. MSStats is a popular tool which has been applied in many contexts and has been extended further from its original roots for SRM/PRM studies to cover TMT approaches and post-translational modifications [83]. The output of MSStats is statistically comprehensive and includes log-fold changes and p-values, results which can be combined with other -omic analyses and implemented in algorithms such as GOstats [84] or GSEA [85]. Thus, MSStats integrates smoothly with other statistical tools, providing numerical quantitation that can be used toward elucidating biological contexts and understandings.

In experiments where complete sequence verification is not necessary and simple verification of protein termini is sufficient, more rapid techniques can be employed. Peptide mass fingerprinting uses the pattern of ion masses from intact peptides to quickly confirm the presence of single proteins (albeit at lower confidence). Another technique is the MALDI In Source Decay (MALDI-ISD) analysis of intact proteins, which generates fragments of the protein termini through laser ablation of a proton-donating matrix and is analyzed using a TOF instrument. Some such spectra can be interpreted manually, but algorithmic tools such as ISDetect [86] can provide objective scoring for terminal verification in a few minutes of total analysis time, providing an alternative to traditional chemical-based approaches such as Edman degradation.

Conclusions and Perspectives

In this chapter, we have sampled the diverse array of mass spectrometric tools available for the characterization of biotherapeutics, underscoring the dynamic and rapidly evolving nature of this field. Mass spectrometry continuously pushes the boundaries of detection and resolution, and when coupled with other "hyphenated" techniques such as capillary electrophoresis (CE), high-performance liquid chromatography (HPLC), and size-exclusion chromatography (SEC), it enables the detailed analysis of complex therapeutic biomolecule mixtures. This integration facilitates the characterization of heterogeneity arising from proteolysis or post-translational modifications.

As biotherapeutics grow increasingly intricate in design, analytical technologies are adapting in tandem. The advent of next-generation TOF and EMR mass spectrometers, for instance, has made it possible to accurately analyze complex moieties that were previously beyond the reach of standard proteomics laboratories, including bispecific antibodies, antibody-drug-conjugates, and complicated proteoforms. Concurrently, software development is keeping pace with these technical advancements, as evidenced by of top-down proteomic analysis tools and the sophisticated algorithms that are enabling DIA.

The symbiotic evolution of biotherapeutic design, mass spectrometric instrumentation, and analytical tools is accelerating at an astonishing rate. Undoubtedly, this trio of design, instrumentation, and analytical capacity will continue, enabling the capabilities of therapeutic recombinant proteins.

Abbreviations

ADC	antibody-drug conjugate
CCS	collisional cross section
CE	capillary electrophoresis
CHCA	alpha-cyano-4-hydroxycinnamic acid
CID	collision-induced dissociation
DAR	drug-to-antibody ratio
DC	direct current
DDA	data-dependent acquisition
DHB	2,5-dihydroxybenzoic acid
DIA	data-independent acquisition
DMS	differential mobility spectrometry
DT-IMS	drift-tube ion mobility spectrometry
ECD	electron capture dissociation
EMR	extended mass range
ESI	electrospray ionization
ETD	electron transfer dissociation
FAIMS	field asymmetric ion mobility spectrometry
FDR	false discovery rate
HPLC	high-performance liquid chromatography
HP-SEC	high-performance size exclusion chromatography

IEC	ion exchange chromatography
IMS	ion mobility spectrometry
IM-MS	ion mobility mass spectrometry
ISD	in-source decay
iTRAQ	multiplexed isobaric tagging technology for relative quantitation
mAb	monoclonal antibody
MALDI	matrix-assisted laser desorption ionization
MALDI-ISD	MALDI In Source Decay
MRM	multiple reaction monitoring
MS	mass spectrometry
MS/MS	tandem mass spectrometry
m/z	mass to charge ratio
PID	photo-induced dissociation
PSM	peptide-spectral match
PTCR	proton transfer charge reduction
QTOF	quadrupole time-of-flight
RF	radio frequency
SEC	size exclusion chromatography
SILAC	stable isotope labeling by amino acids in cell culture
TDP	top-down proteomics
nTDMS	native top-down mass spectrometry
TMT	tandem mass tags
TOF	time of flight
TOF-MS	time of flight mass spectrometry
TW-IMS	traveling-wave ion mobility spectrometry

References

1 Domon, B. and Aebersold, R. (2006). Mass spectrometry and protein analysis. *Science* 312 (5771): 212–217.

2 Carter, P.J. and Rajpal, A. (2022). Designing antibodies as therapeutics. *Cell* 185 (15): 2789–2805.

3 Schachner, L.F., Mullen, C., Phung, W. et al. (2024). Exposing the molecular heterogeneity of glycosylated biotherapeutics. *Nat. Commun.* 15 (1): 3259.

4 Bansal, R., Jha, S.K., and Jha, N.K. (2021). Size-based degradation of therapeutic proteins - mechanisms. *Modell. Contr. Biomol. Concepts* 12 (1): 68–84.

5 Sarvaiya, H., Niu, S., Ladror, D.T. et al. (2023). Bioanalytical strategy for the characterization and bioanalysis of biologics: a global, nonregulated bioanalytical lab perspective. *Bioanalysis* 15 (3): 133–148.

6 Morales, A., Candreva, J., Jayarathne, T. et al. (2023). A comprehensive strategy for the identification of biologics by liquid-chromatography–mass spectrometry for release testing in a regulated environment. *J. Pharm. Biomed. Anal.* 234: 115580.

7 Hillenkamp, F. and Karas, M. (1990). [12] Mass spectrometry of peptides and proteins by matrix-assisted ultraviolet laser desorption/ionization. *Methods Enzymol.* 193: 280–295.

8 Fenn, J.B., Mann, M., Meng, C.K. et al. (1989). Electrospray ionization for mass spectrometry of large biomolecules. *Science* 246 (4926): 64–71.

9 Dodonov, A.F., Chernushevich, I.V., and Laiko, V.V. (1993). Time-of-flight mass spectrometry. *ACS Symp. Ser.* 108–123.

10 Rose, R.J., Damoc, E., Denisov, E. et al. (2012). High-sensitivity orbitrap mass analysis of intact macromolecular assemblies. *Nat. Methods* 9 (11): 1084–1086.

11 Rosati, S., Rose, R.J., Thompson, N.J. et al. (2012). Exploring an orbitrap analyzer for the characterization of intact antibodies by native mass spectrometry. *Angew. Chem. Int. Ed.* 51 (52): 12992–12996.

12 Thompson, N.J., Rosati, S., and Heck, A.J.R. (2014). Performing native mass spectrometry analysis on therapeutic antibodies. *Methods* 65 (1): 11–17.

13 Rosati, S., van den Bremer, E.T., Schuurman, J. et al. (2013). In-depth qualitative and quantitative analysis of composite glycosylation profiles and other micro-heterogeneity on intact monoclonal antibodies by high-resolution native mass spectrometry using a modified Orbitrap. *MAbs* 5 (6): 917–924.

14 de Jong, R.N., Beurskens, F.J., Verploegen, S. et al. (2016). A novel platform for the potentiation of therapeutic antibodies based on antigen-dependent formation of IgG hexamers at the cell surface. *PLoS Biol.* 14 (1): e1002344.

15 Diebolder, C.A., Beurskens, F.J., de Jong, R.N. et al. (2014). Complement is activated by IgG hexamers assembled at the cell surface. *Science* 343 (6176): 1260–1263.

16 Cleary, S.J., Seo, Y., Tian, J.J. et al. (2024). IgG hexamers initiate complement-dependent acute lung injury. *J. Clin. Invest.* 134 (11): e178351.

17 Heck, A.J.R. (2008). Native mass spectrometry: a bridge between interactomics and structural biology. *Nat. Methods* 5 (11): 927–933.

18 Schachner, L.F., Phung, W., Han, G. et al. (2022). High-throughput, quantitative analysis of peptide-exchanged MHCI complexes by native mass spectrometry. *Anal. Chem.* 94 (42): 14593–14602.

19 Jager, S., Cramer, D.A.T., and Heck, A.J.R. (2023). Normal alpha-1-antitrypsin variants display in serum allele-specific protein levels. *J. Proteome Res.* 22 (4): 1331–1338.

20 Hecht, E.S., Mehta, S., Wecksler, A.T. et al. (2022). Insights into ultra-low affinity lipase-antibody non-covalent complex binding mechanisms. *MAbs* 14 (1): 2135183.

21 Campuzano, I.D.G. and Sandoval, W. (2021). Denaturing and native mass spectrometric analytics for biotherapeutic drug discovery research: historical, current, and future personal perspectives. *J. Am. Soc. Mass Spectrom.* 32 (8): 1861–1885.

22 Nagy, G. (2024). High-resolution ion mobility separations coupled to mass spectrometry: what's next? *J. Mass Spectrom.* 59 (5): e5014.

23 Bagal, D., Valliere-Douglass, J.F., Balland, A., and Schnier, P.D. (2010). Resolving disulfide structural isoforms of IgG2 monoclonal antibodies by Ion mobility mass spectrometry. *Anal. Chem.* 82 (16): 6751–6755.

24 Pacholarz, K.J., Peters, S.J., Garlish, R.A. et al. (2016). Molecular insights into the thermal stability of mAbs with variable-temperature ion-mobility mass spectrometry. *ChemBioChem* 17 (1): 46–51.

25 Huang, R.Y.C., Deyanova, E.G., Passmore, D. et al. (2015). Utility of ion mobility mass spectrometry for drug-to-antibody ratio measurements in antibody-drug conjugates. *J. Am. Soc. Mass Spectrom.* 26 (10): 1791–1794.

26 Makey, D.M. and Ruotolo, B.T. (2024). Liquid-phase separations coupled with ion mobility-mass spectrometry for next-generation biopharmaceutical analysis. *Expert Rev. Proteom.* 21 (5–6): 259–270.

27 Stoll, D.R. (2010). Recent progress in online, comprehensive two-dimensional high-performance liquid chromatography for non-proteomic applications. *Anal. Bioanal. Chem.* 397 (3): 979–986.

28 Smith, R.D., Loo, J.A., Edmonds, C.G. et al. (1990). Sensitivity considerations for large molecule detection by capillary electrophoresis-electrospray ionization mass spectrometry. *J. Chromatogr. A* 516 (1): 157–165.

29 Landers, J.P. (2007). *Handbook of Capillary and Microchip Electrophoresis and Associated Microtechniques* (ed. J.P. Landers). CRC Press 1596 p.

30 Gahoual, R., Beck, A., François, Y., and Leize-Wagner, E. (2016). Independent highly sensitive characterization of asparagine deamidation and aspartic acid isomerization by sheathless CZE-ESI-MS/MS. *J. Mass Spectrom.* 51 (2): 150–158.

31 François, Y.N., Biacchi, M., Said, N. et al. (2016). Characterization of cetuximab Fc/2 dimers by off-line CZE-MS. *Anal. Chim. Acta* 908: 168–176.

32 Basu, A., Yang, K., Wang, M. et al. (2006). Structure–function engineering of interferon-β-1b for improving stability, solubility, potency, immunogenicity, and pharmacokinetic properties by site-selective mono-PEGylation. *Bioconjug. Chem.* 17 (3): 618–630.

33 Deeb, S.E., Wätzig, H., El-Hady, D.A. et al. (2016). Recent advances in capillary electrophoretic migration techniques for pharmaceutical analysis (2013–2015). *Electrophoresis* 37 (12): 1591–1608.

34 Ferrige, A.G., Seddon, M.J., Jarvis, S. et al. (1991). Maximum entropy deconvolution in electrospray mass spectrometry. *Rapid Commun. Mass Spectrom.* 5 (8): 374–377.

35 Zhang, Z. and Marshall, A.G. (1998). A universal algorithm for fast and automated charge state deconvolution of electrospray mass-to-charge ratio spectra. *J. Am. Soc. Mass Spectrom.* 9 (3): 225–233.

36 Horn, D.M., Zubarev, R.A., and McLafferty, F.W. (2000). Automated reduction and interpretation of high resolution electrospray mass spectra of large molecules. *J. Am. Soc. Mass Spectrom.* 11 (4): 320–332.

37 Marty, M.T., Baldwin, A.J., Marklund, E.G. et al. (2015). Bayesian deconvolution of mass and ion mobility spectra: from binary interactions to polydisperse ensembles. *Anal. Chem.* 87 (8): 4370–4376.

38 Millán-Martín, S., Carillo, S., Füssl, F. et al. (2021). Optimisation of the use of sliding window deconvolution for comprehensive characterisation of trastuzumab and adalimumab charge variants by native high resolution mass spectrometry. *Eur. J. Pharm. Biopharm.* 158: 83–95.

39 Bern, M., Caval, T., Kil, Y.J. et al. (2018). Parsimonious charge deconvolution for native mass spectrometry. *J. Proteome Res.* 17 (3): 1216–1226.

40 Kostelic, M.M. and Marty, M.T. (2022). Deconvolving native and intact protein mass spectra with UniDec. In: Proteoform Identification: Methods and Protocols. *Methods Mol. Biol.* 2500: 159–180.

41 Phung, W., Bakalarski, C.E., Hinkle, T.B. et al. (2023). UniDec processing pipeline for rapid analysis of biotherapeutic mass spectrometry data. *Anal. Chem.* 95 (30): 11491–11498.

42 Huguet, R., Mullen, C., Srzentić, K. et al. (2019). Proton transfer charge reduction enables high-throughput top-down analysis of large proteoforms. *Anal. Chem.* 91 (24): 15732–15739.

43 Wells, J.M. and McLuckey, S.A. (2005). Collision-induced dissociation (CID) of peptides and proteins. *Methods Enzymol.* 402: 148–185.

44 Brodbelt, J.S. (2014). Photodissociation mass spectrometry: new tools for characterization of biological molecules. *Chem. Soc. Rev.* 43 (8): 2757–2783.

45 Syka, J.E.P., Coon, J.J., Schroeder, M.J. et al. (2004). Peptide and protein sequence analysis by electron

transfer dissociation mass spectrometry. *Proc. Natl. Acad. Sci.* 101 (26): 9528–9533.

46 Zubarev, R.A. (2004). Electron-capture dissociation tandem mass spectrometry. *Curr. Opin. Biotechnol.* 15 (1): 12–16.

47 Zhang, H., Cui, W., and Gross, M.L. (2014). Mass spectrometry for the biophysical characterization of therapeutic monoclonal antibodies. *FEBS Lett.* 588 (2): 308–317.

48 Roepstorff, P. and Fohlman, J. (1984). Proposal for a common nomenclature for sequence ions in mass spectra of peptides. *Biomed. Mass Spectrom.* 11 (11): 601.

49 Johnson, R.S., Martin, S.A., Biemann, K. et al. (1987). Novel fragmentation process of peptides by collision-induced decomposition in a tandem mass spectrometer: differentiation of leucine and isoleucine. *Anal. Chem.* 59 (21): 2621–2625.

50 Catherman, A.D., Skinner, O.S., and Kelleher, N.L. (2014). Top down proteomics: facts and perspectives. *Biochem. Biophys. Res. Commun.* 445 (4): 683–693.

51 Ghosh, G., Shannon, A.E., and Searle, B.C. (2024). Data acquisition approaches for single cell proteomics. *Proteomics* e2400022.

52 Bichmann, L., Gupta, S., and Röst, H. (2024). Peptidomics, methods, and strategies. *Methods Mol. Biol.* 2758: 77–88.

53 Fernández-Costa, C., Martínez-Bartolomé, S., McClatchy, D.B. et al. (2020). Impact of the identification strategy on the reproducibility of the DDA and DIA results. *J. Proteome Res.* 19 (8): 3153–3161.

54 Bakalarski, C.E. and Kirkpatrick, D.S. (2016). A biologist's field guide to multiplexed quantitative proteomics. *Mol. Cell. Proteom.* 15 (5): 1489–1497.

55 Thompson, A., Schäfer, J., Kuhn, K. et al. (2003). Tandem mass tags: a novel quantification strategy for comparative analysis of complex protein mixtures by MS/MS. *Anal. Chem.* 75 (8): 1895–1904.

56 Välikangas, T., Suomi, T., and Elo, L.L. (2018). A systematic evaluation of normalization methods in quantitative label-free proteomics. *Brief. Bioinform.* 19 (1): 1–11.

57 Rozanova, S., Barkovits, K., Nikolov, M. et al. (2021). Quantitative methods in proteomics. *Methods Mol. Biol.* 2228: 85–116.

58 Moulder, R., Goo, Y.A., and Goodlett, D.R. (2016). Quantitative proteomics by mass spectrometry. *Methods Mol. Biol.* 1410: 65–76.

59 Ong, S.E., Blagoev, B., Kratchmarova, I. et al. (2002). Stable isotope labeling by amino acids in cell culture, SILAC, as a simple and accurate approach to expression proteomics*. *Mol. Cell. Proteom.* 1 (5): 376–386.

60 Gygi, S.P., Rist, B., Gerber, S.A. et al. (1999). Quantitative analysis of complex protein mixtures using isotope-coded affinity tags. *Nat. Biotechnol.* 17 (10): 994–999.

61 Dayon, L., Hainard, A., Licker, V. et al. (2008). Relative quantification of proteins in human cerebrospinal fluids by MS/MS using 6-plex isobaric tags. *Anal. Chem.* 80 (8): 2921–2931.

62 Eng, J.K., McCormack, A.L., and Yates, J.R. (1994). An approach to correlate tandem mass spectral data of peptides with amino acid sequences in a protein database. *J. Am. Soc. Mass Spectrom.* 5 (11): 976–989.

63 Eng, J.K., Jahan, T.A., and Hoopmann, M.R. (2013). Comet: an open-source MS/MS sequence database search tool. *Proteomics* 13 (1): 22–24.

64 Perkins, D.N., Pappin, D.J.C., Creasy, D.M., and Cottrell, J.S. (1999). Probability-based protein identification by searching sequence databases using mass spectrometry data. *Electrophoresis* 20 (18): 3551–3567.

65 Cox, J., Neuhauser, N., Michalski, A. et al. (2011). Andromeda: a peptide search engine integrated into the MaxQuant environment. *J. Proteome Res.* 10 (4): 1794–1805.

66 Kim, S. and Pevzner, P.A. (2014). MS-GF+ makes progress towards a universal database search tool for proteomics. *Nat. Commun.* 5 (1): 5277.

67 Zhang, J., Xin, L., Shan, B. et al. (2012). PEAKS DB: de novo sequencing assisted database search for sensitive and accurate peptide identification*. *Mol. Cell. Proteom.* 11 (4): M111.010587.

68 Kong, A.T., Leprevost, F.V., Avtonomov, D.M. et al. (2017). MSFragger: ultrafast and comprehensive peptide identification in mass spectrometry–based proteomics. *Nat. Methods* 14 (5): 513–520.

69 Käll, L., Canterbury, J.D., Weston, J. et al. (2007). Semi-supervised learning for peptide identification from shotgun proteomics datasets. *Nat. Methods* 4 (11): 923–925.

70 da Veiga Leprevost, F., Haynes, S.E., Avtonomov, D.M. et al. (2020). Philosopher: a versatile toolkit for shotgun proteomics data analysis. *Nat. Methods* 17 (9): 869–870.

71 Elias, J.E. and Gygi, S.P. (2007). Target-decoy search strategy for increased confidence in large-scale protein identifications by mass spectrometry. *Nat. Methods* 4 (3): 207–214.

72 Beausoleil, S.A., Villén, J., Gerber, S.A. et al. (2006). A probability-based approach for high-throughput protein phosphorylation analysis and site localization. *Nat. Biotechnol.* 24 (10): 1285–1292.

73 Chalkley, R.J. and Clauser, K.R. (2012). Modification site localization scoring: strategies and performance. *Mol. Cell. Proteom.* 11 (5): 3–14.

74 Wiese, H., Kuhlmann, K., Wiese, S. et al. (2014). Comparison of alternative MS/MS and bioinformatics approaches for confident phosphorylation site localization. *J. Proteome Res.* 13 (2): 1128–1137.

75 Bern, M., Kil, Y.J., and Becker, C. (2012). Byonic: advanced peptide and protein identification software. *Curr. Protoc. Bioinform.* 40 (1): 13.20.1–13.20.14.

76 Tran, N.H., Rahman, M.Z., He, L. et al. (2016). Complete de novo assembly of monoclonal antibody sequences. *Sci. Rep.* 6 (1): 31730.

77 Röst, H.L., Rosenberger, G., Navarro, P. et al. (2014). OpenSWATH enables automated, targeted analysis of data-independent acquisition MS data. *Nat. Biotechnol.* 32 (3): 219–223.

78 Tsou, C.C., Avtonomov, D., Larsen, B. et al. (2015). DIA-umpire: comprehensive computational framework for data-independent acquisition proteomics. *Nat. Methods* 12 (3): 258–264.

79 Demichev, V., Messner, C.B., Vernardis, S.I. et al. (2020). DIA-NN: neural networks and interference correction enable deep proteome coverage in high throughput. *Nat. Methods* 17 (1): 41–44.

80 Bruderer, R., Bernhardt, O.M., Gandhi, T. et al. (2015). Extending the limits of quantitative proteome profiling with data-independent acquisition and application to acetaminophen-treated three-dimensional liver microtissues* [S]. *Mol. Cell. Proteom.* 14 (5): 1400–1410.

81 Yu, F., Teo, G.C., Kong, A.T. et al. (2023). Analysis of DIA proteomics data using MSFragger-DIA and FragPipe computational platform. *Nat. Commun.* 14 (1): 4154.

82 MacLean, B., Tomazela, D.M., Shulman, N. et al. (2010). Skyline: an open source document editor for creating and analyzing targeted proteomics experiments. *Bioinformatics* 26 (7): 966–968.

83 Choi, M., Chang, C.Y., Clough, T. et al. (2014). MSstats: an R package for statistical analysis of quantitative mass spectrometry-based proteomic experiments. *Bioinformatics* 30 (17): 2524–2526.

84 Falcon, S. and Gentleman, R. (2006). Using GOstats to test gene lists for GO term association. *Bioinformatics* 23 (2): 257–258.

85 Subramanian, A., Tamayo, P., Mootha, V.K. et al. (2005). Gene set enrichment analysis: a knowledge-based approach for interpreting genome-wide expression profiles. *Proc. Natl. Acad. Sci.* 102 (43): 15545–15550.

86 Bakalarski, C.E., Gan, Y., Wertz, I. et al. (2016). Rapid, semi-automated protein terminal characterization using ISDetect. *Nat. Biotechnol.* 34 (8): 811–813.

5

Therapeutic Proteins and Hyphenated Methods for Characterization

Dietmar Reusch and Markus Haberger

Roche Diagnostics GmbH, Penzberg, Germany

Introduction

The hyphenation of native protein separation with mass spectrometry is quickly becoming a robust method for the characterization of proteins while maintaining their native states. Reversed-phase chromatography under denaturing conditions coupled to mass spectrometry is not covered in this chapter, as the focus is primarily on the separation of native proteins. This chapter outlines the various available native separation techniques for proteins and discusses practical considerations for coupling these separations with mass spectrometry. The significance of these approaches is highlighted through selected applications that address questions in biomedical and biopharmaceutical research and development. In particular, the chapter focuses on the role of affinity-based separations in simultaneously acquiring structural and functional information about proteins.

Figure 5.1 illustrates various bispecific antibody proteoforms – including aggregates, fragments, and bispecific antibody byproducts – as well as product variants characterized by post-translational modifications (PTMs). These entities have been analyzed using native separation techniques directly coupled with mass spectrometry.

Methods

Size Exclusion Chromatography Mass Spectrometry (SEC-MS)

Bispecific antibodies have gained significant attention in recent years due to their unique ability to simultaneously target two different antigens [1, 2]. However, the complexity of these molecules presents challenges in their characterization. Different technologies are applied to produce bispecific antibodies each with their own advantages and limitations. Mispairing of LCs and HCs can result in bispecific antibody by-products (product-related impurities), in addition, these by-products can also form covalent and non-covalent protein complexes. This chapter aims to provide an overview of SEC-MS as a powerful analytical technique for the structural characterization of bispecific antibodies and their specific byproducts.

SEC operates on the principle that molecules of different sizes will elute through a column packed with porous beads at different rates. The stationary phase consists of these beads, which contain pores of a specific size distribution. Larger molecules cannot enter the pores of the beads and are excluded. These molecules travel the path of least resistance, which is around the outside of the beads, and therefore elute first from the column. The separation is purely based on the hydrodynamic volume of the molecules and not on any chemical interactions with the stationary phase. The resolution of SEC is generally lower than other chromatographic techniques like ion-exchange or reversed-phase chromatography. This means that molecules that are very similar in size may not be effectively separated. Separation is dependent on the used column, flow rates, temperatures, injection volume, and buffer compositions. SEC-MS in recent years is mainly performed using ultra high-performance liquid chromatography (HPLC) systems which improves the separation compared to classical SEC columns. Compared to classical SEC, SEC-MS is operated with a volatile buffer system, here an ammonium acetate buffer system allowing for direct coupling to MS instruments. SEC-MS can be operated without UV detection; however, the usage of UV detection is recommended as the UV data is supportive for the correct interpretation of MS data, for example, in the context of correct quantification of by-products, as small molecules like free LCs will have intrinsically higher ionization efficiencies compared to the antibody investigated.

Characterizing Biotherapeutics: Analytical Methods for Diverse Modalities, First Edition. Edited by Jennie R. Lill and Wendy Sandoval.

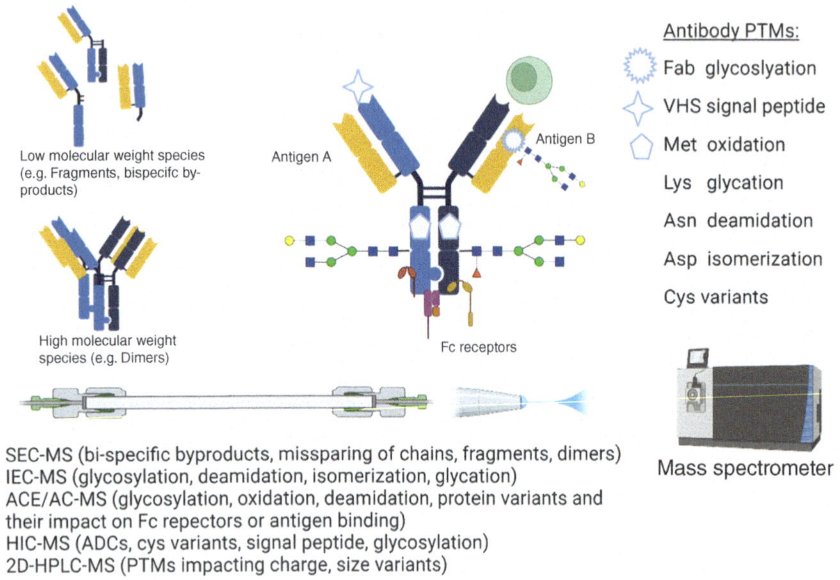

Figure 5.1 Bispecific antibody proteoforms and product variants analyzed by different hyphenated methods for characterization. Source: Created in BioRender. by Haberger, M. (2024) BioRender.com/u26m320.

In general, SEC can effectively resolve aggregates and smaller molecules from the bispecific antibody on their size difference.

Historically, SEC-MALS (multi-angle light scattering) was used for measuring the molecular weight of molecules. However, the mass accuracy of MALS and the resolving power are too low for the characterization of bispecific antibodies by-products because they are often very close in mass (few Dalton, see also Table 5.1) and therefore eluting within size exclusion peak.

Therefore, only high-resolution mass spectrometry using Orbitrap or TOF mass analyzers has the capability to provide the needed accuracy and precise determination of the molecular weight to unambiguously identify bispecific antibodies by-products. Furthermore, modifications of the protein, such as glycosylation or truncations, can be determined by MS.

Figure 5.2 shows a comparison of classical SEC-UV procedure (Figure 5.2a) compared to the native SEC-UV/MS system (Figure 5.2b, UV trace and Figure 5.2c, MS trace) for stability samples (stored for 24 months at 5 °C, red traces) and temperature-stressed material (stored for 3 months at 5°C, blue traces).

Size variants of bispecific antibodies consist of variants which are built during the production process, for example, mispairing of chains, fragments, and aggregates. Proposed assignments of the Native ESI-MS fractions 1–9 (Figure 5.2b) are summarized in Table 5.1.

Aggregates can be covalently and non-covalently linked. As mentioned before, SEC-MS is operated with ammonium acetate in water to enable direct coupling to MS instruments. This buffer system has an additional feature in that it also preserves the native structure and potential non-covalent interactions of the analyzed molecules. This allows for the study of intact complexes and their interactions as they exist in solution. However, even though electro spray ionization is known as gentle ionization method and maintains mild conditions during analysis for biomolecules, fragmentation of molecules can occur during the ESI process and may also depend on the type of instruments used. In particular, non-covalently linked aggregates may tend to fragment. Here, comparison to the UV data can help in data interpretation, for example, if the dimer peak of a molecule is eluting from the SEC column; however, mainly the monomer of the molecule is detected in the MS, indicating fragmentation of the dimer during the MS process.

Haberger et al. [3] applied SEC-MS for the direct characterization of bispecific antibody by-products in different stability samples (5°C versus 40°C) and also investigated the non-covalent interactions of high molecular weight species by the addition of acetonitrile into the buffer system. Similarly, Cramer et al. [4] studied preparative high molecular weight fractions of a trivalent bispecific antibody (2 + 1 Crossmab) demonstrating the resolving power of SEC-MS. Furthermore, Yan et al. [5] used a mixed mode SEC with native mass spectrometry for the sensitive detection and quantitation of homodimer impurities (mispairing of chains) in bispecific IgG using four different bispecific antibody standards. Impressively, they determined the sensitivity for the detection of homodimer impurities with 0.01–0.1% (LOD).

Ren et al. [6] used SEC-MS for the quantitative determination of protein–ligand affinity, they developed a robust platform SEC-MS for measuring non-covalent protein–small molecule interactions.

Table 5.1 Native ESI-UV/MS identification results of CrossMAb size variants detected in stability samples (elution profile shown in Figure 5.3).

SEC-Peak	Protein variant	Description	Charge state	Theoretical molecular mass [Da]	Reference sample [Da]	Stability sample 5°C 24 months [Da]	Stress sample 40°C 3 months [Da]
1	Trimer	Trimer of CrossMAb	35–44	447,146	n.d.	447,651	447,573
2 + 3	Dimer	Dimer 1/2 of CrossMAb	28–36	298,097	298,108	298,131	298,108
4		CrossMAb w/o LCy + LCx/LCy Heterodimer	21–27	172,595$_{(+Cys)}$ 172,782$_{(+GSH)}$	172,611 172,792	172,619 172,800	172,594 172,775
5	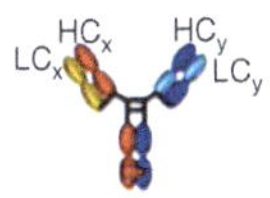	CrossMAb	19–25	149,049	149,043	149,052	149,044
		CrossMAb w/o LC$_y$	19–21	126,918$_{(+Cys)}$ 127,104$_{(+GSH)}$	126,919 127,107	126,933 127,113	126,911 127,100
6	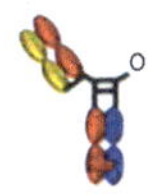	CrossMAb w/o Fab HC$_y$/LC$_y$	16–20	100,327$_{[HCy_240\text{–}241\ H\text{-}T]}$ 100,693$_{[HCy_237\text{–}238\ D\text{-}K]}$ 100,808$_{[HCy_236\text{–}237\ C\text{-}D]}$	100,329 n.d. 100,810	100,334 n.d. 100,810	100,326 100,693 100,806
		CrossMAb w/o Fab HC$_x$/LC$_x$	16–20	101,020$_{[HCx_228\text{–}229\ K\text{-}T]}$ 101,164$_{[HCx_227\text{–}228\ D\text{-}K]}$	n.d. n.d.	n.d. n.d.	101,012 101,165
7		Homodimer LC$_x$/LC$_x$	11–14	46,856	46,856	46,858	46,854
		Heterodimer LC$_x$/LC$_y$	11–14	45,678	45,677	45,679	45,676
8		Fab HC$_y$/LC$_y$	11–14	48,258$_{[HCy_236\text{–}237\ C\text{-}D]}$ 48,355$_{[HCy_237\text{–}238\ D\text{-}K]}$ 48,739$_{[HCy_240\text{–}241\ H\text{-}T]}$	n.d. n.d. n.d.	n.d. n.d. 48,741	48,260 48,356 48,737
9		Fab HC$_x$/LC$_x$	12–14	47,885$_{[HCx_228\text{–}229\ D\text{-}K]}$ 48,029$_{[HCx_227\text{–}228\ K\text{-}T]}$	n.d. n.d.	n.d. 48,033	47,882 48,028

HC$_x$, heavy chain x; HC$_y$, heavy chain y; LC$_x$, light chain x; LC$_y$, light chain y; *, cysteinylation (plus 119 Da) or glutathionylation (plus 305 Da) of heavy chain in position 233 (HC-Cys-233). ° indicates multiple cleavage sites of the upper hinge region, resulting in the corresponding Fc- and Fab-fragments (HC$_x$/LC$_x$ and HC$_y$/LC$_y$) (Peaks 6, 8, and 9). Identity of covalently modified (Peak 4) or truncated CrossMAb variants (Peaks 6–9) were additionally confirmed by LysC peptide mapping; n.d., not detectable.
Source: Table 5.1 published in mAbs. Adapted from Haberger et al. [3].

In conclusion, SEC-MS has emerged as a valuable tool for the comprehensive characterization of bispecific antibodies. This chapter has highlighted the principles, applications, and challenges associated with SEC-MS analysis. By leveraging the capabilities of SEC-MS, researchers and scientists can gain deeper insights into the structural attributes of bispecific antibodies, thus enabling the development of safer and more effective therapeutics.

Ion Exchange Chromatography Mass Spectrometry (IEC-MS)

Ion exchange chromatography (IEC) with UV detection is the standard method for characterizing and testing antibodies. However, the use of non-MS compatible buffers and UV detection only makes peak identification challenging and time consuming, often requiring peak collections followed by buffer exchange steps or complex 2D approaches. This challenge can be overcome in IEC by using MS-friendly buffer systems enabling direct MS coupling.

IEC coupled with mass spectrometry (MS) has emerged as a powerful analytical strategy for the detailed characterization of bispecific antibodies. IEC, known for its high resolution and ability to separate proteins based on their charge properties, provides a robust platform for the initial fractionation of bispecific antibody species. When

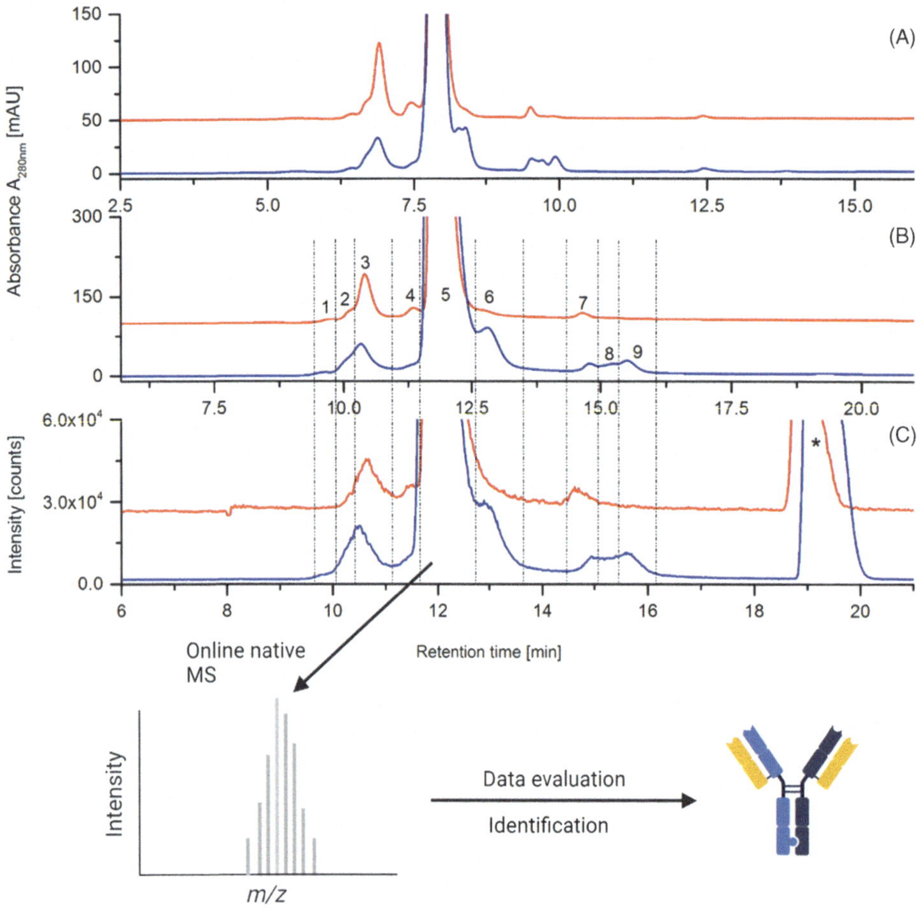

Figure 5.2 Comparison of fast-SEC with UV detection (a) versus native ESI-UV/MS (b, UV trace; c, total ion current chromatogram). Chromatograms show CrossMAb "stability" (red trace, 24 months at 5°C) and temperature-stressed (blue trace, 3 months at 40°C) samples. Sample matrix signals are marked with asterisk (*). Source: Adapted from Haberger et al. 2015.

integrated with the high sensitivity and specificity of MS, this combination allows for comprehensive profiling of the molecular heterogeneity, PTMs, and potential impurities within bispecific antibody preparations.

This chapter delves into the principles and methodologies of IEC and mass spectrometry, elucidating their synergistic application in the characterization of bispecific antibodies. We will explore various IEC techniques, such as cation exchange and anion exchange chromatography, and discuss their optimization for separating complex antibody mixtures. Furthermore, we will examine the role of mass spectrometry in providing detailed structural insights and quantitative analysis, highlighting key advancements in MS technologies that enhance the characterization process.

Through a series of case studies and practical examples, this chapter aims to provide a comprehensive guide for researchers and practitioners in the biopharmaceutical field. By understanding the intricacies of IEC-MS workflows, readers will be better equipped to implement these techniques in their own laboratories, ultimately contributing to the advancement of bispecific antibody therapeutics.

Cation exchange chromatography (CEX) typically involves the separation of molecules based on their charge at a given pH. In this type of chromatography, the stationary phase is negatively charged, meaning it will attract and bind positively charged molecules (cations).

The isoelectric point (pI) of a molecule is the pH at which the molecule carries no net charge. For CEX, the pH of the buffer used should be lower than the isoelectric point of the target molecule. This ensures that the molecule is positively charged and can interact with the negatively charged stationary phase.

The most common type of monoclonal IgG antibodies usually have pIs in the range of 6–9.

For effective CEX, choose a buffer pH that is at least 1–2 pH units below the isoelectric point of the molecule you wish to separate, ensuring that the molecule is positively charged in the process.

To allow for direct MS coupling, volatile buffers are typically chosen; in most cases, ammonium acetate buffer systems are used.

Leblanc et al. [7] developed a mixed mode ion exchange chromatographic method using volatile salts and coupled it online to native mass spectrometry for detailed characterization of monoclonal antibodies charge variants. This approach successfully identified deamidation of the heavy chain as the major degradation pathway in an aged monoclonal antibody, demonstrating the method's effectiveness in analyzing complex charge variant profiles.

Ponniah et al. [8] established an efficient strategy for characterizing recombinant monoclonal antibody charge variants, achieving higher resolution at subunit levels, and localizing potential modifications to specific fragments. Peptide mapping after IdeS digestion identified specific modifications, including aspartate isomerization, asparagine deamidation, and glycation, highlighting the unpredictable chromatographic behavior of antibody charge variants.

Phung et al. [9] discussing the limitations of mass spectrometry in differentiating between correctly paired and light chain-scrambled mispaired antibodies due to their similar biophysical properties. To address this, the study introduces a powerful analytical platform using native weak CEX coupled with high-resolution mass spectrometry, charge variant mass spectrometry (CV-MS), to characterize bispecific and mispaired antibody species. The investigation aims to understand the intrinsic charge, size, and shape differences of these molecules through elution order analysis, molecular modeling, and the hypothesis of localized charge patches offering sufficient micro-variations for resolution and identification.

Shi et al. [10] investigated the use of CEX directly coupled with mass spectrometry (MS) to identify chemical modifications responsible for charge variants in monoclonal antibodies, bispecific antibodies, and Fc-fusion proteins. Additionally, the research explores the use of post-CEX column addition of organic solvent and acid, along with elevated temperatures, to unfold proteins, increase ion intensity, and facilitate top-down analysis. The study demonstrates the utility of the online method in providing site-specific structural information on charge modifications without the need for fraction collection and laborious peptide mapping.

Haberger et al. [11] aim to characterize product variants of a therapeutic T-cell bispecific humanized monoclonal antibody (TCB Mab) and develop an online CEC native electrospray mass spectrometry (CEC-UV-MS) method for direct TCB Mab charge variant monitoring during bioprocess and formulation development. Through offline fractionation and comprehensive analytical testing, the study identified coeluting acid charge variants, such as asparagine deamidation and various glycosylation patterns, including the discovery of a new non-consensus N-glycosylation motif in the heavy chain CDR region. The establishment of an online native CEC-UV-MS method with an ammonium acetate buffer and pH gradient allowed for direct monitoring of TCB Mab charge variants, offering increased throughput, reduced sample handling, and intact protein information.

Schaick et al. [12] described the development of an MS-compatible anion exchange (AEX) method for monitoring multiple quality attributes of the biopharmaceutical erythropoietin (EPO), enabling the assessment of critical quality attributes (CQAs), such as charge heterogeneity, sialic acid content, and the number of N-acetyllactosamine units in a single experiment. The method demonstrated good repeatability and robustness, offering comparable results to common pharmacopeia and standard methods while requiring fewer analytical methods and less sample treatment.

Affinity Separation Mass Spectrometry

Potency testing of proteins is a critical process in biopharmaceutical development, aimed at determining the biological activity of a protein drug to ensure its efficacy and safety. However, this task is fraught with challenges such as the complexity of protein structures, variability in biological assays, and the need for highly sensitive and specific analytical methods to accurately measure protein activity.

Characterization and potency determination of therapeutic protein variants is indeed a time-consuming and challenging step. Typically, variants must first be generated and fractionated to isolate distinct forms. These variants are then characterized using mass spectrometry to determine their molecular properties. Finally, they are analyzed using surface plasmon resonance (SPR) techniques for binding and potency assays to assess their functional efficacy.

However, all these steps can ideally be consolidated into a single experiment using affinity mass spectrometry. This technique streamlines the process by simultaneously characterizing, identifying, and assessing the binding properties and potency of therapeutic protein variants.

The most important biological activities of antibodies and bispecific antibodies include their ability to specifically bind to target antigens, which is crucial for their therapeutic efficacy. Additionally, their interaction with Fc receptors plays a vital role in mediating immune responses, such as antibody-dependent cellular cytotoxicity (ADCC), antibody-dependent cellular phagocytosis (ADCP), and complement activation.

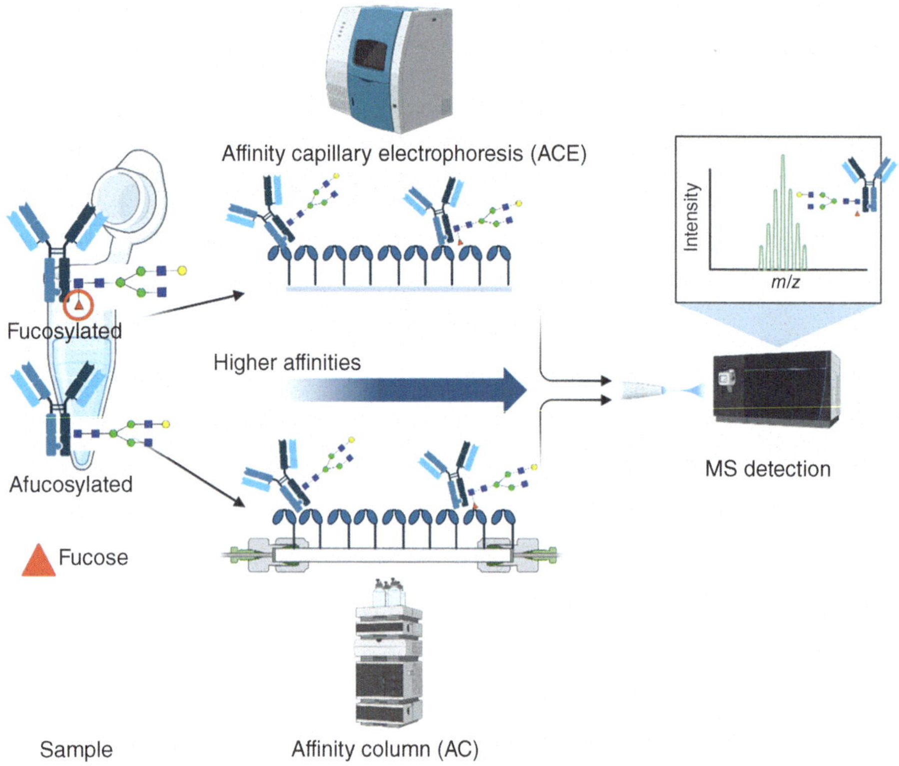

Figure 5.3 Affinity separation using ACE and AC is illustrated by FcγRIIIa affinity testing of a core Fucosylated (lower affinity to FcγRIIIa) and Afucosylated (high affinity to FcγRIIIa) antibody. For simplification, each antibody is shown with only one Fc glycan. Source: Created in BioRender. Haberger, M. (2024) BioRender .com/b18i203.

Affinity mass spectrometry can be conducted using either affinity chromatography mass spectrometry (AC-MS) or affinity capillary electrophoresis mass spectrometry (ACE-MS), refer to Figure 5.3. Each of these methods has its unique advantages and disadvantages, making them suitable for different applications and contexts.

CE-based technology in general offers high-resolution separation and rapid analysis of antibodies, making it suitable for detailed characterization and quality control. ACE coupled with mass spectrometry (MS) requires only minimal sample amounts for both the antibody and its binding partner, and it does not necessitate special capillary preparation, making it both efficient and user-friendly. A major advantage is that a single injection of a 10-μl sample allows for multiple experiments, as only 20 nl is required per experiment. This makes it an ideal technology for the discovery phase. Additionally, ACE-MS theoretically allows for the determination of binding constants. It also offers high-resolution separation and rapid analysis. On the other hand, the limitations include its sensitivity to sample impurities which requires specialized expertise, for example, capillary handling, which might not be readily available in all laboratories. The most compelling advantage of ACE is the quick switch between different receptors or antigens for the sample of interest, which allows for the evaluation of various possible mode of actions for the molecule.

Column-based chromatography allows for the efficient separation and purification of complex mixtures, providing high specificity and reproducibility in the analysis of biomolecules like antibodies. Column-based affinity MS analysis is easy to handle and provides high reproducibility, making it ideal for routine testing and high throughput. It also offers the possibility to collect fractions for further analysis, for example, LC-MS peptide mapping. However, compared to the CE workflows, it requires larger sample amounts and extensive method development. In particular, the preparation of the column with the receptor/antigen of interest involves several steps, including biotinylation, coupling to sepharose beads, and packing of the column. Detailed workflows for column preparations are described e.g. in Thoman et al. [31] and Lippold et al. [16]. Additionally, columns can be prone to clogging or degradation over time (Table 5.2).

The antibody–FcγRIIIa interaction is crucial for immunological responses like ADCC, and differences in antibody glycosylation can significantly affect binding affinity. Affinity liquid chromatography (AC) and ACE with mass spectrometry (MS) can resolve glycoform-specific affinity profiles based on dissociation (AC) or equilibrium (ACE) constants. Gstoettner et al. [13] have shown that both methods can assess mAb–FcγRIIIa interactions in a glycoform-selective manner, with consistent findings

Table 5.2 Difference capabilities between ACE-MS and AC-MS workflows.

Affinity MS method	Sample amount	Receptor amount	Optimal affinity range	Application
ACE-MS	1 ug/μl 10 ul needed for injection, multiple experiments (20 nl per experiment)	Depends on affinities (1 μl volume needed)	Nanomolar to low micromolar	Discovery phase, quick evaluation of different interactions of sample and receptors
AC-MS	Approx. 1 ug–10 μg, 10 μl per injection	0.5 mg/ml of biotinylated constructs for typical column size (inner diameter, 5 mm × length 20 mm)	Independent, can be adjusted through density	Routine testing and high throughput

such as an increased binding for fully afucosylated mAbs and higher galactosylation levels. Additionally, both techniques demonstrated similar binding profiles for the FcγRIIIa F158 allotype compared to the V158 variant, validating their effectiveness in studying antibody glycan interactions.

Affinity Chromatography Mass Spectrometry (AC-MS)

Gahoual et al. [14] reported on the use of FcRn affinity liquid chromatography (LC) coupled with electrospray ionization mass spectrometry (ESI-MS) in native conditions to analyze how modifications affect recombinant monoclonal antibodies' (mAbs) interactions with the immobilized FcRn receptor domain. By optimizing the mobile phase composition, the conditions maintained the proteins in their native state and allowed for sharp pH changes to mimic the IgG Fc domain/FcRn receptor interaction. The study found that oxidized mAb variants had lower retention and decreased affinity for the receptor compared to intact mAbs, demonstrating the methodology's capability to identify and quantify PTMs. This approach significantly reduces the time needed to evaluate the biological relevance of IgG micro heterogeneities, offering valuable insights for biopharmaceutical research and development.

Lippold et al. [15] developed a method linking FcγRIIIa affinity chromatography to mass spectrometry, allowing direct comparison of glycoforms in monoclonal antibodies, thus improving the understanding of glycoform structure–function relationships and their impact on antibody-dependent cell-mediated cytotoxicity (ADCC). This approach aids in evaluating Fc-glycosylation CQAs relevant to ADCC.

Lippold et al. [16] introduced a novel method using CD3 target affinity chromatography coupled with online mass spectrometry (CD3 AC-MS) to separate and identify critically modified proteoforms of T-cell engaging bispecific antibodies (TCBs), aiding in the functional distinction of relevant deamidated and glycosylated variants. This method streamlines the integration of functional and structural characterizations, enhancing the assessment of CQAs for TCBs from early-stage research to late-stage characterization.

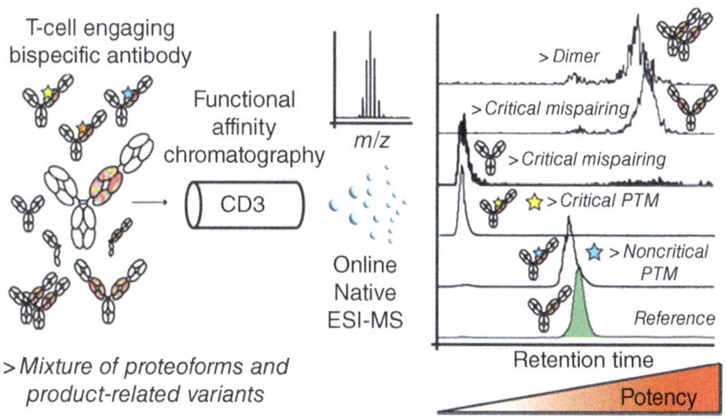

Source: Adapted from Lippold et al. [16].

Affinity Capillary Electrophoresis Mass Spectrometry (ACE-MS)

For the first time, Dominugez et al. [17] introduced ACE coupled with mass spectrometry (MS) as a novel, highly selective tool for assessing protein–protein interactions, demonstrated using the protease inhibitor aprotinin binding to trypsinogen as a model. A trypsinogen sample with several modifications was analyzed using a background electrolyte containing increasing concentrations of aprotinin, and a capillary coating to prevent protein adsorption. The methodology successfully separated trypsinogen variants and revealed a 1:1 binding stoichiometry with aprotinin, allowing for the determination of dissociation constants (Kd's) through shifts in electrophoretic mobility. The ACE-MS technique proved effective for affinity profiling of heterogeneous protein samples, distinguishing components with varying affinities toward aprotinin.

Similar to Gahoul et al. [14] using a FcRn affinity column Gstoettner et al. [18] used mobility shift-ACE–MS to evaluate the binding interactions of monoclonal antibody (mAb) proteoforms with Fc receptors, specifically focusing on the neonatal Fc receptor (FcRn) to determine the impact of oxidation and glycosylation on binding affinities. This high-sensitivity technique provides individual dissociation constant (KD) values and insights into binding stoichiometry, extending its utility to other FcRs and protein interactions for detailed antibody binding studies.

Hydrophobic Interaction Chromatography Mass Spectrometry (HIC-MS)

There are also other methods available combining HPLC hyphenated to a mass spectrometer with a high mass range so that it is able to detect native mass. HIC-MS is a powerful analytical tool that is not widely used yet.

The reason that HIC is not used widely is that normally for separation with HIC buffers with a high salt concentration have to be used. This makes it incompatible for direct hyphenation to a mass spectrometer. However, by using an aqueous mobile phase with a low concentration of a volatile salt (e.g. ammonium acetate), HIC-MS can be used for characterizing monoclonal antibodies (mAbs) and other biotherapeutics under native conditions. HIC itself (without mass spectrometric detection) is the gold standard for determining the drug antibody ratio (DAR) of antibody–drug conjugates (ADCs) that are coupled to a hydrophobic payload. By preserving higher-order structures and noncovalent interactions, not being the case with reversed phase-HPLC (RP-HPLC), HIC-MS can be utilized to elucidate protein structures and PTMs like oxidation and free thiol variants.

As HIC separates biomolecules based on their hydrophobicity, it can be used as an alternative or an orthogonal method for other chromatography modes that are coupled to native mass spectrometry like RP-HPLC-MS, size exclusion MS (SEC-MS), ion exchange-MS (IEC-MS), or AC-MS (affinity-HPLC-MS). It is especially very useful, as stated before for the characterization of ADCs.

Separation principle of HIC is the following: hydrophobic ligand is immobilized on the resin matrix of the column; there is a reversible interaction between the sample (biotherapeutic) and the hydrophobic ligand (highly influenced by the buffers), high salt concentration enhances interaction; by reducing the salt in the buffer with a linear gradient, the interaction is reversed; elution therefore depends on the hydrophobicity of the biotherapeutic (most hydrophobic proteins elute last!).

For being able to hyphenate HIC to native mass spectrometry, a low concentration of volatile salt and sometimes an organic modifier have to be used in the buffers to separate the biotherapeutics (e.g. a monoclonal antibody) and obtain a good MS signal.

Yan et al. in their article "Online coupling of analytical hydrophobic interaction chromatography with native mass spectrometry for the characterization of monoclonal antibodies and related products" [19] successfully combined HIC separation with native MS detection. An innovative makeup and splitting flow strategy (post-column dilution with a makeup flow of water and nanospray electrospray ionization) was employed, enabling the method to tolerate up to 3 M ammonium acetate salt. This resulted in stable electrospray ionization (ESI) over an extended MS analysis time. This allowed the authors to successfully characterize a mixture of monoclonal antibodies, different variants of mAbs resulting from PTMs and Cys-linked ADC mimics, and an analogue of ADCs containing cysteine linkages. They reached a sensitivity of less than 10 μg with a very low sample consumption. The HIC-MS method generated separation profiles highly comparable to conventional HIC-UV methods commonly used in the development of drug. They concluded that this facilitates method comparison and bridging.

Wei et al. stated in their publication "Native Hydrophobic Interaction Chromatography Hyphenated to Mass Spectrometry for Characterization of Monoclonal Antibody Minor Variants" [20] that during HIC, proteins maintain their native folded structure, shielding the free thiols of the constant domain from the stationary phase. Unlike reversed-phase liquid chromatography (RPLC), which exposes all free thiols under denatured conditions, HIC allowed the separation and enrichment of protein variants with free thiol groups in the variable domain, variants that could impact bioactivity.

In addition to free thiol variants, glycation variants were also enriched in HIC. For instance, in the 25+

charge state of m/z, a higher level of +325 Da species was observed. This species could be either 2× glycation or a Valine–Histidine–Serine (VHS) leader sequence at the N-terminus. Considering the low overall glycation level for one of the Mabs and the hydrophilic nature of glycation, it was unlikely that the +325 Da species represented two glycations. Instead, the authors identified it as a VHS segment added to the N-terminus. The VHS-modified variant is typically separated by IEC, indicating accessibility in the native state. The increased hydrophobicity of Valine in the leader sequence aligns with its longer retention on the HIC column.

These glycation and VHS variants were not enriched in either peak of RP-HPLC, highlighting differences in selectivity due to the separation mechanism. These findings demonstrated that HIC-MS provided high-resolution separations of minor variants arising from various PTMs under native conditions.

They could show to be able to do native HIC-MS without using any organic modifier and with a modifier salt concentration under 250 mM. They developed isocratic HIC with an improvement of mAb separation. Their conclusion was that HIC-MS could become a valuable tool for higher-order structure characterization and rapid determination of CQAs of minor protein variants.

Chen et al. [21] state in their paper that recent advancements in top-down proteomics have created a demand for MS-compatible chromatography techniques capable of separating intact proteins using volatile mobile phases. In their study, they introduced a series of more hydrophobic HIC materials that retain proteins using MS-compatible concentrations of ammonium acetate.

These novel HIC materials exhibited characteristics of both conventional HIC and reverse-phase chromatography, functioning as a hybrid approach. Interestingly, the role of the salt appeared to be more about preserving protein structure rather than promoting retention. Online HIC-MS proved feasible for both qualitative and quantitative analyses, as was demonstrated with standard proteins and a complex cell lysate. Notably, the mass spectra of proteins obtained via online HIC-MS displayed low-charge state distributions, consistent with those commonly observed in native mass spectrometry.

Furthermore, HIC-MS enabled the chromatographic separation of proteoforms with minor modifications. Overall, they think that this innovative HIC-MS combination holds promise for advancing top-down proteomics research.

In 2018, Chen et al. also published a paper called "Online Hydrophobic Interaction Chromatography-Mass Spectrometry for the Analysis of intact monoclonal antibodies" [22]. They stated that among various protein chromatography methods, HIC is a popular offline non-denaturing separation technique used for the purification and analysis of mAbs, typically employing non-MS-compatible mobile phases. In their study, they demonstrated the application of direct HIC-MS and HIC-tandem MS (MS/MS) with electron capture dissociation (ECD) for analyzing intact mAbs using quadrupole-time-of-flight (Q-TOF) and Fourier transform ion cyclotron resonance (FTICR) mass spectrometers, respectively. Their method enabled rapid determination of relative hydrophobicity, intact masses, and glycosylation profiles of mAbs, as well as sequence and structural characterization of the complementarity-determining regions in an online configuration. Also these authors concluded that HIC-MS has a high potential for efficient separation and extensive characterization for complex mixtures of monoclonal antibodies.

Two-Dimensional High-Performance Liquid Chromatography-Mass Spectrometry (2D-HPLC-MS)

In addition to using SEC-MS, IEC-MS, HIC-MS, and affinity HPLC-MS, one may also combine two different chromatographic methods and online couple it to the mass spectrometer and measure the native mass. Two-dimensional high-performance liquid chromatography (2D-HPLC) hyphenated to mass spectrometry is an analytical technique that combines two stages of separation to achieve a better resolution and sensitivity of mass spectrometry analysis or to use the second dimension as desalting step so that in the first dimension eluents with a high salt concentration could be used. In principle, the workflow consists of two steps of separation:

Dimension 1: Separation with a liquid chromatography column (e.g. reversed phase (RP), size exclusion (SE), IEC HIC, or AC).

Dimension 2: The effluent from the first dimension is directly coupled to a second chromatography column with a separation mechanism that was not used for Dimension 1 (and the effluent of this column is directly hyphenated to a mass spectrometer).

The objective is to improve the resolution of peaks that are only separated in Dimension 1. It has to be mentioned that the second dimension must be faster than the first. The reason is that there is only one detector for MS analysis. The protein to be measured remains in the native state so that genetic variants and posttranslational modifications can be detected while preserving the three-dimensional structure of the therapeutic.

A review concerning 2D-MS was published by Camperi et al. "Multidimensional LC-MS: the next generation characterization of antibody-based therapeutics by unified online bottom-up, middle-up, and intact approaches" [23].

They highlight the complexity of novel biotherapeutic formats (such as fusion proteins, protein–polymer conjugates, and combination formulations) and the need to enhance the selectivity and resolution of conventional one-dimensional liquid chromatography (1D-LC). Liquid chromatography-mass spectrometry (LC-MS) techniques, including native LC-MS for intact mass analysis and peptide mapping (also known as the bottom-up approach), have shown promise in complementing traditional tools for monoclonal antibody (mAb) characterization.

Two-dimensional liquid chromatography-mass spectrometry (2D-LC-MS) methods have emerged as solutions to address analytical challenges related to novel antibody formats. However, both 1D and 2D-LC-MS methods still require offline sample preparation for in-depth variant characterization at the peptide level. On the other hand multidimensional LC-MS (mD-LC-MS) integrates sample preparation and multi-level analysis (including intact, reduced, middle-up, and peptide levels) within a single chromatographic setup.

The review discusses the pros and cons of mD-LC-MS approaches compared to conventional chromatographic methods (such as 1D-LC-UV for intact proteins and 1D-LC-MS for peptides). Additionally, they explored current trends in antibody characterization using mD-LC-MS beyond the standard 2D-LC-MS approach, sharing their vision of an integrated, multi-level mD-LC-MS characterization platform.

Another review was published by Stoll et al. "Characterization of therapeutic antibodies and related products by two-dimensional liquid chromatography coupled with UV absorbance and mass spectrometric detection" [24]. Given the inherent complexity of complex biomolecules, arising from their size and potential for chemical modifications over time (e.g. due to solution modifications or aggregation), 2D-LC has garnered significant interest as an analytical tool to address the challenges in characterizing these materials. The immediate potential advantages of 2D-LC over conventional 1D-LC in this context include: (i) higher overall resolution, (ii) complementary information obtained from two dimensions of separation in a single analysis, and (iii) enabling indirect coupling of separation modes that are inherently incompatible with mass spectrometric (MS) detection to MS. This is achieved through a more compatible second dimension separation, such as reversed-phase (RP) LC. The authors stated that although the future holds promise for further development in this field, some challenges have already been addressed through new 2D-LC methods. These include characterization of ADCs to understand the extent and localization of small molecule conjugation, detailed study of excipients in protein drug formulations, and detection of host-cell proteins in biotherapeutic molecule preparations.

Grunert et al. stated in their paper "Comprehensive multidimensional liquid chromatography mass spectrometry for the characterization of charge variants of a bispecific antibody" [25] that conventional workflows are time consuming, labor-intensive, and ultimately limited in terms of chromatographic separation for the enhanced structural heterogeneities arising from new antibody formats. To address these challenges, they developed an alternative online characterization strategy for charge variants of a therapeutic bispecific antibody using online middle-down liquid chromatography-mass spectrometry (mD-LC-MS) and bottom-up so-called 4D-LC-MS. With the implemented online mD-LC-MS approach, all medium- and even low-abundant product variants previously identified through offline fraction experiments and liquid chromatography mass spectrometry could be monitored. This automated online mD-LC-MS methodology represents a complementary and partially alternative approach for analytical method validation, via mass spectrometry. It offers several benefits, including increased throughput, reduced sample handling, and combined protein information at both intact protein and peptide levels.

Kumar et al. in their paper "Multiattribute monitoring of aggregates and charge variants of monoclonal antibody through native 2D-SEC-MS-WCX-MS" [26] used a combination of SEC and weak anion exchange chromatography (WCX) for monitoring of aggregates and charge variants of monoclonal antibodies.

They wrote that multi-attribute monitoring allows simultaneous assessment of the attributes using 2D-LC. Typically, only the second-dimension samples are directly analyzed by mass spectrometry in this approach, as the first dimension has limitations regarding direct coupling with mass spectrometry. In the present study, they proposed a novel 2D-SEC-MS/WCX-MS workflow where both dimensions (D1 and D2) of chromatography were directly coupled with a mass spectrometer. This enabled simultaneous analysis of size- and charge-related variants of a monoclonal antibody (mAb) in a native form. Compared to standalone SEC and WCX methods, this approach allowed simultaneous analysis without manual intervention and facilitated the detection of low-abundance variants. Furthermore, this method required 75% less sample material and has a shorter analysis time (25 minutes versus 90 minutes) when analyzing size and charge variants individually. The proposed native 2D-LC-MS workflow was used to analyze a stressed sample of a mAb, revealing the presence of aggregates (8–20%), primarily dimers, in the D1 analysis, and an increase in acidic variants (9–21%) in the D2 analysis.

Rick S. van der Hurk et al. explained in their paper "Comprehensive Two-Dimensional Liquid Chromatography–High-Resolution Mass Spectrometry for Complex Protein Digest Analysis Using Parallel Gradients" [27] why online 2D-HPLC coupled to a high-resolution mass spectrometer has not been widely used to analyze complex protein digests. The reason was the reduced sensitivity due to high 2D flowrates, so flow splitters are needed for coupling to ESI-MS. They compared parallel gradients (2D flow rate 0.7 ml/min) with shifted gradients (2D flow rate 1.4 ml/min). Both methods had high surface coverage and effective peak capacity. But parallel gradients showed higher sensitivity, more identifications of peaks, and increased MS/MS events. The authors even concluded that it outperformed 1D-RPLC showing higher separation performance with limited sensitivity losses.

Farsang et al. concluded in their review "Coupling non-denaturing chromatography to mass spectrometry for the characterization of monoclonal antibodies and related products" [28]: various combinations of ammonium acetate, ammonium formate, and ammonium bicarbonate have been used as mobile phases for the direct coupling of ion exchange chromatography (IEX) and mass spectrometry (MS). Among them, ammonium acetate appears particularly promising, especially when combined with a pH gradient and an ionic strength gradient. In a 2D-LC setup, IEX has been successfully combined with RPLC. This combination not only allows the coupling of IEX with MS but also improves separation efficiency, which can be effectively applied to the analysis of mAb subunits. Another combination is IEX-SEC, which can work with a volatile buffer in the second dimension and maintain non-denaturing conditions throughout the entire analysis. However, further investigations are needed to test this proposed mode of analysis.

For direct SEC-MS analysis, ammonium acetate can be used as an additive. However, due to undesirable secondary interactions between proteins and the packing material or column, this robustly applies only to acidic mAbs. Alternatively, "salt-free" mobile phases can be used, but they may lead to protein denaturation. SEC has been successfully applied in various multidimensional setups (such as SECxRPLC, SECxSEC, and SECxHIC) for mAb and ADC characterization. They found that direct coupling of HIC to MS is more challenging because it relies on protein precipitation and thus requires very high salt concentrations in the mobile phase. Progress has been made to drastically decrease salt concentrations by increasing the hydrophobicity of the stationary phase and adding organic modifiers to the mobile phase. However, some of these methods may need additional optimization before extensive application due to potential denaturation issues.

To summarize, 2D-HPLC-MS could be very helpful for direct characterization of IEC and SEC that is used for release and in cases where one dimension is not enough to separate complex mixtures of biologics.

Considerations

Schaick et al. [29] summarized differences in denatured and native MS workflows. In hyphenated separation MS workflows, volatile electrolytes like ammonium acetate, formate, or bicarbonate buffers are commonly used. Ammonium acetate typically preserves protein conformation during both the separation and ESI process, while formate and bicarbonate may cause protein denaturation. The pH of the separation medium is an important parameter, as it can affect protein activity and prevent unwanted aggregation. The flow rate at which the separation buffer is delivered to the ESI source impacts spray formation, with lower flow rates generally providing higher sensitivity but longer analysis times. Modern ESI sources show optimal ionization performance at flow rates between 0.001 and 0.1 ml/min, but they can accommodate up to 1.0 ml/min. For native LC-MS analysis, flow rates should ideally not exceed 0.4 ml/min, with lower flow rates resulting in better sensitivity but requiring adjustments to the inner diameter of the LC column to maintain separation efficiency.

Prerequisites for native mass spectrometry include instruments capable of detecting MS signals in a high m/z range, coupled with high sensitivity achieved through effective desolvation and efficient ion transfer. Additionally, good resolution is dependent on the declustering of solvent molecules from the sample analyte. Based on our experience, the instruments best suited for native mass spectrometry are those from Thermo and Waters. Specifically, Thermo's EMR and UHMR systems have proven to work perfectly for our intended use.

Quality Control Feasibility

Mass spectrometric methods have a high potential to be included in quality control and to be used as release method for biologics by monitoring several CQAs in one method.

For the time being, the common method is an LC-MS-based method after tryptic digest "Comprehensive multi-attribute method workflow for bio therapeutic characterization and current good manufacturing practices testing" [30]. We think the above-mentioned methods combining native separation with mass spectrometric detection have also a great potential to replace or complement MAM after tryptic digestion. In principle, we would propose

Native SEC-MS or IEC-MS as multi-attribute method, in particular native SEC-MS which can cover quality attributes which are challenging to detect in established MAM approach, like, for example, aggregates, fragments, and PTMs like glycans which are difficult to cope with peptide mapping. The benefit here is that you also can get rid of SEC as release methods. Native separation coupled to MS is less complicated as it involves nearly no sample preparation and should therefore also be easier to validate.

Conclusions

In summary, the native separation techniques presented here, when directly coupled to mass spectrometry, are indispensable tools for the characterization of biologics. The choice of method depends greatly on the nature of the protein and the specific quality attribute being characterized and controlled.

References

1 Labrijn, A.F., Janmaat, M.L., Reichert, J.M., and Parren, P.W.H.I. (2019). Bispecific antibodies: a mechanistic review of the pipeline. *Nat. Rev. Drug Discov.* 18: 585–608.

2 Klein, C., Brinkmann, U., Reichert, J.M., and Kontermann, R.E. (2024). The present and future of bispecific antibodies for cancer therapy. *Nat. Rev. Drug Discov.* 1–19: https://doi.org/10.1038/s41573-024-00896-6.

3 Haberger, M., Leiss, M., Heidenreich, A.K. et al. (2016). Rapid characterization of biotherapeutic proteins by size-exclusion chromatography coupled to native mass spectrometry. *MAbs* 8: 331–339.

4 Cramer, D.A.T., Franc, V., Heidenreich, A.K. et al. (2023). Characterization of high-molecular weight by-products in the production of a trivalent bispecific 2+1 heterodimeric antibody. *MAbs* 15: 2175312.

5 Yan, Y., Xing, T., Wang, S. et al. (2019). Coupling mixed-mode size exclusion chromatography with native mass spectrometry for sensitive detection and quantitation of homodimer impurities in bispecific IgG. *Anal. Chem.* 91: 11417–11424.

6 Ren, C., Bailey, A.O., VanderPorten, E. et al. (2019). Quantitative determination of protein–ligand affinity by size exclusion chromatography directly coupled to high-resolution native mass spectrometry. *Anal. Chem.* 91: 903–911.

7 Leblanc, Y., Ramon, C., Bihoreau, N., and Chevreux, G. (2017). Charge variants characterization of a monoclonal antibody by ion exchange chromatography coupled on-line to native mass spectrometry: Case study after a long-term storage at +5°C. *J. Chromatogr. B* 1048: 130–139.

8 Ponniah, G., Nowak, C., Neill, A., and Liu, H. (2017). Characterization of charge variants of a monoclonal antibody using weak anion exchange chromatography at subunit levels. *Anal. Biochem.* 520: 49–57.

9 Phung, W., Han, G., Polderdijk, S.G.I. et al. (2019). Characterization of bispecific and mispaired IgGs by native charge-variant mass spectrometry. *Int. J. Mass Spectrom.* 446: 116229.

10 Shi, R.L., Xiao, G., Dillon, T.M. et al. (2020). Characterization of therapeutic proteins by cation exchange chromatography-mass spectrometry and top-down analysis. *MAbs* 12: 1739825.

11 Haberger, M., Heidenreich, A.K., Hook, M. et al. (2021). Multiattribute monitoring of antibody charge variants by cation-exchange chromatography coupled to native mass spectrometry. *J. Am. Soc. Mass Spectrom.* 32: 2062–2071.

12 van Schaick, G., Gstöttner, C., Büttner, A. et al. (2021). Anion exchange chromatography – Mass spectrometry for monitoring multiple quality attributes of erythropoietin biopharmaceuticals. *Anal. Chim. Acta* 1143: 166–172.

13 Gstöttner, C., Lippold, S., Hook, M. et al. (2024). Benchmarking glycoform-resolved affinity separation – mass spectrometry assays for studying FcγRIIIa binding. *Front. Immunol.* 15: 1347871.

14 Gahoual, R., Gahoual, R., Heidenreich, A.-K. et al. (2017). Detailed characterization of monoclonal antibody receptor interaction using affinity liquid chromatography hyphenated to native mass spectrometry. *Anal. Chem.* 89: 5404–5412.

15 Lippold, S., Nicolardi, S., Domínguez-Vega, E. et al. (2019). Glycoform-resolved FcγRIIIa affinity chromatography–mass spectrometry. *MAbs* 11: 1191–1196.

16 Lippold, S., Hook, M., Spick, C. et al. (2023). CD3 Target Affinity Chromatography Mass Spectrometry as a New Tool for Function–Structure Characterization of T-Cell Engaging Bispecific Antibody Proteoforms and Product-Related Variants. *Anal. Chem.* 95: 2260–2268.

17 Domínguez-Vega, E., Haselberg, R., Somsen, G.W., and de Jong, G.J. (2015). Simultaneous assessment of protein heterogeneity and affinity by capillary electrophoresis–mass spectrometry. *Anal. Chem.* 87: 8781–8788.

18 Gstöttner, C., Hook, M., Christopeit, T. et al. (2021). Affinity capillary electrophoresis–mass spectrometry as

a tool to unravel proteoform-specific antibody–receptor interactions. *Anal. Chem.* 93: 15133–15141.

19 Yan, Y., Xing, T., Wang, S. et al. (2020). Online coupling of analytical hydrophobic interaction chromatography with native mass spectrometry for the characterization of monoclonal antibodies and related products. *J. Pharm. Biomed. Anal.* 186: 113313.

20 Wei, B., Han, G., Tang, J. et al. (2019). Native hydrophobic interaction chromatography hyphenated to mass spectrometry for characterization of monoclonal antibody minor variants. *Anal. Chem.* 91: 15360–15364.

21 Chen, B., Peng, Y., Valeja, S.G. et al. (2016). Online hydrophobic interaction chromatography–mass spectrometry for top-down proteomics. *Anal. Chem.* 88: 1885–1891.

22 Chen, B., Lin, Z., Alpert, A.J. et al. (2018). Online hydrophobic interaction chromatography–mass spectrometry for the analysis of intact monoclonal antibodies. *Anal. Chem.* 90: 7135–7138.

23 Camperi, J., Goyon, A., Guillarme, D. et al. (2020). Multi-dimensional LC–MS: the next generation characterization of antibody-based therapeutics by unified online bottom-up, middle-up and intact approaches. *Analyst* 146: 747–769.

24 Stoll, D., Danforth, J., Zhang, K., and Beck, A. (2016). Characterization of therapeutic antibodies and related products by two-dimensional liquid chromatography coupled with UV absorbance and mass spectrometric detection. *J. Chromatogr. B* 1032: 51–60.

25 Grunert, I., Heinrich, K., Hingar, M. et al. (2022). Comprehensive multidimensional liquid chromatography–mass spectrometry for the characterization of charge variants of a bispecific antibody. *J. Am. Soc. Mass Spectrom.* 33: 2319–2327.

26 Kumar, S., Savane, T.S., and Rathore, A.S. (2023). Multiattribute monitoring of aggregates and charge variants of monoclonal antibody through native 2D-SEC-MS-WCX-MS. *J. Am. Soc. Mass Spectrom.* 34: 1263–1271.

27 van den Hurk, R.S., Lagerwaard, B., Terlouw, N.J. et al. (2024). Comprehensive two-dimensional liquid chromatography–high-resolution mass spectrometry for complex protein digest analysis using parallel gradients. *Anal. Chem.* 96: 9294–9301.

28 Farsang, E., Guillarme, D., Veuthey, J.L. et al. (2020). Coupling non-denaturing chromatography to mass spectrometry for the characterization of monoclonal antibodies and related products. *J. Pharm. Biomed. Anal.* 185: 113207.

29 van Schaick, G., Haselberg, R., Somsen, G.W. et al. (2022). Studying protein structure and function by native separation–mass spectrometry. *Nat. Rev. Chem.* 6: 215–231.

30 Millán-Martín, S., Jakes, C., Carillo, S. et al. (2023). Comprehensive multi-attribute method workflow for biotherapeutic characterization and current good manufacturing practices testing. *Nat. Protoc.* 18: 1056–1089.

31 Thomann, M., Schlothauer, T., Dashivets, T. et al. (2015). In vitro glycoengineering of IgG1 and its effect on Fc receptor binding and ADCC activity. *PLoS One* 10: e0134949.

6

Methods for Characterization and Optimization of Peptide Therapeutics: Bioanalysis and Biotransformation

Phillip Chu, Suk-Joon Hyung, Robert S. Jones, S. Cyrus Khojasteh, Dennis H. Leung, Bin Ma, Huy Nguyen, Emile Plise, Ola M. Saad, John C. Tran, Qinying Yu, and Xing Zhang

Genentech, Inc., South San Francisco, CA, USA

Introduction

Therapeutic peptides (referred to simply as peptides) have remained a compelling modality, with many successfully marketed drugs [1]. In most instances, the identification of peptides was primarily based on their specific, naturally occurring functions, as opposed to them being tailored for a particular use and therapeutic targets. Peptides hold significant promise as a therapeutic modality, a fact well illustrated by many examples involving endogenous peptides, natural products, and their derivatives. Cyclosporine A, octreotide, oxytocin, vasopressin, calcitonin, and monomethyl auristatin E (MMAE; used as a payload on antibody-drug conjugates) serve as examples across diverse therapeutic areas (Figure 6.1). The majority of these uses cannot be effectively replaced by small molecules or antibodies/biologics due to peptides' unique binding mode. Until recently the mechanisms by which peptides could be deployed and discovered were not amenable to the existing modern drug discovery methods that were developed for small or large molecules. Advances made in the past decade have allowed for the exploration of this new therapeutic area and the generation of modern therapeutic peptides.

Therapeutic Targets

Peptides typically have a molecular weight (MW) that falls between that of small molecules and proteins and usually consist of fewer than 40 amino acids [2]. They occupy a chemical space called beyond the rule of 5 (bRo5) [3]. A significant advantage lies in their ability to bind within regions recognized for protein–protein interactions (PPIs), where peptides can bind and regulate biological activities. PPI occurs in both intracellular and extracellular environments, rendering peptides uniquely qualified due to their expanded binding regions, offering opportunities to facilitate binding [4].

Peptide Technology

The advancement in peptide technology has facilitated the expansion of therapeutic targets, allowing the exploration of new chemical spaces. This encompasses methods such as mRNA display, phage display, ribosome display, and others [5, 6]. Enabling the assessment of PPIs, exploration of large chemical space properties has become feasible through cutting-edge biological methods. These targets can be located in either extracellular or intracellular regions, with intracellular targets specifically requiring adequate cellular permeability [7]. A key enabling technology in advancing the field is peptide RNA or DNA coding, which permits post-potency assessment for reconstructing the peptide sequence involved. Through multiple rounds of selection and amplification, which are considered directed evolution, biomolecules can be optimized to achieve nano- to picomolar binding affinities. This iterative process enables the refinement of biomolecular properties toward specific targets or functions. While potency alone is crucial, combinations with other favorable properties are critical for drug advancement in the clinic [8].

A multitude of elements intersect to enhance the discovery of therapeutic peptides. This encompasses the integration of diverse assays, comprehensive data acquisition, and the utilization of computational tools to construct dashboards conducive to informed and timely

All contributed equally.

Characterizing Biotherapeutics: Analytical Methods for Diverse Modalities, First Edition. Edited by Jennie R. Lill and Wendy Sandoval.

decision-making. These dashboards facilitate the exploration of expansive chemical spaces, enabling researchers to navigate complex molecular landscapes with precision. Furthermore, data mining techniques coupled with a plethora of algorithms, including machine learning, contribute significantly to the advancement of peptide discovery by discerning patterns and optimizing molecular designs.

Building Blocks

Peptides, constructed from amino acids, represent the fundamental building blocks of biological molecules. Recent advancements in peptide synthesis have transcended the limitations of the 20 naturally occurring amino acids, allowing for the incorporation of unnatural amino acids into peptide structures. This expansion of the amino acid repertoire opens up new avenues for peptide engineering, enabling the creation of peptides with enhanced properties and functionalities tailored to specific applications.

Cyclosporine A, derived from the fungus *Tolypocladium inflatum*, is an immunosuppressive drug used to prevent organ transplant rejection. It consists of 11 amino acids and possesses several unique features, as illustrated below (Figure 6.1).

Unnatural Amino Acids (UAA)

UAA are variants not typically found in proteins but can be synthesized or incorporated into peptides. Examples include D-α-amino acids, β-amino acids, amino acids with modified side chains, and those with extended or bulky side chains. These alterations expand the functional diversity of peptides, enabling the design of molecules with unique properties.

N-Methyl Amino Acids (NMe)

NMe are amino acids with a methyl group attached to the nitrogen atom in their backbone. They allow for increased metabolic stability and cellular permeability.

Peptoids

N-substituted glycine is a peptide mimics where the side chains are attached to the backbone nitrogen rather than the alpha carbon.

Overall Shape

Peptides can adopt diverse structural forms, ranging from linear to macrocyclic and other various structural motifs. Linear peptides represent the simplest structure, consisting of a straight chain of amino acids linked by peptide bonds. Cyclic peptides (commonly known as macrocyclic peptides) form closed loops through the linkage of the N- and C-termini or through side chain cross-linkages, providing increased stability and resistance to enzymatic biotransformation. Disulfide-rich peptides, characterized by multiple disulfide bonds, are also found in nature and offer structural rigidity and stability.

Absorption, Distribution, Metabolism, and Excretion (ADME) Properties

While peptides hold significant promise as therapeutics, understanding their ADME characteristics and liabilities is critical for evaluation and refinement. This understanding can inform design strategies, dictate administration routes, and predict potential liabilities and mechanisms of action:

Absorption: Peptides generally may have poor oral bioavailability due to their limited permeability, poor solubility, or susceptibility to biotransformation by proteolytic enzymes in the gastrointestinal tract. This limits their

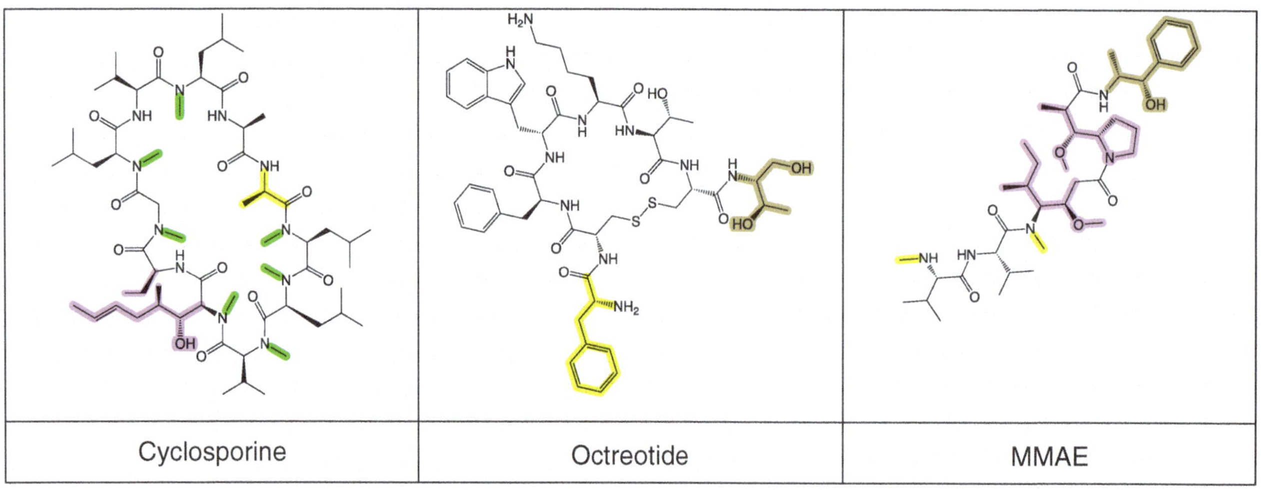

Figure 6.1 The structures of cyclosporine A, ocreotide, and monomethyl auristatin E (MMAE). *N*-methylated amino acids are shown in green, D-amino acids in yellow, endcap in brown, and non-standard amino acids in pink.

uptake into the bloodstream and therefore necessitates the use of alternative methods of administration such as intravenous (either infusion or bolus), subcutaneous (SC), or intramuscular (IM) injections [9]. Understanding factors that impact the bioavailability of peptides can provide insights into attributes to optimize for improved absorption [10].

Distribution: Peptides may face challenges in distribution throughout the body, including crossing biological barriers such as the blood–brain barrier. This can restrict their effectiveness in reaching certain target tissues or organs. Peptide distribution varies depending on molecular size, hydrophilicity, and plasma protein binding. Mechanisms affecting the distribution may include partitioning into organs, permeation of cellular barriers, or binding with drug transport proteins. Studying the distribution behavior of peptides can inform on achieving sufficient therapeutic exposure at target sites [11].

Metabolism: Peptides are susceptible to metabolism by enzymes such as proteases present in various tissues and biological fluids. A rapid rate of metabolism can lead to shortened plasma half-lives and reduced therapeutic efficacy [12].

Excretion: Peptides could be excreted from the body via renal clearance, especially those that are more hydrophilic. Nonetheless, due to their substantial size and lipophilic nature, they can also be eliminated through biliary excretion or alternative pathways.

To address these ADME liabilities, researchers employ various strategies such as structural modifications to enhance stability, conjugation with carrier molecules to improve pharmacokinetic properties, and formulation approaches to facilitate delivery to target tissues. Bioanalytical and biotransformation instruments serve as fundamental tools to perform these ADME evaluations. This book chapter discusses methods ranging from *in vitro* to PK studies that facilitate the evaluation and refinement of peptides.

In Silico to *In Vitro* Measurements

In drug discovery, the assessment of peptides often involves a combination of *in silico* (computational) and *in vitro* methods to evaluate their ADME properties. This integrated strategy encompasses an analysis of various aspects essential for effective deployment in drug development.

Physicochemical Properties

Understanding and predicting the physicochemical properties of molecules is essential during the drug optimization and development stages. As discussed previously, the ADME properties of a molecule are directly influenced by its physicochemical properties, such as solubility and lipophilicity. These properties are commonly predicted for small molecules using *in silico* models and verified with high-throughput assays. Peptides, however, present a unique challenge due to their larger molecular weight, increased number of rotatable bonds, and numerous chiral centers, complicating the prediction of their physicochemical properties. Often, experimental data is required both to confirm these predictions and to refine the models used.

Measuring Solubility

Aqueous solubility is an essential physicochemical property to measure experimentally, as it plays a significant role in absorption and affects downstream assays [13]. Solubility assays are categorized into kinetic and thermodynamic assays. Kinetic solubility assays measure the highest concentration of a peptide (usually from a 10 mM DMSO stock) in a solution prior to protein precipitation and are conducted within 24 hours. Thermodynamic solubility assays, on the other hand, require more material (greater than 1–2 mg) and require the sample to reach equilibrium, necessitating a longer incubation period of over 24 hours. The shake-flask method is widely regarded as the benchmark for measuring aqueous solubility. Assays that utilize samples from DMSO stock solutions should consist of DMSO concentration of less than 5% [14]. To measure solubility, as a general protocol, the sample of interest is first dissolved in the desired buffer. After incubation, samples are prepared for analysis by separating precipitates from the solution using filter plates with a well-defined pore size (Millipore, Multiscreen HTS, 0.45 μm). The concentration of the sample post-incubation is then quantified using HPLC with detectors like UV, mass spectrometry, chemiluminescent nitrogen detection (CLND) [14], or charged aerosol detector (CAD) [15].

Choosing the right aqueous buffer can provide additional insights into the properties of the peptide. Typically, PBS buffer at a physiological pH of 7.4 is used due to its common application in various *in vitro* assays. However, it is also important to consider measuring solubility in assay-specific buffers with the same experimental conditions. Measuring solubility in different buffers provides insight into the assay results, which can be influenced by the buffer's pH, ionic strength, and salt content [16, 17]. Biorelevant buffers, which mimic human intestinal fluids in both fasted and fed states, are also employed to obtain a comprehensive solubility profile of potential drugs [18]. These simulated buffers, such as Fasted State Simulated Intestinal Fluid (FaSSIF) and Fed State Simulated Intestinal Fluid (FeSSIF), offer a more realistic representation of the gastrointestinal environment. The solubility profile

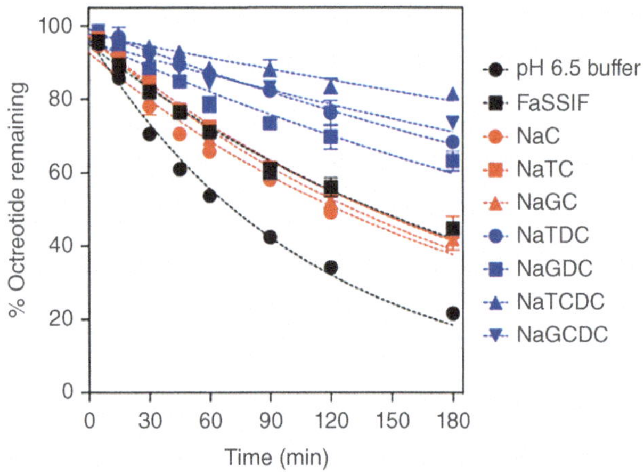

Figure 6.2 Stability of Octreotide in buffers containing different bile salts with pancreatin. The different types of bile salts: sodium taurocholate (NaTC), sodium glycocholate (NaGC), sodium taurodeoxycholate (NaTDC), sodium glycodeoxycholate (NaGDC), sodium taurochenodeoxycholate (NaTCDC), and sodium glycochenodeoxycholate (NaGCDC). Octreotide has the lowest half-life in phosphate buffer at pH = 6.5, is more stable in buffers containing trihydroxy bile salts (red), and is most stable in buffers with dihydroxy bile salts (blue). (Copyright, CC BY-NC-ND 4.0). Source: Reproduced from Dening et al. [19]/with permission of American Chemical Society.

of peptides in these simulated fluids can also provide insight into other inherent properties, such as enzymatic stability. As an example, octreotide, an FDA-approved oral drug (Mycapssa), displayed higher enzymatic stability in FaSSIF (and other buffers containing bile salts, Figure 6.2) compared to just phosphate buffer [19]. Incorporating solubility measurements in various buffers can reveal improved solubility for peptides and potentially information on enzymatic stability, allowing them to be considered for further development.

Measuring Lipophilicity Lipophilicity, generally inversely related to aqueous solubility, is another essential property to predict and measure, especially to optimize peptides for permeability. Although *in silico* models have successfully predicted lipophilicity (e.g. logD at pH 7.4, logP [20]) for typical small molecules, experimental data remain crucial for newer molecular modalities such as cyclic peptides and bRo5 molecules. The traditional shake-flask method is still the gold standard for measuring lipophilicity, where a sample dissolved in DMSO (10 mM) is added into an octanol and PBS mixture (pH adjusted based on pKa for logP, typically pH = 7.4 for logD), mixed, centrifuged, and then the two layers are separated for analysis using mass spectrometry (Triple Quad MS) [21]. To measure lipophilicity for bRo5 molecules, chromatographic methods utilizing reverse-phase HPLC have been developed [22, 23].

Lipophilicity is known to correlate with permeability, but this relationship is less straightforward for larger molecules, which show additional interactions not fully explained by the traditional octanol-PBS system [24]. To better accommodate these properties, octanol has been substituted with 1,9-decadiene [25] and toluene [26]. Both 1,9-decadiene and toluene are more hydrophobic than octanol and lack a hydrogen-bonding hydroxyl group. Measuring the partition coefficient of cyclic peptides in a hydrocarbon-water system like 1,9-decadiene demonstrated a better correlation with PAMPA permeability compared to an octanol–water system (Figure 6.3c and d). Furthermore, measuring logD in toluene promoted peptides to form environment-dependent intermolecular hydrogen bonds (IMHB), providing insight into the molecules' ability to change conformations, known as chameleonicity, which is linked to increased solubility and cell permeability (Figure 6.4c) [26]. To help discover and predict more permeable peptides, new high-throughput assays have been developed to measure the ability of a molecule to possess chameleonic properties.

Recent Advances in Physicochemical Assays

Recent developments in assays measuring physicochemical properties have increasingly focused on identifying molecular chameleons to improve cell permeability. The experimental polar surface area (EPSA) assay, for instance, measures the polar surface area by analyzing the interaction between a peptide and a lipophilic/polar stationary phase ([S]-valine Phenomenex Chirex 3014) in a supercritical CO_2/methanol mixture [27]. This setup creates a low dielectric environment that encourages the formation of intramolecular hydrogen bond (IMHB), ultimately affecting the retention time and an EPSA value is calculated from a set of calibration standards. Utilizing the EPSA assay, Goetz et al. determined that most permeable cyclic peptides had an EPSA of less than 80 (Figure 6.4a) [27].

In addition to the EPSA assay, another chromatography-based assay that has been utilized to elucidate peptide permeability is the immobilized artificial membrane assay (IAM.PC.DD2, Regis Technologies; [28]). With IAM chromatography, the resin in the column contains a layer of covalently bound phospholipids to mimic the lipid bilayer of a cell. Measuring a longer retention time in the IAM column translates to more interaction with the phospholipid layer and is demonstrated to be detrimental to permeability (Figure 6.4b; [29]).

Lastly, the Chamelogk assay is another chromatography-based approach that measures the ability of a peptide to adopt different conformations [30]. Using reverse-phase HPLC with a polystyrene/divinylbenzene reversed-phase

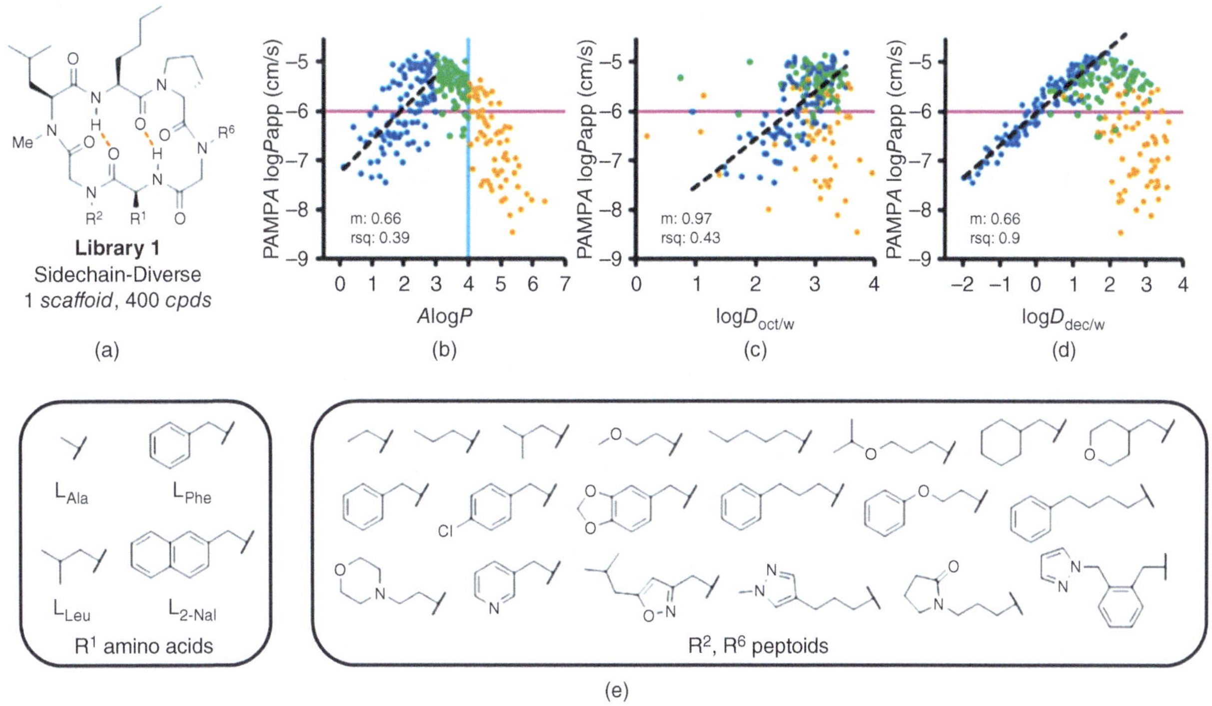

Figure 6.3 PAMPA permeability correlation with different lipophilicity descriptors: AlogP, logD in octanol:water, logD pH = 7.4 measured in a water:1,9-decadiene system. Source: Adapted from Naylor et al. [25].

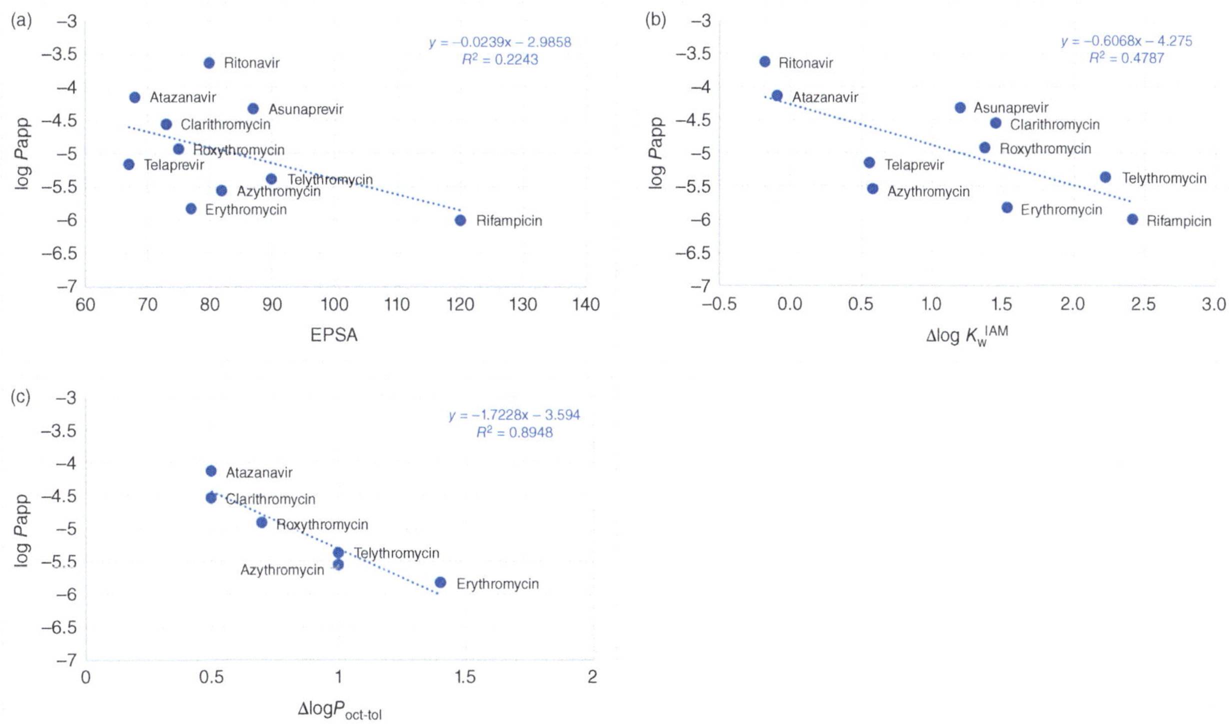

Figure 6.4 Permeability versus (a) EPSA, (b) $\Delta\log K_{W}^{IAM}$, (c) $\Delta\log P_{oct-tol}$. (Copyright, License# 5820951177243). Source: Reproduced from Ermondi et al. [29]/with permission of Elsevier.

column (PLRP-S), retention times are measured at varying concentrations of acetonitrile: 20 mM ammonium acetate pH = 7.0, from 50% to 100% ACN. The difference between the experimentally measured log*k*′ value at 100% acetonitrile and the extrapolated value from lower acetonitrile percentages defines Chamelogk. By measuring Chamelogk, peptides that display chameleonicity can be identified and potentially exhibit high cell permeability and solubility. These assays provide valuable insights into the chameleonic properties of peptides that could potentially enhance their permeability.

In Vitro ADME Assays

Over the past 30 years, *in vitro* ADME assays have evolved significantly and are now standard practice for small molecules. Many of these assays can also be adapted for peptides. Understanding pharmacokinetic (PK) clearance is essential for interpreting the results of *in vitro* assays accurately.

Clearance is the efficiency of drug removal from the body, defined as the volume of fluid from which the drug is completely cleared per unit time. It reflects the rate of change in drug concentration normalized to the concentration at the midpoint of the time interval.

$$CL_{total} = CL_{liver} + CL_{renal} + CL_{others}$$

Total clearance encompasses all mechanisms of drug removal, including metabolism (encompassing liver metabolism) and the excretion of unchanged drug via the kidneys and bile (captured in CL_{liver}). *In vitro* assays are most effective at measuring metabolic turnover.

One other important component is the role that permeability and drug transport play when considering clearance mechanisms. The following are the role permeability plays using Biopharmaceutics Drug Disposition Classification System (BDDCS) considerations [31].

Non-permeable Peptides

These peptides typically have a small volume of distribution. The peptide (depending on the route of administration) could be exposed to a number of peptidases, and the goal is to achieve a reasonable half-life (5–12 hours) by lowering the extent of metabolism. Depending on the peptide's lipophilicity, it may be cleared renally or biliary. More hydrophilic peptides with low plasma protein binding are usually cleared renally, while lipophilic peptides may have significant biliary clearance. Drug transport can play a crucial role in these peptides to change the directionality of the CL pathways.

Permeable Peptide

These peptides typically have a higher volume of distribution and can penetrate different tissues, giving them access to more drug-metabolizing enzymes. Due to their higher lipophilicity, cytochrome P450 enzymes are more involved in their metabolism. For these peptides, renal and biliary clearance is generally lower, with a higher contribution of metabolic clearance to the total clearance.

The strategy in which ADME assays are conducted is determined by numerous factors, including the therapeutic target, target coverage (defined as the concentration and duration needed to prompt efficacy), route of administration, and pertinent species. Delving into assessing ADME and drug properties, the construction of assay cascades for prioritization becomes a key process during drug discovery optimization. The first critical step is the identification of key liabilities (Figure 6.5). Without resolving these liabilities, the ability of the compound to reach its target could be severely compromised. During the early stages, this

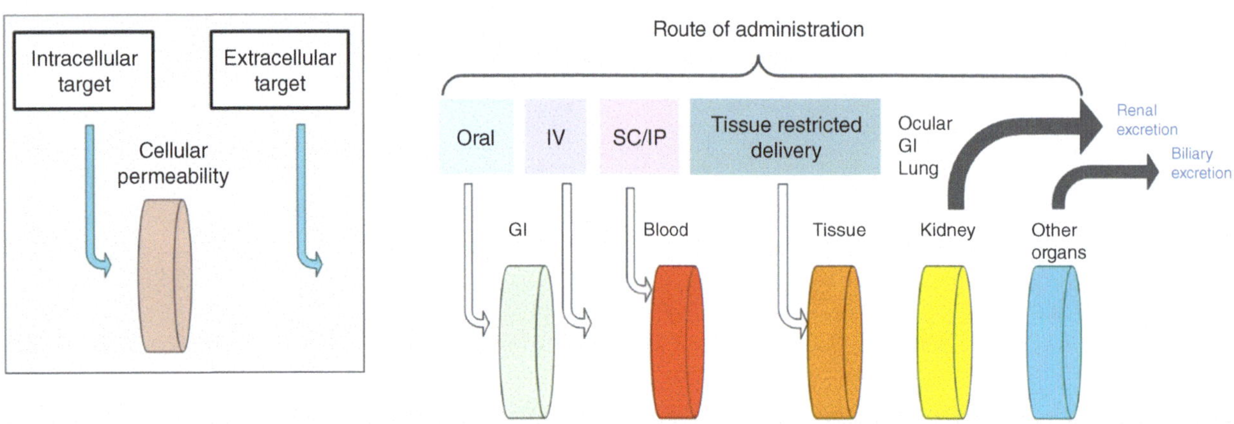

Figure 6.5 The identification of significant ADME liabilities during drug discovery encompasses assessing factors such as target site (intracellular or extracellular), route of administration, permeability barriers, drug transport mechanisms, and metabolism considerations.

focus typically revolves around three aspects: solubility, metabolism, and permeability. In cases where *in silico* models are available, they should be employed prior to the actual synthesis of the compound. However, it should be noted that assays optimized for small molecules might not be the best fit for peptides and further assay optimization may be necessary. Additionally, the threshold used in each assay may require adjustments.

Permeability

It plays a crucial role in the delivery of peptides to their intracellular targets. Essentially, it refers to the ability of a peptide to cross cellular membranes, which is a key determining factor in its efficacy. Higher permeability typically implies better uptake of the peptide by the cell, hence enhancing its bioavailability and therapeutic effect. Therefore, targeting intracellular molecules with peptides often involves the careful design and optimization of permeability.

Metabolism

One class of drug-metabolizing enzymes operating on peptides are peptidases (we consider all hydrolases operating on peptides as peptidases). During drug discovery, one of the goals is to reduce the role of peptidases. These enzymes, which are diverse in nature, facilitate the breakdown of peptide molecules. It should be noted that if the molecule becomes more lipophilic and peptide bonds more stable, other enzymes may be involved. The peptidases are abundant in the liver, which is known for its central role in drug metabolism. However, it's crucial to note that these enzymes are also significantly present in extrahepatic tissues. The distribution of peptidases beyond the liver underscores their potential influence on the metabolic fate of peptides throughout the body. In addition to hydrolysis, oxidation, and reduction are also observed as metabolic pathways of peptides [32]. Though a macrocyclic peptide possesses greater stability than its linear counterpart, it can still be unstable. Once a macrocyclic peptide is metabolized, it may transform into a linear peptide, exposing it to a wider range of enzymes. Hence, metabolite identification is incorporated fairly early in the optimization process of peptide discovery. This early integration can facilitate simple modifications to address possible metabolic liabilities.

Note that numerous other assays are not covered in this chapter. These include, but are not limited to, plasma protein binding assays, which help determine the extent to which a drug binds to proteins in the blood; blood-to-plasma partitioning studies, which assess the distribution of a drug between blood cells and plasma; and assays for drug-transporter interactions, which can identify whether a drug acts as a substrate or inhibitor of specific transport proteins. Additionally, CYP interaction assays may be important for permeable peptides for understanding whether a drug may competitively inhibit or irreversibly inactivate these important metabolic enzymes. Each of these assays provides critical insights into the pharmacokinetics and pharmacodynamics of a drug, contributing to a comprehensive understanding of its behavior in clinical settings.

Tiered Discovery ADME Assays

This could be categorized into Tiers 1, 2, or 3 depending on the initial properties to be optimized. Tier 1 assays are high-throughput tests designed to handle the majority of synthesized compounds, addressing significant and widespread ADME liabilities. In contrast, Tier 2 assays are the derived data from these assays coupled with data from other tests to steer the next phase of compound development. Specific assays either fall in Tier 1 or 2 depending on a compound's identified liabilities [33]. For instance, if peptidase activity is seen as a liability, the related assay would go into Tier 1 to rapidly eliminate highly unstable peptides. If not considered risky, it would fall under Tier 2. Tier 3 assays are specialized, hypothesis-driven studies aimed to pinpoint liabilities missed in previous tiers.

Permeability and Drug Transport Measurements

When assessing the *in vitro* permeability of peptides, it is important to consider several factors. Selecting the right assay to answer the specific question is crucial. For example, if the assessment of oral absorption potential is needed, a Parallel Artificial Membrane Permeability Assay (PAMPA) could be considered. However, cell-based permeability assays such as MDCK, Caco-2, or more recently human intestinal organoids are often preferred. Alternatively, more specialized cellular uptake assays, such as the ChloroAlkane Penetration Assay (CAPA) [34] or NanoClick [35], can be applied to assess cytosolic peptide uptake and accumulation. These assays offer a high-throughput, target-agnostic approach. Semi-quantitative intracellular concentration measurements could be performed in cells expressing the target of interest to determine if therapeutic concentrations have been reached inside the cells by using LC/MS. Lastly, since peptides may have limited permeability, drug transporters can play a significant role in absorption and cellular accumulation. Intestinal efflux transporters such as MDR1 (ABCB1), MRP2 (ABCC2), BCRP (ABCG2), and the PEPT1 (SLC15A1) [36] uptake transporter may be examined in cells over-expressing the desired transporter to determine if the absorption may be reduced or enhanced, respectively.

Table 6.1 Recommended controls for *in vitro* permeability assays.

	PAMPA	MDCK	Caco-2	Intestinal organoids
Low permeability marker	Octreotide, atenolol, nadolol			
High permeability marker	CsA, voclosporin, emodepside, metoprolol, propranolol			
MDR1 positive control	Not needed	Digoxin, loperamide, quinidine		
Monolayer integrity marker	Not needed	Lucifer yellow, Fitc-Dextran, mannitol		

Permeability

PAMPA, MDCK, and Caco-2 permeability protocols used for the assessment of oral absorption of small molecules may be used to examine the permeability of peptides with some modifications. The lower passive permeability of peptides often necessitates longer incubations (3–24 hours, if metabolic stability permits) and sensitive analytical methods (LC/MS) to obtain detectable peptide concentrations in the receiver chamber to calculate an apparent permeability (Papp) value. In addition to longer incubations, permeability assays that include preincubation with compounds to saturate the cells and minimize nonspecific binding to plastic are being successfully employed for cyclic peptides. Recommended controls for *in vitro* permeability assays are captured in Table 6.1. Lipidated peptides, as well as peptides containing multiple *N*-methyl amides or lipophilic moieties, may benefit from having 4% bovine serum albumin (BSA) in the receiver chamber to "pull" the peptide out of the phospholipid phase of PAMPA or cells [37].

PAMPA is a non-cell-based barrier composed of a customizable phospholipid mixture dissolved in an aprotic solvent such as dodecane which occupies the voids in a porous membrane support. The advantages of PAMPA are that it is a high throughput and relatively low-cost method for testing the permeability of peptides. PAMPA is compatible with vigorous stirring and can tolerate relatively high levels of organic solvents such as 5% DMSO to aid solubility. This system lacks endopeptidases present in cells that can potentially metabolize peptides, which can result in an underprediction of permeability. Some disadvantages of PAMPA are that the artificial membrane is approximately 40 μm thick which is 10^3–10^4 times wider than the lipid bilayer of cells. It contains organic solvent and unlike cells lacks tight junctions, organized lipid bilayers, and transporters resulting in a less physiologically relevant system.

Cell-based permeability screens to determine the passive permeability of peptides often use MDCK cells with low expression of the endogenous canine Mdr1 efflux transporter (MDCKII-LE) [38], MDCK with canine Mdr1 gene knocked out (gMDCK) [39], or MDCK/Caco-2 run in the presence of an MDR1 inhibitor such as elacridar.

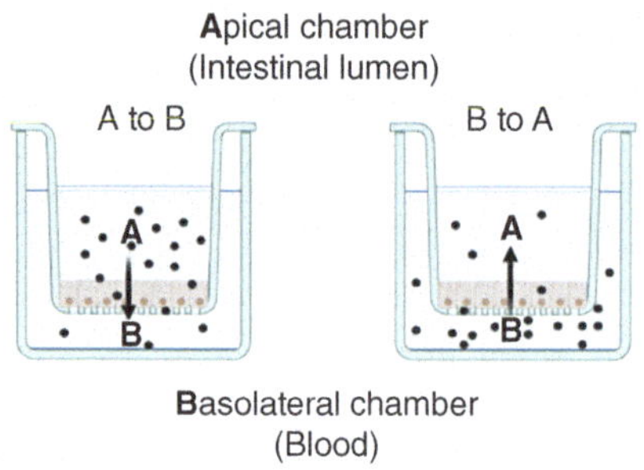

Figure 6.6 Schematic of a bidirectional Transwell permeability assay with the peptide of interest added in the apical donor chamber (A to B) and the basolateral receiver chamber is sampled to determine the rate of permeability (Papp) across the cell monolayer. B to A permeability (Papp) is determined by adding the peptide of interest to the basolateral donor chamber and measuring the rate of permeation into the apical receiver chamber. The efflux ratio, a measure of active efflux is: B to A Papp/A to B Papp. Source: Created with BioRender.com.

Caco-2 cells and organoids (jejunal) grown on Transwell® inserts can be utilized to examine a combination of passive permeability, active uptake, and efflux (Figure 6.6). Distinctly in the case of organoids, CYP-mediated metabolism can be assessed. Interpreting Caco-2 and organoid results can be challenging because it is difficult to determine which transporters and metabolizing enzymes, if any, are affecting the peptide's permeation and to what extent. Permeability protocols with preincubation step [40–42] include an 18–24 h preincubation in cell culture media containing the test peptide. The media containing the peptide is removed, the cells are rinsed and the permeability experiment is initiated by adding fresh peptide in assay buffer for a 1–3 hours incubation to determine the Papp.

Cellular Accumulation

CAPA requires a chloroalkene-modified peptide (ct-peptide) and an engineered cell line expressing a HaloTag facing the cytosol or desired cellular compartment (Figure 6.7). Accumulation is assessed by pulsing with the ct-peptide at multiple concentrations, chased with a ct-dye, and analyzed by flow cytometry. Similarly, NacoClick requires an engineered cell line expressing a NanoLuc-HaloTag fusion protein that is expressed in the cytoplasm and a

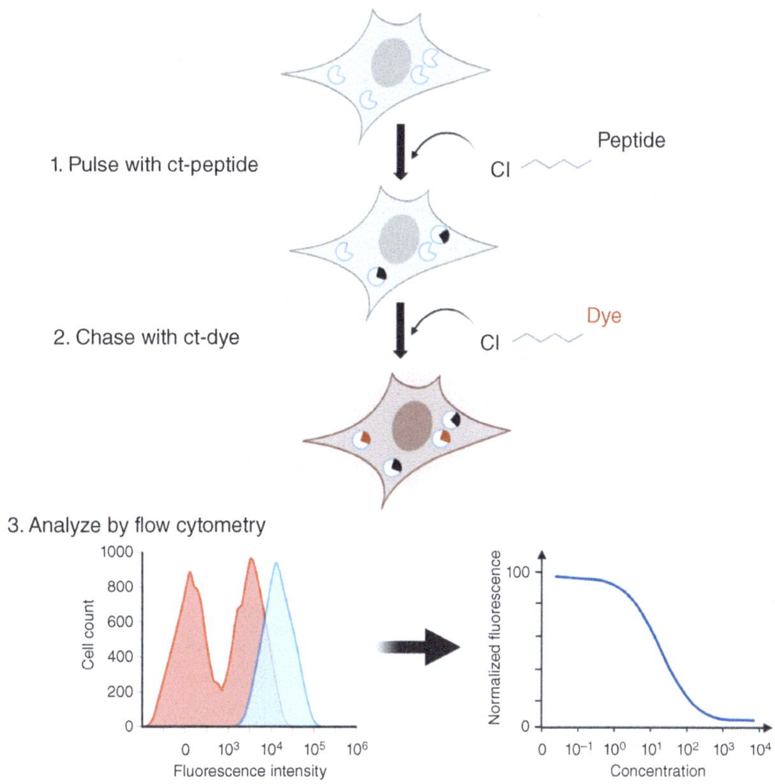

Figure 6.7 Cells expressing HaloTag are pulsed with chloroalkane-tagged molecules (ct-molecule), washed, chased with a chloroalkane-tagged dye, and analyzed by flow cytometry. Source: Adapted from Peraro et al. [34]. Created with BioRender.com.

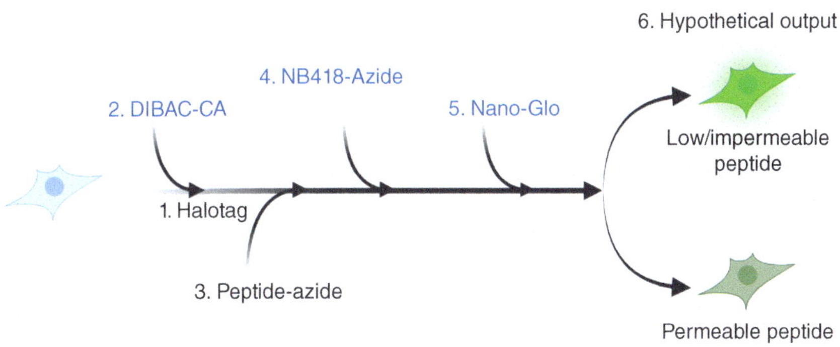

Figure 6.8 1. HeLa cells are transiently transfected with a NanoLuc-HaloTag construct. 2. Cells are incubated with the dibenzoazacyclooctyne–chloroalkane (DIBAC–CA) molecule, and a covalent bond will form between the HaloTag and the chloroalkane linker. 3. The azide-modified peptide is added to the cells and incubated at 37 °C (4–18 hours). If a peptide can traverse the cell membrane and access the cytosol, it will undergo a bioorthogonal strain-promoted click reaction (SPAAC) forming a covalent bond with the DIBAC molecule. 4. Cells are incubated with the NanoBret 618-Azide (NB618-Az) dye molecule which is quantitatively ligated to any unoccupied DIBAC molecules. 5. The NanoLuc substrate is added to cells which yields a NanoLuciferase signal that serves as the BRET energy acceptor. 6. Hypothetical output of NanoClick data. Source: Adapted from Peier et al. [35]. Created with BioRender.com.

peptide conjugated to an azide (az-peptide; Figure 6.8). The dibenzoazacyclooctyne–chloroalkane (DIBAC–CA) covalently binds to the HaloTag expressed in the cells, the az-peptide binds the DIBAC–CA and in turn competes with the azide-dye, minimizing the BRET signal. A cell penetration curve (CP50) can be generated for both assays with the fluorescence signal inversely proportional to the cell penetration of the ct-peptide or az-peptide. Since HaloTags are expressed facing the cytosol for both CAPA and NanoClick, the free intracellular concentration is determined and does not need to be corrected for binding to cell plasma membranes. It is important to note that modifications like a chloroalkane or azide could alter intramolecular hydrogen binding of cyclic peptides which may impact the 3D conformation and its permeability characteristics. Once labeled peptides are proven to accumulate in the CAPA or NanoBret assay, an orthogonal permeability method such as PAMPA or LC/MS-based cell accumulation assay can be used to test the permeability or accumulation of the parent versus the modified peptide. LC/MS-based cellular accumulation assays can be advantageous because engineered cell lines and labeled peptides are not required. The throughput of the LC/MS cell accumulation assay is lower than CAPA and NanoClick but the parent peptide and peptide metabolites can be determined inside the cells. The intracellular concentrations in the LC/MS accumulation assay are total concentrations (the sum of free cytosolic peptides and peptides bound to cellular proteins and membranes). The free intracellular concentration of the peptide can be established by correcting for binding using different methods [43, 44]. The mechanism of cellular accumulation may be multifaceted and complex and all three accumulation assays can be used to elucidate the mechanisms involved. Cellular accumulation rates can be assessed at 4°C versus 37°C to determine if the accumulation is an active or passive process (note the lower temperature inactivates the energy required process [such as ATP] at the same time lower temperature modifies the membrane normal fluidity). The use of inhibitors can be used to probe the involvement of specific endocytosis pathways to gain further knowledge of the accumulation mechanism or dissect the portion of accumulation that is passive compared to what is active [45].

Drug Transport

MDR1, MRP2, and BCRP are promiscuous efflux transporters that are co-localized in the intestine, liver, and kidney. MDR1 can efflux a wide range of substrates including cyclic peptides at least as large as cyclosporin A (CsA). Although MDR1, MRP2, and BCRP share significant substrate overlap, MRP2 generally transports anionic, conjugated endogenous, and xenobiotic substances while BCRP effluxes a wide array of small molecule xenobiotics. MRP2 and BCRP do not transport CsA and since these transporters are not as well studied, it is currently unknown if these transporters play a significant role in the disposition of peptides. A substrate cut-off in terms of size, molecular weight, or number of amino acids has not been reported in the literature for these transporters, most likely due to the complex 3D structures of the peptides and transporters. In order to determine if a peptide is a substrate of MDR1, MRP2, and BCRP a bidirectional permeability assay with cells overexpressing the desired transporter can be employed. Alternatively, if the permeability of the peptide is very low, inside-out vesicles expressing the transporter most is the desired model system because it reduces the analytical detection requirements.

PEPT1 has been reported to uptake substrates in the order of greater than 400 dipeptides and 8,000 tripeptides [46]. Most therapeutic peptides are significantly larger than three peptides, therefore a prodrug approach using a di- or tripeptide handle for a therapeutic peptide can be explored in an effort to increase intestinal absorption.

In Vitro Metabolic Assays

Considering the susceptibility of peptides to biotransformation by peptidases, which are widely distributed throughout the body, it is essential to evaluate the *in vitro* metabolic stability of peptides to rank order and screen for stable lead compounds during the drug discovery stage. Depending on the various routes of administration that may be explored, different matrices have to be considered or in other words, fit-for-purpose metabolic stability assessments are required to answer specific questions (Figure 6.9).

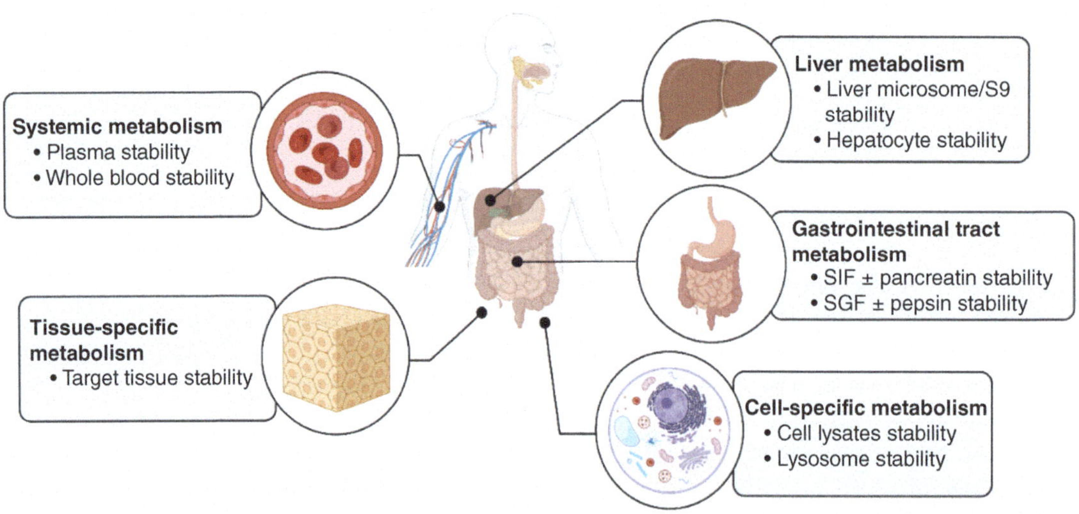

Figure 6.9 Tissues and matrices considered for *in vitro* metabolic stability assays for therapeutic peptides. Source: Created with BioRender.com.

Systemic Metabolism

Intravenous and subcutaneous administration are the common administration routes of peptide therapeutics. These routes have the advantage of avoiding pre-systemic metabolism from hepatic and gastrointestinal enzymes and achieving complete systemic availability of the drug. At the same time, peptides in circulation are susceptible to various drug-metabolizing enzymes, such as peptidases in the blood or present in the tissues. A plasma stability assay is an efficient way to assess such metabolic liability of peptides though it is missing some of peptidases in the blood. Heparinized plasma instead of EDTA-plasma is recommended as the incubation matrix due to the peptidase inhibition activity of EDTA due to chelation. In addition, the relative abundance and substrate specificities of plasma peptidases vary in different species so the selection of plasma from appropriate species needs to be considered when performing stability assays [47]. A consideration of *in vitro* assays is that the enzymes remain active for only a short period, typically less than 24 hours. Additionally, extended incubation times might lead to shifts in the pH of the incubation medium.

Gastrointestinal Tract Metabolism

For orally administered peptides, the presence of high levels of peptidases in the gastrointestinal tract poses a significant challenge. This environment can greatly decrease the oral bioavailability of peptides. Additionally, poor solubility and limited permeability of peptides may further hinder their ability to cross the intestinal mucosa and reach the systemic circulation [48]. There are three types of peptidases: (i) luminally secreted enzymes including pepsins, trypsin, chymotrypsin, elastase, and carboxypeptidase A/B; (ii) brush border membrane-bound enzymes including various carboxypeptidases and aminopeptidases; and (iii) cytoplasmic enzymes [49]. Incubations with simulated gastric fluid (SGF) or intestinal fluid (SIF) mimic the physiological composition of human gastric and intestinal fluids with a similar type and level of surfactants including bile salt and phospholipid such as lecithin [50]. The incubation can be supplemented with pepsin, or pancreatin containing a mixture of trypsin, chymotrypsin, and elastase to evaluate potential peptide instability through the gastrointestinal tract. During experimental setup, the concentration of pancreatin could be diluted to minimize the early rapid rate of metabolism, and thus the disappearance rate of the peptide can be more properly monitored. In the optimization process, as more stable peptides are generated, the concentration of pancreatin could be increased to more physiological levels.

Liver Metabolism

Unlike small molecules, hepatic metabolism may not be a major route of elimination for most peptides, especially for poorly permeable peptides. For peptides with sufficient permeability, access to intracellular drug metabolizing enzymes occurs by combined passive diffusion and/or carrier-mediated uptake mechanism [51]. As peptides become more permeable and stable against peptidases, liver metabolism by enzymes such as cytochrome P450 (CYP) plays a more prominent role. Cyclosporine A, which contains 11 amide bonds, serves as an example of a peptide primarily metabolized in the liver and small intestine by CYP3A4 and CYP3A5, without the involvement of peptidases [52]. Commonly used *in vitro* systems to understand liver metabolism include hepatocytes or subcellular hepatic fractions such as microsomes or S9. Hepatocytes present as a physiologically relevant model, which contains the full complement of hepatic drug-metabolizing enzymes maintained within the intact cell. When using hepatic fractions, it is important to note that while many peptidases are cytosolic and CYP enzymes are in microsomes, liver S9 fractions, which contain both cytosolic and microsomal enzymes, provide a more comprehensive model than microsomes alone. Subcellular fractions can be supplemented with cofactors to investigate beyond peptidases with mainly Phase I with NADPH cofactor as in the case of CYP enzymes. Typical Phase II conjugative enzymes are not involved in the metabolism of peptides.

Cell-Specific Metabolism

During the lead identification stage, biochemical assays are often conducted to assess potency or specificity at a cellular level. In this case, peptide stability in relation to specific cell lines needs to be considered due to the peptidases widely distributed in cells. Peptidases have been found in both cytoplasm within the cell and on the surface of cells, with the catalytic site exposed only at the external surface. Cell lysate has been employed as a good *in vitro* model for stability assay since it is easy to prepare and retains many of the cofactors and other regulatory elements of the cell. Concentration of the cells and incubation time can be further modified based on peptidase activity in the system. For peptides with intracellular targets, it is also beneficial to assess lysosomal stability since this membrane-enclosed organelle contains an array of peptidases capable of breaking down all types of biological polymers including peptides. As an example, commercially available human liver lysosomes are an easy-to-use *in vitro* system. It offers a complete and stoichiometrically correct enrichment of lysosomal contents with lysosomal-specific enzymatic activity and low contaminating activity from mitochondrial enzymes.

Metabolism in Tissue-Specific Homogenates

Peptides can also be evaluated for metabolic stability based on the tissue-specific delivery route and therapeutic target. Intravitreally (IVT) administered ocular peptides are a good example of where this approach may be warranted. Whereas hepatic and gastrointestinal metabolism is broadly studied, ocular drug metabolism, and the impact of local tissue stability are more poorly understood. Metabolic characterization of individual ocular tissues with their enzymatic levels of various peptidases is a useful step in this direction for the development of IVT peptide-based ocular biologics [54]. This can aid in further understanding the biodistribution and fate of the peptide therapeutics but is challenging due to cross-contamination across ocular tissues versus using whole eye homogenates. Physiological and anatomical differences between preclinical species and the human eye can also significantly contribute to the challenges of developing ocular PK models.

Experimental and Analytical Approach

To assess peptide metabolic stability, peptides are incubated with biological matrices (plasma, liver S9, cell lysates, etc.) at 37 °C and at optimal pH depending on specific enzyme activity. In certain instances, glass vials are recommended in the experiment to minimize nonspecific binding between peptides and the surface of plastic wares. However, this consideration of labware material needs to be evaluated in a case-specific manner in order to minimize nonspecific binding, and protein low-binding polypropylene tubes are also successfully used for this purpose. Incubations are quenched at predetermined time points in order to gain a concentration–time profile of the parent compounds. Liquid chromatography online with high-resolution mass spectrometry (LC/HR-MS) is the commonly employed analytical approach benefiting from its advanced specificity, speed, and selectivity. Incubation samples are analyzed by reverse phase columns and the high-resolution MS is usually deployed to collect both MS1 and MS2 data. MS1 spectra with high mass accuracy are utilized to monitor parent compound disappearance as well as capture molecular weight information on the metabolites. The structural information of the peptide fragments will then be confirmed and further elucidated by interpretation of fragment ion and determine the site of metabolism as described in the Biotransformation section.

Bioanalytical

Peptide drugs present a high physicochemical heterogeneity due to the structural modifications that intend to improve the *in vivo* stability and pharmacokinetic properties. As a result, it is challenging to predict their behaviors during bioanalysis. Another bioanalytical challenge is that in many cases, the target shows low abundance in the systemic circulation due to poor bioavailability and short half-life. Therefore, extensive bioanalytical method development is often needed to improve the sensitivity, selectivity, and robustness to support PK investigation of peptides.

During the sample extraction procedure, overcoming the challenges of low recovery and nonspecific binding can facilitate the downstream analysis. Various approaches are explored such as protein precipitation extraction (PPE), solid phase extraction (SPE) [54], liquid–liquid extraction (LLE), electromembrane extraction [55], filtration, and immunopurification [56]. Among these methods, PPE and SPE are the most commonly used and can easily be automated in 96-well format in routine laboratories. PPE in general works for small-size peptides (less than 20 amino acids) provided that the proper precipitating solvent is selected, and in more limited cases can also be applied for larger peptides up to 30–35 amino acids in length-dependent on their physicochemical properties and organic solvent solubility. During the PPE procedure, different organic solvents including methanol, acetonitrile, and acetone can be evaluated and modifiers/additives such as formic acid and trichloroacetic acid can be added to boost the stability, and extraction efficiency and reduce nonspecific binding. PPE is a comprehensive extraction strategy, offering potential post-acquisition exploration for biotransformation or potential biomarkers if an HRMS is used. On the other hand, the SPE approach has been widely adopted for targeted quantitation of peptide drugs with its advantage of great selectivity and wide selection of separation mechanisms; it can also significantly lower the matrix effect compared with the PPE approach. Moreover, the innovative μElution SPE plates [57] are designed for low-volume samples with simplified protocol without evaporation and reconstitution, however, optimization is required to select the best SPE sorbent and optimize the washing/elution conditions to achieve the best recovery, and poor recovery has been reported for peptides with marked hydrophilicity or hydrophobicity. In recent years, microextractions such as electromembrane extraction [58] and solid phase microextraction have gained popularity because of the low consumption of samples and organic solvents, but with limited applications in bioanalysis of peptide drugs due to lack of commercial availability and robustness.

Immunoaffinity Capture

Immunoaffinity (IA) capture can be deployed as a bioanalytical tool for peptide PK and biotransformation analysis. An appropriate capture reagent with high binding affinity provides high specificity and can be leveraged for immunoaffinity enrichment. Two types of reagents can feasibly be used: antibodies generated directly to the peptide or a binding partner of the peptide (typically the extracellular region of a target protein). Various methods of antibody generation have been reported in the literature, but depending on the timeline and resource availability, many may not be feasible [59]. A straightforward route may be to generate polyclonal antibodies via inoculation of rabbits using the peptide of interest directly, or peptides conjugated to carrier proteins, for example, BSA, KLH, and OVA to elicit a sufficient immune response. After ~3–4 months, it is commonly possible to generate milligram levels of antibodies by purifying the rabbit sera. The purity of the antibodies may be improved by purification using columns coated with the peptide antigen of interest while ensuring the affinity of the antibodies, followed by screening with the peptide antigen for implementation in the bioanalytical assay. Alternatively, the binding partner of the peptide can be generated as a recombinant protein. However, such a reagent may be difficult to express or handle due to factors such as aggregation and nonspecific binding to labware, or may not be amenable for the biological system of interest in the case of high level of soluble interference. If a peptide is conjugated (e.g. PEGylated), the additional domain present may also be used as a means of immunoaffinity enrichment, such as with anti-PEG Ab.

Upon immunocapture from the sample to enrich for the therapeutic peptide or metabolites, this can be followed by direct analysis by LC-MS for qualitative biotransformation assessment and/or quantitation or further proteolytic digestion and then analysis of a smaller surrogate analyte for multiple reaction monitoring (MRM) or parallel reaction monitoring (PRM)-based quantification.

LC-MS Conditions

After sample preparation using the aforementioned prefractionation/purification methods, peptide identification and quantification can be performed using various analytical detectors, each with its advantages and drawbacks. For PK studies, commonly used detection methods include UV/fluorescence spectroscopy, ELISA, and mass spectrometry (MS). For biomatrices that are relatively simple or processed through selective affinity purification, subsequent analysis using electrophoresis or chromatography coupled to UV/fluorescence may be used. However, for most complex bio matrices including tissues and blood, usually ELISA or MS is preferred for specificity and sensitivity. In instances where detection antibody reagents required for ELISA are not readily available, MS is most commonly used. Prior to MS, reversed phase chromatography is usually used for front-end separation of peptides because of its high resolution and downstream compatibility. For MS identification, high-resolution instruments including time of flights and orbitraps are commonly used for analyzing peptides from complex matrices. Unambiguous identification can be performed by using tandem mass spectrometry providing peptide sequencing information. In addition, identification can be inferred with high accuracy MS1 detection and with inference through coelution of the analyte with the corresponding stable isotopically labeled peptide and the accurate MS1 measurement of the target peptide. For quantitation, measurement of the intact peptide can be achieved using SIM, and PRM or MRM can also be used to quantitate through product ions. While the latter generally leads to higher sensitivity because of increased selectivity, the sensitivity can be compromised for peptides that do not behave well under collisional energy fragmentation. Unlike UV detection, MS can have response biases amongst different m/z which usually necessitates using calibrant standards and/or isotopic peptides for improving accuracy and precision.

For LC-MS assays, the ideal internal standard (IS) to be used for qualified or validated quantitative assays would be a stable-isotope labeled (SIL) version of the intact peptide therapeutic, whether it be analyzed intact or upon further proteolytic digestion to a smaller size. In the case of an IA-LC-MS-based method, the SIL internal standard (SIL-IS) introduced to the biologic sample provides a means to control for any variability seen in the sample enrichment or immunoaffinity capture steps as well as proceeding sample preparation conditions. However, if the qualification reflects assay reproducibility and robustness, alternative internal standards can be also used.

Biotransformation

In vitro metabolism assays are typically complemented by metabolite identification (MetID) to best understand the biotransformation pathways. Peptide biotransformation allows to complement rates of metabolism by determining the site of metabolism. As mentioned, for peptides the main reaction involves hydrolysis by peptidases and oxidation.

Analytical Methods

The sample process and bioanalysis are shared in the *in vitro* metabolism section. Baseline chromatographic separation of peptide and its metabolites are preferable.

Strategy of analysis: A few general considerations for *in vitro* and *in vivo* MetID studies are below.

In Vitro MetID Selecting the appropriate time point for sample analysis is very important and should be evaluated based on a given peptide's expected stability.

Having a lower concentration of peptides results in more linear kinetic rates.

Multiple matrices may be considered for the assay including for orally administered drugs: FaSSIF + Pancreatin, intestinal S9, plasma stability, and liver S9.

In Vivo MetID Metabolites that are generated *in vitro*, but not observed in circulation primarily because (i) primary metabolites are further metabolized; and (ii) the metabolites are more rapidly cleared from the body.

For more comprehensive characterization, bile-cannulated studies could be performed by collecting both bile and urine.

Similarly, as described for bioanalysis, generally peptides can be purified using organic extraction involving acetonitrile or methanol, at a volume ratio of ~3× relative to the sample. After mixing and centrifugation, the supernatant is typically diluted and extracted for direct analysis via LC-MS.

Immunoaffinity (IA) capture leverages the use of capture reagents with high binding affinity (e.g. anti-peptide antibody, or the target antigen), often immobilized on a solid surface, to selectively bind to and isolate therapeutic peptides from a complex mixture, followed by the utilization of MS as a downstream analysis method to identify and quantify the captured peptides. An IA capture approach requires capture reagents that can bind both the peptide and its metabolites with good affinity to enable unbiased recovery of metabolites. Alternatively, a combination of different capture reagents can also be used to enrich various metabolites depending on which domain of a peptide therapeutic may be modified. The approach is amenable to a range of biological matrices, including serum, plasma, and cell/tissue lysate. The bound therapeutic peptides can be separated from the antibody, often by changing the elution conditions (e.g. pH and ionic strength), to analyze the peptide intact, or be digested into smaller peptides typically using proteases to more specifically localize biotransformation sites. IA combined with MS is a highly specific, selective, and sensitive method for analyzing protein therapeutics and is widely utilized to assess the biotransformation [60, 61].

Radiotracer studies are invaluable for providing insights into biodistribution, metabolism, and excretion. While these studies are commonly employed in the development of small molecules, they present unique challenges when applied to peptides. One significant challenge is the potential need for multiple labels when using ^{14}C, due to potential cleavage sites that can result in the loss of the label [62]. Additionally, the use of ^{125}I as a tracer, particularly when placed on tyrosine residues, can alter the lipophilicity of peptides, potentially affecting their behavior and interactions within biological systems [63]. These changes can complicate the interpretation of tracer study results, as the labeled peptide may not accurately represent the behavior of the native peptide. Therefore, careful consideration and optimization of labeling strategies are essential to ensure that radiotracer studies provide meaningful and accurate data for peptide development.

Types of Metabolic Reactions

Hydrolysis

Proteolytic hydrolysis on the backbone by peptidases is one of the common biotransformation processes. For terminal hydrolysis that leads to truncation, the site of metabolism can be determined from the associated mass shifts from the precursor mass spectra. To predict *in vivo* peptidase activity in blood circulation, serum or plasma incubation is often performed at 37 °C. Note that typically, such systems do not maintain matrix stability past 96 hours resulting in unstable pH and turbidness. For orally delivered peptides that may transit through the gut, enzymatic biotransformation is a likely concern, and protease and pancreatic extract *in vitro* incubation may be performed. Such data may be useful in predicting hydrolysis sites, aiding in the engineering of more stable peptide variants. However, there has been very limited data to suggest good *in vitro–in vivo* correlation for peptidase biotransformation.

Oxidation

More lipophilic and permeable peptides are metabolized through enzymes such as CYP [32].

For peptides containing natural amino acids, oxidation most commonly occurs on residues such as cysteine, methionine, and tryptophan (and to a lesser extent on histidine and tyrosine). Oxidation can be identified in mass spectrometry by the presence of mass shifts of +16 Da (or an integer multiple), +4 Da (or an integer multiple), or combinations of +16 and +4 Da shifts (or an integer multiple). More recently loss of side-chain has also been reported with Carbon–Carbon bond cleavages [65]. Another noteworthy example is NADPH-dependent hydrolysis by cytosolic enzyme(s) [66].

Oxidation can be induced by multiple *in vitro* conditions including oxidizing agents (AAPH, peroxides), ultraviolet (UV) irradiation, and elevated temperatures. Such *in vitro* incubation conditions are commonly used to predict potential liabilities occurring in manufacturing and storage as part of the drug development effort. Incubation can be conducted in biological matrices that mimic the site of action for the therapeutic of interest, not limited to serum and plasma. However, more research will be needed to identify proper predictive *in vitro* conditions for *in vivo* oxidation liabilities since AAPH, peroxide stresses are generally not

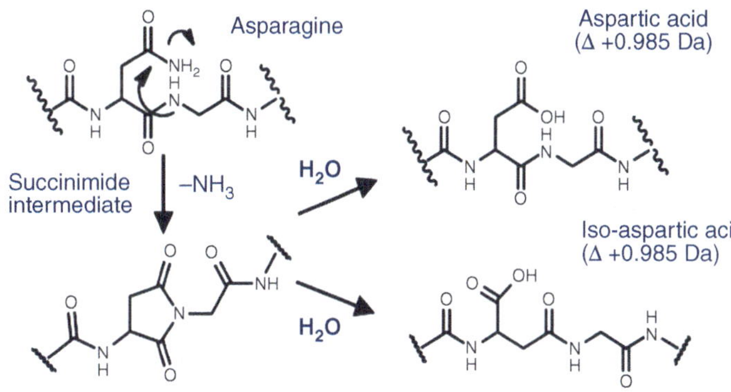

Figure 6.10 The mechanism of deamidation and isomerization begins with a nucleophilic attack by the amide bond amine on the primary amide of the side chain, forming a succinimide intermediate. This intermediate subsequently collapses to yield two acid-containing products. Source : credit [67]/American Chemical Society.

predictive [67]. Considerations for the site of action have been shown to be important, as the environment in the blood does not appear to be oxidizing to peptides, whereas a compartment like ocular tissue may be.

Deamidation and Isomerization

Deamidation is a chemical reaction mechanism that most commonly occurs on the asparagine residue (particularly if preceding a glycine residue). The transformation results in its conversion to aspartic and isoaspartic acid through a succinimide intermediate leading to a detectable mass shift of +0.984 Da (Figure 6.10). Glutamine can also undergo an analogous transformation mechanism but generally at a slower rate.

Isomerization, such as aspartic acid converting to iso-aspartic acid residues, occurs non-enzymatically through the formation of the similar aforementioned cyclic succinimide intermediate. Such biotransformation is most prevalent in DG, DS, and DT motifs. The stable and isomerized peptide species are isobaric and cannot be distinguished in tandem mass spectra using conventional CID or HCD fragmentation. Alternatively, ECD/ETD fragmentation, resulting in signature ions $z-57$ and $c+57$, can be used to identify isomerized residues. Similar to protein therapeutics, deamidation and isomerization on peptides can be generated *in vitro* using incubations under physiologically relevant temperature and pH conditions. Such *in vitro* conditions (e.g. PBS at 37 °C) have been successfully used to predict protein deamidation that is highly translatable to *in vivo* conditions. Source: Chu et al. [67]/ American Chemical Society.

Thiols Oxidation/Reduction

These compounds primarily exist in two forms: thioether and disulfide. Thioethers can be susceptible to oxidation, forming sulfoxides or sulfones through enzymatic, or chemical reactions. Disulfide bonds, on the other hand, are prone to reduction (cleavage) in a reducing environment. Additionally, disulfide bonds can be scrambled, forming new bonds with thiols present in other peptides or proteins.

Conjugated Peptides

Peptides such as pegylated-peptides can also undergo biotransformation or deconjugation of their PEG moieties impacting their physicochemical properties and their overall efficacy and/or safety. PEGylation has been widely used as a drug delivery approach to improve circulating half-life or reduce immunogenicity [68].

Aggregation

Biotransformation typically does not involve aggregation, but it is important to consider that aggregation can occur. High concentrations, lipophilic surfaces, or dithiol cross-linking can lead to peptide aggregation. With the increasing relevance of long-acting delivery of drugs, which typically requires high concentrations and long delivery times, assessment of potential aggregation is increasing in importance. Aggregation may be predicted using incubations at high concentrations in physiological matrices. One such physiologically relevant matrix could include PBS with additional components (e.g. vitreous humor for ocular delivery). As the environment (e.g. pH, chemical composition) within delivery systems may change over time, it is necessary to monitor these changes. High concentrations of drugs often leverage certain pH or other formulation buffer optimizations. However, diffusion of these buffer components out of the device and diffusion of *in vivo* components into the device would change the solubility and other characteristics of the peptide. Therefore, the peptide may precipitate or otherwise act in unexpected ways. Knowing these changes, it would be more straightforward to develop a predictive *in vitro* model by mimicking these conditions. Multiple conditions may be needed to better reflect the formulation changes that can occur *in vivo*. For analytical considerations, complementary techniques to LC/MS, such as CE-SDS or LC-UV (e.g. SEC), are commonly used to detect higher molecular weight species.

Pharmacokinetics of Peptides

Formulation Development

As with other therapeutic modalities, peptides must be formulated as a drug product with the appropriate properties and quality prior to administration *in vivo*. The design of the formulation must be carefully selected in order to optimize the required pharmacokinetic and pharmacodynamic behavior of the peptide with sufficient safety and efficacy within the constraints of the desired route of administration while also maintaining sufficient stability over time. Formulation development can be used to mitigate potential liabilities of the peptide as well as to enhance *in vivo* absorption and exposure.

Formulation development begins first with an understanding of the target dose to be delivered as well as the intended route of administration. The administration route is the primary influence on the extent and rate of drug absorption and, thus, the overall bioavailability and the resulting target dose. Peptides are often administered parenterally through intravenous infusions or injections in order to maximize systemic absorption by avoiding limitations due to poor permeability and pre-systemic metabolism. However, subcutaneous or intramuscular injections have also been explored as they enhance patient convenience and compliance since they can often be self-administered rather than relying upon a clinical setting. Peptides have also been administered via pulmonary delivery, including examples such as inhaled insulin products such as Afrezza®. More recently, orally administered peptides have been successfully developed, including the oral semaglutide product Rybelsus®.

Formulation development then begins with a strong understanding of the physicochemical and ADME properties of the peptide [69, 70]. As noted in 6.0, peptides typically exhibit poor bioavailability due to several factors including poor permeability and solubility as well as stability liabilities. Some of these factors can be mitigated through careful formulation design. These risks are often exacerbated in solution and are thus prevalent in solution formulations typically used for parenteral administration, but similar concerns exist for solid-state formulations such as tablets or capsules for oral administration.

Dose

The target dose of the peptide to be delivered in order to achieve the required pharmacokinetic profile will determine the overall concentration of the formulation. For administration routes with limited dosing volumes such as subcutaneous or intramuscular injections, this can lead to potentially high concentrations of peptide in the formulation, resulting in challenges in solubility, stability, as well as injectability due to high viscosity. Some of these limitations can be addressed via formulation development in the following ways.

Solubility

Solubility is an important parameter for formulation development and plays a critical role in absorption *in vivo*. The solubility of a peptide can be highly dependent on the interaction between its amino acid sequence and its environment, including pH and ionic strength. While a number of computational models exist, aqueous solubility determination remains largely empirical [71, 72]; while, in general, hydrophobic amino acid residues may limit solubility, the secondary structure and conformation of the peptide may change the amount of exposed polar surface area (as noted in chameleonic peptides in Section 2.1). Side chain modification or conjugation may also impact the overall solubility of the peptide. Kinetic solubility in PBS is often obtained early, as described in Section 2. Poor solubility may indicate potential risks to absorption and determining thermodynamic solubility in biorelevant media (e.g. simulated physiological fluids) can be helpful. In the case of intravenous administration, where liquid solution formulations are highly desired, solubility can potentially limit the total dose of the peptide that can be delivered. Fortunately, in many cases, solubility can be enhanced in the formulation. Several factors can be evaluated to improve peptide solubility.

pH

Peptides often contain multiple ionizable functional groups and thus exhibit complex pH-dependent solubility behavior. The isoelectric point (pI) is the pH value of the solution at which the peptide has an overall neutral charge. Thus, in order to increase aqueous solubility, the pH of the formulation should be adjusted away from the pI. However, the acceptable pH range used will be limited by the intended route of administration due to the risk of intolerability at acidic or basic pHs as well as the impact of pH on stability.

Ionic Strength

Since peptides contain many polar functional groups, their solubility can also be dependent on the ionic strength of the solution media. The higher salt content can sometimes mask the polar or ionized functional groups on a peptide and result in lower-than-expected solubility, although this effect can be hard to predict. This effect has been widely observed for proteins, resulting in the commonly described "salting out" or "salting in" solubility behavior [73]. Peptide solubility can also be strongly influenced by the counterion, most commonly observed in the Hofmeister series of counterions and their effect on solubility of proteins [74]. Both the choice of counterion and the overall ionic strength of the peptide formulation can be modified to enhance solubility, although the overall ionic strength of a formulation may be limited by the osmolarity and tonicity ranges acceptable for that route of administration. Injectable formulations should have a tonicity range close to isotonic with bodily fluids to minimize discomfort. For example, an intravenous formulation should be isotonic with plasma (275–295 mOsm/kg). Hyper- or hypotonic formulations may increase the risk of adverse effects and intolerability.

Excipients

When sufficient solubility is not achieved in aqueous solutions despite adjustment of pH or ionic strength, solubilizing excipients may be added [69]. These include surfactants such as polysorbate, complexing agents such as cyclodextrins, water-miscible co-solvents such as polyethylene glycol (PEG), or lipids. The identity and concentration of the excipients may be constrained by the route of administration due to risks of intolerability.

Chemical Stability

When a formulation of a suitable solubility at the desired concentration range is achieved, the short- and the long-term stability of the peptide must be evaluated. Peptides can be susceptible to a number of chemical transformations such as hydrolysis, deamidation, and oxidation, among others, many of which may be sensitive to the pH of the solution. Peptide stability can be determined by incubating the formulation and analyzing by reverse phase high-performance liquid chromatography (RP-HPLC) or capillary electrophoresis (CE) over time. Often, peptide formulations are subjected to accelerated stability conditions such as higher temperatures to understand their potential longer-term stability risk. In some cases, the chemical stability of a peptide can be improved through formulation optimization such as modifying the pH with buffers or adding stabilizing excipients such as antioxidants [75]. A challenge is balancing the effects of pH on stability as well as solubility for a given peptide to give the desired behavior.

While formulation development is primarily focused on optimizing drug product stability for improved shelf-life, in some cases formulation can also improve peptide stability after administration *in vivo* (see next section). For example, for peptides that are sensitive to enzymatic biotransformation *in vivo*, excipients such as pH modifiers or protease inhibitors can be included to provide improved stability to enable improved absorption.

Physical Stability

Peptide activity and potency are often directly linked to not only their primary sequence of amino acids but also to their secondary structure and conformation in solution. In many cases, the secondary structure or conformation of the peptide must be maintained for activity against its target. The physical stability of the peptide must be evaluated in the formulation. Changes in conformation may be detected through circular dichroism (CD) as well as nuclear magnetic resonance spectroscopy (NMR). In some cases, certain amino acid residues may exhibit intrinsic fluorescence which can be used to characterize their conformation in solution.

In addition to changes in their secondary structure, many peptides are also susceptible to oligomerization or aggregation to form high molecular weight species [17]. Peptides are particularly sensitive to aggregation compared to small molecules or even larger therapeutic proteins. Aggregation can occur through a number of different mechanisms including nonspecific hydrophobic interactions or through discrete intermolecular interactions such as disulfide linkages or b-sheet formation in amyloid fibrils. In some cases, aggregation may be reversible or irreversible. Since it is typically the peptide monomer that is active against its target, these changes can reduce the efficacy and potency of the formulation. Aggregation may also reduce absorption. Moreover, aggregation can also pose a significant risk in inducing adverse effects, including immune and foreign body responses. The oligomerization and aggregation state of a peptide can be determined by a number of methods including light scattering as well as size exclusion chromatography. Specific peptide aggregation such as fibrillation can be determined by fluorescence measurements using a reporter dye such as thioflavin T [76]. In addition to aggregation, peptides may also form visible or sub-visible particles as well as other precipitates over time. These effects can challenge the stability and integrity of the formulation.

Aggregation is often correlated with the concentration of the peptide and thus linked to the target dose. For high-concentration peptide formulations, physical stability must be carefully monitored. Mechanical stress should also be evaluated, as excessive agitation or shear forces may also induce aggregation. In some cases, aggregation may be mitigated through formulation optimization such as tuning the pH or adding stabilizing excipients.

Sterilization

Parenteral formulations must be sterile and free of biological contamination and other foreign particles. Terminal sterilization processes may also be employed to ensure the sterility of the final drug product. These can include sterile filtration, steam autoclaves, gamma-ray radiation, and ethylene oxide gas, among others. Endotoxins are a particular type of biological contamination and must be carefully monitored. Endotoxins are lipopolysaccharide remnants found in the cell walls of Gram-negative bacteria. Contamination with endotoxins can lead to severe adverse effects from strong immune responses including inflammation and anaphylaxis. It can be difficult to remove endotoxins from the final formulation through terminal sterilization, so end-to-end aseptic processes are required to minimize their contamination. Drug product formulations intended for multi-use will also require antimicrobial preservatives [77]. However, the chemical and physical stability of the peptide must be evaluated when considering these processes [78].

Container Selection

Finally, the selection of the appropriate container is also important to ensure the stability and quality of the final formulation [75]. Material compatibility studies must be conducted to evaluate the stability of the peptide. For example, plastic or glass surfaces may result in adsorption of the peptide, leading to drug loss. In addition, the volume of the container is also important, as the presence of headspace and the resulting interface between air and liquid can induce peptide aggregation or particle formation. For peptides susceptible to oxidation, purging the container of oxygen and replacing it with an inert gas may be performed. Many syringes may also contain trace amounts of silicone oil as a lubricant, which can induce stability risks as well, which must be evaluated.

The goal of formulation development is achieved when the final drug product can be consistently manufactured with high quality with long long-term shelf-life stability. In the desired drug product, the peptide is formulated in a presentation that can be administered to the patient in a safe and tolerated way while enabling sufficient absorption *in vivo* to achieve the target exposure and efficacy.

Absorption and PK Properties

In general, due to the intrinsic physicochemical and structural properties of peptides, there are many difficulties in optimizing their PK profiles due to common limiting features such as poor absorption, low volume of distribution, as well as rapid protease-mediated elimination; which can all contribute to inadequate (and often sub-therapeutic) exposure, as well as a short elimination half-life in the systemic circulation (Table 6.2). More recently, researchers have attempted to circumvent these challenges by adopting a variety of methods such as using local/extravascular routes of administration, structural modifications (i.e. cyclization, non-natural amino acid substitution, conjugation, etc.), as well as co-formulation approaches (i.e. permeation enhancers). Meibohn and colleagues have previously discussed these attributes in detail in a series of publications related to therapeutic peptide PKPD, and this section seeks to expand upon these concepts as well as provide specific case examples of peptide drugs and the considerations for ADME-related properties [51, 79, 80].

Absorption

As mentioned previously, due to the larger size of peptides relative to traditional small molecules, their absorption is a major challenge when it comes to enabling sufficient bioavailability and systemic exposure to provide therapeutic effect. This is typically reflected in their poor permeability, and sometimes limited aqueous solubility depending on the chemical properties related to ionization (e.g. pKa, salt forms), polarity, and intramolecular interactions (e.g. number of hydrogen bonds, nonpolar groups such as *N*-methylated amino acids), as well as overall hydrodynamic radius and potential for aggregation (e.g. influence of conjugates such as PEG). In particular, for orally administered peptides, these challenges become exacerbated with the presence of hydrolytic enzymes and harsh pH in various compartments of the gastrointestinal tract, which has resulted in poor oral bioavailability (less than 1–2%). Methods to increase oral bioavailability include cyclization [81, 82], coformulation with permeation enhancers (such as SNAC or sodium caprate) [83–85], as well as coadministration with protease inhibitors [86]. More recently, the oral GLP-1 receptor agonist semaglutide was developed for the treatment of Type 2 Diabetes, which was formulated with various doses of SNAC, which resulted in higher exposure when compared to semaglutide without SNAC [87, 88]. However, an increase in the amount of co-administered SNAC did not always result in increased semaglutide exposure suggesting there are limitations for PE coformulation. More recently, these PEs will play a role moving forward for other peptides aiming for oral administration, especially for those that are currently in development for the treatment of obesity and metabolic disorders. For many peptide drugs that are administered via extravascular routes of administration (such as SC and IM), depending on their size, they may be absorbed either into the bloodstream or the lymphatic circulation, or a combination of the two pathways [51]. Depending on the rate-limiting parameter governing the peptide's PK concentration-time profile (i.e. absorption versus elimination rate constant), these molecules can undergo "flip-flop" kinetics where their slowest rate is absorption mediated relative to elimination, especially for peptides undergoing extensive lymphatic absorption (typically larger than 16 kDa), where lymphatic flow rate

Table 6.2 Pharmacokinetic and pharmacokinetic–pharmacodynamic characteristics for select approved therapeutic peptides.

Molecule (reference)	Class	Modification[a)]	Route	MW (kDa)	V_d (l)	$t_{1/2}$ (h)	PK model[b)]	PKPD model[c)]
Bremelanotide [90]	MCR agonist	2	SC	1.03	25.0	2.7	2	ND
Cetrorelix [87]	GnRH antagonist	1	SC	1.43	70 (V_{ss})	12.3	2 or 3	1
Cyclosporine [88, 89]	Immunosuppressant	1	Oral, IV	1.2	63–340	19 (variable)	1	ND
Degarelix [94]	GnRH antagonist	1	SC	1.69	11.4 (V1) 55.7 (V_{ss})	1,272	2 or 3	1
Desmopressin [95]	Synthetic vasopressin	2	IV, IM, SC, nasal, oral	1.07	26.8 (V_{ss})	2	2	1
Exenatide [96]	GLP-1 analogue	1	SC	0.4	28.3	2.4	2	2
Ipamorelin [97]	Growth hormone secretagogue	1	IV	0.71	7.14 (V1) 15.5 (V_{ss})	2	2	1
Leuprorelin [98]	GnRH agonist	1	IM, IV	1.2	27–37	3	ND	ND
Liraglutide [99]	GLP-1 analogue	1	SC	3.75	11.2	13	1	3
Nafarelin [100]	GnRH agonist	1	Nasal	1.32		2.5	1	2
Peginesatide [101]	Erythropoiesis receptor agonist	2, 3	SC, IV	45	3.5	55.3	4	4
Pegcetacoplan [102, 103]	Dual complement C3 and C3b antagonist	2, 3	SC, IVT (ocular)	43	4.0	206	1	2
Romiplostim [104]	TPO analogue	4	SC	60	4.8 (V1)	120–160	2	2, 4
Semaglutide [105, 106]	GLP-1 analogue	1	SC, PO	4.11	7.7	168	2	1
Tesamorelin [107]	Synthetic GHRH	5	SC	5.1	658	0.43–0.63	ND	ND
Tirzeptatide [108]	GIP analogue (GIP/GLP-1 receptor agonist)	1	SC	4.81	10.3	120	2	1
Zilucoplan [109]	Complement C5 antagonist	1, 2	SC	3.56	3.51	172	2	ND

IM, intramuscular; IV, intravenous; GHRH, growth hormone-releasing hormone; GIP, gastric inhibitory polypeptide; GLP, glucagon-like peptide; GnRH, gonadotropin-releasing hormone; MCR, melanocortin receptor; MW, molecular weight; ND, not described; PK, pharmacokinetic; PK–PD, pharmacokinetic–pharmacodynamic; $t_{1/2}$, half-life; TMDD, target-mediated drug disposition; TPO, thrombopoietin; SC, subcutaneous; V1, volume of distribution of central compartment; V_d, volume of distribution; V_{ss} volume of distribution at steady state. The average bodyweight of humans is assumed to be 70 kg where data was reported as l/kg in the literature.

a) Modification: 1 = Amino acid substitution; 2 = Cyclic structure; 3 = PEGylation; 4 = 2 TPO mimetic peptides coupled at each IgG heavy chain; 5 = Terminus modification.

b) PK model: with number of compartments: 1 = one compartment; 2. = two compartment; 3 = three compartment; 4 = One-compartment model with parallel M-M and linear elimination.

c) PKPD models: 1 = Indirect response model; 2 = TMDD; 3 = Glucose-insulin minimal model (indirect with feedback); 4 = Lifespan model with negative feedback regulation.

Source: Table adapted and updated from Diao and Meibohm [51].

and interstitial convective transport are much slower than blood flow and passive diffusion [79, 89]. In contrast, peptide therapeutics administered IV can overcome these barrier challenges by providing high maximum concentrations due to immediate systemic availability. The route of administration and challenges of absorption vary depending on the target localization, desired PK profile, as well as the overall therapeutic strategy and intentions for further development in the clinic.

Distribution

Similar to the challenges presented with the extravascular absorption of peptide therapeutics, there are many complexities involved in the distribution of peptides throughout the body, such as interstitial fluid convection, lymphatic circulation, bulk tissue fluid flow, as well as potential proteins responsible for peptide uptake, transcytosis, and secretion in various organs. The major pathways governing a peptides distribution depend on a variety of factors such as size, ionization state, lipophilicity, as well as propensity to bind to major circulating proteins such as albumin, AAG, or other lipoproteins which can effectively regulate tissue penetration and circulating half-life. Given the bRo5 properties that are characteristic of peptide drugs, the steady-state volume of distribution (V_{ss}) of these molecules is often small and not much larger than the central compartment volume representing plasma and interstitial volume (in humans ~3–8 l) [79].

Metabolism and Excretion

Clearance prediction based on *in vitro* may underpredict *in vivo* CL. In some cases, whole blood, plasma, or serum stability of therapeutically administered peptides may not accurately capture the PK *in vivo* due to disconnects for *in vitro* to *in vivo* extrapolation (also see Section 4).

Acknowledgment

We extend our gratitude to Karen Samy, Shuai Wang, and Matthew Wright for their invaluable contributions to the editing of this book chapter. Their meticulous attention to detail and insightful feedback have greatly enhanced the quality of our work.

References

1 Martinovich, V.P. and Baradzina, K.U. (2022). Peptide hormones in medicine: a 100-year history. *Russ. J. Bioorg. Chem.* 48: 221–232.

2 FDA Guidance for Industry. ANDAs for Certain Highly Purified Synthetic Peptide Drug Products That Refer to Listed Drugs of rDNA Origin. 2021 "FDA considers any alpha amino acid polymer composed of 40 or fewer amino acids to be a peptide, not a protein."

3 Doak, B.C., Zheng, J., Dobritzsch, D., and Kihlberg, J. (2016). How beyond rule of 5 drugs and clinical candidates bind to their targets. *J. Med. Chem.* 59 (6): 2312–2327.

4 Dougherty, P.G., Qian, Z., and Pei, D. (2017). Macrocycles as protein-protein interaction inhibitors. *Biochem. J.* 474 (7): 1109–1125.

5 Wang, L., Wang, N., Zhang, W. et al. (2022). Therapeutic peptides: current applications and future directions. *Signal Transduct. Target. Ther.* 7 (1): 48.

6 You, S., McIntyre, G., and Passioura, T. (2024). The coming of age of cyclic peptide drugs: an update on discovery technologies. *Expert Opin. Drug Discov.* 1-13.

7 Qvit, N. and Rubin, S.J.S. (2018). Cyclic peptides for protein-protein interaction targets. *Curr. Top. Med. Chem.* 18 (7): 525.

8 Walport, L.J., Obexer, R., and Suga, H. (2017). Strategies for transitioning macrocyclic peptides to cell-permeable drug leads. *Curr. Opin. Biotechnol.* 48: 242–250.

9 Ahil, N., Ganesh, A.N., Heusser, C. et al. (2021). Patient-centric design for peptide delivery: trends in routes of administration and advancement in drug delivery technologies. *Med. Drug Discov.* 9: 100079.

10 Tyagi, P., Pechenov, S., and Anand, S.J. (2018). Oral peptide delivery: translational challenges due to physiological effects. *J. Control Release* 287: 167–176.

11 Bolleddula, J., Brady, K., Bruin, G. et al. (2022). Absorption, distribution, metabolism, and excretion of therapeutic proteins: current industry practices and future perspectives. *Drug Metab. Dispos.* 50 (6): 837–845.

12 Barman, P., Joshi, S., Sharma, S. et al. (2023). Strategic approaches to improvise peptide drugs as next generation therapeutics. *Int. J. Pept. Res. Ther.* 29 (4): 61.

13 Martinez, M.N. and Amidon, G.L. (2002). A mechanistic approach to understanding the factors affecting drug absorption: a review of fundamentals. *J. Clin. Pharmacol.* 42: 620–643.

14 Lin, B. and Pease, J.H. (2016). A high throughput solubility assay for drug discovery using microscale shake-flask and rapid UHPLC-UV-CLND quantification. *J. Pharm. Biomed. Anal.* 122: 126.

15 Schilling, K. and Holzgrabe, U. (2020). Recent applications of the charged aerosol detector for liquid chromatography in drug quality control. *J. Chromatography A.* 1619: 460911.

16 Serajuddin, A.T.M. (2007). Salt formation to improve drug solubility. *Adv. Drug Deliv. Rev.* 59 (7): 603–616.

17 Zapadka, K.L., Becher, F.J., Gomes dos Santos, A.L., and Jackson, S.E. (2017). Factors affecting the physical stability (aggregation) of peptide therapeutics. *Interface Focus* 7: 20170030.

18 Galia, E., Nicolaides, E., Hörter, D. et al. (1998). Evaluation of various dissolution media for predicting *in vivo* performance of class I and II drugs. *Pharm. Res.* 15: 698–705.

19 Dening, T.J., Douglas, J.T., and Hageman, M.J. (2021). Do macrocyclic peptide drugs interact with bile salts under simulated gastrointestinal conditions? *Mol. Pharm.* 18 (8): 3086–3098.

20 Win, Z.-M., Cheong, A.M.Y., and Hopkins, W.S. (2023). Using machine learning to predict partition coefficient (logP) and distribution coefficient (logD) with molecular descriptors and liquid chromatography retention time. *J. Chem. Inf. Model.* 63 (7): 1906–1913.

21 Wu, N.P., Wang, W., Gadiagellan, D. et al. (2024). An automated robotic interface for assays: facilitating machine learning in drug discovery by the automation of physicochemical property assays. *ACS Omega* 9 (23): 24948–24958.

22 Ermondi, G., Vallaro, M., Goetz, G. et al. (2019). Experimental lipophilicity for beyond rule of 5 compounds. *Future Drug Discov.* 1 (1).

23 Katz, D., Fike, K., Longenberger, J. et al. (2022). AlphaLogD determination: an optimized reversed-phase liquid chromatography method to measure lipophilicity on neutral and basic small and beyond-rule-of-five compounds. *J. Chromatogr. A* 1674: 463146.

24 Poongavanam, V., Wieske, L.H.E., Peintner, S. et al. (2024). Molecular chameleons in drug discovery. *Nat. Rev. Chem.* 8: 45–60.

25 Naylor, M.R., Ly, A.M., Handford, M.J. et al. (2018). Lipophilic permeability efficiency reconciles the opposing roles of lipophilicity in membrane permeability and aqueous solubility. *J. Med. Chem.* 61 (24): 11169–11182.

26 Shalaeva, M., Caron, G., Abramov, Y.A. et al. (2013). Integrating intramolecular hydrogen bonding (IMHB) considerations in drug discovery using ΔlogP as a tool. *J. Med. Chem.* 56 (12): 4870–4879.

27 Goetz, G.H., Philippe, L., and Shapiro, M.J. (2014). EPSA: a novel supercritical fluid chromatography technique enabling the design of permeable cyclic peptides. *ACS Med. Chem. Lett.* 5 (10): 1167–1172.

28 Valko, K.L., Rava, S., Bunally, S., and Anderson, S. (2020). Revisiting the application of immobilized artificial membrane (IAM) chromatography to estimate *in vivo* distribution properties of drug discovery compounds based on the model of marketed drugs. *ADMET and DMPK* 8 (1): 78–97.

29 Ermondi, G., Vallaro, M., Goetz, G. et al. (2020). Updating the portfolio of physicochemical descriptors related to permeability in the beyond the rule of 5 chemical space. *Eur. J. Pharm. Sci.* 15 (146): 105274.

30 Garcia Jimenez, D., Vallaro, M., Rossi Sebastiano, M. et al. (2023). Chamelogk: a chromatographic chameleonicity quantifier to design orally bioavailable beyond-rule-of-5 drugs. *J. Med. Chem.* 66 (15): 10681–10693.

31 Benet, L.Z. (2010). Predicting drug disposition via application of a biopharmaceutics drug disposition classification system. *Basic Clin. Pharmacol. Toxicol.* 106 (3): 162–167.

32 Guo, Y., Ji, S., Wang, W. et al. (2022). An integrated strategy for assessing the metabolic stability and biotransformation of macrocyclic peptides in drug discovery toward oral delivery. *Anal. Chem.* 94 (4): 2032–2041.

33 Khojasteh, S.C., Wong, H., Zhang, D., and Hop, C.E.C.A. (2022). *Discovery DMPK quick guide: guide to data interpretation and integration*, 1e. 2022 edition. Springer.

34 Peraro, L., Deprey, K.L., Moser, M.K. et al. (2018). Cell penetration profiling using the chloroalkane penetration assay. *J. Am. Chem. Soc.* 140 (36): 11360–11369.

35 Peier, A., Ge, L., Boyer, N. et al. (2021). NanoClick: a high throughput, target-agnostic peptide cell permeability assay. *ACS Chem. Biol.* 16 (2): 293–309.

36 Liu, Z., Wang, C., Liu, Q. et al. (2011). Uptake, transport and regulation of JBP485 by PEPT1 *in vitro* and *in vivo*. *Peptides* 32 (4): 747–754.

37 Saha, P. and Kou, J.H. (2002). Effect of bovine serum albumin on drug permeability estimation across Caco-2 monolayers. *Eur. J. Pharm. Biopharm.* 54 (3): 319–324.

38 Di, L., Whitney-Pickett, C., Umland, J.P. et al. (2011). Development of a new permeability assay using low-efflux MDCKII cells. *J. Pharm. Sci.* 100 (11): 4974–4985.

39 Chen, E.C., Broccatelli, F., Plise, E. et al. (2018). Evaluating the utility of canine Mdr1 knockout Madin-Darby canine kidney I cells in permeability screening and efflux substrate determination. *Mol. Pharm.* 15 (11): 5103–5113.

40 Cui, Y., Desevaux, C., Truebenbach, I. et al. (2021). A bidirectional permeability assay for beyond rule of 5 compounds. *Pharmaceutics* 13 (8): 1146.

41 Muschong, P., Awwad, K., Price, E. et al. (2024). Conquering the beyond rule of five space with an optimized high-throughput Caco-2 assay to close gaps in absorption prediction. *Pharmaceutics* 16 (7): 846.

42 Ohta, A., Tanada, M., Shinohara, S. et al. (2023). Validation of a new methodology to create oral drugs beyond the rule of 5 for intracellular tough targets. *J. Am. Chem. Soc.* 145 (44): 24035–24051.

43 Mateus, A., Matsson, P., and Artursson, P. (2013). Rapid measurement of intracellular unbound drug concentrations. *Mol. Pharm.* 10 (6): 2467–2478.

44 Treyer, A., Walday, S., Boriss, H. et al. (2019). A cell-free approach based on phospholipid characterization for determination of the cell specific unbound drug fraction (fu,cell). *Pharm. Res.* 36 (12): 178.

45 Szewczyk-Roszczenko, O.K., Roszczenko, P., Shmakova, A. et al. (2023). The chemical inhibitors of endocytosis:

from mechanisms to potential clinical applications. *Cells* 12 (18): 2312.
46 Wang, C.Y., Liu, S., Xie, X.N., and Tan, Z.R. (2017). Regulation profile of the intestinal peptide transporter 1 (PepT1). *Drug Des. Devel. Ther.* 11: 3511–3517.
47 Bahar, F.G., Ohura, K., Ogihara, T., and Imai, T. (2012). Species difference of esterase expression and hydrolase activity in plasma. *J. Pharm. Sci.* 101 (10): 3979–3988.
48 Renukuntla, J., Vadlapudi, A.D., Patel, A. et al. (2013). Approaches for enhancing oral bioavailability of peptides and proteins. *Int. J. Pharm.* 447 (1–2): 75–93.
49 Yao, J.F., Yang, H., Zhao, Y.Z., and Xue, M. (2018). Metabolism of peptide drugs and strategies to improve their metabolic stability. *Curr. Drug Metab.* 19 (11): 892–901.
50 Annisa, V., Sulaiman, T.N.S., Nugroho, A.K., and Nugroho, A.E. (2022). Characterization of alginate with natural polymers combination for drug. *Iraqi J Pharm Sci* 31 (2): 150–159.
51 Diao, L. and Meibohm, B. (2013). Pharmacokinetics and pharmacokinetic-pharmacodynamic correlations of therapeutic peptides. *Clin. Pharmacokinet.* 52 (10): 855–868.
52 Kronbach, T., Fischer, V., and Meyer, U.A. (1988). Cyclosporine metabolism in human liver: identification of a cytochrome P-450III gene family as the major cyclosporine-metabolizing enzyme explains interactions of cyclosporine with other drugs. *Clin. Pharmacol. Ther.* 43 (6): 630–635.
53 Hammid, A., Fallon, J.K., Lassila, T. et al. (2022). Activity and expression of carboxylesterases and arylacetamide deacetylase in human ocular tissues. *Drug Metab. Dispos.* 50 (12): 1483–1492.
54 Esposito, S., Vanni, D., Menta, S. et al. (2020). Comparison of different protein precipitation and solid-phase extraction protocols for the study of the catabolism of peptide drugs by LC-HRMS. *J. Pept. Sci.* 26: e3272.
55 Drouin, N., Kubáň, P., Rudaz, S. et al. (2019). Electromembrane extraction: overview of the last decade. *Trends Anal. Chem.* 113: 357–363.
56 Vasicek, L.A., Zhu, X., Spellman, D.S., and Bateman, K.P. (2019). Direct quantitation of therapeutic antibodies for pharmacokinetic studies using Immuno-purification and intact mass analysis. *Bioanalysis* 11 (3): 203–213.
57 Ren, T. et al. (2019). Micro-solid phase extraction and LC–MS3 for the determination of triptorelin in rat plasma and application to a pharmacokinetic study. *J. Pharm. Biomed. Anal.* 166: 13–19.
58 Eie, L.V., Rye, T.K., Hansen, F. et al. (2021). Electromembrane extraction of peptides and amino acids–status and perspectives. *Bioanalysis* 13 (4): 277–289.
59 Roy, A.L., Wilder, E.L., and Anderson, J.L. (2021). Validation of antibodies: lessons learned from the common fund protein capture reagents. *Sci. Adv.* 7 (46): eabl7148.
60 Fourie-O'Donohue, A., Chu, P.Y., Dela Cruz Chuh, J. et al. (2020). Improved translation of stability for conjugated antibodies using an *in vitro* whole blood assay. *MAbs* 12 (1): e1715705.
61 Kaur, S., Bateman, K.P., Glick, J. et al. (2020). IQ consortium perspective: complementary LBA and LC-MS in protein therapeutics bioanalysis and biotransformation assessment. *Bioanalysis* 12 (4): 257–270.
62 Yue, Q., Mulder, T., Rudewicz, P.J. et al. (2013). Evaluation of metabolism and disposition of GDC-0152 in rats using 14C labeling strategy at two different positions: a novel formation of hippuric acid from 4-phenyl-5-amino-1,2,3-thiadiazole. *Drug Metab. Dispos.* 41 (2): 508–517.
63 Wang, M., Defranco, D., Wright, K. et al. (2012). Decreased subcutaneous bioavailability of an oxyntomodulin analog in a controlled release formulation could be caused by skin metabolism in rats. *J. Bioequiv. Bioavail.* 4 (6): 69–77.
64 Qin, X., Wang, Y., Ye, Q. et al. (2024). CYP3A mediates an unusual C(sp2)-C(sp3) bond cleavage via ipso-addition of oxygen in drug metabolism. *Angew. Chem. Int. Ed. Engl.* 63 (23): e202405197.
65 Zhou, J., Qin, X., Zhou, S. et al. (2024). CYP3A-me-diates carbon-carbon bond cleavages in drug metabolism. *Biomolecules* 11: x.
66 Jyrkäs, J. and Tolonen, A. (2021). Hepatic *in vitro* metabolism of peptides; comparison of human liver S9, hepatocytes and Upcyte hepatocytes with cyclosporine a, leuprorelin, desmopressin and cetrorelix as model compounds. *J. Pharm. Biomed. Anal.* 196: 113921.
67 Chu, P.Y., Wang, H., Ross, E. et al. (2023). Translatability of *in vitro* stress for predicting deamidation and oxidation biotransformation on biotherapeutics. *Anal. Chem.* 95 (49): 17957–17961.
68 Gao, Y., Joshi, M., Zhao, Z., and Mitragotri, S. (2023). PEGylated therapeutics in the clinic. *Bioeng Transl Med* 9 (1): e10600.
69 Bak, A., Leung, D., Barrett, S.E. et al. (2014). Physicochemical and formulation developability assessment for therapeutic peptide delivery – a primer. *AAPS J.* 17 (1): 144–155.
70 Di, L. (2014). Strategic approaches to optimizing peptide ADME properties. *AAPS J.* 17 (1): 134–143.
71 Lee, A.C., Harris, J.L., Khanna, K.K., and Hong, J.-H. (2019). A comprehensive review on current advances

in peptide drug development and design. *Int. J. Mol. Sci.* 20 (10): 2383–2401.

72 Oeller, M., Kang, R.J.D., Bolt, H.L. et al. (2023). Sequence-based prediction of the intrinsic solubility of peptides containing non-natural amino acids. *Nature Comm* 14: 7475.

73 Kramer, R.M., Shende, V.R., Motl, N. et al. (2012). Toward a molecular understanding of protein solubility: increased negative surface charge correlates with increased solubility. *Biophys. J.* 102 (8): 1907–1915.

74 Kang, B., Tang, H., Zhao, Z., and Song, S. (2020). Hofmeister series: insights of ion specificity from amphiphilic assembly and interface property. *ACS Omega.* 5 (12): 6229–6239.

75 Nugrahadi, P.P., Hinrichs, W.L.J., Frijlink, H.W. et al. (2023). Designing formulation strategies for enhanced stability of therapeutic peptides in aqueous solutions: a review. *Pharmaceutics* 15 (3): 935.

76 Bolder, S.G., Sagis, L.M., Venema, P., and van der Linden, E. (2007). Thioflavin T and birefringence assays to deter-mine the conversion of proteins into fibrils. *Langmuir* 23 (8): 4144–4147.

77 Stroppel, L., Schultz-Fademrecht, T., Cebulla, M. et al. (2023). Antimicrobial preservatives for protein and peptide formulations: an overview. *Pharmaceutics* 15 (2): 563.

78 Nakamura, K., Kiminami, H., Yamashita, A. et al. (2015). Assessment of the effects of sterilization methods on protein drug stability by elucidating decomposition mechanism and material analysis. *Int. J. Pharm.* 484 (1–2): 51–56.

79 Meibohm, B. (2019). *Pharmaceutical Biotechnology, Fundamentals and Applications*, 5e, 105–137. Dokumen Pub.

80 Tang, L. and Meibohm, B. (2006). *Pharmacokinetics and Pharmacodynamics of Biotech Drugs*, 15–43. Wiley.

81 Merz, M.L., Habeshian, S., Li, B. et al. (2023). De novo development of small cyclic peptides that are orally bioavailable. *Nat. Chem. Biol.* 20 (5): 624–633.

82 Nielsen, D.S., Shepherd, N.E., Xu, W. et al. (2017). Orally absorbed cyclic peptides. *Chem. Rev.* 117 (12): 8094–8128.

83 Kim, D., Jin, L., Park, E.J., and Na, D.H. (2023). Peptide permeation enhancers for improving oral bioavailability of macromolecules. *J. Pharm. Investig.* 53 (1): 59–72.

84 Maher, S., Mrsny, R.J., and Brayden, D.J. (2016). Intestinal permeation enhancers for oral peptide delivery. *Adv. Drug Deliv. Rev.* 106 (Adv. Drug Deliv. Rev. 8 1992): 277–319.

85 Twarog, C., Fattah, S., Heade, J. et al. (2019). Intestinal permeation enhancers for oral delivery of macromolecules: a comparison between salcaprozate sodium (SNAC) and sodium caprate (C10). *Pharmaceutics* 11 (2): 78.

86 Lee, V.H.L. (1990). Protease inhibitors and penetration enhancers as approaches to modify peptide absorption. *J. Control. Release* 13 (2–3): 213–223.

87 Granhall, C., Donsmark, M., Blicher, T.M. et al. (2019). Safety and pharmacokinetics of single and multiple ascending doses of the novel oral human GLP-1 analogue, oral semaglutide, in healthy subjects and subjects with type 2 diabetes. *Clin. Pharmacokinet.* 58 (6): 781–791.

88 Rasmussen, M.F. (2020). The development of oral semaglutide, an oral GLP-1 analog, for the treatment of type 2 diabetes. *Diabetol. Int.* 11 (2): 76–86.

89 Supersaxo, A., Hein, W.R., and Steffen, H. (1990). Effect of molecular weight on the lymphatic absorption of water-soluble compounds following subcutaneous administration. *Pharm. Res.* 7 (2): 167–169.

90 Clayton, A.H., Althof, S.E., Kingsberg, S. et al. (2016). Bremelanotide for female sexual dysfunctions in premenopausal women: a randomized, placebo-controlled dose-finding trial. *Womens Health (Lond).* 12 (3): 325–337.

91 Nagaraja, N.V., Pechstein, B., Erb, K. et al. (2000). Pharmacokinetic and pharmacodynamic modeling of cetrorelix, an LH-RH antagonist, after subcutaneous administration in healthy premenopausal women. *Clin. Pharmacol. Ther.* 68 (6): 617–625.

92 Ptachcinski, R.J., Venkataramanan, R., and Burckart, G.J. (1986). Clinical pharmacokinetics of cyclosporin. *Clin. Pharmacokinet.* 11 (2): 107–132.

93 Yates, C.R., Zhang, W., Song, P. et al. (2003). The effect of CYP3A5 and MDR1 polymorphic expression on cyclosporine oral disposition in renal transplant patients. *J. Clin. Pharmacol.* 43 (6): 555–564.

94 Jadhav, P.R., Agersø, H., Tornøe, C.W., and Gobburu, J.V.S. (2006). Semi-mechanistic pharmacodynamic modeling for degarelix, a novel gonadotropin releasing hormone (GnRH) blocker. *J. Pharmacokinet. Pharmacodyn.* 33 (5): 609–634.

95 Argenti, D., Ireland, D., and Heald, D.L. (2001). A pharmacokinetic and pharmacodynamic comparison of desmopressin administered as whole, chewed and crushed tablets, and as an oral solution. *J. Urol.* 165 (5): 1446–1451.

96 Gao, M., Jin, Y., Tong, Y. et al. (2012). A site-specific PEGylated analog of exendin-4 with improved pharmacokinetics and pharmacodynamics *in vivo*. *J. Pharm. Pharmacol.* 64 (11): 1646–1653.

97 Gobburu, J.V.S., Agersø, H., Jusko, W.J., and Ynddal, L. (1999). Pharmacokinetic-pharmacodynamic modeling

of ipamorelin, a growth hormone releasing peptide, in human volunteers. *Pharm. Res.* 16 (9): 1412–1416.

98 Periti, P., Mazzei, T., and Mini, E. (2002). Clinical pharmacokinetics of depot leuprorelin. *Clin. Pharmacokinet.* 41 (7): 485–504.

99 Agersø, H. and Vicini, P. (2003). Pharmacodynamics of NN2211, a novel long acting GLP-1 derivative. *Eur. J. Pharm. Sci.* 19 (2–3): 141–150.

100 Röblitz, S., Stötzel, C., Deuflhard, P. et al. (2013). A mathematical model of the human menstrual cycle for the administration of GnRH analogues. *J. Theor. Biol.* 321: 8–27.

101 Woo, S., Krzyzanski, W., Duliege, A. et al. (2008). Population pharmacokinetics and pharmacodynamics of peptidic erythropoiesis receptor agonist (ERA) in healthy volunteers. *J. Clin. Pharmacol.* 48 (1): 43–52.

102 Ping, H., Crass, R., Chapel, S., and Ajayi, T. (2023). PB2053: population pharmacokinetics of pegcetacoplan in healthy subjects and patients with paroxysmal nocturnal hemoglobinuria. *Hemasphere* 7 (Suppl): e0771683.

103 de Latour, R.P., Szer, J., Weitz, I.C. et al. (2022). Pegcetacoplan versus eculizumab in patients with paroxysmal nocturnal haemoglobinuria (PEGASUS): 48-week follow-up of a randomised, open-label, phase 3, active-comparator, controlled trial. *Lancet Haematol.* 9 (9): E648–E659.

104 Wang, Y.-M.C., Krzyzanski, W., Doshi, S. et al. (2010). Pharmacodynamics-mediated drug disposition (PDMDD) and precursor pool lifespan model for single dose of romiplostim in healthy subjects. *AAPS J.* 12 (4): 729–740.

105 Overgaard, R.V., Delff, P.H., Petri, K.C.C. et al. (2019). Population pharmacokinetics of semaglutide for type 2 diabetes. *Diabetes Ther.* 10 (2): 649–662.

106 Tham, L.S., Pantalone, K.M., Dungan, K. et al. A model-based simulation of glycaemic control and body weight when switching from semaglutide to 3.0- and 4.5-mg doses of once-weekly dulaglutide. *Diabetes Obes. Metab.* 24 (2): 302–311.

107 Stanley, T.L., Chen, C.Y., Branch, K.L. et al. (2011). Effects of a growth hormone-releasing hormone analog on endogenous gh pulsatility and insulin sensitivity in healthy men. *J. Clin. Endocrinol. Metab.* 96 (1): 150–158.

108 Schneck, K. and Urva, S. (2024). Population pharmacokinetics of the GIP/GLP receptor agonist tirzepatide. *CPT Pharmacometrics Syst. Pharmacol.* 13 (3): 494–503.

109 Howard, J.F. Jr., Bresch, S., Genge, A. et al. (2023). RAISE study team. Safety and efficacy of zilucoplan in patients with generalised myasthenia gravis (RAISE): a randomised, double-blind, placebo-controlled, phase 3 study. *Lancet Neurol.* 22 (5): 395–406.

7

Structural Characterization of Biotherapeutics and New Modalities

Paola Di Lello and Patrick Lupardus

Department of Structural Biology, Genentech, Inc. South San Francisco, CA, USA

Introduction

Historically, drug discovery has predominantly focused on proteins, with efforts centered on identifying and optimizing antibodies or small molecule therapeutics that bind to and modulate the function of key proteins within disease-driving biological pathways. Within this framework, biophysical techniques such as electron microscopy (EM), nuclear magnetic resonance (NMR) spectroscopy, and X-ray crystallography have been indispensable, enabling structure-based drug discovery (SBDD). These methods have been employed across various stages of the early drug discovery process, from target identification, validation, and characterization to epitope mapping and lead optimization.

However, over the past two decades, our understanding of disease biology has significantly advanced, revealing that in certain conditions, targeting crucial proteins modulating dysregulated pathways may not always be the most effective therapeutic strategy. This realization has broadened the spectrum of pharmacological targets, now encompassing not only proteins but also nucleic acids (RNA, DNA) and lipids/membranes. Consequently, the range of therapeutic modalities has expanded beyond small molecules and antibodies to include RNAs (such as aptamers, siRNAs, and antisense oligonucleotides), macrocyclic peptides, DNA (for gene therapy), and cell therapy.

In this evolving and complex landscape, NMR, X-ray crystallography, and EM remain fundamental techniques that guide the drug discovery and development process. Defining the target epitope of a candidate clinical antibody is a critical aspect of validating and characterizing a promising therapeutic antibody. Epitope mapping can elucidate the mechanism of action, decode species cross-reactivity, guide affinity maturation efforts, and protect intellectual property generated during the antibody discovery process. Epitope mapping studies can range from simple competition binding experiments to more sophisticated methods such as alanine scanning mutagenesis, proteolytic mapping, or mass spectrometry-based techniques like hydrogen–deuterium exchange and radiolytic footprinting, which can quickly provide useful epitope data. However, the inherent low resolution of these techniques does not offer a true molecular picture of how the antibody binds its target.

Higher-resolution structural methods such as NMR spectroscopy or X-ray crystallography can provide detailed information about the exact amino acids involved in the interaction, which is crucial for improving the properties of the candidate antibody or understanding how a molecule achieves *in vivo* activity. EM can also provide lower-resolution models of antibody binding quickly and with smaller sample requirements than crystallography or NMR.

In this chapter, we provide an overview of how these three structural methods are applied for epitope mapping and structural characterization of biotherapeutics and new modalities. Additionally, we discuss techniques to express and purify antibody/antigen complexes quickly and efficiently and offer tips for crystallization, structure determination, and NMR analysis. After reading this chapter, one will have the tools and understanding to begin deciphering their chosen epitope at an atomic level.

Antigens, Epitopes, and Paratopes

Given the complicated terminology of the antibody discovery and characterization process, it is important to first clearly define a few terms that describe antibody/target interactions. The first term is *antigen*, which is defined as the target molecule that a given antibody recognizes. In most cases, the antigen is a purified protein used to stimulate an antibody response from, or immunize, a lab

Characterizing Biotherapeutics: Analytical Methods for Diverse Modalities, First Edition. Edited by Jennie R. Lill and Wendy Sandoval.

animal such as a mouse or rabbit. Importantly, antibodies can also be raised against peptides, polysaccharides, or small-molecule haptens. For the purposes of this discussion, we will restrict the definition of an antigen to a peptide or protein. The second term that requires definition is *epitope,* which is the specific part of the antigen that the antibody CDR (complementarity-determining region) loops interact with. Epitopes can be either linear as in an unfolded peptide or three-dimensional and conformationally specific as in the surface of a folded protein. Linear epitopes are generally continuous – meaning that the entire epitope consists of several amino acids in sequence. Three-dimensional epitopes are generally discontinuous – in that amino acids within the epitope can be far apart in sequence order but are arranged closely in the context of the tertiary structure of a folded protein. This is an important concept because most therapeutic antibodies target native, folded protein, and therefore linear epitopes are not preferred. The third term needing definition is *paratope*, or the antibody amino acids that contact the epitope on the antigen. Paratope residues are almost always confined to the hypervariable CDR regions of an antibody, although it is often the case that not all the antibody CDRs will contact the antigen.

Rationale for Structural Characterization of Epitopes

There are a number of reasons a researcher might want to structurally determine an epitope during the antibody development process. One reason is to understand the molecular mechanism by which the antibody has its effect on the target, whether that is inhibition or activation of the target signaling pathway. Yet there are other important reasons to characterize an epitope, including understanding cross-reactivity, to inform engineering of antibody CDRs for affinity maturation, or to understand issues such as CDR cleavage, oxidation, or other modifications that are detrimental to the manufacturing of a therapeutic. In addition, epitope characterization can be extremely important for protection of the intellectual property generated during the development of antibody therapies. In this section, we describe several published examples of structural studies informing the development of a therapeutic antibody.

A common rationale for structural characterization of a therapeutic candidate antibody with its target is to understand the molecular mechanism by which the antibody achieves *in vivo* activity. A noteworthy example of how epitope discovery can help inform mechanism is the case of the HER2-inactivating therapeutics Herceptin (Trastuzumab) and Perjeta (Pertuzumab). The structures of each of these marketed antibodies bound to the extracellular domain of HER2 have been determined by X-ray crystallography. These data have been extremely informative for understanding the mechanisms by which each antibody works *in vivo*. Both molecules block HER2 signaling, but via different mechanisms, with Trastuzumab primarily blocking ligand-independent receptor dimerization, and Pertuzumab sterically blocking ligand-mediated dimerization [1]. Structural analysis revealed that Trastuzumab binds to the membrane proximal EGF domain of HER2 [2], while Pertuzumab binds the receptor dimerization interface in HER2 domains I/II/III [3] (Figure 7.1). Based on the different mechanisms of action and structural characterization that revealed unique binding epitopes, Trastuzumab and Pertuzumab were developed into a combination therapy with better efficacy than the single-agent therapies alone [4].

Understanding mechanisms of cross-reactivity for a candidate antibody is another important reason to characterize an epitope. An excellent example of how structural analysis can help understand cross-reactivity can be found in the characterization of Influenza A neutralizing antibodies. A class of antibodies has been discovered that neutralizes most clinical strains of Influenza A by binding to the stalk region of the viral surface protein hemagglutinin [5–7] (Figure 7.2b–d). The epitope bound by this subset of antibodies is structurally and chemically conserved (Figure 7.2a), likely due to a role in the pH-induced conformational change in the hemagglutinin protein that

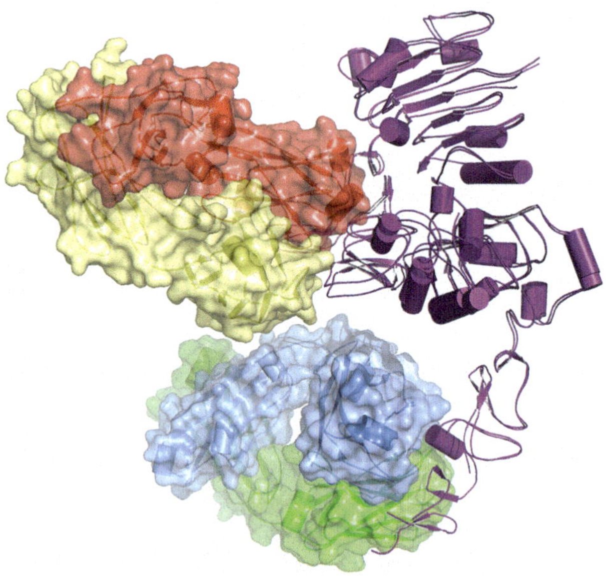

Figure 7.1 Perjeta (red/yellow) and Herceptin (blue/green) bound to the EGF family receptor HER2 (overlaid and colored in purple), illustrating the unique blocking epitopes bound by each of these marketed antibodies. PDB IDs are 1S78 Source: Franklin et al. [3] and 1NZ8 Source: Cho et al. [2].

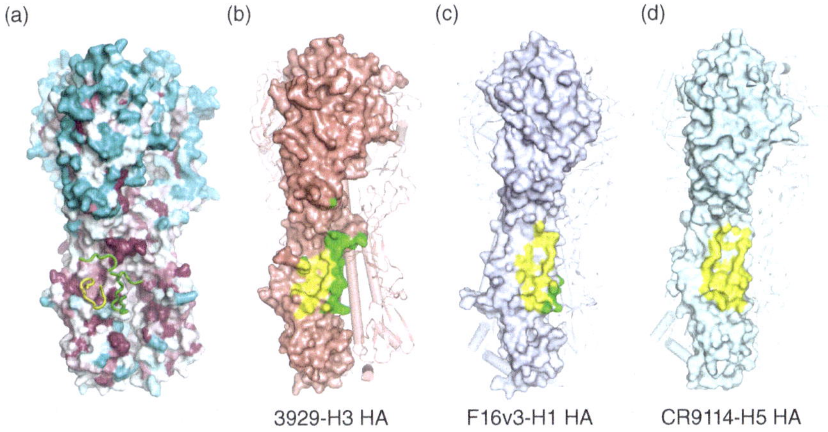

Figure 7.2 A stalk epitope on hemagglutinin (HA) that neutralizes most strains of Influenza A. (a) Consurf analysis showing the conservation of surface residues in an H3 hemagglutinin, displayed on a scale from purple (most conserved) to teal (least conserved). Heavy (yellow) and light (green) chains of the 39.29 antibody Source: Adapted from Nakamura et al. [5] are displayed to show the position of the epitope. (b–d) Epitopes of broadly neutralizing antibodies 39.29 bound to H3 HA Source: Adapted from Nakamura et al. [5] (b), FI6v3 bound to H1 HA Source: Adapted from Corti et al. [6] (c) and CR9114 bound to H5 HA Source: Adapted from Dreyfus et al. [7] (d). Heavy chain epitope is shown as yellow surface and light chain epitope shown as green surface.

occurs upon viral fusion [7]. In this case, structural analysis proposed a mechanism of action for the antibody class, as well as a rationale for cross-reactivity. Validation of this conserved epitope has also provided a target for unique therapeutic scaffolds [8].

Differences in species cross-reactivity are a common problem encountered in the therapeutic antibody discovery process. Practically, it is desirable for a candidate to have cross-reactivity with human, cynomolgus monkey, and mouse at minimum in order to allow for preclinical mouse model utilization as well as pharmacokinetic and toxicology studies in the nonhuman primate cyno species, all with the same antibody. Yet given the sequence diversity between human, nonhuman primates, and mice, it is often difficult to find an antibody with this species cross-reactivity profile, along with all the other desired properties of a therapeutic antibody (pK, stability, solubility, etc.). To help understand differences in species cross-reactivity, a structural approach can be used to identify the epitope and the sequence differences that contribute to differences in binding affinity. Armed with this information, a development team can choose to re-engineer the antibody or pick another candidate for development. An example of structural characterization answering species cross-reactivity questions has been described for VEGF-binding antibodies [9, 10]. This structural study of Avastin and two other human/mouse cross-reactive monoclonal antibodies targeted to VEGF showed that the phage-derived cross-reactive antibodies targeted the receptor binding site on VEGF (Figure 7.3a-c), whereas Avastin bound a nearby epitope that only minimally overlaps with the receptor (Figure 7.3d). The Avastin epitope is poorly conserved between mouse and human, illuminating the reason that Avastin is not cross-reactive with mouse VEGF. Studies such as this can help inform selection of preclinical efficacy and toxicology models, which is critical for efficient development and transition of a molecule into the clinic.

Understanding the structure of a candidate antibody bound to antigen can also be valuable for design of the best possible molecule to take forward into manufacturing and subsequent clinical trials. Once a candidate molecule is decided upon, several "polishing" steps are still needed before the candidate can be manufactured. Antibody affinity, sequence, pharmacokinetics, and effector function properties all may need to be tuned in order to achieve optimal antibody stability and antigen targeting [12]. Importantly, all of these processes can be influenced by an available atomic structure. Knowledge of paratopic residues that contact antigen can accelerate and improve the affinity maturation process [13]. For antibody humanization, a structure-guided approach can be useful, particularly for understanding which human framework to use for grafting CDRs [14, 15]. Crystal structures of antibody Fc interactions with Fc receptors have also revealed ways to engineer and modulate FcRn-mediated recycling and FcγR-mediated effector functions [12].

Lastly, structural approaches can be utilized to improve antibody stability during the manufacturing, formulation, and storage procedures. Antibodies in general are extremely stable *in vitro*, yet can undergo modification during expression and purification steps or over time while in storage. Modifications include peptide cleavage or chemical changes to amino acid sidechains like deamidation,

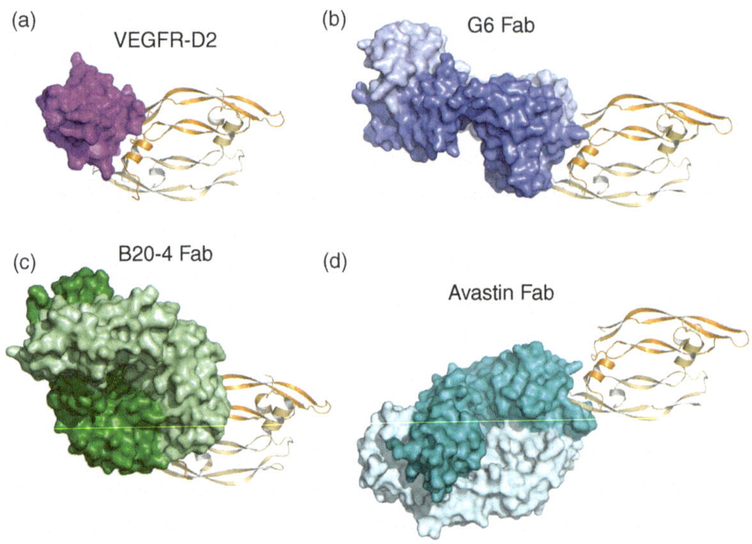

Figure 7.3 Comparison of antibody epitopes on VEGF. Surface representation shown for (a) VEGFR1 (magenta) bound to VEGF (orange) Source: Adapted from Wiesmann et al. [11], (b) the G6 Fab (blue) bound to VEGF Source: Adapted from Fuh et al. [9], (c) the B20–4 Fab (green) bound to VEGF Source: Adapted from Fuh et al. [9], and (d) Avastin Fab bound to VEGF Source: Adapted from Muller et al. [10].

isomerization, and oxidation [16]. These liabilities in antibody sequence can affect both long-term stability and affinity for the target molecule. One example of structural investigation of an antibody modification can be found in a study of Aspartate isomerization in anti-IgE antibodies [17]. A second study utilized structural information to investigate a lysine glycation site near the paratope of a candidate antibody [18]. Structural knowledge can help a development team design changes to the antibody to help mitigate sequence deficiencies.

Choice of Analytical Method for Epitope Mapping

Structural mapping of an antibody epitope has many advantages, but it would be misleading to ignore disadvantages of these methods as well. While an atomic model of the antibody/antigen complex of interest may be nice to have, crystallography, NMR, and EM can be time consuming and expensive endeavors, and sometimes using a structural method is overkill for the question that is to be answered. Before committing resources to a structural effort, it is important to determine whether structural analysis is on the critical path for the development of a molecule. If not, a project may be better served using a simpler method for epitope mapping. If structural analysis is critical, the advent of contract structural biology groups in industry has put these methods within reach of those who do not have the time or budget to set up an in-house structural biology group.

Electron Microscopy for Epitope Analysis

EM has long been an important technique in structural biology, particularly for large protein complexes (i.e. the ribosome) and symmetric protein superstructures such as viral capsids. Technological advances in hardware and software have begun to make EM more accessible to researchers interested in understanding how their antibodies bind to a target protein quickly and with efficient use of sample (reviewed in [19, 20]). A typical single particle EM experiment in negative stain involves placing a dilute sample (ug/ml) onto a carbon grid and staining that grid with a heavy atom such as uranium, which adheres to the carbon grid but leaves the protein unstained (hence the name "negative stain"). EM images of this grid are therefore inverted images in which the electrons transit through the protein but absorbed by the heavy atoms covering the surrounding grid. A number of EM images are taken from a single grid, and individual particles picked and "averaged" in two dimensions by a computer algorithm. These images, called "class averages," can provide a significant amount of information about the epitope of an antibody given a knowledge of what the antigen looks like. Further data collection and analysis can provide a three-dimensional envelope, which can be used for placement of a low-resolution model. For most negative stain EM experiments, a resolution of >15 Å is expected. Yet these data can be collected and analyzed in several days, rather than the weeks to months it might take for crystallography or NMR studies. EM has been used extensively in the study of antiviral antibodies, in particular those against HIV, RSV, and Influenza glycoproteins involved in viral fusion [7, 21, 22].

EM methods have two significant advantages over NMR and crystallography when it comes to epitope determination. The first advantage is that very small samples can yield a structure. Only micrograms of a pure, homogeneous sample are needed, given that the structure is generated from images of individual particles. The second advantage is that EM has the fastest turnaround time. Grid preparation and imaging for negative staining experiments can be done in an afternoon, and data analysis can be completed in a few days. For these reasons, EM can be an extraordinarily powerful way to obtain a quick read on the location of an epitope, in particular for large targets such as viral surface glycoproteins like Influenza hemagglutinin [7]. The major disadvantage of EM is the resolution, which is the lowest among the three methods discussed here. However, detector technology and software have advanced cryo-EM techniques to the point that given the right sample, microscope, and detector, sub-nanometer resolution is achievable [23]. The increased resolution achievable with cryo-EM methods comes at the cost of longer turnaround time. A second disadvantage is the need for the target complex to have a molecular weight greater than 100 kilodaltons (kD), with 200–300 kD being preferred [20]. To mitigate the size limitation, rigid FABs [24, 25] and rigid imaging scaffolds like DARpins (Designed Ankyrin Repeat proteins) [26] have been successfully implemented to increase the overall size of the proteins of interest. These approaches have made possible the determination of high-resolution structures proteins as small as 20–25 kDa at a resolution in the 2.3–3 Å range.

Epitope and Paratope Mapping by NMR

NMR spectroscopy is a biophysical technique that provides information on the structural features and the dynamics properties of a protein in solution. It can also be applied to the study of protein–protein interactions, including the characterization of the binding interface in antibody/antigen complexes. The advantages of using NMR for these types of studies are manifold. First of all, this technique studies proteins in the solution state, thus in an environment where they are most likely to adopt their biologically relevant conformations. Moreover, NMR is quite sensitive to detect weak interactions ($K_d \sim 10^{-6}$–10^{-3} M) that might not be observable by other biophysical techniques. Another advantage of using NMR, especially for epitope mapping, is that one can apply this technique not only to well-folded antigens [27–30] but also to intrinsically disordered proteins (IDP), like in the case of α-synuclein [31], a neuronal protein linked to Parkinson's disease [32], or the Merozoite surface protein 2 (MSP2), an antigen from the malaria parasite *Plasmodium falciparum* [33, 34]

Although NMR is a versatile technique it has limitations, particularly when applied to large molecule-like antibodies. Intact antibodies are not only large in size (~150 kDa) but also quite flexible, two characteristics that make them problematic for NMR studies. Due to their size and dynamic properties, antibodies tumble slowly in solution and this feature gives rise to short relaxation times that, in return, cause significant line broadening of the NMR signals, often beyond detection. This is also observed in the case of antigen/antibody complexes. Formation of a tight complex between a full antibody and a smaller, well-structured antigen will induce a dramatic change in the tumbling rate of the antigen, leading to an overall line broadening of the NMR signals for the antigen, thus making it impossible to distinguish the regions of the antigens that are interacting with the antibody from the regions that are not.

The general strategy to counteract the detrimental effects of short relaxation times, typical of large proteins, is to study fragments rather than full antibodies, to use perdeuterated protein samples [35] and to apply the [^{1}H-^{15}N] TROSY (Transverse Relaxation Optimized SpectroscopY) experiment [36], a variant of the [^{1}H-^{15}N] HSQC (see "Epitope/paratope mapping by chemical shift perturbations"). These approaches, which improve greatly the signal to noise ratio and therefore lead to better sensitivity, have been successfully applied to the structural study of Fv and Fc domains produced in *E. coli* [37], as well as in CHO cells [38]. They also have been shown to be beneficial in mapping the epitope of tight binding antibody fragments to well-folded proteins, like in the case of Canakinumab and Gevokizumab Fabs to interleukin-1β [29]. NMR offers a variety of experiments to map protein-protein binding interfaces [39], however, the most widely applied method is based on chemical shift perturbations (CSPs) mapping using [^{1}H-^{15}N]-HSQC or [^{1}H-^{15}N]-TROSY experiments.

Epitope/Paratope Mapping by Chemical Shift Perturbations (CSPs)

The [^{1}H-^{15}N] HSQC (Heteronuclear Single Quantum Correlation) experiment [40] is one of the key tools in protein NMR. This experiment correlates the resonances of directly bonded ^{1}H and ^{15}N nuclei. For a protein enriched in ^{15}N, one would expect to observe a signal for each of the backbone NH groups (i.e. for every amino acid in the protein, except prolines). The Tryptophan and Arginine side chain HNε groups, as well as the Glutamine and Asparagine side chain NH_2 groups will also be observed in this type of experiment.

Due to the high sensitivity of the ^{1}H and ^{15}N nuclei to the local chemical environment, well-folded proteins in aqueous solution will yield spectra with large dispersion of signals in both ^{1}H and ^{15}N dimensions (Figure 7.4). For these reasons, the two-dimensional [^{1}H-^{15}N]-HSQC spectrum of a protein is an important tool, in fact it provides structural information on a per-residues basis and in the form of a unique "fingerprint" that can be used for epitope and/or paratope mapping. This fingerprint can also be very useful in a variety of other applications, like in assessing and comparing the active conformations of

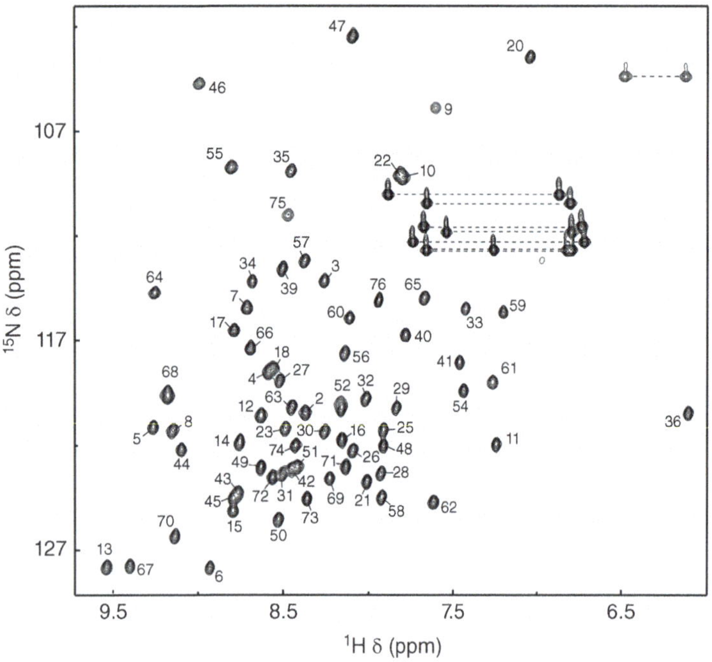

Figure 7.4 2D [^{1}H–^{15}N] HSQC spectrum of uniformly ^{15}N-^{13}C-labeled and 50% deuterated Ubiquitin. Peaks are labeled according to the corresponding amino acid number in the protein sequence. Resonances belonging to the asparagine and glutamine side chain amide protons are connected by dashed lines.

biotherapeutics and biosimilars [41, 42] or in monitoring the effect of mutations and formulation components on the high-order structure (HOS) of protein therapeutics [41].

As noted, the ^{1}H and ^{15}N nuclei are very sensitive to the local chemical environment and this makes them excellent probes to detect perturbations arising from the interactions with any type of ligand. The typical workflow for epitope mapping using [^{1}H-^{15}N]-HSQC experiments consists in recording two spectra of the ^{15}N-labeled antigen, one in the absence and one in the presence of the interacting antibody fragment. Because the antibody is not enriched in ^{15}N, it will not interfere with the detection of the signals from the ^{15}N-labeled antigen. Upon addition of the antibody and formation of the ^{15}N-labeled antigen/antibody complex, select signals of the ^{15}N-labeled antigen will undergo significant changes in position (i.e. ^{1}H and ^{15}N chemical shift) and/or intensity (Figure 7.5a). In general, the antigen residues experiencing the most significant changes upon complex formation will tend to cluster within a distinct surface of the antigen itself, thus defining the binding epitope (Figure 7.5b). Therefore, by comparing the [^{1}H-^{15}N]-HSQC spectra collected for the ^{15}N-labeled antigen in the free and bound forms, one can quickly identify the residues of the antigen involved in the interaction with the antibody.

One aspect to keep in mind when using the chemical shift perturbations approach is that the changes in the NMR signals induced by the binding antibody could also arise from allosteric effects. The nuclei observed by NMR are sensitive to the local chemical environment and this can be perturbed by direct interaction with the antibody surface as well as by local conformational changes. To help distinguish between these two effects one can use additional NMR experiments like the ones based on cross-saturation [44] or deuterium exchange [45]. Of course, one can apply the CSPs method to carry on paratope mapping and this has been shown to work well for a variety of isotopically labeled Fv fragments [34, 43, 46].

Considerations with NMR

Although epitope/paratope mapping by chemical shift perturbations is widely used, it has some drawbacks that need to be carefully considered before embarking on this type of study. An essential requirement of this approach is to have a full or nearly full assignment of the resonances of the isotopically labeled antigen or antibody fragment – in other words, one needs to know which signal belongs to which residue in the protein sequence. This can be achieved through the application of a suite of three-dimensional NMR experiments [47], which often requires production of different protein samples with various labeling schemes, making the process time-consuming if the assignments are not already known. Another aspect to take into consideration is the costs associated with the production of ^{15}N-, ^{13}C-, and ^{2}H-labeled protein samples, which for some expression systems, like insect cells, are still quite expensive. In spite of the disadvantages, this NMR approach can still offer a valuable method to gain low-resolution information on antigen/antibody interactions. For instance, even in the absence of resonance assignment, one could use this technique in a comparative analysis to distinguish

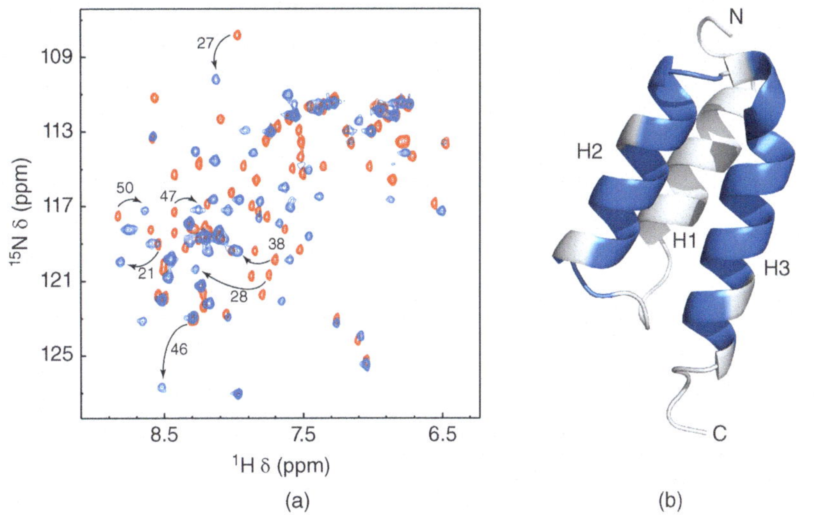

Figure 7.5 Epitope mapping by NMR. (a) Superposition of 2D ^{1}H-^{15}N HSQC spectra of free (red) and Fv-bound ^{15}N-^{13}C-^{2}H-labeled EZ4 a mutant of the *Staphylococcal* protein A (SpA) domain E, in the free (red) and Fv-bound (blue) forms. Peaks undergoing a large change in chemical shifts are labeled. (b) Residues that experience significant chemical shift changes upon complex formation are colored blue on the structure of the E-domain. Source: Adapted from Meininger et al. [43].

between the binding fingerprints of different antibodies on the same antigen.

In the last decade, the technical development of cryogenic probes and the implementation of various methods for fast data acquisition have opened new avenues to tackle more and more complex biological systems, and new NMR approaches have started to emerge that are making possible to obtain structural fingerprints of systems as large as a full IgG1κ antibody, without the need to produce isotopically enriched samples [48].

Epitope Mapping by X-Ray Crystallography

Since the first antibody–antigen complex was crystallized [49], X-ray crystallography has been the gold standard for epitope determination. In a typical crystallography experiment, a complex of highly purified monoclonal Fab fragment and antigen is prepared and concentrated to >10 mg/ml. The concentrated protein solution is set up in drops containing an equal volume of "mother liquor" solutions that contain varied buffers, pH values, precipitants, salts, and additives. These precipitant solutions are organized into 96-well screens (sold by a number of different vendors) and the drops are set in specially designed crystallization plates by a robot liquid handler. Drops are monitored for crystal formation over days to weeks, and once a crystallization condition is found, it is optimized to provide well-diffracting crystals of a size large enough to be harvested for data collection. Diffraction images are then collected by placing the crystal in a focused X-ray beam, from either a commercially available X-ray generator or more often at one of the high-energy synchrotron radiation sources run by a handful of governmental labs around the world. Images are continually collected during rotation of the crystal around a single axis, with anywhere from 45 to 360° of data needed for a complete dataset. The spots found in these images are then indexed based on the crystal lattice properties, with the location and intensity of each spot measured and catalogued in a single data file. It is from this "reduced" data the electron density maps can be obtained. A three-dimensional polypeptide model can then be built into the electron density maps, with the model improving in accuracy as the resolution of the data increases. While a discussion of in-depth methods for X-ray data collection and structure solution are beyond the scope of this chapter, there are many textbooks, reviews, and online resources available for those interested in the details of protein crystallography [50, 51].

Attempting crystallization of an antigen in complex antibody Fab fragment has a number of important benefits. First, Fabs often improve the crystallization properties of an antigen. Monoclonal Fab fragments crystallized readily by themselves and are often used as chaperones to improve the crystallization of difficult structural biology targets [52]. Second, the Fab fragment provides a known structure for overcoming the "phase problem" with X-ray diffraction. Due to the wealth of Fab crystal structures available in the Protein Data Bank (PDB), Fab models from the database can generally be used for molecular replacement into the diffraction data to start the model-building process. Despite these advantages, crystallization is an inherently unpredictable experimental method and sometimes a "perfect" antibody/antigen complex will yield no crystals at all. To improve crystallization properties, several "tricks" can be utilized to produce an antibody/antigen complex with the highest chance of crystallization. We have therefore devoted the next section in this chapter to the protein expression and engineering methods that can be employed to improve crystallization properties of an antibody/antigen complex.

Recombinant Antigen Generation

A critical step in the process of epitope characterization is the production of antigen to be used for structural studies. Often the antigen used for structural characterization is the same antigen used for the antibody discovery campaign (i.e. immunization), without proper consideration of the consequences for structural experiments. Protein reagents perfectly acceptable for immunization, *in vivo* or cell-based experiments are often impractical for use in structural studies. One must think about the structural methods that will be used before making a decision on how to generate the antigen. For example, extracellular proteins expressed in CHO or 293 cells are often used for immunization for reasons such as "native" glycosylation and readily available cell culture capacity. While proteins expressed in mammalian cells with native glycosylation are appropriate for EM, they are only rarely amenable to crystallization or isotopic labeling for NMR studies. It is advisable to spend some time thinking about what experiments will be performed before choosing an expression system, or if possible, trying multiple expression systems to ensure a better chance of success. While the following sections are geared toward generation of antibody/antigen complexes for crystallography, the methods are universal and can be adapted for the structural biologist using NMR or EM for epitope discovery.

E. coli Expression of Antigens

Protein expression in *E. coli* bacteria has been a mainstay of biochemistry and structural biology labs for decades due to a number of advantages that *E. coli* protein production has over more recently developed eukaryotic expression systems [53]. The first advantage is speed - one can start with a cloned construct, transform and grow the bacteria, and purify the protein all within a few days. This turnaround time allows for iterative construct testing and design with a speed impossible for eukaryotic expression systems. The second advantage is cost - *E. coli* expression media and equipment is relatively inexpensive compared to insect or mammalian cell media and the requisite tissue culture facilities. Finally, expression in *E. coli* can be easily optimized to allow the production of proteins that are enriched in ^{15}N, ^{13}C, ^{2}H isotopes, as required for NMR studies [53]. The downsides to *E. coli* expression, which unfortunately loom large, come primarily from a problem that has particular ramifications for the production of protein antigens. Extracellular proteins, which are often the targets of antibody discovery, commonly have folding and post-translational requirements not supported by prokaryotic protein expression machinery [54]. These proteins are usually still produced by the bacterial ribosomal machinery, but instead of being properly folded, they are shunted in an insoluble form to inclusion bodies within the bacterial cytoplasm. While several methods (discussed below) have been developed to overcome these issues, it is common practice in structural biology labs to concurrently test eukaryotic expression systems in order to have more "shots on goal" for successful expression and only use *E. coli* if these early tests show the bacteria make acceptable levels of soluble protein.

Extracellular proteins often contain disulfide bonds that require a proper oxidative environment as well as chaperones to properly form. Therefore, secreted proteins expressed within the *E. coli* cytoplasm are often driven as misfolded intermediates into inclusion bodies. A classical way of dealing with proteins in inclusion bodies is to purify them, denature the inclusion bodies, and then refold at a low concentration in a buffer containing oxidized and reduced glutathione or cysteine as a redox pair [55]. The presence of the redox pair allows reduction of improperly formed disulfide bonds, while proper disulfide bonds remain stably paired. If refolding is successful, it can be a fantastic way to obtain large amounts of antigen for multiple types of structural studies. Unfortunately, refolding can be difficult for many proteins, in particular those dependent on glycosylation for stability.

A second effective method for *E. coli* expression of extracellular proteins is to engineer a prokaryotic secretion signal into the sequence, which targets the protein for transport into the periplasm via the Sec or SRP translocation systems [56]. A protocol specific for periplasmic protein purification can then be used to isolate the protein, which has the added benefit of removing any cytoplasmic components that may interfere with purification. This method works particularly well for antibody fragments [57], but can be tested for any extracellular protein.

A third way to make extracellular proteins in *E. coli* is using a strain or tag that helps with efficient disulfide bond formation in the bacterial cytoplasm. Using the thioredoxin (TrxA protein) fusion tag is one such method, as TrxA can help catalyze disulfide bond formation in the cytoplasm [58]. Another method (which can be combined with the Trx tag) is the use of a modified BL21 strains (Origami or SHuffle) with mutations in thioredoxin reductase (TrxB) and glutathione reductase (Gor) that shift the cytoplasmic environment to more oxidizing conditions [59]. This method has been successfully paired with a low-temperature expression protocol to produce engineered extracellular proteins such as the ligand binding domains from AMPA and NMDA receptors [60, 61].

Insect Cell Expression of Antigens

Over the last several decades insect cell expression using baculovirus as a DNA delivery vector has become one of the most widely used methods for expressing extracellular antigens [53]. This method involves the preparation of a baculovirus transfer vector containing the target gene of interest, followed by cotransfection with a viral bacmid that recombines with transfer vector insert to generate a baculovirus genome containing the desired gene to be expressed. This virus is produced by the host cell and is subject to several rounds of amplification to obtain high titer viral stocks that can be used for large-scale insect cell infection. After a 60–72 hour expression period, the protein is purified either from the cell-free supernatant or the cell paste itself. There are many advantages to using the baculovirus system, including reasonably fast turnaround times, relatively simple cell culture requirements, eukaryotic protein folding and post-translational modification machinery. The use of virus-containing supernatant as the infective agent also makes co-expression of binding partners as simple as adding aliquots of each separate virus to the flask of cells. For the purposes of structural biology, insect cell expression is a desirable system because of its flexibility. Proteins expressed in insect cells can be expressed fully glycosylated and used for EM, or they can be de-glycosylated just as easily for use in crystallization experiments (discussed below). In addition, the advent of insect cell media that contains isotopically labeled amino acids means that insect cell-expressed proteins can be used for NMR experiments [62]. In rare cases, one may find an extracellular protein that does not express or fold properly in insect cells – and in those cases mammalian methods may be required. But for the vast majority of extracellular proteins, the insect cell system provides the simplest means to high-quality protein expression.

Mammalian Expression of Antigens

Mammalian cell protein expression is a highly developed method because of its use in manufacturing of biotherapeutics [53, 63, 64]. The primary cell lines used for large-scale production are Chinese Hamster Ovary (CHO) cells, and properly selected stable cell lines can be engineered to produce gram quantities of protein per liter of cell culture. Unfortunately, the development of cell lines with this kind of yield is resource intensive and takes months of development time. For the purposes of antigen generation for structural biology, expression is usually induced by polyethylenimine (PEI)-mediated transient transfection. Other cell lines can also be used for expression, such as the human embryonic kidney (HEK) 293 cell line or murine myeloma NS0 cell line [60]. All three have their own advantages and disadvantages, which can include surprisingly large differences in expression as well as different glycosylation patterns. A normal workflow for antigen production would take advantage of CHO and 293 expression, followed up by expression in a specialty cell like NS0 if issues arise. While not common, proper expression and folding of a protein from CHO or 293 can be an issue in particular with proteins normally expressed in immune cells.

Another more recent development in mammalian cell expression is the use of baculovirus as a DNA transfection vector. This system, called BACMAM, uses high titer stocks of baculovirus containing a mammalian promoter in front of the gene of interest to deliver DNA into the mammalian cells [65]. Stocks of baculovirus for mammalian expression are produced in the same manner as stocks used for insect cell expression, although much more virus (generally 10-100X) is needed to induce high expression in mammalian cells due to reduced virus tropism for mammalian cells. Fortunately, high titer baculovirus can be made in large amounts in flasks, making the virus preparation a trivial step in the process of BACMAM expression. This method has been used effectively for expression of extracellular and membrane proteins for structural studies [66, 67].

N-Linked Glycosylation

While purified extracellular proteins are often easy to work with because of their "hardiness" (stability at room temperature, no need for detergents or reductants), extracellular proteins nearly always contain at least one (and often many) Asparagine- or N-linked glycosylation sites that can make crystallization extremely difficult. N-Linked glycosylation occurs at Asparagine sites followed by a Serine or Threonine at the +2 position (NxS/T motif) [68] and is catalyzed by glycosylation machinery found in the endoplasmic reticulum, with final processing continuing upon transit into the Golgi apparatus [69]. While each N-linked glycan starts out as a common dolichol-linked precursor oligosaccharide (glucose$_3$-mannose$_9$-*N*-acetylglucosamine$_2$), these initial sugars are trimmed as well as added to by enzymatic machinery that is both species and cell-type specific [68, 69]. For example, CHO cells have different glycosylation patterns than 293S cells, and the pattern of sugars attached to a single protein can also vary. Therefore, glycosylation introduces heterogeneity to the surface of the target protein that often interferes with highly ordered crystal packing required for strong crystal diffraction.

Given the complications that glycosylation creates for the crystallographer, several techniques have been developed to minimize glycosylation on the target protein. First, expression systems where glycosylation can be better controlled like *E. coli* or insect cells are generally more desirable for a crystallization target. *E. coli*, for example, do not contain glycosylation machinery, yet as discussed earlier in the chapter this lack of post-translational processing is often detrimental to expression of normally glycosylated proteins. In insect cells, the simplest method for removal of N-linked glycans is the mutation of the Asparagine in the NxS/T motifs to the related amino acid Glutamine, which blocks transfer of the precursor oligosaccharide to the target protein. Unfortunately, mutation of glycosylation consensus sites often reduces protein expression, and each protein often contains several glycosylated NxS/T motifs. It can be an exponentially complicated effort to find the optimal combination of mutations that gives maximum expression yield with minimum glycosylation. Therefore, much effort in technology development has been put forward to find ways to remove or minimize N-linked glycans in a wholesale manner – methods which are described in the next sections.

A key benefit as well as drawback of expressing extracellular antigens in *E. coli* is the lack of a glycosylation pathway in prokaryotes. This can be a huge benefit for proteins that are well-behaved in the absence of their N-linked glycans, but it can also mean that the *E. coli* system is not a tractable alternative if the protein needs surface sugars for folding or stability. In general, extracellular antigens are best expressed as inclusion bodies and then refolded or secreted into the periplasm. Good examples of refolded proteins used for structural biology are IL-13 [70] and VEGF [71]. Secretion has been successful for T-cell receptors [72] and antibody variable domains [57, 73, 74].

Manipulating N-Linked Glycans on Antigens

For production of highly glycosylated antigens, eukaryotic expression systems are the traditional choice given the presence of appropriate post-translational machinery. Glycans attached to proteins expressed in insect cells are generally 1–6 and 1–3 fucosylated paucimannosidic structures of ~1,000 Da [75]. Glycans attached by the mammalian glycosylation machinery are normally more complex, with bi-antennary and hybrid structures dominating [76]. As previously discussed, the heterogeneity introduced by N-linked oligosaccharides frequently obstructs crystallization [77]. Several methods have been developed for removing N-linked glycans from proteins expressed in eukaryotic cells, which have been successful in a number of studies.

The first method is simple digestion of the sugar off of the protein using a bacterially-derived endoglycosidic enzyme. The Endoglycosidase family of proteins (including EndoH, F1, F2, and F3) digest between the N-linked *N*-acetylglucosamine (GlcNAc) residue and the second GlcNAc residue, leaving the initial GlcNAc attached to the protein [78–80]. The "Endo" proteins have varying specificity for N-linked sugars, with the activity of EndoH and F1 high against high mannose, afucosylated sugars, and EndoF2 and F3 more active against complex bi-antennary structures with or without fucosylation on the core-bound *N*-acteylglucosamine [79]. A cocktail of these enzymes is available for purchase and can sometimes be used for digestion, in particular, if the protein only has one or two glycosylation sites. In addition, there is also the possibility of PNGase digestion, which cuts the N-linked sugar off the Asparagine residue and in the process converts the Asparagine to Aspartate [81]. Importantly, proteins that contain a terminal GlcNAc linked to the Asn glycosylation site are often better behaved than those fully deglycosylated by PNGase or with mutations removing the glycosylation site, likely due to shielding of hydrophobic patches evolved to interact with an attached glycan [77].

An alternate method by which glycosylation can be controlled in eukaryotic expression systems is with small-molecule inhibitors of the enzymatic pathway. Several molecules have been characterized for this purpose, with the endpoint being a recombinant protein that contains afucosylated glycans sensitive to Endoglycosidase H or F1 digestion. The first described method utilized a Lec3.2.8.1 CHO cell line along with the alpha-glucosidase inhibitor *N*-butyldeoxynojirimycin (NB-DNJ) to produce afucosylated glycoproteins for crystallization [82, 83]. Other studies have utilized the mannosidase inhibitors swainsonine or kifunensine to block mannose removal and trap the glycans in an afucosylated state [77]. Swainsonine blocks mannosidase II, resulting in high mannose/hybrid structures that are largely EndoH sensitive [77, 84]. Kifuensine is a potent mannosidase I inhibitor that retains N-linked glycans in a high mannose, EndoH-sensitive state [85]. Proteins expressed in the presence of kifuensine can also be co-expressed with EndoH to allow for glycan processing in the ER. Insect cell expression in the presence of kinfuensine in combination with EndoH coexpression has been used for deglycosylation successfully in a number of cases [86, 87]. On occasion, a protein will not be expressed when co-expressed with EndoH, and in those cases, a rescue strategy can be used to express the protein with kifunensine alone and follow with EndoH or F1 digestion *in vitro* after purification [81].

Another method which has been recently utilized with great success is the use of a 293S cell line that has an inactivating mutation in *N*-acetylglucosaminyltransferase I (GNT1) [88]. Proteins expressed in this cell line only contain high mannose, afucosylated glycans that can be readily digested by EndoH. Coupled with a BACMAM transfection approach, this expression system has been used for generation of crystallization-quality membrane proteins such as the NMDA receptor [89], GPCRs [90], and extracellular receptors that could not be generated in insect cells [91].

Antibody Generation for Crystallography

While intact antibody IgG molecules are often available for structural studies during the development process, the complete IgG containing two Fab fragments connected to an Fc region is not compatible with most structural studies. Full IgGs are often too flexible for crystallization and class averaging in EM studies, and IgGs are generally too large for NMR analysis. Therefore, generation of the isolated Fab fragment is a key step for enabling epitope mapping studies.

The simplest method for obtaining a Fab fragment is by digestion of the intact IgG molecule. Several methods are available for Fab generation, depending on the source of the IgG. Mouse IgG can be efficiently cleaved with Ficin [92] while human IgG can be cleaved with Lys-C protease [93]. As a third alternative, IdeS can be used for generation of Fab′2 fragment, which is a bivalent molecule containing both Fabs connected by uncleaved disulfide-bonded antibody hinge residues [94].

A second method for generation of Fab or single chain Fv (scFv) fragments is by expression of the protein in the periplasm of *E coli* [57, 73]. *E. coli* secretion can provide high yields of the variable domain fragment of interest and has the benefit of not needing a cleavage step. For antibody/antigen complexes with stability issues, co-expression of the Fab with the protein of interest can be tested as a rescue approach. In this case, the antibody heavy and light chains are cloned individually and co-transfected with the antigen construct. This method was used for the RSV F protein in complex with a neutralizing antibody D25 [21]. The advantages of Fab co-expression are twofold – the antigen is stabilized during the expression phase, and only one purification is necessary rather than purification of individual components followed by purification of the antibody/antigen complex.

Crystallization of Antibody/Antigen Complexes

A protein crystallization experiment begins by incubating a molecule of interest (or antibody/antigen complex in this case) in a large number of "precipitant" solutions in an attempt to coax the protein to form an ordered array of molecules, which is held together by noncovalent contacts [50]. The protein is driven from solution into this ordered array, otherwise known as a crystal, the precipitant, which is often a salt or polyethylene glycol (PEG) derivative. Protein crystallization is an inherently stochastic process, involving hundreds of miniature crystallization "experiments" with different precipitants, additives, at a range of pH values performed in a high-throughput manner. Some have called crystallography a "dark art" given the randomness and uncontrolled nature of the crystallization process. Yet a number of tricks can be utilized to improve the chances of successful crystallization of an antibody/antigen complex of interest.

A first suggestion is to work with the highest affinity antibody clone available. A general rule of thumb is that if the antibody has an affinity (KD) weaker than 100 nM, there is a reasonable chance of the formation of crystals of the Fab fragment by itself, without bound antigen. As the affinity improves, there is less free antibody available in solution to crystallize at any one time, raising the odds that any crystals that form are complex crystals. In addition, crystals that form at low pH values (<5) are often Fab crystals, due to a reduction in the affinity of most antibody/antigen interactions at low pH. This leaves more unbound Fab in solution to crystallize. Although all crystals should be checked, it is wise not to assume a crystal grown at low pH is a complex crystal until the structure is solved. A second important step to improve crystallization chances is to ensure high purity of both the Fab and antigen and perform a final purification of the complex using a size exclusion column. This final sizing column is an important quality control step to confirm absolute purity of the sample destined for crystallization as well as confirm formation of a complex between the Fab fragment and antigen. This step also ensures a 1:1 antibody/antigen ratio in the crystallization drop.

If after extensive crystallization trials, no crystals form or X-ray diffraction of the crystals is not adequate for structure solution, several rescue strategies are available to re-invigorate the crystallization process. As discussed earlier in this chapter, deglycosylation of the antigen is very important and can make an enormous difference in the crystallization process. In addition to deglycosylation, one can also produce protein with universally homogenous high-mannose N-linked glycans by expression in the presence of kifunensine or in the GNT1-/- cell line. Proteins with homogeneous, high-mannose sugars are more amenable to crystallization than those with a diversity of N-linked glycans. Another protein modification step that can be very successful is methylation of surface lysines [95].

This method uses dimethylaminoborane and formaldehyde to convert all surface lysines to dimethyl-lysine. This does not alter the charge of the lysine, but the addition of two methyl groups significantly modifies the surface chemistry of both the antibody and antigen. On occasion, this method will break up an antibody/antigen complex, if either the epitope or paratope contains key lysine-mediated interactions, but the methylation reaction can be performed on the Fab or antigen separately, if the complex formation is disrupted.

Consideration for New Modalities and New Targets

Peptides

Over the past few decades, the range of proteins of therapeutic interest has significantly expanded to include intrinsically disordered proteins (IDPs) and proteins involved in protein–protein interactions (PPIs). These types of proteins have traditionally been considered particularly difficult to target by small-molecule inhibitors due to the absence of well-defined druggable pockets (in the case of IDP) and the presence of large, shallow pockets (in the case of PPI targets).

Due to their ability to mimic protein secondary structures and provide large interaction surfaces compared to small molecule inhibitors, peptides (linear, cyclic, disulfide-constrained) are emerging as one of the preferred modalities to modulate the function of these target classes [96–98].

Characterizing the interaction between a peptide and its target at atomic level is a critical step in the drug discovery process and the considerations discussed in the previous sections for EM, NMR, and X-ray crystallography are also relevant in this scenario. It is worth noting that applying NMR for paratope mapping during the early stage of discovery can be highly beneficial. At this stage, the interaction between a peptide binder and its target might be too weak to provide structural information through EM or X-ray crystallography. Therefore, obtaining information on the peptide region that interacts with the target can be extremely useful for a more focused SAR optimization. For linear and disulfide-constrained peptides that contain natural amino acid, it is possible to produce isotopically enriched peptide samples [99] and apply the paratope mapping strategies previously outlined.

Furthermore, NMR provides the opportunity to study the conformational preferences of the unbound peptides in solution, thereby gaining structural and functional insights that can be incorporated into the design of more potent and permeable molecules [100, 101].

RNA

Traditionally, drug discovery has focused on targeting proteins as the main modulators of disease-driving biological pathways. However, in recent years, an increasing variety of therapeutic modalities started to focus on intervening earlier in the central dogma, and RNA has become both a therapeutic modality and a target for inhibition. Pfizer and Moderna COVID-19 vaccines, Moderna/Merck cancer vaccines, and the recent BioNTech/Genentech individualized cancer vaccine for the treatment of PDAC [102] provide good examples of the use of RNA as a therapeutic modality. On the other hand, the inhibition of RNA as a target can be achieved through a variety of modalities: some are RNA-based, like ASOs (e.g. Nusinersen [103]; Casimersen [104]) and siRNAs (e.g. Inclisiran [105]; Patisiran [106]), whereas others are SMIs (e.g. Ribocil [107]; Zotatifin [108, 109]; Risdiplam [110]).

Three-dimensional structural models of RNAs are crucial for the discovery and rational design of therapeutics that modulate RNA function. Historically, structure determination of RNA molecules has been challenging due to their intrinsic flexibility and conformational heterogeneity and, as a result, to date the RCSB protein data bank contains only 1,884 RNA structures, a very small fraction (~0.8%) of the total number of deposited structures.

A variety of computational approaches like phylogenetic analysis and energy minimization are routinely applied to predict RNA secondary structure (i.e. base pairs) with some accuracy [111], but predicting the tertiary structure of RNA molecules remains still challenging [112]. This is mainly due to: (i) the limited experimental data currently available to build the foundations of machine learning approaches and (ii) the inherent flexibility of RNA molecules, which can often adopt different functional conformations like, for instance, in the case of riboswitches.

X-ray crystallography is still widely applied for the structural determination of large RNA, but the flexibility and conformational heterogeneity of these molecules in the apo state, often make it very challenging to crystallize and, thus, obtain structural information.

On the other hand, NMR, has been successfully employed to characterize both structure and dynamics of RNA in solution [113–117] and, to a lesser extent, in solid state [118–120]. However, the application of NMR has been generally limited to small RNA molecules, usually less than 60 nucleotides, because of the spectra complexity that arises as the size of the RNA of interest increases. A variety of strategies like deuteration, nonuniform isotopic labeling, and optimized pulse sequences, as well as the availability of higher magnetic field spectrometers have helped to push the boundary of size limitation [118, 121, 122], but to date, solving RNA structures by NMR still remains a laborious and slow throughput process.

In the last few years, the Cryo-EM "resolution revolution" has provided new opportunities to study RNA-only structures. An important advantage offered by cryo-EM is the possibility to detect and characterize single particles through SPA (single particle analysis), thus capturing the dynamic nature of RNA and unveiling important aspects of the relation between its structure and function. For example, recent cryo-EM studies of the *Tetrahymena* group I ribozyme identified and characterized multiple conformations associated with the intermediate states of the first and second step of the self-splicing process and provided structural and mechanistic insights into this ribozyme activity [123–125].

As seen in the case of proteins, one of the main drawbacks in the application of cryo-EM is the size of the RNA of interest. To mitigate this problem, several strategies have been proposed [126–129]. One such approach, called ROCK (RNA oligomerization-enabled cryo-EM via installing kissing loops), is based on the introduction, into regions of the RNA that are not essential for its function, of RNA sequences that form kissing loops [126]. This allows for the formation of a closed oligomeric ring containing multiple copies of the RNA of interest, thus increasing its overall size, restricting its dynamics and making it more amenable for cryo-EM studies. This strategy has allowed the determination of the *Tetrahymena* group I intron structure (at an overall resolution of 2.98 Å) as well the determination of smaller (<100 kDa) RNAs structures, including the *Fusobacterium nucleatum* FMN riboswitch (~35 kDa) in complex with its ligand (at a resolution of 5.9 Å) [126]. Although promising, the ROCK approach requires careful engineering of the RNA construct for structural studies and poses two major limitations on the applicability of the method: (i) the RNA molecule has to be large enough to have two or more peripheral, nonfunctional helices that could be exploited for the insertion of the kissing loops, (ii) a three-dimensional model (derived from partial structures, computational modeling, etc.) is needed to guide the engineering of the kissing loops.

Additional scaffold-based strategies rely on inserting the RNA sequence of interest into a larger and well-structured RNA. The use of a circularly permuted version of the *Tetrahymena* ribozyme as a scaffold enabled the determination of the structure of RNAs as small as 17 kDa–25 kDa with a resolution ranging from 4.5 Å to 5 Å [127], whereas the use of a group II intron scaffold to display the thiamine pyrophosphate (TPP) riboswitch allowed the determination of this small RNA structure at a 2.5 Å resolution [128].

Similarly to the ROCK approach, the scaffold-based methods also require the presence of at least one nonfunctional stem-loop that can be modified to attach the scaffold. Therefore, it's crucial to have some prior knowledge of the secondary structure and location of the stem-loops in the RNA of interest to guide the design of an optimal construct for structural studies.

As highlighted here, obtaining structural and dynamics information for RNA molecules can be challenging. However, methods like EM, NMR, and X-ray crystallography can provide highly complementary information and, when used in an integrated manner, they can help obtain not only a more accurate depiction of the tridimensional structure of RNA but also a deeper understating of how RNA molecules function [130, 131]. This information is essential for the discovery and rational design of therapeutics that modulate RNA function.

Conclusion

Over the past two decades, the methodologies for structurally mapping antibody epitopes and characterizing new targets and modalities have expanded tremendously. Techniques such as EM, NMR spectroscopy, and X-ray crystallography have become more routine and are now readily accessible to drug development teams eager to understand the interactions between their drugs and their targets.

We anticipate that these methods will become even more accessible in the future, with EM, in particular, experiencing a renaissance and likely becoming an even more integral part of the epitope discovery repertoire. Although not highlighted in this chapter, machine learning is also coming of age, with numerous algorithms being developed to predict biotherapeutic structures or to incorporate synthetic biology, thereby offering a higher level of flexibility in therapeutic molecular design.

Given the critical importance of understanding the mechanism of action and cross-reactivity for potential antibody therapeutics, this progression is highly encouraging for both academic and industrial researchers involved in drug development.

Abbreviations

NMR	nuclear magnetic resonance
EM	electron microscopy
CDR	complementarity-determining region
EGF	epidermal growth factor
PDB	Protein Data Bank

HA	hemagglutinin
FcγR	Fc gamma receptor
Å	Angstrom
μg	microgram
ml	milliliter
kDa	kilodaltons
Da	Daltons
CHO	Chinese Hamster Ovary
HIV	human immunodeficiency virus
RSV	respiratory syncytial virus
TROSY	Transverse Relaxation Optimized SpectroscopY
HSQC	heteronuclear single quantum correlation
CSPs	chemical shift perturbations
HOS	high-order structure
Fv	fragment variable
scFv	single chain fragment variable
Fab	fragment antigen binding
Fc	fragment crystallizable
E.coli	*Escherichia coli*
SRP	signal recognition particle
TrxA	thioredoxin A
TrxB	thioredoxin B
Gor	glutathione reductase
AMPA	α-amino-3-hydroxy-5-methyl-4-isoxazolepropionic acid
NMDA	*N*-Methyl-D-aspartic acid
PEI	polyethylenimine
HEK	human embryonic kidney
Endo	endoglycosidase
GlcNAc	*N*-acetyl glucosamine
Asn	asparagine
NB-DNJ	*N*-butyldeoxynojirimycin
ER	endoplasmic reticulum
GNT1	*N*-acetylglucosaminyltransferase I
GPCR	G-protein-coupled receptor
PEG	polyethylene glycol
ASOs	antisense oligonucleotides
SMIs	small-molecule inhibitors
SAR	structure–activity relationship

References

1 Stern, H.M. (2012). Improving treatment of HER2-positive cancers: opportunities and challenges. *Sci. Transl. Med.* 4 (127): 127rv2.

2 Cho, H.S., Mason, K., Ramyar, K.X. et al. (2003). Structure of the extracellular region of HER2 alone and in complex with the Herceptin fab. *Nature* 421 (6924): 756–760.

3 Franklin, M.C., Carey, K.D., Vajdos, F.F. et al. (2004). Insights into ErbB signaling from the structure of the ErbB2-pertuzumab complex. *Cancer Cell* 5 (4): 317–328.

4 Baselga, J., Cortés, J., Kim, S.B. et al. (2012). Pertuzumab plus trastuzumab plus docetaxel for metastatic breast cancer. *N. Engl. J. Med.* 366 (2): 109–119.

5 Nakamura, G., Chai, N., Park, S. et al. (2013). An in vivo human-plasmablast enrichment technique allows rapid identification of therapeutic influenza a antibodies. *Cell Host Microbe* 14 (1): 93–103.

6 Corti, D., Voss, J., Gamblin, S.J. et al. (2011). A neutralizing antibody selected from plasma cells that binds to group 1 and group 2 influenza a hemagglutinins. *Science* 333 (6044): 850–856.

7 Dreyfus, C., Laursen, N.S., Kwaks, T. et al. (2012). Highly conserved protective epitopes on influenza B viruses. *Science* 337 (6100): 1343–1348.

8 Fleishman, S.J., Whitehead, T.A., Ekiert, D.C. et al. (2011). Computational design of proteins targeting the conserved stem region of influenza hemagglutinin. *Science* 332 (6031): 816–821.

9 Fuh, G., Wu, P., Liang, W.C. et al. (2006). Structure-function studies of two synthetic anti-vascular endothelial growth factor Fabs and comparison with the Avastin Fab. *J. Biol. Chem.* 281 (10): 6625–6631.

10 Muller, Y.A., Chen, Y., Christinger, H.W. et al. (1998). VEGF and the fab fragment of a humanized neutralizing antibody: crystal structure of the complex at 2.4 a resolution and mutational analysis of the interface. *Structure* 6 (9): 1153–1167.

11 Wiesmann, C., Fuh, G., Christinger, H.W. et al. (1997). Crystal structure at 1.7 a resolution of VEGF in complex with domain 2 of the Flt-1 receptor. *Cell* 91 (5): 695–704.

12 Carter, P.J. (2006). Potent antibody therapeutics by design. *Nat. Rev. Immunol.* 6 (5): 343–357.

13 Levin, A.M. and Weiss, G.A. (2006). Optimizing the affinity and specificity of proteins with molecular display. *Mol. BioSyst.* 2 (1): 49–57.

14 Co, M.S. and Queen, C. (1991). Humanized antibodies for therapy. *Nature* 351 (6326): 501–502.

15 Queen, C., Schneider, W.P., Selick, H.E. et al. (1989). A humanized antibody that binds to the interleukin 2 receptor. *Proc. Natl. Acad. Sci. USA* 86 (24): 10029–10033.

16 Wang, W., Singh, S., Zeng, D.L. et al. (2007). Antibody structure, instability, and formulation. *J. Pharm. Sci.* 96 (1): 1–26.

17 Wakankar, A.A., Borchardt, R.T., Eigenbrot, C. et al. (2007). Aspartate isomerization in the complementarity-determining regions of two closely related monoclonal antibodies. *Biochemistry* 46 (6): 1534–1544.

18 Zhang, B., Yang, Y., Yuk, I. et al. (2008). Unveiling a glycation hot spot in a recombinant humanized monoclonal antibody. *Anal. Chem.* 80 (7): 2379–2390.

19 Cheng, Y., Grigorieff, N., Penczek, P.A., and Walz, T. (2015). A primer to single-particle cryo-electron microscopy. *Cell* 161 (3): 438–449.

20 Orlova, E.V. and Saibil, H.R. (2011). Structural analysis of macromolecular assemblies by electron microscopy. *Chem. Rev.* 111 (12): 7710–7748.

21 McLellan, J.S., Chen, M., Joyce, M.G. et al. (2013). Structure-based design of a fusion glycoprotein vaccine for respiratory syncytial virus. *Science* 342 (6158): 592–598.

22 Pejchal, R., Doores, K.J., Walker, L.M. et al. (2011). A potent and broad neutralizing antibody recognizes and penetrates the HIV glycan shield. *Science* 334 (6059): 1097–1103.

23 Cheng, Y. (2015). Single-particle Cryo-EM at crystallographic resolution. *Cell* 161 (3): 450–457.

24 Bailey, L.J., Sheehy, K.M., Dominik, P.K. et al. (2018). Locking the elbow: improved antibody fab fragments as chaperones for structure determination. *J. Mol. Biol.* 430 (3): 337–347.

25 Kung, J.E., Johnson, M.C., Jao, C.C. et al. (2024). Disulfide Constrained Fabs Overcome Target Size Limitation for High-Resolution Single-Particle Cryo-EM. *bioRxiv* 2024.05.10.593593.

26 Castells-Graells, R., Meador, K., Arbing, M.A. et al. (2023). Cryo-EM structure determination of small therapeutic protein targets at 3 A-resolution using a rigid imaging scaffold. *Proc. Natl. Acad. Sci. USA* 120 (37): e2305494120.

27 Grinstead, J.S., Schuman, J.T., and Campbell, A.P. (2003). Epitope mapping of antigenic MUC1 peptides to breast cancer antibody fragment B27.29: a heteronuclear NMR study. *Biochemistry* 42 (48): 14293–14305.

28 Razzera, G., Gadermaier, G., De Paula, V. et al. (2010). Mapping the interactions between a major pollen allergen and human IgE antibodies. *Structure* 18 (8): 1011–1021.

29 Blech, M., Peter, D., Fischer, P. et al. (2013). One target-two different binding modes: structural insights into gevokizumab and canakinumab interactions to interleukin-1beta. *J. Mol. Biol.* 425 (1): 94–111.

30 Blech, M., Seeliger, D., Kistler, B. et al. (2012). Molecular structure of human GM-CSF in complex with a disease-associated anti-human GM-CSF autoantibody and its potential biological implications. *Biochem. J.* 447 (2): 205–215.

31 Sasakawa, H., Sakata, E., Yamaguchi, Y. et al. (2007). Ultra-high field NMR studies of antibody binding and site-specific phosphorylation of alpha-synuclein. *Biochem. Biophys. Res. Commun.* 363 (3): 795–799.

32 Stefanis, L. (2012). Alpha-Synuclein in Parkinson's disease. *Cold Spring Harb. Perspect. Med.* 2 (2): a009399.

33 Adda, C.G., CA, M.R., Reiling, L. et al. (2012). Antigenic characterization of an intrinsically unstructured protein, plasmodium falciparum merozoite surface protein 2. *Infect. Immun.* 80 (12): 4177–4185.

34 Morales, R.A.V., CA, M.R., Seow, J. et al. (2015). Structural basis for epitope masking and strain specificity of a conserved epitope in an intrinsically disordered malaria vaccine candidate. *Sci. Rep.* 5: 10103.

35 Garrett, D.S., Seok, Y.J., Liao, D.I. et al. (1997). Solution Structure of the 30 kDa N-Terminal Domain of Enzyme I of the *Escherichia coli* Phosphoenolpyruvate: Sugar Phosphotransferase System by Multidimensional NMR. *Biochemistry* 36: 2517–2530.

36 Pervushin, K., Riek, R., Wider, G., and Wüthrich, K. (1997). Attenuated T2 relaxation by mutual cancellation of dipole-dipole coupling and chemical shift anisotropy indicates an avenue to NMR structures of very large biological macromolecules in solution. *Proc. Natl. Acad. Sci. USA* 94 (23): 12366–12371.

37 Liu, D., Ren, D., Huang, H. et al. (2008). Structure and stability changes of human IgG1 Fc as a consequence of methionine oxidation. *Biochemistry* 47 (18): 5088–5100.

38 Yamaguchi, Y., Nishimura, M., Nagano, M. et al. (2006). Glycoform-dependent conformational alteration of the Fc region of human immunoglobulin G1 as revealed by NMR spectroscopy. *Biochim. Biophys. Acta* 1760 (4): 693–700.

39 Zuiderweg, E.R. (2002). Mapping protein-protein interactions in solution by NMR spectroscopy. *Biochemistry* 41 (1): 1–7.

40 Bodenhausen, G. and Ruben, D.J. (1980). Natural abundance N-15 Nmr by enhanced Heteronuclear spectroscopy. *Chem. Phys. Lett.* 69 (1): 185–189.

41 Aubin, Y., Hodgson, D.J., Thach, W.B. et al. (2015). Monitoring effects of excipients, formulation parameters and mutations on the high order structure of Filgrastim by NMR. *Pharm. Res.* 32 (10): 3365–3375.

42 Quinternet, M., Starck, J.P., Delsuc, M.A., and Kieffer, B. (2013). Heteronuclear NMR provides an accurate assessment of therapeutic insulin's quality. *J. Pharm. Biomed. Anal.* 78-79: 252–254.

43 Meininger, D.P., Rance, M., Starovasnik, M.A. et al. (2000). Characterization of the binding interface between the E-domain of protein a and an antibody Fv-fragment. *Biochemistry* 39 (6): 1541–1541.

44 Takahashi, H., Nakanishi, T., Kami, K. et al. (2000). A novel NMR method for determining the interfaces of large protein-protein complexes. *Nat. Struct. Biol.* 7 (3): 220–223.

45 Mueller, G.A., Smith, A.M., Chapman, M.D. et al. (2001). Hydrogen exchange nuclear magnetic resonance spectroscopy mapping of antibody epitopes on the house dust mite allergen Der p 2. *J. Biol. Chem.* 276 (12): 9359–9365.

46 Wilkinson, I.C., Hall, C.J., Veverka, V. et al. (2009). High resolution NMR-based model for the structure of a scFv-IL-1beta complex: potential for NMR as a key tool in therapeutic antibody design and development. *J. Biol. Chem.* 284 (46): 31928–31935.

47 Kay, L.E., Ikura, M., Tschudin, R., and Bax, A. (1990). 3-Dimensional triple-resonance NMR-spectroscopy of isotopically enriched proteins. *J. Magn. Reson.* 89 (3): 496–514.

48 Arbogast, L.W., Brinson, R.G., and Marino, J.P. (2015). Mapping monoclonal antibody structure by 2D 13C NMR at natural abundance. *Anal. Chem.* 87 (7): 3556–3561.

49 Amit, A.G., Mariuzza, R.A., Phillips, S.E., and Poljak, R.J. (1985). 3-Dimensional structure of an antigen-antibody complex at 6-a-resolution. *Nature* 313 (5998): 156–158.

50 McPherson, A. and Gavira, J.A. (2014). Introduction to protein crystallization. *Acta Crystallogr. Sect. F Struct. Biol. Cryst. Commun.* 70: 2–20.

51 Shi, Y.G. (2014). A glimpse of structural biology through X-Ray crystallography. *Cell* 159 (5): 995–1014.

52 Bukowska, M.A. and Grütter, M.G. (2013). New concepts and aids to facilitate crystallization. *Curr. Opin. Struct. Biol.* 23 (3): 409–416.

53 Assenberg, R., Wan, P.T., Geisse, S., and Mayr, L.M. (2013). Advances in recombinant protein expression for use in pharmaceutical research. *Curr. Opin. Plant Biol.* 23 (3): 393–402.

54 Baneyx, F. and Mujacic, M. (2004). Recombinant protein folding and misfolding in. *Nat. Biotechnol.* 22 (11): 1399–1408.

55 Burgess, R.R. (2009). Refolding solubilized inclusion body proteins. In: *Guide to Protein Purification*, 2e, vol. 463, 259–282.

56 de Marco, A. (2009). Strategies for successful recombinant expression of disulfide bond-dependent proteins in. *Microb. Cell Factories* 8.

57 Carter, P., Kelley, R.F., Rodrigues, M.L. et al. (1992). High-level *Escherichia coli* expression and production of a bivalent humanized antibody fragment. *Biotechnol.* 10 (2): 163–167.

58 Lavallie, E.R., EA, D.B., Kovacic, S. et al. (1993). A Thioredoxin gene fusion expression system that circumvents inclusion body formation in the *Escherichia coli* cytoplasm. *Biotechnol.* 11 (2): 187–193.

59 Prinz, W.A., Åslund, F., Holmgren, A. et al. (1997). The role of the thioredoxin and glutaredoxin pathways in reducing protein disulfide bonds in the *Escherichia coli* cytoplasm. *J. Biol. Chem.* 272 (25): 15661–15667.

60 Furukawa, H., Singh, S.K., Mancusso, R., and Gouaux, E. (2005). Subunit arrangement and function in NMDA receptors. *Nature* 438 (7065): 185–192.

61 Chen, G.Q., Sun, Y.U., Jin, R., and Gouaux, E. (1998). Probing the ligand binding domain of the GluR2 receptor by proteolysis and deletion mutagenesis defines domain boundaries and yields a crystallizable construct. *Protein Sci.* 7 (12): 2623–2630.

62 Gossert, A.D. and Jahnke, W. (2012). Isotope labeling in insect cells. *Adv. Exp. Med. Biol.* 992: 179–196.

63 Zhu, J.W. (2012). Mammalian cell protein expression for biopharmaceutical production. *Biotechnol. Adv.* 30 (5): 1158–1170.

64 Dalton, A.C. and Barton, W.A. (2014). Over-expression of secreted proteins from mammalian cell lines. *Protein Sci.* 23 (5): 517–525.

65 Boyce, F.M. and Bucher, N.L.R. (1996). Baculovirus-mediated gene transfer into mammalian cells. *Proc. Natl. Acad. Sci. USA* 93 (6): 2348–2352.

66 Dukkipati, A., Park, H.H., Waghray, D. et al. (2008). BacMam system for high-level expression of recombinant soluble and membrane glycoproteins for structural studies. *Protein Expr. Purif.* 62 (2): 160–170.

67 Goehring, A., Lee, C.H., Wang, K.H. et al. (2014). Screening and large-scale expression of membrane proteins in mammalian cells for structural studies. *Nat. Protoc.* 9 (11): 2574–2585.

68 Gahmberg, C.G. and Tolvanen, M. (1996). Why mammalian cell surface proteins are glycoproteins. *Trends Biochem. Sci.* 21 (8): 308–311.

69 Davis, S.J. and Crispin, M. (2011). *Functional and Structural Proteomics of Glycoproteins*, 127–158. Springer.

70 Moy, F.J., Diblasio, E., Wilhelm, J., and Powers, R. (2001). Solution structure of human IL-13 and implication for receptor binding. *J. Mol. Biol.* 310 (1): 219–230.

71 Christinger, H.W., Muller, Y.A., Berleau, L.T. et al. (1996). Crystallization of the receptor binding domain of vascular endothelial growth factor. *Proteins* 26 (3): 353–357.

72 Golden, A., Khandekar, S.S., Osburne, M.S. et al. (1997). High-level production of a secreted, heterodimeric alpha beta murine T-cell receptor in *Escherichia coli*. *J. Immunol. Methods* 206 (1–2): 163–169.

73 Skerra, A. and Pluckthun, A. (1988). Assembly of a functional immunoglobulin Fv fragment in *Escherichia coli*. *Science* 240 (4855): 1038–1041.

74 Bird, R.E., Hardman, K.D., Jacobson, J.W. et al. (1988). Single-chain antigen-binding proteins. *Science* 242 (4877): 423–426.

75 Harrison, R.L. and Jarvis, D.L. (2006). Protein N-glycosylation in the baculovirus-insect cell expression system and engineering of insect cells to produce "mammalianized" recombinant glycoproteins. *Adv. Virus Res.* 68: 159–191.

76 Nettleship, J.E., Assenberg, R., Diprose, J.M. et al. (2010). Recent advances in the production of proteins in insect and mammalian cells for structural biology. *J. Struct. Biol.* 172 (1): 55–65.

77 Chang, V.T., Crispin, M., Aricescu, A.R. et al. (2007). Glycoprotein structural genomics: solving the glycosylation problem. *Structure* 15 (3): 267–273.

78 Trimble, R.B. and Tarentino, A.L. (1991). Identification of distinct endoglycosidase (endo) activities in *Flavobacterium meningosepticum*: endo F1, endo F2, and endo F3. Endo F1 and endo H hydrolyze only high mannose and hybrid glycans. *J. Biol. Chem.* 266 (3): 1646–1651.

79 Tarentino, A.L. and Plummer, T.H. Jr. (1994). Enzymatic deglycosylation of asparagine-linked glycans: purification, properties, and specificity of oligosaccharide-cleaving enzymes from *Flavobacterium meningosepticum*. *Methods Enzymol.* 230: 44–57.

80 Tarentino, A.L. et al. (1993). Multiple endoglycosidase F activities expressed by *Flavobacterium meningosepticum* endoglycosidases F2 and F3. Molecular cloning, primary sequence, and enzyme expression. *J. Biol. Chem.* 268 (13): 9702–9708.

81 Plummer, T.H. Jr. and Tarentino, A.L. (1981). Facile cleavage of complex oligosaccharides from glycopeptides by almond emulsin peptide: N-glycosidase. *J. Biol. Chem.* 256 (20): 10243–10246.

82 Butters, T.D., Sparks, L.M., Harlos, K. et al. (1999). Effects of N-butyldeoxynojirimycin and the Lec3.2.8.1 mutant phenotype on N-glycan processing in Chinese hamster ovary cells: application to glycoprotein crystallization. *Protein Sci.* 8 (8): 1696–1701.

83 Davis, S.J., Puklavec, M.J., Ashford, D.A. et al. (1993). Expression of soluble recombinant glycoproteins with predefined glycosylation: application to the crystallization of the T-cell glycoprotein CD2. *Protein Eng.* 6 (2): 229–232.

84 Crispin, M. et al. (2007). Disruption of alpha-mannosidase processing induces non-canonical hybrid-type glycosylation. *FEBS Lett.* 581 (10): 1963–1968.

85 Elbein, A.D., Tropea, J.E., Mitchell, M., and Kaushal, G.P. (1990). Kifunensine, a potent inhibitor of the glycoprotein processing mannosidase I. *J. Biol. Chem.* 265 (26): 15599–15605.

86 Lupardus, P.J., Birnbaum, M.E., and Garcia, K.C. (2010). Molecular basis for shared cytokine recognition revealed in the structure of an unusually high affinity complex between IL-13 and IL-13Ralpha2. *Structure* 18 (3): 332–342.

87 Lupardus, P.J. and Garcia, K.C. (2008). The structure of interleukin-23 reveals the molecular basis of p40 subunit sharing with interleukin-12. *J. Mol. Biol.* 382 (4): 931–941.

88 Reeves, P.J., Callewaert, N., Contreras, R., and Khorana, H.G. (2002). Structure and function in rhodopsin: high-level expression of rhodopsin with restricted and homogeneous N-glycosylation by a tetracycline-inducible N-acetylglucosaminyltransferase I-negative HEK293S stable mammalian cell line. *Proc. Natl. Acad. Sci. USA* 99 (21): 13419–13424.

89 Lee, C.H., Lü, W., Michel, J.C. et al. (2014). NMDA receptor structures reveal subunit arrangement and pore architecture. *Nature* 511 (7508): 191–197.

90 Burg, J.S., Ingram, J.R., Venkatakrishnan, A.J. et al. (2015). Structural biology. Structural basis for chemokine recognition and activation of a viral G protein-coupled receptor. *Science* 347 (6226): 1113–1117.

91 Ely, L.K., Fischer, S., and Garcia, K.C. (2009). Structural basis of receptor sharing by interleukin 17 cytokines. *Nat. Immunol.* 10 (12): 1245–1251.

92 Mariani, M., Camagna, M., Tarditi, L., and Seccamani, E. (1991). A new enzymatic method to obtain high-yield F(ab)2 suitable for clinical use from mouse IgGl. *Mol. Immunol.* 28 (1–2): 69–77.

93 Gadgil, H.S., Bondarenko, P.V., Pipes, G.D. et al. (2006). Identification of cysteinylation of a free cysteine in the fab region of a recombinant monoclonal IgG1 antibody using Lys-C limited proteolysis coupled with LC/MS analysis. *Anal. Biochem.* 355 (2): 165–174.

94 von Pawel-Rammingen, U., Johansson, B.P., and Bjorck, L. (2002). IdeS, a novel streptococcal cysteine

proteinase with unique specificity for immunoglobulin G. *EMBO J.* 21 (7): 1607–1615.

95 Walter, T.S., Meier, C., Assenberg, R. et al. (2006). Lysine methylation as a routine rescue strategy for protein crystallization. *Structure* 14 (11): 1617–1622.

96 Wang, L., Wang, N., Zhang, W. et al. (2022). Therapeutic peptides: current applications and future directions. *Signal Transduct. Target. Ther.* 7 (1): 48.

97 Johns, D.G., Campeau, L.C., Banka, P. et al. (2023). Orally bioavailable macrocyclic peptide that inhibits binding of PCSK9 to the low density lipoprotein receptor. *Circulation* 148 (2): 144–158.

98 Habazettl, J., Allan, M., Jensen, P.R. et al. (2014). Structural basis and dynamics of multidrug recognition in a minimal bacterial multidrug resistance system. *Proc. Natl. Acad. Sci. USA* 111 (51): E5498–E5507.

99 Hwang, S., Balana, A.T., Martin, B. et al. (2024). Bioproduction platform to generate functionalized disulfide-constrained peptide analogues. *ACS Bio Med Chem Au* 4 (4): 190–203.

100 Rudisser, S.H., Matabaro, E., Sonderegger, L. et al. (2023). Conformations of macrocyclic peptides sampled by nuclear magnetic resonance: models for cell-permeability. *J. Am. Chem. Soc.* 145 (50): 27601–27615.

101 Wang, C.K., Northfield, S.E., Colless, B. et al. (2014). Rational design and synthesis of an orally bioavailable peptide guided by NMR amide temperature coefficients. *Proc. Natl. Acad. Sci. USA* 111 (49): 17504–17509.

102 Rojas, L.A., Sethna, Z., Soares, K.C. et al. (2023). Personalized RNA neoantigen vaccines stimulate T cells in pancreatic cancer. *Nature* 618 (7963): 144–150.

103 Finkel, R.S., Chiriboga, C.A., Vajsar, J. et al. (2016). Treatment of infantile-onset spinal muscular atrophy with nusinersen: a phase 2, open-label, dose-escalation study. *Lancet* 388 (10063): 3017–3026.

104 Vasterling, M.E., Maitski, R.J., Davis, B.A. et al. (2023). AMONDYS 45 (Casimersen), a novel antisense Phosphorodiamidate Morpholino oligomer: clinical considerations for treatment in Duchenne muscular dystrophy. *Cureus* 15 (12): e51237.

105 Ray, K.K., Troquay, R.P., Visseren, F.L. et al. (2023). Long-term efficacy and safety of inclisiran in patients with high cardiovascular risk and elevated LDL cholesterol (ORION-3): results from the 4-year open-label extension of the ORION-1 trial. *Lancet Diabetes Endocrinol.* 11 (2): 109–119.

106 Adams, D., Gonzalez-Duarte, A., O'Riordan, W.D. et al. (2018). Patisiran, an RNAi therapeutic, for hereditary transthyretin amyloidosis. *N. Engl. J. Med.* 379 (1): 11–21.

107 Howe, J.A., Wang, H., Fischmann, T.O. et al. (2015). Selective small-molecule inhibition of an RNA structural element. *Nature* 526 (7575): 672–677.

108 Ernst, J.T., Thompson, P.A., Nilewski, C. et al. (2020). Design of Development Candidate eFT226, a first in class inhibitor of eukaryotic initiation factor 4A RNA helicase. *J. Med. Chem.* 63 (11): 5879–5955.

109 Gerson-Gurwitz, A., Young, N.P., Goel, V.K. et al. (2021). Zotatifin, an eIF4A-selective inhibitor, blocks tumor growth in receptor tyrosine kinase driven tumors. *Front. Oncol.* 11: 766298.

110 Ratni, H., Ebeling, M., Baird, J. et al. (2018). Discovery of Risdiplam, a selective survival of motor Neuron-2 (SMN2) gene splicing modifier for the treatment of spinal muscular atrophy (SMA). *J. Med. Chem.* 61 (15): 6501–6517.

111 Schroeder, S.J. (2018). Challenges and approaches to predicting RNA with multiple functional structures. *RNA* 24 (12): 1615–1624.

112 Schneider, B., Sweeney, B.A., Bateman, A. et al. (2023). When will RNA get its AlphaFold moment? *Nucleic Acids Res.* 51 (18): 9522–9532.

113 Malard, F., Wolter, A.C., Marquevielle, J. et al. (2024). The diversity of splicing modifiers acting on A-1 bulged 5′-splice sites reveals rules for rational drug design. *Nucleic Acids Res.* 52 (8): 4124–4136.

114 Chen, J.L., Taghavi, A., Frank, A.J. et al. (2024). NMR structures of small molecules bound to a model of a CUG RNA repeat expansion. *Bioorg. Med. Chem. Lett.* 111: 129888.

115 Chen, J.L., Zhang, P., Abe, M. et al. (2020). Design, optimization, and study of small molecules that target tau pre-mRNA and affect splicing. *J. Am. Chem. Soc.* 142 (19): 8706–8727.

116 Campagne, S., Boigner, S., Rüdisser, S. et al. (2019). Structural basis of a small molecule targeting RNA for a specific splicing correction. *Nat. Chem. Biol.* 15 (12): 1191–1198.

117 Toews, S., Wacker, A., Faison, E.M. et al. (2024). The 5′-terminal stem-loop RNA element of SARS-CoV-2 features highly dynamic structural elements that are sensitive to differences in cellular pH. *Nucleic Acids Res.* 52 (13): 7971–7986.

118 Marchanka, A., Simon, B., Althoff-Ospelt, G., and Carlomagno, T. (2015). RNA structure determination by solid-state NMR spectroscopy. *Nat. Commun.* 6: 7024.

119 Rout, S.K., Cadalbert, R., Schroder, N. et al. (2023). An analysis of nucleotide-amyloid interactions reveals selective binding to codon-sized RNA. *J. Am. Chem. Soc.* 145 (40): 21915–21924.

120 Ahmed, M., Marchanka, A., and Carlomagno, T. (2020). Structure of a protein-RNA complex by solid-state NMR spectroscopy. *Angew. Chem. Int. Ed. Engl.* 59 (17): 6866–6873.

121 He, J., Liu, X., and Wang, S. (2024). Advances in solid-state NMR methods for studying RNA structures and dynamics. *MRL*.

122 Barnwal, R.P., Yang, F., and Varani, G. (2017). Applications of NMR to structure determination of RNAs large and small. *Arch. Biochem. Biophys.* 628: 42–56.

123 Luo, B.N., Zhang, C., Ling, X. et al. (2023). Cryo-EM reveals dynamics of group I intron self-splicing. *Nat. Catal.* 6 (4): 298.

124 Zhang, X., Li, S., Pintilie, G. et al. (2023). Snapshots of the first-step self-splicing of Tetrahymena ribozyme revealed by cryo-EM. *Nucleic Acids Res.* 51 (3): 1317–1325.

125 Li, S., Palo, M.Z., Zhang, X. et al. (2023). Snapshots of the second-step self-splicing of Tetrahymena ribozyme revealed by cryo-EM. *Nat. Commun.* 14 (1): 1294.

126 Liu, D., Thélot, F.A., Piccirilli, J.A. et al. (2022). Sub-3-a cryo-EM structure of RNA enabled by engineered homomeric self-assembly. *Nat. Methods* 19 (5): 576–585.

127 Langeberg, C.J. and Kieft, J.S. (2023). A generalizable scaffold-based approach for structure determination of RNAs by cryo-EM. *Nucleic Acids Res.* 51 (20): e100.

128 Haack, D.B., Rudolfs, B., Jin, S. et al. (2024). Scaffold-Enabled High-Resolution Cryo-EM Structure Determination of RNA. *bioRxiv*.

129 Sampedro Vallina, N., EK, M.R., Hansen, B.K. et al. (2023). RNA origami scaffolds facilitate cryo-EM characterization of a broccoli-pepper aptamer FRET pair. *Nucleic Acids Res.* 51 (9): 4613–4624.

130 Wang, Y., He, Y., Wang, Y. et al. (2023). Structure of LARP7 protein p65-telomerase RNA complex in telomerase revealed by Cryo-EM and NMR. *J. Mol. Biol.* 435 (11): 168044.

131 Li, S., Su, Z., Lehmann, J. et al. (2019). Structural basis of amino acid surveillance by higher-order tRNA-mRNA interactions. *Nat. Struct. Mol. Biol.* 26 (12): 1094–1105.

8

Modern Methods for Characterizing the Higher-Order Structure (HOS) of Biotherapeutics

Felix Kuhne[1], Rachel P. Liu[2], Lingfei Wang[2], Aaron T. Wecksler[2], Manasi Gaikwad[3], Christine C. Jao[4], Alberto Estevez[5], Roshan M. Regy[6], and Saeed Izadi[6]

[1]*Analytical Characterization, Pharma Technical Development, Roche Diagnostics GmbH, Penzberg, Germany*
[2]*Protein Analytical Chemistry, Genentech Inc., South San Francisco, CA, USA*
[3]*Center for Mass Spectrometry and Optical Spectroscopy (CeMOS), Mannheim University of Applied Sciences, Mannheim, Germany*
[4]*Department of Structural Biology, Genentech Inc., South San Francisco, CA, USA*
[5]*Thomson Instrument Company, CA, Carlsbad, USA*
[6]*Pharmaceutical Development, Genentech Inc., South San Francisco, CA, USA*

Protein Higher-Order Structure (HOS)

Historically, biotherapeutics referred to recombinant peptides and proteins, including cytokines, growth factors, hormones, interferons, monoclonal antibodies (mAbs), and other proteins and peptides. Today, the term also includes viral particles and cells, all of which share the common feature that their efficacy typically depends on specific protein interactions that define the drug's MoA [1–5].

Proteins exhibit structural levels (Figure 8.1), starting with the covalent bonding of amino acids. These peptide bonds are formed by linking the α-carboxyl group of one amino acid to the α-amino group of another. The resulting polypeptide chain comprises the protein backbone, which consists of repeating amide and carbonyl groups, and the individual amino acid or residue side chains. Due to its partial double-bond character, the peptide bond between the carbonyl and amide groups is planar, which restricts its conformational freedom [6, 7]. In contrast, the bonds between the C_α atom and the neighboring carbonyl and amide groups are single bonds, allowing for respective degrees of rotational freedom [7, 8]. This primary structure, defined by the specific sequence of amino acids, ultimately determines how the protein folds into its HOS [9]. Therapeutic proteins are composed of hundreds to thousands of amino acids. An average mAb consists of ~1,300 amino acids with a total molecular weight of ~150 kDa [10].

The local folding of the polypeptide chain into repeating patterns forms the protein's secondary structure (Figure 8.1). The most common secondary structures are the α-helix and β-sheet, both stabilized by hydrogen bonds between the backbone amide and carbonyl groups. Other secondary structure elements include β-turns, ω-loops, and unstructured regions known as random coils. The α-helix is tightly packed with 3.6 amino acids per turn. In contrast, the π-helix contains 4.4 amino acids per turn, and the 3_{10}-helix has 3 amino acids per turn [11]. Certain amino acids, such as valine, isoleucine, threonine, aspartate, asparagine, and serine, can destabilize helical structures. Valine, isoleucine, and threonine destabilize due to branching at C_β atoms, while aspartate, asparagine, and serine interfere through their side chain hydrogen bonds [12]. Additionally, proline is known as a helix breaker because the absence of an amide hydrogen disrupts the helical structure. Glycine has also been reported to promote local helix bending, owing to its lack of a side chain [13]. β-Sheets are similarly stabilized by hydrogen bonding and are formed by parallel and antiparallel β-strands, which can constitute large structural elements within the protein. In mAbs, the secondary structure elements contribute to the formation of antigen-binding sites or paratope regions. Any alteration in these structures could impair the antibody's ability to bind to its target's epitope region, reducing its therapeutic effectiveness [14–17].

Most therapeutic proteins are water soluble, with their thermodynamically stable conformation dictated by the hydrophobic effect. Polar residues are oriented toward the surrounding aqueous solvent, while nonpolar residues are shielded from the solvent and buried within their three-dimensional structure. This three-dimensional conformation, or tertiary structure (Figure 8.1), of a single polypeptide chain facilitates interactions between the side

Characterizing Biotherapeutics: Analytical Methods for Diverse Modalities, First Edition. Edited by Jennie R. Lill and Wendy Sandoval.

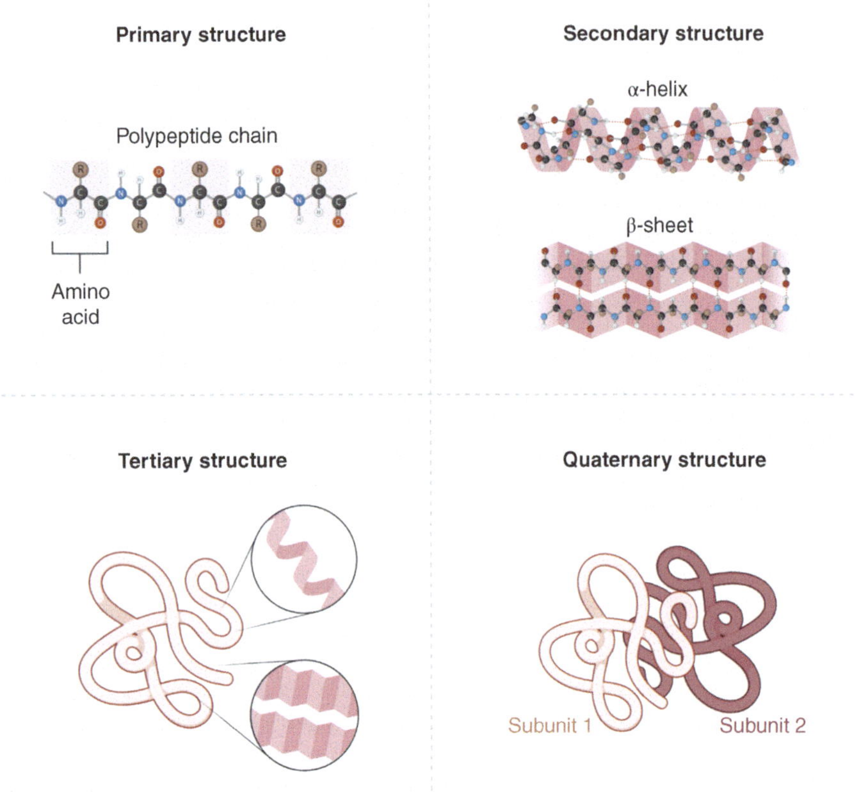

Figure 8.1 Schematic representation of protein structural levels, illustrating primary, secondary, tertiary, and quaternary structures. Source: Created in BioRender. Kuhne, F. (2024). BioRender.com/t58p006. Adapted from "Protein Structure". Retrieved from https://app.biorender.com/biorender-templates.

chains of the amino acids, including hydrogen bonds, ionic bonds, and disulfide bridges [18, 19]. Additionally, tight packing of the protein core is supported by hydrophobic interactions, such as van der Waals forces between nonpolar side chains, which enhance the stability of the protein conformation [20]. The three-dimensional assembly of polypeptide secondary structures forms domains, distinct functional, and structural units within a protein. As such, the tertiary structure is crucial for the protein's biological function, as it defines the spatial arrangement of active sites, binding regions, and other functional domains. In therapeutic proteins, such as enzymes or growth factors, the precise tertiary structure is essential for catalytic activity and receptor binding. For example, the enzyme therapeutic factor VIII requires an intact tertiary structure to properly interact with coagulation factors in the treatment of hemophilia [21] and misfolding or incorrect disulfide bridge formation in mAbs can affect the antibody's ability to bind to its target antigen or lead to increased structural instability, aggregation, and immunogenicity, thereby compromising patient safety [22–25].

For proteins composed of more than one polypeptide chain, the quaternary structure (Figure 8.1) describes the arrangement and interaction of these so-called subunits. The subunits may be identical or different, and their assembly is driven by non-covalent interactions or covalent bonds. For example, antibodies are built of two heavy polypeptide chains and two light polypeptide chains, which are paired together via disulfide bridges. The complementarity-determining regions (CDRs) of therapeutic mAbs, which form the paratope or target-binding region, consist of both heavy- and light-chain domains. Together, these domains create a highly specific binding interface that has been matured to recognize and bind disease target structures with high affinity.

The protein HOS encompasses the secondary, tertiary, and quaternary structures (Figure 8.1). Weiss et al. defined it as all structural elements beyond primary structure, necessary for the protein product to function as intended [26]. The term HOS also covers the dynamic ensemble of protein conformations, which occur over a wide range of time scales, from rapid local motions in picoseconds to large-scale conformational changes in milliseconds, the thermodynamic stability of folded domains, and non-covalent self-association [27]. For example, the so-called breathing motions in proteins refer to relatively slow, large-scale conformational changes that involve the expansion and contraction of specific regions, such as domains or subunits, around a flexible core [28]. These motions allow proteins to adjust their structures for binding events, allosteric regulation, or transitions between different functional states [29]. These dynamic processes are integral to the protein HOS, influencing stability, function, and interactions.

During manufacturing, that is, bioprocessing of recombinant proteins, disruptions, or alterations in the protein structure, whether due to manufacturing processes, formulation changes, or storage conditions, can lead to a loss of function, increased immunogenicity, or aggregation. Variations in bioprocessing can affect cellular metabolism, resulting in altered post-translational modification (PTM), such as glycosylation of the target protein, potentially impacting the protein HOS. For example, different N-glycosylation structures in the antibody fragment crystallizable (Fc) influence antibody effector functions differently through HOS changes [30]. Likewise, chemical modifications induced by the bioprocess or storage conditions can impact the structural integrity of the protein. Deamidation and isomerization at asparagine and aspartate residues as well as oxidation at methionine and tryptophan in or close to the CDRs can significantly influence the target binding behavior of mAbs [31]. Since the CDRs are exposed and flexible regions, the amino acids in these regions are accessible to solvents and thus to modifications. Changes in the HOS can affect their accessibility and susceptibility to modification. Such alterations in the primary sequence can trigger cascading structural changes, leading to aggregation and increasing immunogenicity concerns. Consequently, in the development of biotherapeutics, preserving the integrity of a protein's HOS is crucial for ensuring therapeutic efficacy, safety, and stability.

Importance of HOS Characterization in Biotherapeutic Development

Structural methods support drug design in the early stages of biotherapeutic development. As indicated above, technological advances have led to the ability to generate mAbs with high specificity and affinity toward their targeted antigen. The structural basis for this targeted binding is achieved through the directed design of mAbs, also known as protein or antibody engineering. Display technologies employ large gene libraries, sequenced through next-generation sequencing (NGS), to fast-track the selection of target-binding mAbs by linking genotype and phenotype through surface expression. As such, phage display serves as a crucial technology in mAb affinity maturation [32]. Multiple binders are initially identified and transiently expressed, typically in human embryonic kidney (HEK) cells. These binders then undergo biochemical analysis and developability assessments to identify the most promising candidates. Once an optimal binder is selected via enzyme-linked immunosorbent assay (ELISA), the corresponding genetic constructs are cloned into protein-producing cells, often genetically engineered mammalian cells like Chinese hamster ovary (CHO) cells [33]. As previously described, the structural integrity of proteins significantly influences their functionality and hence efficacy. The functional screening of potential drug candidates is therefore often accompanied by characterization of the precise structural binding interface. The binding interface on the target antigen, known as the epitope, interacts specifically with the antibody's paratope structure, typically composed of the CDRs. Epitopes and paratopes can be linear (continuous) or conformational (discontinuous), with interactions occurring through one-dimensional peptide structures or three-dimensional configurations, respectively (in practice, distinguishing between linear and conformational binding regions is not always straightforward) [34, 35]. Determining the precise epitope enables mechanistic understanding of binding and specificity while minimizing off-target effects. This also aids in evaluating the developability of mAbs by assessing the amino acids in the paratope region for potential modifications and their impact on binding affinity, such as known genetic sequence variants, chemical modifications, or PTMs. Furthermore, intellectual property (IP) related to epitopes and paratopes can be claimed, which is critical for patent protection of biotherapeutics. Epitope mapping, supported by various structural biology methods, can assist in selecting the most promising lead candidates for subsequent bioprocess development and clinical trials. As illustrated by mAbs, epitope mapping across various modalities can be used to understand protein interactions or binding interfaces, MoAs, and to establish IP claims. For example, in addressing the efficiency and safety of gene therapy, antibodies neutralizing adeno-associated viruses (AAVs) could be identified by epitope mapping [36, 37]. Additionally, in early research, epitope mapping plays an important role. For instance, SARS-CoV-2 antibodies in COVID-19 patients could be characterized and classified based on the severity of the disease and their neutralization activity [38].

Developing efficient and robust cell fermentation processes, and highly specific purification of the target protein from thousands of different proteins and other cell components, result in a complex interplay of factors that can potentially affect product quality [39]. Additionally, as outlined in the previous subchapter, the natural structure of proteins is highly heterogeneous, with therapeutic proteins exhibiting natural heterogeneity due to the combination of dozens of possible sequence modifications such as PTMs [40]. This multifactorial interplay of expression systems, process parameters, and protein structures necessitates continuous screening and understanding of their impact on biotherapeutic quality. Robust bioprocesses for producing potent and safe biotherapeutics rely on this understanding to control both process and molecular variability [41]. Therefore, analytical and functional characterization is performed throughout molecule development and shelf life. Among various drug quality factors, biotherapeutic characterization also requires systematic structure-to-function analysis to enable a fundamental understanding of functionality and its dependence on structure [42]. Assessing the protein's HOS is a crucial part of this structure-to-function analysis. While there are many methods for characterizing protein structure, not all are equally suited to describe higher-order structural effects [26, 27]. Well-established spectroscopy methods, such as Circular Dichroism (CD), Raman spectroscopy, and Fourier-transform Infrared (FTIR) spectroscopy, lack the sensitivity and resolution needed to detect fine structural variations. Additionally, while they can capture some dynamic changes, they are limited in providing detailed information on protein dynamics. These limitations hinder a detailed understanding of protein folding, conformational states, and interactions, all crucial for drug and corresponding bioprocess development. In this chapter, we describe a selection of high-resolution structural biology methods capable of, but not limited to, characterizing protein binding interfaces in detail, with a particular emphasis on their application for epitope mapping. The methods presented here offer complementary features, but no single approach can address all the challenges in characterizing protein HOS. Comprehensive characterization of protein HOS requires orthogonal and complementary technologies. Therefore, a corresponding strategy is essential, guided by the specific structural questions and taking into account the strengths and limitations of the available technologies.

For atomic resolution, X-ray crystallography is considered the gold standard. However, it necessitates the crystallization of proteins, posing significant challenges for large protein complexes and flexible regions. In epitope mapping, X-ray crystallography provides precise visualization of antigen–antibody complexes and identifies exact binding residues. However, crystallization conditions may alter the native epitope structures, and the method requires substantial amounts of purified protein. Moreover, X-ray crystallography captures a static snapshot of the protein structure, making it inadequate for representing protein dynamics. In contrast, nuclear magnetic resonance (NMR) is capable of capturing molecular motion, crucial for studying conformational changes, and protein interactions. For epitope mapping, NMR detects subtle changes in protein dynamics and identifies flexible epitopes. Both X-ray crystallography and NMR resolve single residue alterations in protein interactions. However, their technical limitations restrict the analysis of larger proteins. Emerging technologies like cryogenic electron microscopy (cryo-EM) offer near-atomic resolution and can visualize large antigen–antibody complexes, providing high-resolution details of binding interfaces for epitope mapping. When dealing with numerous samples in limited quantities, mass spectrometry (MS) offers an excellent compromise between resolution and throughput. Additionally, MS can handle a broad range of protein sizes. Automation and data interpretation are more straightforward compared to X-ray crystallography and NMR. Hydrogen/deuterium exchange mass spectrometry (H/DX-MS) investigates backbone dynamics and is considered a non-covalent labeling approach. In contrast, covalent labeling approaches, such as hydroxyl radical protein footprinting (HRFP-MS) and diethyl pyrocarbonate (DEPC) labeling MS, provide permanent labels that provide information about side-chain interactions and side-chain solvent accessibility. In epitope mapping, H/DX-MS identifies epitope regions by measuring the decrease in deuterium uptake upon antibody binding. While distinguishing between an actual binding site and a conformational change induced by binding can be challenging with H/DX-MS and other labeling MS methods, combining MS methods with orthogonal approaches like X-ray crystallography often provides valuable insights into the binding interface. A comprehensive overview of the advantages and challenges of the methods described is summarized in Table 8.1. Further technical details and limitations will be discussed in the following method subchapters.

As biotherapeutics become increasingly complex and multifaceted, HOS methods are essential to provide profound insights into the molecular mechanisms of drug structure–function relationships and stability. Anticipated contributions of these methods in bioprocess and biotherapeutic development include formulation development, investigation of aggregation mechanisms, understanding biotherapeutic MoA, assessing and screening CQAs, enabling structure prediction, and collaborating with *in silico* approaches. These methods offer experimental data to validate and refine computational predictions of protein structure and dynamics. This comprehensive analytical characterization using experimental and computational methods enhances our understanding and prediction of protein behavior, facilitating the design of high-quality, safe, and effective next-generation drugs. As these technologies mature, they can support not only general drug discovery research but also drug development within the chemistry, manufacturing, and control (CMC) space.

Table 8.1 Complementarity of structural methods for characterizing protein higher-order structures.

Method	Pros	Cons	Complementarity
H/DX-MS	Rapid analysis High sample-throughput and automation Solution-phase analysis No size restrictions Good sensitivity Captures protein dynamics Compatible with most solution conditions	Low resolution at the amino acid level Resolves H/DX in limited time range Deuterium back-exchange reduces sensitivity Requires low pH and low temperature Unable to differentiate between conformational changes and direct binding events	Information on protein backbone dynamics and solvent accessibility
HRFP-MS	Rapid analysis Solution-phase analysis No size restrictions Irreversible labeling	Stochastic nature impacts reproducibility Requires specific reagents/technologies to generate hydroxyl radicals Buffer/matrix conditions impact radical scavenging Unable to differentiate between conformational changes and direct binding events	Information on side chain solvent accessibility
CL-MS	Rapid analysis Solution-phase analysis No size restrictions Irreversible labeling	Low resolution Requires specific toxic chemical reagents for covalent labeling Alterations in protein structure may occur due to labeling Unable to differentiate between conformational changes and direct binding events	Information on side-chain solvent accessibility
NMR	Atomic resolution Provides information on dynamics, conformation, and interactions Solution-phase analysis	Complex analysis above ~50 kDa protein size Complex sample preparation and data interpretation Requires a high sample amount Complicated peak identification	Spatially resolved information on protein dynamics
X-ray	High atomic resolution Provides detailed three-dimensional structural information	Static structure may not reflect the native structure Time-consuming Challenging to generate diffracting crystals (difficult for large complexes)	Atomic resolution information on the static state of proteins
Cryo-EM	High resolution Works for large complexes	More difficult for smaller proteins Requires a high sample amount Limited to samples purified in solution Buffer/matrix incompatibility considerations	High-resolution information on various static states of proteins

Hydrogen/Deuterium Exchange Mass Spectrometry (H/DX-MS)

The foundation of H/DX lies in the ability of labile amide hydrogens in the protein backbone to exchange with deuterium when the protein is placed in a deuterated solvent, typically heavy water (D_2O). The H/DX phenomenon was first described by Kaj Ulrik Linderstrøm-Lang in the 1950s [43]. He proposed a relationship linking hydrogen exchange rates with the dynamic behavior of protein molecules and nucleic acids, a concept that is still widely used today.

Proteins contain three types of hydrogen: covalently bonded hydrogens, those in exchangeable groups like hydroxyl or amino groups, and amide hydrogens in the protein backbone as an illustration shown in Figure 8.2. Covalently bonded hydrogens do not undergo deuterium exchange. Hydrogens in exchangeable groups on the side chains undergo rapid H/DX and back-exchange. Thus, H/DX focuses specifically on the amide hydrogens, as their exchange rates, with half-lives ranging from approximately 5 seconds to 60 days, can be easily monitored by a liquid chromatography–mass spectrometry (LC–MS) system. These amide hydrogens participate in hydrogen bonding, and their accessibility to solvent reflects the folding and conformation of the protein, providing valuable information about the protein's HOS and dynamics.

The kinetics of H/DX are described by the rate constants of exchange of amide hydrogens in proteins, which are influenced by four main factors: pH, temperature, solvent accessibility, and hydrogen bonding. The rate of exchange is highly dependent on pH, following a bell-shaped curve with a minimum rate roughly at pH 2 – pH 3 [44]. In very acidic or very basic environments, the exchange rates are slower due to the involvement of acid- or base-catalyzed mechanisms. In the neutral pH range, the exchange is large base catalyzed, meaning that hydroxide ions (OH^-) facilitate the exchange. At lower pH, the rate of exchange is slower because there are fewer OH^- available to catalyze the exchange, but it is not completely halted, as acid-catalyzed mechanisms still occur. In practical H/DX experiments, the exchange reaction is often quenched by lowering the pH to around 2.5, reducing the exchange rate for an average backbone amide by 10,000-fold compared to the rate at pH 7.0.

Like most biochemical reactions, H/DX is temperature dependent, with higher temperatures generally leading

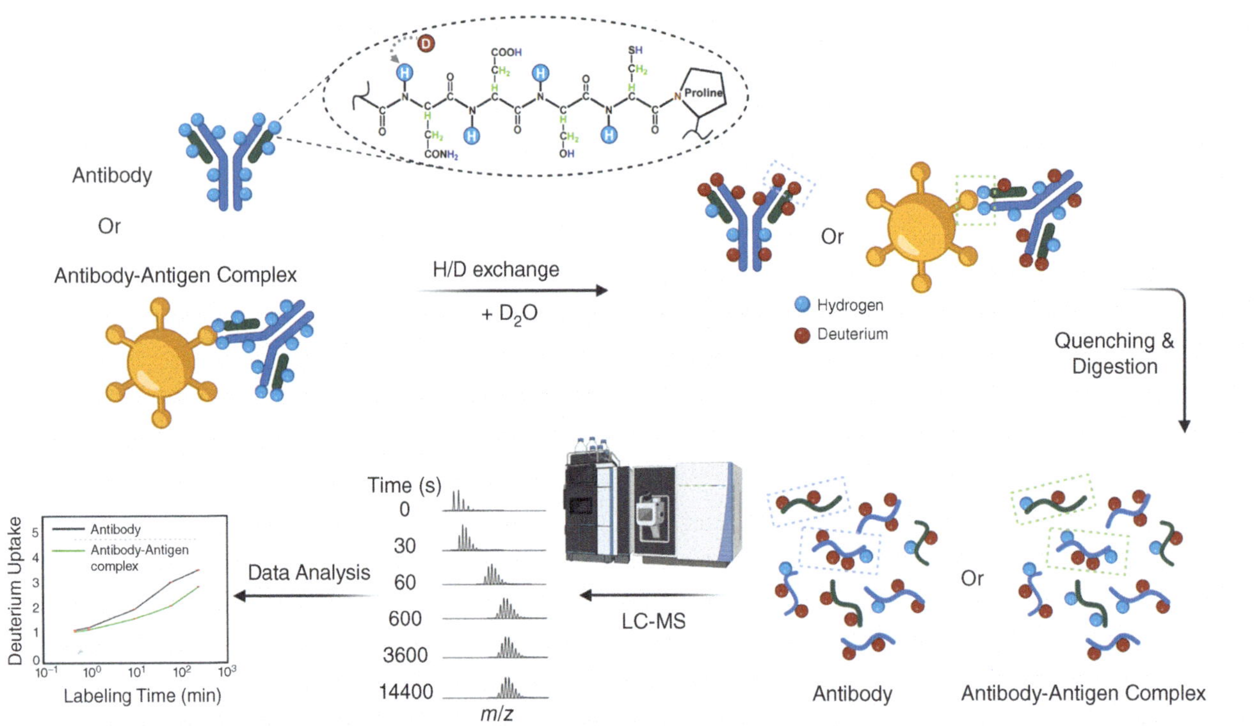

Figure 8.2 A general workflow of hydrogen/deuterium exchange mass spectrometry experiment. Protein samples are incubated in a deuterated buffer, allowing for the exchange of backbone amide hydrogens with deuterium over time. After defined time intervals, the exchange reaction is quenched by lowering the pH and temperature, followed by rapid proteolysis. The proteolytic peptides are analyzed by LC–MS to measure the mass shift resulting from deuterium incorporation. The mass spectra are processed to determine the extent of deuterium uptake for each peptide, providing insights into protein's HOS and dynamics. Source: Created in BioRender. Liu, R. (2024) BioRender.com/r55o093.

to faster exchange rates. Elevated temperatures increase molecular motions, making it easier for amide hydrogens to become accessible to the solvent. In contrast, lower temperatures slow down exchange, which is why many H/DX-MS experiments involve quenching the reaction by reducing the temperature to near-freezing levels, typically around 0°C. This is done to "lock in" the exchange state and prevent back-exchange, where deuterium atoms revert back to hydrogen before analysis.

The protein's conformation and structural dynamics play a crucial role in determining the exchange rates. For regions of the protein that are exposed to the solvent and not involved in strong hydrogen bonds, the exchange occurs relatively quickly. However, in tightly folded regions where the amide hydrogens are involved in stable hydrogen bonds, the exchange occurs more slowly. Overall, the exchange rate of amide hydrogens reflects the structures of proteins. The H/DX reaction is a convoluted process including protein folding and unfolding (closed/open states) as well as the exchange of amid hydrogen atoms with the deuterium atoms. The basic model including EX1 and EX2 kinetics, which continues to serve as a foundational principle for H/DX, was proposed by Kaj Ulrik Linderstrøm-Lang [45]. In the EX2 mechanism, which is the most commonly observed under physiological conditions, the rate of hydrogen exchange is much slower than the rate of protein unfolding and refolding. This means that the amide hydrogen is exposed briefly when protein undergoes local, transient fluctuations, allowing deuterium to replace the hydrogen before the protein refolds. The exchange rate in the EX2 regime is governed by the stability of hydrogen bonds and solvent accessibility, with structured or "closed" regions exchanging slowly and unstructured or "open" regions exchanging rapidly. In contrast, the EX1 mechanism occurs when the rate of unfolding is slower than the rate of exchange. This typically happens in conditions where protein unfolding is slow, and the exchange occurs rapidly once the protein transitions to an open state. In this regime, multiple hydrogens exchange simultaneously upon unfolding, leading to a cooperative exchange event that results in a bimodal distribution of deuterium incorporation. Over many years of research, it has been observed that, under physiological conditions, most proteins undergo rapid local structural fluctuations, leading to EX2 kinetics. In solution, protein predominantly exists in populations where the native, folded, and closed state is favored.

Experimental H/DX-MS Workflow

The detection of H/DX for biological molecules using MS began in the 1990s, taking advantage of the fact that deuterium has roughly twice the mass of hydrogen. Zhang and Smith described a method in which H/DX reactions were monitored through a series of steps, including proteolysis, chromatography separation, and MS analysis [46]. Nowadays, the H/DX workflow begins with incubating the protein in a deuterated (D_2O) buffer, where the exchangeable amide hydrogens in the protein backbone are replaced by deuterium. Most H/DX-MS experiments employ a continuous labeling strategy where the H/D reaction proceeds for various amounts of time from seconds to hours. After a specific labeling period, the reaction is quickly quenched by lowering the pH to approximately 2.5 and the temperature to around 0°C, which slows down any back-exchange of deuterium to hydrogen. Following quenching, the protein is digested into smaller peptides using an acid-resistant protease such as pepsin under cold conditions to preserve the deuterium labeling. The resulting peptides are then separated using ultra-high-performance liquid chromatography (UHPLC) and analyzed by MS to detect mass shifts following deuterium incorporation.

H/DX analysis software can generate time-course deuterium uptake curves, heatmaps by deuterium uptake difference, or butterfly plots, where the x-axis represents the protein sequence and the y-axis represents the level of deuterium incorporation. In a heatmap, the color intensity indicates the level of deuterium uptake at each residue, the hotter colors signifying higher deuterium exchange, and thus greater solvent accessibility in the region of protein structure. Overall, the processed mass spectrometric data can provide a picture of solvent accessibility and structural dynamics at regions of interest.

Applications for H/DX-MS

The use of H/DX-MS in the study of protein HOS offers several advantages over traditional methods such as X-ray crystallography or Cryo-EM. One of the key benefits is that H/DX-MS allows for the study of protein–protein interactions in solutions, close to physiological conditions. This enables the characterization of protein HOS in their native states, capturing dynamic conformational changes that may be missed by techniques requiring crystalline samples. Additionally, the resolution of H/DX-MS has improved significantly with recent advances in instrumentation, more efficient protein digestion, and enhanced data analysis software. High-resolution MS allows for precise detection of deuterium incorporation, enabling more accurate identification of conformational changes. Another key development has been the automation of H/DX workflows. Automated systems allow for more reproducible sample handling, improved throughput, and enhanced data quality. This has been especially beneficial for large-scale epitope mapping projects, where high sample throughput is essential. Automated systems also reduce human error, ensuring more consistent results across experiments. Moreover, H/DX-MS has no limitations in protein size or class, unlike some other structural techniques such as NMR, which can be restricted by protein size. Recent developments in the H/DX-MS field make it possible to address protein problems with larger size and higher complexity. At present, large proteins over 150 kDa such as antibodies have been routinely studied by H/DX-MS.

Despite its broad applicability, H/DX-MS has certain limitations. One challenge is the issue of back-exchange, where deuterium atoms revert to hydrogen during sample processing, potentially reducing the accuracy of detection of deuterium incorporation. Efforts to minimize back-exchange through optimized workflows and experimental conditions are still ongoing, but this remains a technical challenge. Another limitation is that H/DX-MS often requires extensive peptide coverage to map the entire protein surface. Incomplete peptide coverage can lead to gaps in the epitope map, especially in large or highly glycosylated proteins.

H/DX-MS for Epitope Mapping

H/DX-MS is highly sensitive to conformational changes, such as those caused by ligand binding, environmental shifts, or protein–protein interactions. It provides a dynamic picture of protein structures in solution, making it especially valuable for studying the structural integrity and conformational flexibility of biotherapeutics. One of the most critical applications of H/DX-MS in the biopharmaceutical field is epitope mapping, which identifies the regions of an antigen that interacts with an antibody. In order to map the epitopes, H/DX-MS experiments are typically conducted on both the free antigen as a control and the antigen–antibody complex. The data from these two experiments are then compared, and regions of the protein displaying different exchange kinetics are identified as potential epitopes. In a study by Zhang et al., H/DX coupled with FT-ICR MS has been applied to map the epitopes of a large homohexameric food allergen protein, cashew 11S globulin allergen, Ana o 2 (95 kDa) [47]. Ana o 2 is a major cashew nut allergen that may have clinical significance. H/DX-MS data showed that two mAbs, 2B5, and 1F5, can bind to Ana o 2, where mAb 2B5 defines a conformational epitope, and mAb 1F5 defines a linear epitope. Study of overlapping peptides indicated that mAb 1F5 epitope is near the C-terminus of the large subunit composed of the segment from residues 257 to 271 of Ana o 2. In contrast to 1F5, mAb 2B5 binds to the acidic large subunit of Ana o 2 when the large subunit interacts with the small subunit. The cashew large subunit itself does not bind to mAb 2B5, which suggests a conformational epitope. Overall, the epitope mapping results by H/DX-MS agreed with the previous publication by Robotham et al. [48], thereby demonstrating the capability of H/DX-MS epitope mapping of large antigen–antibody complexes for clinical applications.

In the field of cancer immunotherapy, Zhang et al. probed the epitope/paratope of PD-1/Nivolumab by HDX-MS [15]. Programmed cell death protein 1 (PD-1), an antigen co-receptor on cell surfaces, is one of the conspicuous immune checkpoints. Nivolumab is an FDA approved mAb that targets PD-1 and blocks its pathway. Unlike the practical H/DX-MS experiment, the researchers combined H/DX-MS with electron-transfer dissociation (ETD), providing binding-site characterization at residue level. In addition to H/DX-MS, chemical cross-linking also has been explored in the study to provide complementary information on additional epitope/paratope structures, including the BC-loop, the N-terminus of the heavy chain, and a distal conformational change of C'D loop. The integration of H/DX-MS, chemical cross-linking, and docking offers a full picture of PD-1/Nivolumab interactions, which was critical for understanding the antibody's mechanism of action (MoA) and optimizing therapeutic efficacy in the future.

In addition to antibodies, H/DX-MS has been used to study a wide range of drug candidates, from large protein therapeutics to small synthetic peptides. In a study by Puchades et al., H/DX-MS was used to map the epitopes of different drug candidates that target the stem region of Hemagglutinin (HA), the major surface antigen of influenza viruses [49]. The study utilized a semiautomated H/DX-MS method to study HAs and their interaction with drug candidates. H/DX-MS data revealed a major change in the local conformational dynamics of the A helix that has critical mechanistic implications and is associated with broad neutralization, which established a marker for broad anti-HA activity. As such, the findings demonstrate that H/DX-MS can be used to identify promising drug candidates against HA in the early stage of drug discovery.

In addition to epitope mapping, regulatory agencies have increasingly recommended state-of-the-art and orthogonal methodologies be used for demonstrating structural similarity. Houde et al. demonstrated the utility of H/DX-MS in biopharmaceutical comparability studies [50]. Hong et al. applied H/DX-MS for an in-depth comparison of the protein conformations of biosimilars [51]. Both cases show its ability in evaluating similarity in HOS of biotherapeutics.

In summary, H/DX-MS is valuable in the development of mAbs and vaccines, as it provides insights into the structural basis of immune recognition and antibody specificity. Nowadays, with improved instrumentation and

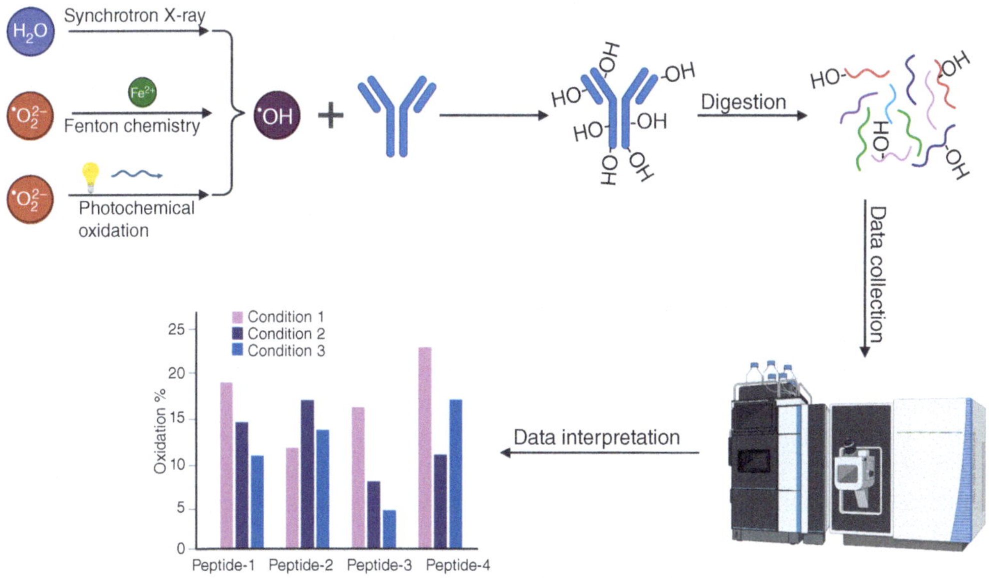

Figure 8.3 The whole workflow of HRPF-MS. Hydroxyl radicals can be generated using various methods and mixed with the sample for the labeling reaction. The labeled sample is subsequently digested into peptides, which are prepared for ultra-performance liquid chromatography – tandem mass spectrometry (UPLC-MS/MS) analysis. The acquired data is then analyzed and mapped against the original sequences of the sample, providing a detailed picture of the oxidation levels at different regions of the protein. The method can offer insights on protein structures and interactions on both peptide and residue levels. Source: Created in BioRender. Wang, L. (2024) BioRender.com/z90q594.

data analysis tools, H/DX-MS has been routinely used in both academia and the biopharmaceutical industry. Combining H/DX MS with complementary structural techniques, such as Cryo-EM and X-ray crystallography, has expanded the applicability of the technique. These integrative approaches provide a more comprehensive view of protein structures by adding high-resolution static data from crystallography or cryo-EM to the dynamic information obtained from H/DX-MS.

Hydroxyl Radical Protein Footprinting Mass Spectrometry (HRPF-MS)

HRPF-MS is a bottom-up mass spectrometry technique in which a target protein is "labeled" with hydroxyl radicals at oxidation-prone amino acid side chains. Hydroxyl radicals are similar in size to that of water molecules and are highly reactive with solvent-exposed amino acids. Thus, the extent of labeling reflects both the reactivity of the amino acid to oxidation, and to the solvent accessibility of the amino acid side chain [52]. Labeled proteins are subsequently subjected to proteolysis, and the resulting peptides are separated and analyzed using LC-MS (Figure 8.3). Quantifying the extent of oxidation on each peptide enables inference regarding the solvent accessibility at the region of the protein in which the peptide is associated with. Therefore, the principle of HRPF is to take advantage of perturbations to a protein structure that impacts the local solvent accessibility. Measuring the extent of oxidation on each peptide in two different states enables the identification of changes in protein structure via the change in solvent accessibility.

Experimental Components of HRPF

Several key experimental factors are crucial to ensuring the success of an HRPF-MS experiment, as summarized in Table 8.2.

Radical Sources

There are primarily two sources for generating hydroxyl radicals. The first method involves irradiating water to produce hydroxyl radicals, which can be performed through electrolysis [53], plasma [54], or X-ray exposure [55, 56]. The second method utilizes hydrogen peroxide as the radical source, which can react with metal ions to produce radicals through Fenton chemistry [57], or can be photolyzed by a UV laser [58] or high-pressure flash-lamp [59]. The latter technique using light irradiation has been known as fast photochemical oxidation of proteins (FPOP).

Scavenger

Scavengers are a component of HRPF that is added to control the lifetime of the radicals to ensure there is no

Table 8.2 Experimental components of HRPF-MS.

Components	Option	Function
Radical source	H_2O/H_2O_2	Radical generation for labeling reaction
Scavenger	Histidine/glycine/methionine/adenine	Controls the lifetime of radicals to prevent over-oxidation
Dosimeter	Leucine Enkephalin/histidine/adenine	Measures radical generation and exposure
Quench	Catalase/DMTU/methionine	Stops reaction and consumes remaining peroxide

over-oxidation of the target protein, or any downstream propagation into other reactive oxygen species (ROS) [60]. In addition, the scavenger is used to ensure that the target protein is only labeled in the native state and not labeled in any subsequent conformation that is a result of the initial labeling process (such as unfolding, aggregation, or conformational changes). Common scavengers include amino acids such as tryptophan, tyrosine, methionine, histidine, and arginine, each with varying reactivities [61]. Molecules that elicit spectroscopic changes upon oxidation, such as Adenine or Alexa Fluor dyes, are also widely used as an alternative scavenger [62].

Dosimeter

Dosimeters are routinely included in HRPF samples to monitor overall effective dose of hydroxyl radical generated during sample labeling [62]. Two common strategies for dosimetry are either the addition of peptides or the use of spectroscopically active molecules. The former strategy has the advantage of using a dosimetry that is more representative of the target protein, but requires offline LC-MS analysis to determine the effective dose of radicals [63]. The latter strategy has the advantage of using changes in absorbance to measure the effective dose of radicals and has been applied to in-line applications [64]. Regardless of the strategy, the use of a dosimeter enables the assurance of reproducible radical generation, reducing sample-to-sample variability.

Quenching

Labeled samples are collected into a quenching solution to halt the excessive labeling reaction by consuming the remaining hydroxyl radicals and hydrogen peroxide. A common quenching solution consists of methionine and catalase. Alternatively, dimethylthiourea (DMTU) can be used as a small molecule quencher for radicals together with methionine [65]. DMTU reduces background noise in peptide mapping data on MS, thereby enhancing the method's sensitivity for complex samples.

Data Interpretation

Data collected from bottom-up MS is analyzed through peptide mapping. The oxidation level is calculated by dividing the amount of oxidized peptide by the total amount of oxidized and corresponding parent peptides. These modified peptides are then mapped against the original sequences of the protein sample. As indicated above, the level of labeling for any given peptide is a direct reflection of the local solvent accessibility and the reactivity of the associated amino acids. The change in solvent accessibility of a peptide when measured in two different states, therefore, provides information about the structural changes that occurred in the peptide. For example, comparing the labeling of a target protein under different conditions (such as buffer conditions, the presence of detergents, and stress conditions) can provide information about the structural changes that have occurred. Likewise, by comparing two interacting proteins in their unbound to their bound states, peptides at the binding interface will exhibit less solvent accessibility and less associated labeling. HRPF can help to characterize binding interfaces and different binding states, making it a valuable tool for studying protein–protein interactions. However, it is challenging to identify binding interfaces solely with HRPF, since it can be difficult to decipher changes in solvent accessibility that are due to a binding interaction or a structural change elicited by a binding interaction. Therefore, it is often used in combination with other structural elucidation methods such as H/DX-MS [66], X-ray crystallography [67], Cryo-EM [68], and NMR [69, 70].

Applications for HRPF-MS

As discussed, HRPF data provides structural information regarding the region-specific solvent accessibility of a target protein, by measuring the extent of labeling for the individual peptides. This technique is best used in a comparative manner to determine changes in solvent accessibility under two different states or conditions. This section will discuss the use of HRPF-MS to identify the binding interface located between protein–protein interactions. The focus of these applications will be on biotherapeutic proteins and their associated binding partners.

HRPF-MS for Epitope Mapping

Therapeutic proteins such as mAbs are effective therapeutic targets due to their highly specific binding interaction with the associated target antigen [71]. The site-specific region on the antigen in which the mAb binds is denoted as the epitope. Understanding the molecular composition of the epitope is critical for therapeutic design and IP. HRPF-MS is an ideal tool for the characterization of antigen epitopes since the technology is typically not limited to the size, shape, or physical properties of the target protein.

One of the first examples of using HRPF-MS for epitope mapping was performed by Zhang et al. in 2017 to map the interface between a therapeutic Fragment antigen binding (Fab) and the target vascular endothelial growth factor (VEGF) [67]. This work identified reduced labeling of both peptides and individual residues, that when mapped onto the available crystal structure, identified the binding interface. However, there were also regions distal from the binding site that showed both increase and decrease in labeling, highlighting the importance of using complementary or orthogonal methods for discerning the difference between a binding site and a structural change.

Li et al. conducted a case study on the epitope mapping of the interleukin-6 receptor (IL-6R) using a combination of multiple MS-based protein footprinting methods, including H/DX-MS, HRPF-MS, and carboxyl group labeling [66]. In this study, FPOP revealed significantly reduced modification levels in peptides from the regions of IL-6R binding to adnectin-1 and adnectin-2. Notably, the combination of FPOP and carboxyl group labeling suggested binding patterns involving rapid dynamics and reorientations of flexible loops, which are typically undetectable by slower footprinting methods like H/DX. The results provided complementary insights, with H/DX-MS mapping the protein backbone and FPOP targeting side chains.

Lin et al. also used a combination of FPOP and carboxyl group labeling to identify two different binding sites on a highly glycosylated antigen utilized in a sandwich ELISA [72]. This work demonstrated that binding of either mAbs to the antigen significantly reduced the solvent accessibility across the antigen, making identifying the binding sites nearly impossible without the use of a combination of methods. This work also paired the bottom-up MS data with negative stain electron microscopy (EM), to enable visualization of the different binding orientations within the sandwich ELISA complex.

Finally, Schick et al. were the first to apply HRPF for epitope mapping of anti-drug antibodies (ADAs) to a clinical bispecific molecule [4]. This work described a bispecific molecule that exhibited unusually high amounts of ADA in clinical trials, not observed in the clinical trials of the respective monovalent parent mAbs. A companion study focusing on bioanalytical characterization of the ADAs suggested that only one of the bispecific arms were being targeted by the ADAs [73], whereas the HRPF-MS data demonstrated conclusively that both bispecific arms were targets of the ADAs. This study highlighted the capability of HRPF-MS to detect complex interactions that may not be detected by other techniques.

The examples provided in this section are intended to demonstrate that HRPF-MS is a powerful bottom-up MS technique for identifying and characterizing important biotherapeutic binding interfaces. However, this technology is often combined with other bottom-up MS techniques, spectroscopic techniques, and visualization HOS techniques such as crystallography and EM [74, 75]. This technique is invaluable for elucidating the structural basis of protein function, aiding in the design and optimization of protein therapeutics, and ensuring the stability and efficacy of biopharmaceuticals. The ability to provide detailed structural information in solutions makes this technology a powerful tool for advancing our understanding of protein HOS in various biological contexts. Further information can be found in these comprehensive reviews [52, 76, 77].

Covalent Labeling Mass Spectrometry (CL-MS)

CL-MS is a chemical labeling technique wherein a labeling reagent will modify the side chains of a protein, which is in turn used to interpret the structural state of the protein. This method involves selecting specific covalent labeling reagents that form stable and irreversible reaction products under experimental conditions. The stability of the labeled product is a significant advantage of CL-MS compared to H/DX-MS, where deuterium labels on the protein can undergo back exchange. To minimize back exchange of labels from the protein, H/DX-MS experiments are conducted at controlled cold temperatures, as detailed in the previous subchapter and in multiple publications [78, 79]. However, this approach incurs additional costs for specialized equipment and trained personnel. In contrast, covalent labeling offers a more stable and less labor-intensive alternative, as the labels bind stably to the side chains of specific amino acids on the protein's surface [80]. Thus, CL-MS provides two advantages: (i) it enables semi-specific targeting of amino acids, allowing for site-specific labeling information, and (ii) it provides insights into changes in protein surface interactions and solvent accessibility rather than the protein backbone dynamics, as seen with H/DX-MS. As a result, covalent labeling is primarily used to investigate alterations in the surface structure of proteins under specific biological or experimental conditions rather than changes in protein dynamics. When comparing altered and unaltered protein structures, differences in covalent label incorporation can be detected via MS, offering a sensitive and accurate readout. The changed label incorporation of proteins in different structural states allows qualitative and quantitative interpretation.

In summary, structural change information is encoded in the mass tags of the covalent labeling reagent. Compared to H/DX-MS, the data analysis is far more straightforward in CL-MS and can be easily adapted to the needs of scientists. Careful sample preparation and mass spectrometric data interpretation are however crucial factors for the success of CL-MS as a structural analytical technique.

CL-MS Methodology and Conditions

During sample preparation of CL-MS, meticulous attention to labeling conditions, such as the concentration of the labeling reagent and maintaining the unaltered structural state of the protein post-labeling, is essential. Common covalent reagents include hydroxyl radicals, fluoroalkyl radicals, carbene, N-hydroxysuccinimide (NHS) ester, and diethylpyrocarbonate (DEPC), as referenced in the literature [80–82]. Martín-Gago and colleagues have also demonstrated the covalent labeling of glutamic acids to probe ligand binding sites [83]. This section will focus on the use and application of DEPC as a covalent labeling reagent as illustrated in Figure 8.4 [84]. DEPC is increasingly valued in structural protein analysis due to its ease of use, in mapping the surface-accessible areas of proteins using standard LC-MS equipment found in most analytical laboratories. DEPC modifies the side chains of amino acids such as lysine, histidine, serine, threonine, tyrosine, free cysteine, as well as the N-termini of proteins. These modifications are detected by MS as a mass shift of +72.021 Da per DEPC label. This +72.021 Da mass-modified product is the most abundant label reaction product, making the mass spectrometric data interpretation straightforward. To increase the sensitivity of label detection, additional modifications, particularly on histidine residues, are well-documented [85]. Another important factor in setting up the covalent labeling experiment is the amount of labeling reagent to be used. Vachet et al. have detailed the reaction kinetics of DEPC labeling, correlating the protein concentration with the appropriate amount of DEPC reagent. Their study concluded that, for a 10 μM protein sample, an 8- to 10-fold molar excess of DEPC reagent reacting for about 5 minutes is optimal for labeling without compromising protein structural integrity [86, 87]. A recent publication further demonstrated that higher DEPC concentrations can enhance sensitivity in structural analysis without affecting protein stability [88]. After the DEPC labeling step, the reaction is quenched with imidazole, allowing the labeled protein to be analyzed by MS either at the intact protein level or at the peptide level [84].

The intact level labeled protein analysis will give the average number of labels per protein sample. One can also attempt to derive detailed information about labeled sites by using top-down sequencing. Given the challenging nature of top-down sequencing as an MS technique, the success of analyzing covalently labeled samples can vary significantly between different proteins. In contrast, bottom-up LC-MS, that is, peptide-level analysis of proteolytically digested, labeled proteins will provide detailed information about the label positions and the relative quantity of labels at each labeled site on the protein. To detect the maximum number of label-modified sites in MS analysis, it is critical to optimize MS settings so that the maximum number of peptides can be fragmented and

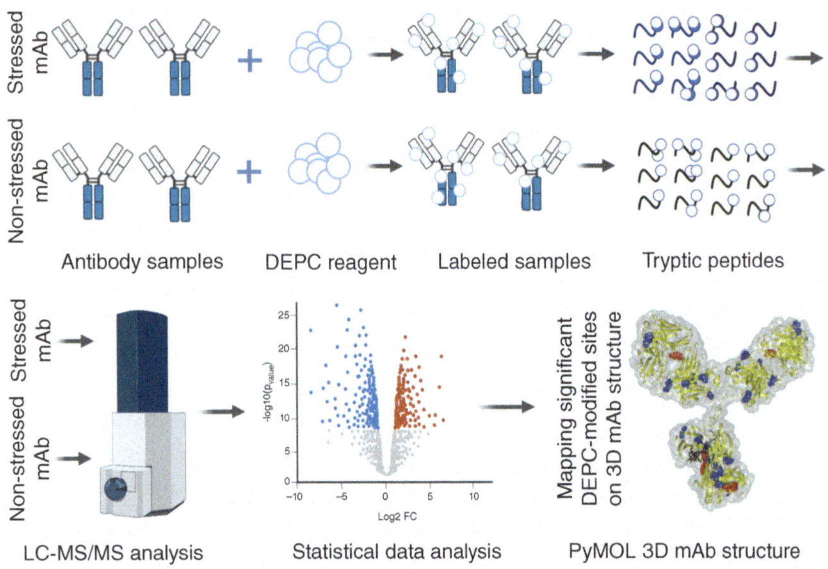

Figure 8.4 Schematic workflow for DEPC-based covalent labeling mass spectrometry workflow. DEPC reagent is allowed to react with the protein samples (in varied structural states) for a limited time and later the reaction is quenched. The labeled samples can be further analyzed at intact or peptide level using mass spectrometry. The tryptic digestion and MS/MS parameters are optimized to enable amino acid-specific detection and quantitative analysis of DEPC-labeled peptides (+72.021 Da mass-modified product). The site and degree of DEPC-modified products are used to infer the structural changes across different protein states. Source: Gaikwad et al. [84]/MDPI/CC BY 4.0.

subsequently detected in the analysis. To confirm the vital information about the label-modified amino acid position per peptide, high-quality tandem MS (MS/MS or MS2) spectra with a comprehensive fragment ion series covering the modified site are needed. Moreover, robust statistical data analysis criteria must be established for quantifying label-modified sites.

Applications for CL-MS

Changes in reactivity of labeling reagents across protein samples indicate the extent of the protein's structural changes [84, 85]. CL-MS is particularly useful for studying protein–protein and protein–ligand interactions, offering a deeper understanding of protein stabilization and molecular pathways [89, 90]. DEPC CL-MS has been utilized in forced degradation studies of recombinant proteins to investigate aggregation-prone areas [91]. The amino acid-specific reactivity can be leveraged to study the structure–function relationship, such as assessing the stability of disulfide bonds at cysteines [84], or differentiating histidine tautomers [92]. Insights gained from these structural studies can inform the design of functional assessment experiments. As a single analytical technique will not satisfy all structural characterization questions, however, CL-MS can be a complementary technique to other structural analytical techniques. It can serve as an initial screening tool in structural protein analysis, with more complex techniques like HDX-MS being used for stringent validation. The complementarity of HDX-MS and CL-MS has been demonstrated in several studies [93–95]. The application of CL-MS in HOS characterization of therapeutic proteins providing reliable and rapid structural information is documented in the literature [91, 93, 96]. Crucial applications of CL-MS in epitope mapping, such as for trastuzumab binding to human epidermal growth factor receptor 2 (HER2) [97], or tumor necrosis factor alpha (TNFα) in complex with various antibodies [98], offer promising potential for expanding the use of this technique by biopharma professionals. The ultimate goal of using CL-MS as a complementary method in HOS characterization of biologics would be to accelerate the process of answering structural questions. If CL-MS proves effective in addressing a broader range of structural inquiries for therapeutic proteins, it could reduce the need for specialized equipment and personnel. However, it is important to note that CL-MS is primarily useful for studying the solvent-accessible surface area of proteins, which depends on factors like the primary sequence, protein structure, microenvironment, and the intrinsic reactivity of the labeling reagent toward specific amino acids. On average, it is reported that CL-MS can provide structural insights for about 30% of a protein [99]. Protein-specific adaptations in experimental conditions may be necessary to improve covalent labeling coverage and detection.

X-Ray Crystallography

X-ray crystallography has long been considered the gold standard for characterizing protein interactions, such as epitope mapping, due to its ability to provide atomic resolution of the binding site. In epitope mapping, both linear epitopes and conformational epitopes can be mapped using crystallography. A linear epitope is composed of amino acids connected to each other in one dimension, while a conformational epitope is composed of residues that come together in three dimensions. By examining the binding region, one can identify the interacting amino acids, for example, from a therapeutic mAb and its target antigen. This information can be used to make rational designs to effect changes in affinity and specificity of the antigen–antibody interaction.

Experimental Workflows and Equipment

A workflow for X-ray crystallography, exemplified by mAb–antigen interaction for epitope mapping, is illustrated in Figure 8.5. After purification of the antigen and the antibody, crystals are grown from protein solutions. These protein crystals are composed of orderly arrays of molecules in three dimensions, also termed lattice. X-ray wavelengths are shorter than bond lengths in proteins. Consequently, crystals are able to diffract X-rays. During a data collection experiment, a crystal is mounted between an X-ray source and detector. When a crystal is bombarded by X-rays, it produces a two-dimensional diffraction pattern on the detector. By measuring the intensities and directions of the diffracted X-rays, computer programs can be used to generate structural models for use in structural analyses such as epitope mapping [100, 101].

There are many methods employed for the discovery of antibodies. Briefly, using purified proteins, immunization campaigns in animals such as mice and rabbits, can be conducted. There are *in vitro* display technologies, such as phage display and yeast surface display, where amino acid sequences from libraries are expressed on the surface of phages and yeast cells. Sequences that bind to the target protein or antigen can be enriched.

Epitope mapping using X-ray crystallography begins with obtaining a monodisperse antibody–antigen complex. In crystallography, formation of ordered arrays of atoms in a crystal lattice is required before data collection can begin. As such, the formation of crystals is the bottleneck in this method. To encourage crystal lattice formation

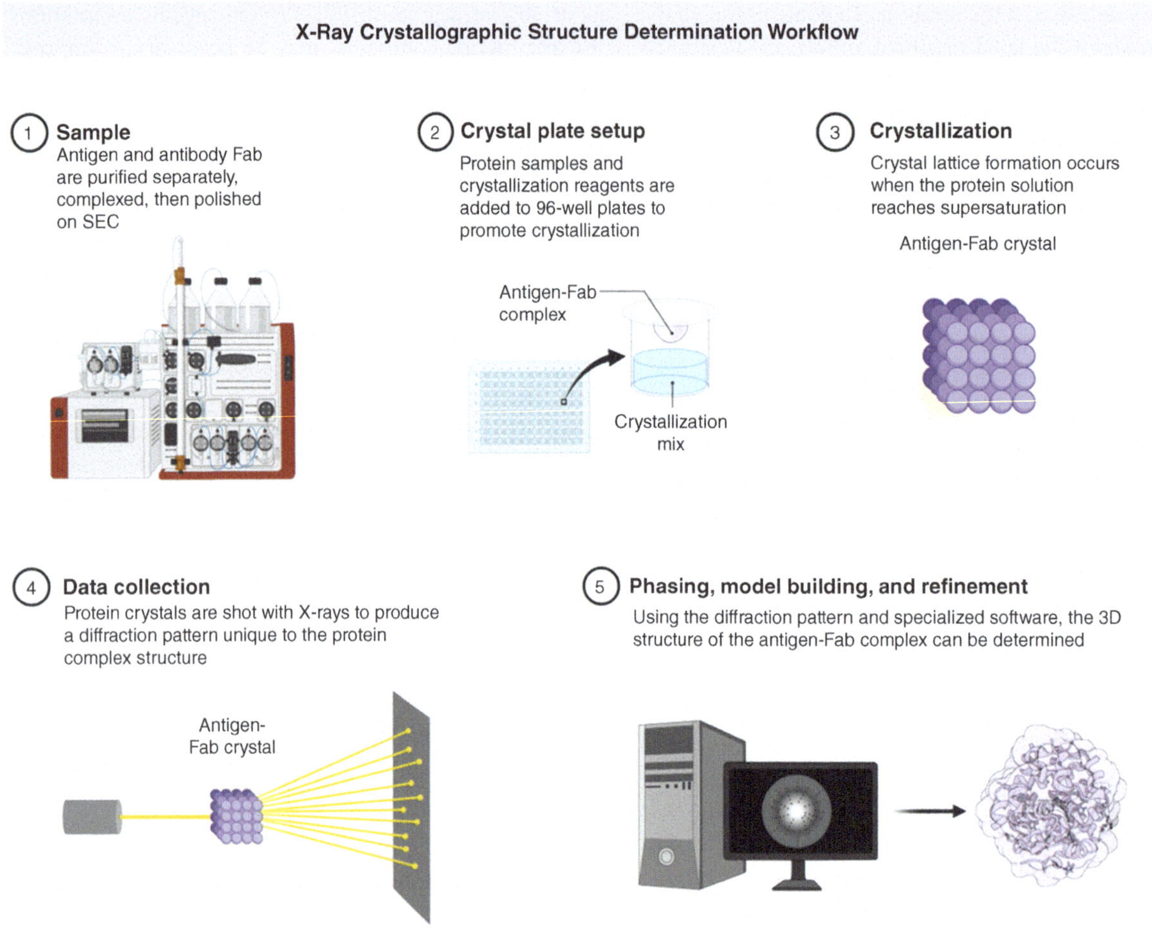

Figure 8.5 X-ray crystallographic structure determination workflow. Source: Created in BioRender. Jao, C. (2024) BioRender.com/t67v024.

in antibody–antigen complexes, several techniques are employed. Initial examination of the antibody sequence shows that there are many regions of flexibility. The full antibody is not used in crystallography experiments. Instead, antibody fragments with fewer flexible regions are employed, such as Fabs, VHHs, and scFvs. Fabs are composed of the variable regions of the heavy and light chains as well as the constant region (~50 kDa), while VHHs contain only the heavy chain (~14 kDa). scFvs are composed of the fusion of the variable domains of the heavy and light chains (~25 kDa). These antibody fragments can be produced in large amounts recombinantly in *Escherichia coli*, which is helpful during crystal trials.

Generally, target proteins with affinity tags like 6x histidine, streptavidin, or FLAG-tag, are purified to homogeneity using affinity chromatography, ion exchange chromatography (IEC), and polished after separation on size exclusion column (SEC). Fabs are similarly purified using affinity to protein A or G and may be polished using SEC. The Fab is added to the target protein at higher than 1x molar excess, allowed to interact at 4°C or room temperature, and then purified on SEC. Fractions which contain both the target protein and Fab are identified by visualization on SDS-PAGE or MS. This sample is concentrated and then used for crystallization trials.

To obtain structural models from X-ray crystallography experiments, protein crystals must first be grown. Growth of well-diffracting crystals is the bottleneck of the technique and is done empirically. Many factors affect the crystallization of proteins, such as pH, temperature, ionic strength, to name a few. Many variables can be sampled during crystal trials. The goal is to find a set of conditions wherein the protein solution is in a supersaturated state. At supersaturation, the proteins can nucleate and come out of solution as crystals. Beyond supersaturation, the proteins will precipitate, which is generally unproductive for crystal growth. The key is to find the balance between supersaturation and precipitation. During crystallization (crystal) trials, the sample is mixed with different buffers to promote crystal lattice formation. Crystal trials can be set up manually or, if available, with the assistance of automated liquid handlers such as Mosquito™. Several concentrations of the complex are tested, starting at 10 mg/mL. Commercially available screens in 96-well blocks, such as those from Hampton Research or Qiagen, among others, are compatible with automation. This allows for testing of 96 different crystallization conditions in one experimental plate. In addition to varying protein concentration, incubation temperature (4°C and 19°C) can also be tested. Sparse matrix screens which cover a wide range of crystallization conditions such as pH, salts, organics, and polymers are used for initial screening. Sitting drop and hanging drop crystallization can also be tested.

Crystal growth is monitored using a microscope or, if available, an automated imaging system such as RockMaker™. Crystals with sharp edges are good candidates for screening. During crystallization experiments, salts from the experimental buffers may crystallize. To eliminate these, crystals can be viewed under a microscope equipped with a UV filter. Because proteins generally contain Tyrosines and Tryptophans, protein crystals will fluoresce when viewed through a UV filter. If UV-positive crystals are identified, but do not have sharp edges, crystal quality can be improved by a finer screen of conditions that promote crystal growth, such as pH. Additives can be screened to evaluate reagents that may improve the initial crystal morphology. If an in-house X-ray generator is available, test shots of the crystals can be performed to screen for diffraction before being sent to a synchrotron for data collection.

At the synchrotron, X-ray radiation is generated when charged particles are accelerated through a magnetic field. Synchrotrons are located in national facilities such as Brookhaven National Laboratory, and Argonne National Laboratory, to name a few. The X-ray beams generated have high intensity and energy, capable of high-resolution structure determination. During data collection, X-rays pass through the protein crystal. The crystal diffracts the X-rays, and diffraction is captured in two dimensions on a detector, the simplest being X-ray film, where spots with different intensities develop after landing on the film. More sensitive detectors capable of photon counting, such as the Dectris Eiger2 detector, are currently in use.

Crystals are composed of a repeating pattern of atoms, and the unit cell is the smallest repeating unit in the crystal. The pattern, or lattice, contains information on the electron density of the atoms in the protein crystal. To determine the three-dimensional arrangement of atoms in the crystal, both the intensity of the signal and phases of the waves are required. X-ray waves may interfere constructively, producing a signal, or destructively, producing no signal on the detector. During data collection, signal intensities, or amplitudes, can be measured, but phase information is lost. To determine the phases, experimental phasing methods using anomalously diffracting atoms like heavy atoms are used [102]. The phase information can be determined computationally if another molecule with a similar structure has been solved using the molecular replacement method [103].

At the synchrotron, there are automated data processing software packages, such as HKL2000 [104], which integrate, scale, and merge intensities from the reflections, to obtain signal amplitudes. Once phases are obtained, either experimentally or computationally, electron density maps are obtained using Phenix, a comprehensive suite of programs for macromolecular structure determination [105]. Programs such as Coot (Crystallographic Object-Oriented Toolkit) [106] can be used to model the electron density map, and PyMol or Chimera can be used for molecular visualization.

X-Ray Crystallography for Epitope Mapping

VEGF or vascular endothelial growth factor is important in the formation of new blood vessels during development, and also from existing ones (angiogenesis) [107]. Dysregulation of angiogenesis has been shown to play a role in diseases such as cancers and diabetic retinopathies [108]. In a murine model, tumor growth suppression was observed after treatment with anti-VEGF antibodies [109]. From this, a humanized version of the antibody was engineered [110, 111]. In order to understand the molecular basis for this inhibition, the VEGF–Fab complex was crystallized [112]. VEGF is a member of the family of cystine knot growth factors and is a homodimeric protein [113]. The structure of the VEGF monomer is similar to other members of the cystine knot growth factors, with a 4-stranded antiparallel β-sheet with the cystine knot at one end. When the monomers of VEGF and VEGF–Fab were superimposed, deviations seen in the structures were very small, and the overall structures are similar [112]. Analysis of the VEGF–Fab binding interface shows that the Fab binds VEGF close to the β-turn between β5 and β6. This is the epitope. On the Fab, the paratope excludes L1 and L2 of the CDRs, while residues in L3, H1, H2, and H3 are involved in binding VEGF. Analysis of the crystal structure identified the structural epitope, a total of 25 residues on the Fab involved in binding to VEGF. Figure 8.6 [112] is a schematic representation of the epitope [112].

Based on this data, characterization of the functional epitope was performed by alanine scanning mutagenesis, where residues of interest are mutated to alanine and then tested in competitive monovalent ELISAs for determination of IC50s (concentration at which 50%

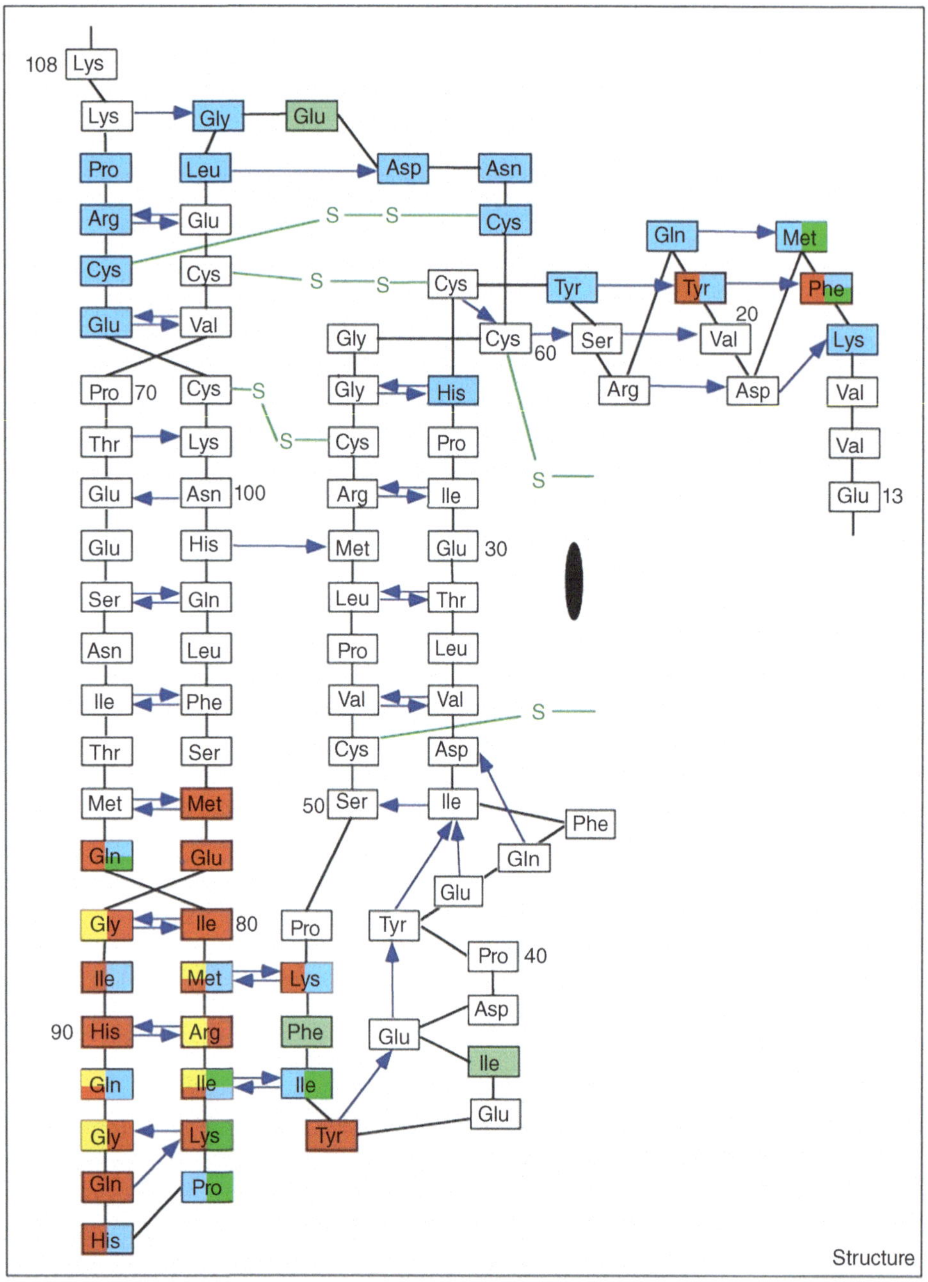

Figure 8.6 The binding epitope of VEGF to the humanized anti-VEGF antibody is schematically shown. Source: [112]/with permission of Elsevier.

binding inhibition is observed). Eight residues on the Fab were identified as critical for binding, and six residues on VEGF. Further, glycine (Gly88) of VEGF makes direct interactions with tyrosine and tryptophan residues in the CDR, identifying this region as important for binding. The CDRs are located on the antibody surface and form a shallow cleft where a linear peptide of VEGF is bound. From the crystal structure, the epitope is formed from a continuous stretch of residues, and not from a conformational epitope from discontinuous residues. As mentioned earlier, when the structure of VEGF alone was superimposed on the VEGF–Fab complex, the structures of VEGF were very similar. The antibody does not induce its neutralizing effect by a conformational change of the receptor binding region in VEGF. An overlap in the epitopes of Fab binding, as well as receptor binding was identified after structural analyses, and the neutralizing effect of Fab binding to VEGF is due to steric hindrance.

In another study, the authors wanted to analyze the molecular basis of the species difference in binding in

Avastin, an antibody approved for treatment of colon cancer [114]. Avastin is the humanized version of an antibody that was developed from mouse hybridoma technology using human VEGF as antigen. This approach prevented the possibility of discovering antibodies that can be used in mouse models, as self-reactive antibodies are not elicited. New antibodies (G6 and B20–4) that bind to both human and murine VEGF were discovered using phage display of synthetic antibody libraries. Alanine scanning mutagenesis was used to map the functional binding sites, while the structural binding sites were determined by X-ray crystallography. When structures of the different VEGF–Fab complexes were superimposed, the overall VEGF structures are very similar to one another. In addition, both the G6 and B20–4 Fabs bind to overlapping epitopes on VEGF, located in the N-terminal region of VEGF. This region is highly conserved and explains why these Fabs bind to both human and murine VEGF. Avastin binding centers around Gly88 in human VEGF. In murine VEGF, the glycine is replaced by a larger residue, serine. When Gly88 is replaced with a Ser in the model of murine VEGF–Avastin complex, a steric clash is observed, resulting in the low binding affinity of Avastin to murine VEGF.

X-ray crystallography, for long the gold standard of epitope mapping, can help us understand the molecular basis of Fab-antigen interactions. The main bottleneck is the crystallization of the complex and the production of diffracting crystals. Cryo-EM, where continual improvements in the technology allow for structure determination of smaller protein complexes and do not require the formation of a crystal lattice, will likely see an increase in epitope mapping.

Cryogenic Electron Microscopy (Cryo-EM)

In 2017, the Nobel Prize was awarded for the technique of Cryo-EM. This ushered in the “Resolution Revolution” which enabled scientists to resolve the atomic structure of proteins, thereby introducing a powerful tool to the structure biology toolbox. Scientific literature has since been filled with images of everything from proteins that cause antibiotic resistance, to the surface of viruses, to endogenous receptors bound to natural and synthesized small molecules. To use this method, highly pure samples are vitrified in ice and kept frozen as they are loaded into very sophisticated microscopes. An image is formed by shooting a coherent beam of electrons at the specimen. The resulting two-dimensional projection is captured using advanced camera sensors. This signal is visualized as a black and white micrograph that contains high-resolution information. At this point, individual particles can be seen, but software is required to do the highly complex task of extracting the correct information from a multitude of micrographs and producing an accurate three-dimensional reconstruction.

The use of Cryo-EM in epitope mapping is one way for biopharma to better understand its current IP as well as future clinical candidates. Epitope mapping in this context is performed by combining an antigen–receptor pair *in vitro* and passing them through a Single Particle Analysis (SPA) workflow to determine the three-dimensional structure of the complex [115]. Once constructed, the three-dimensional map can be analyzed to distinguish how each component interacts with its partner. Depending on the resolution of the resulting map, side chain residues pairings and epitope topography can be determined. This can be done on previously characterized sets of complexes or on a primary complex meant to guide development of next-generation material. In a screen of characterized complexes researchers can closely observe the epitope/paratope interface revealed by the data. This can then be used to structurally describe similarities or variations within a highly curated set. For initial interrogation with a primary complex, the resulting data can allow researchers to model the potential of sequence modifications to enhance or inhibit biological activity, leading to a better understanding of binding properties of mAb candidates.

Key Elements and Workflow of Cryo-EM Epitope Mapping

The workflow for cryo-EM involves several steps that require training and intense attention. From sample purification to intense data processing with specialized software. Thus, it is often carried out by several individuals that are highly trained in a specific aspect of the pipeline. During purification, a maximum soluble concentration is determined as the starting point of a SPA.

To begin the analysis, as illustrated in Figure 8.7, a sample aliquot is applied to a plasma-cleaned copper or gold EM grid that is 3 mm in diameter. The aliquot size can vary around 3–5 µL depending on the constraints dictated by the protein sample. Once deposited onto the grid, a vitrification robot, such as the vitrobot, freezes the grid before allowing the water molecules to align into an ice lattice. An ice lattice would wash out any signal from the target protein rendering the grid useless. The sample is then prepared for loading into an electron microscope where it can stay in a vitrified state for several days or weeks.

The advancements in microscope hardware and software were the major drivers of the Resolution Revolution. Under the guidance of a highly trained microscopist, the camera is calibrated, and grid areas are mapped by software.

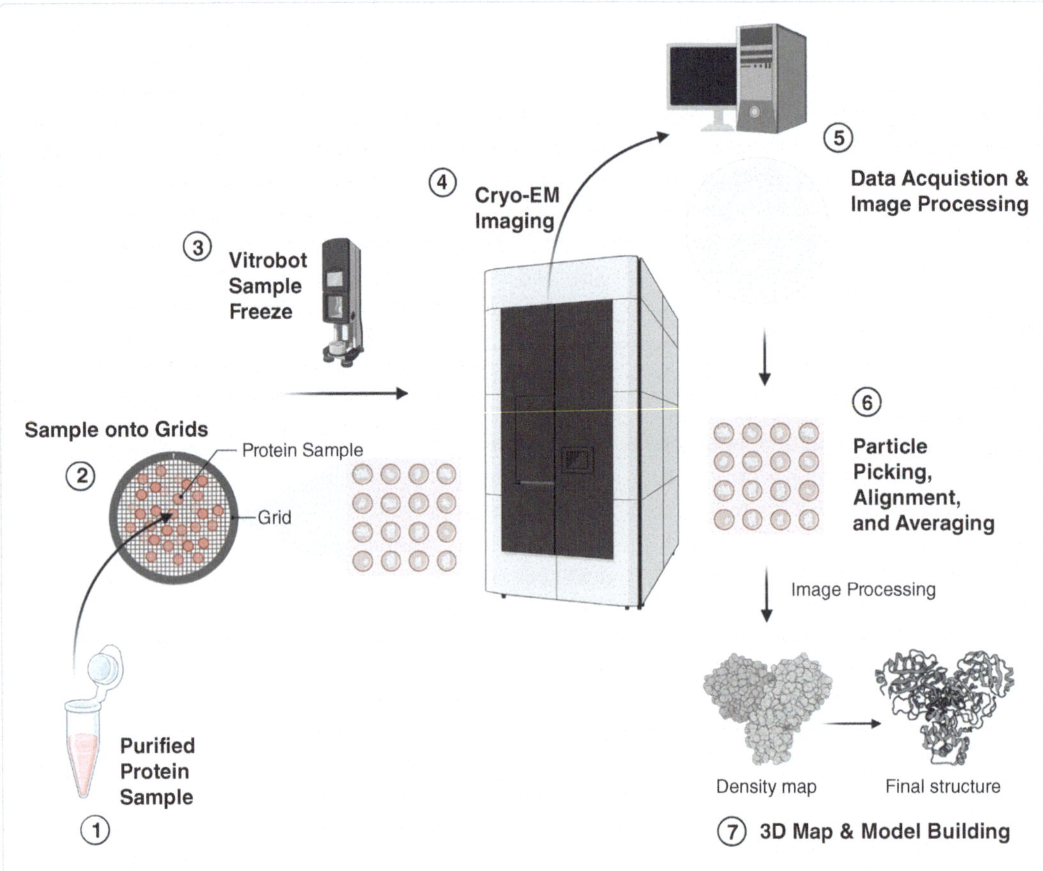

Figure 8.7 General SPA Workflow: 1 – Characterized and highly purified protein, 2 – Deposited onto a Cryo-EM grid, 3 – Then vitrified to a single layer of particles, 4 – Inserted into microscope and 2D micrographs collected, 5 – Initial data processing monitors data collection and preprocesses images for next step, 6 – Particle extraction, 2D classification and dataset refinement, 7 – Creation of three-dimensional maps and models. Source: Created in BioRender. Kuhne, F. (2024) BioRender.com/q54a441.

With only a few micrographs, a quick assessment of sample characteristics can be made such as how many particles can be observed per image, relative ice thickness, and overall potential for high resolution. For setting up a full data collection, software is used to mark hundreds to thousands of holes in the grid that will produce micrographs. Different strategies exist to produce multiple micrographs per hole in order to maximize the use of samples and make automated collection faster.

Once collected, a micrograph is shuttled to the image processing portion of the pipeline. Several image processing software exist, with the most popular being Cryosparc and Relion. They first perform a Contrast Transfer Function to correct inherent imperfections and then determine the resolution limit for each micrograph. Once proper settings have been determined, computationally intensive automated particle extraction and two-dimensional classification begin. This eliminates signals from bad particles and noise from reflections and other factors. The best particles are curated by several rounds of two-dimensional classification before turning to the next step of reconstructing the map in three dimensions. Two-dimensional particles are scored and assigned to a view of an initial map and through iterative scoring at higher and higher resolution until a final map is constructed. This final map can then be exported to analysis software, such as Phenix or ChimeraX. It is at this point that high-resolution maps can be used to describe and map an epitope of interest. In maps with a resolution of less than 3 Å, both tertiary and secondary structures can be recognized. This is especially true if there are pre-existing models available from internal data, public databases like the PDB and EMDB [116, 117], or created through one of the various large language models available for protein structure determination, as will be outlined in the following subchapter. Furthermore, processing modules like 3DFflex allow the signal from a map to be separated into slightly different conformations that may exist as a fluid or stepped transition between states. This greatly increases the understanding of complex affinity and potential allosteric interactions.

As with crystallography, the typical epitope mapping scenario involves Fabs, VHHs, or scFvs with their target antigen. The major reason for this is that a full antibody has several degrees of flexibility between its domains, which prevents downstream reconstruction by software. In structural studies, the antigen becomes the center of mass, with the Fab adding its own signal to the resulting three-dimensional maps. At low resolutions, greater than 4 Å, it is often possible to distinguish where the Fab has bound to its antigen, however, specific information about atomic interactions cannot be observed. With high-resolution data, greater than 3 Å, bond angles and atomic distances can be measured and provide a rich understanding of the binding interactions between each atom in the interacting regions within the complex. Since Cryo-EM does not have the same constraints as crystallography, more than one form or state of the complex can be present in the sample.

Laboratories with previously established purification and characterization workflows are highly amenable to producing good samples for SPA. Those set up for crystallography have a particular advantage due to the high purity and high concentration necessary for that method. In many cases, this allows projects to perform crystallography and cryo-EM in parallel using the same resources and starting material. This can be especially advantageous to projects having difficulty with the precipitation of protein crystals. Most samples brought to high concentrations can tolerate dilution to Cryo-EM working concentrations. Typical concentrations are around 1 mg/mL but can vary significantly. This can enable the investigation of difficult and low expressing proteins. If the sample is well-behaved, high-resolution data can be obtained from as little as 10 uL. Membrane proteins can be reconstituted in a synthetic bilayer of nano discs, amphipols, or low concentrations of detergents. One cannot discount the skilled handling and troubleshooting that takes place during sample preparation for Cryo-EM [118]. Iterative testing and benchmarking are key to setting up an efficient epitope mapping campaign.

One of the major constraints of cryo-EM is the problem of preferred orientation [112]. This occurs when one view of a particle is overrepresented in the data. Thin ice can cause exclusion zones where particles are void or only fit in one orientation. The existence of a surface charge on the protein can also cause it to align in a preferred orientation by sticking to the carbon edge of a hole on the grid, with the air water interface or some other charged component. Interactions with the air water interface are especially problematic. Not only can it cause a preferred orientation, but it can cause protein to denature and add noise to the micrograph. Buffers and additives can be used to mask the charged surface and reduce the preferred orientation. A preferred and therefore overrepresented view of a particle means lack of other orientations critical to reconstructing a good three-dimensional map. The addition of tags and even altering domain sequences can affect the preferred orientation in a sample. Another constraint is that proteins significantly below 50 kDa remain challenging to image at high resolution with standard cryo-EM techniques. Ice thickness is another variable that is difficult to control. Fortunately, grids prepared on a vitrobot often have a gradient of ice thickness across the grid. It is within this gradient that ideal conditions often exist for high resolution.

Fabs as nanobodies can be useful tools for epitope mapping campaigns using SPA. Cryo-EM has a larger effective range than crystallography, with the smallest maps coming from membrane proteins under 60 kDa up to full viral particles. Common nanobody creation based on the camelid Fab of Immunoglobulin G (IgG) can take months to yield a panel of binding partners. Fab maturation campaigns can take equally long but are often in place for high-priority targets. They are created by iteratively selecting for enhanced binding of an engineered receptor panel to a target antigen. This yields a set of Fabs, or Fab-like domains, which can be used for epitope mapping. The nanobody protein acts like a landmark on the antigen. These are commonly used to bulk up low molecular weight target proteins by providing a signal for downstream data processing. The resulting map can be deconvoluted to reveal the precise location of the binding Fabs or nanobodies as well as associated bond pairs that drive the interaction. When compared to apo control data, a nanobody complex can also be used to investigate potential allosteric effects. The potential for complexing with more than one Fab or nanobody can provide added information to an epitope mapping campaign. Strategies have been carried out using streptavidin or green fluorescence protein (GFP) as a landmark or bulking domain to facilitate the characterization of a Fab targeting a different part of the antigen.

In rare cases, looking at the interaction in the context of a full-length antibody can help to understand how the full antibody will behave in the presence of the antigen and can provide added information of higher-order complexing. This information can be crucial when engineering bispecific or modified antibody formats, as it facilitates the understanding of the various potential purification peaks.

With pressure to accelerate the advancement of new potential treatments and the introduction of automated systems, it is possible to carry out a relatively high throughput epitope mapping campaign. New user interfaces like smart EPU lower the complexity of controlling and validating important microscope settings to ensure that the highest quality data is collected. Automated data processing is an additional component that significantly reduces the time required to obtain actionable three-dimensional reconstruction. The latest microscopes, equipped with improved cameras and image processing software, are capable of capturing data at resolutions of less than 3 Å in approximately four hours.

Applications of Cryo-EM

In 2023, a concerted effort started to produce therapeutic antibodies targeting the SARS-CoV-2 spike protein to reduce the burden of the COVID-19 by protecting vulnerable populations. This kickstarted efforts in epitope mapping to understand new potential threats like the heavily mutated "Omicron" B.1.1529 strain of SARS-Cov-1. High-resolution data about an antibody's epitope and MoA may guide the next generation of improved antibody therapeutics. The use of large antibody panels allows researchers to understand the mutations and therefore speed up decision-making in selecting combinations that target nonoverlapping epitopes. In a recent report, it was described how Cryo-EM contributed data that led to an antibody cocktail to reduce the chance of viral escape [119]. This contribution was enabled by a workflow using modern instrumentation and skilled scientists. They established a high-throughput workflow to collect twelve datasets of unique Fab and spike protein complexes in an automated fashion. The workflow involves using Thermo Scientific™ EPU Multigrid Software to automate grid screening on a Thermo Scientific Glacios™ Cryo-TEM. Once captured, data is instantaneously pushed to the processing software Cryosparc. From this screening, 12 grids (one per complex) were set up for automated acquisition on a Thermo Scientific Krios™ G4 Cryo-TEM. A 48-hour automated session produced a set of sub 3 Å Cryo-EM reconstructions that enabled researchers to classify Fabs to their epitope sites that allowed for mapping of epitopes (Figure 8.8).

Another example of epitope mapping is the investigation of malaria during pregnancy [121]. A global health issue caused by *Plasmodium falciparum* parasites. Severe effects arise from the accumulation of infected erythrocytes in the placenta. Here, parasitic erythrocytes adhere to placental chondroitin sulfate A (CS) through VAR2CSA-type *P. falciparum* erythrocyte membrane protein 1 (PfEMP1). This study [121] showed a negative stain and a Cryo-EM structure of PAM1.4, the only known broadly reactive human mAb in complex with VAR2CSA. The resulting data revealed how PAM1.4's broad VAR2CSA reactivity is achieved. Through evolution, multiple conserved residues from different subdomains make up the conformational epitope distant from the CS binding site on the VAR2CSA core structure. Insights from data on complexes such as these may aid the development of a vaccine against placental malaria.

The recent advancements in the speed and robustness of SPA hardware and software will provide yet more precise data to use in modeling, deep learning, and medicinal chemistry efforts. The combination has the potential to deliver new therapeutic antibodies with broad protective capabilities and bring more precise modulation to existing ones. Personalized medicine and gene therapy, where mAbs are developed for a specific patient, can benefit from higher accuracy data gathered more quickly.

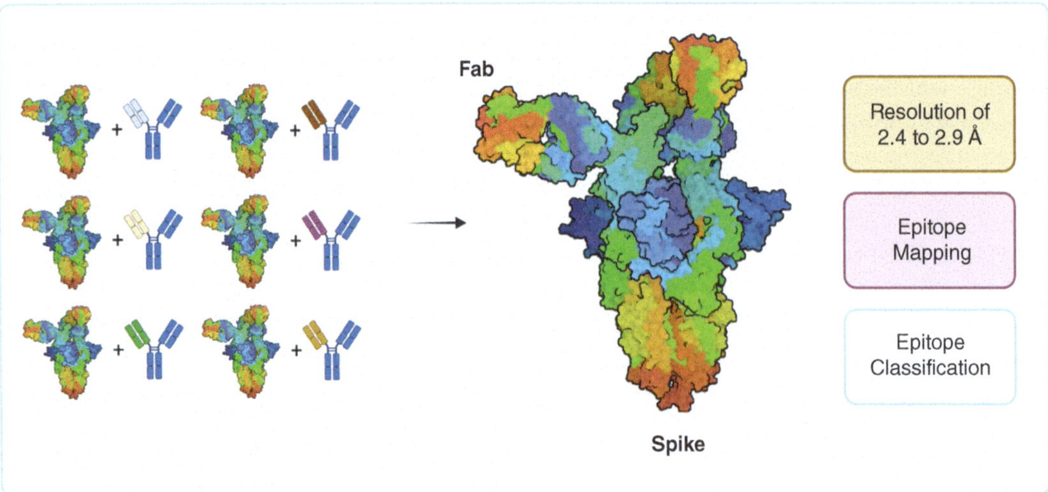

Figure 8.8 A panel of Fabs against the SARS-CoV-2 spike domain was characterized in an epitope mapping experiment. The results enabled researchers to classify each Fab and map its epitope. This method produced <3 Å maps in approximately four hours. Source: Illustration adapted from Drulyte et al. [119]. Illustration shows exemplary PDB ID 7ZJL (https://doi.org/10.2210/pdb7zjl/pdb) Pichkur et al. [120]. Created in BioRender. Kuhne, F. (2024) BioRender.com/d29r494.

Contract research organizations offer rapid epitope mapping services. This opens the door for smaller organizations that have developed robust good antigen and receptor systems, to advance their progress projects with the help of expert microscopists and structural biologists. By avoiding the substantially large investment in the infrastructure necessary for successful in-house Cryo-EM workflows, these organizations can accelerate and diversify their development pipelines.

Multiscale *in-silico* Modeling of Biotherapeutics

Biomolecular simulation has become an invaluable tool in the biotherapeutics field, allowing for the modeling and prediction of biomolecular behavior under various conditions. By capturing the intricate interplay of inter- and intramolecular interactions, the free energy landscape, and conformational changes over time, these simulations offer unique insights into the HOS of biotherapeutics. This chapter explores three primary levels of resolution in biomolecular simulations–quantum mechanical (QM), all-atom molecular dynamics (MD), and coarse-grained (CG) – and discusses their specific applications in studying the HOS of biotherapeutics (Figure 8.9).

Quantum Mechanics

Quantum mechanics is a fundamental theory that describes the behavior of nature at and below the scale of atoms. These calculations reveal critical information about the behavior of electrons within biomolecules, including the distribution of electron density, the nature of chemical bonds, and the energy landscape associated with various conformations. QM methods, such as Density Functional Theory (DFT) and Hartree-Fock (HF) theory, are pivotal in exploring the fundamental interactions that govern molecular behavior. Software tools such as Gaussian [122], VASP [123], ORCA [124], and NWChem [125] are extensively used to carry out DFT and HF calculations. For Quantum Mechanics/Molecular Mechanics (QM/MM) simulations, packages like AMBER [126], GROMACS [127], and CHARMM [128] integrate MD with quantum chemical methods. These tools allow for a multi-scale modeling of chemical reactions, covalent bond formation or breakage, electron transfer processes in proteins.

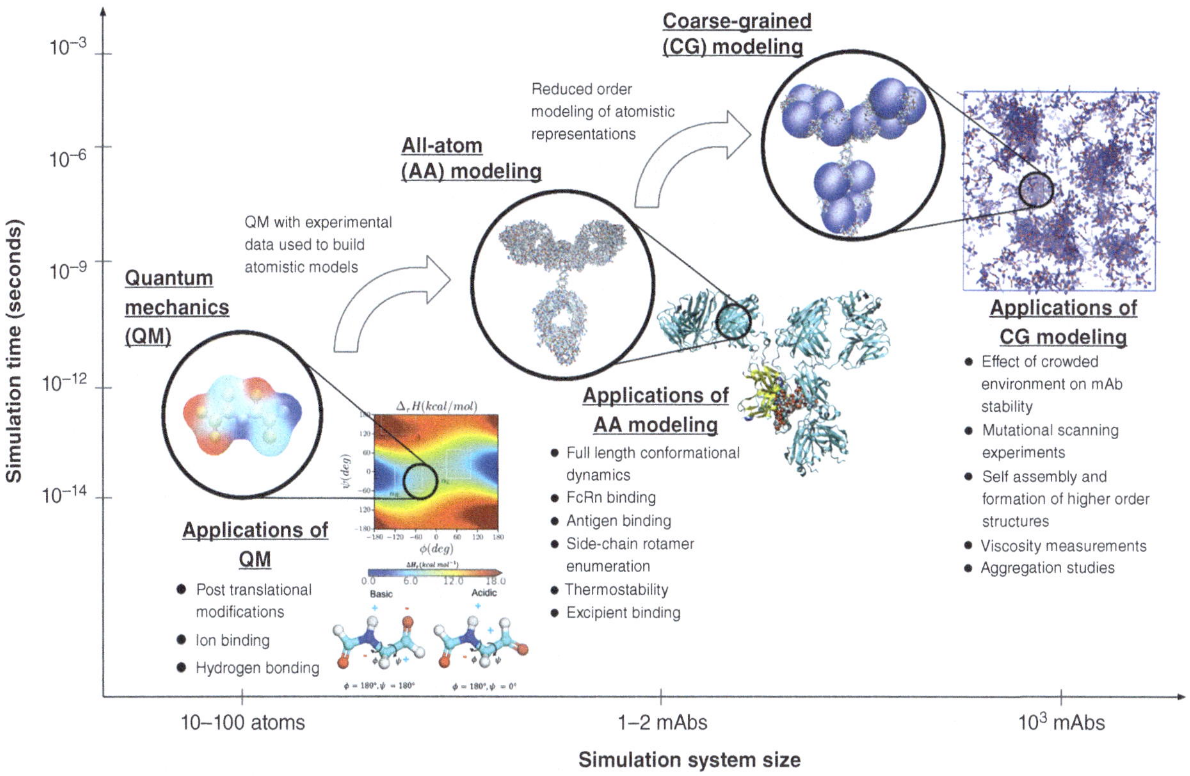

Figure 8.9 The primary levels of resolution in biomolecular simulations (quantum mechanics, all-atom, and coarse-grained) and their respective positions on the simulation time versus system size landscape. Each level of resolution provides unique insights about the HOS of biotherapeutics, especially mAbs and the various phenomena associated with mAbs that affect their potency, stability, and developability [156, 157, 159, 177].

In biotherapeutics, QM simulations are particularly useful for studying the effects of PTM [129–133] on protein structure and dynamics. The nuances of secondary and tertiary structural changes, as well as hydrogen bond formations or breakage, can be modeled using QM, which are essential for stabilizing secondary structures such as α-helices and β-sheets. QM calculations have been used to gain insight into the understanding of the impact of secondary peptide structure and hydration effects on the acidity of N-H backbone amide [132], which affects aspartate (Asp) isomerization and aspargine (Asn) deamidation. Both these PTMs are dependent on the backbone amide acidity, which is modulated by the secondary structure, backbone dihedral angles, as well as hydrogen bond formations or breakage [134, 135]. Effects of solvent and steric hindrance on the reaction rate of Asp/Asn degradation have also been explored using QM/MM [129–133]. Two key factors reducing the degradation propensity are steric hindrance for accessing reactive conformations and displacement of water by less polar side chains in the solvation shell of transition states [129–133].

Ab Initio Quantum Chemistry Calculations and QM/MM Molecular Dynamics Simulations have been used to generate mechanistic Insights into the oxidation of tryptophan by •OH and •OOH radicals [136]. The study found that addition to position 2 of the pyrrole ring is the most favorable site for both radicals. Free energy and QM/MM simulations suggest possible degradation pathways for the adducts. Although inaccuracies in methods limited agreement with experimental product distributions. It was found that local flexibility combined with surface area exposure offered good estimates of oxidative susceptibility in biomolecules.

In another work, QM/MM calculations were used to study the intermolecular interaction between SARS-CoV-2's RNA Binding Domain (RBD) N501 and antibody B38's light chain S30, showing a nearly linear O—H—O hydrogen bond with a 165° angle and a binding strength of −3.32 kcal/mol. MP2 and M06L DFT methods provided accurate results, which could inform COVID-19 vaccine and drug design efforts [137].

Many therapeutic proteins require metal ions for their function and stability. QM methods can provide detailed insights into the coordination geometries and electronic properties of these metal-binding sites, helping to understand how they affect the overall conformation and function of the protein [138]. The impact of metal ions (Fe, Zn, Cu, Mg) on catalytic cycles, drug metabolism, amyloid formation, and nucleic acid synthesis has been studied using QM/MM simulations [139]. Quantum chemical calculations also suggest that phosphorylation of tyrosine, threonine, and serine can lower the optical band gap energy (E_gap), facilitating electronic transitions that easily form or disrupt interactions [140]. These studies offer an atomic and electronic perspective on how phosphorylation promotes conformational and dynamic changes in proteins, which could be valuable for investigating protein function and aiding in protein design.

Limitations

Despite their advantages, QM simulations are computationally expensive, limiting their application to small systems or specific regions of interest within larger biomolecules. Consequently, while QM is invaluable for localized studies and detailed analyses, its use is often restricted to simplified models or specific interactions rather than comprehensive studies of entire biotherapeutic molecules.

All-Atom Modeling

All-atom MD simulations provide a detailed representation of biomolecules by explicitly modeling every atom, allowing for the investigation of atomic-level interactions and conformational dynamics. These simulations leverage classical mechanics and force fields, such as CHARMM [128] or AMBER [126], to calculate forces acting on each atom, thus simulating the motion of biomolecules over time. All-atom simulations are extensively utilized to study various aspects of protein dynamics [141–146], stability, and interactions [141–146], all of which are particularly relevant to HOS characterization. In particular, physics-based force fields can be used to simulate and explore the folding pathways of proteins, the intermediate states, and transition states [141–147], even in a matter of days on commodity hardware [148], depending on their size and topologies.

In the context of epitope mapping, MD simulations have been used to underscore the importance of the CDR loops in the rational design of mAbs and their interactions with antigens [149–151]. It has been shown that the binding pathway in Fab-antigen interactions is governed by a conformation selection mechanism, where the flexible CDR-H3 loop exists as a diverse ensemble in solution, allowing pre-existing conformations to facilitate binding even in the absence of the antigen [151]. Additionally, it has been revealed that specific residues, particularly tyrosine in the heavy chain CDR loops (H2 and H3), play a critical role in antibody recognition, significantly influencing binding free energy. Together, these findings highlight the need to consider both conformational dynamics and the specific contributions of amino acid residues when characterizing antibody–antigen interactions, ultimately enhancing the rational design of more effective antibodies.

MD simulations have also been used to study the pH- and thermally induced fluctuations of a therapeutic antibody Fab [152–154]. Low pH destabilized the constant light (CL) domain, while high temperature affected all domains, particularly the variable light/heavy (VL-VH) and constant light/heavy (CL-CH) interfaces, identifying the CL domain and CL-CH1 interface as key targets for enhancing stability against both stresses [152–154]. In a follow-up study, features from MD simulations were analyzed using machine learning to gain insights into stability mechanisms of Fab A33 and develop models that predict aggregation kinetics across 49 solution conditions, highlighting the impact of formulation factors like pH and temperature on protein stability [153]. Temperature-ramped MD simulations showed that the variable domains of aggregation-prone antibodies tend to unfold and expose aggregation-prone regions (APR) at lower temperatures compared to aggregation-resistant molecules [154].

In the context of antibody developability risk assessment, the molecular surface properties, such as electrostatic charge and hydrophobicity patches, have been used as metrics to define rules of developability [155]. Yet, an accurate calculation of molecular surface properties could be affected by the quality of the predicted structure model and the extent of conformational sampling [156]. Park and Izadi [156] reported systematic shifts in the distribution of surface descriptors depending on the choice of static structure model. These variations led to weak correlations of surface descriptors across different predicted structure models [155, 157]. However, applying MD simulations to the initially predicted structures significantly improved the consistency of developability risk flags based on surface descriptors obtained using different structure models [156]. This shows the importance of simulation-based relaxations to mitigate any potential bias or inaccuracies in the AI-generated structures, such as AlphaFold [158], or antibody-specific structure models such as AB2 [178], DeepAb [179], and Equifold [180].

Atomistic simulations can effectively calculate free energy profiles and barriers related to side-chain and backbone conformational changes that contribute to chemical degradation in antibodies, particularly focusing on aspartate isomerization and asparagine deamidation [133, 159]. For instance, the probability of adopting specific secondary structures, such as right- or left-handed helices and β-sheets, can influence the likelihood of these degradation reactions based on the identity of the $n+1$ residue. Additionally, simulations can quantify the accessibility of aspartate and asparagine side chains for nucleophilic attacks by measuring the distance between the backbone nitrogen and the side chain carbon [133], aiding in the identification of chemical degradation risks. Overall, these insights from MD simulations are valuable for assessing and mitigating chemical instability in antibody therapeutics.

Going beyond Fab arm simulations, the full-length antibodies have been simulated to highlight that the translational and rotational dynamics of Fab arms and their dependence on their interactions with each other and with the fragment-crystallizable (Fc) region, pointing to allosteric interactions between the Fab arms [160]. These simulation data yield novel insights into the inter- and intra-fragment motions of immunoglobulins which could help us better understand the relation between their structure and function. Conformational changes upon glycosylation of the Fc can module the affinity and binding mode of antibody–FcγRIIIa interactions [161]. MD simulations have revealed that the conformations of the Fc domain with and without galactose are dramatically different [162], which can impact the IgG1 hexamerization, thus enhancing C1q avidity and subsequent complement activation.

Limitations

The primary limitation of all-atom simulations is their high computational cost, which restricts the timescales that can be explored. While current advances in computational power have allowed for longer simulations, practical all-atom MD of single proteins of sizes relevant to Biotherapeutics is often limited to several hundred nanoseconds, or a few microseconds. The high computational cost of all-atom MD simulations makes it challenging to study slow conformational changes or large-scale dynamics, such as protein–protein interactions in crowded environments, or many-body protein simulations.

Coarse-Grained Modeling

CG models simplify the representation of biomolecules by aggregating groups of atoms into single interaction sites or beads. This reduction in complexity enables researchers to simulate larger systems over extended timescales, making CG simulations particularly useful for studying collective behaviors and emergent properties in biological systems. CG simulations are advantageous for investigating the HOS of biotherapeutics, in particular to model the self-assembly processes of proteins, shedding light on how these molecules aggregate to form larger complexes.

In the case of mAbs, subcutaneous administration is desirable, but it requires the development of high-concentration formulations. MAbs are massive proteins consisting of ~1,300 amino acids which allows for increased complexity in their interaction network. Hence, developing high-concentration formulations of mAbs requires careful attention to the multitude of intermolecular interactions that can modulate mechanical and chemical properties such as the viscosity, phase separation/aggregation, and degradation in solution. Atomistic modeling provides an in-depth view of the intramolecular interactions of biomolecules. But performing atomistic simulations of large complex molecules like mAbs proves to be quite expensive with currently available computational resources. This complexity is further amplified when we attempt to simulate solutions of mAbs. A CG approach involves simplifying the free energy barriers of a complex system by representing building blocks of large molecules such as amino acids, or protein domains with "beads." The selection and parameterization of the interactions made by different "bead" types is performed carefully to ensure relevant properties of the system are preserved. This allows CG molecular dynamics simulations to capture events that occur over long simulation timescales efficiently. We here discuss coarse-graining techniques which have been successfully used to characterize mAb interactions and their behavior in solution hence showcasing the utility CG modeling brings to investigating HOS of complex biotherapeutics.

A 12-bead and 26-bead domain level CG model was developed for mAbs to capture the effect of charged amino acids on the domain level toward self-association [163]. The beads are placed at the center of mass of the group of residues they are supposed to represent. Each bead has an electrostatic interaction and a van der Waals interaction component. Each bead is assigned a charge equal to the sum of charges of all residues they represent. For the van der Waals interaction, all beads are assigned uniform interaction strengths. MD simulations using this CG approach showed drastic changes in the structures formed by mAbs with varying charge distributions on their domains. Structural differences were shown to correlate with viscosity measured *in vitro* and interactions between the various domains Fab–Fab, Fab–Fc, and Fc–Fc were analyzed to find differences in interactions which lead to varying solution behavior for mAbs. Further studies [164] were conducted to evaluate the impact of charge mutations on the self-association of mAbs and CG models were able to differentiate between the mAbs in qualitative agreement with *in vitro* experiments.

Another domain level CG modeling strategy focused on parameterizing a 12-bead model with short-range electrostatic interactions using fits to small angle X-ray scattering (SAXS) data [165]. Using this model, molecular dynamics simulations were then used to investigate the differences in crucial microstructural properties like radial distribution functions and cluster sizes. The differentiating effects of cosolutes such as NaCl and arginine on the mAb cluster size distribution were also captured via the novel parameterization scheme used for electrostatics. Successful attempts have also been made to include the impact of long-range hydrodynamic interactions on the behavior of mAbs in solution using a 12-bead domain level CG model. This study also focused on measuring transport properties of mAb structures formed in solution and was able to achieve good agreement with experimental data [166]. SAXS-derived CG representations were also used in Monte Carlo simulations of mAbs to faithfully reproduce osmotic virial coefficients from *in vitro* experiments at low salt conditions and later predict the same in high salt conditions providing deeper insights into protein–protein interactions which might lead to aggregation [167]. CG models have also been used to investigate the effect of the anisotropic shape of antibodies on the structures they form in solution. A 3-bead minimal CG model [168] was able to reproduce certain features of the scattering (SAXS) profile of mAbs and to investigate the effect of the anisotropic shape of antibodies on the clusters they make in solution.

Several studies have highlighted the strong correlation between charge properties and hydrophobic character of the domains of a mAb with its self-association in a high-concentration environment. The CG models discussed previously use an averaging scheme to place charges on the beads representing specific domains based on the charged amino acid composition of the sequence of these domains. This interaction was coupled with a uniform van der Waals interaction term to account for interactions between domains apart from ones driven by electrostatic charges such as the hydrophobic character of amino acids. A newer domain level CG modeling approach proposed a novel charge and hydrophobicity parameterization scheme to better capture the electrostatic potential and hydrophobic character observed in the atomistic representation of the antibody [169]. The new charge parameterization uses the optimal point charge approximation theory [170, 171] to arrive at a set of point charges to reproduce low-order molecular multipole moments in the expansion of the original charge distribution. To differentiate between hydrophobic characters of different antibodies, atomistic simulations of variable antibody domains were performed in the presence of a cosolvent, and the distribution of hydrophobic tracer molecules was used to measure or rank the hydrophobic nature of the variable domains which then controls the van der Waals interaction strength of the beads representing these domains. This new CG approach was able to successfully differentiate the viscosity and high-concentration behavior of a set of mAbs with diverse charge distribution and hydrophobicity.

CG models which preserve a sequence level description of mAbs are well suited for evaluating the impact of point mutations on HOS formation in solution. Successful attempts have also been made to parameterize a sequence level CG model which uses a one bead per amino acid level of resolution [172–174]. Both electrostatic interactions and van der Waals interactions are present for all bead types. The basic amino acids lysine and arginine both are assigned a + 1 charge whereas acidic residues like aspartate and glutamic acid are assigned a − 1 charge. The charge on histidine beads is modified between +1 and +0.5 depending on the pH 5 or 6. All beads also interact with a van der Waals interaction using a Lennard Jones potential where the well depth is controlled by the average hydrophobicity of the interacting amino acids. The hydrophobicity of each amino acid is assigned using a previously derived hydrophobicity scale. The CG model discussed here was used in Monte Carlo simulations with the Mayer Overlap sampling technique [175, 176] to estimate the osmotic virial coefficient which provides information about the self-affinity of a mAb. The one bead per amino acid model was able to reproduce binding rates in good agreement with *in vitro* experiments hence showcasing another successful CG modeling approach that can be used to design new mAbs or estimate the impact of mutations.

With this plethora of CG models available to characterize an mAb's interactions and solution behavior, it is possible to build a high-throughput workflow that can be used to assist experimental studies on developability as well as discovery of new mAb molecules. With larger datasets becoming available with wet lab experiments on mAbs, the parameterization of CG models can be further improved using machine learning techniques and further reduce the need for expensive and time-consuming experiments.

Limitations

While CG simulations offer significant computational efficiency, they sacrifice atomic detail and may not accurately represent specific interactions at the atomic level. The mapping of atoms to beads requires careful calibration and validation against experimental data or higher-resolution simulations to ensure reliable results.

Conclusion

Biomolecular simulations, employing various levels of resolution – quantum mechanical, all-atom, and CG – provide essential insights into the HOS of biotherapeutics. Each approach has its strengths and limitations, but together they form a powerful toolkit for understanding the structural and dynamic properties of therapeutic molecules. By leveraging these simulation techniques, researchers can optimize the design of biotherapeutics, enhance their stability and efficacy, and ultimately improve patient outcomes. The continued advancement in computational methods and resources will further enable the exploration of increasingly complex biological systems, paving the way for innovative therapeutic strategies.

References

1 Wang, D., Zhuo, Y., Karfunkle, M. et al. (2021). NMR spectroscopy for protein higher order structure similarity assessment in formulated drug products. *Molecules* 26 (14): 4251.

2 Misra, S.K., Orlando, R., Weinberger, S.R., and Sharp, J.S. (2019). Compensated hydroxyl radical protein footprinting measures buffer and excipient effects on conformation and aggregation in an adalimumab biosimilar. *AAPS J.* 21 (5): 87.

3 Kumar, P., Wang, M., Kumru, O.S. et al. (2023). Correlating physicochemical and biological properties to define critical quality attributes of a rAAV vaccine candidate. *Mol. Ther. – Methods Clin. Dev.* 30: 103–121.

4 Schick, A.J., Lundin, V., Low, J. et al. (2022). Epitope mapping of anti-drug antibodies to a clinical candidate bispecific antibody. *MAbs* 14 (1): 2028337.

5 Biehn, S.E. and Lindert, S. (2021). Accurate protein structure prediction with hydroxyl radical protein footprinting data. *Nat. Commun.* 12 (1): 341.

6 Pauling, L., Corey, R.B., and Branson, H.R. (1951). The structure of proteins: two hydrogen-bonded helical configurations of the polypeptide chain. *Proc. Natl. Acad. Sci.* 37 (4): 205–211.

7 Edison, A.S. (2001). Linus Pauling and the planar peptide bond. *Nat. Struct. Biol.* 8 (3): 201–202.

8 Improta, R., Vitagliano, L., and Esposito, L. (2011). Peptide bond distortions from planarity: new insights from quantum mechanical calculations and peptide/protein crystal structures. *PLoS One* 6 (9): e24533.

9 Anfinsen, C.B. (1973). Principles that govern the folding of protein chains. *Science* 181 (4096): 223–230.

10 Little, M. (2021). *Antibodies for Treating Cancer, Basics, Development, and Applications*, 35–46. Springer.

11 Lord, E.A. (2002). Helical structures: the geometry of protein helices and nanotubes. *Struct. Chem.* 13 (3–4): 305–314.

12 Berg, J.M., Tymoczko, J.L., Gatto jr., G.J., and Stryer, L. (2017). *Biochemistry*, 31–77. New York, NY: W.H. Freeman and Co.

13 Högel, P., Götz, A., Kuhne, F. et al. (2018). Glycine perturbs local and global conformational flexibility of a transmembrane Helix. *Biochemistry* 57 (8): 1326–1337.

14 Opuni, K.F.M., Al-Majdoub, M., Yefremova, Y. et al. (2018). Mass spectrometric epitope mapping. *Mass Spectrom. Rev.* 37 (2): 229–241.

15 Zhang, M.M., Huang, R.Y.C., Beno, B.R. et al. (2020). Epitope and paratope mapping of PD-1/nivolumab by mass spectrometry-based hydrogen–deuterium exchange, cross-linking, and molecular docking. *Anal. Chem.* 92 (13): 9086–9094.

16 Gavade, A., Nagraj, A.K., Patel, R. et al. (2024). Understanding the specific implications of amino acids in the antibody development. *Protein J.* 43 (3): 405–424.

17 Zhang, Z., Yan, Y., Wang, S., and Li, N. (2022). A competitive binding-mass spectrometry strategy for high-throughput evaluation of potential critical quality attributes of therapeutic monoclonal antibodies. *MAbs* 14 (1): 2133674.

18 Widłak, W. (2013). Molecular biology, not only for bioinformaticians. *Lect. Notes Comput. Sci* 15–29.

19 Pace, C.N. and Scholtz, J.M. (1998). A helix propensity scale based on experimental studies of peptides and proteins. *Biophys. J.* 75 (1): 422–427.

20 Almeida, F.C.L., Sanches, K., Pinheiro-Aguiar, R. et al. (2021). Protein surface interactions—theoretical and experimental studies. *Front. Mol. Biosci.* 8: 706002.

21 Gish, J.S., Jarvis, L., Childers, K.C. et al. (2021). Structure of blood coagulation factor VIII in complex with an anti–C1 domain pathogenic antibody inhibitor. *Blood* 137 (21): 2981–2986.

22 Brych, S.R., Gokarn, Y.R., Hultgen, H. et al. (2010). Characterization of antibody aggregation: role of buried, unpaired cysteines in particle formation. *J. Pharm. Sci.* 99 (2): 764–781.

23 Lakbub, J.C., Shipman, J.T., and Desaire, H. (2018). Recent mass spectrometry-based techniques and considerations for disulfide bond characterization in proteins. *Anal. Bioanal. Chem.* 410 (10): 2467–2484.

24 Wang, W., Singh, S., Zeng, D.L. et al. (2007). Antibody structure, instability, and formulation. *J. Pharm. Sci.* 96 (1): 1–26. Available from: http://www.ncbi.nlm.nih.gov/pubmed/16998873.

25 Liu, H. and May, K. (2012). Disulfide bond structures of IgG molecules. *MAbs* 4 (1): 17–23.

26 Weiss, W.F., Gabrielson, J.P., Al-Azzam, W. et al. (2016). Technical decision making with higher order structure data: perspectives on higher order structure characterization from the biopharmaceutical industry. *J. Pharm. Sci.* 105 (12): 3465–3470.

27 Gabrielson, J.P. and Weiss, W.F. (2015). Technical decision-making with higher order structure data: starting a new dialogue. *J. Pharm. Sci.* 104 (4): 1240–1245.

28 Makowski, L., Rodi, D.J., Mandava, S. et al. (2008). Molecular crowding inhibits intramolecular breathing motions in proteins. *J. Mol. Biol.* 375 (2): 529–546.

29 Stank, A., Kokh, D.B., Fuller, J.C., and Wade, R.C. (2016). Protein binding pocket dynamics. *Acc. Chem. Res.* 49 (5): 809–815.

30 Kuhne, F., Bonnington, L., Malik, S. et al. (2019). The impact of immunoglobulin G1 fc sialylation on backbone amide H/D exchange. *Antibodies* 8 (4): 49.

31 Markus, H., Katrin, B., Katharina, D. et al. (2014). Assessment of chemical modifications of sites in the CDRs of recombinant antibodies. *MAbs* 6 (2): 327–339.

32 Lu, R.M., Hwang, Y.C., Liu, I.J. et al. (2020). Development of therapeutic antibodies for the treatment of diseases. *J. Biomed. Sci.* 27 (1): 1.

33 Goulet, D.R. and Atkins, W.M. (2019). Considerations for the design of antibody-based therapeutics. *J. Pharm. Sci.* 109 (1): 74–103.

34 Nilvebrant, J. and Rockberg, J. (2018). Epitope mapping protocols. *Methods Mol. Biol.* 1785: 1–10.

35 Liu, X.R., Huang, R.Y.C., Zhao, F. et al. (2022). Advances in mass spectrometry-based epitope mapping of protein therapeutics. *J. Pharm. Biomed. Anal.* 215: 114754.

36 Emmanuel, S.N., Smith, J.K., Hsi, J. et al. (2021). Structurally mapping antigenic epitopes of adeno-associated virus 9: development of antibody escape variants. *J. Virol.* 96 (3): e01251–e01221.

37 Piletska, E., Veron, P., Bertin, B. et al. (2023). Analysis of adeno-associated virus serotype 8 (AAV8)-antibody complexes using epitope mapping by molecular imprinting leads to the identification of Fab peptides that potentially evade AAV8 neutralisation. *Nanomed: Nanotechnol. Biol. Med.* 52: 102691.

38 Haynes, W.A., Kamath, K., Bozekowski, J. et al. (2021). High-resolution epitope mapping and characterization of SARS-CoV-2 antibodies in large cohorts of subjects with COVID-19. *Commun. Biol.* 4 (1): 1317.

39 Steven, K. and Patrick, S. (2006). Current and future issues in the manufacturing and development of monoclonal antibodies. *Adv. Drug Deliv. Rev.* 58 (5–6): 707–722.

40 Alain, B. and Hongcheng, L. (2019). Macro- and micro-heterogeneity of natural and recombinant IgG antibodies. *Antibodies* 8 (1): 18.

41 Vulto, A.G. and Jaquez, O.A. (2017). The process defines the product: what really matters in biosimilar design and production? *Rheumatology* 56 (suppl_4): iv14–iv29.

42 Chen, G., Tao, L., and Li, Z. (2022). Recent advancements in mass spectrometry for higher order structure characterization of protein therapeutics. *Drug Discov. Today* 27 (1): 196–206.

43 Hvidt A, Linderstrom-Lang K. (1954). Exchange of hydrogen atoms in insulin with deuterium atoms in aqueous solutions. *Biochim. Biophys. Acta.* 14 (4): 574-5. https://doi.org/10.1016/0006-3002(54)90241-3. PMID: 13198919.

44 Englander, S.W., Mayne, L., Kan, Z.Y., and Hu, W. (2016). Protein folding-how and why: by hydrogen exchange, fragment separation, and mass spectrometry. *Annu. Rev. Biophys.* 45: 135–152. https://doi.org/10.1146/annurev-biophys-062215-011121. Epub 2016 Apr 27. PMID: 27145881; PMCID: PMC5588872.

45 Hvidt, A. and Linderstorm-Lang, K. (1955). The kinetics of the deuterium exchange of insulin with D_2O; an amendment. *Biochim. Biophys. Acta* 16 (1): 168–169. https://doi.org/10.1016/0006-3002(55)90200-6. PMID: 14363248.

46 Zhang, Z., Post, C.B., and Smith, D.L. (1996). Amide hydrogen exchange determined by mass spectrometry: application to rabbit muscle aldolase. *Biochemistry* 35 (3): 779–791.

47 Zhang, Q., Willison, L.N., Tripathi, P. et al. (2011). Epitope mapping of a 95 kDa antigen in complex with antibody by solution-phase amide backbone hydrogen/deuterium exchange monitored by Fourier transform ion cyclotron resonance mass spectrometry. *Anal. Chem.* 83 (18): 7129–7136.

48 Robotham, J.M., Xia, L., Willison, L.N. et al. (2010). Characterization of a cashew allergen, 11S globulin (ana o 2), conformational epitope. *Mol. Immunol.* 47 (9): 1830–1838.

49 Puchades, C., Kűkrer, B., Diefenbach, O. et al. (2019). Epitope mapping of diverse influenza hemagglutinin drug candidates using HDX-MS. *Sci. Rep.* 9 (1): 4735.

50 Houde, D., Berkowitz, S.A., and Engen, J.R. (2011). The utility of hydrogen/deuterium exchange mass spectrometry in biopharmaceutical comparability studies. *J. Pharm. Sci.* 100 (6): 2071–2086.

51 Hong, J., Lee, Y., Lee, C. et al. (2016). Physicochemical and biological characterization of SB2, a biosimilar of Remicade® (infliximab). *MAbs* 9 (2): 364–382.

52 McKenzie-Coe, A., Montes, N.S., and Jones, L.M. (2022). Hydroxyl radical protein footprinting: a mass spectrometry-based structural method for studying the higher order structure of proteins. *Chem. Rev.* 122 (8): 7532–7561.

53 Monroe, E.B. and Heien, M.L. (2013). Electrochemical generation of hydroxyl radicals for examining protein structure. *Anal. Chem.* 85 (13): 6185–6189.

54 Minkoff, B.B., Blatz, J.M., Choudhury, F.A. et al. (2017). Plasma-generated OH radical production for analyzing three-dimensional structure in protein therapeutics. *Sci. Rep.* 7 (1): 12946.

55 Gupta, S., Celestre, R., Petzold, C.J. et al. (2014). Development of a microsecond X-ray protein footprinting facility at the advanced light source. *J. Synchrotron Radiat.* 21 (4): 690–699.

56 Goldsmith, S.C., Guan, J.Q., Almo, S.C., and Chance, M.R. (2001). Synchrotron protein footprinting: a technique to investigate protein-protein interactions. *J. Biomol. Struct. Dyn.* 19 (3): 405–418.

57 Shcherbakova, I., Mitra, S., Beer, R.H., and Brenowitz, M. (2006). Fast fenton footprinting: a laboratory-based method for the time-resolved analysis of DNA, RNA and proteins. *Nucleic Acids Res.* 34 (6): e48.

58 Hambly, D.M. and Gross, M.L. (2005). Laser flash photolysis of hydrogen peroxide to oxidize protein solvent-accessible residues on the microsecond timescale. *J. Am. Soc. Mass Spectrom.* 16 (12): 2057–2063.

59 Weinberger, S.R., Chea, E.E., Sharp, J.S., and Misra, S.K. (2021). Laser-free hydroxyl radical protein footprinting to perform higher order structural analysis of proteins. *J. Vis. Exp.* 172.

60 Gau, B.C., Sharp, J.S., Rempel, D.L., and Gross, M.L. (2009). Fast photochemical oxidation of protein footprints faster than protein unfolding. *Anal. Chem.* 81 (16): 6563–6571.

61 Dean, R.T., Fu, S., Stocker, R., and Davies, M.J. (1997). Biochemistry and pathology of radical-mediated protein oxidation. *Biochem. J.* 324 (1): 1–18.

62 Xie, B. and Sharp, J.S. (2015). Hydroxyl radical dosimetry for high flux hydroxyl radical protein footprinting applications using a simple optical detection method. *Anal. Chem.* 87 (21): 10719–10723.

63 Garcia, N.K., Sreedhara, A., Deperalta, G., and Wecksler, A.T. (2020). Optimizing hydroxyl radical footprinting analysis of biotherapeutics using internal standard dosimetry. *J. Am. Soc. Mass Spectrom.* 31 (7): 1563–1571.

64 Sharp, J.S., Chea, E.E., Misra, S.K. et al. (2021). Flash oxidation (FOX) system: a novel laser-free fast photochemical oxidation protein footprinting platform. *J. Am. Soc. Mass Spectrom.* 32 (7): 1601–1609.

65 Shami, A.A., Misra, S.K., Jones, L.M., and Sharp, J.S. (2023). Dimethylthiourea as a quencher in hydroxyl radical protein footprinting experiments. *J. Am. Soc. Mass Spectrom.* 34 (12): 2864–2867.

66 Li, K.S., Chen, G., Mo, J. et al. (2017). Orthogonal mass spectrometry-based footprinting for epitope mapping and structural characterization: the IL-6 receptor upon binding of protein therapeutics. *Anal. Chem.* 89 (14): 7742–7749.

67 Zhang, Y., Wecksler, A.T., Molina, P. et al. (2017). Mapping the binding interface of VEGF and a monoclonal antibody Fab-1 fragment with fast photochemical oxidation of proteins (FPOP) and mass spectrometry. *J. Am. Soc. Mass Spectrom.* 28 (5): 850–858.

68 Wang, K., Dagil, R., Lavstsen, T. et al. (2021). Cryo-EM reveals the architecture of placental malaria VAR2CSA and provides molecular insight into chondroitin sulfate binding. *Nat. Commun.* 12 (1): 2956.

69 Peng, Y., Cao, S., Kiselar, J. et al. (2019). A metastable contact and structural disorder in the estrogen receptor transactivation domain. *Structure* 27 (2): 229–240.e4.

70 Wang, X., Watson, C., Sharp, J.S. et al. (2011). Oligomeric structure of the chemokine CCL5/RANTES from NMR, MS, and SAXS data. *Structure* 19 (8): 1138–1148.

71 Shepard, H.M., Phillips, G.L., Thanos, C.D., and Feldmann, M. (2017). Developments in therapy with monoclonal antibodies and related proteins. *Clin. Med.* 17 (3): 220–232.

72 Lin, M., Krawitz, D., Callahan, M.D. et al. (2018). Characterization of ELISA antibody-antigen interaction using footprinting-mass spectrometry and negative staining transmission electron microscopy. *J. Am. Soc. Mass Spectrom.* 29 (5): 961–971.

73 Peng, K., Siradze, K., and Fischer, S.K. (2021). Characterization of robust immune responses to a bispecific antibody, a novel class of antibody therapeutics. *Bioanalysis* 13 (4): 239–252.

74 Olson, L.J., Misra, S.K., Ishihara, M. et al. (2020). Allosteric regulation of lysosomal enzyme recognition by the cation-independent mannose 6-phosphate receptor. *Commun. Biol.* 3 (1): 498.

75 Niu, B., Appleby, T.C., Wang, R. et al. (2020). Protein footprinting and X-ray crystallography reveal the interaction of PD-L1 and a macrocyclic peptide. *Biochemistry* 59 (4): 541–551.

76 Ralston, C.Y. and Sharp, J.S. (2022). Structural investigation of therapeutic antibodies using hydroxyl radical protein footprinting methods. *Antibodies* 11 (4): 71.

77 Xu, G. and Chance, M.R. (2007). Hydroxyl radical-mediated modification of proteins as probes for structural proteomics. *Chem. Rev.* 107 (8): 3514–3543.

78 Pan, J., Zhang, S., Parker, C.E., and Borchers, C.H. (2014). Subzero temperature chromatography and top-down mass spectrometry for protein higher-order structure characterization: method validation and application to therapeutic antibodies. *J. Am. Chem. Soc.* 136 (37): 13065–13071. Available from: http://www.ncbi.nlm.nih.gov/pubmed/25152011.

79 Oganesyan, I., Lento, C., and Wilson, D.J. (2018). Contemporary hydrogen deuterium exchange mass spectrometry. *Methods* 144: 27–42.

80 Limpikirati, P., Liu, T., and Vachet, R.W. (2018). Covalent labeling-mass spectrometry with non-specific reagents for studying protein structure and interactions. *Methods* 144: 79–93.

81 Barth, M., Bender, J., Kundlacz, T., and Schmidt, C. (2020). Evaluation of NHS-acetate and DEPC labelling for determination of solvent accessible amino acid residues in protein complexes. *J. Proteome* 222: 103793.

82 Manzi, L., Barrow, A.S., Hopper, J.T.S. et al. (2017). Carbene footprinting reveals binding interfaces of a multimeric membrane-spanning protein. *Angew. Chem.* 129 (47): 15069–15073.

83 Martín-Gago, P., Fansa, E.K., Winzker, M. et al. (2017). Covalent protein labeling at glutamic acids. *Cell Chem. Biol.* 24 (5): 589–597.e5.

84 Gaikwad, M., Richter, F., Götz, R. et al. (2023). Site-specific structural changes in long-term-stressed monoclonal antibody revealed with DEPC covalent-labeling and quantitative mass spectrometry. *Pharmaceuticals* 16 (10): 1418.

85 Mendoza, V.L. and Vachet, R.W. (2009). Probing protein structure by amino acid-specific covalent labeling and mass spectrometry. *Mass Spectrom. Rev.* 28 (5): 785–815.

86 Mendoza, V.L. and Vachet, R.W. (2008). Protein surface mapping using diethylpyrocarbonate with mass spectrometric detection. *Anal. Chem.* 80 (8): 2895–2904.

87 Pan, X., Limpikirati, P., Chen, H. et al. (2020). Higher-order structure influences the kinetics of diethylpyrocarbonate covalent labeling of proteins. *J. Am. Soc. Mass Spectrom.* 31 (3): 658–665.

88 Kirsch, Z.J. and Vachet, R.W. (2024). Proteins can withstand more extensive labeling while providing accurate structural information in covalent labeling-mass spectrometry. *J. Am. Soc. Mass Spectrom.* 35 (5): 1030–1039.

89 Dhingra, K., Gudhka, R.B., and Cramer, S.M. (2023). Evaluation of preferred binding regions on ubiquitin and IgG1 FC for interacting with multimodal cation exchange resins using DEPC labeling/mass spectrometry. *Biotechnol. Bioeng.* 120 (6): 1592–1604.

90 Kirsch, Z.J., Blake, J.M., Huynh, U. et al. (2023). Membrane protein binding interactions studied in live cells via diethylpyrocarbonate covalent labeling mass spectrometry. *Anal. Chem.* 95 (18): 7178–7185.

91 Limpikirati, P., Hale, J.E., Hazelbaker, M. et al. (2019). Covalent labeling and mass spectrometry reveal subtle

higher order structural changes for antibody therapeutics. *MAbs* 11 (3): 463–476.
92 Pan, X., Kirsch, Z.J., and Vachet, R.W. (2022). Distinguishing histidine tautomers in proteins using covalent labeling-mass spectrometry. *Anal. Chem.* 94 (2): 1003–1010.
93 Tremblay, C.Y., Limpikirati, P., and Vachet, R.W. (2021). Complementary structural information for stressed antibodies from hydrogen–deuterium exchange and covalent labeling mass spectrometry. *J. Am. Soc. Mass Spectrom.* 32 (5): 1237–1248.
94 Liu, T., Limpikirati, P., and Vachet, R.W. (2019). Synergistic structural information from covalent labeling and hydrogen–deuterium exchange mass spectrometry for protein–ligand interactions. *Anal. Chem.* 91 (23): 15248–15254.
95 Tremblay, C.Y., Kirsch, Z.J., and Vachet, R.W. (2022). Complementary structural information for antibody–antigen complexes from hydrogen–deuterium exchange and covalent labeling mass spectrometry. *J. Am. Soc. Mass Spectrom.* 33 (7): 1303–1314.
96 Garcia, N.K., Deperalta, G., and Wecksler, A.T. (2019). Current trends in biotherapeutic higher order structure characterization by irreversible covalent footprinting mass spectrometry. *Protein Pept. Lett.* 26 (1): 35–43.
97 Fojtík, L., Kalaninová, Z., Fiala, J. et al. (2024). Structural characterization of monoclonal antibodies and epitope mapping by FFAP footprinting. *Anal. Chem.* 96 (19): 7386–7393.
98 Tremblay, C.Y., Kirsch, Z.J., and Vachet, R.W. (2022). Epitope mapping with diethylpyrocarbonate covalent labeling-mass spectrometry. *Anal. Chem.* 94 (2): 1052–1059.
99 Zhou, Y. and Vachet, R.W. (2012). Diethylpyrocarbonate labeling for the structural analysis of proteins: label scrambling in solution and how to avoid it. *J. Am. Soc. Mass Spectrom.* 23 (5): 899–907.
100 McPherson, A. (2010). *Introduction to Macromolecular Crystallography*, 1–18. Wiley.
101 Rhodes, G. (2006). *Crystallography Made Crystal Clear*, 3e, 7–30. Academic Press.
102 Hendrickson, W.A. (2023). Facing the phase problem. *IUCrJ.* 10 (Pt 5): 521–543.
103 Taylor, G. (2003). The phase problem. *Acta Crystallogr. Sect. D* 59 (11): 1881–1890.
104 Otwinowski, Z. and Minor, W. (1997). Processing of X-ray diffraction data collected in oscillation mode. *Methods Enzymol.* 276: 307–326.
105 Liebschner, D., Afonine, P.V., Baker, M.L. et al. (2019). Macromolecular structure determination using X-rays, neutrons and electrons: recent developments in Phenix. *Acta Crystallogr. Sect. D* 75 (Pt 10): 861–877.
106 Emsley, P., Lohkamp, B., Scott, W.G., and Cowtan, K. (2010). Features and development of coot. *Acta Crystallogr. Sect. D: Biol. Crystallogr.* 66 (4): 486–501.
107 Ferrara, N. (1995). The role of vascular endothelial growth factor in pathological angiogenesis. *Breast Cancer Res. Treat.* 36 (2): 127–137.
108 Folkman, J. (1995). Angiogenesis in cancer, vascular, rheumatoid and other disease. *Nat. Med.* 1 (1): 27–30.
109 Kim, K.J., Li, B., Winer, J. et al. (1993). Inhibition of vascular endothelial growth factor-induced angiogenesis suppresses tumour growth in vivo. *Nature* 362 (6423): 841–844.
110 Baca, M., Presta, L.G., O'Connor, S.J., and Wells, J.A. (1997). Antibody humanization using monovalent phage display. *J. Biol. Chem.* 272 (16): 10678–10684.
111 Presta, L.G., Chen, H., O'Connor, S.J. et al. (1997). Humanization of an anti-vascular endothelial growth factor monoclonal antibody for the therapy of solid tumors and other disorders. *Cancer Res.* 57 (20): 4593–4599.
112 Muller, Y.A., Chen, Y., Christinger, H.W. et al. (1998). VEGF and the fab fragment of a humanized neutralizing antibody: crystal structure of the complex at 2.4 å resolution and mutational analysis of the interface. *Structure* 6 (9): 1153–1167.
113 Sun, P.D. and Davies, D.R. (1995). The cystine-knot growth-factor superfamily. *Annu. Rev. Biophys. Biomol. Struct.* 24 (1): 269–292.
114 Fuh, G., Wu, P., Liang, W.C. et al. (2006). Structure-function studies of two synthetic anti-vascular endothelial growth factor Fabs and comparison with the AvastinTM fab. *J. Biol. Chem.* 281 (10): 6625–6631.
115 Lyumkis, D. (2019). Challenges and opportunities in cryo-EM single-particle analysis. *J. Biol. Chem.* 294 (13): 5181–5197.
116 EMDB (2024). EMDB < Home. www.ebi.ac.uk/emdb (accessed 17 September 2024).
117 RCSB (2024). RCSB PDB: Homepage. https://www.rcsb.org (accessed 17 September 2024).
118 Weissenberger, G., Henderikx, R.J.M., and Peters, P.J. (2021). Understanding the invisible hands of sample preparation for cryo-EM. *Nat. Methods* 18 (5): 463–471.
119 Drulyte I, Koester S, Lundberg D, Eng K, Malyutin A, McGrath A, et al. (2022). High-Throughput Cryo-EM Epitope Mapping of SARSCoV-2 Spike Protein Antibodies Using EPU Multigrid. https://assets.thermofisher.com/TFS-Assets/MSD/Reference-Materials/cryo-em-epitope-mapping-wp0031.pdf?cid=CMP-05320-M2V8&utm_source=Comms-Blog&utm_medium=EM%20Blog&utm_campaign=CMP-05320-M2V8 (accessed 15 November 2024).

120 Pichkur, E.B., Lyukmanova, E.N., and Shenkarev, Z.O. (2023). *Delta SARS-CoV-2 Spike Protein in Complex with REGN10987 Fab Homologue*. Protein Data Bank.

121 Raghavan, S.S.R., Dagil, R., Lopez-Perez, M. et al. (2022). Cryo-EM reveals the conformational epitope of human monoclonal antibody PAM1.4 broadly reacting with polymorphic malarial protein VAR2CSA. *PLoS Pathog.* 18 (11): e1010924.

122 Frisch, M.J., Trucks, G.W., Schlegel, H.B., et al. (2016). Gaussian, Inc., Wallingford, CT.

123 Kresse, G. and Hafner, J. (1993). Ab initio molecular dynamics for liquid metals. *Phys. Rev. B.* 47: 558–561. Available from: https://link.aps.org/doi/10.1103/PhysRevB.47.558.

124 Neese, F., Wennmohs, F., Becker, U., and Riplinger, C. (2020). The ORCA quantum chemistry program package. *J. Chem. Phys.* 152 (22): 224108. Available from: https://doi.org/10.1063/5.0004608.

125 Aprà, E., Bylaska, E.J., de Jong, W.A. et al. (2020). NWChem: past, present, and future. *J. Chem. Phys.* 152 (18): 184102.

126 Case, D.A., Cheatham, T.E. III, Darden, T. et al. (2005). The Amber biomolecular simulation programs. *J. Comput. Chem.* 26 (16): 1668–1688. Available from: https://doi.org/10.1002/jcc.20290.

127 Abraham, M.J., Murtola, T., Schulz, R. et al. (2015). GROMACS: high performance molecular simulations through multi-level parallelism from laptops to supercomputers. *SoftwareX* 1 (1–2): 19–25. Available from: https://www.sciencedirect.com/science/article/pii/S2352711015000059.

128 Brooks, B.R., Brooks, C.L. 3rd, Mackerell, A.D.J. et al. (2009). CHARMM: the biomolecular simulation program. *J. Comput. Chem.* 30: 1545–1614.

129 Takahashi, O., Kirikoshi, R., and Manabe, N. (2014). Roles of intramolecular and intermolecular hydrogen bonding in a three-water-assisted mechanism of succinimide formation from aspartic acid residues. *Molecules* 19 (8): 11440–11452.

130 Catak, S., Monard, G., Aviyente, V., and Ruiz-López, M.F. (2006). Reaction mechanism of deamidation of asparaginyl residues in peptides: effect of solvent molecules †. *J. Phys. Chem. A* 110 (27): 8354–8365.

131 Catak, S., Monard, G., Aviyente, V., and Ruiz-López, M.F. (2009). Deamidation of asparagine residues: direct hydrolysis versus succinimide-mediated deamidation mechanisms. *J. Phys. Chem. A* 113 (6): 1111–1120.

132 Oliva, A., Henry, B., and Ruiz-López, M.F. (2013). Insights on peptide backbone N–H acidity: structure of anions, hydration effects. *Chem. Phys. Lett.* 561: 153–158.

133 Plotnikov, N.V., Singh, S.K., Rouse, J.C., and Kumar, S. (2017). Quantifying the risks of asparagine deamidation and aspartate isomerization in biopharmaceuticals by computing reaction free-energy surfaces. *J. Phys. Chem. B* 121 (4): 719–730.

134 Radkiewicz, J.L., Zipse, H., Clarke, S., and Houk, K.N. (2001). Neighboring side chain effects on asparaginyl and aspartyl degradation: an Ab initio study of the relationship between peptide conformation and backbone NH acidity. *J. Am. Chem. Soc.* 123 (15): 3499–3506. Available from: https://doi.org/10.1021/ja0026814.

135 Radkiewicz, J.L., Zipse, H., Clarke, S., and Houk, K.N. (1996). Accelerated racemization of aspartic acid and asparagine residues via succinimide intermediates: an ab initio theoretical exploration of mechanism. *J. Am. Chem. Soc.* 118 (38): 9148–9155. Available from: https://doi.org/10.1021/ja953505b.

136 Wood, G.P.F., Sreedhara, A., Moore, J.M. et al. (2016). Mechanistic insights into radical-mediated oxidation of tryptophan from ab initio quantum chemistry calculations and QM/MM molecular dynamics simulations. *J. Phys. Chem. A* 120 (18): 2926–2939.

137 Liu, Y., Sulaiman, H.F., Johnson, B.R. et al. (2023). QM/MM study of N501 involved intermolecular interaction between SARS-CoV-2 receptor binding domain and antibody of human origin. *Comput. Biol. Chem.* 102: 107810.

138 Jing, Z., Liu, C., Qi, R., and Ren, P. (2018). Many-body effect determines the selectivity for Ca2+ and Mg2+ in proteins. *Proc. Natl. Acad. Sci.* 115 (32): E7495–E7501.

139 Vidossich, P. and Magistrato, A. (2014). QM/MM molecular dynamics studies of metal binding proteins. *Biomolecules* 4 (3): 616–645.

140 Han, W., Zhu, J., Wang, S., and Xu, D. (2017). Understanding the phosphorylation mechanism by using quantum chemical calculations and molecular dynamics simulations. *J. Phys. Chem. B* 121 (15): 3565–3573.

141 Lindorff-Larsen, K., Trbovic, N., Maragakis, P. et al. (2012). Structure and dynamics of an unfolded protein examined by molecular dynamics simulation. *J. Am. Chem. Soc.* 134 (8): 3787–3791.

142 Rocco, A.G., Mollica, L., Ricchiuto, P. et al. (2008). Characterization of the protein unfolding processes induced by urea and temperature. *Biophys. J.* 94 (6): 2241–2251.

143 Salimi, N.L., Ho, B., and Agard, D.A. (2010). Unfolding simulations reveal the mechanism of extreme unfolding cooperativity in the kinetically stable α-lytic protease. *PLoS Comput. Biol.* 6 (2): e1000689.

144 Settanni, G. and Fersht, A.R. (2008). High temperature unfolding simulations of the TRPZ1 peptide. *Biophys. J.* 94 (11): 4444–4453.

145 Collu, F., Spiga, E., Chakroun, N. et al. (2018). Probing the early stages of prion protein (PrP) aggregation with atomistic molecular dynamics simulations. *Chem. Commun.* 54 (57): 8007–8010.

146 Childers, M.C. and Daggett, V. (2018). Validating molecular dynamics simulations against experimental observables in light of underlying conformational ensembles. *J. Phys. Chem. B* 122 (26): 6673–6689.

147 Lindorff-Larsen, K., Piana, S., Dror, R.O., and Shaw, D.E. (2011). How fast-folding proteins fold. *Science* 334 (6055): 517–520. Available from: https://doi.org/10.1126/science.1208351.

148 Nguyen, H., Maier, J., Huang, H. et al. (2014). Folding simulations for proteins with diverse topologies are accessible in days with a physics-based force field and implicit solvent. *J. Am. Chem. Soc.* 136 (40): 13959–13962. Available from: https://doi.org/10.1021/ja5032776.

149 Osajima, T. and Hoshino, T. (2016). Roles of the respective loops at complementarity determining region on the antigen-antibody recognition. *Comput. Biol. Chem.* 1 (64): 368–383. Available from: https://www.sciencedirect.com/science/article/pii/S1476927116301773.

150 Yamashita, T. (2023). Molecular dynamics simulation for investigating antigen-antibody interaction. *Methods Mol. Biol.* 2552: 101–107.

151 Fernández-Quintero, M.L., Kraml, J., Georges, G., and Liedl, K.R. (2019). CDR-H3 loop ensemble in solution – conformational selection upon antibody binding. *MAbs* 11: 1077–1088.

152 Zhang, C., Codina, N., Tang, J. et al. (2021). Comparison of the pH- and thermally-induced fluctuations of a therapeutic antibody fab fragment by molecular dynamics simulation. *Comput. Struct. Biotechnol. J.* 19: 2726–2741.

153 Wang, Y., Williams, H.D., Dikicioglu, D., and Dalby, P.A. (2024). Predictive model building for aggregation kinetics based on molecular dynamics simulations of an antibody fragment. *Mol. Pharm.* 21 (11): 5827–5841.

154 Berner, C., Menzen, T., Winter, G., and Svilenov, H.L. (2021). Combining unfolding reversibility studies and molecular dynamics simulations to select aggregation-resistant antibodies. *Mol. Pharm.* 18 (6): 2242–2253.

155 Raybould, M.I.J., Turnbull, O.M., Suter, A. et al. (2024). Contextualising the developability risk of antibodies with lambda light chains using enhanced therapeutic antibody profiling. *Commun. Biol.* 7 (1): 62.

156 Park, E. and Izadi, S. (2024). Molecular surface descriptors to predict antibody developability: sensitivity to parameters, structure models, and conformational sampling. *MAbs* 16 (1): 2362788.

157 Raybould, M.I.J., Marks, C., Krawczyk, K. et al. (2019). Five computational developability guidelines for therapeutic antibody profiling. *Proc. Natl. Acad. Sci.* 116 (10): 4025–4030.

158 Jumper, J., Evans, R., Pritzel, A. et al. (2021). Highly accurate protein structure prediction with AlphaFold. *Nature* 596 (7873): 583–589. Available from: https://doi.org/10.1038/s41586-021-03819-2.

159 Irudayanathan, F.J., Zarzar, J., Lin, J., and Izadi, S. (2022). Deciphering deamidation and isomerization in therapeutic proteins: effect of neighboring residue. *MAbs* 14 (1): 2143006.

160 Natesan, R. and Agrawal, N.J. (2022). Non-covalent fc-fab interactions significantly alter internal dynamics of an IgG1 antibody. *Sci. Rep.* 12 (1): 9321.

161 Sun, Y., Izadi, S., Callahan, M. et al. (2021). Antibody–receptor interactions mediate antibody-dependent cellular cytotoxicity. *J. Biol. Chem.* 297 (1): 100826.

162 Wei, B., Gao, X., Cadang, L. et al. (2021). Fc galactosylation follows consecutive reaction kinetics and enhances immunoglobulin G hexamerization for complement activation. *MAbs* 13 (1): 1893427.

163 Chaudhri, A., Zarraga, I.E., Kamerzell, T.J. et al. (2012). Coarse-grained modeling of the self-association of therapeutic monoclonal antibodies. *J. Phys. Chem. B* 116 (28): 8045–8057.

164 Chaudhri, A., Zarraga, I.E., Yadav, S. et al. (2013). The role of amino acid sequence in the self-association of therapeutic monoclonal antibodies: insights from coarse-grained modeling. *J. Phys. Chem. B* 117 (5): 1269–1279.

165 Dear, B.J., Bollinger, J.A., Chowdhury, A. et al. (2019). X-ray scattering and coarse-grained simulations for clustering and interactions of monoclonal antibodies at high concentrations. *J. Phys. Chem. B* 123 (25): 5274–5290.

166 Wang, G., Varga, Z., Hofmann, J. et al. (2018). Structure and relaxation in solutions of monoclonal antibodies. *J. Phys. Chem. B* 122 (11): 2867–2880.

167 Mahapatra, S., Polimeni, M., Gentiluomo, L. et al. (2022). Self-interactions of two monoclonal antibodies: small-angle X-ray scattering, light scattering, and coarse-grained Modeling. *Mol. Pharm.* 19 (2): 508–519.

168 Corbett, D., Hebditch, M., Keeling, R. et al. (2017). Coarse-grained modeling of antibodies from small-angle scattering profiles. *J. Phys. Chem. B* 121 (35): 8276–8290.

169 Izadi, S., Patapoff, T.W., and Walters, B.T. (2020). Multiscale coarse-grained approach to investigate Self-Association of Antibodies. *Biophys. J.* 118 (11): 2741–2754.

170 Anandakrishnan, R., Baker, C., Izadi, S., and Onufriev, A.V. (2013). Point charges optimally placed to represent the multipole expansion of charge distributions. *PLoS One* 8 (7): e67715.

171 Izadi, S., Anandakrishnan, R., and Onufriev, A.V. (2016). Implicit solvent model for million-atom atomistic simulations: insights into the organization of 30-nm chromatin Fiber. *J. Chem. Theory Comput.* 12 (12): 5946–5959.

172 Grünberger, A., Lai, P.K., Blanco, M.A., and Roberts, C.J. (2013). Coarse-grained modeling of protein second osmotic virial coefficients: sterics and short-ranged attractions. *J. Phys. Chem. B* 117 (3): 763–770.

173 Calero-Rubio, C., Saluja, A., and Roberts, C.J. (2016). Coarse-grained antibody models for "weak" protein–protein interactions from low to high concentrations. *J. Phys. Chem. B* 120 (27): 6592–6605.

174 Janke, J.J., Starr, C.G., Kingsbury, J.S. et al. (2024). Computational screening for mAb colloidal stability with coarse-grained, molecular-scale simulations. *J. Phys. Chem. B.* 128 (6): 1515–1526.

175 Benjamin, K.M., Singh, J.K., Schultz, A.J., and Kofke, D.A. (2007). Higher-order virial coefficients of water models. *J. Phys. Chem. B* 111 (39): 11463–11473.

176 Singh, J.K. and Kofke, D.A. (2004). Mayer sampling: calculation of cluster integrals using free-energy perturbation methods. *Phys. Rev. Lett.* 92: 220601. Available from: https://link.aps.org/doi/10.1103/PhysRevLett.92.220601.

177 Park, E. and Izadi, S. (2024). Molecular surface descriptors to predict antibody developability: sensitivity to parameters, structure models, and conformational sampling. *mAbs* 16 (1). https://doi.org/10.1080/19420862.2024.2362788.

178 Raybould, M.I.J., Marks, C., Krawczyk, K. et al. (2019). Five computational developability guidelines for therapeutic antibody profiling. *Proc. Natl. Acad. Sci. U.S.A.* 116 (10): 4025–4030. https://doi.org/10.1073/pnas.1810576116.

179 Norman, R.A., Ambrosetti, F., Bonvin, A.M.J.J. et al. (2020). Computational approaches to therapeutic antibody design: established methods and emerging trends. *Brief. Bioinf.* 21 (5): 1549–1567. https://doi.org/10.1093/bib/bbz095.

180 Irudayanathan, F.J., Zarzar, J., Lin, J., and Izadi, S. Deciphering deamidation and isomerization in therapeutic proteins: Effect of neighboring residue. *MAbs* 14 (1): 2143006.

181 Abanades, B., Wong, W.K., Boyles, F. et al. (2023). ImmuneBuilder: deep-learning models for predicting the structures of immune proteins. *Commun. Biol.* 6 (1): 575. https://doi.org/10.1038/s42003-023-04927-7.

182 Ruffolo, J.A., Sulam, J., and Gray, J.J. (2022). Antibody structure prediction using interpretable deep learning. *Patterns* 3 (2): 100406. https://doi.org/10.1016/j.patter.2021.100406.

183 Lee, J.H., Yadollahpour, P., Watkins, A. et al. (2022). EquiFold: protein structure prediction with a novel coarse-grained structure representation. *bioRxiv* https://www.biorxiv.org/content/10.1101/2022.10.07.511322v1.

9

Target Binding Assays for Biotherapeutics

Eric Janezic[1] *and Xiangdan Wang*[2]

[1]*Biochemical and Cellular Pharmacology, Genentech Inc., South San Francisco, CA, USA*
[2]*Bioanalytical Sciences, Genentech Inc., South San Francisco, CA, USA*

Introduction

Verification of target engagement is a critical step in the identification of biotherapeutics. Single-concentration enzyme-linked immunosorbent assay (ELISA) and fluorescence-activated cell sorting (FACS)-based assays are commonly performed to confirm the drug–target interaction. While these assays are useful to triage binders and non-binders, they do not provide quantitative information on the strength of the interaction. Understanding the biophysical and biochemical parameters that dictate the strength of these interactions is required for lead selection, molecular engineering, and dose selection for pharmacokinetic/pharmacodynamic modeling.

Binding affinity reflects the strength of the binding interaction between a biomolecule and its binding partner. Affinity is typically represented by the equilibrium dissociation constant (K_D). The smaller the K_D value, the stronger the binding affinity. For a simple binding interaction of a therapeutic drug (D) to its target (T) that forms the complex (DT) at a 1:1 stoichiometry ratio, assuming the interaction is reversible, the reaction scheme is represented as:

$$D + T \underset{k_d}{\overset{k_a}{\rightleftharpoons}} \mathrm{DT} \tag{9.1}$$

where k_a is the association rate constant, describing how fast the drug and its target bind to each other, and k_d is the dissociation rate constant, describing how fast the complex dissociates. The formation of the complex at any given time can be described as:

$$\frac{d[\mathrm{DT}]}{dt} = k_a \times [D] \times [T] - k_d \times [\mathrm{DT}] \tag{9.2}$$

where $[D]$ is the concentration of free drug, $[T]$ is the concentration of free target, and [DT] is the concentration of the complex. When the reaction reaches equilibrium, the rate at which the complexes are formed is equal to the rate at which they dissociate, this results in:

$$K_D = \frac{[D][T]}{[\mathrm{DT}]} = \frac{k_d}{k_a} \tag{9.3}$$

Based on Equation (9.3), binding affinity can be determined either from the ratio of free drug and target to bound complex at equilibrium or from the ratio between kinetic rate constants. Interactions with the same binding affinity could have different kinetic profiles. Studies have demonstrated that kinetics can provide unique information for understanding the biological process behind an interaction and might be useful for the design and optimization of therapeutics [1, 2].

Assays to Determine Affinity of Biotherapeutics

AV Hill was the first to report that the receptor–ligand interaction is reversible and follows the Law of Mass Action [3]. Further work by Langmuir [4], Michaelis and Menten [5], and Clark [6] validated and expanded on these findings. While these works laid the foundation for understanding how drugs interact with their receptors, it was not until 1947 that the first method to measure binding affinity of ligands was developed [7]. This indirect method, known as Schild regression analysis, allows the calculation of an antagonist's affinity from functional assays. The next major breakthrough in the field came from Paton and Rang, who performed the first assays to directly measure the affinities of drug–target interactions using radioactivity [8] followed by analysis from George Scatchard [9]. All of these landmark methods were originally developed for studying small molecules. The first assays to measure the binding affinities of biotherapeutics were not developed

Characterizing Biotherapeutics: Analytical Methods for Diverse Modalities, First Edition. Edited by Jennie R. Lill and Wendy Sandoval.

Table 9.1 Common technologies to determine affinity of biotherapeutics to their targets.

Technology	Target (protein or cell based)	Capable of providing kinetic information (k_a/k_d)	Equilibrium required?	Labeling required?	Biological relevance of target protein	Throughput
Surface plasma resonance (SPR)	Protein based	Yes	No	No	+/−[a]	High
Biolayer interferometry (BLI)	Protein based	Yes	No	No	+/−[a]	High
switchSENSE	Protein based	Yes	No	Yes	+/−[a]	Mid
Isothermal titration calorimetry (ITC)	Protein based	No	Yes	No	+/−[a]	Low
Microscale thermophoresis (MST)	Protein based	No	Yes	Yes	+/−[a]	Mid
Flow-induced dispersion analysis (FIDA)	Protein based	No	Yes	Yes	+/−[a]	Mid
Kinetic exclusion assay	Protein or cell based	Yes	Yes	No	+/−[a]	Mid
Saturation assay	Cell based	No	Yes	Yes	+	Low
Competition assay	Cell based	No	Yes	Yes	+	Low
LigandTracer	Cell based	Yes	No	Yes	+	Low
HeliXcyto	Cell based	Yes	No	Yes	+	Mid
SPRm200	Cell based	Yes	No	No	+	Low
Pre-equilibrium kinetic exclusion assay	Cell based	Yes	No	No	+	High

a) Not always feasible to produce recombinant proteins.

until the end of the twentieth century [10, 11]. Since then, the number of technologies to study these interactions has expanded significantly (Table 9.1) and relies either on purified/recombinant target proteins (protein-based) or on cells expressing the receptor of interest (cell-based).

Protein-Based Assays

Common technologies that utilize purified or recombinant target proteins for affinity measurement include surface plasmon resonance (SPR) (Biacore [12], Carterra [13], Sierra [14], Nicoya [15]), biolayer interferometry (Octet [16], Gator [17]), KinExA [18, 19], isothermal titration calorimetry (ITC) [20], MicroScale Thermophoresis (MST) [21], switchSENSE [22], and flow-induced dispersion analysis (FIDA) [23]. Among these methods, SPR assays are the most frequently used. Since the first SPR-based Biacore instrument was launched, many researchers have applied SPR assays to characterize biotherapeutics binding to their targets during early or advanced stages of biotherapeutic development. In this section, we will focus on commonly used SPR affinity assays and general considerations when conducting these experiments.

SPR Assays for Affinity Measurement

The SPR phenomenon occurs when polarized light is directed toward a thin metal layer (such as gold) at the interface between media with different refractive indices.

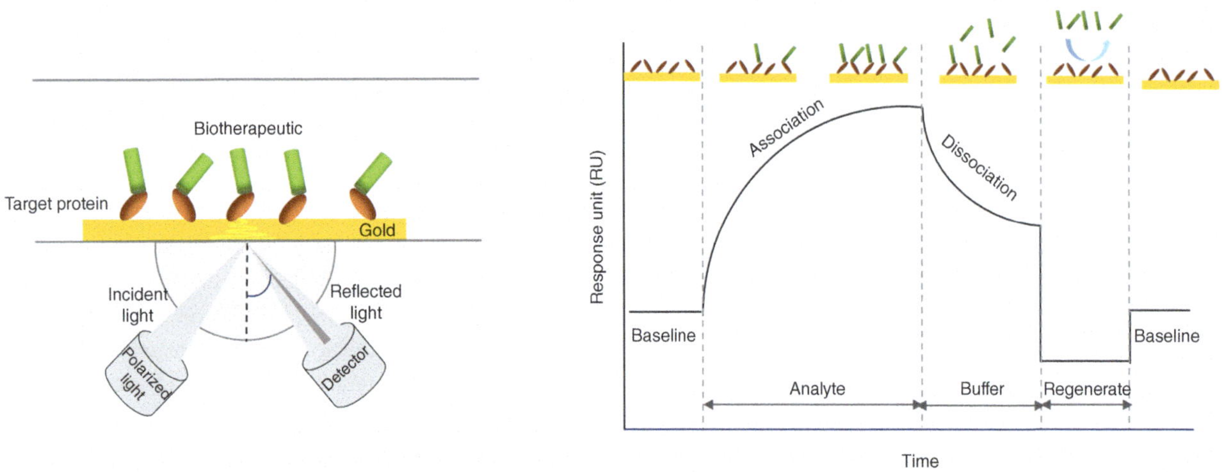

Figure 9.1 A biotherapeutic protein binding to its target on the surface of a sensor chip in an SPR setup.

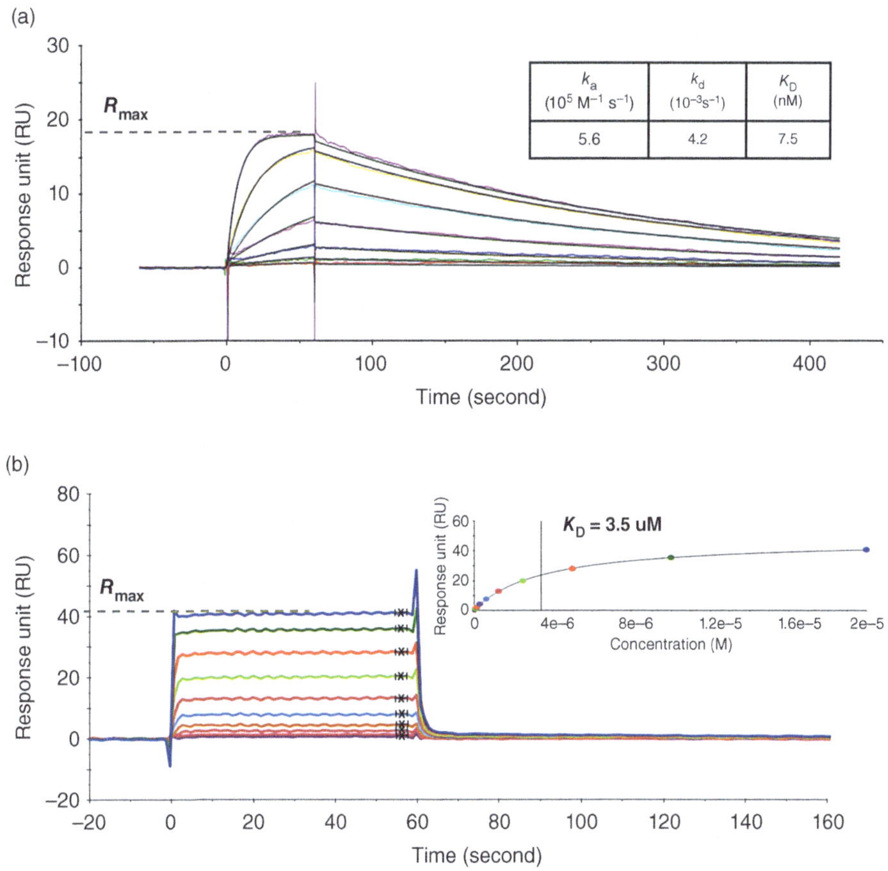

Figure 9.2 Examples of binding affinity of therapeutics to their targets determined from (a) kinetic or (b) equilibrium (steady-state) analysis in SPR assays. Colored lines represent experimental data corresponding to different analyte concentrations, and black lines are the fit models. Experimentally determined R_{max} values are shown. Source: Adapted from Wang et al. [24, 25].

In a typical SPR assay, target molecules are immobilized on a sensor chip surface followed by injection of the other binding partner, or analyte, allowing the interaction to take place (Figure 9.1). During the interaction, the refractive index of medium close to the gold surface increases due to the formation (association) of therapeutic-target complexes, resulting in a refractive index angle shift. The angle change is recorded in real time as an increase in binding response unit (RU) over time in a sensorgram. The increase of binding response level gradually slows down as more complexes are formed with less binding sites available on the surface. After the injection of the analyte stops, only buffer flows over the surface and the binding response decreases due to dissociation of the biotherapeutic molecules from the bound complexes. The sensor chip can be regenerated by removing bound therapeutic molecules from the surface with a regeneration solution, allowing the surface to be ready for the next interaction cycle. The binding data can be fitted to an appropriate binding model, enabling the determination of the affinity from either kinetic (Figure 9.2a) or equilibrium (steady-state) analysis (Figure 9.2b).

As discussed in the Introduction, binding affinity can be derived as the ratio between kinetic rate constants, k_d and k_a. The rate of the formation of a drug–target complex can be described by Equation (9.2). In an SPR assay setup, Equation (9.2) can be expressed as

$$\frac{dR}{dt} = k_a C(R_{max} - R) - k_d R \tag{9.4}$$

where R is the SPR signal at any given time t, dR/dt is the change of the SPR signal, C is the concentration of analyte (it can be either the drug or the target), and R_{max} is the maximum binding response. The dissociation rate can be described by

$$\frac{dR}{dt} = -k_d R \tag{9.5}$$

Kinetic rate constants can be derived by fitting experimental data to the model equations that describe binding interactions.

Though kinetic analysis is the most commonly used, it is not always feasible. For example, the interaction kinetics may not be able to be resolved for weak interactions with very fast dissociation rates. In these cases, the binding affinity can be determined by steady-state analysis, and the K_D value can be obtained from a plot of equilibrium response (R_{eq}) versus analyte concentration (C):

$$R_{eq} = \frac{CR_{max}}{C + K_D} \tag{9.6}$$

The requirement of applying the steady-state affinity analysis to an interaction is that the binding curves of analyte at all concentrations must reach equilibrium. If equilibrium is not reached at some tested concentrations, the K_D value will be overestimated.

Experiments to determine binding kinetics and affinity can be performed by using multi-cycle kinetics (MCK) or single-cycle kinetics (SCK) approaches (Figure 9.3). MCK involves injecting multiple cycles of analyte at different concentrations over the ligand, and surface regeneration is needed after each cycle. MCK provides great flexibility in experimental design and data analysis and is the most frequently used approach. SCK involves sequential injections of analyte with five increasing concentrations over the ligand in one cycle without the need to regenerate the surface, and the dissociation of the highest concentration is monitored and used for data analysis. Although developed later and less commonly used than MCK, SCK analysis provides all the information necessary to calculate binding kinetic and affinity data associated with an interaction [26, 27] and has been useful for interactions in which it is challenging to remove bound analyte or when regeneration conditions are detrimental to the ligand on a sensor chip [28].

Compared to technologies that only measure binding affinity of molecular interactions at equilibrium such as ITC, MST, and FIDA, SPR assay can provide real-time kinetics and affinity information. SPR is highly sensitive to refractive index changes caused by binding events happening at the sensor chip surface and is suitable for studying most biomolecular interactions with a wide range of K_D values from millimolar (mM) to picomolar (pM). SPR assays in general are easy to set up, consume low sample volumes, and do not require labeling of molecules. The measurements can be carried out with crude samples, such as cell culture medium or biological matrices, without the need for laborious sample preparation steps. Most available SPR platforms have a high degree of automation, and there are several systems with high-throughput capability such as Biacore 8K, Carterra, and Sierra. These features make SPR assays suitable for characterizing binding properties of biotherapeutics including new modalities with complex molecular structures such as bispecific antibodies [29, 30] and PEGylated multimeric antibodies [24, 31].

Considerations When Conducting SPR Assays

To determine reliable kinetic and affinity parameters, careful experimental design and execution are required. Binding affinity is influenced by non-covalent interactions such as electrostatic interactions, hydrogen bond, hydrophobic interactions, and van der Waals forces. Many factors such as sensor chip types, target protein immobilization approaches, reagents, and assay conditions (i.e. flow rate, ligand immobilization levels, temperature, buffer composition, pH, etc.) can affect the data quality and accuracy of the calculated K_D in an SPR assay.

Sensor Chip Selection

There are a wide range of commercially available SPR sensor chips. The choice of sensor chip selection in an SPR assay depends on multiple factors including the application requirements, the properties of tested molecules, and operational considerations. Cytiva provides a variety of Biacore sensor chips with different features that are suitable for various applications. Xantec also manufactures Biacore compatible sensor chips as well as sensor chips for other SPR platforms such as Carterra and Sierra. The

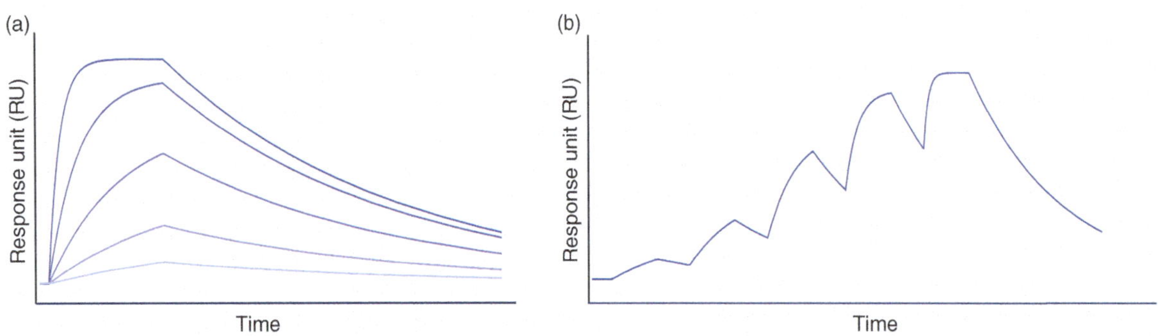

Figure 9.3 Simulated sensorgrams of experiments performed using (a) multi-cycle kinetics (MCK) or (b) single-cycle kinetics (SCK) approaches. Source: Adapted from Biacore™ Simul8 (Cytiva).

surface of most available Biacore sensor chips is covered by a layer of carboxymethyl (CM) dextran matrix. The dextran matrix provides a hydrophilic surface suitable for most solution-based biomolecular interactions. Additionally, the flexible hydrogel is intended to create a "solution-like" environment allowing attached molecules to have relatively free movement within the surface layer. For the most widely used sensor chip, CM5, the dextran layer on the surface is approximately 100 nm thick. The high density of CM dextran provides high surface capacity for immobilization and makes sensor chip CM5 suitable for most applications [32]. For analyte molecules that are highly positively charged, nonspecific binding to the surface might be a concern, and sensor chips with a lower degree of carboxylation such as CM4 and CM3 can be considered. An alternative surface to reduce nonspecific binding is sensor chip C1, which has a flat surface with significantly lower amount of carboxyl groups but no dextran matrix.

Immobilization of Ligand

Ligands can be attached to the sensor chip by either direct immobilization or indirect capture methods. The direct immobilization method covalently couples a protein to the dextran matrix via specific reactive groups such as amine, thiol, or aldehyde groups and does not require additional steps to modify the ligand protein. Amine coupling conjugates the protein to the chip via primary amine groups in lysine or N-terminal residues and is the most often used chemistry. The immobilization processes consume very low amounts of ligand proteins, and the immobilized proteins are generally stable on the surface if regeneration conditions are optimized. One main drawback of the direct immobilization method is that the coupling sites are random, which can impact the binding properties of the protein if the modified reactive groups are in proximity to the binding site. Therefore, it is always a good practice to assess the activity of a ligand after it has been covalently immobilized on a sensor chip.

An alternative approach to circumvent the drawback of the direct immobilization method is to indirectly capture a ligand on the surface of a sensor chip. This approach relies on the specific interaction between a ligand and a capture molecule that is covalently immobilized on a sensor chip using the direct immobilization method described above. Compared with the direct immobilization method, the indirect capture method offers several advantages: binding properties of ligand molecules can be better preserved, ligand molecules can be oriented in a more uniform manner on the surface, and interference from impurities in ligand proteins can be minimized. Because of these advantages, the indirect capture format is often used in SPR analysis to evaluate interactions between therapeutics and their binding partners including challenging transmembrane proteins [27, 33–35]. For example, Griffiths et al. successfully developed an SPR assay using the indirect capture format to characterize interactions between human single-domain antibodies and chemokine receptor CXCR4 expressed on lipoparticles [35]. Common capture reagents include antibodies against specific epitopes or tags, Protein A for antibodies, and streptavidin or neutravidin for biotinylated molecules. To generate reliable binding affinity data using the indirect capture format, the interaction between the capture reagent and the ligand must be tight enough to ensure a stable baseline of the ligand during an analysis cycle. Therefore, it is critical to monitor the baseline stability of a ligand after it is captured on a sensor chip. If a significant amount of ligand molecules dissociates from the surface, alternative capture reagents need to be explored and tested [34]. It is worth noting that the observed stable baseline may not always be indicative of a tight interaction between the capture reagent and the ligand, but rather a result of rebinding of ligand molecules on the surface [36]. The rebinding of the captured ligand might be disrupted by the ligand–analyte interaction and in turn affects the measured kinetics and affinity data. The recommended best practice is to assess the binding affinity of a capture reagent to the ligand and to immobilize the high-affinity capture reagent on sensor chips at high densities. Additionally, caution should be taken when selecting a capture reagent to ensure that it does not interfere with the ligand binding to its target via steric hindrance or by altering the binding properties [34, 36].

Reagents

Reliable affinity determination is heavily dependent on the quality of reagents used in SPR assays. Reagents suitable for other assays may not always be of sufficient quality for SPR analysis. Additionally, reagent quality may change during storage conditions, and it is important to assess the quality of reagents immediately before they are used in SPR assays and not rely on previous characterization results [24]. This practice is especially important for reagents that are aggregation prone, have gone through multiple freeze–thaw cycles, or are generated for the first time with unknown stabilities. Besides characterizing reagents by traditional analytical tools such as mass spectrometry, sodium dodecyl sulfate polyacrylamide gel electrophoresis (SDS-PAGE), and size-exclusion chromatography with multi-angle light scattering (SEC-MALS), it is highly desirable to assess binding activity of proteins when conducting SPR assays with appropriate assay format. The binding activity of ligand can be evaluated based on the ratio between experimentally determined R_{max} and theoretically calculated R_{max}. The expected ligand binding activity for direct immobilization methods typically ranges from 50% to 80% and is greater than 90% for indirect capturing method [37]. Poor ligand binding activity can be an indication of low active concentration associated with a ligand protein or caused by modified binding properties of a protein introduced by the immobilization process [38]. Under these circumstances, higher-quality ligand protein or an alternative immobilization method should be considered.

Mass Transport Limitations

In a surface-based fluidic platform such as the SPR system, analyte molecules need to be transported from the bulk solution to the sensor surface before they can associate with ligand molecules on the surface. If the transport rate is slower than the association rate, mass transport will be a limiting factor and can impact the determined binding kinetic data. Using high flow rates (≥30 μL/min) and immobilizing the ligand at low densities that result in R_{max} values no greater than 50 – 150 RU can help reduce mass transport limitations (MTLs) [39]. If MTLs cannot be completely avoided, such as in the case of interactions with very fast k_a values (>10^7 $M^{-1}s^{-1}$), the impact of partial MTLs can be corrected during data analysis by using a fitting model with MTLs accommodated [40, 41].

Multivalent Binding

In an SPR assay, multivalent binding of one analyte molecule to multiple ligand molecules can cause the measured dissociation rate to appear slower than expected and affect the determined affinity [25]. In addition, multivalent binding interactions are often heterogeneous, which can add challenges to data analysis. It is generally recommended to immobilize or capture multivalent molecules such as antibodies on sensor chips to avoid multivalent binding. In situations where multivalent binding to the ligand cannot be avoided due to reagent limitations or where both binding partners are multivalent, careful experimental design such as immobilizing or capturing ligand at very low densities and/or using a C1 chip can be applied to reduce multivalent interaction [24]. Alternatively, Interaction Map (IM), an algorithm developed by Ridgeview, can be utilized to analyze heterogeneous binding data associated with multivalent binding. This method assumes that the binding of an analyte to a heterogeneous mixture of targets can be expressed as a sum of interactions where each interaction is represented in an on–off map by its specific pair of recognition ($\log(k_a)$) and stability ($\log(k_d)$) coordinates and a color corresponding to the calculated contribution (Figure 9.4) [42, 43]. Wang et al. applied IM to analyze SPR binding data between anti-CD20 antibodies and CD20 expressed on extracellular vesicles [27]. In this example, both the analyte (anti-CD20 antibodies) and the ligand (CD20) are multimeric, and multivalent binding was expected. However, IM analysis revealed that binding mechanisms and kinetics differ between antibodies and identified a unique Fab–Fab interaction that mediates the interaction of rituximab and CD20 dimers.

Temperature

Temperature can also affect the kinetic rate constants of an interaction and cause the derived binding affinity to be temperature dependent [44, 45]. Experiments to determine binding kinetics and affinity should be performed at a relevant temperature. For most biotherapeutics binding to their targets, measures at 37°C are more appropriate and should be considered unless stability of the proteins at higher temperature is a concern.

Cell-Based Affinity assays

Protein-based methods to assess the strength of the interaction between a biomolecule and its target receptor are heavily utilized throughout the drug discovery and development process. However, with the increasing complexity of biomolecules (i.e. bispecific and multispecific antibodies, protein conjugates, etc.) and receptor targets, these

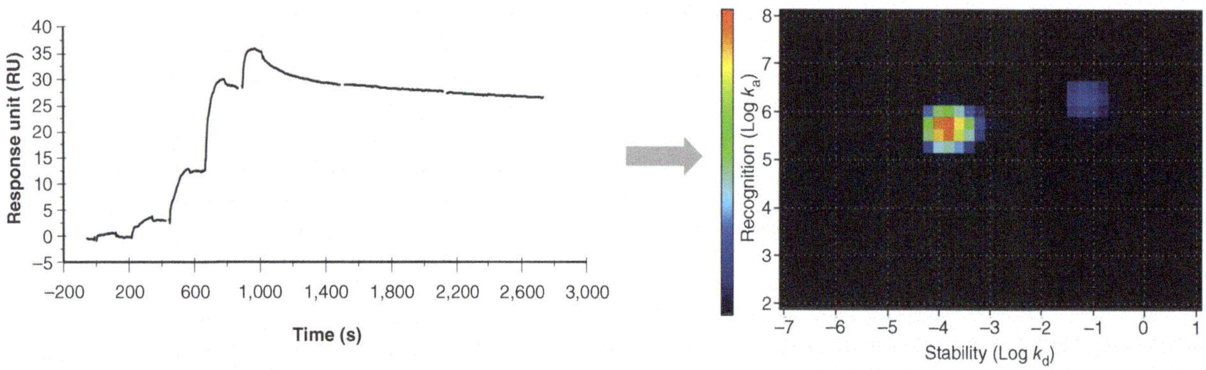

Figure 9.4 Representative SPR sensorgram of Obinutuzumab binding to CD20 expressed on extracellular vesicles using single-cycle kinetics (left) and corresponding Interaction Map (right). Source: Adapted from Wang et al. [27].

methods may not always be suitable [46]. For example, understanding the role of avidity can be important for selecting the lead candidate of complex biomolecules; and in the case of complex membrane targets, such as GPCRs, producing well-behaved antigens can be technically challenging. In these cases, cell-based affinity methods are often implemented. They can be classified into two categories: equilibrium and pre-equilibrium. In this section, we will discuss methods within both categories and provide general considerations and guidelines for selecting the most appropriate assay.

Equilibrium Assays

Equilibrium binding assays are the most commonly used cell-based binding assays and require the interaction to reach equilibrium to determine an accurate K_D. Most equilibrium binding experiments require labeling of the biotherapeutic using radioisotopes [47], fluorophores [11], or protein tags [48]. The two main types of equilibrium binding assays are saturation and competition experiments.

Saturation Assays

Saturation binding studies are performed by incubating a fixed number of cells against a titration of the labeled drug (Figure 9.5a) until equilibrium is reached. The cells are then washed to remove the unbound drug, and the amount of bound drug is measured. The signal represents the total binding (the sum of the specific and nonspecific interactions) and can be plotted according to the total drug concentration. The data is fit using Equation (9.7):

$$Y = \frac{B_{max} \times [D]_{total}}{K_D + [D]_{total}} + (NS \times [D]_{total}) + \text{Background} \tag{9.7}$$

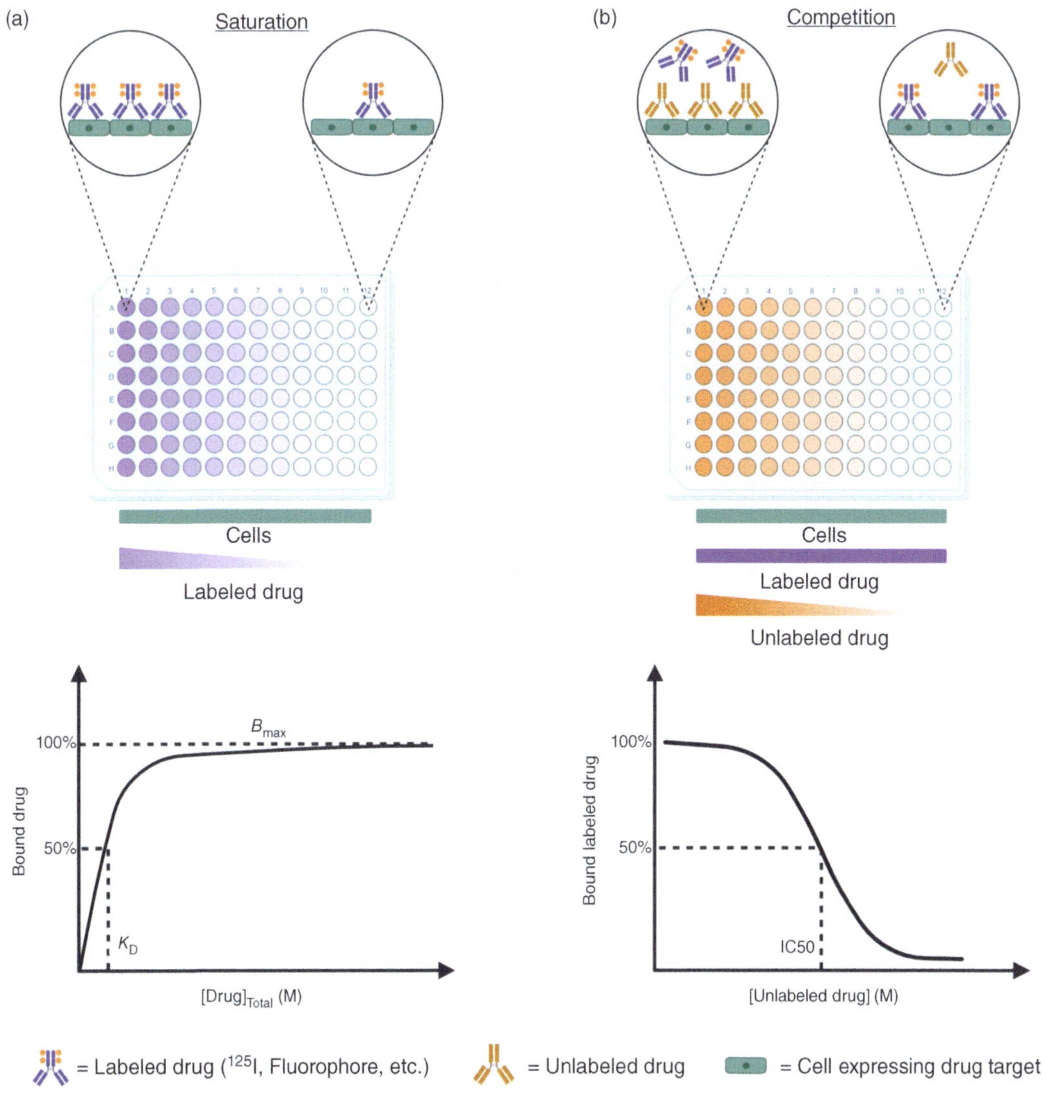

Figure 9.5 Schematic of (a) saturation and (b) competition cell-based equilibrium assays. Source: Created with BioRender.com.

where B_{max} is the maximum binding site density, K_D is the affinity of the labeled molecule, NS is the slope of the nonspecific signal, and Background is the background signal with no labeled molecule present. When using whole cells, the receptor copy number per cell can be calculated from the B_{max} value if the degree of labeling per drug molecule is known. The degree of labeling is obtained by determining the specific activity (CPM/fmol) of the radiolabeled biotherapeutics or the molar ratio of fluorophore to protein for fluorescence-based assays.

Nonspecific binding is defined as interactions that are not mediated by the target receptor (i.e. cell membranes, assay plate, membrane filter, etc.). It results in a non-saturable signal linearly proportional to the drug concentration in contrast to the specific binding, which can be represented using a saturation curve. The nonspecific binding can be determined by incubating the cells with a mixture of a titration of labeled drug with a fixed concentration of unlabeled drug in excess. The high concentration of unlabeled drug outcompetes the labeled drug for the specific binding sites leaving only the signal from the nonspecific interactions of the labeled drug. The nonspecific signal can be fit with a simple linear equation and can be subtracted from the total binding at each concentration within the titration to obtain the specific binding curve [47]. These data can be fit using Equation (9.8):

$$Y = \frac{B_{max} \times [D]_{Total}}{K_D + [D]_{Total}} \tag{9.8}$$

It is important to note that the equations above assume that the total bound drug is less than 10% of the total drug added to the reaction. A higher percentage would result in ligand depletion, which is observed when the target receptor is expressed at high levels and for very tight binders. Though there are equations that can account for ligand depletion, it is preferred to optimize the experimental conditions to prevent it. The most common approach to overcome ligand depletion is to increase the volume of the reaction and/or decrease the number of cells used.

Competition Assays

Competition binding assays measure the ability of an unlabeled biotherapeutic to displace a labeled molecule (Figure 9.5b). The labeled molecule can be a tool molecule or can be identical to the unlabeled molecule. To perform these assays, cells are incubated with a fixed concentration of labeled molecule less than K_D and a titration of unlabeled drug. The concentration of unlabeled drug that is able to displace 50% of the labeled molecule is called the IC50. This value is dependent on the concentration of labeled ligand that is used in an individual experiment and is useful to understand the relative interaction strength of biotherapeutics. While knowing the IC50 may be useful, it does not provide the affinity of the unlabeled molecule. To calculate the affinity of the unlabeled molecule, or K_i, one must first know the K_D of the labeled antibody and use the Cheng–Prusoff correction [49]. Though one may assume the K_D and K_i should be the same for a given molecule, this is often not the case for biotherapeutics, likely due to changes introduced by the labeling process. Therefore, it is recommended to perform both a saturation and a competition assay to determine the true affinity of a therapeutic. Competition assays can also suffer from ligand depletion as discussed above.

Kinetic Exclusion Assays

Kinetic exclusion assays are an alternative method to quantitate the receptor density and affinity of biotherapeutics to cells. Kinetic exclusion assays determine the K_D by quantifying the amount of unbound drug in solution at equilibrium (Figure 9.6) using a sandwich immunoassay. The solution containing free drug is quickly passed over a column containing the immobilized target. Because the drug concentration is fixed across the titration series, the amount of captured drug is directly proportional to free drug and the K_D can be calculated from the resulting isotherm. This method was first described for soluble proteins [19], and later adapted to cells by Xi and colleagues [50] using Sapidyne Instruments's KinExA. Gyrolab xP and xPand instruments can also be used to calculate affinities using this method [51, 52]. The principle of the assay is the same regardless of the instrument being used.

For the purpose of this section, we will primarily be focusing on adapting this assay for cells. When implementing these assays for cells, it is necessary to separate the media containing the unbound biotherapeutic from the cells first to prevent clogging of the instrument. For antibody therapeutics, it is common to use biotinylated or fluorescently labeled anti-Fc antibodies as capture and detection reagents, respectively. However, it may be advantageous to use the antigen as the capture reagent if it can be easily purified and biotinylated. The capture and detection pairing may need to be determined empirically for the specific interaction being studied.

These assays are typically performed in two steps. First, a fixed concentration of drug above the predicted K_D value is incubated across a titration of cells until equilibrium is reached. Because the drug concentration is above the K_D, the curve is considered stoichiometrically controlled [53]. This stoichiometrically controlled curve

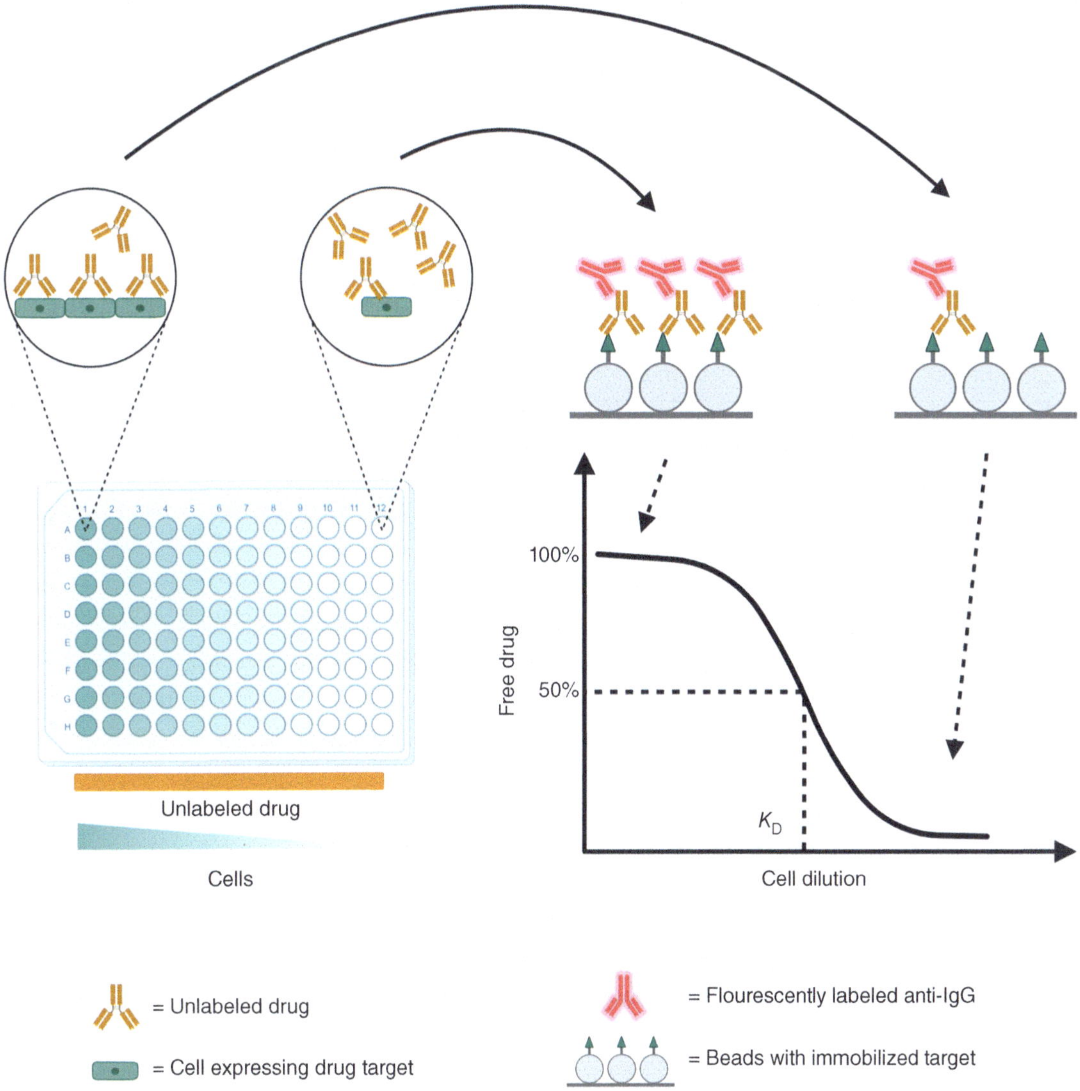

Figure 9.6 Principles of kinetic exclusion assay to determine binding affinity of therapeutics on cells. Source: Created with BioRender.com.

is used to determine the receptor density. Next, the assay is repeated but the antibody concentration is fixed below the predicted affinity in order to ensure that the reaction is K_D controlled. These two steps may also be performed in tandem to control receptor density differences between passages. The two curves can be fit using a global analysis to improve the robustness of the K_D determination. Interestingly, this does not appear to affect the receptor density calculations [54].

Kinetic exclusion assays have distinct considerations and advantages and disadvantages compared to more traditional saturation and competition assays. The main advantage of these experiments is the reliance on ligand depletion in order to calculate the affinity from the concentration of free drug [19]. This makes these assays exceptionally powerful for very tight binders where ligand depletion can be hard to overcome. For example, this method has been used to characterize femtomolar affinity interactions [55]. In order to reliably measure the concentration of free biotherapeutic for these strong interactions, the assay conditions must be sensitive enough to detect concentrations less than K_D. The time required to reach equilibrium should also be considered and may be restrictively long, which may impact cell viability. Finally, this technique can require very high cell densities in order to deplete the free ligand. This is a particular issue for cell lines or primary cells with low receptor concentration, where the cell concentration required to completely deplete the ligand may be incompatible with keeping the cells in solution.

Pre-equilibrium Assays

Equilibrium methods are routinely utilized for cell-based affinity measurements due to the ease of implementation. In the case of very high-affinity interactions, the time to reach equilibrium may become a limitation as it will significantly compromise cell viability. Additionally, equilibrium assays can only determine the K_D and do not provide kinetic information, such as k_a and k_d. Knowledge of these parameters can oftentimes be informative for understanding a biotherapeutic's mechanism of action. To address these needs, cell-based real-time (Figure 9.7a) and time-point pre-equilibrium binding assays (Figure 9.7b) have been developed.

Real-Time Binding Assays

Real-time binding assays are characterized by the ability to measure the association and dissociation of biotherapeutics to cells in real time, similar to SPR. These methods typically involve immobilizing the target cells by culturing the cells on the assay plate [56, 57] or using customized cell cages [58]. Once immobilized, the cells are exposed to increasing concentrations of the biotherapeutic for a set amount of time followed by an antibody-free buffer. The association and dissociation of the antibody to the cells are monitored in real time by tracking changes in fluorescence or radioactive emission. From these real-time binding curves, the association and dissociation rate constants and the affinity can be calculated.

In addition to calculating the binding kinetics of simple 1:1 interactions, real-time binding assays are used to understand more complex binding profiles [42, 59]. Similar to SPR, these heterogeneous binding traces can be further evaluated using IM analysis. Using this analysis, Bondza and colleagues found that Type I anti-CD20 antibodies have concentration-dependent binding profiles, but Type II anti-CD20 antibodies bind almost exclusively in a bivalent manner, regardless of the concentration [59]. Interestingly, the percent of bivalent binding was inversely related to antibody concentration for both of the Type II molecules.

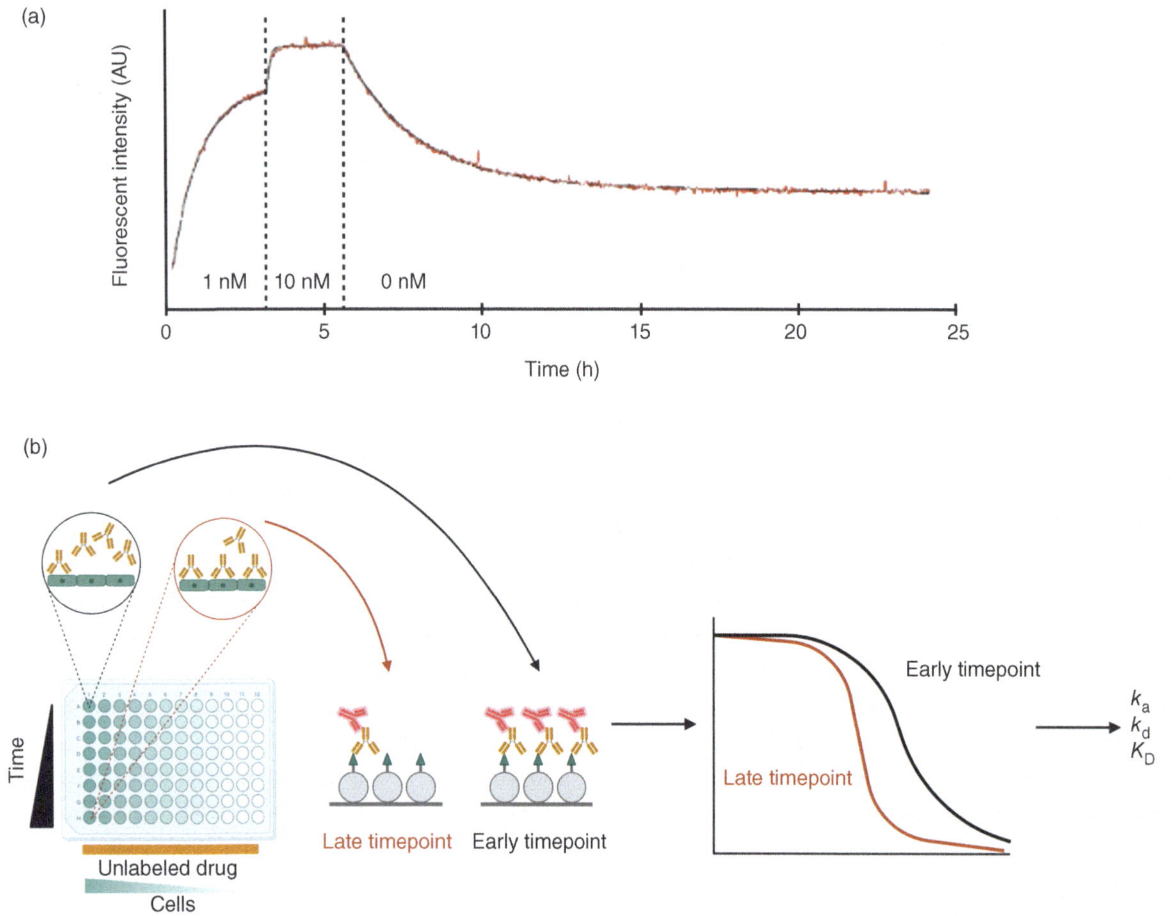

Figure 9.7 Pre-equilibrium cell-based assays. (a) An example of a real-time binding trace (red line) from a biotherapeutic binding to cells expressing the target receptor. The data was fit with a complex binding model (black line). (b) Workflow of a time-point pre-equilibrium kinetic exclusion assay.

Bivalent binding is the most common example of heterogeneous binding profiles for antibody-based therapeutics; however, it is not the only type of complex binding that can be analyzed. Endogenous protein–ligand interactions have also been characterized with these types of assays. For example, a recent study used LigandTracer assays to show that the chemokine, CCL1, binds to the G protein-coupled receptor, CCR8, in a two-step mechanism – engagement with one binding site induces a conformational shift in CCR8 that exposes a second binding site [60].This study also used real-time quenching assays [61] to show that an anti-CCR8 antibody can interact with CCL1-occupied receptors. This data was able to explain the potential mechanism of action and affinity–function disconnect of this particular antibody.

Real-time binding assays offer great insight into the binding mechanisms of biotherapeutics. These assays not only provide kinetic rate constants but also give insight into the potential heterogeneity of binding that may be missed with equilibrium assays. While we focused on LigandTracer instruments in this section, we should note that there are other commercially available products to assess binding in real time to cells. These instruments include Dynamic Biosensors's heliXcyto [58], Attana's Cell 250 QCM instruments [62], and the SPRm 200 instrument from Biosensing Instrument [63].

Time-Point Pre-equilibrium Assays

While real-time binding assays allow for the in-depth analysis of biotherapeutic binding kinetics, the current methods require specialized equipment and are often low throughout. Alternatively, kinetic rate constants can be determined by sampling the binding reaction at different time points before reaching equilibrium to assess the amount of drug bound over time. These assays typically require labeling the antibody with a fluorophore or a radioactive isotope [64]. Label-free methods to determine the binding kinetics have also been developed for kinetic exclusion assays [52, 65, 66].

Typical time point pre-equilibrium assays are performed in two steps. First, the equilibrium K_D is determined using one of the techniques discussed previously. Next, the association rate of the ligand to the receptor is determined by incubating two or three non-saturating concentrations of the labeled drug with a fixed number of cells and measuring the amount of bound drug over time. The association of the ligand to the receptor occurs rapidly initially but slows down as binding sites become occupied. Therefore, it is good practice to sample the amount of bound drug more frequently at early time points. This data can be fit to a one-phase association pseudo-first-order equation:

$$Y = Y_0 + \left(\left(\text{Plateau} - Y_0\right) \times \left(1 - e^{-k_{obs} \times t}\right)\right) \qquad (9.9)$$

where Y_0 is the signal at time 0, Plateau is the asymptote, t is the time of incubation in seconds, and k_{obs} is the observed or apparent association rate in s^{-1}. The calculated k_{obs} and drug concentration share a linear relationship and can be defined as:

$$k_{obs} = k_a \times [D] + k_d \qquad (9.10)$$

Therefore, the association and dissociation rate constants can be determined by simply plotting the observed association rate on the y-axis against ligand concentration on the x-axis. The slope of this line will provide k_a and k_d is the y-intercept. However, the accuracy of k_d determined by this method is often not accurate, and it is recommended to calculate the dissociation rate constant using Equation (9.3) [67].

This method to calculate the kinetic rate constants can become quite labor intensive and often requires multiple experiments. To overcome these limitations, Janezic et al. developed a kinetic exclusion assay to determine the binding rate constants simultaneously [52]. This assay is performed similar to the kinetic exclusion assays described above, but the concentration of unbound antibody across the cell titration is measured at different time points and the data is fitted with a previously described complex equation [67]. The strength of this method over other time-point pre-equilibrium methods is the improved accuracy of both k_a and k_d. Furthermore, 30 ligand–receptor interactions can be assessed in a single experiment using the Gyrolab platform. This is the highest throughput cell-based pre-equilibrium assay to date. KinExA platforms can also be used for this assay, but the throughput will be decreased.

Conclusions

The binding affinity is a quantitative measurement that defines the strength of molecular interactions, while the kinetic rate constants define the speed at which the binding partners associate or dissociate. These measurements are often informative for selecting and engineering clinical candidates when discovering novel biotherapeutic molecules, and the data are used to predict potency [68, 69], determine in vivo dosing and biodistribution [70], and define mechanisms of action [59]. A number of methods have been developed to measure the parameters that dictate molecular interactions. Protein- and cell-based assays are often combined using a fit-for-purpose strategy throughout the discovery process to select and characterize the final clinical candidate because of their complementary features [24]. For example, when developing multispecific biotherapeutics against targets on the same cell, it may be useful to determine the K_D of each arm individually with SPR and cell-based methods for the finalized molecule [56, 71–73]. Additionally, SPR assays are suitable for interactions with well-behaved target proteins and may be utilized at earlier stages in the screening process due to the ease of setting up the assays and the high-throughput potential, and cell-based assays can be used at later stages when selecting in vivo dosing or to assess the antigen in a native environment. The use of multiple orthogonal methods is likely to increase as the complexity of biotherapeutics and their targets increases [46].

References

1 Berezov, A., Zhang, H.T., Greene, M.I., and Murali, R. (2001). Disabling erbB receptors with rationally designed exocyclic mimetics of antibodies: structure-function analysis. *J. Med. Chem.* 44 (16): 2565–2574.

2 Copeland, R.A., Pompliano, D.L., and Meek, T.D. (2006). Drug-target residence time and its implications for lead optimization. *Nat. Rev. Drug Discov.* 5 (9): 730–739.

3 Hill, A.V. (1909). The mode of action of nicotine and curari, determined by the form of the contraction curve and the method of temperature coefficients. *J. Physiol.* 39 (5): 361–373.

4 Langmuir, I. (1918). The adsorption of gases on plane surfaces of glass, mica and platinum. *J. Am. Chem. Soc.* 40 (9): 1361–1403.

5 Michaelis, L. and Menten, M.L. (2007). Die Kinetik der Invertinwirkung. *Biochem. Z.* 49: 333–369.

6 Clark, A.J. (1926). The reaction between acetyl choline and muscle cells. *J. Physiol.* 61 (4): 530–546.

7 Schild, H.O. (1947). pA, a new scale for the measurement of drug antagonism. *Br. J. Pharmacol. Chemother.* 2 (3): 189–206.

8 Paton, W.D. and Rang, H.P. (1965). The uptake of atropine and related drugs by intestinal smooth muscle of the Guinea-pig in relation to acetylcholine receptors. *Proc. R. Soc. Lond. B Biol. Sci.* 163: 1–44.

9 Scatchard, G. (1949). The attractions of proteins for small molecules and ions. *Ann. N. Y. Acad. Sci.* 51 (4): 660–672.

10 Karlsson, R., Michaelsson, A., and Mattsson, L. (1991). Kinetic analysis of monoclonal antibody-antigen interactions with a new biosensor based analytical system. *J. Immunol. Methods* 145 (1–2): 229–240.

11 Sklar, L.A., Finney, D.A., Oades, Z.G. et al. (1984). The dynamics of ligand-receptor interactions. Real-time analyses of association, dissociation, and internalization of an N-formyl peptide and its receptors on the human neutrophil. *J. Biol. Chem.* 259 (9): 5661–5669.

12 Torreri, P., Ceccarini, M., Macioce, P., and Petrucci, T.C. (2005). Biomolecular interactions by surface plasmon resonance technology. *Ann. Ist. Super. Sanita* 41 (4): 437–441.

13 Brown, M.E., Bedinger, D., Lilov, A. et al. (2020). Assessing the binding properties of the anti-PD-1 antibody landscape using label-free biosensors. *PLoS One* 15 (3): e0229206.

14 Kamat, V. and Rafique, A. (2017). Exploring sensitivity & throughput of a parallel flow SPRi biosensor for characterization of antibody-antigen interaction. *Anal. Biochem.* 525: 8–22.

15 Hanson, E.K. and Whelan, R.J. (2023). Application of the Nicoya OpenSPR to studies of biomolecular binding: a review of the literature from 2016 to 2022. *Sensors (Basel)* 23 (10).

16 Yang, D., Singh, A., Wu, H., and Kroe-Barrett, R. (2017). Determination of high-affinity antibody-antigen binding kinetics using four biosensor platforms. *J. Vis. Exp.* 122.

17 Li, J.F., He, L., Deng, Y.Q. et al. (2021). Generation and characterization of a nanobody against SARS-CoV. *Virol. Sin.* 36 (6): 1484–1491.

18 Darling, R.J. and Brault, P.A. (2004). Kinetic exclusion assay technology: characterization of molecular interactions. *Assay Drug Dev. Technol.* 2 (6): 647–657.

19 Blake, R.C. 2nd, Pavlov, A.R., and Blake, D.A. (1999). Automated kinetic exclusion assays to quantify protein binding interactions in homogeneous solution. *Anal. Biochem.* 272 (2): 123–134.

20 Ladbury, J.E. and Chowdhry, B.Z. (1996). Sensing the heat: the application of isothermal titration calorimetry to thermodynamic studies of biomolecular interactions. *Chem. Biol.* 3 (10): 791–801.

21 Magnez, R., Thiroux, B., Taront, S. et al. (2017). PD-1/PD-L1 binding studies using microscale thermophoresis. *Sci. Rep.* 7 (1): 17623.

22 Daub, H., Traxler, L., Ismajli, F. et al. (2020). The trimer to monomer transition of tumor necrosis factor-alpha is a dynamic process that is significantly altered by therapeutic antibodies. *Sci. Rep.* 10 (1): 9265.

23 Jensen, H. and Ostergaard, J. (2010). Flow induced dispersion analysis quantifies noncovalent interactions in nanoliter samples. *J. Am. Chem. Soc.* 132 (12): 4070–4071.

24 Wang, X., Phan, M.M., Li, J. et al. (2020). Molecular interaction characterization strategies for the

development of new biotherapeutic antibody modalities. *Antibodies (Basel)* 9 (2): 7.

25 Wang, X., McKay, P., Yee, L.T. et al. (2017). Impact of SPR biosensor assay configuration on antibody: neonatal fc receptor binding data. *MAbs* 9 (2): 319–332.

26 Rich, R.L., Papalia, G.A., Flynn, P.J. et al. (2009). A global benchmark study using affinity-based biosensors. *Anal. Biochem.* 386 (2): 194–216.

27 Wang, X., Phan, M.M., Sun, Y. et al. (2022). Development of an SPR-based binding assay for characterization of anti-CD20 antibodies to CD20 expressed on extracellular vesicles. *Anal. Biochem.* 646: 114635.

28 Karlsson, R., Katsamba, P.S., Nordin, H. et al. (2006). Analyzing a kinetic titration series using affinity biosensors. *Anal. Biochem.* 349 (1): 136–147.

29 Ovacik, A.M., Li, J., Lemper, M. et al. (2019). Single cell-produced and in vitro-assembled anti-FcRH5/CD3 T-cell dependent bispecific antibodies have similar in vitro and in vivo properties. *MAbs* 11 (2): 422–433.

30 Staflin, K., Zuch de Zafra, C.L., Schutt, L.K. et al. (2020). Target arm affinities determine preclinical efficacy and safety of anti-HER2/CD3 bispecific antibody. *JCI Insight* 5 (7): e133757.

31 Agard, N.J., Zhang, G., Ridgeway, J. et al. (2022). Direct Tie2 agonists stabilize vasculature for the treatment of diabetic macular edema. *Transl. Vis. Sci. Technol.* 11 (10): 27.

32 Biacore *Sensor Surface Handbook: BR-1005-71 AB 05/2008*.

33 Igawa, T., Tsunoda, H., Tachibana, T. et al. (2010). Reduced elimination of IgG antibodies by engineering the variable region. *Protein Eng. Des. Sel.* 23 (5): 385–392.

34 Yang, J., Wang, X., Fuh, G. et al. (2014). Comparison of binding characteristics and in vitro activities of three inhibitors of vascular endothelial growth factor a. *Mol. Pharm.* 11 (10): 3421–3430.

35 Griffiths, K., Dolezal, O., Cao, B. et al. (2016). I-bodies, human single domain antibodies that antagonize chemokine receptor CXCR4. *J. Biol. Chem.* 291 (24): 12641–12657.

36 Kamat, V., Rafique, A., Huang, T. et al. (2020). The impact of different human IgG capture molecules on the kinetics analysis of antibody-antigen interaction. *Anal. Biochem.* 593: 113580.

37 Nikolovska-Coleska, Z. (2015). Studying protein-protein interactions using surface plasmon resonance. *Methods Mol. Biol.* 1278: 109–138.

38 Karlsson, R. (2016). Biosensor binding data and its applicability to the determination of active concentration. *Biophys. Rev.* 8 (4): 347–358.

39 Jason-Moller, L., Murphy, M., and Bruno, J. (2006). Overview of biacore systems and their applications. *Curr. Protoc. Protein Sci.* Chapter 19:Unit 19 3.

40 Karlsson, R. (1999). Affinity analysis of non-steady-state data obtained under mass transport limited conditions using BIAcore technology. *J. Mol. Recognit.* 12 (5): 285–292.

41 Myszka, D.G., He, X., Dembo, M. et al. (1998). Extending the range of rate constants available from BIACORE: interpreting mass transport-influenced binding data. *Biophys. J.* 75 (2): 583–594.

42 Altschuh, D., Bjorkelund, H., Strandgard, J. et al. (2012). Deciphering complex protein interaction kinetics using interaction map. *Biochem. Biophys. Res. Commun.* 428 (1): 74–79.

43 Bjorkelund, H., Gedda, L., Barta, P. et al. (2011). Gefitinib induces epidermal growth factor receptor dimers which alters the interaction characteristics with (1)(2)(5)I-EGF. *PLoS One* 6 (9): e24739.

44 Johnstone, R.W., Andrew, S.M., Hogarth, M.P. et al. (1990). The effect of temperature on the binding kinetics and equilibrium constants of monoclonal antibodies to cell surface antigens. *Mol. Immunol.* 27 (4): 327–333.

45 Shopes, B. (1995). Temperature-dependent binding of IgG1 to a human high affinity fc receptor. *Mol. Immunol.* 32 (5): 375–378.

46 Carter, P.J. and Lazar, G.A. (2018). Next generation antibody drugs: pursuit of the 'high-hanging fruit'. *Nat. Rev. Drug Discov.* 17 (3): 197–223.

47 Cardarelli, P.M., Quinn, M., Buckman, D. et al. (2002). Binding to CD20 by anti-B1 antibody or F(ab')(2) is sufficient for induction of apoptosis in B-cell lines. *Cancer Immunol. Immunother.* 51 (1): 15–24.

48 Dela Cruz Chuh, J., Go, M., Chen, Y. et al. (2021). Preclinical optimization of Ly6E-targeted ADCs for increased durability and efficacy of anti-tumor response. *MAbs* 13 (1): 1862452.

49 Cheng, Y. and Prusoff, W.H. (1973). Relationship between the inhibition constant (K1) and the concentration of inhibitor which causes 50 per cent inhibition (I50) of an enzymatic reaction. *Biochem. Pharmacol.* 22 (23): 3099–3108.

50 Xie, L., Mark Jones, R., Glass, T.R. et al. (2005). Measurement of the functional affinity constant of a monoclonal antibody for cell surface receptors using kinetic exclusion fluorescence immunoassay. *J. Immunol. Methods* 304 (1–2): 1–14.

51 Dai, Z., Juneja, J., Schneeweis, L. et al. (2020). Application of the gyrolab microfluidic platform to measure

picomolar affinity of a PD-L1-binding adnectin radioligand for positron emission tomography. *Biotechniques* 69 (3): 200–205.

52 Janezic, E.M., Doan, A., Mai, E. et al. (2023). A novel, label-free, pre-equilibrium assay to determine the association and dissociation rate constants of therapeutic antibodies on living cells. *Br. J. Pharmacol.* 181 (20): 3836–3855.

53 Drake, A.W. and Klakamp, S.L. (2007). A rigorous multiple independent binding site model for determining cell-based equilibrium dissociation constants. *J. Immunol. Methods* 318 (1–2): 147–152.

54 Vanhove, M. (2021). A method to assess the robustness of complex mathematical models used for quantitative interpretation of experimental data by nonlinear regression analysis: application to high-affinity binding models. *Eur. Biophys. J.* 50 (7): 1045–1054.

55 Erasmus, M.F., Dovner, M., Ferrara, F. et al. (2023). Determining the affinities of high-affinity antibodies using KinExA and surface plasmon resonance. *MAbs* 15 (1): 2291209.

56 Bjorke, H. and Andersson, K. (2006). Measuring the affinity of a radioligand with its receptor using a rotating cell dish with in situ reference area. *Appl. Radiat. Isot.* 64 (1): 32–37.

57 Bondza, S., Foy, E., Brooks, J. et al. (2017). Real-time characterization of antibody binding to receptors on living immune cells. *Front. Immunol.* 8: 455.

58 Harwardt, J., Geyer, F.K., Schoenfeld, K. et al. (2024). Balancing the affinity and tumor cell binding of a two-in-one antibody simultaneously targeting EGFR and PD-L1. *Antibodies (Basel)* 13 (2): 36.

59 Bondza, S., Ten Broeke, T., Nestor, M. et al. (2020). Bivalent binding on cells varies between anti-CD20 antibodies and is dose-dependent. *MAbs* 12 (1): 1792673.

60 Sun, D., Sun, Y., Janezic, E. et al. (2023). Structural basis of antibody inhibition and chemokine activation of the human CC chemokine receptor 8. *Nat. Commun.* 14 (1): 7940.

61 Bondza, S., Bjorkelund, H., Nestor, M. et al. (2017). Novel real-time proximity assay for characterizing multiple receptor interactions on living cells. *Anal. Chem.* 89 (24): 13212–13218.

62 Peiris, D., Spector, A.F., Lomax-Browne, H. et al. (2017). Cellular glycosylation affects herceptin binding and sensitivity of breast cancer cells to doxorubicin and growth factors. *Sci. Rep.* 7: 43006.

63 Wang, W., Yang, Y., Wang, S. et al. (2012). Label-free measuring and mapping of binding kinetics of membrane proteins in single living cells. *Nat. Chem.* 4 (10): 846–853.

64 Spiegelberg, D., Stenberg, J., Richalet, P., and Vanhove, M. (2021). K(D) determination from time-resolved experiments on live cells with LigandTracer and reconciliation with end-point flow cytometry measurements. *Eur. Biophys. J.* 50 (7): 979–991.

65 Hardy, F., Djavadi-Ohaniance, L., and Goldberg, M.E. (1997). Measurement of antibody/antigen association rate constants in solution by a method based on the enzyme-linked immunosorbent assay. *J. Immunol. Methods* 200 (1–2): 155–159.

66 Rathanaswami, P., Roalstad, S., Roskos, L. et al. (2005). Demonstration of an in vivo generated sub-picomolar affinity fully human monoclonal antibody to interleukin-8. *Biochem. Biophys. Res. Commun.* 334 (4): 1004–1013.

67 Vanhove, E. and Vanhove, M. (2018). Affinity determination of biomolecules: a kinetic model for the analysis of pre-equilibrium titration curves. *Eur. Biophys. J.* 47 (8): 961–966.

68 Tang, Y., Lou, J., Alpaugh, R.K. et al. (2007). Regulation of antibody-dependent cellular cytotoxicity by IgG intrinsic and apparent affinity for target antigen. *J. Immunol.* 179 (5): 2815–2823.

69 Velders, M.P., van Rhijn, C.M., Oskam, E. et al. (1998). The impact of antigen density and antibody affinity on antibody-dependent cellular cytotoxicity: relevance for immunotherapy of carcinomas. *Br. J. Cancer* 78 (4): 478–483.

70 Mandikian, D., Takahashi, N., Lo, A.A. et al. (2018). Relative target affinities of T-cell-dependent bispecific antibodies determine biodistribution in a solid tumor mouse model. *Mol. Cancer Ther.* 17 (4): 776–785.

71 Jarantow, S.W., Bushey, B.S., Pardinas, J.R. et al. (2015). Impact of cell-surface antigen expression on target engagement and function of an epidermal growth factor receptor x c-MET bispecific antibody. *J. Biol. Chem.* 290 (41): 24689–24704.

72 Marei, H., Tsai, W.K., Kee, Y.S. et al. (2022). Antibody targeting of E3 ubiquitin ligases for receptor degradation. *Nature* 610 (7930): 182–189.

73 Moores, S.L., Chiu, M.L., Bushey, B.S. et al. (2016). A novel bispecific antibody targeting EGFR and cMet is effective against EGFR inhibitor-resistant lung tumors. *Cancer Res.* 76 (13): 3942–3953.

10

Functional Assays for Screening Protein Biotherapeutics

Maureen Beresini and Kelly Loyet

Department of Biochemical and Cellular Pharmacology, Genentech Inc., South San Francisco, CA, USA

Introduction

The development and implementation of functional assays for screening protein biotherapeutics hold crucial importance in their discovery, optimization, and characterization. Functional assays are quantitative in vitro procedures that characterize the biological activity of candidate molecules in terms of the magnitude of their effects and/or the potency with which these effects are achieved. They may also provide insights into candidate mechanisms of action. Each assay assesses how well a candidate biotherapeutic works and is specifically designed to determine its involvement in molecular interactions that are important for a particular biological process, signaling pathway, or enzymatic activity. Biochemical functional assays can be used to assess effects on molecular interactions or enzyme activity. Cellular functional assays interrogate effects on biological processes and/or interactions in cells. Whether one is screening protein biotherapeutics of various modalities including monoclonal antibodies, bi- or multi-specific antibodies, fusion proteins, peptides, or antibody–drug conjugates (ADCs), the need for comprehensive functional screening methodologies weighs heavily. The selection of the right lead therapeutic relies on the nature and combination of functional assays used to screen the candidates. Thus, one must ask the right questions and design the assay screening cascade with the goals of the therapeutic in mind. This chapter explores the complex landscape of functional assays, highlighting the basic principles in the design of both biochemical and cell-based functional assays as well as the fundamentals of building an assay cascade and validating assay performance for screening diverse protein biotherapeutics.

Development of a Screening Cascade

Effective screening and nomination of lead biotherapeutics relies on well-designed functional assays. Functional assays go beyond examining the target-binding properties that are tested in early screening assays. They play a seminal role, serving as critical tools for identifying lead candidates with the desired biological activities. Prior to starting a campaign to select a lead protein biotherapeutic, the first step is to design a screening cascade. A screening cascade is a map that guides the decision-making process to filter the optimal from the nonoptimal biotherapeutic candidates and rank them for specific properties. Designing a screening cascade that is both relevant and effective in identifying lead biotherapeutics demands an understanding of the underlying target biology. Researchers can tailor the screening cascade to specifically interrogate aspects critical to the target's function and modulation.

There are generally three tiers to a screening cascade (see Figure 10.1 for a general antibody screening example). The first tier contains the screening assays that interrogate target binding; it focuses on specificity, affinity, and epitope binning and may also examine species-cross-reactivity. These assays are for primary screening but are not functional assays per se. A biotherapeutic protein binding to a target does not always predict function. Although when a biotherapeutic binds to the epitope relevant to the function, for example, at the region of interface with a target ligand or receptor one intends to block, the highest affinity candidate is often the most potent. In the case of therapeutic antibodies [1], first, primary antigen-binding screening assays are performed on thousands of antibodies minimally purified and in small amounts. Commonly

Characterizing Biotherapeutics: Analytical Methods for Diverse Modalities, First Edition. Edited by Jennie R. Lill and Wendy Sandoval.

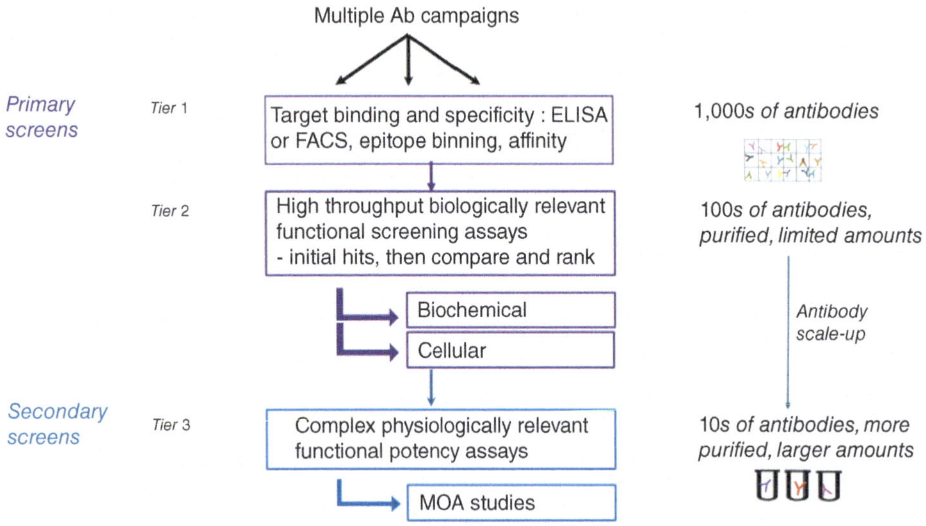

Figure 10.1 Screening cascade to find the lead antibody candidates. This general cascade is for an antibody campaign. Tier 1 and tier 2 (purple) are the primary screens. Tier 1 screening assays interrogate antibody candidate target binding and target specificity and may test thousands of antibodies. Tier 2 functional screening assays identify initial hits from hundreds of antibodies that are somewhat purified and rank them with high-throughput biologically relevant biochemical and/or cellular assays. Tier 3 assays are secondary screens (blue) and are complex physiologically relevant functional potency assays that test or confirm the mechanism of action (MOA) of lead antibodies that are more purified and in larger amounts.

used technologies for therapeutic antibody generation include hybridoma [2], B-cell cloning [3], phage [4], [5], yeast [6], or ribosome display [7]. For the latter, selection may be accomplished via panning with antigen-coated wells or vials [8]. Tier 1 hybridoma and B-cell screening technologies may include enzyme-linked immunosorbent assays (ELISA) and flow cytometry (or flow-assisted cell sorting, FACS) target binding specificity assays to select antibodies that are specific to the target and do not bind homologous proteins. A previous review described additional homogeneous, heterogeneous, and cell-based assay technologies that have been used successfully to screen for the binding of candidate antibodies [9]. Cell capture has also been described for this purpose [10]. Oftentimes, monoclonal antibodies are initially selected by ELISA with a single antigen coated per plate, typically using 384-well microtiter plates. Negative control plates are coated with single antigens such as BSA or irrelevant proteins with the same purification tag. A single (neat) antibody concentration is tested and a cutoff point is generally set for ~3–fivefold over the background. FACS screening is performed via 96-well or 384-well plates with antigen-positive and -negative cells from the same parent line using an isotype immunoglobulin (IgG) as a negative control, testing a single antibody concentration. At the ELISA and FACS screening stage, the monoclonal antibodies (mAbs) are characterized for molecule selectivity and/or cross-reactivity by examining binding to various isoforms and family members; the animal species selectivity and/or cross-reactivity may also be tested by looking at antigens of different species such as mouse, monkey, and human. To further characterize binding, affinity for the target may be determined by surface plasmon resonance or biolayer interferometry, either of which may also be used for the characterization of the antibody–antigen interaction by epitope binning [11]. Epitope binning allows the categorization of candidate biotherapeutic antibodies by whether or not they bind the same epitope, or antigen area. This step needs smaller amounts of antibodies and may be performed at the same time as the tier 2 assays. SPR typically requires higher amounts of candidate molecules and thus may be performed at the same time as the tier 3 assays, although more recently HTS SPR methods have evolved for both affinity and epitope binning [12].

The scale of the screen depends on the antibody generation species and platform. For rodent hybridoma, we expect hundreds to a thousand antibodies to be purified from hybridoma supernatants within a timeframe of 3–5 months post-immunization. Generation of recombinant antibodies is performed by sequencing DNA inserts encoding heavy and light antibody chains, sub-cloning them into a mammalian expression vector, then transfection into a cell line such as CHO for production as full-length antibodies, and followed by purification. Incorporating automation into the tier 1 antigen-binding assays allows ease and speed during this discovery phase [13]. For a rabbit B-cell culture [14], following selection, we expect tens to hundreds of antibodies that are recombinantly expressed within a timeframe of 3–4 months post-immunization. A novel example of efficient human antibody generation includes the activation of human peripheral blood mononuclear cells from vaccinated donors and enrichment in an antigen-specific manner to produce many antigen-specific plasmablasts; with this method, four broadly neutralizing influenza A antibodies were discovered following screening of 840 human antibodies [15]. Large-scale information on antibody repertoire diversity can be obtained through high-throughput sequencing [16]. The combination of high-throughput sequencing and high-throughput screening (HTS) may be used to directly discover and engineer new therapeutic antibodies [17].

The screening and discovery processes for antibodies versus non-antibody protein biotherapeutics have notable similarities and differences. While antibody screening cascades typically involve the identification and selection of mAbs, non-antibody protein biotherapeutic screening cascades encompass a broader range of protein-based therapeutics, such as fusion proteins, enzymes, and cytokines. Recent efforts to design proteins de novo have sought to create small, stable proteins with shapes specifically tailored to bind therapeutic targets [18]. Unlike the broad antibody generation and screening processes, which typically involve creating large libraries through hybridoma technology, B-cell cloning, or library selection to generate thousands of candidates, non-antibody biotherapeutics based on known entities focus on rational molecule design guided by structural studies and engineering efforts [19]. These efforts are mainly focused on modulating desirable properties such as stability, pharmacokinetics, gain or loss of potency, and lack of toxicity. They replace the broad candidate generation approach with more targeted strategies, potentially resulting in smaller numbers of initial candidates to screen, although there may be multiple iterations of targeted optimization. The types of assays for non-antibody protein biotherapeutics will be dictated by the types of functionalities beyond the endogenous sequence, added via modification of the protein. Modifications of protein biotherapeutics via derivatization include glycoengineering for enhanced potency, PEGylation for extended half-life or decreased immunogenicity, addition of an Fc region for increased half-life, and alkylation or acylation for stability.

In addition to modification, three different strategies are often employed in protein engineering efforts to improve biotherapeutic properties: knowledge-based mutagenesis, computational protein design, and directed evolution [20]. Long-acting glucagon-like peptide-1 (GLP-1) analogs, liraglutide, and semaglutide are examples of protein biotherapeutics that were modified through rational design with fatty acids for albumin binding, thus allowing greater stability and longer biotherapeutic half-life [21]. The design of these GLP-1 biotherapeutics relied on structural knowledge to optimize their potencies and stabilities. Non-antibody protein biotherapeutics can also be generated from peptide and protein scaffold libraries such as affibody [22, 23], anticalin® [24, 25], and thermostable designed ankyrin repeat proteins (DARPins) [26], among others [27, 28], and their main applications have been in the treatment and diagnosis of cancer and inflammatory diseases [29]. The screening cascades for non-antibody scaffold protein biotherapeutics are quite similar to those of antibodies in terms of assay composition as well as in generation and quantities of candidate molecules screened. For each of these types of biotherapeutics, lead candidate selection relies on interrogating multiple aspects of the therapeutically relevant biological function in cells. Such biotherapeutics may also be characterized biochemically if the nature of the target is amenable. When optimizing the broad range of molecule properties relevant to therapeutic administration, it is essential to use functional assays to ensure that the biological activity is adequately retained.

The purpose of the functional assay is to test whether the biotherapeutic mechanism of action (MOA) is selectively on target through the characterization of a biologically or physiologically relevant aspect of its function. Primary and secondary functional screening assays differ in their throughput and complexity. In the second tier of the screening cascade (Figure 10.1) lie the primary functional screening assays, which may be biochemical assays, cell-based assays, or a combination of both. The nature of the biochemical or cell-based assays is discussed in the next two sections. These assays should be high throughput in order to screen hundreds (and in some cases thousands) of candidate protein biotherapeutics, yet still capture the biological process of interest. Typically these samples are produced from one-milliliter cell culture and have undergone a one-column purification process, so they are relatively low in concentration (typically 100 µl of 100–200 µg/ml) and low in purity. Biologically relevant assays represent a delicate balance between achieving high throughput and ease of implementation on the one hand and capturing the complexity and relevance of biological systems on the other. Balancing these factors requires careful consideration of assay design, validation, and optimization in the context of a deep understanding of the biological mechanisms so that the assays adequately reflect the subtleties of *in vivo* systems. Therefore, the challenge lies in devising assays that leverage advanced technologies and automation while preserving biological relevance to enable the rapid identification of lead biotherapeutics without sacrificing biological fidelity [30]. Achieving this balance is crucial for maximizing the translational potential of preclinical research and expediting the development of clinically meaningful therapeutics.

In order to efficiently screen a high number of candidates in the primary functional assay (Figure 10.1, tier 2), the first primary functional assay is often performed with a single concentration, in duplicate. Depending on the throughput of the assay and the essential information needed, instead of a single concentration in duplicate, two to three different concentrations in singlicate may be preferred. This alternative workflow may be desirable if the assay shows a biphasic response such that any single concentration may give a false-negative response. Alternatively, testing low versus high concentrations may be integral to the screening strategy for a specific function related to the safety of the potential biotherapeutic [31], for example, the desire to have a potent biotherapeutic (test low concentration, readout #1) paired with the desire to reduce toxicity at higher concentrations (test high concentration, readout #2). In addition, if the throughput of the primary functional assay and sample number allow, assaying a few concentrations of the sample may enable early ranking of the hits. Following hit selection, confirmation and ranking of the hits with a full titration should be performed. It is likely that the amount of sample is not sufficient to perform a full titration, so a larger-scale expression of the protein is implemented on the hits, typically a 30 ml expression with a two-column purification, yielding milligrams of higher purity protein at greater than 1 mg/ml concentration. It is not unusual for hits to be eliminated from further consideration at this stage, because an impurity in the initial small-scale purification caused a false-positive result. The larger-scale material may first be used for confirmation and ranking in the primary functional assay, and then used for MOA studies in the secondary functional assay.

Highly valued secondary functional assays are content and information rich with an emphasis on physiological relevance [9]. In the third tier (Figure 10.1), there are the complex physiologically relevant potency assays used to deeply investigate the MOA of a smaller number of lead biotherapeutic candidates. These assays may have a lower throughput because it is often only necessary to interrogate a small number (ten or below) of biotherapeutic candidates. In addition, these assays may require a larger amount of the sample and with a higher purity. A higher purity is necessary for many cell-based assays so that impurities such as endotoxins do not lead to false positives or false negatives. Closely mimicking the physiological context in which the target operates is the goal when selecting or developing these types of assays. Doing so will enhance the likelihood of characterizing biotherapeutics that exert meaningful therapeutic effects. Moreover, an in-depth understanding of target biology allows for the incorporation of relevant disease models or cellular systems into the screening cascade, further increasing the odds of success in the characterization process by simulating the conditions encountered in pathological states. Ultimately, the relevance of the screening cascade hinges on its ability to accurately capture and pinpoint the relevant aspects of target biology, guiding the identification of lead candidates poised to address unmet medical needs with precision and good efficacy.

Beyond identifying active biotherapeutics and assessing the magnitudes of their effects, another key objective of functional assay cascades is to verify that candidate molecules demonstrate the desired MOA. To do this, the assays must establish that the candidates are acting on the target, are acting on the pathway, and have the intended function. Target engagement is often confirmed in tier 1 and tier 2 assays that demonstrate binding to the target antigen on a solid phase, in solution, and/or on the surface of cells (where relevant). In tier 3, target engagement may be interrogated more deeply on a limited number of candidates by examining the kinetics of binding to the target as presented on cells. Techniques such as microfluidics-based immunoassays (e.g. Gyros technology [32]) or fluorescent tracer binding assays [33] may be used to determine K_D values and/or on- and off-rates for candidate binding to the target protein (see Chapter 15 of this volume for more detail). The assessment of the binding to the target protein on cells in the absence of fixation more faithfully replicates the cellular context encountered *in vivo*. Furthermore, when binding to surface receptors is studied with live cells, the target protein may move freely within the membrane and cluster or interact with other proteins. This provides a more physiologically relevant representation of the dynamics of biotherapeutic binding and may promote a better understanding of the observed functional effects. Beyond target engagement, on-pathway activity can be demonstrated with mechanistic assays that interrogate specific elements in the relevant cellular process. For example, on a signaling pathway, the phosphorylation state of a kinase target may be interrogated, the production of a second messenger such as cAMP may be measured, or the levels of gene expression or protein production may be assessed. If signaling pathways diverge post-target, it may be valuable to investigate both arms of the pathway, particularly if the preference is to target one arm specifically. Confirmation of modulation of pathway activity is critical for ensuring that the observed phenotypic effects are occurring through the intended mechanisms. Lastly, phenotypic assays are essential for confirming that the candidate biotherapeutics have the expected biological activities in cellular systems. Examples of phenotypic assays include cell proliferation, tumor cell killing, and cell differentiation. These assays should demonstrate functions that are as close as possible to the desired *in vivo* effects, and as such, physiological relevance is of utmost importance in the design of the phenotypic assays in the screening cascade. Because the number of molecules is reduced as candidates progress through the cascade, assays of greater complexity may be strategically applied to facilitate rigorous function-based selections of lead biotherapeutics.

Inherent in some biotherapeutic modalities is a complexity that significantly influences the design and execution of functional assay screening cascades. Biotherapeutics containing multiple functionalities, such as multi-specific targeted immune cell engager therapeutics that bridge immune effector cells to their target [34], demand more elaborate screening methodologies compared to simpler constructs like monovalent fusion proteins or Fabs. The development of assays capable of interrogating each functional component is necessitated by the multifaceted nature of these advanced modalities. Thus, the screening cascade must be tailored to address the nuanced interactions and mechanisms intrinsic to the complex biotherapeutic design. Consequently, a thorough evaluation can be made of their therapeutic potential and suitability for further development. See Figure 10.2 for a general example of tier 2 and tier 3 functional assays for a bivalent antibody biotherapeutic. Another example of tier 1 and tier 2 assays examines a protein biotherapeutic that targets an enzyme (see Figure 10.3). In this case, possibly no single ideal tier 2 assay exists. However, a deeper understanding of these complex biotherapeutics is achieved by integrating a variety of readouts which facilitates decision-making and accelerates the identification of promising candidates for clinical development.

A Note on Full and Partial Activity

In functional screening assays, understanding the nuances between partial and complete inhibition, as well as the extent of agonist activation compared to the natural ligand response, is critical for evaluating the efficacy and specificity of biotherapeutic candidates. Partial inhibition

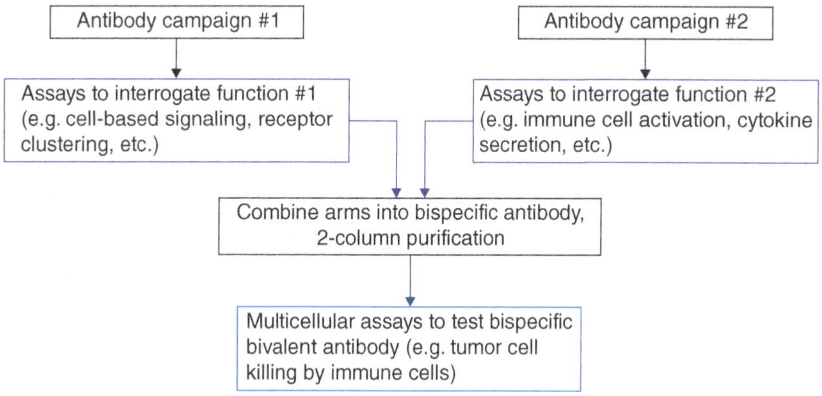

Figure 10.2 Example of screening cascade tier 2 and tier 3 assays for a bispecific biotherapeutic antibody campaign. This example shows tier 2 (purple) primary functional assays for two different antibody campaigns. The lead antibodies from the respective campaigns are combined into bivalent antibody candidates to be tested in tier 3 (blue) secondary functional assays.

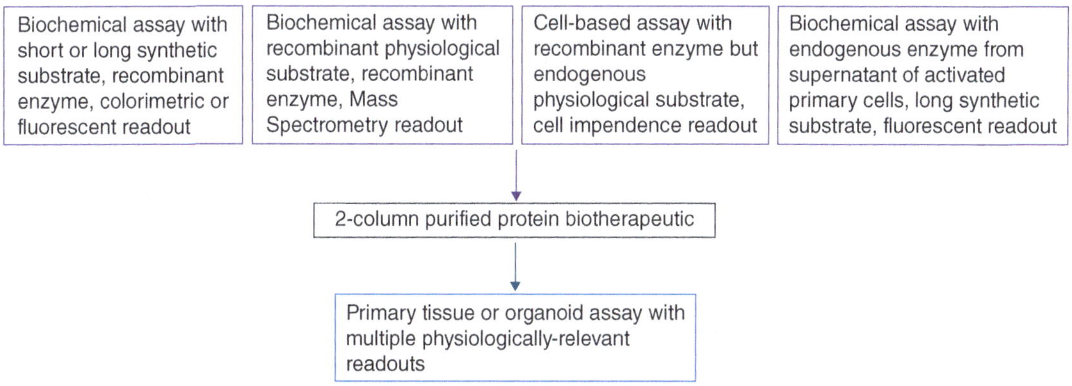

Figure 10.3 Example of screening cascade tier 2 and tier 3 assays for a protease-targeting protein biotherapeutic. Tier 2 assays (purple) illustrate various high-throughput biochemical and cell-based primary functional screening assays. Tier 3 assay (blue) represents a low-throughput more complex physiologically relevant secondary functional screening assay for a small number of lead candidates.

refers to a scenario in which the biotherapeutic molecule only partially blocks the activity of the target molecule, resulting in a submaximal response compared to complete inhibition, where the target activity is fully abolished. Partial inhibition may suggest an antagonist with partial agonism or a biotherapeutic candidate with limited potency that reflects either a suboptimal MOA or binding epitope, potentially necessitating further optimization. On the other hand, agonist activation involves stimulating the target molecule to elicit a response akin to its natural ligand. Assessing agonist activation in comparison to the response induced by the natural ligand provides insights into the biotherapeutic's ability to mimic, enhance, or attenuate the physiological effects of the endogenous ligand. The protein biotherapeutic Fc.IL-2 mutein is an example of an engineered cytokine with weaker agonism compared to wild-type IL-2 [35]. In this case, the reduced agonism is beneficial in enhancing Treg cell selectivity, reducing activation of effector cells, and improving pharmacokinetics by reducing receptor-mediated clearance. Discriminating between partial and complete inhibition, as well as the extent of agonist activation versus natural ligand response, aids in identifying biotherapeutic leads with optimal potency, efficacy, and specificity for subsequent development and clinical translation. The use of assay controls (to be discussed in Section "Assay Controls and Performance Characterization") promotes the distinction between full and partial activity of candidate biotherapeutics.

Biochemical Assays for Cellular Receptor-Binding or Extracellular Targets

Biochemical assays represent a cornerstone in the functional evaluation of biotherapeutics targeting cellular receptors or extracellular entities, facilitating the critical assessment of their efficacy and specificity to modulate biologically relevant protein/ligand interactions and/or enzyme activities. These assays employ a myriad of techniques, including ELISA, fluorescence-based assays, and proximity-based assays, to name a few. Biochemical assays are generally amenable to high-throughput applications and have the benefit of being constructed with well-defined components relative to cellular assays, and their performance is typically more robust. Moreover, assay development and application are often more straightforward. Importantly, these assays play an indispensable role in screening candidate molecules at the level of the target protein, ensuring the selection of potent therapeutics with on-target activity, thus advancing the landscape of precision medicine and therapeutic interventions.

Ligand or Receptor-Binding or Blocking

A central focus of biochemical functional assays for screening biotherapeutics is to assess ligand-binding or receptor-binding interactions, as well as the potential blocking effects exerted by these therapeutic agents. These assays are designed to elucidate the functional consequences of biotherapeutics interacting with their intended targets, whether they be cell surface receptors or extracellular ligands. A well-known example of a biotherapeutic that blocks a ligand is adalimumab, the first phage-derived human monoclonal antibody approved for therapy. It binds to and blocks tumor necrosis factor (TNF), preventing binding to its receptors and underlying its success in the treatment of arthritis and other inflammatory diseases [36]. Functional assays provide a platform to investigate the ability of biotherapeutics to block receptor–ligand interactions, thereby modulating downstream signaling pathways and biological responses. By dissecting these molecular interactions, biochemical functional assays play a key role in the early stages of biotherapeutic development. An example of a biochemical functional assay for screening biotherapeutics is an ELISA in which one component, such as the extracellular domain of a receptor, is coated on the plate and the therapeutic candidates are chosen by their ability to block the binding of a ligand. This example uses a competition ELISA with the extracellular domain (ECD) coated on the plate and a constant concentration, equal to the EC_{70}–EC_{80}, of tagged ligand added to the plate with a pre-incubation of the candidates either with the ECD or the ligand, depending on the target, and detection by an antibody to the tag on the ligand. Other assay technologies commonly used for biochemical protein–ligand binding assays include TR-FRET (time-resolved fluorescence resonance energy transfer) and AlphaScreen. These assay technologies differ from ELISA in that they are homogeneous assays, and as such, are higher throughput, easier to perform, and amenable to miniaturization.

Enzyme Inhibition with a Focus on Protease Cleavage Inhibition

In the realm of biochemical functional assays for screening biotherapeutics, the assessment of enzyme inhibition, including protease cleavage inhibition, stands as a crucial aspect. These assays aim to evaluate the ability of biotherapeutics to modulate enzymatic activity. One example of this is in the context of proteases that play pivotal roles in various physiological processes. Employing techniques such as fluorogenic or chromogenic substrate assays, FRET assays, or ELISA, researchers quantitatively measure the extent to which biotherapeutics impede protease cleavage events. By elucidating the kinetics (V_{max}, substrate K_m, inhibitor K_I) and mechanisms of enzyme inhibition (competitive, noncompetitive, or uncompetitive), these assays provide insights into the potential therapeutic efficacy of biotherapeutics to modulate protease-mediated pathways associated with diseases such as cancer, inflammation, and viral infections. Additionally, biochemical functional assays enable the screening of biotherapeutic candidates for their potency and selectivity in inhibiting the targeted proteases over related proteases, aiding in the identification and optimization of molecules with promising therapeutic profiles. An example of this type of screening campaign yielded a noncompetitive inhibitory antibody against human β-tryptase, a secretory granule protease in human lung mast cells [37]. Following immunization of mice, rabbits, and rats with monomeric and tetrameric β1-tryptase as well as panning with a human antibody phage library, a chromogenic peptide substrate assay was used to screen for antibodies that had >90% inhibition of tryptase cleavage activity. Selection of the top clone was based on its ability to inhibit human βI-, βII-, and βIII-tryptases, cross-reactivity with cynomolgus monkey tryptase to enable safety and preclinical pharmacokinetic studies, as well as inhibition by both fragment antigen-binding and IgG formats of the antibody. This allosteric anti-tryptase antibody, 31A, was a clinical candidate for severe asthma treatment. Another example is the selection of murine and human antibodies against Kallikrein-related peptidases KLK5 or KLK7 [38]. Antibodies against KLK5 were selected using a Boc-VPR-AMC direct fluorogenic substrate, and antibodies against KLK7 were selected using a Mca-RPKPVE-Nva-WRK(Dnp)-NH2 quenched fluorogenic substrate. Preclinical mouse models of Netherton Syndrome (NS) and atopic dermatitis (AD) demonstrated that combined antibody blockade of KLK5 and KLK7 improved disease severity. Thus, a humanized hKLK5/7 bispecific antibody was generated based on biochemical and human primary keratinocyte assays, and it holds promise for the treatment of NS and AD in humans. Historically, proteases are the enzyme class commonly targeted for inhibition by protein biotherapeutics because of their pathological roles in disease conditions and accessibility in the extracellular environment. In another type of protein therapeutic, the enzyme itself acts therapeutic with applications in cardiovascular disease, cancer treatment, and rare diseases as well as in enzyme replacement therapies [39]. Such enzyme therapeutics must be suitable for *in vivo* administration, and this may require modifications of the protein to enhance stability, solubility, and pharmacokinetic characteristics as well as to reduce immunogenicity. Throughout such modifications, characterization of the enzyme activity in biochemical assays is beneficial to ensure that the enzyme function, substrate specificity, and stability are appropriately optimized for maximal therapeutic benefit.

Assay Complexity – Endogenous Versus Purified Protein Assays

The utilization of endogenous proteins or biofluids, such as serum, in a biochemical assay, may provide a significant enhancement in the relevance and accuracy of an assay compared to an assay using only purified proteins. Oftentimes, biochemical functional assays utilize purified proteins or nonphysiological peptide substrates. Endogenous proteins, naturally present within biological systems, offer a more authentic representation of the complex interactions occurring within living organisms. One may consider using the endogenous substrate of an enzyme and utilizing mass spectrometry for the readout instead of a fluorogenic peptidic substrate with a spectrometric readout. As an example, antagonistic antibodies to the protease bone morphogenic protein 1 (BMP1), were discovered using a heterogeneous 66 kDa procollagen substrate and peptide-based mass spectrometry readout [40]. Biofluids such as serum encapsulate a myriad of proteins, metabolites, and other biomolecules, reflecting the dynamic state of an organism's physiology. An example of using serum in a biochemical assay is a complement activity ELISA [41, 42] that uses serum complement proteins to reconstitute the complement terminal complex on the surface of a microtiter plate; this could be used for screening complement inhibitor biotherapeutics [43]. This approach not only mirrors the *in vivo* environment more closely but also considers the potential influence of various factors present in the biological milieu, thus offering a more comprehensive understanding of biochemical processes. This also leads to more complexity and the need for careful characterization of reagents since different lots of serum may have different properties. By incorporating endogenous proteins or biofluids into assays, researchers may gain valuable insights into the intricacies of biological systems, fostering advancements in diagnostics, drug development, and biomedical research.

Cellular Assays for a Cell Signaling Readout

Although both biochemical assays with purified targets and cell-based assays play critical roles in screening cascades [44], cellular assays allow for the interrogation of targets and pathways in a more physiological setting and are ideal for identifying lead biotherapeutics targeting cell signaling pathways [45, 46]. These assays enable the evaluation of agonist and antagonist biotherapeutics by assessing their effects on cellular responses mediated through specific signaling cascades. Utilizing reporter cell systems equipped with reporter genes such as luciferase, secreted alkaline phosphatase, or GFP allows for the detection of signaling pathway activation or inhibition. Endogenous target cell lines, primary cells, organoids, and patient-derived cell systems also serve as valuable platforms to mimic physiological conditions and disease states, enhancing the relevance and translatability of the assay results. Potential readouts in native cells include phosphorylation of a protein in the signaling pathway or detection of the production of a second messenger. These cellular assays that probe cell signaling not only provide insights into the efficacy and potency of potential therapeutics but also contribute significantly to the understanding of complex signaling networks underlying various diseases, facilitating the development of novel treatments.

Agonist Versus Antagonist

A powerful and accessible biotherapeutic MOA is to either block or initiate a signaling pathway cascade. A typical biotherapeutic antagonist will block either the ligand from binding to a cell-surface receptor or the receptor at the cell surface from binding to its ligand. Anti-tumor necrosis factor (TNF) biotherapeutics have been clinically successful in treating inflammatory and autoimmune disorders such as psoriasis, inflammatory bowel disease, and arthritis and act by blocking the TNF ligand, although recent efforts have focused on receptor selective intervention using antibodies or selective ligands [47]. A biochemical assay may directly test candidates for the ability to block binding. The researcher is cautioned that this screening method may lead to false-positive or false-negative results. By using a cell-based assay with the relevant signaling pathway readout, additional information to confirm antagonism may be obtained. For example, if a therapeutic candidate with a receptor target blocks the ligand binding but also acts as an agonist of the signaling molecule, potentially by causing the association of inactive monomeric receptor subunits into an active homodimer, the binding ELISA method would fail to inform of this undesired agonist activity and the researcher may obtain a false positive for the desired antagonist activity. In addition, if the ECD is not properly folded or is incorrectly modified post-translationally, it may not be a physiologically relevant proxy for endogenous cell-receptor or ligand-blocking activity, and the researcher may obtain either a false-negative or false-positive result.

Receptor Clustering

Arranging signaling molecules into clusters or complexes boosts the efficiency, fidelity, and robustness of cellular signaling thus overcoming the random macromolecular interactions that occur through spatial confinement [48]. The activation of immune cells is a common example of this phenomenon [49, 50], and many disease-relevant cytokines also signal through receptor heterodimerization [51]. This feature of signaling has inspired the emergence of various novel protein biotherapeutic modalities such as bispecific T-cell engagers [52] that trigger T-cell receptor (TCR) clustering via the formation of an immunological synapse. Recently an i-shaped antibody (iAbs) engineering approach has been described that manipulates the geometry, which is the distance and orientation, of target receptor engagement [53]. An example of an agonist biotherapeutic that signals through receptor clustering is a multivalent PEG-Fab conjugate, Tie2.1-hexamer that oligomerizes Tie2 to induce receptor activation [54]. For this novel modality, primary functional screening was performed with phage-generated human antibodies. The candidate mAbs were crosslinked with polyclonal goat anti-human IgG for testing agonist activity in primary human umbilical cord endothelial cells (HUVECs) using a phospho-Ak strain transforming (Akt) assay. The phospho-Akt assay monitored signal transduction through a membrane-proximal marker. Due to the fundamental role of multimerization and/or clustering in normal biological processes as well as in the MOA of certain protein biotherapeutics, it is essential that functional assays be designed to enable and employ this capability where relevant in their biological readouts.

Signaling Assays

Unlike antagonist screening, it would be difficult to design a biochemical functional assay for agonist biotherapeutic screening. By nature of the MOA, an agonist requires a cell-signaling pathway readout. Some options may include a membrane-proximal readout, measurement of endogenous early gene expression [55], nuclear translocation of a pathway component, or a reporter cell readout. Often, the available reagent tools dictate which type of assay may accelerate discovery. Membrane-proximal readouts in a cell signaling functional assay refer to measurements taken at or near the cell membrane to assess the initial events triggered by signaling molecules. These readouts often involve monitoring the activation or recruitment of proteins, such as kinases or adaptor molecules, at the membrane following receptor stimulation. Fluorescence microscopy techniques, proximity assays, and immunoprecipitation coupled with Western blotting are commonly employed to detect changes in protein localization, interaction, or phosphorylation status. Immunoprecipitation coupled with Western blotting is not high throughput, so direct measurement following cell lysis of phosphorylation status via technologies such as electrochemiluminescence, AlphaLisa, ELISA, or TR-FRET allows the researcher to screen many candidates. A proximity assay is a laboratory technique used to detect interactions between molecules that are in close physical proximity to each other within a cellular context. These assays are based on the principle that when two molecules are close enough, they can generate a signal that indicates their proximity. Proximity assays often utilize various methods such as FRET, bioluminescence resonance energy transfer (BRET), enzyme fragment complementation (ECF) [56], or proximity ligation assays (PLA). In FRET and BRET, the energy transfer between a donor molecule associated with one protein and an acceptor molecule associated with its binding partner occurs when they are within a certain distance, leading to measurable changes in fluorescence or luminescence. PLA, on the other hand, involves the ligation of DNA probes attached to the interacting molecules followed by signal amplification, allowing for the visualization of molecular interactions with high specificity and sensitivity. These assays are valuable tools for studying protein–protein interactions, protein conformational changes, and spatial organization within cells. By focusing on events occurring in close proximity to the membrane, researchers can gain insights into the early stages of signal transduction cascades, providing valuable information about the dynamics and specificity of cellular responses to extracellular cues.

Image-based readouts for cell-based functional assay screening of biotherapeutics encompass a diverse array of techniques that leverage microscopy and image analysis to assess cellular responses. One common approach involves monitoring changes in cell morphology or phenotype upon treatment with biotherapeutics, which can include alterations in cell shape, size, granularity, or subcellular structures such as organelles or cytoskeletal components. Fluorescence-based assays are frequently employed, where fluorescently labeled probes or antibodies are used to visualize specific cellular features or biomolecules of interest, enabling the quantification of protein localization, expression levels, or post-translational modifications. Additionally, live-cell imaging techniques enable real-time monitoring of dynamic cellular processes such as cell migration, proliferation, apoptosis, or signaling pathway activation in response to biotherapeutic treatment. A high-content, quantitative, live-cell imaging assay was previously used to measure the rate and extent of megakaryocyte maturation and proplatelet production and to establish the mechanism by which trastuzumab emtansine (T-DM1), an antibody-drug conjugate, affects platelet production [57]. Another example is the use of a real-time image-based efferocytosis assay to discover anti-MerTK antibodies that block tumor-associated macrophage-mediated efferocytosis [58]. High-content screening platforms further enhance the throughput and multiplexing capabilities of image-based assays, allowing for the simultaneous assessment of multiple cellular parameters within individual cells or across entire populations. Nuclear translocation assays are experimental techniques used to study the movement of molecules from the cytoplasm into the nucleus of a cell. This movement is often critical for regulating cellular processes such as gene expression, DNA repair, and signal transduction. In these assays, researchers typically label the molecule of interest with a fluorescent marker, such as green fluorescent protein (GFP) or a fluorescent dye, or use immunofluorescence to detect the molecule of interest with a labeled antibody. For the former method, following the introduction of this labeled molecule into cells, the localization of the protein can be monitored over time using fluorescence microscopy.

Impedance-based readouts are a powerful technique used for cell-based functional screening of biotherapeutics [59, 60]. This approach measures changes in electrical impedance resulting from cellular responses to biotherapeutics in real time. Cells are cultured on electrodes embedded in the bottom of microtiter wells and form a confluent monolayer. As cells adhere and proliferate, they create an electrical barrier, which impedes the flow of electrical current between the electrodes. When cells change morphology, adhesion, viability, or proliferation in response to biotherapeutics, this impedance barrier is altered, leading to changes in the electrical impedance measured by the system. By monitoring these impedance changes over time, researchers can assess various cellular behaviors, including cell attachment, spreading, migration, proliferation, and cytotoxicity induced by biotherapeutic agents. Impedance-based assays offer several advantages, including label-free detection, real-time monitoring, high sensitivity, and compatibility with HTS in 96- and 384-well plates. Additionally, impedance-based readouts provide quantitative endpoint or kinetic data, enabling the characterization of both short-term and long-term effects of biotherapeutics on cellular physiology.

Multiplexed readouts in cell-based assays involve the simultaneous measurement of multiple parameters within the same sample or experimental system. This approach provides a comprehensive view of cellular responses to biotherapeutics, allowing researchers to assess multiple aspects of cellular physiology or signaling pathways in a single experiment. Multiplexed assays can incorporate various types of readouts, including fluorescence-based, luminescence-based, impedance-based, or label-free detection methods, depending on the specific parameters being measured. For example, fluorescence-based multiplexing might involve the use of different fluorescent labels or probes targeting distinct cellular components, such as proteins, organelles, or biomolecules, allowing simultaneous visualization and quantification of multiple cellular features. Similarly, multiplexed assays can combine different signaling pathway reporters or biomarkers to assess the activation or inhibition of multiple pathways in response to biotherapeutic treatment. High-content imaging and flow cytometry platforms further enhance multiplexing capabilities by enabling the simultaneous assessment of numerous cellular parameters within individual cells or across entire cell populations. Overall, multiplexed readouts in cell-based assays offer a comprehensive and efficient approach for characterizing the functional effects of biotherapeutics on cellular behavior, facilitating the identification of lead molecules and the elucidation of underlying mechanisms of action.

Engineered Expression Reporter Cell Systems

Engineered expression cellular reporter systems are invaluable tools for functional screening assays aimed at identifying biotherapeutic agonists or antagonists. These systems typically involve cells that have been modified to express a reporter gene under the control of a specific promoter or response element that is activated upon binding of the biotherapeutic or its target. For agonist screening, these reporter cells can be designed to express the reporter gene when the biotherapeutic binds to its receptor, thereby activating downstream signaling pathways. By measuring the expression level of the reporter gene, researchers can quantitatively assess the potency and efficacy of potential agonists in stimulating the desired cellular response. The use of reporter cells for the discovery of bispecific T-cell engager (BiTE) therapeutics is exemplified with a novel high-throughput, single-cell-based functional screening assay that employs a droplet microfluidic-based system to compartmentalize individual bispecific antibody-producing cells with co-encapsulated cell reporters and to interrogate for both T-cell receptor activation and target cell binding [61]. For antagonist screening, the biotherapeutic and a signal activator such as the receptor ligand, are added to the reporter cells. The successful biotherapeutic upon binding the target molecule represses the expression of the reporter gene by inhibiting the signaling pathway. One example of this is a reporter-based bioactivity assay for anti-VEGF based biologic agents that uses a HEK293 cell line stably expressing vascular endothelial growth factor receptor 2 (VEGFR-2) and a luciferase reporter under the control of nuclear factor activated T-cell (NFAT) response elements [62]. This assay enabled VEGF antagonists to be identified by their ability to block the VEGFR-2 signaling pathway which resulted in decreased expression of the luciferase reporter. These engineered expression reporter cell systems offer several advantages for functional screening assays. They provide a sensitive and quantitative readout of cellular responses, allowing for the rapid identification of biotherapeutic candidates with desired agonist or antagonist activities. Additionally, these assays can be adapted to high-throughput formats, enabling the screening of large libraries to identify lead candidates for further development.

While engineered expression reporter cell systems offer numerous advantages for functional assay screening of biotherapeutics, they also come with several potential disadvantages. Engineered reporter systems may not fully recapitulate the complex signaling networks and cellular microenvironments found *in vivo*. As a result, the cellular responses observed in these systems may not always accurately reflect the physiological effects of the biotherapeutic. Reporter assays rely on the activity of a specific promoter or response element, which may be influenced by factors unrelated to the intended target or pathway. This can lead to false positives or negatives, where candidate molecules are incorrectly identified as agonists or antagonists due to off-target effects or experimental artifacts. The choice of cell type for the reporter system can significantly impact assay results, as different cell types may exhibit variable responses to the same biotherapeutic; biologically relevant molecules and signaling pathways must be present in the chosen cell line in order to recapitulate the functional response targeted by the biotherapeutic [63]. Additionally, the availability of suitable cell lines for specific targets or pathways may be limited, constraining the applicability of this assay format. Engineered reporter assays often require extensive optimization to ensure robust and reliable performance. Factors such as reporter gene selection, promoter strength, and assay conditions must be carefully optimized to achieve optimal assay sensitivity, specificity, and reproducibility. Additionally, engineered expression reporter cell systems may introduce artifacts due to the manipulation of the genetic background or the addition of exogenous elements. These artifacts can skew the results and lead to inaccurate assessments of the biotherapeutic's efficacy or toxicity. Finally, engineered expression reporter cell systems may have a vastly different target receptor copy number and different post-translational modifications, and may not fully replicate the complexities of the native environment. To better represent the relevant biological context in engineered cells, CRISPR/Cas9 can be used to insert a small reporter tag into an endogenous protein locus. For example, this approach has been used to introduce an 11-amino acid HiBiT tag into a variety of proteins to enable monitoring of the protein levels through enzyme fragment complementation [64]. Overall, while engineered expression reporter cell systems are valuable tools for functional assay screening of biotherapeutics, researchers must carefully consider the potential disadvantages and employ appropriate controls and validation strategies to ensure the reliability and relevance of assay results.

Endogenously Expressing Cell Systems

While engineered expression reporter cell systems can offer certain advantages such as higher throughput or greater sensitivity, using cell lines with endogenously expressed targets may be preferred in some cases for functional assay screening of biotherapeutics due to their physiological relevance, avoidance of artifacts, and enhanced predictive validity. Cell lines with endogenously expressed targets mimic the natural physiological conditions more closely. This can provide more accurate insights into how the biotherapeutic will behave *in vivo*. To provide a direct measurement of the endogenous gene expression in native cells, the mRNA level may be determined by quantitative PCR (qPCR). For this purpose, the mRNA is extracted from the cells and transcribed to cDNA using specific primers. Then the cDNA is amplified and fluorescent oligonucleotide probes are used to quantitate its amplification in real time. qPCR is a highly sensitive and highly specific technology that has been broadly applied in drug discovery [65]. Favorably, it is amenable to multiplexing. Essential to the specificity of qPCR is the careful design of primers and probes; however, advances in design software have greatly aided this process. Equally critical for the generation of robust results are careful sample preparation and rigorous quality control including normalization to appropriately selected housekeeping genes and the use of proper controls such as no-template controls to identify potential contamination. Although qPCR may be more technically demanding, with lower throughput and higher cost than reporter gene assays, its applicability to cell lines with endogenously expressed targets means that the results are more likely to reflect the response of native tissues or organs to the biotherapeutic. This enhances the predictive validity of the assay and increases the likelihood that the activity obtained *in vitro* will translate to meaningful outcomes *in vivo*. Although using cell lines with endogenously expressed targets simplifies the experimental system and reduces the likelihood of some assay design complications, it is not without its own complexities and care must be taken in both assay design and execution.

Cellular Assay Systems of Increasing Complexity

In general, immortalized cell lines and cancer cell lines are relatively easy to work with: their cultures can be scaled up for screening purposes, and they typically exhibit good reproducibility in their application. These characteristics make them well-suited for use in many cellular assays, for example, initial screening assays and mechanistic assays. However, they are often grown within artificial environments or on nonphysiological substrates and the media may provide an inaccurate representation of key physiological conditions such as normal glucose, vitamin, and amino acid levels [66]. When the properties of an assay using a single cell line are insufficient for addressing the biology under investigation, the use of a more complex and/or specialized cell system should be explored.

Primary Cell Assays

Primary cell assays involve the use of cells directly isolated from tissues or organs of human or animal origin, maintaining their native physiological characteristics and microenvironment. These assays offer a more biologically relevant model for screening biotherapeutics compared to immortalized cell lines. Primary cells retain their tissue-specific functions, cell–cell interactions, and signaling pathways, providing insights into the complex mechanisms underlying drug responses. Assays with primary cells enable the evaluation of biotherapeutic candidate effects on diverse cellular processes, including proliferation, differentiation, apoptosis, migration, and secretion of cytokines or growth factors. Primary cells cover the spectrum of disease-relevant tissues for biotherapeutic targets; these include immune cells, epithelial cells (e.g. primary human bronchial epithelial cells or primary human renal epithelial cells), endothelial cells (e.g. HUVECs and primary human microvascular endothelial cells), fibroblasts, muscle cells, neurons, hepatocytes, adipocytes, chondrocytes, osteoblasts, keratinocytes, glial cells, and mesenchymal stem cells. Cells from internal organs are challenging to obtain while cells from blood, skin, or umbilical cord are relatively easy to obtain because they come from less invasive procedures or are from discarded tissues. Primary immune cells are easily obtained and separated from peripheral blood or cord blood and are often required for screening immune-related biotherapeutics when immortalized cell lines do not faithfully represent the immune cell biology. One assay example is the use of antibody-dependent cell-mediated cytotoxicity

(ADCC) assay to evaluate anti-tumor immunotherapy biotherapeutic products bearing an Fc region [67]; the Fc region of a tumor-specific antibody recruits Fc receptors on human primary natural killer (NK) cells that in turn exert cytotoxic activity towards the tumor cell. In addition to targeted antibodies, immune checkpoint inhibitors [68] and immune cell engagers [69] are other categories of immunotherapy protein biotherapeutics. Given the role of cytokines as crucial regulators of the immune system, engineered cytokines [70] are being developed for a variety of immune-related disorders such as cancer, autoimmune disease, and infection. All such immune-related biotherapeutics benefit from the inclusion of cell-based assays using primary cells in their discovery and optimization.

Nonetheless, the use of primary cells for screening biotherapeutics can pose significant challenges. Primary cells can pose safety issues because they are derived from living tissues, which can potentially carry pathogens, thus their handling and disposal require stringent biosafety measures to prevent exposure to hazardous biological materials. In addition, primary cells exhibit inherent biological variability between donors, batches, and tissue sources, which can complicate data interpretation and reproducibility across experiments. Often it is necessary to obtain a large single donor batch of cells (i.e. by leukapheresis for primary blood cells) and prepare frozen stocks for consistent results over time. If frozen stocks are not utilized, primary cells have a finite lifespan in culture and may undergo senescence or phenotypic changes over time, limiting their utility for long-term or repeated screening assays. Additionally, primary cells typically have slower growth rates and lower proliferation capacities compared to immortalized cell lines, which can impact assay throughput and scalability. Primary cell cultures may contain heterogeneous cell populations, including contaminating cell types or subpopulations with varying phenotypes or responses to biotherapeutics. Achieving consistent purity and homogeneity of primary cell cultures for HTS can be technically demanding and may require additional purification steps or cell sorting techniques. Primary cell culture-based assays can be more resource intensive and expensive compared to assays using immortalized cell lines due to the need for specialized culture media, equipment, and expertise. HTS with primary cells may require significant investments in infrastructure, automation, and personnel training. Despite their challenges in culture maintenance and limited proliferative capacity, primary cell assays remain indispensable tools in drug discovery and development, facilitating the identification of promising biotherapeutic candidates and the translation of preclinical findings to clinical applications.

Co-cultured Cell Systems

In addition to ligand–receptor interactions, cell signaling can also be initiated by or result from the direct interaction of a cell with its neighboring cell. Co-cultured cell-based assay systems involve the cultivation of multiple cell types together in the same culture environment to mimic the complex interactions that occur within tissues or organs *in vivo*. These systems are particularly valuable for screening biotherapeutics because they enable the assessment of the biotherapeutic effects on cell–cell interactions, paracrine signaling, and multicellular responses. The source of co-cultured cells may be from cell lines and/or primary cells. One common application of co-cultured cell-based assays is in the study of immune responses to biotherapeutics. Some examples of co-culture models include inflammatory biology, neovascularization, and tumor microenvironments [71]. By co-culturing immune cell types, such as T cells, B cells, dendritic cells, and macrophages, with target cells or tissues, researchers can evaluate the immunomodulatory effects of biotherapeutics, including cytokine release, immune cell activation, and antigen presentation. The application of a 3D cell co-culture to replicate the tumor microenvironment is an example of a potential physiologically relevant functional assay for the discovery of novel cancer immunotherapies [72]. Overall, when complex biological processes cannot be adequately captured in single-cell culture models, co-cultured cell-based assay systems provide a more physiologically relevant and holistic approach to screening biotherapeutics.

Organoid and Patient-Derived Cell Systems

Organoid and patient-derived cell systems offer significant advantages in terms of physiological relevance and disease modeling. Bioengineered human disease models hold great promise for drug discovery and development although model validation and scalable production are future milestones that may facilitate increased implementation in preclinical research [73]. Organoids are three-dimensional structures derived from stem cells or tissue samples that recapitulate some aspects of organ architecture and function. *In vitro*-differentiated human pluripotent stem cells offer an assay system that may better recapitulate human physiology, although implementation of these systems has been slower than anticipated [74]. Human-induced pluripotent stem cells (iPSCs) are a promising tool for screening biotherapeutics but require constant attention with media changes every 24 hours; automation of this type of cell-based system may help minimize technical errors and provide more flexibility for researchers [75, 76]. Patient-derived cell systems involve the isolation and culture of cells directly obtained from patient samples, such as tumor biopsies or blood specimens. Patient-derived organoids (PDOs), mimicking tumor structures and harboring cancer stem cells, are valuable tools for personalized medicine. A study conducted on a large colorectal cancer PDO biobank and healthy colonic mucosa samples screened over 500 bispecific antibodies (bAbs) targeting WNT and receptor tyrosine kinase (RTK) pathways; this effort led to the development of MCLA-158, a bAb that selectively degrades epidermal growth factor receptor in cancer stem cells, demonstrating therapeutic efficacy against KRAS-mutant colorectal cancers and other epithelial cancer types in preclinical models [77].

While these models hold great promise for personalized medicine and preclinical drug testing, their application in HTS settings has been limited by several factors. Organoids and patient-derived cell systems exhibit inherent biological complexity and variability, both within and between samples. This heterogeneity can complicate assay standardization, reproducibility, and data interpretation in high-throughput settings, where consistency and reliability are paramount. Organoid and patient-derived cell cultures often require specialized culture conditions and manual handling techniques, making them less amenable to high-throughput automation and miniaturization. Achieving the throughput necessary for large-scale screening campaigns may be technically challenging and resource-intensive. Systems such as the human gut-on-a-chip [78] and the human small airway-on-a-chip [79] may be able to recapitulate complex features of disease but use is restricted to lead optimization instead of screening. Organoids and patient-derived cell cultures typically have longer culture times compared to conventional cell lines, requiring extended incubation periods to allow for proper differentiation, maturation, and formation of complex tissue structures. High-throughput assays with these models may necessitate prolonged experimental durations, further complicating workflow optimization and throughput. The use of organoids and patient-derived cell systems in drug screening raises ethical and regulatory concerns related to patient consent, privacy, and sample sourcing. Compliance with ethical guidelines and regulatory requirements for working with human-derived materials, including validation of assay reproducibility and accuracy, adds complexity and time to the screening process.

Despite these challenges, efforts are underway to adapt organoid and patient-derived cell systems for high-throughput functional assays, leveraging advances in automation, microfluidics, and imaging technologies. These efforts aim to improve assay throughput, reproducibility, and scalability while retaining the physiological relevance and disease-modeling capabilities of these models. As technology continues to evolve and methodologies are refined, the use of organoid and patient-derived cell systems in HTS of biotherapeutics may become more feasible, offering enhanced predictive value and personalized treatment approaches in drug discovery and development.

Fundamentals of Assay Development, Validation, and Implementation

Assay Components

The availability, quality, and composition of the assay components play vital roles in promoting the successful application of functional assays for screening biotherapeutics. Key components include the target proteins and/or cells that present the target molecules. Proteins required for biochemical assays such as enzyme activity or receptor/ligand interaction assays may include the target protein, binding partners, and substrates. Cellular assays may require a protein (possibly the target of the therapeutic) to stimulate a response in cells. Most often, recombinant proteins are used. Among the considerations for the choice of protein construct are construct boundaries, tags, and fusions. These properties may facilitate the expression and purification of the protein in addition to benefiting the configuration of and performance in the assay. The expression system (e.g. *Escherichia coli*, baculovirus, CHO cells, and NSO cells) may also play a pivotal role in optimizing protein expression and ensuring that the protein has the proper post-translational modifications where relevant. Protein reagents should be characterized for identity, purity, concentration, and biological activity. Selection of cells for use in a functional assay likewise requires multiple considerations. Biological relevance is foremost, taking into consideration phenotypic characteristics and the presence of mutations. The specific role of the assay in the assay cascade may also influence the choice of species as well as the use of cell lines or primary cells. It may be necessary to genetically engineer the cells, for example, to introduce tags on specific proteins, to construct reporters of gene expression, or to generate isogenic cell lines. All engineered modifications should be verified by PCR and/or sequencing, and the identities of cell lines should be authenticated by STR profiling [80]. Other assay components include detection reagents, cell media, and assay buffers. The detection reagents range broadly across various types of assays. Antibodies are often essential reagents in many functional assays. In biochemical assays, they may be used for binding to and recognizing the target protein, its binding partners, or the product of an enzyme reaction in such assay formats as ELISA, TR-FRET, or fluorescence polarization. In cellular assays, antibodies may serve to characterize signaling pathways or cellular responses by methods such as immunoassays, immunofluorescent imaging, or FACS. To address the diverse needs of the broad array of functional assays, an extensive range of high-quality commercial reagents are available. However, there are some instances in which commercially available reagents fall short, and the need for bespoke reagents arises. Despite the additional time investment, tailored reagents can enhance assay accuracy and potentially yield superior lead biotherapeutics. Suboptimal reagents (e.g. unstable, impure, nonspecific, etc.) can skew assay results, jeopardizing the reliability of the screening process and leading to inaccurate conclusions such as false positives and/or false negatives. Consequently, diligent selection and optimization of the assay components is imperative for advancing the discovery and development of effective biotherapeutics.

Assay Development Basics

Functional assay development can take many directions depending on the specific nature of the assay. Most functional assays for biotherapeutic screening are conducted in multi-well plates. Following the preliminary selection of key assay components and the detection technology, fundamental parameters for optimization that are common to most multi-well assays include the well configuration and material composition of the assay plates, the total assay volume, how the components are staged into the assay, and the composition of the assay buffer or medium. The concentrations of individual assay components and incubation times are among the most critical factors influencing the magnitude of the assay window, which is the difference between the maximal and minimal signals in the assay. To ensure robust assay performance, reagent stability and signal stability should be monitored for compatibility with the time period of a typical assay run. The composition of the assay buffer in biochemical assays may influence both assay window and reagent stability and can be optimized to maximize these assay characteristics. Other factors to consider for enzyme activity assays are the percent conversion of substrate, the substrate Km, reaction linearity with time and enzyme concentration, and whether a stop reagent is needed. Additional considerations for cellular assays include cell culture model (e.g. 2D, 3D, co-culture, etc.), cell density, serum conditions, treatment conditions, and cell passage number restrictions. More detailed contemplations regarding assay configuration are provided above in the specific assay sections. There are many factors to define and optimize when developing a functional assay. These factors often influence each other, so it is essential to optimize multiple factors in parallel and/or to optimize conditions iteratively. As such, assay development is an ideal target for the application of statistical design of experiments (DoE) [81]. Notably, the use of this approach for assay optimization is increasing due to improvements in software tools [82].

Assay Controls and Performance Characterization

Assay controls (Table 10.1) are essential for defining the assay signal range and normalizing the data to control for variation across plates, runs, reagent lots, and researchers. In an inhibition assay, the positive control produces a basal or background signal and represents the 100% inhibition condition, and the negative control produces a maximal signal and is used to define the no inhibition condition. To best represent the expected pharmacology, the positive control should be a known inhibitor at a concentration yielding maximal inhibition. If a known inhibitor is not available, then a key assay component may be omitted as an alternate approach to simulate complete inhibition. The negative control may be the buffer used for the preparation of the samples or a known inactive molecule representing the same modality as the test samples. In an activation assay, the positive control is defined by using a known activator (when available) that produces a "maximal" signal or 100% response, and the negative control represents the no activation condition and produces a minimal or basal signal. The concentration of the positive control is set at a level that produces maximal activation, and the negative control is similar to that used for inhibition assays. The terms MIN and MAX are often used to refer to control conditions producing the minimal and maximal signal magnitude, respectively, regardless of whether the control is positive or negative control.

A two-point method is typically used to normalize the data. The MIN and MAX controls are run in multiple replicates for this purpose, and means or medians of the replicates are calculated. The use of median values is favored to make the normalization more robust by reducing its sensitivity to potential outlying control values. The normalized

Table 10.1 Controls for assay normalization.

Type of assay	Type of control	Response	Signal magnitude
Inhibition	Positive	100% inhibition	MIN
	Negative	0% inhibition	MAX
Activation	Positive	100% activation	MAX
	Negative	0% activation	MIN

values for a given sample represent the percent inhibition or activation for that sample and may be used to assess its activity relative to that of the controls and other samples. Additionally, normalized results facilitate the evaluation of reproducibility across assay plates or assay runs.

$$\text{Normalized\%Inhibition} = \frac{\text{Median}_{\text{MAX}} - \text{Median}_{\text{SAMPLE}}}{\text{Median}_{\text{MAX}} - \text{Median}_{\text{MIN}}} \times 100$$

$$\text{Normalized\%Activation} = \frac{\text{Median}_{\text{SAMPLE}} - \text{Median}_{\text{MIN}}}{\text{Median}_{\text{MAX}} - \text{Median}_{\text{MIN}}} \times 100$$

In addition to their use for normalization, positive and negative assay controls enable a means for characterizing assay performance through the calculation of several control-based statistics on a per-plate basis. Key statistics used for this purpose are assay window, signal/background, and Z'. In addition to the assay window that provides the absolute magnitude of the difference between the positive and negative controls, the signal/background provides a relative measure of the difference between these two controls by dividing the MAX signal by the MIN signal. There is not *a priori* a specific magnitude to recommend for the assay window or signal/background, as the target values are assay dependent. Nevertheless, larger magnitudes within a given assay generally translate to better assay performance. As such, these two parameters are important to monitor and optimize during assay development. Tracking these statistics during the routine application of an assay is valuable for verifying consistency and identifying potential discrepancies. Z' is a widely used, dimensionless statistic that assesses the size of the assay window relative to the noise [83]. It assumes a normal distribution of the results, and the use of larger numbers of replicates improves its accuracy. As originally conceived, Z' incorporates into a single statistic standard deviations (SD) and mean values for both controls. It can be made robust by using medians in place of means and median absolute deviations (MAD) in place of SD. The use of such a robust Z' or RZ' is favored over the elimination of individual control data points as a means of addressing outlying control values.

$$RZ' = 1 - \frac{3 \times \text{MAD}_{\text{MAX}} + 3 \times \text{MAD}_{\text{MIN}}}{|\text{Median}_{\text{MAX}} - \text{Median}_{\text{MIN}}|}$$

The range of possible RZ' (or Z') values runs from $-\infty$ to 1. A value between 0.5 and 1 indicates a large separation band and represents the ideal range to target during the development of an assay as well as in routine operation. For some more complex cellular assays, RZ' values down to 0.4 may still provide acceptable results for biotherapeutic screening. Z' or RZ' should be evaluated periodically throughout assay development to ensure that assay performance is improving. Determination of the assay window or signal/background and the Z' or RZ' is an essential component of assay validation. Additionally, the continued application of these statistics as part of ongoing performance monitoring after assay implementation is vital to ensure that the assay quality is compatible with robust identification and ranking of active molecules. At the conclusion of assay development, acceptance criteria for these performance statistics should be defined and documented in the written assay protocol.

Characterizing the Activity of Biotherapeutics

The activity of potential biotherapeutics in an assay can be determined using one or a limited number of concentrations of the test molecule to evaluate percent inhibition or percent activation. This approach is often used for early tier 2 functional assays in which the foremost task is to reliably identify functionally active molecules while minimizing false-positive and false-negative results. Alternatively, biotherapeutics may be more fully characterized by assessing activity across a range of concentrations to enable the generation of dose–response curves and calculation of IC_{50} or EC_{50} values, which are the concentrations at which the molecules inhibit or activate, respectively, to 50% of the maximum level. Minimum significant ratio (MSR) provides a measure of the reproducibility of IC_{50} or EC_{50} determinations in a given assay [84]. It is a statistic that can be used to determine the smallest ratio between IC_{50} or EC_{50} values of two samples that can be distinguished statistically. Additionally, it enables comparisons of the relative precision of potency estimates across assays. A solid target for the MSR of an assay is <3.

In the case of functional assays, assay sensitivity refers to the limit of the assay in discriminating among highly potent molecules. The sensitivity of biochemical assays is limited to IC_{50} or EC_{50} values of greater than half the concentration of the target protein. For very potent biotherapeutics, this may present a severe limitation. However, in some cases, assay sensitivity may be extended to approximately 1/100 of the target protein concentration using mathematical and operational approaches [85, 86]. Often, cellular assays provide a more sensitive means for distinguishing molecules with potencies at sub-nanomolar levels. However, for complex assays or those with indeterminate target concentrations, it may not be possible to determine the assay sensitivities non-empirically. When many of the most potent molecules appear to be equivalently potent, the assay is likely not sensitive enough to differentiate leads. In this case, the assay should be re-optimized or a new assay developed for increased sensitivity to enable ranking of the candidate potencies. Notably, the inherent variability within an assay will not allow the distinction of leads with potency differences that are less than the MSR. If leads are indistinguishable in terms of potency, then orthogonal means of differentiation such as sequence uniqueness, developability, binding affinity, epitope binning, or a combination of these approaches may be necessary [87].

Reference Molecules and Benchmarking

In addition to positive and negative controls, a reference molecule also serves as a crucial control for a biotherapeutics discovery effort. Ideally, the reference molecule is of the same modality as the test samples, but this is not always possible. Reference molecules may be sourced internally or from commercial vendors specializing in reference standards for biotherapeutics. In screening, the potency of the sample may be evaluated by relative comparison to the potency of a reference standard [88]. Such a molecule will be run alongside samples in the same fashion in each plate in every run of an assay in practice. It will also be used for assessing assay robustness and reliability in terms of the reproducibility of potency determinations and can be used for determining the MSR. Reference controls with well-characterized pharmacological activities that set minimum standards for the activity and potency may also function as benchmark molecules with characteristics that potential biotherapeutics must meet or exceed to be considered viable candidate molecules. By employing rigorous benchmarking and control strategies, researchers can enhance the reliability and validity of functional assays, facilitating the identification of biotherapeutics with desirable therapeutic characteristics for further development and clinical translation.

Assay Validation

After the assay is developed and is approaching implementation for screening, its performance should be characterized by conducting a set of validation studies. These studies include a plate uniformity test to rigorously evaluate assay performance based on positive and negative assay controls. Additionally, a reference control at a concentration targeting 50% effect may be used but is not required. A key feature of uniformity testing is that it will also enable the identification of plate effects such as edge effects and signal drift. Assay validation strategies have been well documented for HTS of small molecules [89–91], and similar approaches can be applied for assays of biotherapeutics. Minimally, the validation studies should include positive and negative controls alternating by column across an assay plate run on each of three days (Figure 10.4a). The RZ′ (or Z') and the signal/background (or assay window) should be calculated for each plate. If a reference molecule is used, it should be similarly rotated into every third column across the plate, and the % effect should be determined. All parameters should meet acceptance criteria that are defined in the assay protocol. If insufficient data has been collected to date for the establishment of these criteria, then the assay validation data may be used for this purpose. Additionally, the normalized data should be plotted as a heat map per plate (Figure 10.4b) and as a scatter plot across all plates (Figure 10.4c [by well] and 10.4d [by column as relevant in this example]). These visualizations facilitate the identification of trends, for example, loss of activity across plates or runs, edge effects, or signal drift. Potential causes of edge effects include evaporation, temperature gradients, or CO_2 gradients. They are more frequently encountered in cellular assays and may be remedied by avoiding the use of the affected outer wells, building this restriction into the plate layout, and defining it in the assay protocol. Signal drift may also be caused by similar sources, and additionally, by liquid handling, plate handling, poor plate construction, or plate reader variability. The potential causes of drift should be investigated and mitigated before proceeding to the screen. Figure 10.4 provides an example of a plate uniformity test that demonstrates the presence of signal drift as evident in a decrease from left to right in the normalized activity for the positive control in this activation assay.

When encountering poor assay quality, it is recommended that the assay developer carefully examine the data and consider all biological and technical factors. Besides well location-based variability, many factors may negatively affect the overall assay quality. It would be difficult to enumerate all the factors here. Nevertheless, a common culprit if performing a primary cell-based assay may include donor-to-donor variability. In this case, it is important to determine if cells from multiple donors are needed. For other cell-based assays, the cell source and quality may affect the output of the assay, in particular, the cell passage number and whether a live passage or a frozen assay-ready stock is used. The performance and responsiveness of cells may also contribute to assay variability, so it is critical to use good cell culture practices and establish standard operating procedures (SOPs) for handling cultures [92]. Manual pipetting techniques differ between analysts and may increase variability [93], so automated liquid handlers are often preferred. On the other hand, automated liquid handlers may introduce variability if they are not properly maintained or are not operating correctly. Cell dispensing, whether automated or manual, may also

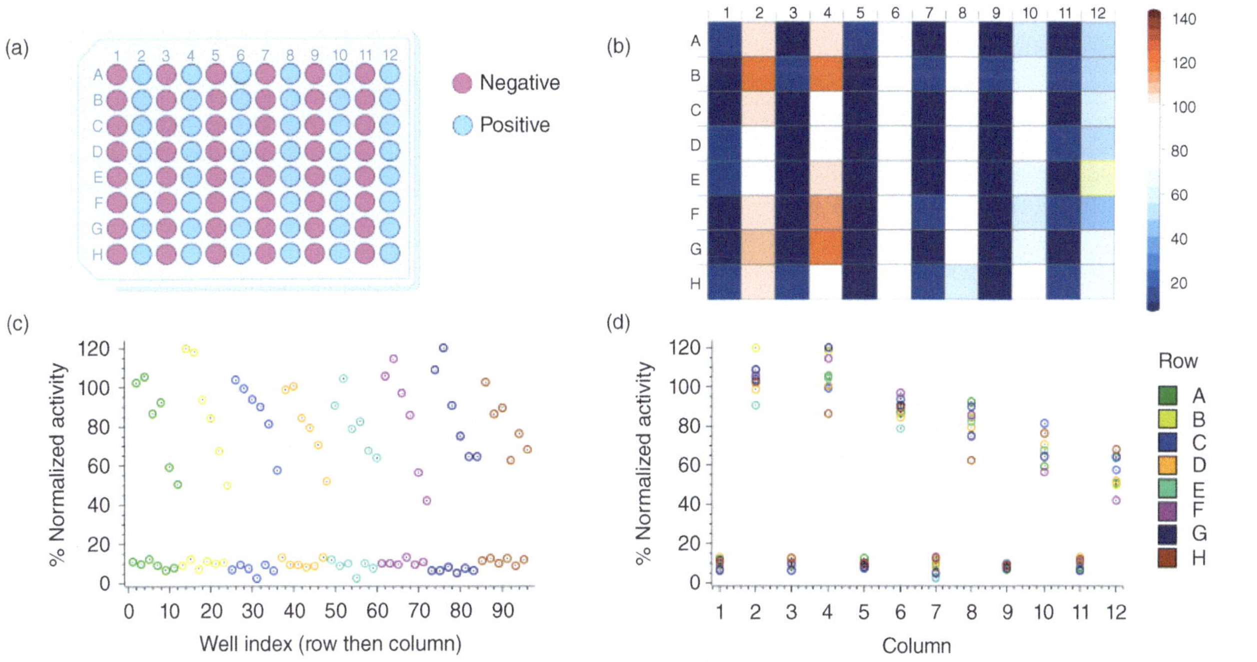

Figure 10.4 Application and analysis of assay controls to assess assay uniformity across the plate. This example of one plate of a 96-well activation assay with two controls (positive and negative) shows assay drift across the plate as evident in a decrease in the magnitude of the positive control signal from column 2 to column 12. (a) Alternating layout of assay controls by column in assay plate. (b) Normalized assay results are depicted as a heat map of activity (100% = maximum activity; 0% = minimum activity). (c, d) Normalized activity is depicted as scatterplots by well (c) and by column (d) with points colored by row on the assay plate.

lead to variability if care is not taken to ensure that the cells are being equally distributed. As discussed above, reagent quality may also affect the quality of the assay. Below we discuss the effect of the sample.

Suitability of the Sample for Screening

Various factors impact confidence in assay results including the sample buffer. The sample to be screened may contain a high salt concentration (due to the purification process) that is incompatible with the assay. During screening assay development, it is crucial to test the effects of the sample buffer on the assay. It may be difficult to precisely replicate the composition of the sample buffer that results from a small-scale purification process (i.e. 1 ml) which lends itself to variable buffer composition. In that case, a workaround is to obtain mock samples from an unrelated screening campaign in which the purified protein is not expected to interact in the assay. If it is determined that the sample buffer causes interference in the assay, it is recommended that the incoming screening samples undergo a buffer exchange step using the desired buffer.

Just as the purity of assay reagents affects the assay quality, the purity of the samples as well as the other key quality attributes of the candidate biotherapeutics may affect the outcome of the functional screening assay and lead to a lower confidence in the biotherapeutic hits. The key quality attributes of a biotherapeutic, such as an antibody [94], may be assessed analytically [95], although the quantity of purified samples generally precludes this type of analysis during the early screening phases. The purity of samples subjected to functional screening assays profoundly impacts the reliability and confidence in identifying biotherapeutic hits. Sample quality typically increases as samples are scaled up and progressed through the screening cascade, so earlier screening assays may be particularly subject to issues related to sample purity. Contaminants, impurities, or aggregates present in the samples can interfere with the assay readout, yielding misleading results and potentially misidentifying active samples as false negatives or inactive samples as viable biotherapeutic candidates. These impurities may interact with assay components, leading to nonspecific binding or inhibitory effects, complicating the process of selecting promising lead candidates for further development. Therefore, ensuring the purity of samples is paramount to obtaining accurate and trustworthy outcomes from functional screening assays, enabling confident decision-making in the pursuit of effective biotherapeutics. Additionally, a pure sample also allows an accurate determination of the concentration, which enables sample normalization and proper ranking of the hit potencies.

Assay Implementation

When preparing for assay implementation, the assay developer should establish plate layouts that include positive and negative controls as well as reference controls (where available). For a primary screen, it is possible that only one concentration (neat or diluted) is tested to determine the top hits. These hits will then proceed to a secondary screen in the same assay, often with an eight-point serial dilution to further rank the candidates by IC_{50} or EC_{50} values. The researcher may also select multiple concentrations or a combination of one and multiple concentrations for their primary screen, based on the project needs, assay capacity, and material availability. Plate layouts must consider these needs and can be flexible; however, the use of automated sample preparation protocols may place some constraints. In the final preparation for a screen, it is ideal to run a set of test molecules (if available) with the planned plate layouts and control schemes (single concentration and/or dose–response format) for final validation of assay quality including reproducibility across a range of molecule potencies. At least two independent assay runs should be performed. If a scatterplot of the result values from one run versus the other shows acceptable reproducibility (all values within ~threefold of the fitted line), then the assay is ready to use. As a standard practice, control values and plate statistics should be tracked throughout the screening process and monitored for deviations. Failure of an assay plate to meet the acceptance criteria defined in the assay protocol should result in a rerun of the samples in the failed assay plate.

In planning for the screening campaign, it is paramount that researchers generating the samples and those running the functional screening assays are well aligned on the considerations and parameters needed for success. Considerations for transfer of samples require clear communication on the timing of sample arrival, numbers of samples, type of labware, layout of samples, and concentration and amount of sample. The preparation and assay of samples should also incorporate FAIR (Findability, Accessibility, Interoperability, and Reusability) principles of data management [96]. This includes the use of barcoded plates or tubes that contain registered samples to be evaluated in registered assays. Ideally, sample check-in, execution of automated workflows (assignment of protocols, worklist generation, and sample tracking), and data analysis are all coordinated through a LIMS or other integrated data management system.

Concluding Remarks

Attempting to replicate the complexity of the human body entirely within *in vitro* assays is an unrealistic endeavor. However, the pursuit of physiologically relevant assays serves as a vital bridge between isolated functional activities and their biological significance. By starting with the desired pharmacological activity and working backward, functional assays provide a systematic approach to evaluating biotherapeutic candidates, ranking their potencies, and ultimately selecting the most promising leads. These assays strive to translate disease biology insights into a comprehensive pharmacological understanding, capturing the important interactions and mechanisms crucial for the development of novel transformative medicines. The goal is to not only identify biotherapeutics with the desired pharmacological effects but also elucidate their modes of action within the context of complex biological systems, laying the groundwork for the development of targeted therapies that precisely and effectively address the underlying mechanisms of disease. Through this iterative process of screening and selection, physiologically relevant functional assays play a central role in translating scientific insights into tangible advancements in medicine, offering new avenues for the treatment of diverse diseases and improving patient outcomes.

In conclusion, functional assays play a pivotal role in the screening and characterization of biotherapeutics, offering valuable insights into their efficacy and mechanisms of action. From biochemical assays for extracellular targets to traditional cell-based assays using immortalized cell lines to advanced organoid and patient-derived cell systems, a diverse array of experimental approaches exists to interrogate the functional effects of biotherapeutics on cellular physiology and disease processes. By leveraging innovative technologies such as high-content screening platforms, multiplexed readouts, and microfluidic systems, researchers can enhance assay throughput, sensitivity, and relevance, paving the way for more efficient drug discovery and development pipelines. While challenges remain in assay standardization, scalability, and validation, continued advancements in assay methodologies, automation, and data analytics promise to address these limitations and further propel the field forward. Finally, functional assays provide indispensable tools for identifying lead biotherapeutics, elucidating drug mechanisms, and ultimately translating promising therapeutic candidates from bench to bedside, shaping the future of precision medicine, and personalized healthcare.

References

1 Lu, R.M., Hwang, Y.C., Liu, I.J. et al. (2020). Development of therapeutic antibodies for the treatment of diseases. *J. Biomed. Sci.* 27 (1): 1.

2 Parray, H.A., Shukla, S., Samal, S. et al. (2020). Hybridoma technology a versatile method for isolation of monoclonal antibodies, its applicability across species, limitations, advancement and future perspectives. *Int. Immunopharmacol.* 85: 106639.

3 Pedrioli, A. and Oxenius, A. (2021). Single B cell technologies for monoclonal antibody discovery. *Trends Immunol.* 42 (12): 1143–1158.

4 Singh, V., Garg, S., Raj, N. et al. (2022). Protocol for high throughput screening of antibody phage libraries. *BIO-Protoc.* 12 (12): e4450.

5 Zambrano, N., Froechlich, G., Lazarevic, D. et al. (2022). High-throughput monoclonal antibody discovery from phage libraries: challenging the current preclinical pipeline to keep the pace with the increasing mAb demand. *Cancer* 14 (5): 1325.

6 Sheehan, J. and Marasco, W.A. (2015). Phage and yeast display. *Microbiol. Spectr.* 3 (1): AID-0028-2014.

7 Kunamneni, A., Ogaugwu, C., Bradfute, S., and Durvasula, R. (2020). Ribosome display technology: applications in disease diagnosis and control. *Antibodies* 9 (3): 28.

8 Ledsgaard, L., Kilstrup, M., Karatt-Vellatt, A. et al. (2018). Basics of antibody phage display technology. *Toxins* 10 (6): 236.

9 Cariuk, P., Gardener, M.J., and Vaughan, T.J. (2013). Evolution of biologics screening technologies. *Pharmaceuticals* 6 (5): 681–688.

10 Sherman, D.J., Kenanova, V.E., Lepin, E.J. et al. (2010). A differential cell capture assay for evaluating antibody interactions with cell surface targets. *Anal. Biochem.* 401 (2): 173–181.

11 Noy-Porat, T., Alcalay, R., Mechaly, A. et al. (2021). Characterization of antibody-antigen interactions using biolayer interferometry. *STAR Protoc.* 2 (4): 100836.

12 Matsunaga, R., Ujiie, K., Inagaki, M. et al. (2023). High-throughput analysis system of interaction kinetics for data-driven antibody design. *Sci. Rep.* 13 (1): 19417.

13 Duo, J., Dong, H., DeSilva, B., and Zhang, Y.J. (2013). A generic template for automated bioanalytical ligand-binding assays using modular robotic scripts in support of discovery biotherapeutic programs. *Bioanalysis* 5 (14): 1735–1750.

14 Zhang, Z., Liu, H., Guan, Q. et al. (2017). Advances in the isolation of specific monoclonal rabbit antibodies. *Front. Immunol.* 8: 494.

15 Nakamura, G., Chai, N., Park, S. et al. (2013). An in vivo human-Plasmablast enrichment technique allows rapid identification of therapeutic influenza A antibodies. *Cell Host Microbe* 14 (1): 93–103.

16 Parola, C., Neumeier, D., and Reddy, S.T. (2018). Integrating high-throughput screening and sequencing for monoclonal antibody discovery and engineering. *Immunology* 153 (1): 31–41.

17 Porebski, B.T., Balmforth, M., Browne, G. et al. (2024). Rapid discovery of high-affinity antibodies via massively parallel sequencing, ribosome display and affinity screening. *Nat. Biomed. Eng.* 8 (3): 214–232.

18 Chevalier, A., Silva, D.A., Rocklin, G.J. et al. (2017). Massively parallel de novo protein design for targeted therapeutics. *Nature* 550 (7674): 74–79.

19 Ebrahimi, S.B. and Samanta, D. (2023). Engineering protein-based therapeutics through structural and chemical design. *Nat. Commun.* 14 (1): 2411.

20 Tobin, P., Richards, D., Callender, R., and Wilson, C. (2015). Protein engineering: a new frontier for biological therapeutics. *Curr. Drug Metab.* 15 (7): 743–756.

21 Knudsen, L.B. and Lau, J. (2019). The discovery and development of liraglutide and semaglutide. *Front. Endocrinol.* 10: 155.

22 Frejd, F.Y. and Kim, K.T. (2017). Affibody molecules as engineered protein drugs. *Exp. Mol. Med.* 49 (3): e306–e306.

23 Ståhl, S., Gräslund, T., Karlström, A.E. et al. (2017). Affibody molecules in biotechnological and medical applications. *Trends Biotechnol.* 35 (8): 691–712.

24 Rothe, C. and Skerra, A. (2018). Anticalin® proteins as therapeutic agents in human diseases. *BioDrugs* 32 (3): 233–243.

25 Morales-Kastresana, A., Siegemund, M., Haak, S. et al. (2022). Anticalin®-based therapeutics: expanding new frontiers in drug development. *Int. Rev. Cell Mol. Biol.* 369: 89–106.

26 Schilling, J., Jost, C., Ilie, I.M. et al. (2022). Thermostable designed ankyrin repeat proteins (DARPins) as building blocks for innovative drugs. *J. Biol. Chem.* 298 (1): 101403.

27 Luo, R., Liu, H., and Cheng, Z. (2022). Protein scaffolds: antibody alternatives for cancer diagnosis and therapy. *RSC Chem. Biol.* 3 (7): 830–847.

28 Gebauer, M. and Skerra, A. (2020). Engineered protein scaffolds as next-generation therapeutics. *Annu. Rev. Pharmacol. Toxicol.* 60 (1): 391–415.

29 Škrlec, K., Štrukelj, B., and Berlec, A. (2015). Non-immunoglobulin scaffolds: a focus on their targets. *Trends Biotechnol.* 33 (7): 408–418.

30 Mayr, L.M. and Fuerst, P. (2008). The future of high-throughput screening. *SLAS Discov.* 13 (6): 443–448.

31 Danilenko, D.M. and Wang, H. (2012). The Yin and Yang of immunomodulatory biologics: assessing the delicate balance between benefit and risk. *Toxicol. Pathol.* 40 (2): 272–287.

32 Dai, Z., Juneja, J., Schneeweis, L. et al. (2020). Application of the gyrolab microfluidic platform to measure picomolar affinity of a PD-L1-binding adnectin™ radioligand for positron emission tomography. *BioTechniques* 69 (3): 200–205.

33 Bondza, S., Foy, E., Brooks, J. et al. (2017). Real-time characterization of antibody binding to receptors on living immune cells. *Front. Immunol.* 8: 455.

34 Rolin, C., Zimmer, J., and Seguin-Devaux, C. (2024). Bridging the gap with multispecific immune cell engagers in cancer and infectious diseases. *Cell. Mol. Immunol.* 21 (7): 643–661.

35 Khoryati, L., Pham, M.N., Sherve, M. et al. (2020). An IL-2 mutein engineered to promote expansion of regulatory T cells arrests ongoing autoimmunity in mice. *Sci. Immunol.* 5 (50): eaba5264.

36 Frenzel, A., Schirrmann, T., and Hust, M. (2016). Phage display-derived human antibodies in clinical development and therapy. *MAbs* 8 (7): 1177–1194.

37 Maun, H.R., Jackman, J.K., Choy, D.F. et al. (2019). An allosteric anti-tryptase antibody for the treatment of mast cell-mediated severe asthma. *Cell* 179 (2): 417–431.e19.

38 Chavarria-Smith, J., Chiu, C.P.C., Jackman, J.K. et al. (2022). Dual antibody inhibition of KLK5 and KLK7 for Netherton syndrome and atopic dermatitis. *Sci. Transl. Med.* 14 (675): eabp9159.

39 Tandon, S., Sharma, A., Singh, S. et al. (2021). Therapeutic enzymes: discoveries, production and applications. *J. Drug Deliv. Sci. Technol.* 63(J. Bacteriol. 154 1 1983): 102455.

40 Wu, C., Cook, R., Wu, P. et al. (2022). Using a peptide-based mass spectrometry approach to quantitate proteolysis of an intact heterogeneous procollagen substrate by BMP1 for antagonistic antibody screening. *Anal. Bioanal. Chem.* 414 (22): 6601–6610.

41 Roos, A. and Wieslander, J. (2013). Evaluation of complement function by ELISA. *Methods Mol. Biology Clifton N J.* 1100: 11–23.

42 Wu, M., Li, H., Zhang, Y., and Chen, D. (2015). Development of a C3c-based ELISA method for the determination of anti-complementary potency of Bupleurum polysaccharides. *Acta Pharm. Sin. B* 5 (4): 316–322.

43 Garred, P., Tenner, A.J., and Mollnes, T.E. (2021). Therapeutic targeting of the complement system: from rare diseases to pandemics. *Pharmacol. Rev.* 73 (2): 792–827.

44 Moore, K. and Rees, S. (2001). Cell-based versus isolated target screening: how lucky do you feel? *J. Biomol. Screen.* 6 (2): 69–74.

45 Fen, W., Sicen, W., and Xilan, G. (2021). A review for cell-based screening methods in drug discovery. *Biophys. Rep.* 7 (6): 504–516.

46 Michelini, E., Cevenini, L., Mezzanotte, L. et al. (2010). Cell-based assays: fuelling drug discovery. *Anal. Bioanal. Chem.* 398 (1): 227–238.

47 Fischer, R., Kontermann, R.E., and Pfizenmaier, K. (2020). Selective targeting of TNF receptors as a novel therapeutic approach. *Front. Cell Dev. Biol.* 8: 401.

48 Cebecauer, M., Spitaler, M., Sergé, A., and Magee, A.I. (2010). Signalling complexes and clusters: functional advantages and methodological hurdles. *J. Cell Sci.* 123 (3): 309–320.

49 Grakoui, A., Bromley, S.K., Sumen, C. et al. (1999). The immunological synapse: a molecular machine controlling T cell activation. *Science* 285 (5425): 221–227.

50 Li, M. and Yu, Y. (2021). Innate immune receptor clustering and its role in immune regulation. *J. Cell Sci.* 134 (4): jcs249318.

51 Spangler, J.B., Moraga, I., Mendoza, J.L., and Garcia, K.C. (2014). Insights into cytokine–receptor interactions from cytokine engineering. *Annu. Rev. Immunol.* 33 (1): 1–29.

52 Tian, Z., Liu, M., Zhang, Y., and Wang, X. (2021). Bispecific T cell engagers: an emerging therapy for management of hematologic malignancies. *J. Hematol. Oncol.* 14 (1): 75.

53 Romei, M.G., Leonard, B., Katz, Z.B. et al. (2024). I-shaped antibody engineering enables conformational tuning of biotherapeutic receptor agonists. *Nat. Commun.* 15 (1): 642.

54 Agard, N.J., Zhang, G., Ridgeway, J. et al. (2022). Direct Tie2 agonists stabilize vasculature for the treatment of diabetic macular Edema. *Transl. Vis. Sci. Technol.* 11 (10): 27.

55 Moore, M., Ferguson, J., and Burns, C. (2014). Applications of cell-based bioassays measuring the induced expression of endogenous genes. *Bioanalysis* 6 (11): 1563–1574.

56 Eglen, R.M. (2002). Enzyme fragment complementation: a flexible high throughput screening assay technology. *Assay Drug Dev. Technol.* 1 (1): 97–104.

57 Thon, J.N., Devine, M.T., Begonja, A.J. et al. (2012). High-content live-cell imaging assay used to establish mechanism of trastuzumab emtansine (T-DM1)–mediated inhibition of platelet production. *Blood* 120 (10): 1975–1984.

58 Bravo, D.D., Shi, Y., Sheu, A. et al. (2023). A real-time image-based Efferocytosis assay for the discovery of functionally inhibitory anti-MerTK antibodies. *J. Immunol.* 210 (8): 1166–1176.

59 McGuinness, R. (2007). Impedance-based cellular assay technologies: recent advances, future promise. *Curr. Opin. Pharmacol.* 7 (5): 535–540.

60 Xi, B., Yu, N., Wang, X. et al. (2008). The application of cell-based label-free technology in drug discovery. *Biotechnol. J.* 3 (4): 484–495.

61 Segaliny, A.I., Jayaraman, J., Chen, X. et al. (2023). A high throughput bispecific antibody discovery pipeline. *Commun. Biol.* 6 (1): 380.

62 Wang, L., Xu, G.L., Gao, K. et al. (2016). Development of a robust reporter-based assay for the bioactivity determination of anti-VEGF therapeutic antibodies. *J. Pharm. Biomed. Anal.* 125: 212–218.

63 Grailer, J., Moravec, R.A., Cheng, Z.J. et al. (2020). Immuno-oncology, cellular and translational approaches. *Methods Pharmacol. Toxicol.* 131–156.

64 Schwinn, M.K., Machleidt, T., Zimmerman, K. et al. (2018). CRISPR-mediated tagging of endogenous proteins with a luminescent peptide. *ACS Chem. Biol.* 13 (2): 467–474.

65 Derveaux, S., Vandesompele, J., and Hellemans, J. (2010). How to do successful gene expression analysis using real-time PCR. *Methods* 50 (4): 227–230.

66 Horvath, P., Aulner, N., Bickle, M. et al. (2016). Screening out irrelevant cell-based models of disease. *Nat. Rev. Drug Discov.* 15 (11): 751–769.

67 González-González, E., Camacho-Sandoval, R., Jiménez-Uribe, A. et al. (2019). Validation of an ADCC assay using human primary natural killer cells to evaluate biotherapeutic products bearing an fc region. *J. Immunol. Methods* 464: 87–94.

68 Alturki, N.A. (2023). Review of the immune checkpoint inhibitors in the context of cancer treatment. *J. Clin. Med.* 12 (13): 4301.

69 Fenis, A., Demaria, O., Gauthier, L. et al. (2024). New immune cell engagers for cancer immunotherapy. *Nat. Rev. Immunol.* 24 (7): 471–486.

70 Deckers, J., Anbergen, T., Hokke, A.M. et al. (2023). Engineering cytokine therapeutics. *Nat. Rev. Bioeng.* 1 (4): 286–303.

71 Berg, E.L., Hsu, Y.C., and Lee, J.A. (2014). Consideration of the cellular microenvironment: physiologically relevant co-culture systems in drug discovery. *Adv. Drug Deliv. Rev.* 69: 190–204.

72 Lee, J.Y. and Chaudhuri, O. (2021). Modeling the tumor immune microenvironment for drug discovery using 3D culture. *APL Bioeng.* 5 (1): 010903.

73 Loewa, A., Feng, J.J., and Hedtrich, S. (2023). Human disease models in drug development. *Nat. Rev. Bioeng.* 1 (8): 545–559.

74 Engle, S.J. and Puppala, D. (2013). Integrating human pluripotent stem cells into drug development. *Cell Stem Cell* 12 (6): 669–677.

75 Namatame, I., Ishii, K., Shin, T. et al. (2023). Screening station, a novel laboratory automation system for physiologically relevant cell-based assays. *SLAS Technol.* 28 (5): 351–360.

76 Bassil, R., Shields, K., Granger, K. et al. (2021). Improved modeling of human AD with an automated culturing platform for iPSC neurons, astrocytes and microglia. *Nat. Commun.* 12 (1): 5220.

77 Herpers, B., Eppink, B., James, M.I. et al. (2022). Functional patient-derived organoid screenings identify MCLA-158 as a therapeutic EGFR × LGR5 bispecific antibody with efficacy in epithelial tumors. *Nat. Can.* 3 (4): 418–436.

78 Kim, H.J., Li, H., Collins, J.J., and Ingber, D.E. (2016). Contributions of microbiome and mechanical deformation to intestinal bacterial overgrowth and inflammation in a human gut-on-a-chip. *Proc. Natl. Acad. Sci.* 113 (1): E7–E15.

79 Benam, K.H., Villenave, R., Lucchesi, C. et al. (2016). Small airway-on-a-chip enables analysis of human lung inflammation and drug responses in vitro. *Nat. Methods* 13 (2): 151–157.

80 Almeida, C.T.K. (2023). Authentication of human and mouse cell lines by short tandem repeat (STR) DNA genotype analysis abstract introduction. In: *Authentication of Human and Mouse Cell Lines by Short Tandem Repeat (STR) DNA Genotype Analysis [Internet]* (ed. S. Markossian, A. Grossman, M. Arkin, et al.). Bethesda (MD): Eli Lilly & Company and the National Center for Advancing Translational Sciences Available from: https://www.ncbi.nlm.nih.gov/books/NBK144066.

81 Onyeogaziri, F.C. and Papaneophytou, C. (2019). A general guide for the optimization of enzyme assay conditions using the design of experiments approach. *SLAS Discov.* 24 (5): 587–596.

82 Raniga, K., Stebbeds, W., Shivalingam, A. et al. (2023). Design of experiments for the automated development of a multicellular cardiac model for high-throughput screening. *SLAS Discov.* 28 (8): 410–417.

83 Zhang, J.H., Chung, T.D.Y., and Oldenburg, K.R. (1999). A simple statistical parameter for use in evaluation and

validation of high throughput screening assays. *SLAS Discov.* 4 (2): 67–73.

84 Eastwood, B.J., Chesterfield, A.K., Wolff, M.C., and Felder, C.C. (2013). *Pharmaceutical Sciences Encyclopedia*, 1–22. Wiley.

85 Copeland, R.A. (2013). *Evaluation of Enzyme Inhibitors in Drug Discovery: A Guide for Medicinal Chemists and Pharmacologists*. Wiley.

86 Copeland, R.A. (2023). *Enzymes: A Practical Introduction to Structure, Mechanism, and Data Analysis*, 3e. Wiley.

87 Ditto, N.T. and Brooks, B.D. (2016). The emerging role of biosensor-based epitope binning and mapping in antibody-based drug discovery. *Expert Opin. Drug Discov.* 11 (10): 925–937.

88 Faya, P., Borer, M.W., Griffiths, K.L., and Parekh, B.S. (2020). Potency assignment of biotherapeutic reference standards. *J. Pharm. Biomed. Anal.* 191: 113577.

89 Murray, D. and Wigglesworth, M. (2017). High throughput screening methods: evolution and refinement. *Chem. Biol.* 1–15.

90 Iversen, P.W., Beck, B., Chen, Y.F. et al. (2012). HTS assay validation abstract 1. Overview. In: *HTS Assay Validation [Internet]* (ed. S. Markossian, A. Grossman, M. Arkin, et al.). Bethesda (MD): Eli Lilly & Company and the National Center for Advancing Translational Sciences Available from: https://www.ncbi.nlm.nih.gov/books/NBK83783.

91 Chai, S.C., Goktug, A.N., and Chen, T. (2015). Assay validation in high throughput screening - from concept to application. In: *Drug Discovery and Development – from Molecules to Medicine* (ed. O. Vallisuta and S. Olimat). InTech.

92 Riss, T.L., Moravec, R.A., Duellman, S.J., and Niles, A.L. (2021). Treating cells as reagents to design reproducible assays. *SLAS Discov.* 26 (10): 1256–1267.

93 Pandya, K., Ray, C.A., Brunner, L. et al. (2010). Strategies to minimize variability and bias associated with manual pipetting in ligand binding assays to assure data quality of protein therapeutic quantification. *J. Pharm. Biomed. Anal.* 53 (3): 623–630.

94 Alt, N., Zhang, T.Y., Motchnik, P. et al. (2016). Determination of critical quality attributes for monoclonal antibodies using quality by design principles. *Biologicals* 44 (5): 291–305.

95 Alhazmi, H.A. and Albratty, M. (2023). Analytical techniques for the characterization and quantification of monoclonal antibodies. *Pharmaceuticals* 16 (2): 291.

96 Wilkinson, M.D., Dumontier, M., IJ, A. et al. (2016). The FAIR guiding principles for scientific data management and stewardship. *Sci. Data.* 3 (1): 160018.

11

Immunogenicity Profiling of Biotherapeutics (Wiley Book – Bioanalytical Characterization of New Therapeutic Modalities)

Qui T. Phung[1,*], *M. Violet Lee*[2], *Ola Saad*[2], *Sivan Cohen*[2], *Yinyin Li*[3], *Mercedesz Balasz*[3], *Jieming Chen*[4], *Zhaojun Yin*[2], *Zhenru Zhou*[1], *and Jennie R. Lill*[1,*]

[1]*Proteomic and Genomic Technologies, Genentech, South San Francisco, CA, USA*
[2]*Bioanalytical Sciences, Genentech, Inc., South San Francisco, CA, USA*
[3]*Biochemical and Cellular Pharmacology, Genentech, South San Francisco, CA, USA*
[4]*gRED Computational Sciences, Genentech, South San Francisco, CA, USA*

Introduction

Recombinant proteins have been successfully administered as therapeutics since 1982 when recombinant human (rh) insulin was initially approved for human use in type I diabetes [1]. Several different classes of recombinant therapeutic proteins are now approved for use [2–4], including enzyme replacement therapies, growth factors, cytokines, fusion proteins, clotting factors, monoclonal antibodies, antibody fragments, and fusion proteins. Over the last 25 years, antibody therapeutic development in particular has moved at a rapid pace with formats becoming more sophisticated, and many now being modified with other moieties such as chemotherapeutic agents. Currently, more than 180 antibodies are approved in one or more countries [5, 6].

Biopharmaceuticals such as therapeutic monoclonal antibodies (mAbs) are often complex with a high degree of structural organization. In efforts to reduce the risk of immunogenicity, humanized or fully human antibodies are often preferred over chimeric or germ-line mouse-derived molecules [7, 8]. These mAbs are often subject to post-translational modifications (PTMs), which play an important role in their functionality. However, these PTMs, for instance, glycosylation in the case of antibodies, can contribute to micro-heterogeneity within the therapeutic protein [9, 10] making its bioanalytical analysis complex, and downstream *in vivo* effects less predictable. Furthermore, they can undergo biotransformations such as oxidation and deamidation during the different stages of the drug production lifecycle which could impact therapeutic efficacy, safety, or stability. To reduce biotransformation liabilities, amino acid level changes are sometimes incorporated but with unintentional, unexpected immunogenic risk to the re-engineered protein. In recent decades, advancements in protein engineering have revolutionized the development of bispecific antibodies (bsAb) that can bind to two different antigen moieties [11, 12] and antibody–drug conjugates (ADCs) [13–18] as therapeutics to treat diseases such as but not limited to cancer, autoimmune and inflammatory diseases. The bioengineered molecular improvements to the immunoglobulins to yield increased specificity, stability, and more tolerability *in vivo* can sometimes also inadvertently introduce sequence-level immunogenic risks.

A protein therapeutic might contain multiple B and/or T-cell epitopes, which could trigger the formation of a heterogeneous mixture of anti-drug antibodies (ADAs) due to an immune response when administered to a patient. These ADAs can bind to the target protein with varying affinities. In some cases, even fully humanized antibodies might not be completely tolerated by patients, due to the unique properties of the peptide sequences within complementarity-determining regions (CDRs). This could potentially trigger an immune response as these sequences might be perceived as "foreign" by the patient's immune system. In addition, product quality attributes including aggregation, can drive a relatively non-immunogenic drug to induce antibodies even if there is high sequence homology. These product quality attributes can act as adjuvants.

The clinical impact of a polyclonal ADA response is highly variable, not only across time within the same individual but also among different patients. Consequently, this presents a significant challenge for the successful

*Joint corresponding authors.

Characterizing Biotherapeutics: Analytical Methods for Diverse Modalities, First Edition. Edited by Jennie R. Lill and Wendy Sandoval.

application of protein therapeutics. To compound these concerns, biosimilars are increasingly gaining relevance in modern disease intervention and offer a wider variety of therapeutic options to doctors and patients. They are gaining momentum due to their potential to significantly reduce healthcare expenditure due to their lower cost of manufacturing compared to the original biologics. However, their entrance into the market also comes with a range of challenges related to safety profiling and risk assessment. A primary concern is establishing the safety and efficacy of biosimilars and given their increasingly complex molecular structures and the nature of their production processes, slight variations can occur between different batches of the same product and between biosimilars and reference products [19]. Thus, a detailed comparison to assess biosimilarity in terms of safety, purity, and potency is paramount to ensure therapeutic equivalence.

Pharmacovigilance, or the process of collecting, detecting, assessing, monitoring, and preventing adverse effects with pharmaceutical products, plays a crucial role in managing the risks associated with biosimilars. Proper pharmacovigilance can help detect unanticipated adverse reactions, understand risk factors, and contribute to risk management plans.

Immunogenicity, or the ability of a particular substance to provoke an immune response in the body, is another significant aspect of risk assessment. Any biotherapeutic or biosimilars may trigger unwanted immune responses, which could be harmful. Therefore, assessing the immunogenic potential of these drugs remains a crucial step in ensuring their safety.

There are many strategies to de-risk the potential of generating ADAs. Table 11.1 summarizes each of these risks and potential mitigation strategies.

By implementing these strategies, researchers, and pharmaceutical developers aim to create safer, more effective biotherapeutic products that are less likely to trigger unwanted immune responses in patients. Additionally, it is imperative to be cognizant that the product, the patient, and disease-specific factors can all contribute to and influence immunogenicity.

The development of biotherapeutics is a multi-billion dollar business and significant work goes into the design, preclinical, and clinical testing of every protein drug. Therefore where possible, researchers aim to mitigate risk associated with timelines and expenditure by ensuring only the most efficacious, safe, and robust biotherapeutics are brought into the clinic.

In this chapter, we discuss how ADAs can be detrimental both to the safety and efficacy of the drug and deleterious to the health of the patient, as well as the various bioanalytical and computational approaches available to mitigate these risks.

Table 11.1 Strategies to de-risk the immunogenic potential of biotherapeutics.

Strategy	Description
Humanization	Modify proteins derived from nonhuman sources to make them more similar to endogenous human proteins. This includes humanizing antibodies to reduce the presence of foreign epitopes.
De-immunization	Identify T-cell and B-cell epitopes that can potentially elicit an immune response using techniques such as MAPPs data analysis, and then modify these epitopes (e.g. through point mutations) to reduce immunogenicity while retaining functionality.
Glycosylation	Adding or modifying glycan structures on proteins can enhance their stability and reduce immunogenicity by masking epitope recognition. This is however not a very attractive strategy as this adds a lot of CMC complexity in characterizing glycans at new sites. Glycans are themselves problematic, for example, variable domain glycans in cetuximab triggered anaphylaxis in some patients.
Improving protein folding and stability	Properly folded proteins are less likely to expose hydrophobic regions that can be recognized as foreign. Ensuring correct folding and stability may reduce aggregation and denaturation, diminishing the immunogenic potential.
Administration and formulation strategies	Adjusting the route of administration or the formulation to avoid components that can boost the immune response (e.g. avoiding adjuvant-like excipients). ADA rates for subcutaneous administration are commonly similar to IV with some exceptions [20].
Tolerance induction strategies	Techniques such as inducing specific immune tolerance to the therapeutic protein can be explored, using methods such as oral tolerance or co-administration with specific immunosuppressant, for example, methotrexate for anti-TNFs
Regulatory and monitoring strategies	Implementing rigorous quality controls to ensure batch-to-batch consistency and monitoring patients for anti-drug antibodies (ADAs) to understand and manage immunogenic responses appropriately.

The Anti-drug Antibody Response

ADAs are produced when the immune system identifies a biotherapeutic as an external threat and thus mounts an immune reaction against it (Figure 11.1). This response is part of the adaptive immune reaction, but it is likely that the innate immune system is also implicated in the initiation of this process. In this context, the exposure to a biotherapeutic leads to an immune response, where antigen-presenting cells (APCs), such as DCs, engulf the biotherapeutic, internalize it, and perform endocytic processing, which is then presented by the major histocompatibility complex (MHC) class II molecule where antigen fragments are subsequently translocated to the cell surface where they can interact with circulating CD4+ T cells [21].

Upon recognition of the peptide–MHC complex by a naive helper T cell, a series of events is triggered that involves the activation of CD4+ helper T cells, secretion of cytokines, and activation of B cells downstream. This chain of T-cell-dependent B-cell activations eventually leads to plasma cell secretion of ADAs as well as the generation of memory B cells.

Because the process of class-switching, affinity maturation, and the generation of long-lived memory B and plasma cells secreting ADA necessitates the assistance of CD4+ T cells, there is significant interest in exploring the mechanisms that initiate the generation of T-cell-dependent antibody responses.

T-cell epitopes are peptide fragments of antigens (such as biotherapeutics) that are processed and presented on the cell surface to T-cell receptors (TCRs) as part of a peptide–MHCII complex. It is important to characterize T-cell epitopes since the generation of a mature immune response relies on the assistance of CD4+ T cells. Due to their linear nature, T-cell epitopes can be sequenced and re-engineered more readily compared to B-cell epitopes, B-cell epitopes are currently difficult to predict computationally. Also, experimentally mapping B-cell epitopes from animal studies has questionable relevance to clinical immunogenicity. For the goal of mitigating immunogenic

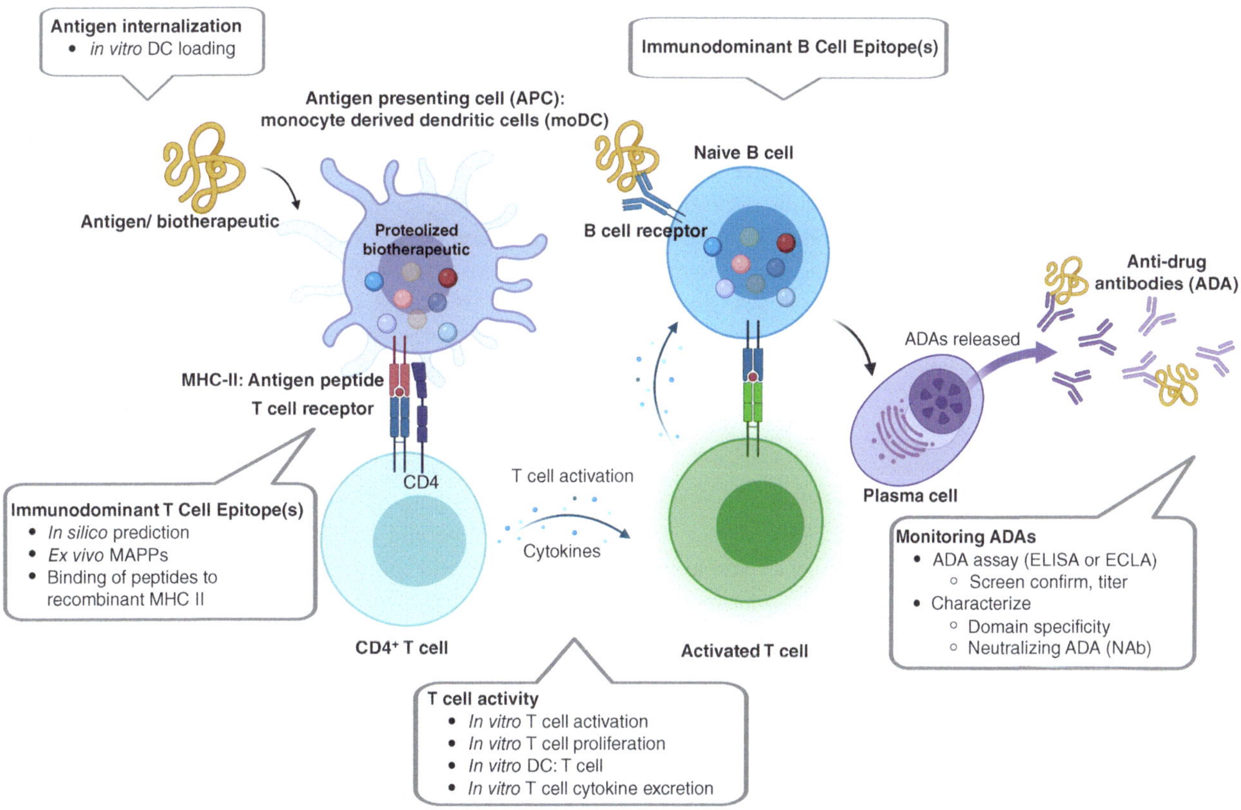

Figure 11.1 Overview of the key steps in the generation of a T-cell-dependent ADA response to a protein therapeutic. Methods used in this study to evaluate the immunogenicity risk of monospecific and bispecific antibodies are highlighted: *in silico* prediction of peptides presented by MHC II, DC internalization, T-cell proliferation, and MAPPs. Source: Tsai et al. [88]/with permission of Taylor & Francis.

risks, T-cell epitopes are particularly appealing for mapping, given their tendency to stay static during an ADA response progression, with typically multiple T-cell epitopes defined per human leukocyte antigen (HLA) type for every biotherapeutic component.

In contrast, a B-cell-mediated immune response to a biotherapeutic can lead to the expansion and diversification of the B-cell population against the biotherapeutic over time, thereby resulting in a more heterogeneous ADA response, which exhibits a broader specificity that can bind different structures on the biotherapeutic. This phenomenon, known as "epitope spreading," is frequently observed in both B-cell and T-cell epitopes [22, 23]. For further information on B-cell epitope mapping and associated *in silico* predictive tools, the study by Potocnakova et al. is recommended [24].

T-cell epitopes, in general, are better understood and suited for analytical characterization and re-engineering than B-cell epitopes, given the vital role T cells play in the formation of durable immune responses. As a result, recent research efforts have been dedicated primarily to designing algorithms and performing *in vitro* assays to analyze T-cell epitopes, leading to a "T-cell-centric" approach adopted by the biopharmaceutical industry for reducing immunogenicity. Nevertheless, it should be noted that B cells can be activated directly without the aid of T cells. For example, recurring viral envelope epitopes or highly structured repetitive self-antigens can cross-link B-cell receptors, thus triggering an Immunoglobulin M (IgM) response independent of T cells [25, 26]. Thus, as more is learned about the ADA response to therapeutic proteins, continued exploration of the role of B-cell epitopes will be important.

Clinical Monitoring of ADAs

The formation of ADAs is routinely monitored in clinical studies involving protein therapeutics. This is essential because all therapeutic proteins can elicit immune responses, which may affect pharmacokinetics, pharmacodynamics, safety, and efficacy of the molecule to various degrees. As monitoring and characterizing immunogenicity is also a regulatory requirement, numerous guidelines from the FDA, European Medicines Agency, World Health Organization, and industry "White Papers" have been written to provide a summary of best practices for evaluating immunogenicity [19, 20].

The extent of monitoring and characterization of immune responses should be determined based on a risk-based assessment. New therapeutic modalities with multiple components or extensive engineering (e.g. fusion proteins, fabylated PEGs, stabilized peptides, ADCs, and AAVs) may pose higher immunogenicity risks for patients, necessitating more frequent and/or detailed ADA testing and characterization. Initial risk-based assessments should be re-evaluated and may need to be modified as clinical data accrues.

Immunogenicity Sample Analysis Strategy

Immunogenicity sample testing typically follows a tiered-based approach for ADA sample analysis [27]. This includes a screening assay, a confirmatory assay, a titering assay (to measure the magnitude of the ADA response), and characterization assay(s) including *in vitro* neutralizing ability of the ADA or domain mapping.

Screening assays for the presence of ADA predominantly use a bridging assay format, such as an enzyme-linked immunosorbent assay (ELISA) or electrochemiluminescence assay (ECLA). This format allows two labeled drug molecules to be bridged by an ADA. The ADA assay is quasi-quantitative, applying a cut point to differentiate ADA-positive samples from negative ones. Typically, a surrogate positive control generated from immunized animals is used to assess the assay's sensitivity.

In some instances, the ADA sample is pre-incubated with both reagents. In others, it is first captured by the drug attached to the surface of a plate and then detected by a second antibody. Sample pre-treatments, such as acid dissociation, can also be incorporated to overcome drug interference in ADA detection. However, these pre-treatments could alter ADA properties, affecting their detection in assays [28].

Confirmation, the second step of the analysis strategy, is conducted on all screen-positive samples using a drug-inhibition test. This test confirms the specificity of the antibody signal and distinguishes between true- and false-positive samples. Only samples that are positive in both the screening and confirmatory assays are reported as ADA positive and considered for further assessment, including titer and characterization.

Each confirmed ADA-positive sample is then assessed for the magnitude (titer) of the ADA response, as well as the *in vitro* neutralizing ability of the ADA (NAb), especially in late-stage clinical trials. ADA-positive samples in a study may undergo further analysis in context-dependent characterization assays.

The decision on which additional characterization assays to use, and when to implement them, depends on the clinical phase, indication, and specifics of the molecule. This should be described in the testing strategy. Further characterizations, though less common, may include isotype determinations, relative binding affinity, complement activating ability of the ADA, or cross-reactivity with endogenous proteins. These are considered on a case-by-case basis for a molecule and potentially indication-specific basis, based on the overall risk-based assessment of the program [27, 29].

ADA Assay Development Challenges and Considerations for New Modalities

ADA assays pose unique challenges in their development, particularly concerning sensitivity, drug tolerance, cross-reactivity, interference, and reference standardization. Complex modalities can introduce additional challenges to ADA assay development and immunogenicity assessment. Peptide-based therapeutics, for instance, can present several challenges due to their size and structures. Some of the challenges include the generation of labeled reagents, where additional conjugation sites of the label (e.g. biotin and digoxigenin) may be difficult to generate given the molecule structure or sequence. When peptide-based therapeutics are conjugated as well, this challenge becomes even more pronounced. Over-conjugation of the peptide reagent can also introduce steric hindrance, potentially affecting ADA binding. Positive control antibodies may also not react strongly to the peptide or may be difficult to generate.

In the cases of complex multicomponent drug modalities, such as Fc-fusion proteins, bispecifics, or other conjugates, the possibility of immunogenicity may be higher as these protein scaffolds may present neo-epitopes in their molecular structure. For instance, in a monoclonal antibody (mAb) therapeutic, ADAs could develop in different regions of the antibody, such as the CDR or the Fc region. In the case of an ADC, however, ADAs could also develop to the linker or the cytotoxic drug components. To adequately assess immunogenicity instead of using multiple ADA screening assays, additional characterization assessments may be conducted for domain-mapping to identify where most of the ADA response is focused. This can be achieved by incorporating competitive-binding experiments into the screening and confirmatory assay strategy [30].

With adeno-associated virus (AAVs) modalities, in addition to assessing ADA and NAb potential to the transgene product, additional assays may be developed to assess preexisting and new ADA responses to the AAV capsid as well. These bioanalytical methods to measure the clinical humoral immune response are similar in immunoassay format to those used for conventional biotherapeutics' immunogenicity assessment. While less common, cellular immunogenicity assessment is also important to consider for cell and gene therapies. The bioanalytical methods and assays used for these modalities include ELISpot, which is considered the standard due to the assay reproducibility, flow cytometry-based assays, and other bioanalytical platforms under evaluation for multiplexing and improved sensitivity [29, 31, 32]. These challenges for new modalities necessitate careful consideration and optimization for each program during ADA assay development to ensure accurate and reliable immunogenicity assessment.

Integrating Immunogenicity Assessments, Harmonization, and Incorporation of Clinically Relevant Information into Drug Product Labeling

Immunogenicity assessment approaches should be product and case specific, considering therapeutic risk. With reporting, terminology in this area can sometimes be incongruous. For example, in the field of clotting factor and enzyme replacement, the term "inhibitor" has conventionally been used to denote a neutralizing ADA (NAb) response. The recent introduction of "harmonized" abbreviations and definitions should facilitate the standardization of immunogenicity documentation [21, 22]. These include the development of ADA status descriptors for study samples, subjects, and populations.

It is also important to note that ADA assays for different therapeutics cannot be directly compared as results depend on individual assay characteristics including drug tolerance and sensitivity in addition to sampling strategies used. Harmonized ADA standards and assay development methods that could potentially enable such future comparisons are being evaluated but are not available or routinely employed [33]. Furthermore, immunogenicity is context dependent, meaning the same protein therapeutic may induce the formation of ADAs at different rates across different patient populations.

Finally, an integrated summary of immunogenicity from registration-directed studies is typically included in a Biological License Application/Common Technical Document filing, with a summarized version also incorporated in the product insert. Recent guidance from health authorities further recommends incorporation of the clinical relevance information of any observed ADA on PK, efficacy, and safety into product labels [34]. This aims to enable health practitioners to more easily access and understand this information, aiding in managing patient care.

The Clinical Impact of an ADA

ADA responses to a therapeutic protein in a treated patient can vary, with some having no clinical impact while others leading to severe or frequent sequelae. Adverse clinical effects are most often associated with high titer indicating sustained ADA responses that can disrupt the pharmacokinetics (PK), efficacy, potency, and safety of the biotherapeutic [35–37].

Unfavorable immune reactions can significantly affect a protein therapeutic program either before or after licensure. These responses can interfere with the effectiveness of therapies in patients. Patients may develop ADA responses that alter the clearance of a biotherapeutic, shortening or extending its PK half-life, and are labeled as clearing or sustaining ADAs, respectively. A survey of immunogenicity data in the United States revealed around 42% of medicinal products reported clearing antibodies development, while about 19% reported sustaining antibodies development [38].

ADAs that accelerate clearance can be functionally neutralizing. NAb can bind to a biotherapeutic's active site, diminishing its effectiveness. However, not all ADAs are neutralizing. Some can cross-react with the endogenously produced protein, resulting in a functionality loss. A prominent example is recombinant hemophilia A and B, with approximately 30% and 5% of patients developing neutralizing antibodies to recombinant factor VIII (FVIII) and FIX, respectively, which can limit the effectiveness of the blood clotting factor infusion [39, 40]. NAbs can also be challenging for enzyme replacement therapies, as observed in treatments for Pompe disease, a lysosomal storage disorder. Although replacement therapy dramatically improves survival rates and quality of life, a subset of these patients develop neutralizing antibodies hindering treatment efficacy and safety [41, 42].

The development of auto-reactive CD4+ T cells varies in healthy individuals compared to patients with autoimmune diseases or other health conditions. While the "Major Histocompatibility Complex-Associated Peptide Proteomics" (MAPPs) technology, discussed later in this chapter, can identify naturally processed and presented MHC II peptides, including a subset of immunodominant T-cell epitopes of endogenous and therapeutic proteins, other elements of the immune system also need to be considered to fully comprehend why some patients develop ADAs, while others do not.

Therapeutic mAbs targeting cell surface receptors or secreted signaling molecules are also prone to eliciting immune responses. Although original rodent-based therapeutics were significantly immunogenic to treated patients, the next generation of therapeutic antibodies was designed with as few nonhuman or substituted amino acid residues as possible to reduce the chance of ADA responses [43, 44]. Even fully human antibody therapeutics may provoke unwanted immune responses in some patients [38].

Assays Employed to Assess Immunogenicity Risk

Here, we discuss the various bioanalytical tools developed and currently utilized by the biopharmaceutical industry to assess the immunogenicity potential of protein therapeutics. Several *in silico*, *in vitro*, and *ex vivo* models are available for immunogenicity "risk ranking" [15, 38]. *In silico* prediction tools use the algorithmic processing power to assess the immunogenic potential at the amino acid sequence level of protein therapeutics. *In vitro* assays such as T-cell proliferation and T-cell activation can help determine relative CD4+ T-cell immunogenic response upon treatment of DCs with a biotherapeutic when compared against positive controls of therapeutics of known immunogenicity or clinical ADA incidence rate. In this similar suite of T-cell assays, DC:T-cell co-culture assays can be employed to evaluate immunomodulators. Dendritic cell internalization assay is commonly used as an indicator of a biotherapeutic's immunogenicity by measuring the degree of uptake of the protein drug by APCs [45]. To identify potential immunogenic T-cell epitopes, a mass spectrometric technique known as "Major Histocompatibility Complex-Associated Peptide Proteomics" or MAPPs is employed. Unlike *in silico* prediction, the unique power of MAPPs is their ability to identify the epitope sequences that are processed and presented on the cell surface as part of the peptide–MHC complex.

In Silico Approaches for Immunogenicity Risk Assessment

The induction of CD4 T-cell immunogenicity can be affected by many factors [46]. Advancements in machine learning have brought about various improvements in direct computational modeling of immunogenicity [47, 48]. However, it still remains a challenging task, primarily due to a dearth of appropriate immunogenicity training data, and identification of features for the models to learn. Consequently, a main focus in computational T-cell epitope prediction has been to develop strategies to examine and predict correlations of immunogenicity (instead of immunogenicity *per se*), so that collectively, we might be able to enrich for true T-cell epitopes.

In this section, we will examine two related, but conceptually different, strategies. One approach searches computationally for peptides that are more likely to be presented by the HLA II molecules on the cell surface. The premise is that these peptides should then be more likely to interact with the TCR and evoke an immunogenic response. Conversely, a second strategy identifies candidates that are less likely to be immunogenic. The idea is that candidates that are more similar to the human proteome, that is, more "human-like," should be more tolerated by a healthy immune system.

In Silico Peptide–MHC Presentation Prediction Algorithms

While peptide presentation is not a singularly sufficient factor for immunogenicity, it is a necessary prerequisite for the downstream generation of CD4 T-cell response [49, 50]. Hence, one way to enrich T-cell epitopes is to identify potential candidates that can be derived from presented peptides. However, historically, most *in silico* algorithms have focused on predicting pMHC candidates from various aspects upstream of antigen presentation on the cell surface, for example, peptide–MHC binding, peptide processing, and peptide–MHC stability.

More recently, the confluence of various technological advancements has enabled better prediction for peptide–MHC II (pMHCII) presentation on the cell surface [51–53]. Chiefly, the massive use of high throughput mass spectrometry (such as in MAPPS and other large compendia) has generated voluminous amounts of eluted MHC ligands, often on the order of 100–1,000s of peptides per HLA allotype [51–53]. This is coupled with cell engineering techniques to create cell lines that harbor only single HLA allotypes [53, 54], further allowing training algorithms to better learn the "rules" of antigen presentation for each allotype. Subsequently, new advancements in machine learning, particularly deep learning approaches, can then leverage these deep and rich ligandomes in order to maximize their predictive prowess [55–58]. In the context of ADA immunogenicity risk assessment, many studies have applied pMHCII presentation prediction algorithms.

For example, the ISPRI workflow has been widely used to screen for pMHCII candidates that are more likely to bind and/or present in various preclinical [59, 60] and clinical studies [61, 62]. The workflow relies heavily on empirically determined position-specific scoring matrices (PSSMs), which are used sequentially to screen for possible candidates [63]. At ISPRI's core, EpiMatrix is a PSSM that was derived from 9mer peptide cores of the pMHCII complexes. Cohen et al. identified and compared binding cores between mono- and bispecific variants of antibody-based therapeutics using NetMHCIIpan3.2 and ISPRI. They found that bispecific antibodies generally appear to have higher epitope counts than the monospecific variant [60].

The perceptron-based NetMHCIIpan series of tools is also highly regarded. The NetMHCIIpan tool, as well as its underlying NNAlign algorithm, has evolved and expanded its applications over time, such as (i) the inclusion of more and diverse ligandome data to augment binding data to improve prediction of presented binders across HLA-DR, DP, and DQ MHC II genes, and (ii) having the ability to deconvolute multiple alleles in their model training, which allowed them to include MAPPS data since each donor in MAPPS can possess up to 12 MHC II alleles. They have further contextualized and adapted their NNalign algorithm (NNAlign_MAC) for the task of ADA immunogenicity risk assessment. They then validated NNAlign_MAC performance using infliximab and rituximab MAPPS and EliSpot experiments and showed that the new algorithm could better capture presented peptides and immunogenic epitopes for DR allotypes than the original NetMHCIIpan [64].

Humanization

Peptides derived from human native sequences are most likely to either fail to bind to MHC II or evade recognition by reactive T cells, making them non-immunogenic [65]. This tolerance phenomenon occurs because T cells that recognize host peptides that strongly bind to MHC II are deleted during T-cell maturation in the thymus or anergized in the peripheral immune system. Thus, self-proteins are considered less likely to provoke an immune response compared to foreign proteins [66]. It is commonly believed that protein therapeutics exhibiting high immunogenicity will contain a high number of T-cell epitopes, while those with low immunogenicity tend to exhibit a low number of such epitopes [67]. Since protein therapeutics containing nonhuman components tend to have more T-cell epitopes compared to fully human proteins, one of the effective tools to reduce immunogenicity risk is to identify and remove foreign sequences, a process known as humanization [7]. Numerous *in silico* tools have been developed to evaluate the human-likeness of protein therapeutics in order to assess their immunogenicity risk. One effective tool for distinguishing between human and nonhuman antibodies is based on pairwise sequence identity between the test sample and a set of reference human sequences. This "humanness score" demonstrated a direct correlation between higher human-like scores and reduced immunogenicity of antibodies. Furthermore, it suggests that the process of humanization can effectively mitigate the risk of immunogenicity, to a similar degree as fully human antibodies [68]. The recent growth of publicly available antibody-specific sequence, structure, and experimental databases such as TABS and Thera-SAbDab has allowed for better development of bioinformatics tools to facilitate the analysis and design of antibodies and evaluate their human-likeness [69]. Recently, the Observed Antibody Space (OAS) database, which offers NGS sequence data on

B-cell receptors from 80 studies, was used for data mining and the development of machine-learning classifiers that can distinguish between human and nonhuman antibody variable domains [70]. Using this computational humanization tool and analyzing 217 antibodies, Marks *et al.* were able to show that high humanness scores correlated with low immunogenicity and *vice versa* [71].

Methods for antibody humanization based on the comparison of 9-mer segments of the tested antibody with human germline genes such as human string content (HSC) may better identify foreign peptides at the level of potential MHC/T-cell epitopes [65]. Based on this approach, Prihoda et al. developed the OASis model, which compares and scores the human prevalence of any 9-mer peptide in a tested antibody. These scores were shown to correlate with the level of clinical immunogenicity [72]. Interestingly, using NetMHCIIpan 3.2 alongside OASis did not result in a better correlation of the model with clinical immunogenicity compared to using OASis alone. More recently, we demonstrated an immunogenicity risk-scoring method based on both peptide presentation prediction and self-peptides removal. Employing this approach, we calculated the total potential T-cell epitopes predicted by NetMHCIIpan-4.0 and Graph-pMHC, and then removed peptides that scored high with OASis. Our findings revealed a superior performance of this model in correlating with clinical immunogenicity compared to OASis alone when testing a dataset containing 109 monoclonal human and humanized antibodies [73], suggesting that integrating a human-like score into peptide presentation prediction algorithms could significantly enhance their ability to predict the immunogenicity risk of protein therapeutics.

Challenges and Additional Thoughts

In silico algorithms, if and when performed, can help to identify potential immunogenic epitopes, and/or more definitive non-immunogenic ones. Such results can aid in optimizing antibody sequences in biotherapeutics, thereby reducing the need to run more expensive and time-consuming cell cultures or animal experiments. However, there are some caveats in these *in silico* approaches in the context of ADA immunogenicity risk assessment that may limit some of their utility, and at the same time, leave room for more improvements to come.

Data

A fundamental tenet in machine learning is the infamous computer science concept of "Garbage in, garbage out." Machine learning performance is closely tied to the nature and quality of the training, test, and validation data. Each model is only as good as the data that is used to train it. Additionally, how it is tested and validated needs to reflect the reality of its application to gauge performance in a way that is relevant. Hence, it is important to employ datasets that best reflect the application scenario. Concurrently, it is also imperative to understand its limitations. For example, most pMHCII algorithms previously focused on predictions from HLA-DR allotypes [74, 75], because most of the public databases contain primarily HLA-DR data. Large single-allele DP and DQ ligandome datasets were only available in very recent years [51, 53]. The advent of better *in silico* approaches will be highly impacted by the development of high throughput technologies that generate data in high volume and diversity for more effective training of the models.

Context

Most peptides bound to MHC II molecules do not trigger TCR-mediated binding and/or recognition. As such, prediction of pMHCII binding and/or presentation serves as an inadequate proxy for actual immunogenicity, with possibly a substantial number of false positives and negatives. A possible next step might be to better develop tools to predict the downstream interaction between pMHC-II complexes and TCRs. Various computational tools have been developed [76]. However, these tools are heavily limited by the quality and availability of paired TCR sequencing data (TCR α and β) and the limited set of peptides included in the training data. However, the impact of such tools on ADA immunogenicity risk assessment is largely unknown at this point. The development of technologies for high-throughput paired sequencing of TCRs with known peptide–MHC-II targets may help to introduce more improvements, and subsequent examination of relevance.

Algorithm

Each computational approach is imperfect, but some will be more useful when employed appropriately. Fundamentally, this can be achieved by understanding their pros and cons, especially in the context of their applications. For example, deep learning approaches can usually obtain better performance than traditional machine learning methodologies because they can often grasp nonlinear relationships within the training data. However, they usually require large amounts of data to uncover patterns, and typically have less interpretability of the results. For example, PSSMs are used in the ISPRI workflow. They capitalize on the information offered by the contributions of each individual amino acid at each position independently. They are highly interpretable and easy to compute. However, they usually cannot capture complex dependencies and relationships *between* the positions. Approaches for assessing humanness are usually very intuitive and highly interpretable. On the other hand, it might be hard to understand the factors that contribute to the predictive power of deep learning approaches, even though the latter might typically have superior performance.

B-Cell Epitopes

ADAs are B-cell-mediated responses that are typically triggered by T-cell responses. A big challenge is that B-cell epitopes could be linear or conformational. There are multiple tools to predict linear B-cell epitopes. However, since more than 90% of B-cell epitopes are believed to be conformational, the prediction ability of those algorithms was shown to be poor [77, 78]. Conformational B-cell epitopes include residues that may not be close to each other in the primary linear sequence but close together because of protein folding, polypeptide folding in the three-dimensional structure. *In silico* prediction of conformational B-cell epitopes based on an algorithm trained on X-ray structures of antibody–antigen complexes have been shown to perform better than algorithms trained solely on linear sequences. Yet, the overall performance of *in silico* B-cell epitope prediction is generally low and will have to be improved for the prospective design of protein therapeutics with low B-cell epitope counts [79, 80].

Finally, multiple factors may influence the immunogenicity potential of protein therapeutics, such as the drug's formulation and stability, dosage, and the patient's genetic makeup and immune status. Product quality attributes such as aggregation, glycosylation, host cell proteins (HCP), and excipient formulation that can change the protein therapeutic structure or induce biotransformations could act as an adjuvant to increase the immune response [81]. Some of these can be picked up by *in vitro* testing such as the DC internalization assay. Many computational methods have been developed to examine various aspects and causes of immunogenicity, such as homology modeling [82], and protein aggregation [83, 84]. Perhaps an *in silico* approach that can unify many of these features can provide a more comprehensive computational assessment of ADA immunogenicity. To that end, further benchmarking of multiple tools with the appropriate "ground truth" may be needed to more clearly assess relevance, performance, and impact of the tools and features on ADA immunogenicity.

CD4 immunogenicity is highly complex and heterogeneous and implicates many aspects of human immunology. Improvements are continually being made in computational approaches for the prediction of CD4 T-cell immunogenicity and its correlates [56, 85], since they have a variety of applications beyond ADA immunogenicity risk assessment, such as neoepitope discovery and vaccine development. Hence, it may be beneficial to the community at large to rally efforts in examining, more systematically, the performance and utility of cross-disciplinary but relevant *in silico* approaches in the specific application of ADA immunogenicity assessment.

T-Cell and Other Cellular Assays

Preclinical *in vitro* assays are developed to determine the likelihood of a biotherapeutic triggering CD4+ T-cell responses, which serves as an indicator for immunogenicity. These methods typically investigate the impact of biotherapeutics on T-cell activation, proliferation, and cytokine secretion from cultured peripheral blood mononuclear cells (PBMCs) from healthy donors. The suite of assays includes T-cell activation, T-cell proliferation, and DC:T-cell co-culture assays, particularly for immunomodulators.

T-Cell Proliferation Assay

T helper cells are instrumental in triggering the immunogenic response to biotherapeutics, making their proliferation a potentially valuable biomarker for *in vitro* immunogenicity assays targeting monoclonal antibody (mAb) pharmaceuticals. To assess the immunogenic potential of biotherapeutics, multiple T-cell proliferation assays have been developed. These assays provide crucial insights into how biotherapeutics may elicit immune responses, facilitating the development and selection of therapeutic agents with reduced immunogenic risks. An *in vitro* enriched CD4+ T cells: PBMC assay was shown to evaluate several critical parameters that predict the immunogenicity potential of biotherapeutics [86]. This assay assessed the frequency of responding donors, the magnitude of their response, and the proliferation of antigen-specific T cells, which was quantified by measuring the incorporation of ^{3}[H]-thymidine. Subsequently, another research group refined the assessment approach by developing a 7-day PBMC-based CD4+ T-cell proliferation assay as measured with BrdU-positive percentage by flow cytometry. This assay setting includes the depletion of CD8+ T cells and CD25+ T cells [87], which was intended to enhance the interactions between APCs and CD4+ T cells, as well as to amplify T-cell responses specifically to factor VIII biotherapeutics, to achieve higher sensitivity in detecting immunogenic responses. In this assay, a stimulation index (SI) was calculated as the frequency of proliferating CD4+ T cell in each mAb-treated group/proliferating CD4+ T cell in the untreated group. Samples from 218 donors treated with etanercept with low clinic ADA incidence were measured to determine a positive threshold level. Data showed

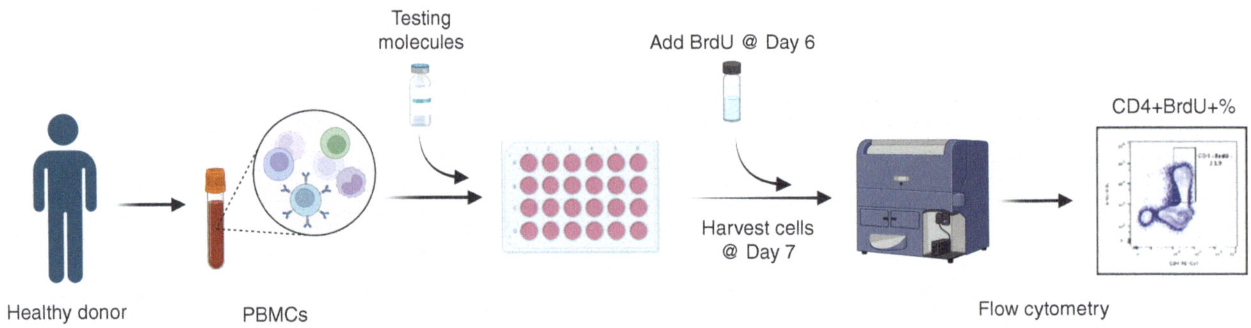

Figure 11.2 PBMCs from healthy donors were stimulated with testing molecules. After culturing for 6 days, cells were then treated with BrdU reagent for another day, and harvested on day 7. CD4+ T-cell proliferation was monitored with BrdU percentage by flow cytometry. Source: Adapted from BioRender.com.

that a T-cell proliferation assay involving 40 donors could provide statistically significant differences when comparing mAbs with low versus high immunogenicity. In addition, we have utilized a very similar T-cell proliferation assay by monitoring BrdU-based CD4+ T-cell proliferation from 72 healthy donors with flow cytometry (Figure 11.2), together with other *in vitro* assays, to assess the immunogenicity potential of knobs-into-holes bispecific IgG1 antibodies [88]. After comparison with low and high ADA benchmarks, there is low preclinical immunogenicity risk from using KIH mutations and low to intermediate risk for Fab charge-pair mutations. Onartuzumab, the clinical stage one-armed KIH antibody, exhibited a low T-cell proliferation SI, which was not significantly different from the Avastin low ADA benchmark but as compared to the positive controls KLH, Boco, and ATR107 (Figure 11.3).

Overall, the T-cell proliferation assay serves as a valuable tool for screening and ranking biotherapeutic candidates, though estimating clinical ADA frequency accurately still poses a challenge. Some clinical parameters such as the drug's mechanism of action (MOA), patient population, dosing frequency, strength, and route can influence immunogenicity. For example, in a patient population, autoimmune disease patients are much more likely to raise an immune response against therapeutic protein than oncology patients, especially if they are immunosuppressed.

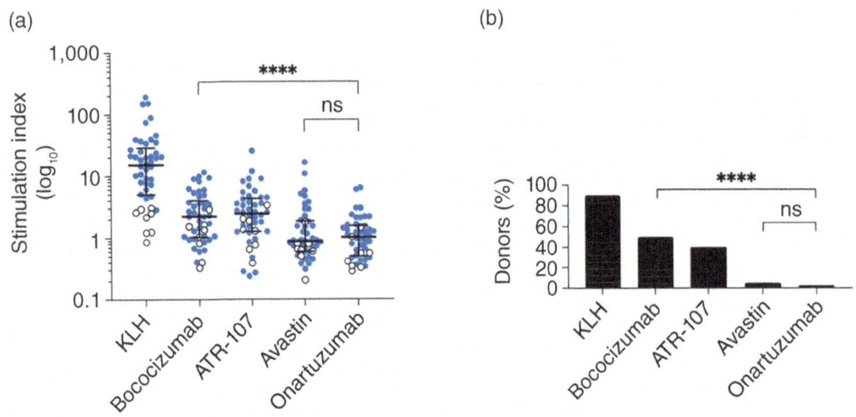

Figure 11.3 *In vitro* BrdU-based T-cell proliferation assay for immunogenicity risk assessment. Human donor-derived PBMCs ($n = 72$) were co-eluted with clinical-grade onartuzumab. (a) Stimulation index (SI) is defined here as the ratio of the percentage of treated $BrdU^+CD4^+CD3^+$ T cells to untreated $BrdU^+CD4^+CD3^+$ T cells. Blue and open circles represent donors that were positive or negative, respectively, in response to KLH, and all donors were used for the following calculations. (b) The median and interquartile range are overlaid on the donor data points (black bars). The percentage of positive donors out of total number of donors for each sample is shown. A positive response is defined here by using a cut-point value corresponding to the 95% quantile of the SI measurements of Avastin (reference antibody with low incidence of clinical ADAs). Donors with a SI greater than this cut-point are considered positive donors. The *P*-values (Fisher's exact test) are shown: ns: not significant and ****: $P \leq 0.0001$. Source: Tsai et al. [88]/with permission of Taylor & Francis.

T-Cell Cytokine Secretion Assay

CD4+ T-cell cytokine secretion assays are essential for evaluating the immunogenicity risk of biotherapeutics. These assays provide insights into how T cells respond to a therapeutic agent by producing specific cytokines. The early phases of T-cell responses can generally be outlined in three stages [89, 90]. Initially, T cells engage with APCs, which prompts the production of cytokines such as interleukin (IL)-2 and interferon (IFN)-γ. This crucial interaction precedes the third phase, characterized by rapid T-cell proliferation. Given this sequence of T-cell responses, assessing cytokine release – happening before cell proliferation – can greatly reduce culture times, streamlining the immunogenicity assessment process. Importantly, IL-2 and IFN-γ, secreted by T cells in response to therapeutic antibodies, are critical endpoints that can be measured using ELISpot and ELISA in immunogenicity assessment. PBMC from up to 50 healthy donors are usually monitored at the T-cell proliferation phase with repeated antigen stimulation, which also requires a long culture time [91]. A recent study showed a rapid *in vitro* immunogenicity risk assessment tool for biotherapeutics has been developed by detecting IL-2 secretion from CD4+ T cell as early as day 3 after stimulation [92]. CD8-CD25- T cells were seeded at the 24-well plate and stimulated with tested biotherapeutic for 3 days. IL-2 secreting CD4+ T cells were then measured by flow cytometry. Seven therapeutic antibodies, each with a known incidence of clinically detected ADAs, were tested to assess the frequency of IL-2 secreting CD4+ T cells. The results showed that the frequency of these cells was consistent with the clinical immunogenicity profiles of the antibodies. One major advantage of this assay is that it can be completed within three days, significantly enhancing throughput by allowing multiple samples to be measured in a shorter cycle than previously possible. This rapid turn-around time facilitates the ranking and selection of therapeutic antibody candidates, aiming to identify those with low immunogenic potential for future clinical use.

T-Cell Activation Assay

Several cell-based assays have been developed to examine T-cell responses as predictors of clinical immunogenicity [93]. One tool to assess the immunogenicity risk of biotherapeutics *in vitro* is to evaluate changes in T-cell activation markers. A previous study examining 14 biotherapeutics with varying clinical immunogenicity found a correlation between immunogenicity and the ability to elicit CD4+ T-cell responses. This was measured by the proportion of CD4+ T cells positive for CD134 or CD137 using flow cytometry following a two-day stimulation of PBMCs. Biotherapeutics with high rates of ADA exhibited higher proportions of CD4+ T cells positive for CD134, CD137, or both compared to those with low clinical immunogenicity or unstimulated cells. These markers may be useful in preclinical screening to identify biotherapeutics with a high probability of leading to higher ADA in the clinic [94].

In comparison to the T-cell proliferation assay, this CD134/CD137 T-cell activation assay requires a short duration of cell culture time and allows for rapid evaluation of multiple biotherapeutics. Moreover, additional markers can be incorporated into the flow cytometry panel to further identify and characterize T-cell populations, potentially enhancing the assay's specificity.

This assay utilized a cohort of PBMCs from 40 healthy donors to ensure representation of the HLA-II diversity of the worldwide population. Additionally, the PBMCs used were isolated from healthy donors and are considered naïve, having never been exposed to the tested biotherapeutic. It is important to note that overall immune system function, response, and other patient-specific factors may significantly influence ADA development. Therefore, PBMCs from healthy donors may exhibit different responses to a tested biotherapeutic compared to those obtained from patients [95]. Another limitation arises in evaluating molecules with mechanism of action (MOA) that can interfere with assay outcomes, such as TNFα blockers. Integration of this assay with different outputs to assess immunogenicity risk, such as *in silico* analysis and DC internalization, may be necessary to provide a more comprehensive interpretation of the results [82].

DC:T-Cell Co-culture Assay

While T-cell activation and proliferation assays are valuable for assessing preclinical immunogenicity risk for many biotherapeutics, they present challenges when used to evaluate immunomodulators. These substances can interfere with the assays, complicating the accurate assessment of immunogenic responses. This situation underscores the need for specifically tailored techniques that can effectively address the unique properties of immunomodulators. One potential solution could be the DC:T-cell co-culture assay, which may offer a more reliable method to overcome these difficulties. A recent study described and characterized a dendritic cell and CD4+ T-cell co-culture assay [96]. This assay involves the co-culture of monocyte-derived dendritic cells (moDCs) and autologous CD4+ T cells, with the inclusion of a re-stimulation step intended to boost the sensitivity of the assay (Figure 11.4). Initially, immature DCs were loaded with each test protein along with a DC maturation cocktail. The activation of the DCs was verified by assessing the expression of key surface markers at both

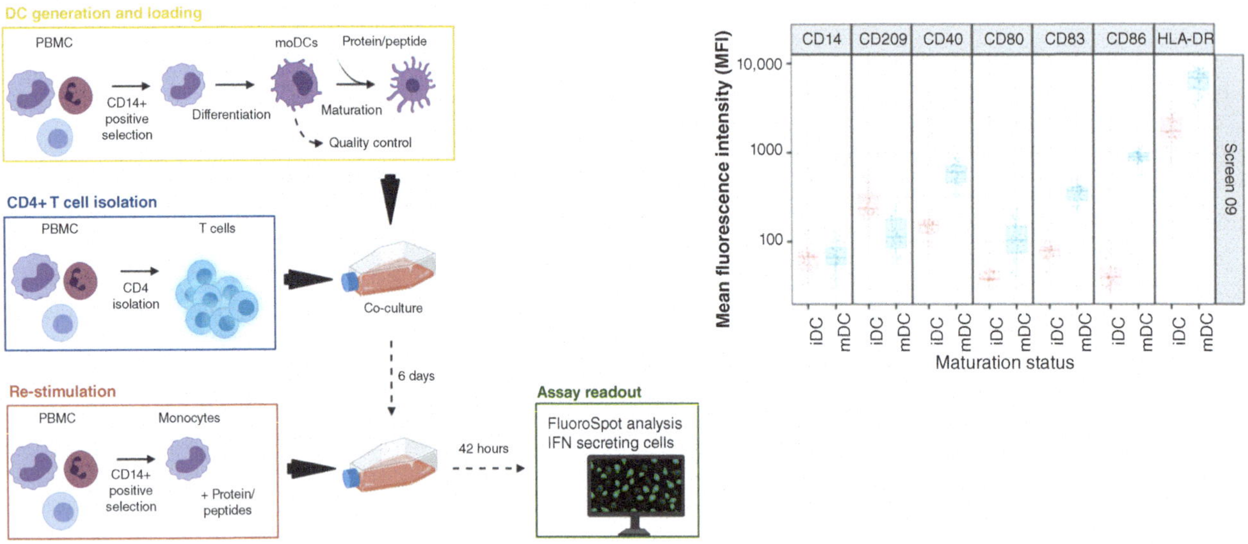

Figure 11.4 The DC:CD4+ T-cell restimulation assay. (a) Experimental setup of the DC:CD4+ re-stimulation assay. The assay starts with the isolation of monocytes from healthy donor PBMCs, followed by the loading of the protein of interest and maturation of the monocyte-derived dendritic cells (moDCs). Autologous CD4+ T cells are isolated and co-cultured with the loaded moDCs. After an incubation of 6 days, freshly isolated monocytes are challenged with the same protein and added to the co-culture for an additional 42 hours before analyzing the production of IFN-γ by FluoroSpot. (b) DCs were characterized by the expression of the following cell surface markers: CD11c, CD14, CD80, CD83, CD86, CD209, and HLA-DR before and after DC activation by addition of TNFα and IL-1β to ensure good cell fitness. Source: Pharmaceutics 2022/MDPI/CC BY 4.0.

the immature and mature stages via flow cytometry. Before the interaction with T cells, mature dendritic cells (mDCs) undergo a thorough washing procedure to remove any residual test proteins, thus minimizing potential interference with the CD4+ T cells. After a 6-day co-culture period, autologous monocytes were isolated from PBMCs and loaded with the initially selected protein. Following another thorough washing to remove any excess protein, these monocytes were then co-cultured with CD4+ T cells for a restimulation phase. The presence of IFN-γ secreting cells during this stage was monitored using FluoroSpot analysis to assess the specific immune response elicited by the interaction between the monocytes and T cells. The authors evaluated the predictive capability of this assay by comparing its results with the clinical ADA rate for 24 marketed antibodies. The outcome showed an accuracy rate of 83%, indicating substantial alignment between the assay predictions and observed clinical immunogenicity.

The implementation of this assay, especially when applied to immunomodulators, as part of a comprehensive risk assessment, has the potential to significantly enhance the robustness and informativeness of immunogenicity risk assessments in the early stages of drug development. This more detailed analysis can aid in better identifying and mitigating potential immunogenic risks before progressing to later stages of development and clinical trials.

DC Internalization Assay

Antigen internalization and processing by professional APCs, mainly DCs, are essential to initiate an immune response and generate ADAs against biotherapeutics. In general, the DCs have two distinct phenotypes, immature DCs and mature DCs. The immature DCs are highly endocytic and very efficient in antigen capture and accumulation, while the mature DCs are poor at antigen uptake but good at processing and presenting antigen to T cells. The immature DCs use several different mechanisms to uptake antigens, including macropinocytosis, phagocytosis, receptor-mediated endocytosis, and autophagy. The internalized antigen is delivered to a multivesicular late endosomal–lysosomal antigen-processing compartment, also known as the MHC class II compartment (MIIC), where the antigen is degraded into short peptides and loaded onto MHC-II molecules. A higher internalization might contribute to an increased number of peptide–MHC-II complexes presented on the surface of DCs and stronger activation of cognate CD4T cells. Previous studies have revealed that the amount of internalization by DC is correlated with the immunogenicity risk of test antibodies [45, 97]. Therefore, the DC internalization assay has been recognized as an important tool complementary to many other *in vitro* immunogenicity risk assessment assays. A combined analysis of DC internalization and *in silico* prediction data is often performed to corroborate results from the MAPPs and/or T-cell assay. In addition, the DC internalization assay could be used for T-cell engagers and immunomodulators such as TNF antagonist or checkpoint inhibitors [98, 99], which might interfere with the readout of T-cell activation or proliferation assays due to their mechanism of actions.

The factors that impact the internalization of biotherapeutics by DCs include physicochemical properties, target receptor expression level, and turnover rates. The positive charge patch [100] or aggregates [101] have been shown to enhance the DC internalization probably due to the interaction with the negatively charged cell membrane or Fcγ receptors, respectively. The immature DCs are typically used to assess the internalization of biotherapeutics due to their efficiency in antigen capture. It should be noted that the target receptor expression level could vary significantly after DC maturation. Although the mature DCs have much-reduced macropinocytosis, the receptor-mediated endocytosis is still maintained. For example, immature DCs treated with LPS showed much higher internalization of adalimumab compared with immature DCs due to the increased expression of membrane-bound TNFα [99]. Evidence showed increased DC maturation in rheumatoid arthritis (RA) patients, especially in the inflamed RA synovial tissues [102]. In addition to membrane-bound targets, the soluble multimeric targets such as TNFα or IL17 might also contribute to increased internalization by forming immune complexes [103].

The human monocyte-derived DCs, also known as inflammatory DCs, are widely used to evaluate the internalization of biotherapeutics as they can be easily obtained in large quantities from human peripheral blood mononuclear cells (PBMC) with a well-established protocol [104–106]. Upon isolation and generation of DCs, the internalization of biotherapeutics can be measured using two different techniques. In the first approach, Melendez *et al.* used a fluorescent secondary antibody specific to the Fab or Fc domain of antibody therapeutics (unlabeled approach) to study 11 antibody therapeutics with a known ADA incidence rate [45]. Here, both surface staining and total staining are performed for each sample. The internalization index can be determined by subtracting the external staining from the total staining. In a panel of anti-PCSK9 antibodies with varying degrees of clinical immunogenicity, the authors observed a correlation between greater measurement of DC internalization for test biotherapeutics with a higher incidence rate of ADA (bococizumab, 48% ADA in the clinic) as compared to those with lower ADA incidence [107]. The advantage of this approach is that no modification is required and therefore no risk of changing the physicochemical properties of the test molecules. Since only the intact molecule can be detected, the timing of measuring internalization is important as this technique does not measure the accumulation of internalized molecules. The other approach uses a pH-sensitive dye or FRET technique to label the test molecules, which provide increased fluorescence when the internalized molecules are delivered to the low-pH endosomal/lysosomal compartments or degraded in the lysosome (labeled approach) [97, 107]. This method measures the accumulation of internalized molecules as the dyes used in the labeling typically do not leak from cells. It is also more relevant to the antigen processing compared with the unlabeled approach. However, the cost of labeling, either directly or indirectly, is the possibility of changing the physicochemical properties of test molecules, by introducing additional charges or increasing aggregations. Therefore, the internalization of biotherapeutics should be evaluated on a case-by-case basis. The suitability of dye labeling, target internalization, level and structure of soluble targets, and disease condition are important considerations. Additionally, a caveat to this assay is the level of stimulation required for the assay may be more robust and result in a higher level of activation than would be experienced by patients.

MHC II Immunopeptidome

In the mid-1980s, several studies showed that short proteolytically processed peptides displayed by MHC class I and II induced an immune response [108, 109]. This could be either via cytotoxic T cells (for MHC I) or through helper T-cell-mediated responses (MHC II). With the development of new mass spectrometric ionization techniques during the late 1980s and early 1990s, the characterization of both synthetic and naturally derived MHC class I and II peptides increased dramatically by mass spectrometry, in addition to conventional Edman degradation. MHC II peptides are more typically involved in the immunogenicity of biotherapeutics and its associated ADA response. MHC class II peptides primarily originate from bacterial proteins or proteins absorbed through endocytosis or from surrounding cellular debris [110]. In the context of biotherapeutics, these might also come from intravenous or subcutaneous administrations. Our understanding of these processes has expanded over the past decades, with breakthroughs by researchers like Grey et al. who began investigating the structural stipulations and biological implications of peptides-MHC interactions [111]. They demonstrated that the peptides-MHC binding is reasonably flexible, although specific structural features of peptides are crucial for their binding to certain MHC alleles. Each

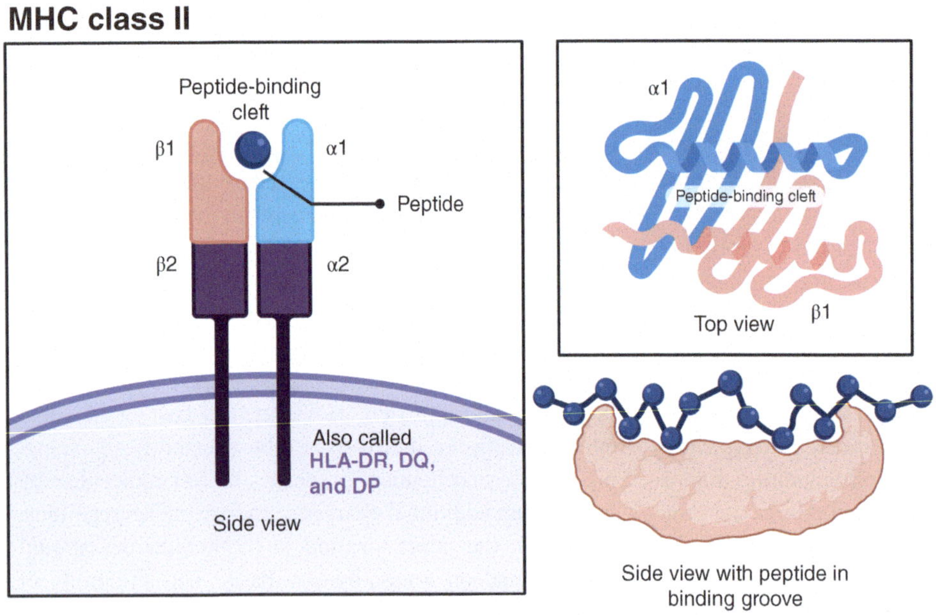

Figure 11.5 Peptide binding properties of MHC II molecules. Left panel shows MHC protein α and β chains. Top right panel shows top view of α1/β1 binding groove. The lower right panel shows a schematic view of the binding site, with four anchor residues. Source: MHC class II, (2024)/Biology Notes Online.

molecule possesses two membrane-proximal immunoglobulin domains which provide structural support to the peptide binding unit which have an average length of 15 residues, and four anchor residues (Figure 11.5). The binding groove is open on both ends of these highly polymorphic MHC II proteins, allowing peptides of varying lengths to be presented [112]. MHC II presented peptides' C-termini are not structurally constrained by their interaction with the binding site, resulting in peptides of variable lengths.

Lippolis et al. conducted a thorough analysis of the heterogeneity of peptides linked with HLA-DR*0401 MHC II peptides using mass spectrometry. Over 700 potential peptides were characterized in a single analysis; revealing that, out of those, 142 peptides consisted of 11 nested sets, each comprising of 3–36 distinct peptides. Peptides that were made up of 12 or fewer residues were found to exist only in very low abundance, despite predictions of them being able to fill the HLA-DR*0401 molecule entirely. On the other hand, the relative abundance of longer species suggested that these are likely to be determined by proteolytic events following MHC binding, which eventually define the final structure of most class II-associated peptides. In this context, this study suggested that C-terminal residues of these peptides are largely influenced by the action of specific peptidases that cleave at specific/preferred amino acids, while the position of amino termini seems to be decided more by the proximity to the class II MHC-binding site.

Our understanding of epitope processing and the features of the resulting peptides across a variety of MHC II alleles has expanded significantly over recent years, thanks to the sequencing of an increasingly large number of MHC II peptides under a wide range of biological conditions [113–117]. This has helped gain a clearer understanding of the biochemical and biophysical processes that govern MHC II presentations, and also understand the role immunodominant epitopes play in varying autoimmune diseases [118, 119] together with numerous allergens [120]. As a result, different therapeutic strategies have been developed to induce tolerance against such epitopes with the aim of preventing patients from generating an immune response to a certain set of MHC II epitopes. The IEDB database serves as a comprehensive resource for this field, as it catalogs thousands of MHC I and II epitopes associated with infectious disease, allergy, transplant, and autoimmunity [55].

In recent years, advancements in cancer immunotherapies have elevated the interest and increased publications describing analytical methods for optimizing the detection of and computational algorithms for the prediction of MHC class I and II peptides. Research interest has expanded beyond how MHC II peptides play a role in disease progression or possible interventions, including the identification of antigens linked to noninfectious or autoimmune-associated illnesses such as cancer. Within the last decade, there has been an increase in interest toward developing techniques for studying the immunogenicity of protein therapeutics in preclinical and clinical settings. In the following section, we discuss an *ex vivo* mass spec-based technique called MHC-associated peptide proteomics (MAPPs) for the detection and identification of T-cell epitopes presented in the peptide–MHC complex.

MAPPs Analysis

The terms "immunopeptidomics" and MHC "ligandomics" have originated over the years in order to describe the features of MHC I and II peptide collections. However, the term MAPPs is uniquely utilized to describe a procedure that involves pulsing an APCs with a biotherapeutic, performing immuno-affinity capture of the resulting peptide–MHC II complexes, and subsequently analyzing these peptides using mass spectrometry. This term was initially introduced by Rohn et al. and quickly gained popularity within the biotechnology and bio-pharmaceutical sectors [121]. Now, it is used by organizations such as the ABIRISK consortium and in documents related to FDA inquiries. Notably, the term MAPPs enables distinguishing between a generic approach of examining all MHC ligands/peptides and a highly specific approach oriented toward a selected subset of MHC II peptides that are derived from endocytosed and proteolytically processed biotherapeutics that are then presented.

The MAPPs procedure adheres to the fundamental outline presented in Figure 11.6. The assay requires a minimum of 2-5X10E6 APCs. Acquiring primary mature APCs like DCs or macrophages is difficult, therefore, researchers generally derive differentiated monocytes from peripheral blood, mostly from healthy donors, to acquire an adequate quantity for the assay. Monocytes from healthy human donor peripheral blood mononuclear cells (PBMCs) are isolated using CD14+ magnetic resin (or a similar method). These monocytes are then incubated for about 5 days with recombinant human granulocyte-macrophage colony-stimulating factor (GM-CSF) and interleukin-4 cytokines (IL-4), promoting differentiation into immature DCs. These DCs are then subjected to 100 µg/mL protein therapeutic and matured using either tumor necrosis factor (TNF) or lipopolysaccharide (LPS) for 24 hours. Despite the commonality of using these maturation agents, their downstream effects are notably different. TNF, a protein, can influence the repertoire of both self-peptides and biotherapeutic peptides presented as potential T-cell epitopes via MHC II. In contrast, LPS, a phospholipid,

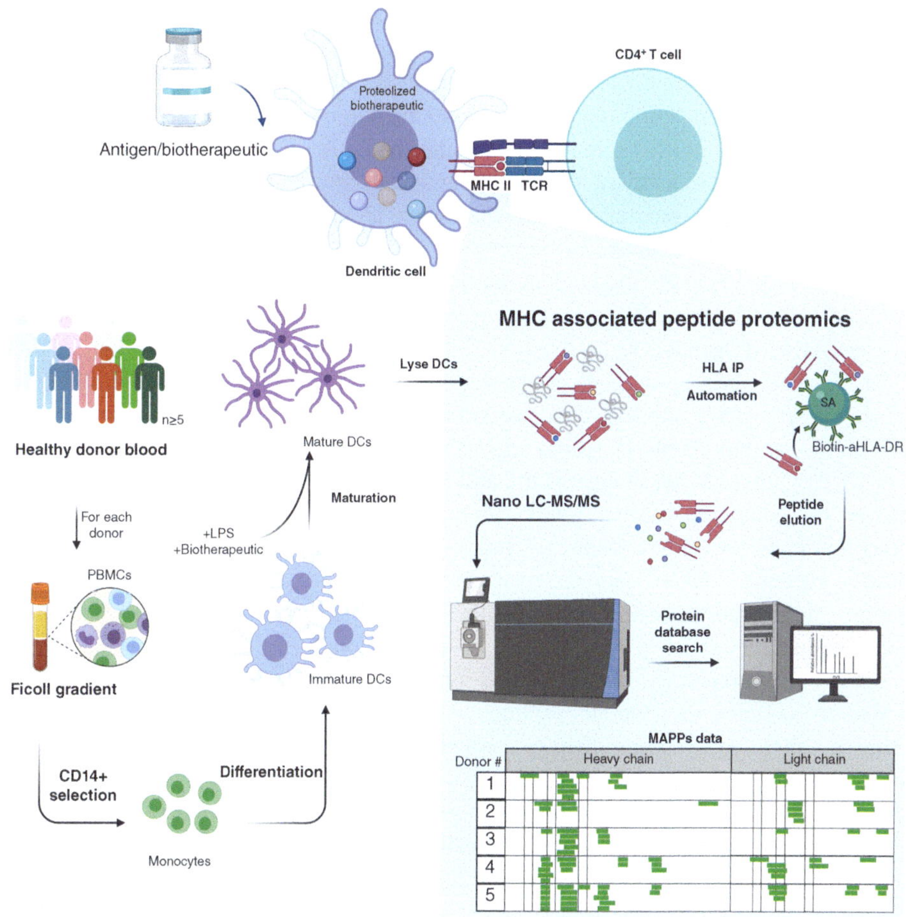

Figure 11.6 The isolation and differentiation of monocytes from PBMCs into immature DCs is accomplished via incubation with GM-CSF and IL-4. DCs are stimulated with protein therapeutic and matured using either TNF or LPS for 24 hours. Subsequent to priming and pulsing with the biotherapeutic, cells are pelleted and rinsed with a physiologically relevant buffer such as PBS. DC cell pellets are lysed and exposed to MHC II HLA specific antibodies. After multiple washes, MHC II peptides are acidically eluted, separated by high-performance liquid chromatography (HPLC), and analyzed via mass spectrometry. The resultant data undergo a subsequent bioinformatics analysis.

will not be directly presented to the immune system via MHC II, though it could trigger an inflammatory reaction via pyroptosis resulting in changes in the genes expressed by the APC. Furthermore, when performing MAPPs on anti-TNF antibodies it is advisable to avoid cell priming with TNF. After the biotherapeutic is pulsed and primed, cells are subsequently centrifuged into a pellet, rinsed with an appropriate buffer like phosphate-buffered saline, and then preserved in a frozen state until future analysis is required.

To profile peptides presented by MHC class II, derived from biotherapeutics, cell pellets are lysed in the presence of a cocktail of protease inhibitors. These lysates are subsequently incubated with MHC II HLA-specific antibodies (i.e. pan anti-HLA-DR, DP, and DQ) that are affixed to a resin or beads, consequently enabling efficient capture of the MHC II-peptide complexes. After incubating with the resin-bound antibodies, the antibody/HLA allele-specific complex undergoes several washing steps to remove any nonspecific interacting peptides, proteins, excipients, or other contaminants. The MHC II peptides are then subjected to mild acid-elution, separated using high-performance liquid chromatography, and analyzed either offline or in-line via mass spectrometry. With sensitivity, scan speed, and mass accuracy vastly improved within the last two decades, mass spectrometric analysis can yield significant data from as low as one million cells, provided the biotherapeutic under study contains potential T-cell epitopes. With technological advancements in mass spectrometric sensitivity, fast scanning speed, and orthogonal gas phase separation; high-confidence identifications of immunopeptides can be achieved with both high-resolution precursor and tandem mass spectral scans [122, 123]. This data can determine the coverage of a wide range of MHC II ligands.

To elucidate the sequence of presented MHC II peptides, tandem mass spectral data are processed using proteomic protein database search algorithms such as PEAKS db [124, 125], MS Fragger [126], and Mascot [127]. These are compared against public protein databases preconditioned for human taxonomy and appended with protein sequences taken from the investigated biotherapeutic. Immunopeptidome characterization is extremely computationally intensive as compared to conventional "bottoms-up" proteomics analysis where enzyme(s) are added to samples to generate peptides with expected cleavage sites, mass spectrometric compatible length, and ionization efficiency [128]. In contrast, MHC peptide identification requires interrogating a greater search space to include (endogenously proteolytically processed) non-enzymatic specificity queries [116, 129]. Validation of the identified immuno-peptides by MAPPs can be achieved by comparing the peptide sequences observed by mass spectrometry to those determined using a complementary tool, as described later. The field of MHC peptide profiling via mass spectrometry has been propelled forward by the precise adjustments made to the mass spectrometric instrumentation to produce optimal spectra and by the development of our understanding of the informatics tools that help decipher peptide sequence information. Additionally, the exceptional sensitivity, selectivity, and scan speed offered by contemporary mass spectrometers are allowing researchers to very efficiently characterize MHC II peptides, even from a minimal number of cells.

MAPPs Automation

Historically, MAPPs methods nonspecifically conjugated capture antibodies, such as a pan anti-HLA specific antibody (DR, DP, or DQ), directly to beads or resin [130–132] through primary amines (NHS). The immunoaffinity procedures were typically manual, resulting in lower throughput and less robustness. Nonspecific conjugation may interfere with optimal MHC capture if, for instance, the binding site on the antibody was blocked or the antibody conjugated in an orientation on the bead or resin that limited the binding of MHC. The automation of MAPPs employs the high-affinity and high-specificity of biotin and streptavidin [133]. MAPPs automation utilizes a streptavidin–biotin spacer between the bead and the capture antibody. This strategy minimizes steric hindrance, thus enabling more efficient capture of HLA complexes. Due to biotin's small size, it is less likely to disturb the function of the molecule to which it is conjugated.

Previous methods were typically developed for research purposes and not required to be as rigorously evaluated for robustness and high throughput to the extent required for drug development. The field also lacked technical reports and standardized protocols that would allow meaningful comparison of results across different laboratories. Recent works by [131, 133] have systematically evaluated and demonstrated improved robustness and well-characterized MAPPs methods for implementation in drug development by incorporating biotin/streptavidin immunoaffinity capture with automation.

Automating MAPPs provides four main advantages over a manual workflow: allows for greater analysis throughput, greater day-to-day and analyst-to-analyst reproducibility, minimizes sample handling thus increasing confidence in the detection and identification of biotransformations, and overall assay robustness that affords confidence in the epitopes identified. In MAPPs assessments where DCs from donors that contain HLA alleles representative of the population [134] are combined with automated MAPPs method, as few as 5 donors, but ideally 10, are needed to be representative of a biotherapeutic's epitope presentation profile [133].

The median length of peptides identified are ~16 amino acids long. The putative biotherapeutic specific T-cell epitope(s) peptide(s) can be visualized plotted in a qualitative (Figure 11.6) and quantitative [135] manner onto a heat map. Across the top represents the residues of the biotherapeutic. In the case of antibody-based therapeutics, the HC and LC sequences along with their respective CDRs are annotated. Each block corresponds to a unique peptide sequence identified. Some peptides will align in a core region and share a consensus sequence but have ragged N- and C-termini; this group of peptides is referred to as a cluster [136]. Higher peptide and cluster presentation trends with the clinical immunogenicity rate of the biotherapeutic. Conversely, molecules with lower immunogenicity present fewer peptides and clusters. Thus, the number of peptides and clusters identified may potentially be an indicative readout of a molecule's risk for immunogenicity. Based on the sequence identification, the potential immunogenic epitope(s) can be tolerized or re-engineered to de-immunize the biotherapeutic molecule.

Pros and Cons of MAPPs

MAPPs exploit the combination of highly selective and sensitive antibodies against various class II alleles and the recent advancements in mass spectrometry. Coupling this with genotyping data and a panel of donors representing the population's major allelic diversity, MAPPs aid in identifying protein therapeutic sequences with potential immunogenicity. Early immunogenic potential assessments during preclinical development can contribute to improved therapeutic success.

However, the MAPPs method has limitations: the peptide repertoire presented by MHC II on *in vitro* generated DCs from monocytes might not match physiologically relevant *in vivo* sourced DCs, macrophages, or other APCs. Additionally, MHC II presentation of a peptide does not guarantee its recognition by a CD4+ T cell or induce an immune response. Immune-tolerized epitopes may also be presented and detected by MAPPs. Several factors such as T-cell clonal selection play a role in differing "self" or "foreign" peptides, and MHC II presentation does not predict immunogenic potential.

Furthermore, profiling MHC II peptides can be challenging due to their physicochemical properties. They are often heterogeneous and because these peptides are created endogenously, not using commercial proteolytic enzymes such as trypsin, they may lack basic residues that aid in ionization or fragmentation during tandem mass spectrometry (MS/MS). Additionally, biopharmaceuticals can also harbor PTMs such as glycosylation, oxidation, and deamidation to name a few. The standard MAPPs workflow can identify native, non-modified T-cell epitopes; however, routinely detecting epitopes harboring PTMs can be challenging. Some labs have shown detection of post-translationally modified MHC II peptide [10, 137, 138] is feasible but depending on the stoichiometry of the modified to unmodified immunopeptidome in the sample, additional method optimization to enrich for low abundance modified MHC presented peptides may be necessary. Another caveat is minimizing sample handling and processing to allow differentiation between endogenous PTMs versus those introduced by the sample processing.

MHC class II complexes are heterodimers formed of alpha and beta chains, each consisting of two domains. These complexes are encoded by highly polymorphic HLA-DR, HLA-DP, and HLA-DQ genes. The HLA-DR alpha chain is highly conserved, whereas the others are more polymorphic [139]. The allelic diversity in the human population due to the polymorphic nature of the HLA-DR, HLA-DP, and HLA-DQ amplifies the complexity of the immunopeptidome. In this era of NGS and genomic sequencing, the number of known MHC-II alleles has grown steadily for over 20 years and reached >10,000 in 2022 [140].

Given that HLA-DR alpha chain is the most highly conserved, as compared to HLA-DP and HLA-DQ, the pan anti-HLA-DR antibodies (e.g. L243 and G46–6) that are used, typically target the alpha chain and work well for efficiently capturing the majority of HLA-DR complexes. However, HLA-DP and HLA-DQ have differing expression levels from HLA-DR [141]. Both HLA-DP and HLA-DQ are believed to have a role in biotherapeutic immunogenicity [142]. While current available pan-MHC II, HLA-DP, and HLA-DR antibodies are lacking in specificity and selectivity, Hartman et al. demonstrate a mixed IP strategy for the additional identification and characterization of HLA-DP and HLA-DQ-specific epitopes [143]. With greater interest in developing a more comprehensive MAPPs method to profile T-cell epitopes presented by all class II alleles, further investigation is warranted into peptide presentation associated with HLA-DP and HLA-DQ in order to enhance our understanding of biotherapeutic immunogenicity.

While MAPPs underrepresents the number of peptides displayed by MHC II, current *in silico* T-cell epitope prediction algorithms tend to over-predict. This is due to limited training datasets, biases in previously cataloged MHC II peptide sequences, and the failure of not all predicted peptides from a protein to be generated during processing and presentation. Despite MAPPs limitations, this tool is still useful for predicting T-cell epitopes that could lead to an ADA response. It is worth noting, however, that these predictions are not a sure guarantee of a peptide not being presented, detected, or being immunogenic, and MAPPs peptides identified from native constant domains of IgG are likely to be tolerized.

False-negative results are a common occurrence and should be factored in when evaluating potential immunogenicity. When MAPPs results conflict with results from other assays, the conflicts should be considered. The use of this tool to mitigate a potential ADA response becomes important in such scenarios.

Data Visualization and Curation

Data processing and visualization tools play an important role in facilitating the interpretation of MAPPs results and showcasing the assay's strength in risk assessment of drug-induced immunogenicity. A robust and reliable data analysis pipeline is needed to present complex mass spectrometric data sets. Previous studies on a panel of marketed monoclonal antibodies demonstrated that molecules with a rather low clinical immunogenicity rate on average showed lower numbers of presented sequence regions and low T-cell response rate in T-cell activation assay, and monoclonal antibodies that showed increased immunogenicity also showed elevated numbers of presented sequence regions and T-cell response rate [136]. This study indicated that the number of clusters, which are defined as one or more peptide length variants occurring within a sequence region that share the same MHC II binding core, is useful information to rank similar biotherapeutic candidates. Therefore, cluster maps of T-cell epitopes observed across donors are determined by counting the number of different or total clusters identified in a donor set for the whole molecule [132, 144].

MAPPs Visualization

Many different computational tools have been developed for the rapid and spatial visualization of cluster maps. For example, PepMapViz, a custom R-package script developed and used by our lab to seamlessly map, identified peptides from cross-software mass spectrometry results to protein sequences. Highlighting distinct domains and regions of interest and accentuating mutations and post-translational modifications, all while enabling comparison across diverse experimental conditions. The process includes peptide filtering to a length range of 11–30 amino acids, peptide mapping to protein sequence, annotating biotherapeutic regions, calculation of peptide intensity and then generating the heatmap spanning across the amino acid sequence of the protein therapeutic colored by the peptide intensity [88]. Another MAPPs data processing pipeline, dataMAPPs, can also be used including quality control, signal normalization, sequence mapping, and data visualization [135]. In addition, dataMAPPs feature in recording peptides' associated peak area for enabling advanced quality control and normalization procedures.

Data Curation

Each assay mentioned throughout this chapter for immunogenicity risk assessment has inherent limitations that may lead to false-positive or -negative predictions [82, 145]. Data from each assay should be considered independently and integrated together for an overall immunogenicity risk assessment. Therefore, building a central hub for both experimental results and metadata related to biotherapeutic immunogenicity is recommended. Still harmonizing and standardizing all the assays remains challenging, and the selection of assays should be driven by the nature of the data acquired in the study and its intended application in assessing immunogenicity risk [145].

Engineering Best Practices Post-immunogenicity Hotspot Data Elucidation

Data from the various immunogenicity de-risking assays mentioned in this chapter alone can provide some but limited insight into the immunogenicity risk of a protein therapeutic; however, by looking at all the data holistically, it could shed more light on where the immunogenic hotspot may occur [88]. By corroborating T-cell proliferation assay results, MAPPs data of similar HLA matching donors, and comparing *in silico* predicted versus experimentally observed T-cell epitopes, it is possible to narrow and identify potentially immunogenic epitopes. The hotspots can be genetically engineered out and replaced with residues that can be checked by *in silico* prediction for immunogenicity risk level. The newly re-designed therapeutic can be subjected to another round of de-risking assays and the process is repeated until a protein candidate is maximally de-immunized. Additionally, these moieties will also need to be screened for affinity, biologic potency, and developability including chemical hotspots, that is, multiparameter optimization. By subjecting preclinical drug candidates through this immunogenicity de-risking strategy, researchers can design biotherapeutics that minimize the presence of potentially immunogenic epitopes before taking them through costly clinical development.

This approach helps in altering or removing specific amino acid sequences in the therapeutic protein to reduce its immunogenicity risk, maintaining therapeutic efficacy while minimizing adverse immune reactions. While it is theoretically possible to engineer out every potentially immunogenic T-cell epitope, several cost factors need to be considered: time, resources, and introduction of changes with unforeseen impact on the drug's physicochemical attributes and mode of action. Another consideration is the potential consequences of ADA and how it is factored into the strategic decision. If the consequences of an immune response are very low, is it worth the effort to de-immunize.

Conclusion

The field of immunogenicity de-risking has made significant strides in recent years, yet there remains a substantial amount of work to fully understand which tools are most effective for accurately predicting ADA formation. Additionally, prioritizing these tools based on their ability to quickly, sensitively, and reproducibly identify immunogenic "hot spots" in biotherapeutics is still a challenge.

Immunogenicity remains a critical concern in the development of biotherapeutics, impacting safety, efficacy, and cost. Various factors, including the physical properties of the formulation material, aggregation, and non-sequence-related parameters, may contribute to ADA induction. However, CD4+ T-cell epitope generation is the primary driver of immunogenicity *in vivo*. As more data becomes available, it is evident that no single test can adequately monitor ADA risk *in vitro*. Instead, a combination of whole protein assays and peptide-specific identification methods, such as *in silico* prediction and MAPPs analysis, is necessary to pinpoint immunogenic hotspots and inform re-engineering strategies.

A challenge in the field is the disconnect between data from clinical studies and its application to the available assays for ADA prediction. Some ADA incidences may target patient cohorts with specific HLA II alleles. Despite the value of MHC profiling in patients receiving clinical biotherapeutics, it has not been standardized yet. Another obvious next step in the methods for profiling immunogenicity is the incorporation of machine learning algorithms to better predict and alleviate ADAs. This will require data to be collected in a manner that can feed into such machine-learning approaches but it is already on the horizon and will allow a leap forward in the field. Such data would be invaluable for future investigations into the correlation between HLA type, genotypic profiles, and ADA outcomes.

Abbreviations

ADA	anti-drug antibody
MAPPs	major histocompatibility complex-associated peptide proteomics
MHC I	major histocompatibility class I
MHC II	major histocompatibility class II
Rh	recombinant human
GM-CSF	granulocyte-macrophage colony-stimulating factor
IL-4	interleukin-4 cytokine
DC	dendritic cell
mAbs	monoclonal antibodies
HLA	human leukocyte antigen
HLA-DR	human leukocyte antigen – antigen D related
APC	antigen-presenting cells
PK	pharmacokinetics
PD	pharmacodynamics
FDA	Food and Drug Administration
APC	antigen-presenting cell
NAb	neutralizing anti-drug antibodies
$CD4^+$ T-cell	cluster of differentiation 4-positive thymus cell
CD11c	cluster of differentiation 11c
CD14	cluster of differentiation 14
CD80	cluster of differentiation 80
CD83	cluster of differentiation 83
CD86	cluster of differentiation 86
CD209	cluster of differentiation 209
PCSK9	proprotein convertase subtilisin/Kexin type 9
MS	precursor mass spectrum
MS/MS	tandem mass spectrum
TNF	tumor necrosis factor
LPS	lipopolysaccharide
PBS	phosphate-buffered saline
IEDB	immune epitope database
HPLC	high-performance liquid chromatography
PTMs	post-translational modifications
IgG	immunoglobulin G
IgM	immunoglobulin M
NHS	N-Hydroxysuccinimide
NCBI	National Center for Biotechnology Information
CDR	complementarity determining region
Ova	ovalbumin
ADC	antibody–drug conjugate
KLH	keyhole limpet hemocyanin
bsAb	bispecific antibody
AAV	adeno-associated virus
pMHCII	peptide–MHC II
MOA	mechanism of action

References

1 Fineberg, S.E., Kawabata, T.T., Finco-Kent, D. et al. (2007). Immunological responses to exogenous insulin. *Endocr. Rev.* 28 (6): 625–652.

2 Carter, P.J. and Lazar, G.A. (2018). Next generation antibody drugs: pursuit of the "high-hanging fruit". *Nat. Rev. Drug Discov.* 17 (3): 197–223.

3 Whyte, M.P., Valdes, R., Ryan, L.M., and McAlister, W.H. (1982). Infantile hypophosphatasia: enzyme replacement therapy by intravenous infusion of alkaline phosphatase-rich plasma from patients with Paget bone disease. *J. Pediatr.* 101 (3): 379–386.

4 Hofmann, C., Jakob, C., Seefried, F. et al. (2015). Recombinant enzyme replacement therapy in hypophosphatasia. *Subcell. Biochem.*

5 Mullard, A. (2021). FDA approves 100th monoclonal antibody product. *Nat. Rev. Drug Discov.* 20 (7): 491–495.

6 (2024) The Antibody Society. Therapeutic monoclonal antibodies approved or in regulatory review. www.antibodysociety.org/antibody-therapeutics-product-data (accessed 03 May 2024).

7 Riechmann, L., Clark, M., Waldmann, H., and Winter, G. (1988). Reshaping human antibodies for therapy. *Nature* 332 (6162): 323–327.

8 Hwang, W.Y.K. and Foote, J. (2005). Immunogenicity of engineered antibodies. *Methods* 36 (1): 3–10.

9 Kozlowski, S. and Swann, P. (2006). Current and future issues in the manufacturing and development of monoclonal antibodies. *Adv. Drug Deliv. Rev.* 58 (5–6): 707–722.

10 Kuriakose, A., Chirmule, N., and Nair, P. (2016). Immunogenicity of biotherapeutics: causes and association with posttranslational modifications. *J Immunol Res* 2016: 1298473.

11 Brinkmann, U. and Kontermann, R.E. (2017). The making of bispecific antibodies. *mAbs* 9 (2): 182–212.

12 Labrijn, A.F., Janmaat, M.L., Reichert, J.M., and Parren, P.W.H.I. (2019). Bispecific antibodies: a mechanistic review of the pipeline. *Nat. Rev. Drug Discov.* 18 (8): 585–608.

13 Sievers, E.L. and Senter, P.D. (2013). Antibody-drug conjugates in cancer therapy. *Medicine* 64 (1): 15–29.

14 Sehn, L.H., Herrera, A.F., Flowers, C.R. et al. (2020). Polatuzumab vedotin in relapsed or refractory diffuse large B-*cell* lymphoma. *J. Clin. Oncol.* 38 (2): 155–165.

15 Verma, S., Miles, D., Gianni, L. et al. (2012). Trastuzumab emtansine for HER2-positive advanced breast cancer. *N. Engl. J. Med.* 367 (19): 1783–1791.

16 Kaur, S. (2013). *Bioanalysis* special focus issue on antibody–drug conjugates. *Bioanalysis* 5 (9): 981–983.

17 Gorovits, B., Alley, S.C., Bilic, S. et al. (2013). Bioanalysis of antibody–drug conjugates: American Association of Pharmaceutical Scientists Antibody–Drug Conjugate Working Group position paper. *Bioanalysis* 5 (9): 997–1006.

18 Dumontet, C., Reichert, J.M., Senter, P.D. et al. (2023). Antibody–drug conjugates come of age in oncology. *Nat. Rev. Drug Discov.* 22 (8): 641–661.

19 Detrez, I. and Gils, A. (2018). Risk assessment and monitoring of antibody responses to biosimilars in chronic inflammatory diseases. *Curr. Pharm. Des.* 23 (44): 6739–6745.

20 Davis, J.D., Padros, M.B., Conrado, D.J. et al. (2024). Subcutaneous administration of monoclonal antibodies: pharmacology, delivery, immunogenicity, and learnings from applications to clinical development. *Clin. Pharmacol. Ther.* 115 (3): 422–439.

21 Baker, M., Reynolds, H.M., Lumicisi, B., and Bryson, C.J. (2010). Immunogenicity of protein therapeutics: the key causes, consequences and challenges. *SelfNonself* 1 (4): 314–322.

22 Cornaby, C., Gibbons, L., Mayhew, V. et al. (2015). B cell epitope spreading: mechanisms and contribution to autoimmune diseases. *Immunol. Lett.* 163 (1): 56–68.

23 Vanderlugt, C.J. and Miller, S.D. (1996). Epitope spreading. *Curr. Opin. Immunol.* 8 (6): 831–836.

24 Potocnakova, L., Bhide, M., and Pulzova, L.B. (2016). An introduction to B-cell epitope mapping and in silico epitope prediction. *J Immunol Res* 2016: 6760830.

25 Bachmann, M.F. and Zinkernagel, R.M. (1997). Neutralizing antiviral B cell responses. *Annu. Rev. Immunol.* 15 (1): 235–270.

26 Fehr, T., Bachmann, M.F., Bucher, E. et al. (1997). Role of repetitive antigen patterns for Induction of antibodies against antibodies. *J. Exp. Med.* 185 (10): 1785–1792.

27 Koren, E., Smith, H.W., Shores, E. et al. (2008). Recommendations on risk-based strategies for detection and characterization of antibodies against biotechnology products. *J. Immunol. Methods* 333 (1–2): 1–9.

28 Du, J., Yang, Y., Zhu, L. et al. (2023). Method validation of a bridging immunoassay in combination with acid-dissociation and bead treatment for detection of anti-drug antibody. *Heliyon* 9 (3): e13999.

29 Yang, T., Braun, M., Lembke, W. et al. (2022). Immunogenicity assessment of AAV-based gene therapies: an IQ consortium industry white paper. *Mol. Ther. – Methods Clin. Dev.* 26: 471–494.

30 Carrasco-Triguero, M., Yi, J.-H., Dere, R. et al. (2013). Immunogenicity assays for antibody–drug conjugates:

case study with ado-trastuzumab emtansine. *Bioanalysis* 5 (9): 1007–1023.

31 Long, B.R., Veron, P., Kuranda, K. et al. (2021). Early phase clinical immunogenicity of valoctocogene roxaparvovec, an AAV5-mediated gene therapy for hemophilia A. *Mol. Ther.* 29 (2): 597–610.

32 Mora, J., Palmer, R., Wagner, L. et al. (2024). 2023 white paper on recent issues in bioanalysis: ISR for ADA assays, the rise of dPCR vs qPCR, international reference standards for vaccine assays, anti-AAV TAb post-dose assessment, nanoString validation, ELISpot as gold standard (part 3 – recommendations on gene therapy, cell therapy, vaccines immunogenicity & technologies; iotherapeutics Immunogenicity & Risk Assessment; ADA/NAb assay/reporting harmonization). *Bioanalysis* 16 (7): 77–119.

33 Pöhler, A., Emrich, T., Jordan, G. et al. (2022). Comparison of assay formats used for the detection of pre-existing anti-drug antibodies against monoclonal antibodies. *Bioanalysis* 14 (13): 923–933.

34 FDA (2022) Immunogenicity information in human prescription therapeutic protein and select drug product labeling – content and format. *FDA Draft Guidance for Industry*, (FDA-2021-D-1041).

35 Krishna, M. and Nadler, S.G. (2016). Immunogenicity to biotherapeutics – the role of anti-drug immune complexes. *Front. Immunol.* 7: 21.

36 Zhong, Z.D., Clements-Egan, A., Gorovits, B. et al. (2017). Drug target interference in immunogenicity assays: recommendations and mitigation strategies. *AAPS J.* 19 (6): 1564–1575.

37 Calabresi, P.A., Giovannoni, G., Confavreux, C. et al. (2007). The incidence and significance of anti-natalizumab antibodies. *Neurology* 69 (14): 1391–1403.

38 Wang, Y.-M.C., Wang, J., Hon, Y.Y. et al. (2016). Evaluating and reporting the immunogenicity impacts for biological products—a clinical pharmacology perspective. *AAPS J.* 18 (2): 395–403.

39 Ettinger, R.A., Liberman, J.A., Gunasekera, D. et al. (2018). FVIII proteins with a modified immunodominant T-cell epitope exhibit reduced immunogenicity and normal FVIII activity. *Blood Adv.* 2 (4): 309–322.

40 Lusher, J.M. (2000). Hemophilia treatment. Factor VIII inhibitors with recombinant products: prospective clinical trials. *Haematologica* 85 (10 Suppl): 2–5; discussion 5-6.

41 Banugaria, S.G., Prater, S.N., Ng, Y.-K. et al. (2011). The impact of antibodies on clinical outcomes in diseases treated with therapeutic protein: lessons learned from infantile Pompe disease. *Genet. Med.* 13 (8): 729–736.

42 Casadevall, N., Cournoyer, D., Marsh, J. et al. (2004). Recommendations on haematological criteria for the diagnosis of epoetin-induced pure red cell aplasia. *Eur. J. Haematol.* 73 (6): 389–396.

43 Welt, S., Divgi, C.R., Real, F.X. et al. (1990). Quantitative analysis of antibody localization in human metastatic colon cancer: a phase I study of monoclonal antibody A33. *J. Clin. Oncol.* 8 (11): 1894–1906.

44 Hale, G., Clark, M.R., Marcus, R. et al. (1988). Remission induction in non-hodgkin lymphoma with reshaped human monoclonal antibody campath-1h. *Lancet* 332 (8625): 1394–1399.

45 Melendez, R., Ordonia, B., Guerrero, J. et al. (2022). Introducing dendritic cell antibody internalization as an immunogenicity risk assessment tool. *Bioanalysis* 14 (10): 703–713.

46 Schellekens, H. (2002). Bioequivalence and the immunogenicity of biopharmaceuticals. *Nat. Rev. Drug Discov.* 1 (6): 457–462.

47 Peters, B., Nielsen, M., and Sette, A. (2020). T cell epitope predictions. *Annu. Rev. Immunol.* 38 (1): 1–23.

48 Gartner, J.J., Parkhurst, M.R., Gros, A. et al. (2021). A machine learning model for ranking candidate HLA class I neoantigens based on known neoepitopes from multiple human tumor types. *Nat. Can.* 2 (5): 563–574.

49 Kim, H.-J. and Cantor, H. (2014). CD4 T-cell subsets and tumor *immunity*: the helpful and the not-so-helpful. *Cancer Immunol. Res.* 2 (2): 91–98.

50 Swain, S.L., McKinstry, K.K., and Strutt, T.M. (2012). Expanding roles for CD4+ T cells in immunity to viruses. *Nat. Rev. Immunol.* 12 (2): 136–148.

51 Ramarathinam, S.H., Ho, B.K., Dudek, N.L., and Purcell, A.W. (2021). HLA class II immunopeptidomics reveals that co-inherited HLA-allotypes within an extended haplotype can improve proteome coverage for immunosurveillance. *Proteomics* 21 (17–18): e2000160.

52 Marcu, A., Bichmann, L., Kuchenbecker, L. et al. (2021). HLA ligand atlas: a benign reference of HLA-presented peptides to improve T-cell-based cancer immunotherapy. *J. Immunother. Cancer* 9 (4): e002071.

53 Stražar, M., Park, J., Abelin, J.G. et al. (2023). HLA-II immunopeptidome profiling and deep learning reveal features of antigenicity to inform antigen discovery. *Immunity* 56 (7): 1681–1698.e13.

54 Abelin, J.G., Harjanto, D., Malloy, M. et al. (2019). Defining HLA-II ligand processing and binding rules with mass spectrometry enhances cancer epitope prediction. *Immunity* 51 (4): 766–779.e17.

55 Vita, R., Mahajan, S., Overton, J.A. et al. (2019). The immune epitope database (IEDB): 2018 update. *Nucleic Acids Res.* 47 (D1): D339–D343.

56 Yogurtcu, O.N., Sauna, Z.E., McGill, J.R. et al. (2019). TCPro: an *in silico* risk assessment tool for biotherapeutic protein immunogenicity. *AAPS J.* 21 (5): 96.
57 Groot, A.S.D., Moise, L., Terry, F. et al. (2020). Better epitope discovery, precision immune engineering, and accelerated vaccine design using Immunoinformatics tools. *Front. Immunol.* 11: 442.
58 Chen, B., Khodadoust, M.S., Olsson, N. et al. (2019). Predicting HLA class II antigen presentation through integrated deep learning. *Nat. Biotechnol.* 37 (11): 1332–1343.
59 Mattei, A.E., Gutierrez, A.H., Seshadri, S. et al. (2024). *In silico* methods for immunogenicity risk assessment and human homology screening for therapeutic antibodies. *MAbs* 16 (1): 2333729.
60 Cohen, S., Chung, S., Spiess, C. et al. (2021). An integrated approach for characterizing immunogenic responses toward a bispecific antibody. *MAbs* 13 (1): 1944017.
61 Mufarrege, E.F., Giorgetti, S., Etcheverrigaray, M. et al. (2017). De-immunized and functional therapeutic (DeFT) versions of a long lasting recombinant alpha interferon for antiviral therapy. *Clin. Immunol.* 176: 31–41.
62 Koren, E., Groot, A.S.D., Jawa, V. et al. (2007). Clinical validation of the "*in silico*" prediction of immunogenicity of a human recombinant therapeutic protein. *Clin. Immunol.* 124 (1): 26–32.
63 Reche, P.A. and Reinherz, E.L. (2007). Immunoinformatics, predicting immunogenicity *in silico*. *Methods Mol. Biol.* 409: 185–200.
64 Barra, C., Ackaert, C., Reynisson, B. et al. (2020). Immunopeptidomic data integration to artificial neural networks enhances protein-drug immunogenicity prediction. *Front. Immunol.* 11: 1304.
65 Lazar, G.A., Desjarlais, J.R., Jacinto, J. et al. (2007). A molecular immunology approach to antibody humanization and functional optimization. *Mol. Immunol.* 44 (8): 1986–1998.
66 Yin, L., Dai, S., Clayton, G. et al. (2013). Recognition of self and altered self by T cells in autoimmunity and allergy. *Protein Cell* 4 (1): 8–16.
67 Groot, A.S.D. and Moise, L. (2007). Prediction of immunogenicity for therapeutic proteins: state of the art. *Curr. Opin. Drug Discov. Dev.* 10 (3): 332–340.
68 Gao, S.H., Huang, K., Tu, H., and Adler, A.S. (2013). Monoclonal antibody humanness score and its applications. *BMC Biotechnol.* 13 (1): 55.
69 Norman, R.A., Ambrosetti, F., Bonvin, A.M.J.J. et al. (2020). Computational approaches to therapeutic antibody design: established methods and emerging trends. *Brief. Bioinform.* 21 (5): 1549–1567.
70 Kovaltsuk, A., Leem, J., Kelm, S. et al. (2018). Observed antibody space: a resource for data mining next-generation sequencing of antibody repertoires. *J. Immunol.* 201 (8): 2502–2509.
71 Marks, C., Hummer, A.M., Chin, M., and Deane, C.M. (2021). Humanization of antibodies using a machine learning approach on large-scale repertoire data. *Bioinformatics* 37 (22): 4041–4047.
72 Prihoda, D., Maamary, J., Waight, A. et al. (2022). BioPhi: a platform for antibody design, humanization, and humanness evaluation based on natural antibody repertoires and deep learning. *MAbs* 14 (1): 2020203.
73 Thrift, W.J., Perera, J., Cohen, S. et al. (2024). Graph-pMHC: graph neural network approach to MHC class II peptide presentation and antibody immunogenicity. *Brief. Bioinform.* 25 (3): bbae123.
74 Nielsen, M., Lundegaard, C., and Lund, O. (2007). Prediction of MHC class II binding affinity using SMM-align, a novel stabilization matrix alignment method. *BMC Bioinform.* 8 (1): 238.
75 Nielsen, M., Lund, O., Buus, S., and Lundegaard, C. (2010). MHC class II epitope predictive algorithms. *Immunology* 130 (3): 319–328.
76 Deng, L., Ly, C., Abdollahi, S. et al. (2023). Performance comparison of TCR-pMHC prediction tools reveals a strong data dependency. *Front. Immunol.* 14: 1128326.
77 EL-Manzalawy, Y., Dobbs, D., and Honavar, V.G. (2016). Prediction of protein secondary structure. *Methods Mol. Biol.* 1484: 255–264.
78 Ansari, H.R. and Raghava, G.P. (2010). Identification of conformational B-cell epitopes in an antigen from its primary sequence. *Immunome Res.* 6 (1): 6.
79 Ras-Carmona, A., Lehmann, A.A., Lehmann, P.V., and Reche, P.A. (2022). Prediction of B cell epitopes in proteins using a novel sequence similarity-based method. *Sci. Rep.* 12 (1): 13739.
80 Cia, G., Pucci, F., and Rooman, M. (2023). Critical review of conformational B-cell epitope prediction methods. *Brief. Bioinform.* 24 (1): bbac567.
81 Wen, Y. and Jawa, V. (2021). The impact of product and process related critical quality attributes on immunogenicity and adverse immunological effects of biotherapeutics. *J. Pharm. Sci.* 110 (3): 1025–1041.
82 Harris, C.T. and Cohen, S. (2024). Reducing immunogenicity by design: approaches to minimize immunogenicity of monoclonal antibodies. *BioDrugs* 38 (2): 205–226.
83 Pham, N.B. and Meng, W.S. (2020). Protein aggregation and immunogenicity of biotherapeutics. *Int. J. Pharm.* 585: 119523.

84 Moussa, E.M., Panchal, J.P., Moorthy, B.S. et al. (2016). Immunogenicity of therapeutic protein aggregates. *J. Pharm. Sci.* 105 (2): 417–430.

85 Racle, J., Michaux, J., Rockinger, G.A. et al. (2019). Robust prediction of HLA class II epitopes by deep motif deconvolution of immunopeptidomes. *Nat. Biotechnol.* 37 (11): 1283–1286.

86 Schultz, H.S., Reedtz-Runge, S.L., Bäckström, B.T. et al. (2017). Quantitative analysis of the CD4+ T cell response to therapeutic antibodies in healthy donors using a novel T cell:PBMC assay. *PLoS One* 12 (5): e0178544.

87 Ito, S., Ikuno, T., Mishima, M. et al. (2019). In vitro human helper T-cell assay to screen antibody drug candidates for immunogenicity. *J. Immunotoxicol.* 16 (1): 125–132.

88 Tsai, W.-T.K., Li, Y., Yin, Z. et al. (2024). Nonclinical immunogenicity risk assessment for knobs-into-holes bispecific IgG1 antibodies. *MAbs* 16 (1): 2362789.

89 Mempel, T.R., Henrickson, S.E., and von Andrian, U.H. (2004). T-cell priming by dendritic cells in lymph nodes occurs in three distinct phases. *Nature* 427 (6970): 154–159.

90 Wullner, D., Zhou, L., Bramhall, E. et al. (2010). Considerations for optimization and validation of an in vitro PBMC derived T cell assay for immunogenicity prediction of biotherapeutics. *Clin. Immunol.* 137 (1): 5–14.

91 Joubert, M.K., Deshpande, M., Yang, J. et al. (2016). Use of in vitro assays to assess immunogenicity risk of antibody-based biotherapeutics. *PLoS One* 11 (8): e0159328.

92 Arata, Y., Motoyama, S., Yano, M. et al. (2023). Rapid in vitro assessment of the immunogenicity potential of engineered antibody therapeutics through detection of CD4+ T cell interleukin-2 secretion. *MAbs* 15 (1): 2253570.

93 Cohen, S. and Chung, S. (2021). In vitro immunogenicity prediction: bridging between innate and adaptive immunity. *Bioanalysis* 13 (13): 1071–1081.

94 Cohen, S., Myneni, S., Batt, A. et al. (2021). Immunogenicity risk assessment for biotherapeutics through in vitro detection of CD134 and CD137 on T helper cells. *MAbs* 13 (1): 1898831.

95 Duke, B.R. and Mitra-Kaushik, S. (2020). Current in vitro assays for prediction of T cell mediated immunogenicity of biotherapeutics and manufacturing impurities. *J. Pharm. Innov.* 15 (2): 202–218.

96 Siegel, M., Steiner, G., Franssen, L.C. et al. (2022). Validation of a dendritic cell and CD4+ T cell restimulation assay contributing to the immunogenicity risk evaluation of biotherapeutics. *Pharmaceutics* 14 (12): 2672.

97 Wen, Y., Cahya, S., Zeng, W. et al. (2020). Development of a FRET-based assay for analysis of mAbs internalization and processing by dendritic cells in preclinical immunogenicity risk assessment. *AAPS J.* 22 (3): 68.

98 Xue, L., Hickling, T., Song, R. et al. (2016). Contribution of enhanced engagement of antigen presentation machinery to the clinical immunogenicity of a human interleukin (IL)-21 receptor-blocking therapeutic antibody. *Clin. Exp. Immunol.* 183 (1): 102–113.

99 Deora, A., Hegde, S., Lee, J. et al. (2017). Transmembrane TNF-dependent uptake of anti-TNF antibodies. *MAbs* 9 (4): 680–695.

100 Wen, Y., Waltman, A., Han, H., and Collier, J.H. (2016). Switching the immunogenicity of peptide assemblies using surface properties. *ACS Nano* 10 (10): 9274–9286.

101 Ahmadi, M., Bryson, C.J., Cloake, E.A. et al. (2015). Small amounts of sub-visible aggregates enhance the immunogenic potential of monoclonal antibody therapeutics. *Pharm. Res.* 32 (4): 1383–1394.

102 Santiago-Schwarz, F., Anand, P., Liu, S., and Carsons, S.E. (2001). Dendritic cells (DCs) in rheumatoid arthritis (RA): progenitor cells and soluble factors contained in RA synovial fluid yield a subset of myeloid DCs that preferentially activate Th1 inflammatory-type responses. *J. Immunol.* 167 (3): 1758–1768.

103 Kroenke, M.A., Barger, T.E., Hu, J. et al. (2021). Immune complex formation is associated with loss of tolerance and an antibody response to both drug and target. *Front. Immunol.* 12: 782788.

104 Nair, S., Archer, G.E., and Tedder, T.F. (2012). Isolation and generation of human dendritic cells. *Curr. Protoc. Immunol.* 99 (1): 7.32.1–7.32.23.

105 Vissers, M.C.M., Jester, S.A., and Fantone, J.C. (1988). Rapid purification of human peripheral blood monocytes by centrifugation through ficoll-hypaque and sepracell-MN. *J. Immunol. Methods* 110 (2): 203–207.

106 Ferrante, A. and Thong, Y.H. (1980). Optimal conditions for simultaneous purification of mononuclear and polymorphonuclear leucocytes from human blood by the hypaque-ficoll method. *J. Immunol. Methods* 36 (2): 109–117.

107 Ridker, P.M., Tardif, J.-C., Amarenco, P. et al. (2017). Lipid-reduction variability and antidrug-antibody formation with Bococizumab. *N. Engl. J. Med.* 376 (16): 1517–1526.

108 Bjorkman, P.J., Saper, M.A., Samraoui, B. et al. (1987). Structure of the human class I histocompatibility antigen, HLA-A2. *Nature* 329 (6139): 506–512.

109 Takahashi, H., Houghten, R., Putney, S.D. et al. (1989). Structural requirements for class I MHC molecule-mediated antigen presentation and cytotoxic T cell recognition of an immunodominant determinant of the human immunodeficiency virus envelope protein. *J. Exp. Med.* 170 (6): 2023–2035.

110 Mantegazza, A.R., Magalhaes, J.G., Amigorena, S., and Marks, M.S. (2013). Presentation of phagocytosed antigens by MHC class I and II. *Traffic* 14 (2): 135–152.

111 Grey, H.M., Buus, S., Colon, S. et al. (1989). Structural requirements and biological significance of interactions between peptides and the major histocompatibility complex. *Philos. Trans. R. Soc. Lond. B, biol. Sci.* 323 (1217): 545–552.

112 Brown, J.H., Jardetzky, T.S., Gorga, J.C. et al. (1993). Three-dimensional structure of the human class II histocompatibility antigen HLA-DR1. *Nature* 364 (6432): 33–39.

113 Nag, B., Mukku, P.V., Arimilli, S. et al. (1994). Antigenic peptide binding to MHC class II molecules at increased peptide concentrations. *Mol. Immunol.* 31 (15): 1161–1168.

114 Gugasyan, R., Vidavsky, I., Nelson, C.A. et al. (1998). Isolation and quantitation of a minor determinant of hen egg white lysozyme bound to I-Ak by using peptide-specific immunoaffinity. *J. Immunol.* 161 (11): 6074–6083.

115 Dick, R.L., Aldrich, C., Jameson, C.S. et al. (1994). Proteolytic processing of ovalbumin and beta-galactosidase by the proteasome to a yield antigenic peptides. *J. Immunol.* 152 (8): 3884–3894.

116 Hunt, D.F., Henderson, R.A., Shabanowitz, J. et al. (1992). Characterization of peptides bound to the class I MHC molecule HLA-A2.1 by mass spectrometry. *Science* 255 (5049): 1261–1263.

117 Bakke, O. and Dobberstein, B. (1990). MHC class II-associated invariant chain contains a sorting signal for endosomal compartments. *Cell* 63 (4): 707–716.

118 Nishimura, Y., Oiso, M., Fujisao, S. et al. (1998). Peptide-based molecular analyses of HLA class II-associated susceptibility to autoimmune diseases. *Int. Rev. Immunol.* 17 (5–6): 229–262.

119 Lepage, V., Lamm, L.U., and Charron, D. (1993). Molecular aspects of HLA class ii and some autoimmune diseases. *Int. J. Immunogenet.* 20 (3): 153–164.

120 Christie, J.F., Dunbar, B., and Kennedy, M.W. (2008). The ABA-1 allergen of the nematode *Ascaris suum*: epitope stability, mass spectrometry, and N-terminal sequence comparison with its homologue in *Toxocara canis*. *Clin. Exp. Immunol.* 92 (1): 125–132.

121 Röhn, T.A., Reitz, A., Paschen, A. et al. (2005). A novel strategy for the discovery of MHC class II–restricted tumor antigens: identification of a melanotransferrin helper T-cell epitope. *Cancer Res.* 65 (21): 10068–10078.

122 Purcell, A.W., Ramarathinam, S.H., and Ternette, N. (2019). Mass spectrometry–based identification of MHC-bound peptides for immunopeptidomics. *Nat. Protoc.* 14 (6): 1687–1707.

123 Gravel, N.H., Nelde, A., Bauer, J. et al. (2023). TOFIMS mass spectrometry-based immunopeptidomics refines tumor antigen identification. *Nat. Commun.* 14 (1): 7472.

124 Tran, N.H., Zhang, X., Xin, L. et al. (2017). De novo peptide sequencing by deep learning. *Proc. Natl. Acad. Sci.* 114 (31): 8247–8252.

125 Tran, N.H., Qiao, R., Xin, L. et al. (2019). Deep learning enables de novo peptide sequencing from data-independent-acquisition mass spectrometry. *Nat. Methods* 16 (1): 63–66.

126 Kong, A.T., Leprevost, F.V., Avtonomov, D.M. et al. (2017). MSFragger: ultrafast and comprehensive peptide identification in mass spectrometry–based proteomics. *Nat. Methods* 14 (5): 513–520.

127 Perkins, D.N., Pappin, D.J.C., Creasy, D.M., and Cottrell, J.S. (1999). Probability-based protein identification by searching sequence databases using mass spectrometry data. *Electrophoresis* 20 (18): 3551–3567.

128 Zhang, Y., Fonslow, B.R., Shan, B. et al. (2013). Protein analysis by shotgun/bottom-up proteomics. *Chem. Rev.* 113 (4): 2343–2394.

129 Lippolis, J.D., White, F.M., Marto, J.A. et al. (2002). Analysis of MHC class II antigen processing by quantitation of peptides that constitute nested sets. *J. Immunol.* 169 (9): 5089–5097.

130 Sekiguchi, N., Kubo, C., Takahashi, A. et al. (2018). MHC-associated peptide proteomics enabling highly sensitive detection of immunogenic sequences for the development of therapeutic antibodies with low immunogenicity. *MAbs* 10 (8): 1168–1181.

131 Ianni, A.D., Fraone, T., Balestra, P. et al. (2024). Assessing MAPPs assay as a tool to predict the immunogenicity potential of protein therapeutics. *Life Sci. Alliance* 7 (1): e202302095.

132 Karle, A.C. (2020). Applying MAPPs assays to assess drug immunogenicity. *Front. Immunol.* 11: 698.

133 Lee, M.V., Saad, O.M., Wong, S. et al. (2023). Development of a semi-automated MHC-associated peptide proteomics (MAPPs) method using streptavidin bead-based immunoaffinity capture and nano LC-MS/MS to support immunogenicity risk assessment in drug development. *Front. Immunol.* 14: 1295285.

134 Paul, S., Arlehamn, C.S.L., Scriba, T.J. et al. (2015). Development and validation of a broad scheme for

prediction of HLA class II restricted T cell epitopes. *J. Immunol. Methods* 422: 28–34.

135 Steiner, G., Marban-Doran, C., Langer, J. et al. (2020). Enabling routine MHC-II-associated peptide proteomics for risk assessment of drug-induced immunogenicity. *J. Proteome Res.* 19 (9): 3792–3806.

136 Karle, A., Spindeldreher, S., and Kolbinger, F. (2016). Secukinumab, a novel anti–IL-17A antibody, shows low immunogenicity potential in human in vitro assays comparable to other marketed biotherapeutics with low clinical immunogenicity. *MAbs* 8 (3): 536–550.

137 Karle, A.C., Oostingh, G.J., Mutschlechner, S. et al. (2012). Nitration of the pollen allergen bet v 1.0101 enhances the presentation of bet v 1-derived peptides by HLA-DR on human dendritic cells. *PLoS One* 7 (2): e31483.

138 Depontieu, F.R., Qian, J., Zarling, A.L. et al. (2009). Identification of tumor-associated, MHC class II-restricted phosphopeptides as targets for immunotherapy. *Proc. Natl. Acad. Sci.* 106 (29): 12073–12078.

139 Unanue, E.R., Turk, V., and Neefjes, J. (2015). Variations in MHC class II antigen processing and presentation in health and disease. *Annu. Rev. Immunol.* 34 (1): 1–33.

140 Barker, D.J., Maccari, G., Georgiou, X. et al. (2022). The IPD-IMGT/HLA database. *Nucleic Acids Res.* 51 (D1): D1053–D1060.

141 Yamamoto, F., Suzuki, S., Mizutani, A. et al. (2020). Capturing differential allele-level expression and genotypes of all classical HLA loci and haplotypes by a new capture RNA-Seq method. *Front. Immunol.* 11: 941.

142 Jawa, V., Cousens, L.P., Awwad, M. et al. (2013). T-cell dependent immunogenicity of protein therapeutics: preclinical assessment and mitigation. *Clin. Immunol.* 149 (3): 534–555.

143 Hartman, K., Steiner, G., Siegel, M. et al. (2023). Expanding the MAPPs assay to accommodate MHC-II Pan receptors for improved predictability of potential T cell epitopes. *Biology* 12 (9): 1265.

144 Rombach-Riegraf, V., Karle, A.C., Wolf, B. et al. (2014). Aggregation of human recombinant monoclonal antibodies influences the capacity of dendritic cells to stimulate adaptive T-cell responses in vitro. *PLoS One* 9 (1): e86322.

145 Ducret, A., Ackaert, C., Bessa, J. et al. (2022). Assay format diversity in pre-clinical immunogenicity risk assessment: toward a possible harmonization of antigenicity assays. *MAbs* 14 (1): 1993522.

12

High-Throughput Biophysical Assays for Developability Assessment

Emma Pelegri-O'Day[1] and Nithya Srinivasan[2]**

[1]Amgen Research, Amgen Inc, South San Francisco, CA, USA
[2]Amgen Research, Amgen Inc, Thousand Oaks, CA, USA

Introduction

Biological therapeutics such as monoclonal antibodies, bispecific antibodies, and antibody-drug conjugations (ADCs), are an important category of innovative drugs and represent one of the fastest growing segments in the pharmaceuticals market [1]. Although the growth of biologics has increased significantly in the past decade, the overall rate of success for advancing a biotherapeutic from early screening to commercial regulatory approval remains very low [2, 3]. The low success rate is likely driven by a combination of factors such as poor efficacy in clinical trials, safety concerns, or manufacturability and quality issues with the drug. Given the high costs associated with the development of biologics and the competitive industry landscape, it is critical to have a holistic strategy to select the right candidate and to de-risk liabilities that may impact advancement to clinic and ultimately to the market.

The focus during the early screening stages of a biologics development process is typically on the mechanism of action and biological function. However, as end-to-end product development timelines shrink, an early developability assessment of candidates becomes a critical consideration to accelerate timelines. Developability refers to the attributes of a biological candidate that define its manufacturability, safety, and efficacy, which will ultimately enable advancement to clinical trials and regulatory approval [4–7]. Typical considerations for a developability assessment include ensuring the titer and yield of the manufacturing process, selecting an appropriate formulation to ensure long-term stability, ensuring acceptable product quality throughout the manufacturing process, and identifying an appropriate delivery mechanism for the drug. Since early screening panels are often large and not generated using representative manufacturing processes, it is essential to possess a robust fit-for-purpose set of high-throughput assays that reliably predict downstream success and consume extremely low amounts of material.

Previous excellent reviews on this topic have focused on the analytical techniques to evaluate developability [5, 8, 9]. This book chapter will take a different view on developability by focusing on critical quality attributes that can typically be measured via multiple analytical techniques. The chapter will describe the advantages and drawbacks of each assay to assess the suitability of an assay to measure a particular attribute while contextualizing considerations such as sample size and throughput. Recent advancements in the field of high-throughput analytics will be discussed, with an emphasis on the biophysical attributes being measured such as identity, aggregation, clipping, hydrophobicity, and chemical modifications (Figure 12.1). The overall intent of this chapter is to provide readers with a comprehensive set of modality-agnostic attributes and associated techniques to reliably evaluate the developability of early-stage candidates in a high-throughput manner.

High-Throughput Intact Mass Spectrometry

When assessing biotherapeutics for their developability characteristics, a critical assay in the toolbox is intact mass spectrometry (MS), both to confirm the identity and purity of the sample of interest. Traditional approaches have used reverse-phase LC separations coupled to electrospray ionization (ESI)-MS; these methods are ubiquitous across the biopharmaceutical industry [11]. However, even with the use of ultra-high pressure liquid chromatography (UPLC)

* Both authors contributed equally.

Characterizing Biotherapeutics: Analytical Methods for Diverse Modalities, First Edition. Edited by Jennie R. Lill and Wendy Sandoval.

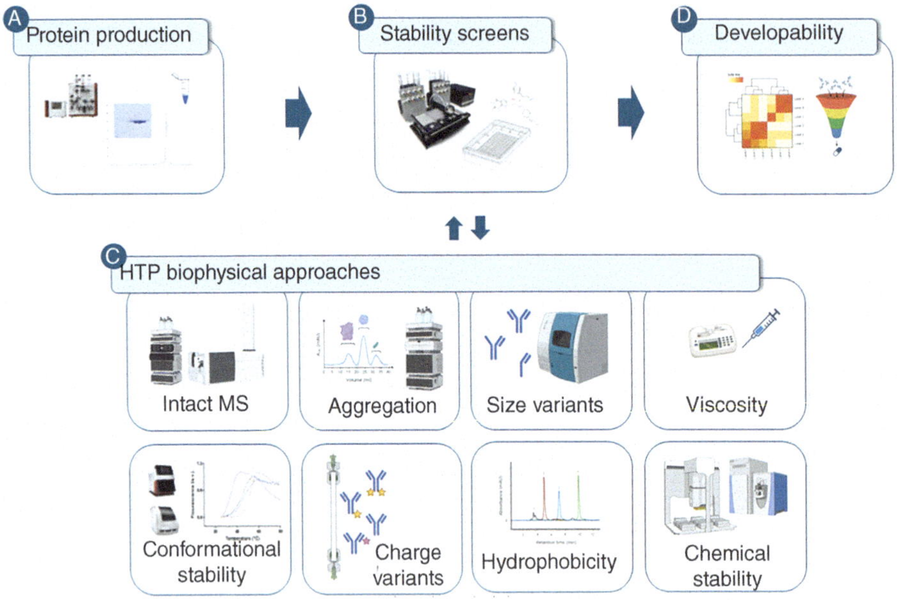

Figure 12.1 Workflow for incorporation of high-throughput biophysical assays into a process for assessing molecule developability. (a) Purification of proteins using standard or high-throughput techniques. (b) Preparation of accelerated stability screens. (c) Critical HTP biophysical approaches mentioned in this chapter to measure stability-indicating attributes, such as identity by intact MS, aggregation, size variants, viscosity, conformational stability, charge variants, hydrophobicity, and chemical stability. (d) Compilation of analytical results into developability assessments or scores for biotherapeutic selection. Source: Kwan et al. [10] / Frontiers Media S.A. / CC BY 4.0.

systems, these separations are typically 5–7 minutes in length [12]. In addition to reverse-phase LC separations, there are also many reports of other LC techniques applied to MS hyphenation, for instance, hydrophobic interaction chromatography (HIC), size-exclusion chromatography (SEC), and cationic exchange chromatography (CEX). More detail for these hyphenation techniques will be discussed in subsequent sections of this chapter.

A good alternative for high-throughput intact mass analysis is to use solid phase extraction (SPE)-MS. SPE-MS uses a trap, wash, and elution strategy to remove salts, then couples to an electrospray ion source for analysis. Using an Agilent RapidFire (RF)-MS system, Campuzano et al. first demonstrated the use of SPE-MS on proteins to quantify the extent of small-molecule modification of a therapeutic target with an analysis time of ~20 s/sample [13]. Subsequent work by Sawyer and coworkers showed that RF-MS could also be used for therapeutic antibodies (>140 kDa) and was applied toward cell line development, quantifying the extent of mispairing in 248 bispecific antibody samples in 2 hours (Figure 12.2) [14]. Importantly, the role of intact MS was critical in this application to accurately characterize the bispecific mispairing in this sample set and subsequently identify optimal cell line clones. Campuzano and coworkers also showed that RF-MS could be applied toward multiple diverse therapeutic and target protein formats including membrane proteins, protein–polymer conjugates, complex bispecifics, and nucleic acid-based modalities with accurate MWs consistent with those reported by other intact MS measurement techniques [15]. For all SPE-MS technologies, it is important to optimize wash and elute conditions to avoid adducts which may make accurate MS identification difficult.

Other technologies which function on this time scale include the SampleStream, which uses a membrane-based separation to desalt and concentrate samples at rates of ~15 s/sample under both native and denaturing conditions, with comparable performance to LC–MS [16]. For further throughput enhancements, a prospective option is acoustic ejection MS (AEMS), in which nanoliter-size droplets are generated and transferred via a carrier liquid into an ESI ion source [17]. An initial report has demonstrated the application of AEMS toward the intact analysis of the NIST mAb with an analysis speed of 1 s/sample [18]. However, more data would need to be generated to prove that AEMS is a reliable and robust HT method for biopharmaceuticals.

Together with improving the throughput of MS data acquisition, it is equally critical to develop workflows to increase the throughput of MS data processing. These processes should work for a diversity of molecule formats, including mAbs, bispecifics, or even highly polydisperse samples such as ADCs or PEGylated biomolecules [19]. For instance, Ziegengeist and coworkers have described a high-throughput workflow capable of identifying and quantifying mispaired species [20]. Waldenmaier et al. have implemented an automated system

Figure 12.2 Application of RF-MS for the identification of mispaired bispecific antibody species in a clone selection workflow. (a) RF-MS workflow using a trap, wash and elution strategy for removing salts from bispecific antibody samples. (b) Extracted ion chromatogram (EIC) showing the high throughput of RF-MS analysis. (c) Deconvoluted mass spectra of a bispecific antibody sample with annotated mispairing as identified by RF-MS. Source: Sawyer et al. [14] / National Academy of Sciences / CC BY 4.0.

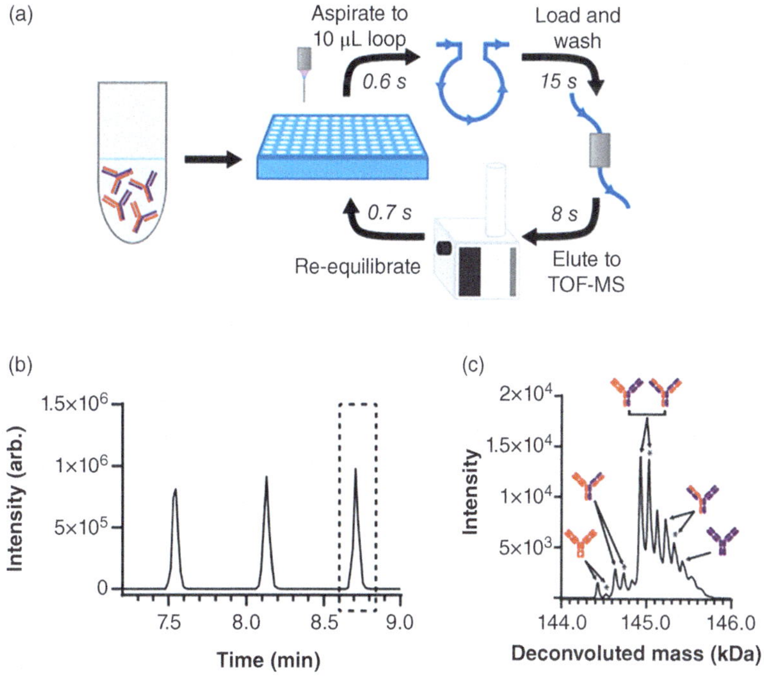

for MS characterization compiling data from multiple instruments into a single cloud data platform [21]. Phung and coworkers have described the use of an open-source deconvolution pipeline for high-throughput intact analysis [22]. A combination of wet lab methods and software support is critical for the full implementation of high-throughput MS analysis.

Aggregation

Protein aggregation is a critical product quality attribute during the development of biological therapeutics. Since aggregation can lead to undesirable outcomes such as particulation or precipitation and could induce immunogenic responses from patients, this attribute is very closely monitored across the lifecycle of the molecule. During molecular assessment, the aggregate levels of candidates under a variety of stress conditions such as thermal and light stress are often key determinants of lead selection. As development proceeds, the aggregates are further structurally characterized to yield a more complete picture of the risks associated with aggregate formation.

It is essential to have a suite of high-throughput assays to rapidly screen large panels of molecules during the lead selection process (Figure 12.3). SEC remains the gold standard for detecting and quantifying aggregation [24, 25]. Recent advances in column technology have facilitated the development of SEC methods with short run times, enabling high-throughput workflows to be established at the early lead screening stages [5, 26, 27]. Reliable peak separation and quantitation are key factors to assess the suitability of an SEC method for a particular molecule. A limitation of SEC is a relatively long run time that may not be amenable to large panel sizes. Fluorescence-based aggregation analysis workflows attempt to overcome this obstacle by leveraging a dye such as Sypro Orange that binds to hydrophobic patches exposed on antibody aggregates and undergoes aggregation-induced emission when excited [28, 29]. While Sypro Orange appears to be specific for structurally altered proteins, in 2002 scientists showed that by using high-throughput, TIRF-based photobleaching experiments on Thioflavin T-incubated aggregates, it was possible to detect lower-order oligomeric species [30]. Recent advances in microplate technology can further enhance the throughput of fluorescence-based aggregation detection workflows [31]. Using specially designed 384-well plates that are preloaded with dried capture beads, Perrou et al. demonstrated that the plate-based assay readout correlates well with aggregation levels detected by SEC.

Dynamic light scattering (DLS) is a quick, label-free, and nondestructive assay to understand size for many biologics including peptides, proteins, and viruses that lends itself seamlessly to high-throughput optimization. The hydrodynamic size (radius) of the protein and the polydispersity readout are good indicators of the stability in the chemical and physical conditions (temperature) used [32]. Using microliter volumes and plate-based approaches, behavior of proteins in the presence of ligands, substrates, additives, and buffers can now be determined in parallel [33–35].

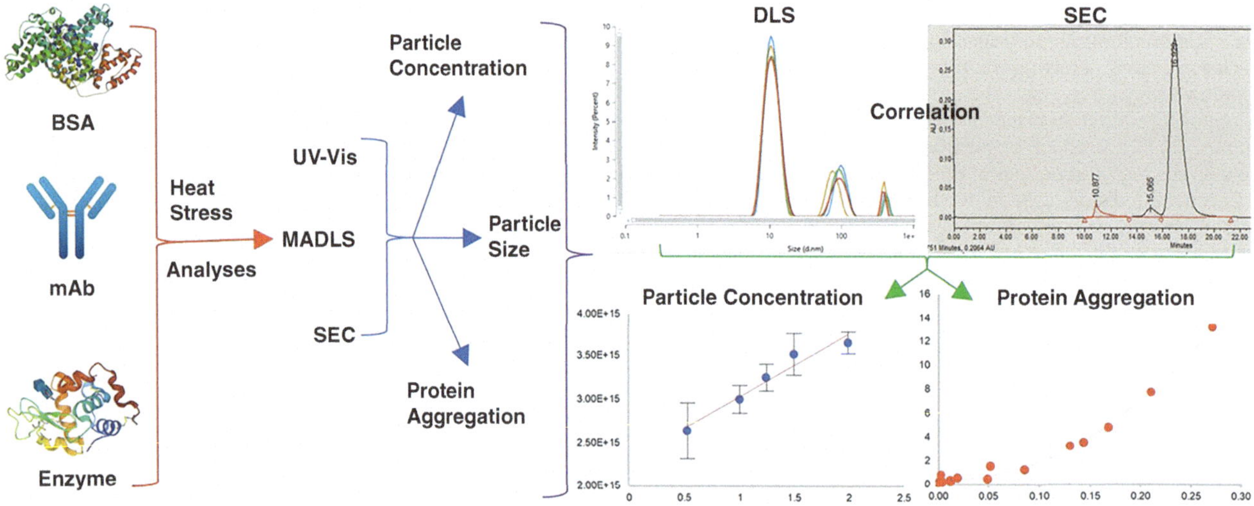

Figure 12.3 Overview of biophysical assays to monitor aggregation in biotherapeutics. Source: Sharma et al. [23] / Springer Nature.

Kwan 2022 described using DLS to monitor degradation or changes in aggregation by measuring real-time changes over time and thereby identifying twelve excipients that inhibit aggregation of the target protein [10]. High-throughput DLS measurements also find applications in quality control labs to determine sample quality and compare production batches [35–38].

In recent years, mass photometry has emerged as a convenient tool for providing rapid insights into the aggregation profile therapeutic candidates. It relies on the changes in local refractive index when a molecule binds to a glass surface [39]. Monitoring these single molecule binding events with optimized image contrast enables very accurate quantification of the reflectivity and therefore exceptional mass accuracy and resolution. MP has been demonstrated to have superior resolution to negative stain EM when assessing the integrity and stability of large molecular complexes [40].

A highly useful approach that has been recently employed to gain deeper insight into the structural aspects of therapeutics is the hyphenation of native MS with separation techniques such as SEC [41, 42]. Using 100 mM ammonium acetate (pH 6.8) as the mobile phase, Ehkirch et al. demonstrated the ability of SEC-MS to both identify and quantify degradation products of pharmaceutical monoclonal antibodies (mAbs) [43]. Along similar lines, Haberger et al. used SEC–MS to assess bispecific antibodies for product-related impurities arising from light or heavy chain mispairing, and aggregates during accelerated stability studies [44]. Although SEC-MS is currently not mature for high-throughput applications, the recent advances demonstrate potential for deployment in developability screening in the near future [41, 45, 46].

Size Variants

Size variants and fragments of therapeutic proteins are also critical attributes that are measured to monitor a molecule's integrity and purity. This section covers rapid analytical methods to monitor and identify molecule clipping and partial species. The industry standard for quantifying clipped species and partial antibody fragments has been capillary electrophoresis-sodium dodecyl sulfate (CE-SDS), which can be run on a number of commercial instruments and provides high resolution with relatively low sample consumption [47]. CE-SDS is run under denaturing conditions and is typically used as a complementary method to SEC, specifically to quantify species that are smaller than the desired biotherapeutic. Most industrial CE-SDS methods are low throughput, with a cycle time of 50–60 minutes where capillary rinsing takes up to 50% of that total time. Instrument vendors have recently released systems capable of improving this throughput. For instance, Sciex released a multicapillary system in 2021 containing 8 capillaries; while the overall separation time does not change, the throughput is increased by 8x. Other products have focused on reduction of separation time; the ProteinSimple Turbo cartridge offers a 4–5x improvement in speed by using a shorter separation capillary and a higher separation field [48].

While it is the gold standard for clip quantification, CE-SDS cannot be used for clip identification. The buffer composition is not MS compatible, and currently commercial systems capable of fractionation are not available. The most common approach to identify unknown peaks is for scientists to fractionate using a different separation mode (often SEC or RPLC). The isolated fraction may then be split, with one portion analyzed for identity (rpLC–MS or MS/MS) and the other portion repeated by CE-SDS to confirm the migration time. This method was successfully used by Wang and coworkers to identify fragments in an IgG1 mAb, running online RP-HPLC hyphenated to intact mass and top-down MS/MS to confirm fragment identity, and then performing off-line RP fractionation followed by CE-SDS of each fraction to assign the identified species to a peak in the resulting electropherogram [49]. Fragment identification has also been achieved by directly correlating CE-SDS migration times of low molecular weight species with the molecular weight of identified clips as measured by rpLC–MS [50].

A common high-throughput alternative to CE-SDS is to pursue microchip capillary electrophoresis (MCE), which is well-established for multiple antibody product quality assays [51]. MCE injections can be performed in less than 60 seconds of analysis time, which is a significant improvement over their conventional capillary counterparts. Studies on the current model, the LabChip GXII, reported comparable resolution to conventional capillary assays after method optimization [52, 53]. For high-throughput applications, MCE is critical to provide rapid quantification of LMW species, as a complementary purity metric to SEC. Similar to CE-SDS, care should be taken that the sample preparation conditions do not affect the reported sample purities (e.g. unanticipated fragmentation for non-reduced samples, or incomplete reduction for reduced samples).

Reverse-phase liquid chromatography-mass spectrometry (rpLC–MS) may also be used to both quantify and identify the site of biotherapeutic clipping [12]. rpLC–MS has been used to identify clip sites in the heavy chain observed during the cell culture development of a glycosylated mAb [54]. rpLC–MS has also been used to identify three cleavage sites within a mAb heavy chain after accelerated incubation [55]. MS-based clipping analyses are especially beneficial in that they provide identification of biotherapeutic clip sites and may offer opportunities to reduce those liabilities through protein engineering.

Hydrophobicity

Hydrophobicity is a key attribute that informs on the propensity of the molecule to self-associate and is often an indicator of success during the molecule assessment stage. The primary assay to evaluate the hydrophobicity of a biologic is HIC [56, 57]. The technique can provide insight into structurally different conformers that are driven by hydrophobic interactions as well as the aggregation-prone regions resulting from surface exposure of hydrophobic patches [58, 59]. Increased retention times on a hydrophobic matrix during chromatography has been shown to correlate with a propensity to precipitate in a set of monoclonal antibodies [60]. HIC has been used widely in the high-throughput landscape for developability assessments [4, 5] as well as for development of in silico models to predict viscosity and mouse clearance [6].

A drawback of HIC is its low throughput that is often not amenable to processing the large numbers of candidates that are commonly screened during antibody discovery. An alternative, higher throughput, assay is based on evaluating antibody self-association at high salt concentrations using affinity-capture self-interaction nanoparticle spectroscopy (AC-SINS). Estep 2015 demonstrated that hydrophobic mAbs, as identified by the HIC retention time, generally show significant self-association at low to moderate ammonium sulfate concentrations, while hydrophilic mAbs typically show self-association only at high ammonium sulfate concentrations [61]. To further improve throughput and ease of executions, Meissner 2023 has developed an easy-to-automate microplate-based workflow [62]. 24 mAbs were tested using chemically defined beads to detect the binding to hydrophobic surfaces and were compared to published HIC retention times. The bead-based assay was able to successfully identify all mAbs with high hydrophobicity and displayed a broad dynamic range in terms of the assay concentration.

Hyphenation of HIC with mass spectrometry has been explored as a way to identify conformational changes, PTMS such as oxidation, glycosylation, drug to antibody ratio, and free thiol variants of the mAbs [63, 64]. In the past, HIC was not considered an MS-compatible method because the mobile phase contains nonvolatile buffer salts at high concentrations [65]; however, the use of a HIC column with longer surface carbon chain length and higher hydrophobicity allows for good separation using MS-compatible acetate buffers [66, 67]. Such advances that blend protein compatibility and the high resolution of HIC with the ever-increasing capabilities of mass spectrometry merit close monitoring for future incorporation into developability packages.

Charge Variants

Molecule charge variants are key attributes that reflect the heterogeneity of the molecule and may affect potency. There are two standard methods for measuring and identifying acidic and basic molecule variants: capillary isoelectric focusing (cIEF) and ion exchange chromatography (IEX). cIEF is a denaturing method wherein analytes are first separated based on isoelectric point in the focusing step, then mobilized across a detector for quantification.

This is a time-intensive process that is also susceptible to loss of resolution during the mobilization step. Imaged cIEF does not require the mobilization step and instead quantifies variants using whole-column imaging, which reduces the amount of time to both develop and run charge variant analysis assays [68]. However, icIEF still requires 5–15 min/sample, which is significant and often precludes its application for large biomolecule panels.

Capillary zone electrophoresis (CZE) is another CE-based separation technique that may be used to quantify charge variants. CZE separates analytes based on electrophoretic mobilities and may be used with coated or uncoated capillaries; a rapid method has been published describing rapid CZE analysis (11 minutes) of mAbs in uncoated silica gel capillaries using a dynamic coating [69]. Microfluidic chips may also be used to reduce sample requirements and improve throughput. Han and coworkers demonstrated the use of the LabChip GXII to perform high-throughput (80 s/sample) charge variant analyses of mAbs [69]. Wheeler et al. later applied a similar microchip zone electrophoresis (MZE) method on a LabChip GXII toward the quantification of known charge variants (C-terminal lysine modification and deamidation), as well as clone selection (Figure 12.4) [70].

Electrophoresis-based charge variant detection methods may also be hyphenated to MS, thereby providing both charge variant quantification and identification. When hyphenating a CE system to MS, it is critical to maintain separation and resolution as much as possible while enabling MS compatibility. Two recent commercial systems have demonstrated significant advances in this area. The ZipChip is a commercialized CE-ESI-integrated device which has resolved charge variants resulting from incomplete C-terminal lysine processing and from varying levels of sialylation [71, 72]. Native CZE-MS methods using the commercialized device have demonstrated high sensitivity, good fidelity with established charge variant techniques, and LOD as low as 0.01 ng on-chip [73, 74]. Microfluidic CE-MS offers significant benefits due to its low sample consumption and rapid separations. However, it is often more difficult to obtain reproducible migration times compared to LC-based separations [75]. Commercial systems from Sciex and Advanced Electrophoresis Solutions have been developed which enable online icIEF-MS charge variant analysis [76–78]. While the separation time is still too long to be truly high-throughput, significant progress has been made. Wu and coworkers compared online icIEF-MS using the CE Infinite device with a hyphenated CEX-MS method and found icIEF-MS to exhibit improved separation, sensitivity, and accuracy, although the CEX-MS offered shorter run-times and higher throughput [78].

IEX may be used with cationic or anionic resins and may be used with salt- or pH-based separation. In salt-based gradients, the pH is held constant and a gradient of increasing ionic strength is applied. In contrast, for pH-based separations, the pH is increased over the course of the analytical method, and species of interest will elute at their isoelectric point (pI). pH-based separations typically offer higher resolution than salt-based separations and require less sample-specific method optimization [79]. The two gradient types may also be combined to keep ionic strength constant throughout the run and improve resolution [80]. The throughput of CEX has been improved through the development of ultrashort (5–50 mm) columns for CEX analyses. Good charge variant resolution in less than 2 minutes was obtained for subunit analysis of therapeutically relevant samples using a 15 × 2.1 mm column and a pH-based gradient [81].

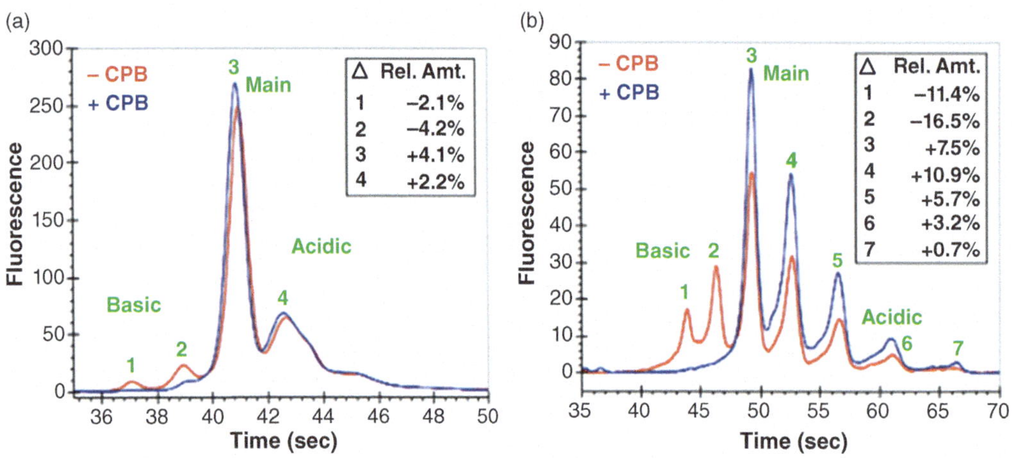

Figure 12.4 High-throughput MZE separation of (a) mAb A and (b) mAb C before (−CPB) and after (+CPB) carboxypeptidase digestion to fully remove C-terminal lysine residues. Inset shows a change in relative amount of each peak. Source: Wheeler et al. [70] / with permission of American Chemical Society.

Ion exchange chromatography may also be hyphenated to MS; this approach requires the identification of MS-compatible buffer systems that still enable high-resolution chromatographic separations [82]. Ma and coworkers have reported an online CEX-MS method capable of chromatographically resolving charge variants for mAb samples with pIs ranging from 6.4 to 9.2 [83]. To achieve improved chromatographic performance, they performed a separation combining both pH and salt gradients. While less common, AEX may also be coupled to MS for characterization of charge variants; Liu and coworkers have shown that this is useful for IgG4 molecules, which typically have lower pIs than IgG1 molecules [84].

Viscosity

Self-association-driven attributes include colloidal stability and viscosity. Here, we discuss several high-throughput methods that predict successful manufacturability and the utility of these tools in the molecule assessment space. The industry standard for viscosity measurement of biopharmaceuticals uses cone-and-plate measurements using a rheometer [85]. Cone-and-plate measurements are not parallelized, require manual cleaning between measurements, and require >10 mg of protein per high-concentration measurement, all of which preclude their use in high-throughput developability assays. There are two main commercial microfluidic systems that have been developed to improve the throughput and decrease the sample requirements of high-concentration viscosity measurements. The Initium was the first system; released in 2016 by RheoSense, it can inject samples directly from a 96-well microplate which facilitates walk-away analysis. However, samples are still analyzed 1-by-1, and the system requires a robust cleaning method between injections to avoid sample carryover, leading to long overall analysis times. The HoneyBun is another system, released in 2022 by Unchained Labs. While analysts must physically transfer samples from a microplate or other container into the analytical capillary; the HoneyBun does offer parallelized analysis, which decreases overall time.

Other methods have been developed to measure viscosity of biopharmaceutical samples in microplate formats. One approach toward measuring viscosity is to add a reporter material to a high-concentration protein solution and then measure the effects of the solution viscosity on the reporter material. For instance, Razinkov and coworkers developed a high-throughput viscosity assay by adding 150 nm polystyrene beads to protein solutions and measuring the apparent radius (R) of the beads using a DLS system with a microplate reader [86]. Beads in viscous samples resulted in an increase in the apparent bead radius as measured by the autocorrelation function, enabling measurement of viscosity. Good correlations between DLS viscosity and cone-and-plate viscosity were observed over a range of concentrations and formulations with two mAb samples.

Despite significant improvements in increasing throughput and decreasing sample need for viscosity measurements, their application at the screening stage is limited due to the need for high-concentration (150 ± 10 mg/ml) material for measurements. There is therefore significant interest in identifying surrogate low-concentration assays that measure the underlying molecular attributes contributing to antibody viscosity at high concentration. One example is AC-SINS, which captures antibodies onto gold beads, and then measures the resulting shift in plasmon wavelength for self-interactive antibody samples [87, 88]. Multiple studies from the biopharmaceutical industry have been reported using AC-SINS as a high-throughput surrogate to guide and evaluate the effect of viscosity-reduction engineering [89, 90]. However, a downside of AC-SINS is the need to prepare fresh lots of the gold-antibody reagent for each panel analysis, introducing the potential for lot-to-lot variability with corresponding variability in sample measurements.

Another promising surrogate assay is clone self-interaction by biolayer interferometry (CSI-BLI) [91]. Using an Octet instrument, samples are first immobilized on a surface, and then tested for self-binding. CSI-BLI output has been shown to be correlative with AC-SINS and aggregation-related outputs (percent monomer, purification titer) [92], as well as correlative with assays measuring cross-interaction [4]. The diffusion coefficient (kD) of low-concentration mAb samples as measured by DLS has also been shown to correlate with direct measurement of high-concentration mAb viscosity [7, 93, 94]. Wei and coworkers recently demonstrated that the addition of salt to these low-concentration kD measurements suppresses electrostatic repulsion between mAb molecules at formulation pH, enabling the measurement of short-range self-association and improving the correlation between kD and viscosity compared to kD measured in the absence of salt (Figure 12.5) [95].

Antibody self-association may also lead to other colloidal stability issues, including precipitation, phase-induced separation, or low solubility. In many cases, direct observation of these undesirable attributes is not possible until proteins are scaled up at high concentrations. In addition to the self-interaction assays described above, a common method of measuring colloidal stability in protein solutions is to induce precipitation by the addition of poly(ethylene glycol) (PEG) [96, 97]. The addition of PEG reduces the solubility of the product mAbs, decreasing

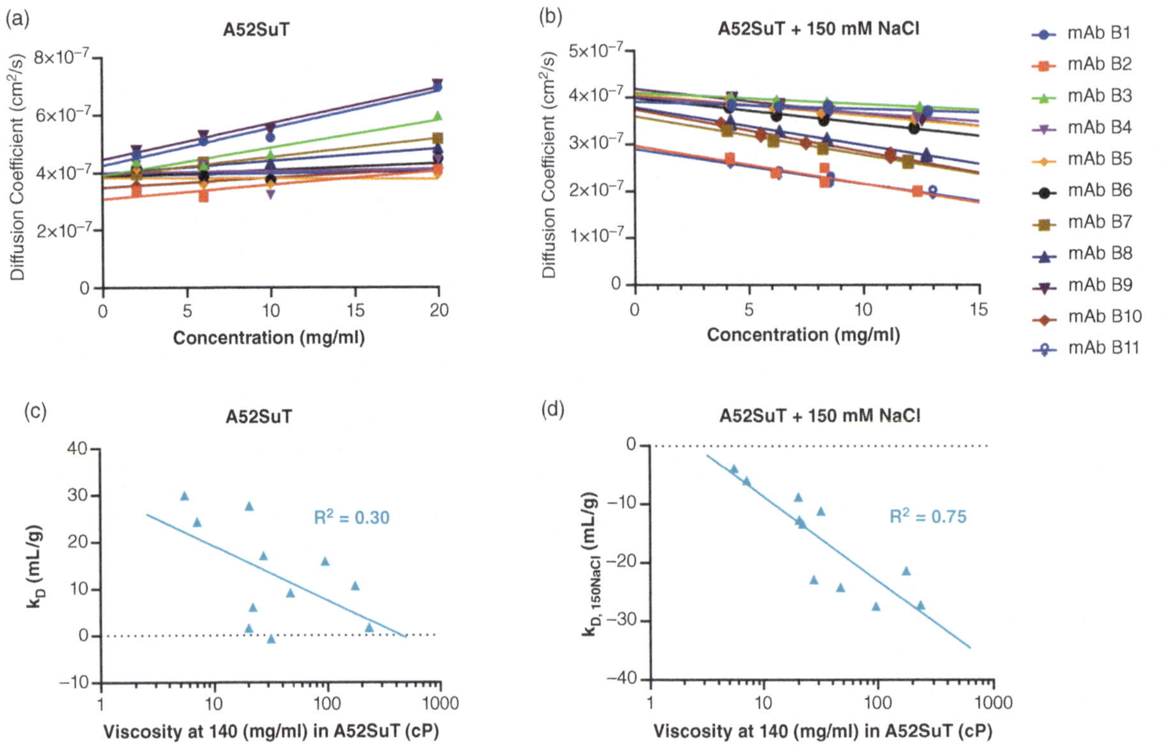

Figure 12.5 Improved correlation between k_D and viscosity with the addition of 150 mM NaCl for a panel of 11 IgG1 mAbs. (a, b) DLS k_D plots of the mAbs in formulation buffer alone (10 mM acetate, 6% sucrose, pH 5.2, 0.01% PS80, A52SuT) and with the addition of 150 mM NaCl. (c, d) Correlation scatter plots between viscosity at 140 mg/ml and measured k_D ($n = 3$). Source: Wei et al. [95] / with permission of American Chemical Society.

the amount of material required. Chai and coworkers reported a high-throughput adaptation of these earlier works and demonstrated its utility in identifying optimal formulations [98].

Conformational Stability

The functional activity of most therapeutic proteins depends on the integrity of their three-dimensional structures. Conformational stability provides insight into molecular behavior under accelerated stress conditions and can be a good predictor of long-term stability. Differential scanning fluorimetry (DSF) is a key tool to interrogate the inherent conformational stability of a biologic. The assay measures changes in the fluorescence as the conformation of the protein changes during thermal unfolding as the temperature is ramped up. This technique provides biophysical properties such as protein unfolding transitions (T_m and T_{onset}), and the transition associated with aggregation (T_{agg}) to inform on conformational and colloidal stability. Conventionally, the T_m was measured in the presence of a fluorescent dye, such as SYPRO orange or CPM, that preferentially binds to unfolded proteins [99]. A major advantage of the dye-based methods is that they lend themselves to facile plate-based, high-throughput optimization.

The current industry standard for a dye-free alternative is nano-DSF, which is a capillary-based assay that can measure 48 samples in parallel within 90 minutes. NanoDSF-based thermostability screening approach is widely used to screen and select the optimal buffer conditions for liquid formulation of antibodies [100, 101]. For instance, Kim 2022 demonstrated that sodium acetate buffer at pH 4.6 is better suited for IgG formulations in comparison to sodium citrate [102]. NanoDSF is also employed as part of a high-throughput developability workflow during lead discovery campaigns to inform early stages of candidate selection [5, 103]. Bailly 2020 showed that the differences in T_m/T_{agg} correlated well with aggregation propensity during expression and accelerated stability, as well as particle formation [5]. With the availability of an autosampler, which is an all-inclusive system with robotic arm, computer, and monitor, it is now able to analyze up to 1,536 samples for hands-free characterization of thermal unfolding and chemical denaturation in a single nanoDSF experiment. Further innovation in the form of capillary-free, dye-free nanoDSF simplifying

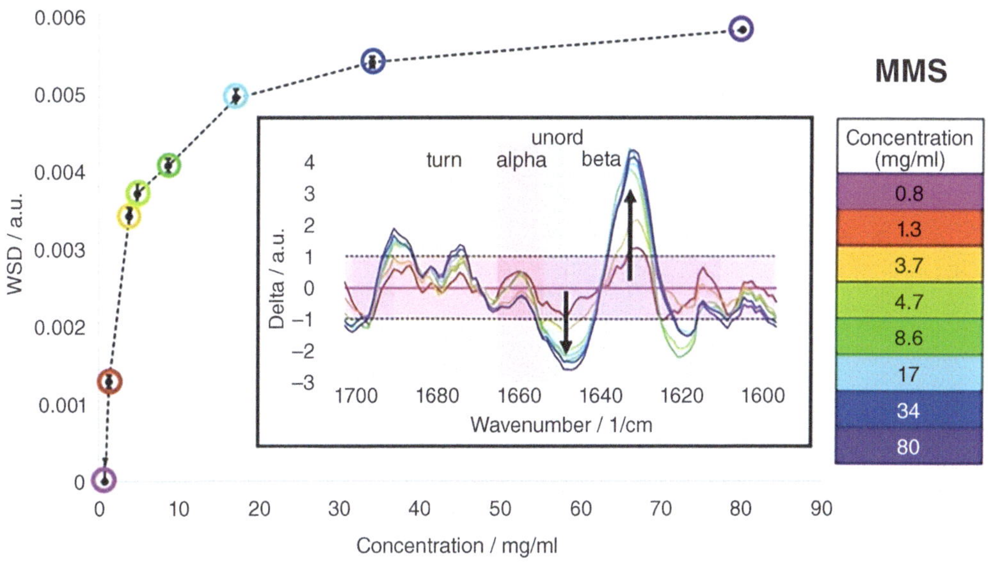

Figure 12.6 The MMS weighted spectral difference (WSD) across the measured concentration range between 0.8 and 80 mg/ml shows that the changes observed as a function of concentration predominantly occur in the beta-sheet and unordered range. Source: RedShiftBio [106] / AZoNetwork.

sample preparation into a single microplate should aid in throughput improvements for lead selection.

Other conventional methods routinely used to monitor the secondary structure of proteins and assess conformational stability include Fourier transform infrared spectroscopy (FTIR) and circular dichroism (CD), but both suffer from several limitations such as specific concentration requirements, low sensitivity, inconsistent reproducibility, or poor excipient incompatibility [104, 105]. These limitations have been addressed by microfluidic modulation spectroscopy (MMS), an automated infrared technology. Figure 12.6 shows the MMS spectra recorded at different concentrations between 0.8 and 80 mg/ml. The changes predominantly occurring due to the beta-sheet and the unordered region can be observed in the range of 1 and 20 mg/ml. MMS offers a superior limit of quantitation for determining structural impurity at 3.2%, which is significantly better than that of FTIR at 23% [104, 107]. This work and other studies [108, 109] highlight the advancement of MMS as a highly sensitive technique to detect small changes of protein structures due to aggregation or misfolding.

Chemical Stability Assessment

In addition to physical instability challenges such as aggregation and precipitation, biological therapeutics are susceptible to a myriad of chemical degradation pathways that are a major concern throughout the product life cycle. These pathways can give rise to chemical modifications such as deamidation, isomerization, oxidation, and glycation that may alter the biological activity of the therapeutic, potentially cause immunogenicity, impact pharmacokinetics, and may induce further degradation [110]. Chemical modifications to amino acid residues in the complementarity-determining regions (CDRs), can especially affect the antibody's binding affinity, effector function, or the pharmacokinetics (PK) of therapeutic IgGs, potentially leading to reduced efficacy [111]. Furthermore, chemically degraded biologics can be prone to additional physical instabilities, underscoring that both phenomena are closely intertwined and interdependent. Given the mechanistically diverse nature of chemical degradation and its criticality, having robust analytical methods for quantifying chemical liabilities is paramount during product development.

Multi-attribute methods based on liquid chromatography–mass spectrometry have emerged as an important technology that can identify and quantify multiple product quality attributes critical to developability and manufacturability assessments [112–114]. A high throughput means of executing peptide mapping analysis for chemical liability identification includes a robust multi-instrument, integrated automated system designed to execute multiple processes in series. These include microplate-enabled, liquid handling robotics for protein digestion, high-resolution separation techniques such as LC, MS/MS together with an automated data processing pipeline (Figure 12.7).

The high value proposition of deploying chemical liability analysis earlier in the pipeline for developability assessments has necessitated higher throughput and

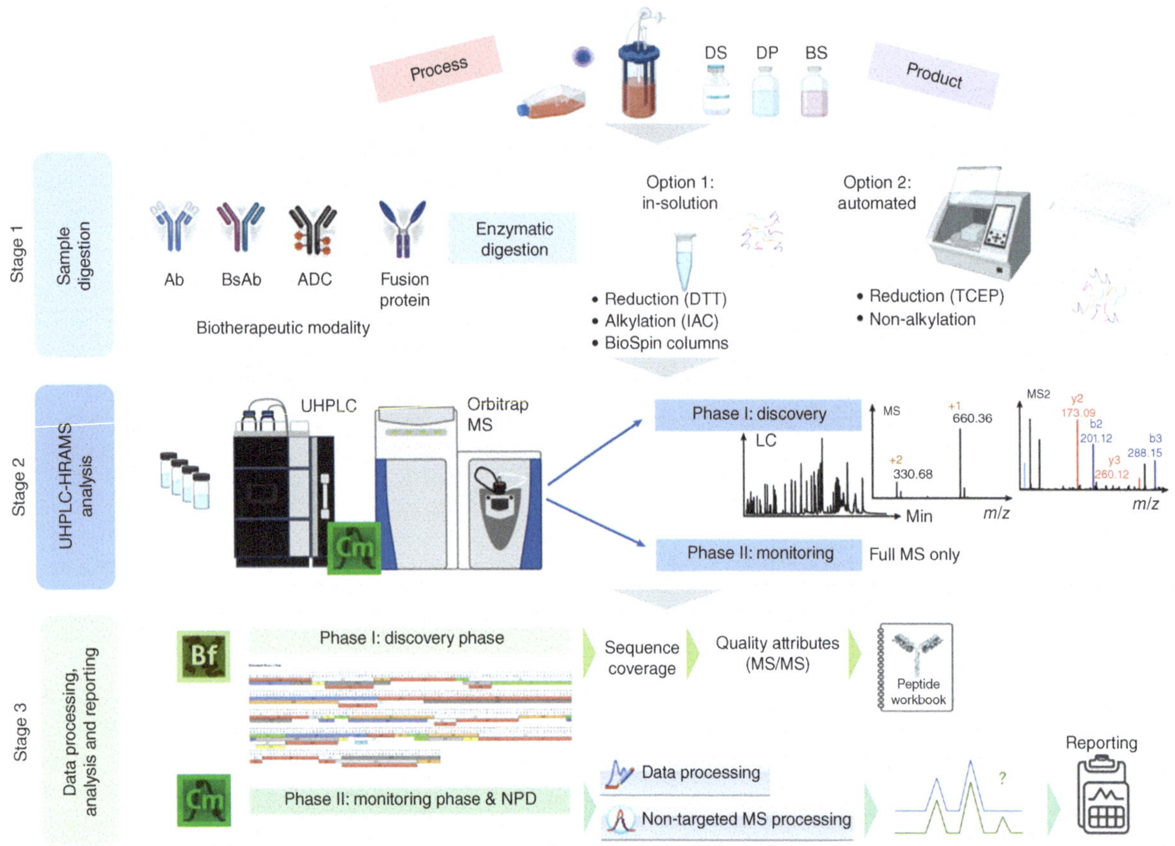

Figure 12.7 An overview of an end-to-end MAM set up that includes liquid handling robotics for protein digestion, high-resolution separation techniques such as LC and MS/MS together with an automated data processing pipeline. Source: Millán-Martín et al. [115] / Taylor & Francis / CC BY 4.0.

improved laboratory operational efficiency of MS techniques to characterize molecules. Richardson 2011 reported an automated digestion protocol including protein denaturation, reduction, alkylation, and enzymatic digestion on a commonly used HPLC autosampler [116]. The protocol uses a series of dilutions, instead of an on-column buffer exchange to reduce the adverse effects of denaturants and interfering reagents. Further work by the same group describes a hands-free automation of peptide mapping using standard HPLC hardware [117]. Other liquid handlers have been described for automated peptide digestion. Chelius 2008 describes a workflow on Tecan using a custom-designed, 96-well desalting plate filled with silica-based desalting medium [118]. AssayMAP, developed by Agilent (Santa Clara, CA), is another semi-automated platform used for analytical-scale protein digest preparation that works in conjunction with the Bravo liquid handler. A hybrid workstation, capable of a completely automated workflow for proteomics sample preparation, has been described here [119, 120]. Qian 2021 describes a "walk-away" automated procedure using the Hamilton STAR liquid handling platform together with a 10 K MWCO microdialysis plate for sample clean-up in an automated process [121].

LC-based separation of digested peptides in line with MS/MS is the method of first intent for chemical liability analysis. The typical gradient times are on the order of 90 minutes with some as long as 190 minutes [122]. Several studies on optimization of the column, the LC method used for separation, and MS-related parameters have been reported [123, 124]. A comprehensive review of these aspects is beyond the scope of this chapter and can be found in the published literature [115, 125, 126].

In addition to LC-based separation of digested peptides, multiplexed CE has been increasingly employed in peptide analysis due to its rapid yet efficient separation, low sample requirements, and compatibility with biological materials [127, 128]. A high degree of separation efficiency was achieved using six different separation buffers in a 96-capillary array CE system [129] and further improved by He and colleagues in 2002 [130]. Model proteins were digested with three different enzymes (trypsin, chymotrypsin, and pepsin), and the digests were separated with eight CZE buffers and two MEKC buffers in order to achieve a comprehensive peptide map. A chip-based capillary electrophoresis mass spectrometry (CZE-MS) was used as an integrated microfluidic platform to demonstrate rapid (<10 minutes) resolution of isoAsp from Asn- and Asp-containing peptides over a range of peptide sizes and hydrophobicity [131]. Low-level nanomolar quantitation for the peptides was demonstrated when resolution was achieved.

Waldenmeir et al. describe the integration of the different high-throughput approaches to provide a unified solution for MS analysis for biotherapeutics [21]. Samples are purified from clarified supernatant digested with trypsin and chymotrypsin and loaded onto LC–MS systems completely handsfree with subsequent data automatically processed and reported. Automation is further enabled through the application of embedded workflows in the data analysis software. The capabilities include identification of post-translational modification to identify potential liabilities present in CDRs as well as sequence variant analysis, and host cell protein contamination analysis and evaluating molecule liabilities. The large volume of MS data produced by the system is processed into useful reports for manual review and curation by analysts. The work describes the end-to-end, modular automation of a complex, multistep workflow that demonstrates the potential for further "lab of the future" applications.

Leveraging AI/ML for Molecule Assessment

As the number of success criteria that a therapeutic must meet for advancement increases, the associated analytical package also increases proportionally, necessitating an alternate approach to predict success. Artificial intelligence has the potential to fill the gap in the existing wet lab workflows, especially in cases where analytics are still relatively low throughput [132]. For example, significant efforts have focused on the development of viscosity prediction algorithms, since it is highly material-intensive when measured by cone-and-plate rheometry [85]. While some models exclusively rely on calculated features (e.g. sequence or structure), some use experimental output from low-concentration surrogate assays to improve model accuracy. Mock and coworkers found that accuracy for a viscosity prediction model was improved when combining DLS and AC-SINS experimental outputs with structure-based features [6]. This model was employed to deselect candidates in an engineering campaign to successfully reduce the viscosity of a therapeutic candidate [133]. Other examples have focused on determining rules and descriptors for classifying high- and low-viscosity samples. For instance, Lai and coworkers developed a decision tree model to determine the effect of net charge and hydrophobic residues on measured high-concentration viscosity [134]. Makowski and coworkers have also demonstrated how a decision tree model can influence engineering decisions to successfully reduce antibody viscosity [90].

Chemical liability quantification is another wet lab assay which is time and resource intensive, and therefore efforts have been invested into developing predictive models. Some efforts have focused on identifying in silico features as predictors of susceptibility toward chemical modification. For instance, Agrawal et al. found that the calculated average water correlation number (as calculated using molecular dynamics) correlates well with experimental measurements of hydrogen peroxide-induced methionine oxidation [135]. However, most work has found that machine learning models outperform predictions based on single features. Sankar and coworkers have reported that a random forest-based model to predict chemically induced oxidation of antibody methionine residues outperformed a classifier using solvent accessible surface area (SASA) alone [136]. A random forest model which predicts photooxidation of methionine residues has also been successfully developed [137].

Predictive models for other important biotherapeutic attributes have also been developed. Using amino acid sequence information and buffer conditions such as pH and salt concentration Gentiluomo et al. predict the biophysical properties such as melting temperature (T_m), diffusion interaction parameter (kD), and aggregation onset temperature (T_{agg})] of mAbs [138]. Thermal stability is an attribute of interest that is often a focus of predictive approaches. The ability to predict long-term stability using accelerated stability studies and measurable biophysical parameters such as T_m and T_{agg} in different formulations has been demonstrated [139]. Along the same lines, the intrinsic propensity for aggregation of mAbs has been predicted using sequence-derived physicochemical properties [140].

While each assay as performed may address a single attribute, or multiple attributes, with interactions and correlations between results, it is also important to develop systems to holistically compile and evaluate the assay results. Hedbitch et al. used the dataset published by Jain et al., which assessed 12 different assays across 137 antibodies, to develop ML models for all the 12 biophysical properties using only their sequence-based descriptors as input [4, 141].

The current process of acquiring data to train these models is still labor-intensive. To expand and accelerate the potential of AI/ML tools, it is imperative that ML methods are uniformly applied and combined with consistent data storage. Federated learning is a potential option for pooling data in order to train more accurate models, while still keeping sequence and other proprietary information confidential [142].

Future of Developability Assessments

The assays mentioned in this chapter address a variety of biophysical attributes that are important to assessing developability. A critical aspect of implementing high-throughput assays includes automation such as liquid handlers and integrated robotic automation systems, which should be used to prevent sample handling errors and to maximize "walkaway time." Analytical methods such as the ones mentioned above should be standardized as much as possible, and any deviations from procedure should be well-documented. Additionally, automated analysis and reporting tools are required to process and manage the large number of generated data points. Data should be stored in queryable databases and should be stored with as much metadata as possible, enabling future data consumers to fully understand how an experiment was carried out, including both data acquisition and processing.

Although wet lab analytical efforts offer robust insights into the biophysical characteristics of a biologic, it is important to recognize that successful commercialization is a highly complex multivariate problem that requires judicious optimization. The design space for protein sequences is large, and each molecule could be formulated with multiple combinations of buffer salts and excipients thereby exponentially increasing the size of the panel. Therefore, in addition to identifying experimental assays, it is also important to develop systems to holistically compile and evaluate the assay results. For instance, Gentiluomo and coworkers have reported a series of weighted Stability Risk Scores, with separate readouts for computed properties, and low, medium, and high levels of experimental analytical effort [143]. Other work has been done to benchmark and understand the characteristics of developable biotherapeutics, for instance, the Therapeutic Antibody Profiler [144].

As more effort is put into decreasing product development timelines and improving the success of clinical trials, the structure and analytical assays that make up developability assessments will also continue to evolve. Overall, we envision that the evolution of experimental techniques and advances in ML methods have the potential to reduce the experimental resource burden (Figure 12.8). Concomitantly, the cost of biotherapeutics development will likely decrease due to the shrinking of product development timelines and higher success rate in clinical trials.

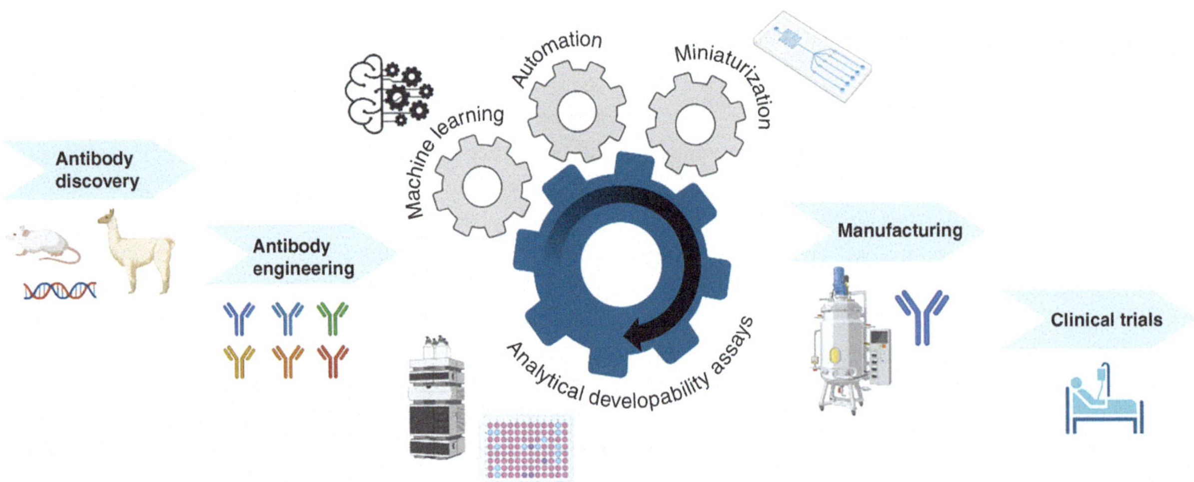

Figure 12.8 Schematic overview of the antibody discovery, engineering and manufacturing process for the development of biologics. Machine learning, automation, and miniaturization can be applied during the developability assessment process to increase throughput and reduce costs and resources. Source: Fernández-Quintero et al. [145]/Taylor & Francis/CC BY 4.0.

References

1 Deshaies, R.J. (2020). Multispecific drugs herald a new era of biopharmaceutical innovation. *Nature* 580 (7803): 329–338.

2 Mullard, A. (2016). Parsing clinical success rates. *Nat. Rev. Drug Discov.* 15 (7): 447–448.

3 Sun, D., Gao, W., Hu, H., and Zhou, S. (2022). Why 90% of clinical drug development fails and how to improve it? *Acta Pharm. Sin. B* 12 (7): 3049–3062.

4 Jain, T., Sun, T., Durand, S. et al. (2017). Biophysical properties of the clinical-stage antibody landscape. *Proc. Natl. Acad. Sci.* 114 (5): 944. https://doi.org/10.1073/pnas.1616408114.

5 Bailly, M., Mieczkowski, C., Juan, V. et al. (2020). Predicting Antibody Developability Profiles Through Early Stage Discovery Screening. *MAbs* 12 (1): 1743053. https://doi.org/10.1080/19420862.2020.1743053.

6 Mock, M., Jacobitz, A.W., Langmead, C.J. et al. (2023). Development of in silico models to predict viscosity and mouse clearance using a comprehensive analytical data set collected on 83 scaffold-consistent monoclonal antibodies. *MAbs* 15 (1): 2256745. https://doi.org/10.1080/19420862.2023.2256745.

7 Kingsbury, J.S., Saini, A., Auclair, S.M. et al. (2020). A single molecular descriptor to predict solution behavior of therapeutic antibodies. *Sci. Adv.* 6 (32): eabb0372.

8 Xu, Y., Wang, D., Mason, B. et al. (2019; Taylor & Francis). Structure, heterogeneity and developability assessment of therapeutic antibodies. *MAbs* 11: 239–264.

9 Zhang, W., Wang, H., Feng, N. et al. (2023). Developability assessment at early-stage discovery to enable development of antibody-derived therapeutics. *Antib. Ther.* 6 (1): 13–29.

10 Kwan, T.O., Kolek, S.A., Danson, A.E. et al. (2022). Measuring protein aggregation and stability using high-throughput biophysical approaches. *Front. Mol. Biosci.* 9: 890862.

11 Campuzano, I.D.G. and Sandoval, W. (2021). Denaturing and Native Mass Spectrometric Analytics for Biotherapeutic Drug Discovery Research: Historical, Current, and Future Personal Perspectives. *J. Am. Soc. Mass Spectrom.* 32 (8): 1861–1885. https://doi.org/10.1021/jasms.1c00036.

12 Dillon, T.M., Bondarenko, P.V., Rehder, D.S. et al. (2006). Optimization of a reversed-phase high-performance liquid chromatography/mass spectrometry method for characterizing recombinant antibody heterogeneity and stability. *J. Chromatogr. A* 1120 (1–2): 112–120. https://doi.org/10.1016/j.chroma.2006.01.016 From NLM.

13 Campuzano, I.D., San Miguel, T., Rowe, T. et al. (2016). High-throughput mass spectrometric analysis of covalent protein-inhibitor adducts for the discovery of irreversible inhibitors: a complete workflow. *J. Biomol. Screen.* 21 (2): 136–144.

14 Sawyer, W.S., Srikumar, N., Carver, J. et al. (2020). High-throughput antibody screening from complex matrices using intact protein electrospray mass spectrometry. *Proc. Natl. Acad. Sci.* 117 (18): 9851–9856. https://doi.org/10.1073/pnas.1917383117.

15 Campuzano, I.D., Pelegri-O'Day, E.M., Srinivasan, N. et al. (2022). High-throughput Mass spectrometry for biopharma: a universal modality and target Independent analytical method for accurate biomolecule characterization. *J. Am. Soc. Mass Spectrom.* 33 (11): 2191–2198.

16 Park, H.-M., Winton, V.J., Drader, J.J. et al. (2019). Novel interface for high-throughput analysis of biotherapeutics by electrospray mass spectrometry. *Anal. Chem.* 92 (2): 2186–2193.

17 Zhang, H., Liu, C., Hua, W. et al. (2021). Acoustic ejection mass spectrometry for high-throughput analysis. *Anal. Chem.* 93 (31): 10850–10861.

18 Zacharias, A.O., Liu, C., VanAernum, Z.L. et al. (2023). Ultrahigh-Throughput Intact Protein Analysis with Acoustic Ejection Mass Spectrometry. *J. Am. Soc. Mass Spectrom.* 34 (1): 4–9. https://doi.org/10.1021/jasms.2c00276.

19 Campuzano, I.D., Robinson, J.H., Hui, J.O. et al. (2019). Native and denaturing MS protein deconvolution for biopharma: monoclonal antibodies and antibody–drug conjugates to polydisperse membrane proteins and beyond. *Anal. Chem.* 91 (15): 9472–9480.

20 Ziegengeist, T., Orth, J., Kroll, K. et al. (2023). High-Throughput and Format-Agnostic Mispairing Assay for Multispecific Antibodies Using Intact Mass Spectrometry. *Anal. Chem.* 95 (27): 10265–10278. https://doi.org/10.1021/acs.analchem.3c00742.

21 Waldenmaier, H.E., Gorre, E., Poltash, M.L. et al. (2023). "Lab of the Future"—Today: fully automated system for high-throughput mass spectrometry analysis of biotherapeutics. *J. Am. Soc. Mass Spectrom.* 34 (6): 1073–1085. https://doi.org/10.1021/jasms.3c00036.

22 Phung, W., Bakalarski, C.E., Hinkle, T.B. et al. (2023). UniDec processing pipeline for rapid analysis of

biotherapeutic mass spectrometry data. *Anal. Chem.* 95 (30): 11491–11498. https://doi.org/10.1021/acs.analchem.3c02010.

23 Sharma, A., Beirne, J., Khamar, D. et al. (2023). Evaluation and screening of biopharmaceuticals using multi-angle dynamic light scattering. *AAPS PharmSciTech* 24 (4): 84.

24 Hawe, A., Friess, W., Sutter, M., and Jiskoot, W. (2008). Online fluorescent dye detection method for the characterization of immunoglobulin G aggregation by size exclusion chromatography and asymmetrical flow field flow fractionation. *Anal. Biochem.* 378 (2): 115–122.

25 Philo, J.S. (2009). A critical review of methods for size characterization of non-particulate protein aggregates. *Curr. Pharm. Biotechnol.* 10 (4): 359–372.

26 Fekete, S. and Guillarme, D. (2023). Ultra-short columns for the chromatographic analysis of large molecules. *J. Chromatogr. A* 1706: 464285. https://doi.org/10.1016/j.chroma.2023.464285.

27 Mou, X., Yang, X., Li, H. et al. (2014). A high throughput ultra performance size exclusion chromatography assay for the analysis of aggregates and fragments of monoclonal antibodies. *Pharm. Bioprocess.* 2 (2): 141–156.

28 He, F., Phan, D.H., Hogan, S. et al. (2010). Detection of IgG aggregation by a high throughput method based on extrinsic fluorescence. *J. Pharm. Sci.* 99 (6): 2598–2608.

29 Liau, B. (2020) High-Throughput Fluorescence Based mAbs Aggregation Analysis Workflow. *Agilent Application Note* 2020, (5994-1745EN).

30 Tahirbegi, B., Magness, A.J., Piersimoni, M.E. et al. (2022). Toward high-throughput oligomer detection and classification for early-stage aggregation of amyloidogenic protein. *Front. Chem.* 10: 967882.

31 Perrou, A., Meissner, C., Bussat, R. et al. (2022). Rapid assessment of aggregation of biologics in different matrices and from upstream and downstream development process stages. *27th Meeting European Society for Animal Cell Technology, Lisbon 2022.*

32 Shiba, K., Niidome, T., Katoh, E. et al. (2010). Polydispersity as a parameter for indicating the thermal stability of proteins by dynamic light scattering. *Anal. Sci.* 26 (6): 659–663.

33 Meyer, A., Dierks, K., Hussein, R. et al. (2015). Systematic analysis of protein–detergent complexes applying dynamic light scattering to optimize solutions for crystallization trials. *Acta Crystallogr. Sect. F Struct. Biol. Cryst. Commun.* 71 (1): 75–81.

34 Kwan, T.O., Reis, R., Siligardi, G. et al. (2019). Selection of biophysical methods for characterisation of membrane proteins. *Int. J. Mol. Sci.* 20 (10): 2605.

35 Dauer, K., Pfeiffer-Marek, S., Kamm, W., and Wagner, K.G. (2021). Microwell plate-based dynamic light scattering as a high-throughput characterization tool in biopharmaceutical development. *Int. J. Pharm.* 13 (2): 172.

36 Yu, Z., Reid, J.C., and Yang, Y.-P. (2013). Utilizing dynamic light scattering as a process analytical technology for protein folding and aggregation monitoring in vaccine manufacturing. *J. Pharm. Sci.* 102 (12): 4284–4290.

37 Hulse, W.L., Gray, J., and Forbes, R.T. (2013). Evaluating the inter and intra batch variability of protein aggregation behaviour using Taylor dispersion analysis and dynamic light scattering. *Int. J. Pharm.* 453 (2): 351–357.

38 Bhirde, A.A., Chiang, M.-J., Venna, R. et al. (2018). High-throughput in-use and stress size stability screening of protein therapeutics using algorithm-driven dynamic light scattering. *J. Pharm. Sci.* 107 (8): 2055–2062.

39 Cole, D., Young, G., Weigel, A. et al. (2017). Label-free single-molecule imaging with numerical-aperture-shaped interferometric scattering microscopy. *ACS photonics* 4 (2): 211–216.

40 Sonn-Segev, A., Belacic, K., Bodrug, T. et al. (2020). Quantifying the heterogeneity of macromolecular machines by mass photometry. *Nat. Commun.* 11 (1): 1772.

41 Deslignière, E., Ley, M., Bourguet, M. et al. (2021). Pushing the limits of native MS: online SEC-native MS for structural biology applications. *Int. J. Mass Spectrom.* 461: 116502. https://doi.org/10.1016/j.ijms.2020.116502.

42 Jooß, K. and Somsen, G. *Native Separation-Mass Spectrometry in Biopharmaceutical Analysis.* 2023.

43 Ehkirch, A., Hernandez-Alba, O., Colas, O. et al. (2018). Hyphenation of size exclusion chromatography to native ion mobility mass spectrometry for the analytical characterization of therapeutic antibodies and related products. *J. Chromatogr. B* 1086: 176–183.

44 Haberger, M., Leiss, M., Heidenreich, A.-K. et al. (2016; Taylor & Francis). Rapid characterization of biotherapeutic proteins by size-exclusion chromatography coupled to native mass spectrometry. *MAbs* 8: 331–339.

45 Lambiase, G., Inman, S.E., Muroni, M. et al. (2022). High-throughput multiplex analysis of mAb aggregates and charge variants by automated two-dimensional size exclusion-cation exchange chromatography coupled to mass spectrometry. *J. Chromatogr. A* 1670: 462944.

46 Juliano, B.R., Keating, J.W., Li, H.W. et al. (2023). Development of an automated, high-throughput

methodology for native mass spectrometry and collision-induced unfolding. *Anal. Chem.* 95 (45): 16717–16724.

47 Kaur, H., Beckman, J., Zhang, Y. et al. (2021). Capillary electrophoresis and the biopharmaceutical industry: therapeutic protein analysis and characterization. *TrAC Trends Anal. Chem.* 144: 116407. https://doi.org/10.1016/j.trac.2021.116407.

48 Minkner, R., Burger, U., Malz, L. et al. (2023). Efficient and reliable CE-SDS separations by the high-speed turbo CE-SDS® cartridge. *J. Liq. Chromatogr. Relat. Technol.* 46 (16–20): 293–301. https://doi.org/10.1080/10826076.2023.2269233.

49 Wang, W.-H., Cheung-Lau, J., Chen, Y. et al. (2019). Specific and high-resolution identification of monoclonal antibody fragments detected by capillary electrophoresis–sodium dodecyl sulfate using reversed-phase HPLC with top-down mass spectrometry analysis. *MAbs* 11 (7): 1233–1244. https://doi.org/10.1080/19420862.2019.1646554.

50 Langlois, N.I., Evans, K.C., Dillon, T.M. et al. (2023). Characterization of IgG1 low molecular weight variants by correlating RPLC-MS with a CE-SDS migration time shift assay. In: *CE Pharm 2023*. Philadelphia, PA.

51 Chen, X., Tang, K., Lee, M., and Flynn, G.C. (2008). Microchip assays for screening monoclonal antibody product quality. *Electrophoresis* 29 (24): 4993–5002.

52 Kahle, J., Maul, K.J., and Wätzig, H. (2018). The next generation of capillary electrophoresis instruments: performance of CE-SDS protein analysis. *Electrophoresis* 39 (2): 311–325. https://doi.org/10.1002/elps.201700278 (accessed 2024/04/27).

53 Cai, H., Song, Y., Zhang, J. et al. (2016). Optimization of microchip-based electrophoresis for monoclonal antibody product quality analysis revealed needs for extra surfactants during denaturation. *J. Pharm. Biomed. Anal.* 120: 46–56.

54 Ivleva, V.B., Schneck, N.A., Gollapudi, D. et al. (2018). Investigation of Sequence Clipping and Structural Heterogeneity of an HIV Broadly Neutralizing Antibody by a Comprehensive LC-MS Analysis. *J. Am. Soc. Mass Spectrom.* 29 (7): 1512–1523. https://doi.org/10.1007/s13361-018-1968-0.

55 Kerr, R.A., Keire, D.A., and Ye, H. (2019). The impact of standard accelerated stability conditions on antibody higher order structure as assessed by mass spectrometry. *MAbs* 11 (5): 930–941. https://doi.org/10.1080/19420862.2019.1599632.

56 Haverick, M., Mengisen, S., Shameem, M., and Ambrogelly, A. (2014). Separation of mAbs molecular variants by analytical hydrophobic interaction chromatography HPLC: overview and applications. *MAbs* 6 (4): 852–858. https://doi.org/10.4161/mabs.28693 From NLM.

57 Fekete, S., Veuthey, J.L., Beck, A., and Guillarme, D. (2016). Hydrophobic interaction chromatography for the characterization of monoclonal antibodies and related products. *J. Pharm. Biomed. Anal.* 130: 3–18. https://doi.org/10.1016/j.jpba.2016.04.004 From NLM.

58 Janson, J.-C. (2011). *Protein Purification: Principles, High Resolution Methods, and Applications*. John Wiley & Sons.

59 Queiroz, J.A., Tomaz, C.T., and Cabral, J.M. (2001). Hydrophobic interaction chromatography of proteins. *J. Biotechnol.* 87 (2): 143–159. https://doi.org/10.1016/s0168-1656(01)00237-1 From NLM.

60 Kohli, N., Jain, N., Geddie, M.L. et al. (2015). A novel screening method to assess developability of antibody-like molecules. *MAbs* 7 (4): 752–758. https://doi.org/10.1080/19420862.2015.1048410. PubMed.

61 Estep, P., Caffry, I., Yu, Y. et al. (2015). An alternative assay to hydrophobic interaction chromatography for high-throughput characterization of monoclonal antibodies. *MAbs* 7 (3): 553–561. https://doi.org/10.1080/19420862.2015.1016694. PubMed.

62 Meissner, C., Wosnitza, C., Perrou, A., Giehring, S. (2024). Developability assessment for nonspecificity and polyspecificity in high-throughput bead-based assays in microplates. *PEGS Boston Summit* 2024.

63 Saikusa, K., Kinumi, T., and Kato, M. (2022). Development of native mass spectrometry with nanoelectrospray ionization coupled to size exclusion chromatography for proteins. *Rapid Commun. Mass Spectrom.* 36 (21): e9395.

64 Gilardoni, E. and Regazzoni, L. (2022). Liquid phase separation techniques for the characterization of monoclonal antibodies and bioconjugates. *J. Chromatogr. Open* 2: 100034.

65 Yan, Y., Xing, T., Wang, S. et al. (2020). Online coupling of analytical hydrophobic interaction chromatography with native mass spectrometry for the characterization of monoclonal antibodies and related products. *J. Pharm. Biomed. Anal.* 186: 113313. https://doi.org/10.1016/j.jpba.2020.113313.

66 Schneck, N.A., Khatri, K., Maust, M.D. et al. (2022). Separation techniques for intact antibody analysis by mass spectrometry. *J. Liq. Chromatogr. Relat. Technol.* 45 (17–20): 271–283.

67 Wei, B., Han, G., Tang, J. et al. (2019). Native Hydrophobic Interaction Chromatography Hyphenated to Mass Spectrometry for Characterization of Monoclonal Antibody Minor Variants. *Anal. Chem.* 91 (24): 15360–15364. https://doi.org/10.1021/acs.analchem.9b04467 From NLM.

68 Salas-Solano, O., Kennel, B., Park, S.S. et al. (2012). Robustness of iCIEF methodology for the analysis of monoclonal antibodies: an interlaboratory study. *J. Sep. Sci.* 35 (22): 3124–3129. https://doi.org/10.1002/jssc.201200633 (accessed 2024/05/19).

69 He, Y., Isele, C., Hou, W., and Ruesch, M. (2011). Rapid analysis of charge variants of monoclonal antibodies with capillary zone electrophoresis in dynamically coated fused-silica capillary. *J. Sep. Sci.* 34 (5): 548–555.

70 Wheeler, T.D., Sun, J.L., Pleiner, S. et al. (2014). Microchip zone electrophoresis for high-throughput analysis of monoclonal antibody charge variants. *Anal. Chem.* 86 (11): 5416–5424. https://doi.org/10.1021/ac500497n.

71 Redman, E.A., Batz, N.G., Mellors, J.S., and Ramsey, J.M. (2015). Integrated microfluidic capillary electrophoresis-electrospray ionization devices with online MS detection for the separation and characterization of intact monoclonal antibody variants. *Anal. Chem.* 87 (4): 2264–2272. https://doi.org/10.1021/ac503964j.

72 Füssl, F., Trappe, A., Carillo, S. et al. (2020). Comparative elucidation of cetuximab heterogeneity on the intact protein level by cation exchange chromatography and capillary electrophoresis coupled to mass spectrometry. *Anal. Chem.* 92 (7): 5431–5438.

73 Wu, Z., Wang, H., Wu, J. et al. (2023). High-sensitivity and high-resolution therapeutic antibody charge variant and impurity characterization by microfluidic native capillary electrophoresis-mass spectrometry. *J. Pharm. Biomed. Anal.* 223: 115147. https://doi.org/10.1016/j.jpba.2022.115147.

74 Cao, L., Fabry, D., and Lan, K. (2021). Rapid and comprehensive monoclonal antibody Characterization using microfluidic CE-MS. *J. Pharm. Biomed. Anal.* 204: 114251. https://doi.org/10.1016/j.jpba.2021.114251.

75 Naghdi, E., Moran, G.E., Reinau, M.E. et al. (2023). Concepts and recent advances in microchip electrophoresis coupled to mass spectrometry: technologies and applications. *Electrophoresis* 44 (1–2): 246–267.

76 He, X., ElNaggar, M., Ostrowski, M.A. et al. (2022). Evaluation of an icIEF-MS system for comparable charge variant analysis of biotherapeutics with rapid peak identification by mass spectrometry. *Electrophoresis* 43 (11): 1215–1222.

77 Schlecht, J., Moritz, B., Kiessig, S., and Neusüß, C. (2023). Characterization of therapeutic mAb charge heterogeneity by iCIEF coupled to mass spectrometry (iCIEF–MS). *Electrophoresis* 44 (5–6): 540–548.

78 Wu, G., Yu, C., Wang, W. et al. (2023). Mass spectrometry-based charge heterogeneity characterization of therapeutic mAbs with imaged capillary isoelectric focusing and ion-exchange chromatography as separation techniques. *Anal. Chem.* 95 (4): 2548–2560.

79 Farnan, D. and Moreno, G.T. (2009). Multiproduct high-resolution monoclonal antibody charge variant separations by pH gradient ion-exchange chromatography. *Anal. Chem.* 81 (21): 8846–8857.

80 Zhang, L., Patapoff, T., Farnan, D., and Zhang, B. (2013). Improving pH gradient cation-exchange chromatography of monoclonal antibodies by controlling ionic strength. *J. Chromatogr. A* 1272: 56–64.

81 Navarro-Huerta, J.A., Murisier, A., Nguyen, J.M. et al. (2021). Ultra-short ion-exchange columns for fast charge variants analysis of therapeutic proteins. *J. Chromatogr. A* 1657: 462568.

82 Bailey, A.O., Han, G., Phung, W. et al. (2018). Charge variant native mass spectrometry benefits mass precision and dynamic range of monoclonal antibody intact mass analysis. *MAbs* 10 (8): 1214–1225. https://doi.org/10.1080/19420862.2018.1521131.

83 Ma, F., Raoufi, F., Bailly, M.A. et al. (2020). Hyphenation of strong cation exchange chromatography to native mass spectrometry for high throughput online characterization of charge heterogeneity of therapeutic monoclonal antibodies. *MAbs* 12 (1): 1763762. https://doi.org/10.1080/19420862.2020.1763762.

84 Liu, A.P., Yan, Y., Wang, S., and Li, N. (2022). Coupling Anion Exchange Chromatography with Native Mass Spectrometry for Charge Heterogeneity Characterization of Monoclonal Antibodies. *Anal. Chem.* 94 (16): 6355–6362. https://doi.org/10.1021/acs.analchem.2c00707.

85 Burckbuchler, V., Mekhloufi, G., Giteau, A.P. et al. (2010). Rheological and syringeability properties of highly concentrated human polyclonal immunoglobulin solutions. *Eur. J. Pharm. Biopharm.* 76 (3): 351–356. https://doi.org/10.1016/j.ejpb.2010.08.002.

86 He, F., Becker, G.W., Litowski, J.R. et al. (2010). High-throughput dynamic light scattering method for measuring viscosity of concentrated protein solutions. *Anal. Biochem.* 399 (1): 141–143. https://doi.org/10.1016/j.ab.2009.12.003.

87 Sule, S.V., Dickinson, C.D., Lu, J. et al. (2013). Rapid analysis of antibody self-association in complex mixtures using immunogold conjugates. *Mol. Pharm.* 10 (4): 1322–1331. https://doi.org/10.1021/mp300524x.

88 Liu, Y., Caffry, I., Wu, J. et al. (2014). High-throughput screening for developability during early-stage antibody

discovery using self-interaction nanoparticle spectroscopy. *MAbs* 6 (2): 483–492. https://doi.org/10.4161/mabs.27431 PubMed.

89 Wu, J., Schultz, J.S., Weldon, C.L. et al. (2015). Discovery of highly soluble antibodies prior to purification using affinity-capture self-interaction nanoparticle spectroscopy. *Protein Eng. Des. Sel.* 28 (10): 403–414. https://doi.org/10.1093/protein/gzv045 (acccessed 2/27/2020).

90 Makowski, E.K., Chen, H.-T., Wang, T. et al. (2024). Reduction of monoclonal antibody viscosity using interpretable machine learning. *MAbs* 16 (1): 2303781. https://doi.org/10.1080/19420862.2024.2303781.

91 Sun, T., Reid, F., Liu, Y. et al. (2013). High throughput detection of antibody self-interaction by bio-layer interferometry. *MAbs* 5 (6): 838–841. https://doi.org/10.4161/mabs.26186.

92 Kelly, R.L., Sun, T., Jain, T. et al. (2015). High throughput cross-interaction measures for human IgG1 antibodies correlate with clearance rates in mice. *MAbs* 7 (4): 770–777. https://doi.org/10.1080/19420862.2015.1043503.

93 Connolly, B.D., Petry, C., Yadav, S. et al. (2012). Weak interactions govern the viscosity of concentrated antibody solutions: high-throughput analysis using the diffusion interaction parameter. *Biophys. J.* 103 (1): 69–78.

94 Yadav, S., Shire, S.J., and Kalonia, D.S. (2012). Viscosity behavior of high-concentration monoclonal antibody solutions: correlation with interaction parameter and electroviscous effects. *J. Pharm. Sci.* 101 (3): 998–1011.

95 Wei, Y., Qi, W., Maglalang, E. et al. (2023). Improved diffusion interaction parameter measurement to predict the viscosity of concentrated mAb solutions. *Mol. Pharm.* 20 (12): 6420–6428. https://doi.org/10.1021/acs.molpharmaceut.3c00797.

96 Li, L., Kantor, A., and Warne, N. (2013). Application of a PEG precipitation method for solubility screening: a tool for developing high protein concentration formulations. *Protein Sci.* : a publication of the Protein Society 22 (8): 1118–1123. https://doi.org/10.1002/pro.2289 From NLM.

97 Gibson, T.J., McCarty, K., McFadyen, I.J. et al. (2011). Application of a high-throughput screening procedure with PEG-induced precipitation to compare relative protein solubility during formulation development with IgG1 monoclonal antibodies. *J. Pharm. Sci.* 100 (3): 1009–1021. https://doi.org/10.1002/jps.22350 From NLM.

98 Chai, Q., Shih, J., Weldon, C. et al. (2019). Development of a high-throughput solubility screening assay for use in antibody discovery. *MAbs* 11 (4): 747–756. https://doi.org/10.1080/19420862.2019.1589851.

99 Niesen, F.H., Berglund, H., and Vedadi, M. (2007). The use of differential scanning fluorimetry to detect ligand interactions that promote protein stability. *Nat. Protoc.* 2 (9): 2212–2221.

100 Breitsprecher, D., Glücklich, N., Hawe, A., Menzen, T. (2017). Comparison of nanoDSF and µDSC for thermal stability assessment during biopharmaceutical formulation development. *NanoTemper Application Note* 2017, (NT-PR-006-01).

101 Ristic, M., Rosa, N., Seabrook, S.A., and Newman, J. (2015). Formulation screening by differential scanning fluorimetry: how often does it work? *Acta Crystallogr F Struct Biol Commun* 71 (Pt 10): 1359–1364. https://doi.org/10.1107/s2053230x15012662 From NLM.

102 Kim, S.H., Yoo, H.J., Park, E.J., and Na, D.H. (2021). Nano differential scanning fluorimetry-based thermal stability screening and optimal buffer selection for immunoglobulin G. *J. Pharm.* 15 (1): 29.

103 Magnusson, A.O., Szekrenyi, A., Joosten, H.J. et al. (2019). nanoDSF as screening tool for enzyme libraries and biotechnology development. *FEBS J.* 286 (1): 184–204.

104 Ma, T.M., Huang, R.H., Collins, V.I. et al. (2023). Advancement of microfluidic modulation spectroscopy as a highly sensitive protein characterization technique for the detection of small structural changes. *Anal. Biochem.* 683: 115350.

105 Pelton, J.T. and McLean, L.R. (2000). Spectroscopic methods for analysis of protein secondary structure. *Anal. Biochem.* 277 (2): 167–176. https://doi.org/10.1006/abio.1999.4320.

106 RedShiftBio. (2024) Advancing protein structure analysis: MMS versus CD spectroscopy in detecting in-situ structural changes. https://www.news-medical.net/whitepaper/20240215/Advancing-Protein-Structure-Analysis-MMS-Versus-CD-Spectroscopy-in-Detecting-In-Situ-Structural-Changes.aspx

107 Kendrick, B.S., Gabrielson, J.P., Solsberg, C.W. et al. (2020). Determining spectroscopic quantitation limits for misfolded structures. *J. Pharm. Sci.* 109 (1): 933–936.

108 Batabyal, D., Wang, L., Zonderman, J., and Wikström, M. (2020). Shaping IR spectroscopy into a powerful tool for biopharma characterizations. *Biopharm Int* 33: 42–47.

109 Liu, L.L., Wang, L., Zonderman, J. et al. (2020). Automated, high-throughput infrared spectroscopy for secondary structure analysis of protein biopharmaceuticals. *J. Pharm. Sci.* 109 (10): 3223–3230.

110 Nowak, C., Cheung, J.K., Dellatore, S.M. et al. (2017). Forced degradation of recombinant monoclonal antibodies: a practical guide. *MAbs* 9 (8): 1217–1230. https://doi.org/10.1080/19420862.2017.1368602.

111 Beck, A. and Liu, H. (2019). Macro-and micro-heterogeneity of natural and recombinant IgG antibodies. *Antibodies* 8 (1): 18.

112 Rogers, R.S., Abernathy, M., Richardson, D.D. et al. (2018). A view on the importance of "multi-attribute method" for measuring purity of biopharmaceuticals and improving overall control strategy. *AAPS J.* 20: 1–8.

113 Rogers, R.S., Nightlinger, N.S., Livingston, B. et al. (2015). Development of a quantitative mass spectrometry multi-attribute method for characterization, quality control testing and disposition of biologics. *MAbs* 7 (5): 881–890. https://doi.org/10.1080/19420862.2015.1069454 From NLM.

114 Rogstad, S., Yan, H., Wang, X. et al. (2019). Multi-attribute method for quality control of therapeutic proteins. *Anal. Chem.* 91 (22): 14170–14177.

115 Millán-Martín, S., Jakes, C., Carillo, S. et al. Multi-Attribute Method (MAM): an Emerging Analytical Workflow for Biopharmaceutical Characterization, Batch Release and cGMP Purity Testing at the Peptide and Intact Protein Level. *Crit. Rev. Anal. Chem.* 1–18. doi: 10.1080/10408347.2023.2238058.

116 Richardson, J., Shah, B., Xiao, G. et al. (2011). Automated in-solution protein digestion using a commonly available high-performance liquid chromatography autosampler. *Anal. Biochem.* 411 (2): 284–291.

117 Richardson, J. and Zhang, Z. (2023). Fully Unattended Online Protein Digestion and LC–MS Peptide Mapping. *Anal. Chem.* 95 (42): 15514–15521.

118 Chelius, D., Xiao, G., Nichols, A.C. et al. (2008). Automated tryptic digestion procedure for HPLC/MS/MS peptide mapping of immunoglobulin gamma antibodies in pharmaceutics. *J. Pharm. Biomed. Anal.* 47 (2): 285–294.

119 Johansen, E. and Skilton, S. J. (2017). Automated High Throughput Trypsin Digests for Highly Reproducible Peptide Map Analysis. *Sciex Application Note* 2017, (RUO-MKT-02-6976-C).

120 Yang, H., Koza, S. M. and Yu, Y. Q. (2023). Automated High-Throughput LC-MS Focused Peptide Mapping of Monoclonal Antibodies in Microbioreactor Samples. *Waters Application Note* 2023, (720007885).

121 Qian, C., Niu, B., Jimenez, R.B. et al. (2021). Fully automated peptide mapping multi-attribute method by liquid chromatography–mass spectrometry with robotic liquid handling system. *J. Pharm. Biomed. Anal.* 198: 113988.

122 Evans, A.R., Capaldi, M.T., Goparaju, G. et al. (2019; Taylor & Francis). Using bispecific antibodies in forced degradation studies to analyze the structure–function relationships of symmetrically and asymmetrically modified antibodies. *MAbs* 11: 1101–1112.

123 Kristensen, D.B., Ørgaard, M., Sloth, T.M. et al. (2022). Optimized multi-attribute method workflow addressing missed cleavages and chromatographic tailing/carry-over of hydrophobic peptides. *Anal. Chem.* 94 (49): 17195–17204. https://doi.org/10.1021/acs.analchem.2c03820.

124 Dick, L.W. Jr., Mahon, D., Qiu, D., and Cheng, K.C. (2009). Peptide mapping of therapeutic monoclonal antibodies: improvements for increased speed and fewer artifacts. *J. Chromatogr. B* 877 (3): 230–236. https://doi.org/10.1016/j.jchromb.2008.12.009 From NLM.

125 Graf, T., Heinrich, K., Grunert, I. et al. (2020). Recent advances in LC–MS based characterization of protein-based bio-therapeutics–mastering analytical challenges posed by the increasing format complexity. *J. Pharm. Biomed. Anal.* 186: 113251.

126 Yang, H. and Zubarev, R.A. (2010). Mass spectrometric analysis of asparagine deamidation and aspartate isomerization in polypeptides. *Electrophoresis* 31 (11): 1764–1772.

127 Gomes, F.P. and Yates, J.R. III, (2019). Recent trends of capillary electrophoresis-mass spectrometry in proteomics research. *Mass Spectrom. Rev.* 38 (6): 445–460.

128 Štěpánová, S. and Kašička, V. (2019). Recent developments and applications of capillary and microchip electrophoresis in proteomics and peptidomics (2015–mid 2018). *J. Sep. Sci.* 42 (1): 398–414.

129 Kang, S.H., Gong, X., and Yeung, E.S. (2000). High-throughput comprehensive peptide mapping of proteins by multiplexed capillary electrophoresis. *Anal. Chem.* 72 (14): 3014–3021.

130 He, Y., Zhong, W., and Yeung, E.S. (2002). Multiplexed on-column protein digestion and capillary electrophoresis for high-throughput comprehensive peptide mapping. *J. Chromatogr. B* 782 (1): 331–341. https://doi.org/10.1016/S1570-023202)00698-0.

131 Dykstra, A.B., Flick, T.G., Lee, B. et al. (2021). Chip-Based Capillary Zone Electrophoresis Mass Spectrometry for Rapid Resolution and Quantitation of Critical Quality Attributes in Protein Biotherapeutics. *J. Am. Soc. Mass Spectrom.* 32 (8): 1952–1963. https://doi.org/10.1021/jasms.0c00424.

132 Khetan, R., Curtis, R., Deane, C.M. et al. (2022). Current advances in biopharmaceutical informatics: guidelines, impact and challenges in the computational

developability assessment of antibody therapeutics. *MAbs* 14 (1): 2020082. https://doi.org/10.1080/19420862.2021.2020082.

133 Estes, B., Jain, M., Jia, L. et al. (2024). Sequence-Based Viscosity Prediction for Rapid Antibody Engineering. *Biomol.* 14 (6): 617.

134 Lai, P.-K., Fernando, A., Cloutier, T.K. et al. (2021). Machine learning applied to determine the molecular descriptors responsible for the viscosity behavior of concentrated therapeutic antibodies. *Mol. Pharm.* 18 (3): 1167–1175.

135 Agrawal, N.J., Dykstra, A., Yang, J. et al. (2018). Prediction of the hydrogen peroxide–induced methionine oxidation propensity in monoclonal antibodies. *J. Pharm. Sci.* 107 (5): 1282–1289.

136 Sankar, K., Hoi, K.H., Yin, Y. et al. (2018). Prediction of methionine oxidation risk in monoclonal antibodies using a machine learning method. *MAbs* 10 (8): 1281–1290. https://doi.org/10.1080/19420862.2018.1518887 From NLM.

137 Delmar, J.A., Buehler, E., Chetty, A.K. et al. (2021). Machine learning prediction of methionine and tryptophan photooxidation susceptibility. *Mol Ther Methods Clin Dev* 21: 466–477. https://doi.org/10.1016/j.omtm.2021.03.023 From NLM.

138 Gentiluomo, L., Roessner, D., Augustijn, D. et al. (2019). Application of interpretable artificial neural networks to early monoclonal antibodies development. *Eur. J. Pharm. Biopharm.* 141: 81–89.

139 Gentiluomo, L., Roessner, D., and Frieß, W. (2020). Application of machine learning to predict monomer retention of therapeutic proteins after long term storage. *Int. J. Pharm.* 577: 119039.

140 Obrezanova, O., Arnell, A., De La Cuesta, R.G. et al. (2015; Taylor & Francis). Aggregation risk prediction for antibodies and its application to biotherapeutic development. *MAbs* 7: 352–363.

141 Hebditch, M. and Warwicker, J. (2019). Charge and hydrophobicity are key features in sequence-trained machine learning models for predicting the biophysical properties of clinical-stage antibodies. *PeerJ* 7: e8199.

142 Mock, M., Edavettal, S., Langmead, C., and Russell, A. (2023). AI can help to speed up drug discovery - but only if we give it the right data. *Nature* 621 (7979): 467–470. https://doi.org/10.1038/d41586-023-02896-9 From NLM.

143 Gentiluomo, L., Svilenov, H.L., Augustijn, D. et al. (2020). Advancing Therapeutic Protein Discovery and Development through Comprehensive Computational and Biophysical Characterization. *Mol. Pharm.* 17 (2): 426–440. https://doi.org/10.1021/acs.molpharmaceut.9b00852.

144 Raybould, M.I., Marks, C., Krawczyk, K. et al. (2019). Five computational developability guidelines for therapeutic antibody profiling. *Proc. Natl. Acad. Sci.* 116 (10): 4025–4030.

145 Fernández-Quintero, M.L., Ljungars, A., Waibl, F. et al. (2023). Assessing developability early in the discovery process for novel biologics. *MAbs* 15 (1): 2171248. https://doi.org/10.1080/19420862.2023.2171248.

13

Establishing Analytical Methods for RNA-Based Therapeutics Characterization

Julien Camperi[1], Emily Freund[2], Carolina Galan[2], and Axel Guilbaud[3]

[1] *Cell Therapy Engineering and Development, Genentech, South San Francisco, CA, USA*
[2] *Department of Molecular Biology, Genentech, South San Francisco, CA, USA*
[3] *Protein Analytical Chemistry, Genentech, South San Francisco, CA, USA*

Introduction to RNA-Based Therapeutics

Using *in vitro* transcribed (IVT) messenger RNA (mRNA) molecules, scientists can deliver specific instructions to cells, directing them to trigger desired immune responses or produce therapeutic proteins using their natural machinery. mRNA-based therapies are a significant advance in drug development, representing the third major milestone after chemical and protein-based drugs [1]. There are currently two approved mRNA-based therapeutics: the Sars-CoV-2 vaccines BNT 162b2 (BioNTech/Pfizer) and mRNA 1273 (Moderna) [2, 3]. The outstanding success of mRNA vaccines against COVID-19 has paved the way for the emergence of mRNA as a new class of drugs. There are several other mRNA-based therapeutics targeting additional diseases such as cardiovascular, cancer, rabies, and cystic fibrosis currently in human clinical trials [1, 4].

The use of RNA as a therapeutic agent offers a number of distinct advantages, including increased safety, efficiency, and versatility [5]. In comparison to DNA drugs, RNA cannot integrate into the genome, minimizing the risk of insertional mutagenesis. Moreover, RNA therapeutics have transient activity and can be metabolized by the human body. The delivery systems of RNA therapeutics are also straightforward compared to DNA-based approaches, as they do not need to enter the nucleus to exert their function. Researchers are working to exploit these advantages of IVT mRNA. Recent advances have laid the foundations for a wide range of potential therapeutic applications by overcoming the challenges inherent in this class of drugs, particularly with regard to the control of mRNA translation efficiency, stability, and reactogenicity.

Applications of RNA-Based Drugs

Current clinical efforts are focused on protein production, protein replacement therapies, mRNA-based vaccines, genome editing, and cell therapies [6] (Figure 13.1). Messenger RNA supports transient and spatially controlled protein expression and when paired with targeted delivery strategies enables replacement of defective proteins in rare genetic diseases. In addition, mRNA can also be used to express factors that modulate defective biological pathways in common diseases. In regard to vaccines, mRNA is specifically well-suited as a therapeutic. As soon as a viral genome sequence is determined, an mRNA can be rapidly designed to encode antigens that prompt an adaptive immune response. In addition, the single-stranded mRNA molecule itself can activate the innate immune response delivering adjuvant activity that contributes to vaccine efficacy. In summary, protein-coding mRNAs can be used for target gene modification (e.g. Cas9 protein expression), antigen presentation, functional protein expression, or as starting material for the manufacture of T-cell therapy products.

For clinical use, the manufacture of high-quality IVT mRNA molecules is both critical and complex. It involves fine-tuning a series of factors, including codon optimizing the initial genetic sequences, ensuring the quality of the linearized plasmids used as the IVT templates, and identifying optimal conditions for the IVT reaction. This ensures the final integrity of the mRNA and, ultimately, its efficiency in expression. In addition, optimization of the mRNA chemistry, such as the 5′ cap, modified nucleosides, and the non-coding regulatory elements (e.g. 5′ and 3′ UTRs) and poly(A) tail length, is crucial to achieve high

Characterizing Biotherapeutics: Analytical Methods for Diverse Modalities, First Edition. Edited by Jennie R. Lill and Wendy Sandoval.

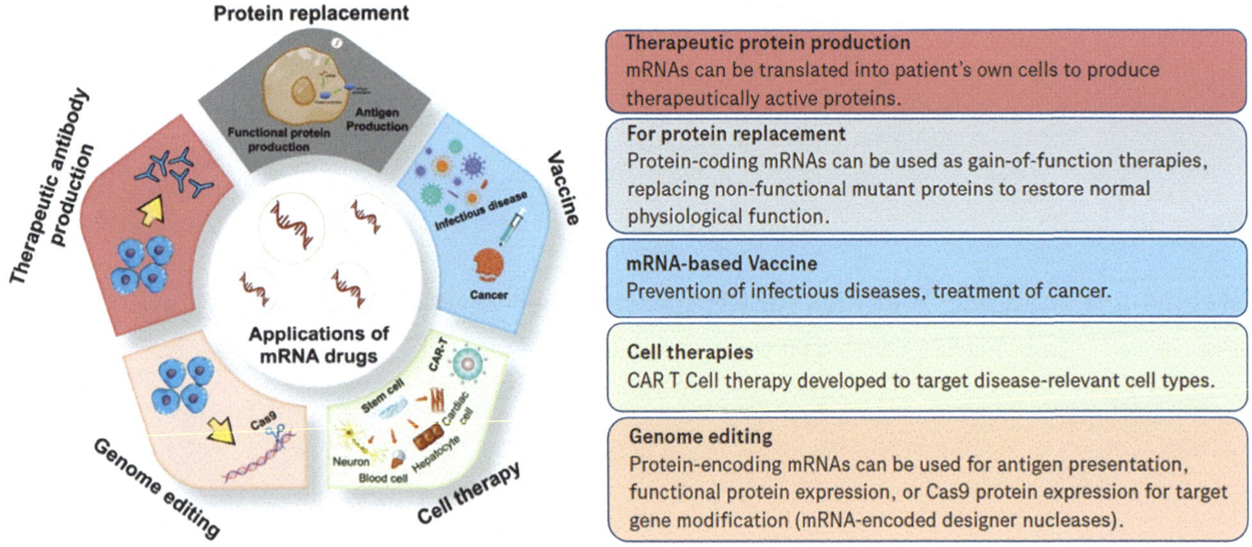

Figure 13.1 Applications of RNA-based drugs including therapeutic protein production, protein replacement, mRNA vaccines, cell and genome-editing applications. Source: Wang et al. [7]/Springer Nature/CC BY 4.0.

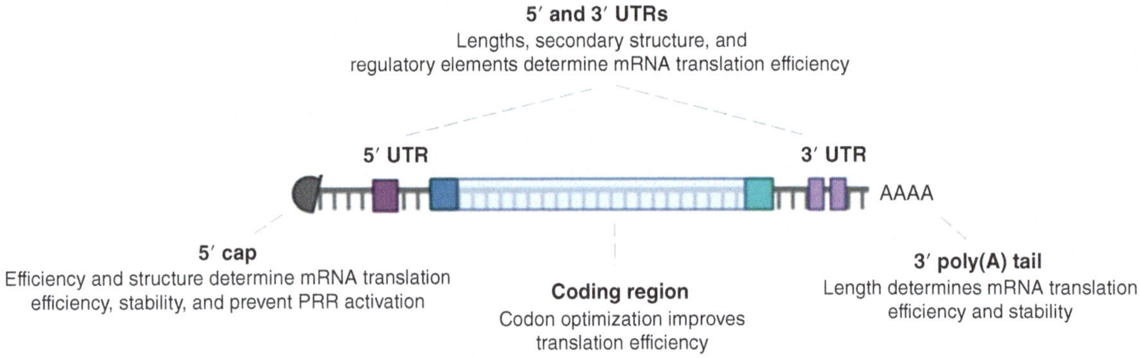

Figure 13.2 Structure and function of messenger RNA.

expression efficiency, stability, low immunogenicity and minimal toxicity in the resulting mRNA molecule. As shown in Figure 13.2, an mRNA therapeutic has five major regions: (i) the 5′ cap [8]; (ii) the 5′ untranslated region (UTR) [9, 10]; (iii) the coding region [11]; (iv) a 3′ UTR [12] and; (v) a 3′ poly(A) tail [13, 14].

Reliable and consistent *in vitro* mRNA synthesis depends firstly on the quality of the starting double-stranded DNA (dsDNA) template and then the IVT reaction conditions. Plasmids encoding a DNA-dependent RNA polymerase promoter (e.g. T7) followed by all critical mRNA components, including the 5′ and 3′ UTRs, the coding region and, ideally, an encoded poly(A) tail sequence, can be used directly as templates for IVT. In this case, plasmid linearization is performed prior to IVT to avoid the chance of the transcription enzyme (e.g. T7 RNA polymerase) reading through termination signals in the circular plasmid. Alternatively, the plasmids can be the substrate in a PCR reaction to produce a dsDNA product encoding the mRNA that is then used as the template in the IVT reaction. A template composed of an efficiently linearized plasmid or dsDNA PCR product with minimal impurities (i.e. host genomic DNA, RNA, and proteins) ensures mRNA integrity. Additionally, considerations that help ensure plasmid template quality for IVT or PCR include a single linearized DNA band on an agarose gel, high-supercoiled ratio, and a sequence-verified poly(A) tail.

The IVT mRNA purity directly affects translation efficiency. Therefore, several components introduced or formed in the IVT reaction (Figure 13.3), such as the DNA template, free unincorporated nucleotides, residual enzymes, short abortive RNA products, and dsRNA, are targeted for removal. Frequently, a combination of

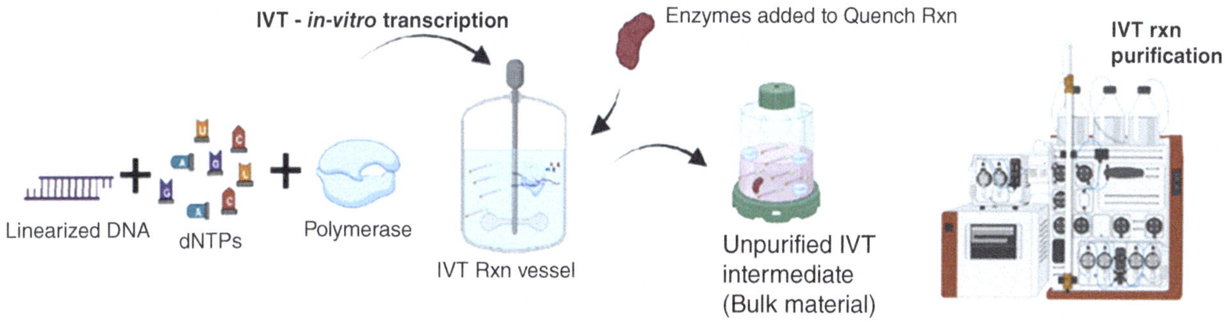

Figure 13.3 *In vitro* transcription (IVT) mRNA manufacturing process.

purification methods is used, including DNase treatment to remove the residual DNA templates and chromatographic techniques to isolate the full-length synthetic mRNA from other impurities [15–19]. Beyond effects on mRNA stability, translation efficiency, removing impurities such as dsRNA is critical to reducing the immunogenicity of the final IVT mRNA product [11].

The unique characteristics of IVT mRNA molecules make a comprehensive analysis of its final purity and predicted efficiency multifaceted and challenging. RNA is a large, single-stranded nucleic acid that exerts a strong negative charge, is highly susceptible to degradation by nucleases, is prone to structural heterogeneity, and is subject to a variety of chemical modifications [20]. In addition, the functional efficacy of mRNA-based therapies depends not only on the sequence integrity but also on correct secondary and tertiary structures, appropriate capping and adequate polyadenylation.

Capping is an absolute requirement to achieve mRNA stability. Uncapped IVT mRNA is targeted for degradation by cellular exonucleases, such as RNases [8]. Notably, some cap structures protect IVT mRNA from decapping cellular enzymes and promote translation by recruiting factors which aid in ribosome recruitment and binding. The specific cap structure also affects translation efficiency in a cell type-dependent manner. As reported by others, depending on the cell type, there may be some benefit to using Cap1 or the more highly modified Cap2 in IVT mRNA [21]. The cap structures each shield the mRNA from recognition by cellular sensors which activate the innate immune response and inhibit translation. Next to the cap structure, the poly(A) tail is critical since it serves different functional roles in mRNA splicing, stability, transport, and translation. For translation initiation, the 3′ poly(A) tail synergizes with the 5′ cap through associated proteins promoting mRNA circularization and ribosomal recruitment. For stabilization, interaction with poly(A)-binding protein PABPC prevents deadenylation and degradation of the transcript. As each aspect of the mRNA is impactful in its functionality, a comprehensive analytical strategy is imperative to ensure the safety, efficacy, and consistency of mRNA-based products throughout all stages of development of any RNA-based applications.

Quality Attributes and Techniques for the Characterization of mRNA

In developing an analytical framework for mRNA quality, it is imperative to include an assessment of all relevant quality attributes. Figure 13.4 delineates the myriad of quality attributes of mRNA and lists the techniques that can be used to characterize each one.

RNA Content

Measuring the content or amount of RNA present in a sample is commonly done with UV spectroscopy [22]. This technique is often preferred for its simplicity and rapid execution. The nucleic acid content is determined by simply monitoring the absorbance at a wavelength of 260 nm. To then evaluate the purity of the sample, the ratio of absorbance at 260 nm to that at 280 nm, which is influenced primarily by the presence of protein contaminants, is calculated. High-purity mRNA samples are generally characterized by a 260:280 ratio of 2. This ratio serves as a quick and effective gauge of simple purity as deviations often indicate unacceptable levels of protein contamination or other impurities.

Despite its convenience, the simplicity of the UV spectroscopy method comes with inherent limitations that necessitate careful management. It is crucial to treat samples with DNase prior to analysis to ensure the quantification pertains exclusively to RNA, thereby eliminating any residual DNA template that might have persisted from the upstream manufacturing process. Additionally, several extrinsic factors can skew the UV absorbance readings, potentially compromising the accuracy of the results.

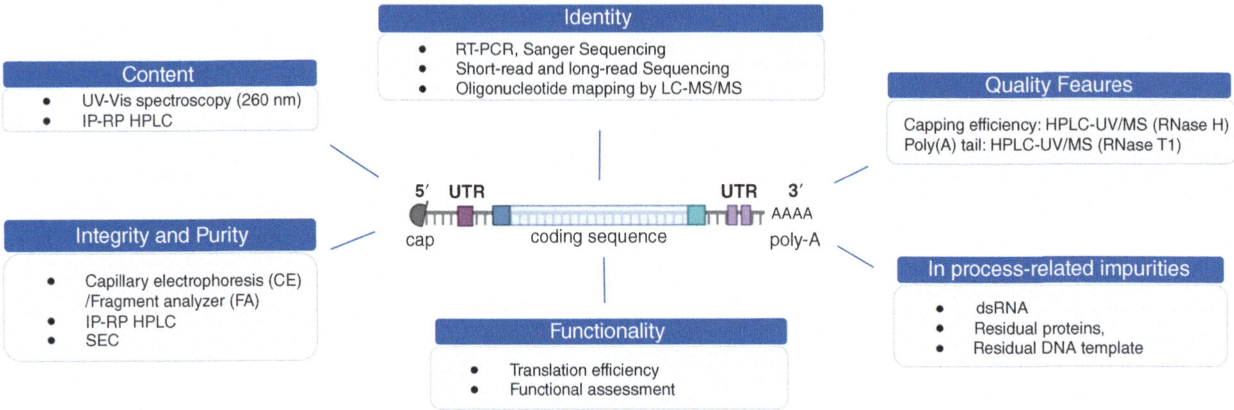

Figure 13.4 List of quality attributes and associated techniques for mRNA.

These include, but are not limited to, the material and optical quality of the cuvettes used, the temperature during measurement, and the pH level of the solution. Extreme pH can affect nucleic acid structure and absorbance characteristics [23]. Meticulous control of these parameters is therefore essential to prevent technical artifacts that could lead to erroneous conclusions and unnecessary investigative procedures. Understanding and mitigating these limitations enhance the reliability of UV spectroscopy in mRNA analysis, making it a valuable tool in the field of molecular biology and genomics.

Identity

RNA identity analysis ensures that the intended protein is produced and is therefore critical to drug development and subsequent regulatory filings. A widely used method is to perform reverse transcription polymerase chain reaction (RT-PCR) on the IVT-produced mRNA. This technique converts RNA into complementary DNA (cDNA). The cDNA can then be subjected to Sanger sequencing. Sanger sequencing on cDNA involves several key steps to ensure accurate determination of the nucleotide sequence. During sequencing, chain-terminating dideoxynucleotides are incorporated into the growing DNA strand, resulting in fragments of varying lengths. These fragments are then separated by capillary electrophoresis, allowing the sequence to be read and analyzed. The data analysis is straightforward, and it provides read lengths of up to ~800 base pairs making it very useful for ensuring that the correct DNA template was used in the IVT reaction.

However, Sanger sequencing also has its drawbacks. Its sensitivity is relatively low, with a detection limit of around 15–20%, which may be insufficient to detect low-frequency variants. In addition, it requires the production of cDNA and, to determine the full identity of longer mRNA sequences (>1 kb), multiple reads must be performed on the entire sequence. This means designing primers that "walk" along the cDNA to cover its entire length.

While Sanger sequencing can be used for verifying the sequence of cDNA, advanced methods to analyze mRNA have been developed using short-read (Illumina) and long-read (Oxford Nanopore Technologies [ONT] and PacBio) sequencing [24]. These methods can be used to measure additional key mRNA quality features, including sequence identity, but also integrity, purity, poly(A) tail length, detection of nucleoside modifications, as well as RNA structure analysis. Here we detail identity as sequence, detection of nucleoside modifications, and RNA structure (while integrity, purity, and poly(A) tail length are described in later sections).

Illumina sequencing provides high-throughput, reliable short reads that can be aligned to reference sequences to confirm mRNA identity with high confidence as the chemistry is designed to achieve high accuracy and quality. ONT has direct RNA sequencing capabilities, which circumvents the need for cDNA synthesis [25, 26]. This is especially advantageous as it preserves the native RNA molecule, allowing for mRNA identity confirmation as well as detection of nucleoside modifications and providing a broader understanding of the mRNA sequence and underlying structure. For example, direct RNA sequencing with ONT, combined with advancements in software, enables the site-specific detection of relevant modified nucleosides like pseudouridine and *N1*-methylpseudouridine (m1Ψ) [27–31]. Additionally, ONT sequencing can be paired with chemical probing methods to perform RNA structure analysis. SHAPE-MaP (selective 2′-hydroxyl acylation analyzed by primer extension and mutational profiling) [32] has been adapted to nanopore sequencing (nanoSHAPE) using a 2′-hydroxyl reactive compound (1-acetylimidazole) that does not interfere with translocation through the nanopore to chemically modify an RNA of interest. The chemically modified RNA of interest is sequenced alongside the unmodified control RNA and differences in RNA sequencing translocation rates are used to computationally derive a secondary structure. This technique offers full-length structure at single-molecule resolution using nanopore sequencing [33].

PacBio sequencing combines long-read capabilities with high accuracy, making it highly effective for precise and detailed mRNA identity analysis by sequencing full-length cDNA. Compared to short-read sequencing technologies like Illumina, both ONT and PacBio have a lower throughput limiting the number of samples that can be processed simultaneously [34].

Leveraging these workflows provides crucial insights for the development and quality control of mRNA-based therapeutics. In fact, an all-in-one workflow, VAX-seq has detailed how nanopore long-read sequencing can effectively measure mRNA quality attributes (sequence identity, integrity, and contamination) [35]. Without a doubt, more workflows will emerge as sequencing coupled with advanced bioinformatics tools can provide comprehensive insight into mRNA-based therapeutics.

As an alternative to sequencing-based methods, oligonucleotide mapping workflows involving mRNA digestion by RNAse (mainly RNase T1), followed by a liquid chromatography–tandem mass spectrometry (LC-MS/MS) analysis has attracted interest for mRNA characterization [36–38]. Gao et al. demonstrated the value of such a workflow to support the development of Comirnaty, the world's first commercial mRNA vaccine that immunizes against the SARS-CoV-2 virus [36] (Figure 13.5). This approach ensures the quality, safety, and efficacy of mRNA vaccines by providing: confirmation of construct identity and primary structure. Like sequencing-based methods, oligonucleotide mapping can also provide additional information, including confirmation of structural features such as the presence of a 5′ cap and a 3′ polyA tail.

RNA Integrity and Purity

Ensuring the production of high-quality, reproducible RNA is pivotal for the success of therapeutic mRNA applications, as the purity of IVT mRNA directly affects protein translation efficiency. Impurities, such as abortive transcripts, mRNA with partial poly(A) tail, or aggregates can lead to reduced protein expression, low efficacy, and potentially be stimulatory activating the innate immune system. Consequently, developing robust tests for assessing RNA purity is essential.

Analytical methods, such as capillary electrophoresis (CE) and high-performance liquid chromatography (HPLC), have become indispensable tools in this regard. These technologies enable the detection and quantification of various forms of impurities, including size and structural variants, the identification of which are critical to predicting the functional integrity of mRNA.

Capillary Electrophoresis (CE)

Electrophoresis techniques – particularly agarose gel and the more sophisticated capillary gel electrophoresis (CGE) – play a central role in assessing RNA integrity and purity [39, 40]. These techniques rely on the differential migration of analytes based on their hydrodynamic radius (size) through a sieving gel. In the example of CGE, the negatively charged RNA molecules migrate through a narrow capillary filled with a gel buffer solution, under the influence of an electric field. This enables precise and effective size-based separation of RNA, allowing determination of the percentage of an IVT RNA sample purity representing the intended RNA species [41].

CGE is applicable to RNA molecules up to 4,500 nucleotides (nts) in length, with satisfactory resolution. This is valuable as studies have demonstrated that small impurities often consist of RNA variants with different poly(A) tail lengths, resulting from aborted additions or partial hydrolysis [42]. These can be visualized, when CGE

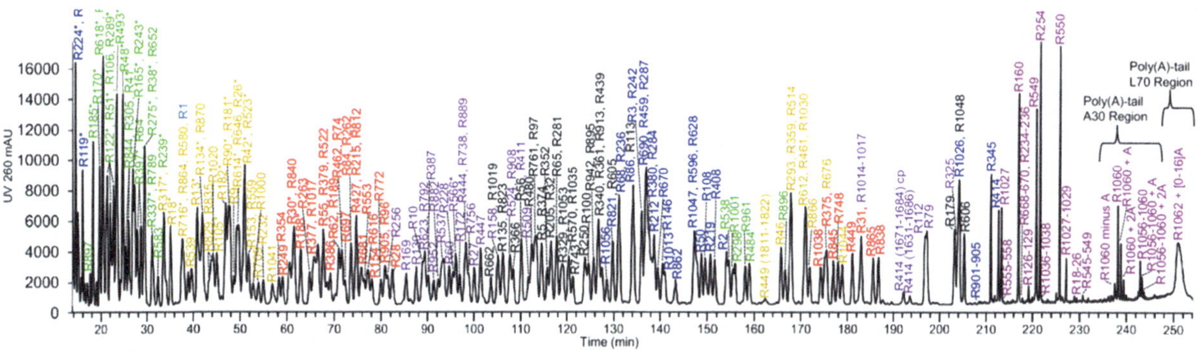

Figure 13.5 RNAse T1 Oligonucleotide mapping LC-MS/MS profile of the SARS-CoV-2 mRNA vaccine. Sequence coverage of 100% was obtained, enabling the confirmation of the identity of the mRNA including the presence of structural features such as 5′ cap and 3′ poly(A) tail. 5′ cap fragment is not shown here. Source: Gau et al. [36]/Springer Nature/CC BY 4.0

is performed on RNA with and without poly(A) tails. This method typically involves using high concentrations of denaturing agents such as formamide or urea [39, 40]. These agents facilitate the linearization of RNA, thereby enabling the efficient separation and migration of peaks. Under these stringent denaturing conditions, the content of aggregates detected in CGE can be identified as covalent dsRNA by-product, because the denaturing agents disrupt non-covalent base-pairing and base-stacking interactions [42]. Despite its great potential for achieving detailed separation of RNA size variants, CGE is characterized by relatively long separation times (between 40 and 60 minutes), limiting its wider adoption in RNA analysis.

To streamline analysis, fragment analysis (FA) is a valuable alternative, consisting of automated parallel capillary electrophoresis for reliable mRNA quality control (QC). Common QC bottlenecks are solved by automating key steps such as gel loading and sample injection, boosting laboratory efficiency. This makes FA the method of choice for determining mRNA purity in RNA-based drug development area [43].

Further advancements have led to the exploration of microcapillary electrophoresis (mCE), a miniaturized version of CE, particularly applicable for analyzing RNA in lipid nanoparticle vaccines [44]. mCE offers a shorter turnaround time, acceptable resolution, and the potential for increased sample throughput. Comparative studies between conventional CGE and mCE focusing on the assessment of RNA purity under denaturing conditions have shown that mCE produces consistent results regarding RNA purity, small impurities, intact RNA, and aggregate content [42]. These findings indicate that mCE is equally reliable and reproducible as traditional CGE, with the benefit of a significantly reduced run time (1.5 minutes for mCE compared to 60 minutes for CGE). Moreover, mCE provides the added advantage of analyzing RNA samples under both denaturing and native conditions. This versatility allows for the monitoring of both covalent and non-covalent aggregates. Consequently, mCE represents a promising advancement in RNA integrity and purity assessment, offering similar separation efficiency and data quality as conventional CGE but with enhanced speed and versatility.

Ion Pair-Reverse-Phase (IP-RP) HPLC

IP-RP HPLC has emerged as a pivotal technique for purity determination of IVT mRNAs [45, 46]. Given the polyanionic nature of nucleic acids, conventional RP chromatographic mode often falls short due to insufficient interactions between the analyte and the stationary phase. The IP-RP HPLC mode addresses this challenge by incorporating ion-pairing agents, which enhance the retention and separation of mRNA [47].

The methodology typically involves high column temperatures and the use of ion-pairing agents in association with organic solvents, in particular methanol [48]. The choice of ion-pairing agents and column settings has been fine-tuned through extensive research. For instance, studies have shown that using triethylammonium acetate (TEAA) at high temperatures can optimize the resolution of RNA ladders ranging from 155 to 1,770 nucleotides, which is critical for reducing confounding intermolecular and intramolecular RNA interactions [49].

Recent advancements in IP-RP HPLC have focused on improving separation efficiency, sensitivity, applicability, and turnaround times. For example, the use of octadecyl-based RPLC columns with superwide pores has shown promise in achieving higher separation for long mRNAs [50]. Moreover, incorporating alkylamines as ion-pairing agents and fluoroalcohols as counterions facilitates mass spectrometry (MS) integration, providing enhanced analytical capabilities over traditional UV detection methods [51].

IP-RP HPLC offers not only desirable selectivity for RNA analysis in terms of size but also sequence specificity, even under denaturing conditions involving high temperatures and organic solvents [52].

Similar to the rigorous characterization processes applied to monoclonal antibodies, various complimentary analytical assays – including but not limited to IP-RP HPLC – are necessary to uncover the different types of mRNA impurities, such as size and structural variants. Several companies have successfully applied IP-RP HPLC in their workflows to monitor mRNA heterogeneity and stability [49]. Notably, these methods involve the use of specialized columns and optimized mobile phases, demonstrating significant improvements in run times and resolution.

Other promising approaches, such as the use of superficially porous particles and polystyrene divinylbenzene polymeric columns, have also shown efficacy in separating large RNAs containing up to 6,000–8,000 bases [53, 54]. This ongoing innovation underscores the importance of developing robust, reliable technologies to ensure the production of high-purity mRNA, thereby enhancing the overall success and therapeutic efficacy of mRNA-based treatments.

Recent studies have highlighted the value of using orthogonal methods to assess mRNA purity, as each method can help distinguish different types of mRNA impurities. As an example, the complementarity between IP-RP HPLC and CGE is illustrated in Figure 13.6, where mRNAs with different poly(A) tail lengths are mainly

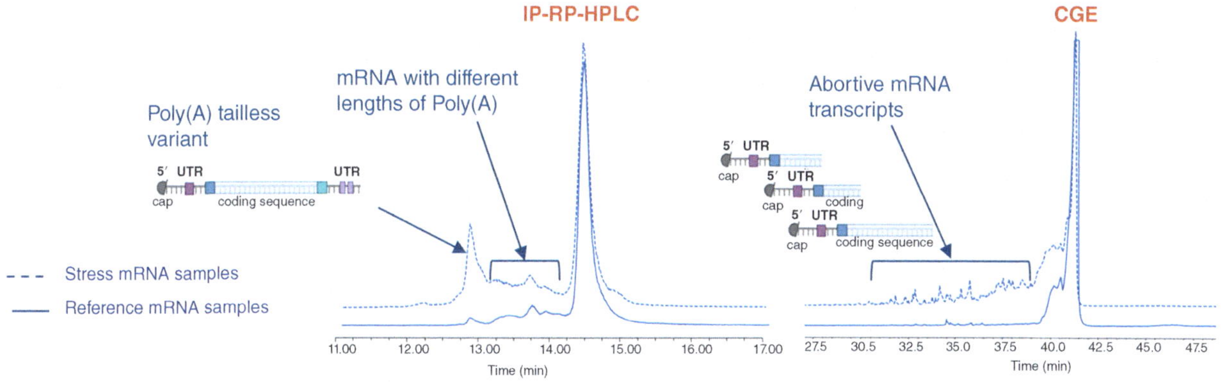

Figure 13.6 Importance of using orthogonal methods to assess mRNA. Similar reference and stress (heat stress) mRNA samples were analyzed by both IP-RP HPLC (left side) and CGE (right side). Impurities are shown.

observed by IP-RP HPLC, while CGE enabled the detection of aborted mRNA transcripts.

Size-Exclusion Chromatography (SEC)

Despite the solid array of analytical tools already available, questions persist as to the impurities present in the new mRNA vaccines. One area of concern is the presence of dsRNA aggregates and their potential impact on the efficacy and safety of these vaccines. Specifications for mRNA vaccines must include a clear accounting of these aggregates, to ensure transparency and address any potential implications they may have on vaccine performance.

SEC is a widely used method for analyzing monoclonal antibody aggregates by separating analytes according to size; however, its use remains limited for mRNA analysis. The main reason is due to the larger size of nucleic acids, which makes it difficult to separate aggregates by SEC due to the limited pore size of conventional SEC columns. To address this problem, efforts are currently underway to develop SEC columns with ultra-large pores, which enable efficient separation of mRNA aggregates and lipid nanoparticles [55, 56]. These studies have demonstrated the ability of SEC to explore the structural characteristics, purity, and aggregation states of mRNA, as well as offering the possibility of RNA purification. Its application can be further extended to include molecular weight (MW) determination and shape estimation when multi-angle light scattering (MALS) and online dynamic light scattering (DLS), are associated [57]. SEC-MALS should be considered as an important analytical tool for mRNA analysis, as it can provide biophysical and structural information that is otherwise difficult to obtain due to the large size and high charge density of mRNA.

Sequencing

Beyond its clear benefits to RNA identity, sequencing-based methods can be used as complementary strategies to assess RNA integrity and purity to verify mRNA-based therapeutics (Figure 13.7). Illumina sequencing provides high-throughput, accurate short reads that can be aligned to reference sequences to confirm mRNA identity with high confidence [58, 59]. These aligned short reads can be used to infer mRNA integrity by assembling overlapping reads to build a coverage map. Upon assembly, the coverage map can be interpreted whereby uniform coverage indicates higher RNA integrity while a 3′ or 5′ coverage bias can be indicative of RNA degradation or abortive transcription. Illumina sequencing can also provide insight into the purity of mRNA samples by mapping reads to potential contaminating sequences (e.g. plasmid backbone) [60]. This allows for the assessment of mRNA purity by calculating the proportion of total reads that align to the mRNA payload of interest versus those that align to contaminant sequences. Quantitative measurements such as this can help ensure that the mRNA sample is free from significant contamination, which is crucial for the reliability and efficacy of mRNA-based therapeutics.

To evaluate RNA integrity using long-read ONT, RNA is prepared with a cDNA library preparation workflow by adding adapters to the 5′ and 3′ ends of the mRNA [35, 61, 62]. Subsequent reads that include both flanking adapters allow one to distinguish full-length mRNA molecules from truncated mRNA molecules, providing quantitative information about 5′ and 3′ truncations (whether due to hydrolysis, degradation, or abortive transcription). Further analysis of the length of mRNA fragments can provide information about the proportion of full-length mRNA to shorter fragments, creating a distribution profile. This profile allows individual peaks to be analyzed to determine problematic regions with single base pair resolution in a PCR-free format.

PacBio sequencing follows a similar workflow as ONT in that RNA is first converted into full-length cDNA and prepared for sequencing using PacBio's library preparation

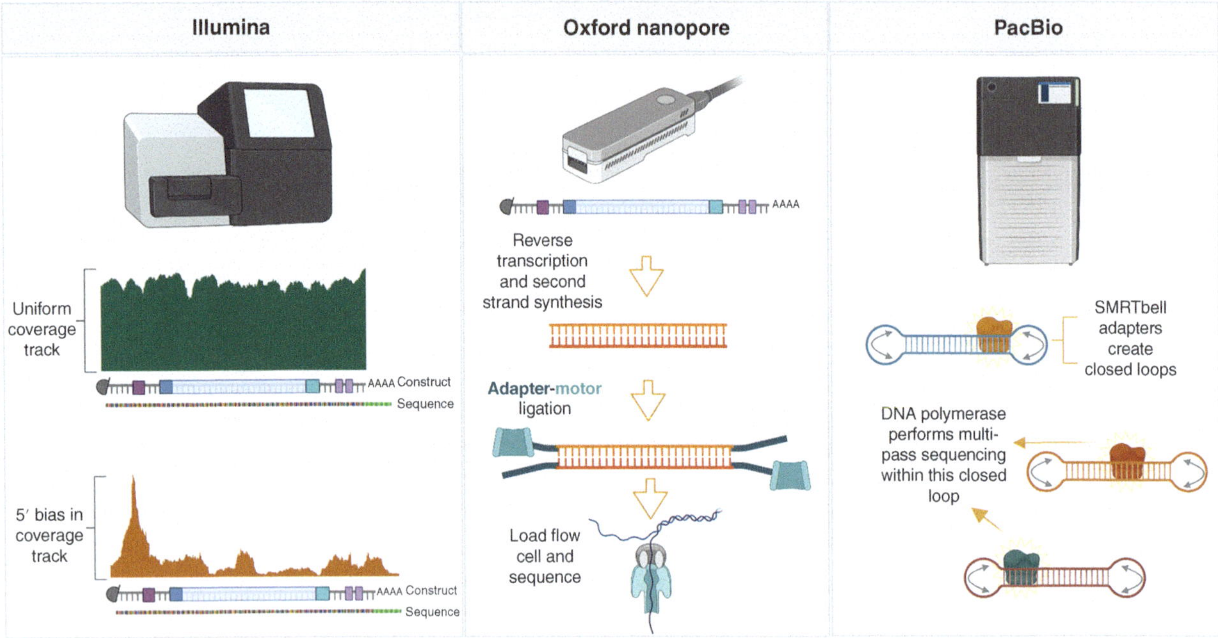

Figure 13.7 Overview of sequencing applications for mRNA vaccine QC using Illumina, Oxford Nanopore, and PacBio. Illumina coverage plots can be used to infer integrity where a coverage plot with uniform coverage indicates higher RNA integrity while a 5′ bias indicates a degradation or abortive transcription (representative plots shown). Oxford nanopore overview highlighting cDNA library prep. PacBio leverages SMRTbell adapters ligated to both ends of the cDNA molecules to create a closed loop and DNA polymerase performs multi-pass sequencing with this closed loop.

protocols. This involves adding proprietary SMRTbell adapters to both ends of the cDNA molecules to create a closed loop. PacBio's unique sequencing technology whereby DNA is capped on both sides with hairpin adapters creates a circular template for polymerase allowing repeated sequencing of the same molecule generating highly accurate circular consensus sequencing (CCS) [63] and PacBio's unique chemistry makes it particularly suitable for resolving complex regions or repetitive sequences ensuring a detailed and precise analysis of RNA full length, degradation, and truncation events [34].

mRNA Quality Features

Capping Efficiency Measurement

Capping efficiency is a crucial quality attribute because the 5′ cap structure of mRNA plays a vital role in the stability, translation efficiency, and overall efficacy of the mRNA [8]. Proper capping is essential for efficient ribosome recognition and protection from exonuclease degradation, which directly impacts the therapeutic effectiveness and safety of mRNA-based treatments and vaccines. Therefore, it is critical to accurately measure and ensure high capping efficiency for the development and quality control of mRNA therapeutics.

RNA cap analysis involves several steps: generation of short, predefined fragments from the 5′ end of mRNA molecules using RNase H, DNAzyme or ribozyme, enrichment of the 5′ cleavage products and analysis by LC-MS [64].

Significant attention has been given to the use of RNase H in cap analysis. This approach uses RNase H cleavage probe that is complementary to the 5′ end of the mRNA. The cleavage probe consists of an RNA sequence that contains 2′-O-methyl modifications, except at the 3′-end cleavage site that contains four to six deoxyribose (DNA) nucleotides. The terminal stretch of DNA directs the RNase H to cleave the mRNA at a specific location as the 2′-O-methyl-modified RNA is not recognized as a cleavage site by RNase H. The probe must be long enough (15–30 nt) to avoid random hybridization but still generate uniform fragments for better detection sensitivity. The RNase H-based cap isolation process involves three main steps: denaturing the IVT mRNA at 80–90 °C to remove secondary structures, hybridizing the probe at a temperature below its melting temperature, and digesting the RNA–DNA hybrid RNase H at 37 °C. Another report demonstrated an isothermal process using thermostable RNase H, where hybridization and partial unwinding of the mRNA occurred at 50°C [65]. The ratio of capped to uncapped IVT mRNA can be determined by LC-MS, as the longer 3′-end elutes later than the 5′-end after digestion. Depending on the IVT mRNA and the RNase H cleavage probe,

LC conditions may need optimization for the complete separation of 3′- and 5′-species. To enhance sensitivity, the 5′ end can be purified before LC-MS analysis. Beverly et al. demonstrated that using a biotinylated DNA–RNA chimera oligo, the 5′ RNase H cleavage fragments could be readily isolated for qualitative and quantitative analyses using LC-MS [66].

One of the drawbacks of RNase H is that the cleavage site is not unique, as other cleavage products of different lengths can also be produced, making the method not ideal for quantification purposes. A recent paper reported the use of a biosensor that detects both mRNA overlap and integrity [67]. However, the biosensor method only allows changes in capping levels to be detected in the range of at least 20%, making this approach difficult to quantify with precision. A more recent study by Vlatkovic et al. demonstrated the value of designing custom ribozymes (Rz) that specifically cleave IVT mRNA at a single position near the 5′ end, releasing short 5′-capped or uncapped cleavage products in the 10–30 nt range [68]. Defined 5′ cleavage products cut by ribozyme from capped mRNA differ by one nucleoside in length, in particular the structure of the cap itself, compared with cleavage products cut from uncapped RNA. RNA fragments from the cleavage reaction can be analyzed directly by denaturing PAGE or further purified using a silica-based column and analyzed by LC-MS, allowing determination of methylation status or minor capped products.

Poly(A) Tail Length and Distribution

In addition to the cap structure, the poly(A) tail has also gained significant attention in IVT mRNA analysis. The poly(A) tail enhances mRNA stability, influences nuclear export, and plays a critical role in the protein translation process by promoting ribosome binding. Variation in poly(A) tail length can affect mRNA half-life and translation efficiency, thereby impacting protein synthesis [69]. Better understanding these variations can help in optimizing mRNA constructs for therapeutic application by ensuring consistent expression and stability of the mRNA. For these reasons, it is crucial to monitor poly(A) tail length and distribution for the development and quality control of mRNA-based therapies and vaccines.

Due to the limitation of current molecular biology techniques, LC-MS has become the preferred method for poly(A) tail analysis [70]. Traditional methods like deep sequencing struggle with long repetitive sequences, particularly when the IVT mRNA has a well-defined poly(A) tail length encoded in the DNA template. Moreover, when poly(A) tail is added at the posttranscriptional level, this often leads to different poly(A) tail lengths, making it necessary to determine the average poly(A) tail length–a task well suited to LC-MS. For this approach, poly(A) tails are cleaved from IVT mRNA using the enzyme RNAse T1 and analyzed by IP-RP LC-MS, which provides information on tail length at single-nucleotide resolution (Figure 13.8). If the mRNA resulting from digestion interferes with the elution of the poly(A) tail species, it may be necessary to purify the poly(A) tail mixture prior to LC-MS analysis. To this end, after digestion with RNase T1, poly(A) tails can be isolated by hybridizing them to oligo(dT)25-coated magnetic beads. Following denaturation of the oligo(dT)25:poly(A) tail duplex, the length distribution of denatured poly(A) tails can be determined by IP-RP LC-MS. In a recent report, the development of several HPLC methods for measurement of mRNA poly(A) tail length and heterogeneity was reported [71]. The authors showed the interest of the SEC to measure the average length of a poly(A) tail, while IP-RP HPLC was more suitable for size heterogeneity, offering resolution of poly(A) tail oligonucleotide variants up to 150 nucleotides long (Figure 13.8).

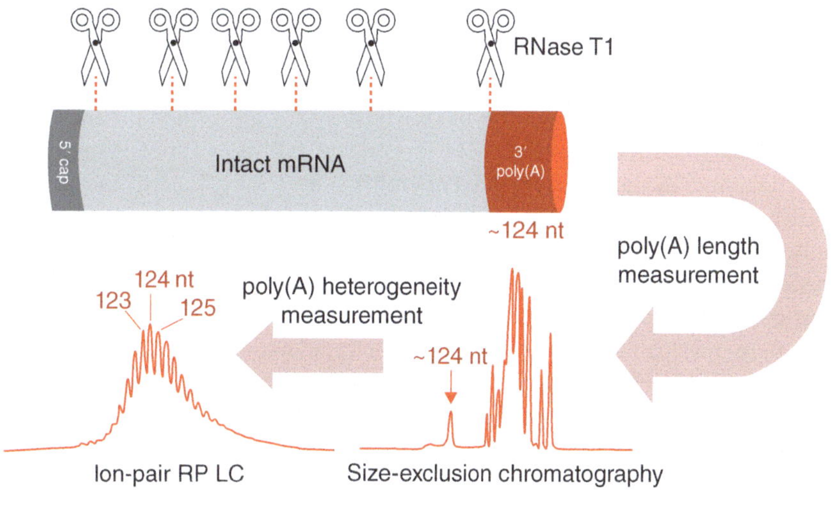

Figure 13.8 Analytical HPLC-based methods for measurement of mRNA poly(A) tail length and heterogeneity after RNase T1 digestion. Poly(A) tail was first cleaved from mRNA with the RNase T1 enzyme. Average length of a release poly(A) tail was analyzed by SEC, while size heterogeneity can be estimated with P RP HPLC. Source: Gilar et al. [71]/ American Chemical Society/CC BY-NC-ND 4.0.

Recently, long-read sequencing technologies, such as PacBio HiFi sequencing, can be highly useful for monitoring poly(A) tail length in IVT mRNA by providing the ability to sequence long and repetitive regions accurately. Unlike short-read sequencing, which can face challenges in accurately calling low-complexity sequences and has difficulty in alignment of the poly(A) tail due to its repetitive nature, long-read sequencing can capture the entire length of the poly(A) tail in a single, continuous read. This sequencing approach paired with specialized algorithms allows for precise determination of the poly(A) tail length and distribution [72].

Functionality Testing

Functionality testing to ensure that synthesized mRNA can be efficiently translated into a protein that performs its intended function is a crucial step in developing mRNA-based therapeutics and vaccines. Assays for this purpose can focus solely on translation efficiency and these commonly read out as yield and/or protein product quality. Additionally, assays can be designed to also evaluate the functional activity of the expressed protein, and these are tailored to its intended role under physiological conditions.

Translation Level Assessment

Assays designed to measure the translation efficiency of mRNA vary widely, ranging from cell-free systems that use recombinant proteins to cell-based assays. Cell-based assays can be more complicated, but they have the benefit of evaluating protein translation levels in the context of cellular factors.

Protein Synthesis Using Recombinant Elements (PURE) Systems

The PURE system is a fully reconstituted translation system made from purified components required for protein synthesis [73]. These include ribosomes, tRNAs, translation factors, aminoacyl tRNA synthetases, and energy regeneration components. This system bypasses the complexities and variabilities of cell lysates, providing a highly controlled environment to evaluate mRNA functionality. By introducing the IVT mRNA into the PURE system, researchers can directly assess the level of full-length protein produced. The simplicity of the PURE system also allows for easy manipulation and troubleshooting, making it a robust method for functional analysis and therefore invaluable for basic research and industrial applications [74, 75]. However, while this platform can reliably detect translation issues derived from the mRNA sequence, it will not detect additional challenges to protein production introduced by cellular factors.

In Vitro Translation Assays

Compared to the PUR system, *in vitro* translation assays employ cellular lysates such as Rabbit Reticulocyte Lysate (RRL), Wheat Germ Extract (WGE), or HeLa Cell Lysate. These systems derive all the necessary components for translation, including ribosomes, tRNAs, and initiation factors from the cellular context of a lysate allowing them to be potentially more biologically relevant than the PURE system [76–78]. The readout of mRNA translation efficiency from this assay can be achieved in multiple ways. By incorporating luciferase reporter genes within the mRNA sequence, researchers can monitor translation efficiency through luminescence measurements, providing a direct and quantitative readout of translation activity [79]. Alternatively, if researchers need to assess the production of a specific sequence without the inclusion of a reporter sequence, western blot analysis can be employed to track expression levels. In this approach the protein products of the *in vitro* translation reaction are run on an SDS-PAGE gel, where they are denatured and resolved by molecular weight. Following separation, proteins are transferred onto a membrane, such as nitrocellulose or PVDF. The membrane is incubated with a specific primary antibody that recognizes the target protein, followed by a secondary antibody linked to an enzyme or fluorescent label. This amplification allows for visualization using chemiluminescent or fluorescent detection methods. By analyzing the resulting bands, researchers can assess the presence and abundance of the full-length protein. The presence of bands corresponding to smaller proteins indicates truncated protein products allowing this assay to also report on the purity of the protein product. It is important to note that while powerful, western blots require the use of an antibody specific to your protein of interest which can be limiting. Alternatively, incorporating fluorescently labeled amino acids, such as lysines, into the *in vitro* translation reaction allows for the visualization of all protein products on the gel without the use of an antibody. By knowing the expected size of your target protein, you can determine its abundance relative to any prematurely terminated protein products. In summary, *in vitro* translation assays can provide insight into the efficiency of translation and help clarify the production levels of one's desired sequence.

Ribo-seq

Ribosome profiling (Ribo-seq) is a cutting-edge cell-based technique that can interrogate specific sequences of RNA at the nucleotide level for their translation efficiency by providing a snapshot of ribosome positions on mRNA during protein synthesis [80, 81]. The workflow begins with delivering mRNA into cells, followed by the isolation of ribosome-protected mRNA fragments generated by treating the cells with ribonucleases. The fragments are then purified and converted into a sequencing library. High-throughput sequencing enables the mapping of these fragments back to the reference genome, revealing both the abundance and translation efficiency of specific transcripts. Ribo-seq can provide a higher-resolution look at protein translation by identifying sequences within the mRNA that exhibit lower translational efficiency. This data may allow researchers to redesign their mRNAs for higher and more uniform translation across the entire sequence, optimizing protein production. Ribo-seq is increasingly recognized as a vital tool in molecular biology and genomics.

ELISA (Enzyme-Linked Immunosorbent Assay)

ELISA is another cell-based technique that can be used to quantify the level of protein produced by an mRNA delivered both *in vitro* in a cell culture system or *in vivo* in an animal model [82]. The assay involves capturing the target protein on a microplate using specific antibodies, followed by detection with enzyme-linked secondary antibodies [83]. The enzyme catalyzes a colorimetric or luminescent reaction, which can be quantified using a plate reader. The protein levels measured by ELISA correlate with the translation efficiency of the mRNA, providing a sensitive and specific readout of mRNA functionality. Importantly, the assay can be used to monitor protein produced by mRNA intracellularly by probing cellular lysate or secreted proteins by assaying supernatants. Additionally, the system allows for the assessment of secreted protein levels *in vivo* by measuring serum samples.

Reporter Tether Assays

Tether assays involve appending an easily measurable protein product such as a fluorescent protein, enzyme or small tag onto the desired mRNA sequence to monitor translation. A commonly used example of this is the HiBiT system, which provides a versatile tool for detecting mRNA translation *in vitro* in cell culture systems or *in vivo*. A small peptide termed the HiBIT is encoded with the desired sequence and when translated interacts with a complementary NanoBiT component that is added exogenously to produce a functional luciferase enzyme [84]. The luminescence signal of the luciferase can be quantified with high sensitivity, oftentimes even greater than that of the ELISA assay. Similar to the ELISA assay, it allows real-time monitoring of mRNA translation both *in vitro* both within lysed cells or through the harvested supernatant or *in vivo* from the serum of animal models. Importantly the HiBIT tag can be placed at the start, middle, or end of the mRNA to allow for assessment of translation throughout the sequence. An experimental setup analogous to the HiBIT workflow can be used with any tag or sequence that produces a protein that can be detected through imaging or chemiluminescent methods.

Fluorescence *In Situ* Hybridization

Techniques to visualize mRNA, such as Fluorescence *In Situ* Hybridization (FISH), involve designing fluorescently labeled probes complementary to portions of the mRNA sequence. These probes are then delivered to cells harboring the mRNA and allowed to hybridize. The annealed probes can then be detected using fluorescence microscopy indicating the amount and localization of specific mRNA species. Single-molecule fluorescence microscopy extends this approach by tracking individual mRNA molecules and their translation events in real time. This high-resolution technique provides insights into the dynamics of mRNA translation and its spatial distribution within cells, offering detailed information on mRNA functionality and translational regulation [85, 86].

Functionality Assessment

The primary function of an mRNA is to carry the code needed to direct translation of a specific protein. As discussed in the last section this can be assessed independent of that protein's activity. However, testing the functionality of the encoded protein is also critical as it verifies whether the mRNA sequence is adequately designed to produce the intended protein capable of performing its specific function. This adds complexity to the analysis as each protein must have an assay designed to evaluate its specific function.

Cas9 mRNA Functionality

A common example of tailoring an assay to assess a specific protein is when one wants to determine the functionality of a genome-engineering enzyme such as Cas9. This involves deploying the Cas9-expressing mRNA in gene-editing experiments. Researchers transfect cells with Cas9 mRNA and specific guide RNAs (gRNAs) that direct Cas9 to targeted genomic loci. The efficiency of gene editing is subsequently evaluated using techniques like T7 endonuclease assay [87], antibody-based detection of the targeted protein, next-generation sequencing, or reporter assays. By measuring the frequency and precision of the introduced edits, researchers can determine the functional activity of the Cas9 produced by the mRNA. This method is crucial for applications in gene therapy, genomic research, and biotechnology. The efficiency and precision of gene edits directly correlate with the functionality or ability of the mRNA to robustly express the Cas9 protein, making this method critical for validating mRNA used in gene-editing applications and therapeutic strategies.

Bone Marrow-Derived Macrophages (BDMCs)

Murine cells are extremely sensitive to dsRNA contaminants, likely due the level of certain cellular sensors [88]. Therefore, a useful system for testing the level of dsRNA produced in an IVT mRNA prep is by delivering the RNA to mouse cells *in vitro* and assessing expression and monitoring cytokine production (e.g. IL-6, TNF-α). Elevated immune responses paired with reduced translation of the mRNA indicate the presence of immunogenic contaminants or problematic mRNA sequences. This assessment helps ensure the IVT mRNA does not provoke unintended immune reactions, which is critical for therapeutic applications. This approach is crucial for ensuring the safety and non-immunogenic nature of mRNA therapeutics, particularly for clinical applications where immune activation must be minimized.

In Process-Related Impurities

Impurities such as dsRNA, residual proteins, and plasmid DNA can be still found after the purification process. Eliminating these impurities in the IVT mRNA manufacturing process is crucial for ensuring the quality, efficacy, and safety of mRNA-based therapeutics.

For example, there are certain cell types like primary immune cells that are very sensitive to dsRNA as these are recognized as foreign by specific receptors triggering an immune response. Residual proteins from the manufacturing process can interfere with cellular translation machinery, reducing the efficiency of mRNA translation into the desired protein. Finally, residual DNA is another critical impurity that must be addressed, as traces of template DNA can lead to unintended biological effects. The cumulative presence of these impurities can trigger unwanted immune responses, as cells recognize the contaminants as foreign. This can result in inflammation, while also causing reduced or erroneous protein synthesis. It is clear that robust analytical methods are needed to track the level of these impurities paired with rigorous purification processes to produce high-quality mRNA therapeutics.

Double-Stranded RNA

Double-stranded RNA can inadvertently form during the IVT process, and effectively controlling its presence is essential for producing a high-quality mRNA product. Residual dsRNA remaining in the final IVT produced RNA can be sensed by cellular receptors leading to adverse immune responses and also potentially compromising therapeutic efficacy by triggering a reduction in translation [89]. To mitigate dsRNA formation, several strategies can be implemented during and after IVT synthesis. One approach is the use of RNase III or dsRNA-specific nucleases to selectively degrade dsRNA while sparing the desired single-stranded RNA [90]. Furthermore, the incorporation of modified nucleotides (e.g. *N1*-methyl-pseudouridine (m1Ψ)) during transcription can reduce RNA duplex formation [91]. More recently there has been an effort to identify mutations in the T7 RNA polymerases that result in reduced dsRNA formed during transcription. When used in IVT reactions, these mutants result in reduced dsRNA by-products without a significant loss in yield making them an attractive solution [92].

In addition to enzymatic treatments, purification techniques are employed to separate and remove dsRNA. Cellulose purification of mRNA is an effective method for isolating messenger RNA from dsRNA contaminants. This process typically involves binding mRNA to a cellulose matrix under conditions that favor the selective adsorption of single-stranded RNA. The sample is passed through a column filled with cellulose, where mRNA binds due to its structural properties. dsRNA, which has a more stable, double-helical structure, does not adhere as strongly and is washed away during subsequent elution steps. Finally, the purified mRNA can be eluted from the column, resulting in a high-quality preparation suitable for downstream applications such as cloning, sequencing, or expression analysis. More advanced methods such as HPLC, particularly ion-exchange or SEC modes, can also be used to differentiate between ssRNA and dsRNA based on their physical and chemical properties. Affinity chromatography utilizing columns with ligands that specifically bind dsRNA can further enhance the purification process by selectively removing dsRNA contaminants.

To quantitatively measure any residual dsRNA to ensure its thorough removal, several detection methods can be employed. Dot or slot blot assays involve immobilizing RNA samples on a membrane followed by probing with dsRNA-specific antibodies (J2 antibody). Detection is then achieved using chemiluminescent or colorimetric substrates that provide visual confirmation of the dsRNA and its quantification can be achieved through the establishment of a standard curve. Similarly, ELISA uses the same dsRNA-specific antibodies to bind residual dsRNA [93]. The bound antibodies are then detected using enzyme-linked secondary antibodies, providing a quantitative measure of dsRNA through a colorimetric or fluorescent readout [94]. Importantly, the J2 antibody can only detect dsRNA longer than ~40 bp, requiring the use of alternative strategies such as the BMDC assay discussed above to detect the level of all species of dsRNA in a sample.

PAGE can also be utilized for this purpose. Native PAGE, in particular, separates RNA molecules based on their size and structure without denaturing them, allowing the distinct separation of dsRNA from ssRNA. Specific stains such as acridine orange [95] or northern blotting can be utilized post-separation for visualization.

Certain HPLC-based methods, such as anion-exchange chromatography, can separate dsRNA based on its negative charge density, providing high sensitivity and resolution for detecting dsRNA contaminants. NGS also offers a platform for comprehensive identification and quantification of dsRNA in an mRNA product. Once the contaminating dsRNA is purified or enriched, the dsRNA can be reverse transcribed and converted to cDNA using random primers for downstream library preparation and sequenced with Illumina instruments. Notably, most dsRNA sequencing is in the context of viral surveillance [96] and infection [97] as this method is relatively complex and data intensive, albeit with great potential to offer detailed insights into the identity of dsRNA as an IVT process impurity.

The combination of enzymatic degradation, chromatographic separation, and sensitive detection assays establishes a robust framework for achieving and verifying the removal of dsRNA, meeting stringent regulatory standards for use in therapeutics.

Residual Proteins

Beyond dsRNA, another impurity that must be removed to achieve a high purity and safe final product are residual proteins. Residual proteins can originate from various sources, such as the enzymes used during transcription and purification processes. To remove these proteins, several purification techniques can be employed. Affinity chromatography, utilizing matrices that specifically bind to proteins of interest, can selectively isolate and remove these contaminants. Additional purification steps, like ion-exchange chromatography and SEC, can further ensure the removal of residual proteins by separating them based on their charge and size differences.

To confirm the effectiveness of protein removal and ensure minimal residual protein content, highly sensitive and specific detection methods are adopted. One of the primary techniques is the ELISA, which uses antibodies specific to the residual proteins to detect and quantify their presence. Additionally, gel electrophoresis techniques such as SDS-PAGE, followed by protein staining, can be used to visualize and quantify proteins, particularly when combined with densitometry analysis. Recently, a fluorescence-based assay protein kit (nanorange) has been introduced, enabling a very sensitive and easy detection of protein content, with detection as low as 10 ng/ml of total protein content in solution. For collecting more information, mass spectrometry should be considered, offering a precise identification and quantification of each protein.

Residual DNA Template

During the IVT process for mRNA manufacturing, residual DNA templates can remain, and eliminating them is a crucial step to ensure high purity of the final mRNA product. One common approach to remove DNA templates is enzymatic digestion using DNases, such as DNase I, which cleaves DNA into smaller fragments that can be separated from the mRNA product during purification steps like chromatography or ultrafiltration. After enzymatic treatment, the mRNA product typically undergoes additional purification steps, including affinity purification and SEC, to further reduce any residual DNA.

To confirm the effectiveness of the DNA removal process and ensure the absence of residual DNA, highly sensitive and specific detection methods are employed. Quantitative PCR (qPCR) is one of the most widely used techniques due to its high sensitivity and ability to quantitatively measure even trace amounts of DNA. The process involves using specific primers that amplify only the DNA sequences of interest, enabling precise quantification. For even higher sensitivity and absolute quantification, digital PCR (dPCR) can be utilized, providing detailed information about the residual DNA content. These methods, when combined, ensure the robustness of the purification process and the highest standards of mRNA product quality.

Conclusion

The characterization of RNA-based therapeutics is a multifaceted endeavor requiring a suite of advanced analytical methods to ensure the purity, integrity, and overall quality of mRNA products. Techniques such as CGE, IP-RP HPLC, and SEC are pivotal in assessing mRNA purity and integrity, offering high-resolution separation and quantification of various RNA species, though each comes with its specific limitations. For instance, while CGE provides detailed size separation, it is time intensive, and IP-RPLC requires careful optimization of ion-pairing agents and column conditions to avoid nonspecific adsorption issues. The content analysis of mRNA often utilizes UV absorbance to quantify RNA concentration, but this method can be compromised by contaminants and requires complementary techniques for validation. Identity confirmation is typically performed using RT-PCR and RNA sequencing, which offer high specificity and accuracy but can be hampered by template degradation and sequence complexity. Assessing critical quality attributes such as capping efficiency and poly(A) tail length and distribution is crucial, as these factors directly impact mRNA stability and protein translation efficiency. Techniques like ELISA, dot blot, and cell-based assays are employed to gauge the functionality of mRNA, but these bioassays can be influenced by variability in cellular responses and reagent quality. Handling and identifying impurities, including dsRNA, residual proteins, and residual DNA, present significant challenges. Methods such as ELISA and dot blot provide sensitive detection of dsRNA, while residual proteins and DNA are typically quantified through fluorescence/mass spectrometry and qPCR, respectively.

The main limitations of these analytical methods include their susceptibility to contamination, the need for extensive method optimization, and the potential for incomplete impurity removal. Moreover, the inherent heterogeneity in mRNA vaccines, induced by variations such as poly(A) tail length, must be carefully monitored to ensure consistent therapeutic efficacy.

Looking ahead, the safety and efficacy of mRNA-based therapeutics hinge on continuous improvement in the IVT mRNA manufacturing process. This includes refining purification steps to eliminate contaminants, optimizing analytical methods to reduce analysis time and increase sensitivity, and developing robust bioassays to better predict *in vivo* functionality. Future directions must also focus on standardizing analytical protocols to enhance reproducibility and regulatory compliance, paving the way for the widespread adoption of mRNA therapeutics.

In summary, while significant strides have been made in the analytical characterization of RNA-based therapeutics, ongoing innovation and method refinement are essential in overcoming existing challenges. These efforts are crucial to ensuring the development of safe, effective, and high-quality mRNA therapeutics capable of meeting the demands of modern medicine.

References

1 Qin, S., Tang, X., Chen, Y. et al. (2022). mRNA-based therapeutics: powerful and versatile tools to combat diseases. *Signal Transduct. Target. Ther.* 7: 166.

2 Jeeva, S., Kim, K.-H., Shin, C.H. et al. (2021). An update on mRNA-based viral vaccines. *Vaccine* 9: 965.

3 Poh, C.M., Zheng, J., Channappanavar, R. et al. (2020). Multiplex screening assay for identifying cytotoxic CD8+ T cell epitopes. *Front. Immunol.* 11: 400.

4 Zogg, H., Singh, R., and Ro, S. (2022). Current advances in RNA therapeutics for human diseases. *Int. J. Mol. Sci.* 23: 2736.

5 Sparmann, A. and Vogel, J. (2023). RNA -based medicine: from molecular mechanisms to therapy. *EMBO J.* 42: e114760.

6 Kowalski, P.S., Rudra, A., Miao, L., and Anderson, D.G. (2019). Delivering the messenger: advances in technologies for therapeutic mRNA delivery. *Mol. Ther.* 27: 710–728.

7 Wang, Y.-S., Kumari, M., Chen, G.-H. et al. (2023). mRNA-based vaccines and therapeutics: an in-depth survey of current and upcoming clinical applications. *J. Biomed. Sci.* 30: 84.

8 Ramanathan, A., Robb, G.B., and Chan, S.-H. (2016). mRNA capping: biological functions and applications. *Nucleic Acids Res.* 44: 7511–7526.

9 Kozak, M. and Shatkin, A.J. (1978). Identification of features in 5′ terminal fragments from reovirus mRNA which are important for ribosome binding. *Cell* 13: 201–212.

10 Kozak, M. (1987). At least six nucleotides preceding the AUG initiator codon enhance translation in mammalian cells. *J. Mol. Biol.* 196: 947–950.

11 Karikó, K., Muramatsu, H., Ludwig, J., and Weissman, D. (2011). Generating the optimal mRNA for therapy: HPLC purification eliminates immune activation and improves translation of nucleoside-modified, protein-encoding mRNA. *Nucleic Acids Res.* 39: e142–e142.

12 Kwon, H., Kim, M., Seo, Y. et al. (2018). Emergence of synthetic mRNA: in vitro synthesis of mRNA and its applications in regenerative medicine. *Biomaterials* 156: 172–193.

13 Weissman, D. (2015). mRNA transcript therapy. *Expert Rev. Vaccines* 14: 265–281.

14 Goldstrohm, A.C. and Wickens, M. (2008). Multifunctional deadenylase complexes diversify mRNA control. *Nat. Rev. Mol. Cell Biol.* 9: 337–344.

15 Korenč, M., Mencin, N., Puc, J. et al. (2021). Chromatographic purification with CIMmultus™ oligo dT increases mRNA stability. *CGTI* 7: 1207–1216.

16 Jacobsen, N. (2004). Direct isolation of poly(A)+ RNA from 4 M guanidine thiocyanate-lysed cell extracts using locked nucleic acid-oligo(T) capture. *Nucleic Acids Res.* 32: e64.

17 Green, M.R. and Sambrook, J. (2019). Isolation of poly(A) + messenger RNA using magnetic oligo(dT) beads. *Cold Spring Harb. Protoc.* 2019 (10): pdb.prot101733.

18 Pereira, M.J.B., Behera, V., and Walter, N.G. (2010). Nondenaturing purification of co-transcriptionally folded RNA avoids common folding heterogeneity. *PLoS One* 5: e12953.

19 Baiersdörfer, M., Boros, G., Muramatsu, H. et al. (2019). A facile method for the removal of dsRNA contaminant from in vitro-transcribed mRNA. *Mol Ther - Nucleic Acids.* 15: 26–35.

20 Guo, P., Coban, O., Snead, N.M. et al. (2010). Engineering RNA for targeted siRNA delivery and medical application. *Adv. Drug Deliv. Rev.* 62: 650–666.

21 Inagaki, M., Abe, N., Li, Z. et al. (2023). Cap analogs with a hydrophobic photocleavable tag enable facile purification of fully capped mRNA with various cap structures. *Nat. Commun.* 14: 2657.

22 Okamoto, T. and Okabe, S. (2000). Ultraviolet absorbance at 260 and 280 nm in RNA measurement is dependent on measurement solution. *Int J Mol Med* (cited 2024 Sep 30); Available from: http://www.spandidos-publications.com/10.3892/ijmm.5.6.657.

23 Wilfinger, W.W., Mackey, K., and Chomczynski, P. (1997). Effect of pH and ionic strength on the spectrophotometric assessment of nucleic acid purity. *BioTechniques* 22: 474–481.

24 Stark, R., Grzelak, M., and Hadfield, J. (2019). RNA sequencing: the teenage years. *Nat. Rev. Genet.* 20: 631–656.

25 Garalde, D.R., Snell, E.A., Jachimowicz, D. et al. (2018). Highly parallel direct RNA sequencing on an array of nanopores. *Nat. Methods* 15: 201–206.

26 Jain, M., Abu-Shumays, R., Olsen, H.E., and Akeson, M. (2022). Advances in nanopore direct RNA sequencing. *Nat. Methods* 19: 1160–1164.

27 Fleming, R. (2020). ADC Analysis by Hydrophobic Interaction Chromatography. In: *Antib-Drug Conjug* (ed. L.N. Tumey), 147–161. New York, NY: Springer US (cited 2020 Apr 8). Available from: http://link.springer.com/10.1007/978-1-4939-9929-3_10.

28 Zhang, Y., Yan, H., Wei, Z. et al. (2024). NanoMUD: profiling of pseudouridine and N1-methylpseudouridine using Oxford Nanopore direct RNA sequencing. *Int. J. Biol. Macromol.* 270: 132433.

29 Tavakoli, S., Nabizadeh, M., Makhamreh, A. et al. (2023). Semi-quantitative detection of pseudouridine modifications and type I/II hypermodifications in human mRNAs using direct long-read sequencing. *Nat. Commun.* 14: 334.

30 Huang, S., Zhang, W., Katanski, C.D. et al. (2021). Interferon inducible pseudouridine modification in human mRNA by quantitative nanopore profiling. *Genome Biol.* 22: 330.

31 Begik, O., Lucas, M.C., Pryszcz, L.P. et al. (2021). Quantitative profiling of pseudouridylation dynamics in native RNAs with nanopore sequencing. *Nat. Biotechnol.* 39: 1278–1291.

32 Smola, M.J., Rice, G.M., Busan, S. et al. (2015). Selective 2′-hydroxyl acylation analyzed by primer extension and mutational profiling (SHAPE-MaP) for direct, versatile and accurate RNA structure analysis. *Nat. Protoc.* 10: 1643–1669.

33 Stephenson, W., Razaghi, R., Busan, S. et al. (2022). Direct detection of RNA modifications and structure using single-molecule nanopore sequencing. *Cell Genom.* 2: 100097.

34 Pardo-Palacios, F.J., Wang, D., Reese, F. et al. (2024). Systematic assessment of long-read RNA-seq methods for transcript identification and quantification. *Nat. Methods* 21: 1349–1363.

35 Gunter, H.M., Idrisoglu, S., Singh, S. et al. (2023). mRNA vaccine quality analysis using RNA sequencing. *Nat. Commun.* 14: 5663.

36 Gau, B.C., Dawdy, A.W., Wang, H.L. et al. (2023). Oligonucleotide mapping via mass spectrometry to enable comprehensive primary structure characterization of an mRNA vaccine against SARS-CoV-2. *Sci. Rep.* 13: 9038.

37 Jiang, T., Yu, N., Kim, J. et al. (2019). Oligonucleotide sequence mapping of large therapeutic mRNAs via parallel ribonuclease digestions and LC-MS/MS. *Anal. Chem.* 91: 8500–8506.

38 Lobue, P.A., Jora, M., Addepalli, B., and Limbach, P.A. (2019). Oligonucleotide analysis by hydrophilic interaction liquid chromatography-mass spectrometry in the absence of ion-pair reagents. *J. Chromatogr. A* 1595: 39–48.

39 Lu, T., Klein, L.J., Ha, S., and Rustandi, R.R. (2020). High-resolution capillary electrophoresis separation of large RNA under non-aqueous conditions. *J. Chromatogr. A* 1618: 460875.

40 Skeidsvoll, J. and Ueland, P.M. (1996). Analysis of RNA by capillary electrophoresis. *Electrophoresis* 17: 1512–1517.

41 Guimaraes, G.J., Kim, J., and Bartlett, M.G. (2023). Characterization of mRNA therapeutics. *Mass Spectrom. Rev.* mas.21856.

42 Camperi, J., Lippold, S., Ayalew, L. et al. (2024). Comprehensive impurity profiling of mRNA: evaluating current technologies and advanced analytical techniques. *Anal. Chem.* 96: 3886–3897.

43 Warzak, D.A., Pike, W.A., and Luttgeharm, K.D. (2023). Capillary electrophoresis methods for determining the IVT mRNA critical quality attributes of size and purity. *SLAS Technol.* 28: 369–374.

44 Raffaele, J., Loughney, J.W., and Rustandi, R.R. (2022). Development of a microchip capillary electrophoresis method for determination of the purity and integrity of mRNA in lipid nanoparticle vaccines. *Electrophoresis* 43: 1101–1106.

45 Sutton, J.M., Guimaraes, G.J., Annavarapu, V. et al. (2020). Current state of oligonucleotide characterization using liquid chromatography–mass spectrometry: insight into critical issues. *J. Am. Soc. Mass Spectrom.* 31: 1775–1782.

46 Basiri, B. and Bartlett, M.G. (2014). LC–MS of oligonucleotides: applications in biomedical research. *Bioanalysis* 6: 1525–1542.

47 Kadlecová, Z., Kalíková, K., Tesařová, E., and Gilar, M. (2022). Phosphorothioate oligonucleotides separation in ion-pairing reversed-phase liquid chromatography: effect of ion-pairing system. *J. Chromatogr. A* 1676: 463201.

48 Wei, B., Wang, J., Cadang, L. et al. (2022). Development of an ion pairing reversed-phase liquid chromatography-mass spectrometry method for characterization of clustered regularly interspaced short palindromic repeats guide ribonucleic acid. *J. Chromatogr. A* 1665: 462839.

49 Azarani, A. (2001). RNA analysis by ion-pair reversed-phase high performance liquid chromatography. *Nucleic Acids Res.* 29: 7e–7e.

50 Kuwayama, T., Ozaki, M., Shimotsuma, M., and Hirose, T. (2023). Separation of long-stranded RNAs by RP-HPLC using an octadecyl-based column with super-wide pores. *Anal. Sci.* 39: 417–425.

51 Guimaraes, G.J., Saad, J.G., Annavarapu, V., and Bartlett, M.G. (2023). Mobile phase aging and its impact on electrospray ionization of oligonucleotides. *J. Am. Soc. Mass Spectrom.* 34: 2691–2699.

52 Fekete, S., Doneanu, C., Addepalli, B. et al. (2023). Challenges and emerging trends in liquid chromatography-based analyses of mRNA pharmaceuticals. *J. Pharm. Biomed. Anal.* 224: 115174.

53 Close, E.D., Nwokeoji, A.O., Milton, D. et al. (2016). Nucleic acid separations using superficially porous silica particles. *J. Chromatogr. A* 1440: 135–144.

54 Yamauchi, Y., Taoka, M., Nobe, Y. et al. (2013). Denaturing reversed phase liquid chromatographic separation of non-coding ribonucleic acids on macro-porous polystyrene-divinylbenzene resins. *J. Chromatogr. A* 1312: 87–92.

55 D'Atri, V., Lardeux, H., Goyon, A. et al. (2024). Optimizing messenger RNA analysis using ultra-wide pore size exclusion chromatography columns. *Int. J. Mol. Sci.* 25: 6254.

56 Goyon, A., Tang, S., Fekete, S. et al. (2023). Separation of plasmid DNA topological forms, messenger RNA, and lipid nanoparticle aggregates using an ultrawide pore size exclusion chromatography column. *Anal. Chem.* 95: 15017–15024.

57 Jia, X., Liu, Y., Wagner, A.M. et al. (2021). Enabling online determination of the size-dependent RNA content of lipid nanoparticle-based RNA formulations. *J. Chromatogr. B* 1186: 123015.

58 Chao, H.-P., Chen, Y., Takata, Y. et al. (2019). Systematic evaluation of RNA-Seq preparation protocol performance. *BMC Genomics* 20: 571.

59 Pease, J. and Sooknanan, R. (2012). A rapid, directional RNA-seq library preparation workflow for Illumina® sequencing. *Nat. Methods* 9: i–ii.

60 McKernan K, Helbert Y, Kane LT, McLaughlin S. (2023) Sequencing of bivalent Moderna and Pfizer mRNA vaccines reveals nanogram to microgram quantities of expression vector dsDNA per dose. (cited 2024 Oct 1). Available from: https://osf.io/b9t7m

61 Wang, Y., Zhao, Y., Bollas, A. et al. (2021). Nanopore sequencing technology, bioinformatics and applications. *Nat. Biotechnol.* 39: 1348–1365.

62 Soneson, C., Yao, Y., Bratus-Neuenschwander, A. et al. (2019). A comprehensive examination of nanopore native RNA sequencing for characterization of complex transcriptomes. *Nat. Commun.* 10: 3359.

63 Rhoads, A. and Au, K.F. (2015). PacBio sequencing and its applications. *Genomics Proteomics Bioinformatics* 13: 278–289.

64 Beverly, M., Dell, A., Parmar, P., and Houghton, L. (2016). Label-free analysis of mRNA capping efficiency using RNase H probes and LC-MS. *Anal. Bioanal. Chem.* 408: 5021–5030.

65 Morreel, K., t'Kindt, R., Debyser, G. et al. (2022). Diving into the structural details of in vitro transcribed mRNA using liquid chromatography–mass spectrometry-based oligonucleotide profiling. *LCGC Europe* 220–236.

66 Beverly, M., Hagen, C., and Slack, O. (2018). Poly a tail length analysis of in vitro transcribed mRNA by LC-MS. *Anal. Bioanal. Chem.* 410: 1667–1677.

67 Moya-Ramírez, I., Bouton, C., Kontoravdi, C., and Polizzi, K. (2020). High resolution biosensor to test the capping level and integrity of mRNAs. *Nucleic Acids Res.* 48: e129–e129.

68 Vlatkovic, I., Ludwig, J., Boros, G. et al. (2022). Ribozyme assays to quantify the capping efficiency of in vitro-transcribed mRNA. *Pharm.* 14: 328.

69 Jalkanen, A.L., Coleman, S.J., and Wilusz, J. (2014). Determinants and implications of mRNA poly(a) tail size – does this protein make my tail look big? *Semin. Cell Dev. Biol.* 34: 24–32.

70 Brouze, A., Krawczyk, P.S., Dziembowski, A., and Mroczek, S. (2023). Measuring the tail: methods for poly(A) tail profiling. *WIREs RNA* 14: e1737.

71 Gilar, M., Doneanu, C., and Gaye, M.M. (2023). Liquid chromatography methods for analysis of mRNA poly(a) tail length and heterogeneity. *Anal. Chem.* 95: 14308–14316.

72 Legnini, I., Alles, J., Karaiskos, N. et al. (2019). FLAM-seq: full-length mRNA sequencing reveals principles of poly(A) tail length control. *Nat. Methods* 16: 879–886.

73 Shimizu, Y., Inoue, A., Tomari, Y. et al. (2001). Cell-free translation reconstituted with purified components. *Nat. Biotechnol.* 19: 751–755.

74 Pardee, K., Slomovic, S., Nguyen, P.Q. et al. (2016). Portable, on-demand biomolecular manufacturing. *Cell* 167: 248–259.e12.

75 Whittaker, J.W. (2013). Cell-free protein synthesis: the state of the art. *Biotechnol. Lett.* 35: 143–152.

76 Zemella, A., Thoring, L., Hoffmeister, C., and Kubick, S. (2015). Cell-free protein synthesis: pros and cons of prokaryotic and eukaryotic systems. *ChemBioChem* 16: 2420–2431.

77 Carlson, E.D., Gan, R., Hodgman, C.E., and Jewett, M.C. (2012). Cell-free protein synthesis: applications come of age. *Biotechnol. Adv.* 30: 1185–1194.

78 Endo, Y. and Sawasaki, T. (2006). Cell-free expression systems for eukaryotic protein production. *Curr. Opin. Biotechnol.* 17: 373–380.

79 Smale, S.T. (2010). *Luciferase Assay*, vol. 2010, pdb.prot5421. Cold Spring Harb Protoc.

80 Ingolia, N.T., Hussmann, J.A., and Weissman, J.S. (2019). Ribosome profiling: global views of translation. *Cold Spring Harb. Perspect. Biol.* 11: a032698.

81 Ingolia, N.T., Ghaemmaghami, S., Newman, J.R.S., and Weissman, J.S. (2009). Genome-wide analysis in vivo of translation with nucleotide resolution using ribosome profiling. *Science* 324: 218–223.

82 Engvall, E. and Perlmann, P. (1971). Enzyme-linked immunosorbent assay (ELISA) quantitative assay of immunoglobulin G. *Immunochemistry* 8: 871–874.

83 Lequin, R.M. (2005). Enzyme immunoassay (EIA)/enzyme-linked immunosorbent assay (ELISA). *Clin. Chem.* 51: 2415–2418.

84 Oh-hashi, K., Furuta, E., Fujimura, K., and Hirata, Y. (2017). Application of a novel HiBiT peptide tag for monitoring ATF4 protein expression in Neuro2a cells. *Biochem Biophys Rep* 12: 40–45.

85 Buxbaum, A.R., Haimovich, G., and Singer, R.H. (2015). In the right place at the right time: visualizing and understanding mRNA localization. *Nat. Rev. Mol. Cell Biol.* 16: 95–109.

86 Swanger, S.A., Bassell, G.J., and Gross, C. (2011). High-resolution fluorescence in situ hybridization to detect mRNAs in neuronal compartments in vitro and in vivo. In: *RNA Detect Vis* (ed. J.E. Gerst), 103–123. Totowa, NJ: Humana Press (cited 2024 Oct 1). Available from: http://link.springer.com/10.1007/978-1-61779-005-8_7.

87 Bloom, K., Ely, A., and Arbuthnot, P. (2017). A T7 endonuclease I assay to detect talen-mediated targeted mutation of HBV cccDNA. In: *Hepat B Virus* (ed. H. Guo and A. Cuconati), 85–95. New York, NY: Springer New York (cited 2024 Oct 1). Available from: http://link.springer.com/10.1007/978-1-4939-6700-1_8.

88 Chen, Y.G. and Hur, S. (2022). Cellular origins of dsRNA, their recognition and consequences. *Nat. Rev. Mol. Cell Biol.* 23: 286–301.

89 Mu, X. and Hur, S. (2021). Immunogenicity of *in vitro* -transcribed RNA. *Acc. Chem. Res.* 54: 4012–4023.

90 Nicholson, A.W. (2014). Ribonuclease III mechanisms of double-stranded RNA cleavage. *WIREs RNA* 5: 31–48.

91 Mu, X., Greenwald, E., Ahmad, S., and Hur, S. (2018). An origin of the immunogenicity of in vitro transcribed RNA. *Nucleic Acids Res.* 46: 5239–5249.

92 Dousis, A., Ravichandran, K., Hobert, E.M. et al. (2023). An engineered T7 RNA polymerase that produces mRNA free of immunostimulatory byproducts. *Nat. Biotechnol.* 41: 560–568.

93 De Faria, I.J.S., Imler, J.-L., and Marques, J.T. (2023). Protocol for the analysis of double-stranded RNAs in virus-infected insect cells using anti-dsRNA antibodies. *STAR Protocols* 4: 102033.

94 Aramburu, J., Navas-Castillo, J., Moreno, P., and Cambra, M. (1991). Detection of double-stranded RNA by ELISA and dot immunobinding assay using an antiserum to synthetic polynucleotides. *J. Virol. Methods* 33: 1–11.

95 Ma, T., Zhao, Y., and Cheng, X. (2024). Detection of dsRNA by Acridine Orange staining. In: *Double-Stranded RNA* (ed. X. Cheng and G. Wu), 7–12. New York, NY: Springer US (cited 2024 Oct 1). Available from: https://link.springer.com/10.1007/978-1-0716-3702-9_2.

96 Izumi, T., Morioka, Y., Urayama, S. et al. (2021). DsRNA sequencing for RNA virus surveillance using human clinical samples. *Viruses* 13: 1310.

97 Decker, C.J., Steiner, H.R., Hoon-Hanks, L.L. et al. (2019). dsRNA-Seq: identification of viral infection by purifying and sequencing dsRNA. *Viruses* 11: 943.

14

Analytical Methods Shaping Cell Therapy Characterization

Amritha Lewis, Sudeshna Sadhu, Rebecca Melmon, Rahamthulla Shaik, Egemen Tutuncuoglu, and Amy Gilbert

Adaptive Biotechnologies, Seattle, WA, USA

Introduction

Harnessing the potential of living cells to restore tissue function or modulate immune responses has come to light as a promising new class of therapeutics over the last decade to treat diseases ranging from cancer to autoimmune disorders [1–3]. While cell therapy has emerged as a new drug modality with the recent success of chimeric antigen receptor (CAR) T cells in hematological malignancies, the concept of cell therapy dates back centuries with early examples including blood transfusions for treating hemorrhage and bone marrow transplants for hematopoietic disorders. The study of HIV deepened our understanding of T-cell biology, the most common type of cell used as a therapeutic and facilitated the development of techniques to engineer T cells, such as engineering CARs using viruses. More recent advances in stem cell biology, gene editing technologies, and immunotherapy have revolutionized the field, paving the way for the development of novel cell-based therapeutics with unprecedented precision and efficacy. Analytical methods play a crucial role in every stage of the cell therapy workflow, from research and development to manufacturing and clinical monitoring. These methods encompass a diverse range of techniques, including flow cytometry, gene expression analysis, functional cell-based assays, and omics technologies, each offering unique insights into the biology and behavior of therapeutic cells. Prior to embarking on clinical trials, rigorous preclinical characterization of cell therapy products is imperative to ensure safety and efficacy. Central to this endeavor are characterization assays to measure the expression of immune receptors engineered on the cell surface and functional assays which encompass a suite of techniques aimed at evaluating the functional attributes of T-cell candidate products. These assays, spanning from in vitro proliferation assays to later in vivo efficacy studies, provide valuable insights into the therapeutic potential of T-cell therapies, guiding further optimization and development.

As cell therapies transition from preclinical studies to clinical trials, comprehensive immune monitoring becomes essential for evaluating treatment outcomes and understanding the complex interactions between administered T cells and the host immune system. Monitoring T-cell persistence, functionality, and host immune responses in clinical settings offers critical insights into treatment efficacy and safety profiles. By leveraging advanced analytical techniques, such as flow cytometry and next-generation sequencing, we can decipher the dynamic interplay between T cells and the immune milieu, paving the way for optimized therapeutic interventions and improved patient outcomes.

One of the key challenges in cell therapy development lies in the analytical characterization of living cells and their inherent biological variability. From an analytical perspective, there is inherent variability in starting material for autologous products, limited material available for testing, lack of appropriate reference standards, and undefined or multiple mechanisms of actions that make potency assessments in vitro complex. The field faces several challenges that require careful consideration. As the field matures, analytical sciences will continue to evolve, providing tools to address these challenges and enhance the quality and safety of cell therapies.

This chapter delves into analytical techniques crucial for assessing cell therapies, with a primary focus on methods used for T-cell therapies for the treatment of cancer. By starting with the preclinical characterization of potency through the monitoring of immune responses during clinical trials, we can enhance our understanding of T-cell

Characterizing Biotherapeutics: Analytical Methods for Diverse Modalities, First Edition. Edited by Jennie R. Lill and Wendy Sandoval.

biology, optimize therapeutic strategies, and ultimately improve patient outcomes in the flourishing field of cell therapy.

Fundamentals of T-Cell Immunology in Cell Therapy

The role of immune cells, namely T cells, in preventing or treating neoplastic disease has been described since the early nineteenth century. Starting from Colley's toxins used to treat sarcomas to the more recent understanding that T cells are capable of recognizing cancer neoantigens expressed in tumors resulting in an immune response that can often eradicate a tumor, T-cell-based therapies have now shown immense promise in clinical settings [4, 5]. The therapies range from checkpoint blockade that removes the "brakes" imposed on T-cell activity by the tumor, adoptive cell therapy (ACT) that involves transferring T cells into a tumor, CAR T cells that are engineered to target tumor-specific antigens and cancer vaccines used to boost immune responses to cancer-specific antigens.

T cells develop in the thymus, a small organ above the heart, from T-cell progenitor cells that migrate to the thymus from the bone marrow. Within the thymus, these T-cell progenitors undergo a process of random VDJ rearrangement that generates a T-cell receptor (TCR) that is selected by positive selection to be restricted to the host's HLA, and self-reactive TCRs are deleted from the immune repertoire by negative selection. This results in a highly diverse array of TCRs that exit the thymus and circulate in the peripheral blood and tissues looking for their cognate antigen. When a T cell interacts with its cognate antigen, these resting or naive T cells become activated. Activated T cells begin to rapidly proliferate and expand and produce cytokines such as interleukin (IL)-2, IL-4, and IL-7 to support their expansion (Figure 14.1).

These T cells also start to express markers of activation such as CD69 and CD137 following interaction with cognate antigens that are routinely used to track T-cell activation in preclinical and clinical settings. These activated "effector" T cells also produce cytokines and cytolytic enzymes such as interferon gamma (IFNγ), granzyme B, and tumor necrosis factor alpha (TNFα), among others, that are cytolytic and kill cells expressing the target antigen. Once this phase of effector T-cell killing is complete, the T cells shrink to form a memory population that exists long term in the blood, constantly surveilling for repeated threats from the antigen. In this way, T cells are strongly poised for continual immune surveillance against not just foreign pathogens but also cancer antigens, keeping neoplastic growths under control and contributing to their clearance if they do occur.

Measuring Immune Receptor Expression

One of the primary mechanisms of action of cell therapy products is the targeted depletion of a cell population such as cancer cells. Recognition of targeted cells by the T cell either is mediated through endogenous immune receptors, such as the expression of neoantigen TCRs for TIL therapies, or requires the engineering of tumor-reactive immune receptors. These tumor-reactive immune receptors can be engineered in the form of introducing an engineered

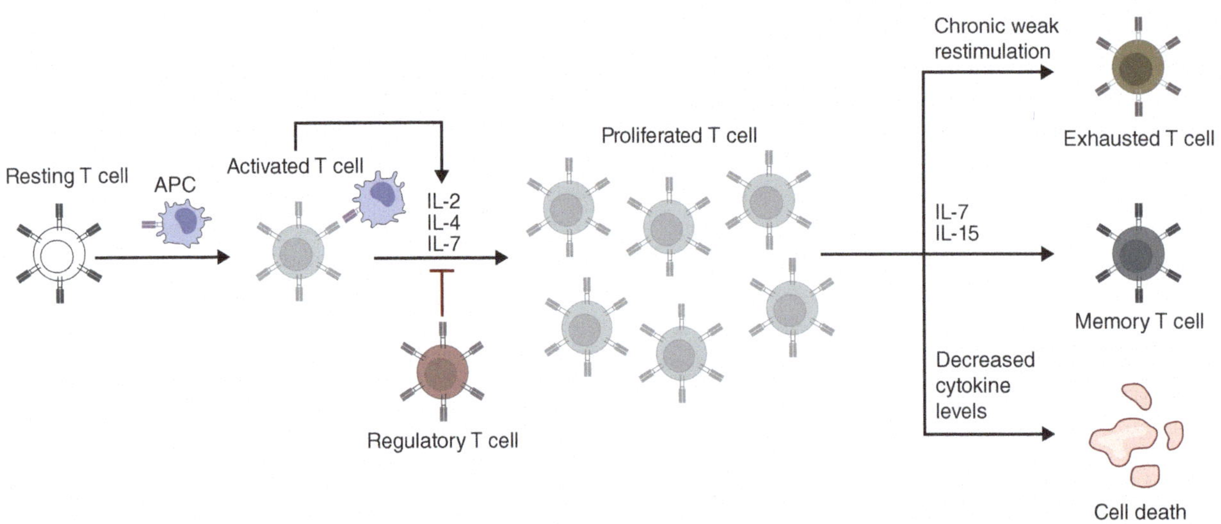

Figure 14.1 T-cell activation by antigen presenting cells. The image outlines the process of T-cell activation by antigen presenting cells and some of the different subsequent cell fates that occur after proliferation. Source: Adapted from BioRender.com.

TCR to recognize intracellular antigens or through the introduction of a CAR, which is a hybrid antibody and TCR molecule that recognizes cell surface proteins. In this section, we will highlight tools used in the T-cell therapy field to measure the expression of tumor-reactive immune receptors, which is important not only to understand if engineering was successful but also because the number of immune receptors on a per cell basis can impact the potency of the cells.

T cells express a clonal and antigen-specific receptor called T-cell receptor (TCR) on their surface that recognizes and binds to an antigenic peptide–MHC (p-MHC) complex. TCR cell surface expression is crucial for the initiation of an adaptive immune cell response; therefore, monitoring TCRs provides valuable insight in assessing the effectiveness of T-cell-based therapeutics. A possible analysis method is to directly use soluble monomeric p-MHC coupled to a detector molecule such as a fluorophore to assess antigen-specific T-cell response. However, unlike high-affinity antibodies, antigen-specific TCRs have low affinity to p-MHC complexes, and the dissociation constant (Kd) ranges from 1 μM to 100 μM [6]. This makes the direct usage of monomeric p-MHC ineffective for measuring an antigen-specific TCR-mediated response. In 1996, Altman et al. introduced MHC tetramers, where they complexed several p-MHC molecules on a fluorescently labeled backbone, thereby increasing the p-MHC avidity for their cognate TCR. This helped in direct evaluation of antigen-specific T cells ex vivo by flow cytometry. MHC tetramers are the most common tool to investigate the expression of antigen-specific TCRs in the T-cell therapy field [7].

Several multimers have been developed since the groundbreaking discovery of MHC tetramers by Altman and colleagues. Besides the frequently used tetramers, there are different types of irreversible (pentamers and dextramers) and reversible (streptamers and NTAmers) multimers described below. Tetramers are the first MHC class I multimers developed in 1996. Four p-MHC complexes are conjugated to an avidin or a non-glycosylated analogue of avidin namely streptavidin. The backbone has very high affinity for biotin and comprises four biotin binding pockets where biotinylated p-MHC complex can be bound. Both avidin and streptavidin bind with biotinylated p-MHC complex, leading to the formation of a stable tetramer. The backbone is tagged with a fluorescent molecule which allows detection of antigen-specific TCR by staining.

In the early 2000s, pentamers were developed, comprising five fluorescent or biotin-tagged p-MHC complexes multimerized on a self-assembling coiled-coil domain. Pentamers have high avidity interaction with TCRs since all the five p-MHC complexes face the same direction, unlike tetramers. Streptamers were developed around 2002 and were the first commercially available reversible multimers. In this approach, a short amino acid sequence called Twin-Strep-tag is fused to p-MHC. This, in turn, binds to a variant of streptavidin called Strep-tactin. The reversibility of streptamers is achieved by the addition of biotin, which has higher affinity for Strep-tactin compared to Strep-tagged p-MHC. Low-frequency antigen-specific T-cell clones could be reliably detected by reversible streptamers. In addition, since streptamers are reversible, the label-free cells could be utilized for functional assays or can be expanded to perform downstream analysis in the context of cell therapy. Dextramers allow detection of low-affinity antigen-specific T cells. Several fluorescently conjugated p-MHC complexes are bound to a dextran fiber backbone, allowing the detection of T-cell responses. Lastly, NTAmers are a type of noncommercial multimer, where the p-MHC complex is multimerized via oligohistidine-tag (His-tag) on Ni2+-nitrilotriacetic acid (NTA) moiety. This is a reversible multimer, wherein the addition of imidazole weakens the binding and dissolves the multimerization [8].

The key steps in manufacturing MHC class I tetramers include folding, purification, biotinylation, and tetramerization (Table 14.1). MHC tetramers are a valuable tool to assess the expression of antigen-specific TCRs but have some limitations. First, synthesizing MHC tetramers is labor intensive, has batch-to-batch variability, and is not cost-effective. Moreover, considering the number of unique peptide–MHC-TCR interactions, only one peptide–MHC tetramer can recognize its respective TCR, thereby reducing flexibility for its application.

In the preclinical development of CAR T-cell products, protein L is a common reagent to detect the cell surface expression of CARs on transduced lymphocytes by flow

Table 14.1 How are MHC class I tetramers generated?

Step 1: The heavy chain of an MHC molecule containing the bacterial BirA-recognition site is synthesized. It is folded along with MHC molecule light chain, beta2 microglobulin, and peptide of interest to form p-MHC monomers. The formation of a p-MHC monomer is monitored by high-performance liquid chromatography (HPLC).
Step 2: Each refolded p-MHC monomer is biotinylated by BirA enzyme at the C terminus of the MHC heavy chain and the monomeric p-MHC complex is purified by column chromatography.
Step 3: Subsequently, in the presence of fluorochrome labeled-streptavidin, containing four biotin binding per molecule, four purified biotinylated p-MHC monomers are linked together to form tetramers. This allows four TCR to bind simultaneously with MHC tetramers, thereby creating a stronger p-MHC–TCR interaction compared to a single p-MHC:TCR [9].

cytometry. Protein L is an immunoglobulin (Ig) binding protein derived from bacterial surface protein that specifically binds to variable (kappa) light chains of all IgGs and Fab fragments without interfering with the antigen binding site. Moreover, it also binds to single-chain antibody fragments, making it a convenient reagent to measure the expression of various types of CARs on the surface of transduced lymphocytes [10]. Additional approaches for measuring CAR expression on the surface of T cells can include recombinant proteins of the targeted antigen or the addition of a tag in the N terminal region of the CAR, such as FLAG or histidine (HIS). Disadvantages of these approaches can include a lack of detection of CAR in the presence of tumor or soluble antigens or impacting the expression of the receptor. Another tool to identify and quantify CAR T cells in patients is using an anti-idiotype antibody. This typically comes after the identification of a clinical lead candidate. In a study by Jena B et al., the researchers measured CAR T cells using an anti-idiotype monoclonal antibody that binds to the unique scFv expressed as part of the CAR on the T cells. This antibody specifically recognizes the idiotype of the CD19 CAR, as measured by flow cytometry, allowing for the detection and quantification of CAR T cells in patient PBMCs [11].

The biophysical properties, such as affinity and avidity of the immune receptor to its cognate antigen, can influence the functional response of cells. CAR or TCR affinity can be measured using recombinant scFv proteins or soluble TCRs and is measured by surface plasmon resonance (SPR) and bio-layer interferometry (BLI), where they measure equilibrium constant by monitoring real-time binding kinetics between two biomolecules. CAR avidity can be measured in vitro by exposing transduced T cells to varying amounts of antigens, and cells with high functional avidity respond to the lowest antigen dose. For example, in a proof-of-concept study, Greenman et al. assessed avidity, affinity, and antigen density of CAR binding [12]. They developed an experimental system consisting of several B-cell hybridomas secreting antibodies against melanoma tumor-associated antigen Tyr/HLA-A2 to generate a series of CARs against Tyr/HLA-A2 with varying affinities. They tested the monovalent affinity of each Fab construct using SPR. They next constructed CARs by converting the Fab into scFv, transduced and expressed on CD8+ T cells, and measured avidity by assessing the expression of Tyr/HLA-A2 multimer binding. Moreover, molecular techniques like quantitative polymerase chain reaction (qPCR), RNA sequencing, and integration site analysis monitor CAR transduction, expression, and in vivo CAR T-cell expansion kinetics.

Measuring the immune receptor expression on cells is a part of the product identity assay often used for product release, and expression levels can potentially make dramatic differences in clinical outcomes. Additionally, immune receptor expression measurements can be used to calculate cell dosage, which relies on understanding the percentage of engineered cells in the total cell product. Although evaluating TCRs and CAR expression on engineered T cells is a crucial first step in characterization, assessing the functional potency determines its effectiveness as a cell therapy product. The next section will elucidate various cell-based functional assays used in evaluating the potency of engineered cell therapy products.

Functional Cell-Based Assays for *In Vitro* Characterization

Preclinical evaluation of engineered cell therapy products is crucial to assess their mechanisms of action, such as anti-tumor efficacy, leading to desired clinical outcomes. There is a wide range of assays to measure T-cell function and anti-tumor efficacy in preclinical settings, including antigen-specific T-cell-mediated cytotoxicity, cytokine production, proliferation, and the functional avidity of the TCR (Table 14.2).

Cytotoxicity assays are the standard method used to measure the tumor-killing potential of effector T cells. The most

Table 14.2 *In vitro* characterization assays to measure cell function.

Principal measurement	Assay	Readout	References
Cytotoxicity	^{51}Cr Release	Radioactivity (^{51}Cr Release)	[13]
Cytotoxicity	Bioluminescence	Luciferase activity	[14]
Cytotoxicity	Live cell imaging assay	Cell death by imaging	[15]
Cytotoxicity	Impedance	Cell detachment	[16]
Cytotoxicity	Flow cytometry	Live/dead staining, phenotype	[17]
Cytokine production	Multiple ELISAs (Ella/Luminex)	Automated ELISA measuring effector molecules	[18]
Cell proliferation	Flow cytometry-based assay	CFSE tracking dye	[19]
TCR functional avidity	ELISA	IFN-γ release	[20]

common methods for determining T-cell-mediated cytotoxicity include the Chromium-51 (^{51}Cr) release assay, luminescence-based luciferase assay, flow cytometry-based assays, and label-free impedance-based assays. Considerations for assay selection for functional characterization are described below, and some inter-assay comparisons with the same CAR T have also been performed [13].

The ^{51}Cr release assay is a limited, low-throughput assay to measure cytotoxicity, with measurement usually restricted to a single time point (4-hour endpoint assay), limiting the possibility of repeated cytotoxicity measurements over time. This is one of the oldest techniques used to measure cell death, and outside of needing a gamma counter, no specialized equipment is required. The assay has a low signal-to-background ratio and requires the use of radioactive isotopes. The luciferase-based cytotoxicity assay is a simple assay that measures the bioluminescence activity of luciferase without the need to work with radioisotopes such as ^{51}Cr [18]. However, this assay's limitations include the requirement to engineer target cells to express luciferase and its short half-life. Luciferase release assays also only offer an endpoint assay without any additional kinetics.

The live cell imaging-based cytotoxicity (Incucyte) assay measures cytotoxicity based on the detection of cell permeability, a hallmark of cell death. A DNA-binding fluorescent dye is used to measure the level of cytotoxicity [15]. It has the primary advantage that you can measure the kinetics of T-cell-mediated cell death. While not offering real-time kinetics, a flow cytometry-based cytotoxicity assay has the advantage of distinguishing heterogeneous cell populations when stained with specific fluorescent antibodies. The high throughput, multiparametric flow cytometry assay developed by Martinez et al. can measure cytotoxicity and cell phenotype in a single assay [17]. Lastly, the impedance-based assay is a high-throughput method to measure cytotoxicity in real time. This assay is based on the difference in impedance levels of electric current flow between electrodes in the presence and absence of target cell attachment. The change in impedance is recorded as a "cell index" when target tumor cell detachment occurs due to effector cell-mediated cytotoxicity. For non-adherent cells, coating the plate wells with extracellular matrix proteins (antibodies) is required to allow proper attachment of the target cells to the wells on the plate. This assay can measure cytotoxicity at various effector-to-target (E:T) ratios over an extended period [16]. All three of these techniques offer the ability to remove or directly characterize T cells from the assays to better understand the "state" of the cell product, such as expression of activation or exhaustion markers such as PD1, TIM3, and LAG3 on the cell surface.

Antigen-specific recognition and binding of T cells to tumor cells result in T-cell activation, leading to proliferation and production of effector cytokines. Cytokine production ability is measured by determining the cytokine profile of T cells using multiplex ELISA assays (Luminex/ELLA). Some of the most common cytokines measured in these assays include IFN-γ, TNF-α, granzyme B, and IL-2 [18]. For TCR-based cell therapy products, TCR functional avidity, or antigen sensitivity assay measures the ability of TCRs to recognize cognate peptides, which corresponds with the amount of IFN-γ produced in a co-culture assay [20]. This is an additional characterization assay often used for TCR T-cell products.

T-cell expansion and persistence have been shown to correlate with clinical outcomes [21], and in vitro models to assess if either scFv binder or engineering approaches may provide valuable research tools. T-cell proliferation can be assessed by staining with a tracking dye such as CFSE [19]. Additional elegant in vitro systems have been developed to evaluate the proliferation and persistence of engineered cells, such as serial stimulation assays, which allow researchers to understand how repeated activation and restimulation of the T cells can impact cell function [22]. Transitioning from in vitro functional characterization, we now delve into understanding the state of the T cells mediating the functional response to better understand their fitness and differentiation, which is an important consideration for designing and tailoring cell therapy products for optimal function [23].

Cell Phenotyping

One of the unique challenges in the field of cell therapy is the inherent variability of the cell product derived from living cells. In the autologous (using patient's own cells) or healthy donor allogeneic (using cells from a different individual) setting, the cell product is created from a blood draw, capturing the immune repertoire at a moment in time which may vary in its composition. The characterization of T-cell phenotype is essential for the design and functional properties of cell therapy products. Phenotypic analysis provides insights into the identity, differentiation status, and activation state of T cells, enabling the tailoring of therapeutic strategies and may predict treatment outcomes [23–25].

A comprehensive phenotyping approach encompasses the evaluation of surface markers, intracellular proteins, and functional markers to delineate T-cell subsets and their immunological properties. Surface marker expression profiling by flow cytometry allows for the identification and quantification of distinct T-cell subsets based on the

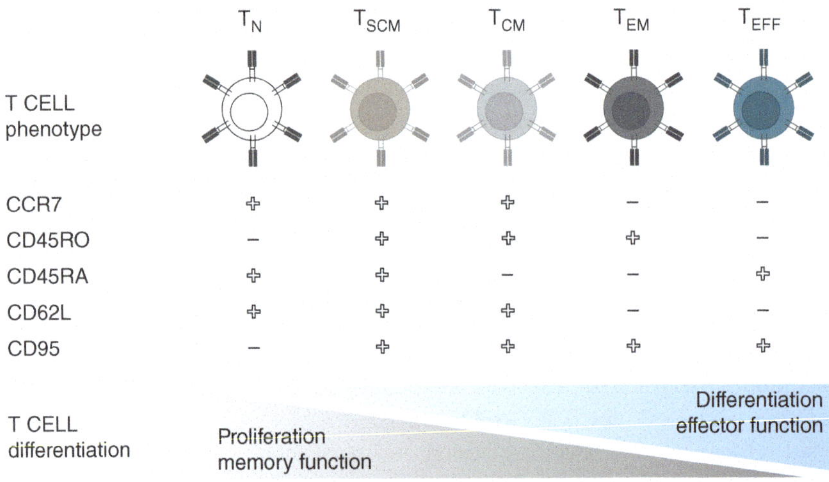

Figure 14.2 T-cell differentiation subsets and associated cell surface markers. The schematic depicts the complex process of T-cell maturation and specialization in immune responses.

presence or absence of specific cell surface markers. Commonly used markers include CCR7, CD45RO, CD45RA, CD62L, and CD95 (Figure 14.2), which distinguish between naive (TN), effector, memory, and regulatory T-cell subsets [26]. Flow cytometry is the cornerstone technique for surface marker analysis, offering high-throughput, multiparametric assessment of T-cell phenotypes. Careful panel design should be considered for multiparameter flow cytometry, including selection of fluorophore, positive and negative controls for receptor detection, as well as utilizing a fluorescence minus one (FMO) control.

The differentiation status of T cells to improve proliferative potential of a cell over time and understanding of other T-cell phenotypes in the product can be important in understanding treatment responses. For example, single-cell proteomic profiling has revealed an association with the presence of CAR Treg cells on day 7 post-infusion to be associated with progressive disease [25, 27] and long-term persistence of CD4+ CAR T has been associated with long-term remissions [28].

In addition to potency assays, understanding the populations of T cells by phenotyping is a critical component of cell product characterization, providing valuable insights into the composition, functionality, and therapeutic potential of T-cell-based therapies. By leveraging advanced analytical techniques, such as flow cytometry and single cell profiling, we can comprehensively assess T-cell phenotypes and tailor therapeutic interventions to maximize clinical efficacy.

Immune Monitoring: BCR and TCR Sequencing

To recognize a wide variety of potential antigens, T cells have a well-diversified repertoire of receptors. This diversity is created by the combinatorial rearrangement of receptor gene segments. Information on receptor sequence and clonality can provide key understanding and knowledge of the adaptive immune response. However, due to the heterogeneity of TCR sequences, a common primer cannot be used to amplify all receptors. To simultaneously amplify multiple DNA sequences of TCRs, a multiplex PCR reaction is used with a mixture of primers targeting the variable and joining segments. Amplification bias can be a barrier in PCR reactions, where the efficiency of amplification varies from one sequence to another, resulting in inaccurate readouts of receptor clonality. To mitigate this issue, a repertoire of synthetic immune receptors can be employed. These synthetic receptors are used to recognize and identify biases in amplification and remove them during analysis. This unbiased multiplex PCR assay is called immunoSEQ [29].

ImmunoSEQ can be used to monitor the efficacy of a cell therapy product in patients during clinical trials, providing valuable secondary endpoint data and a more in-depth understanding of the kinetics, persistence, and effectiveness of the infused cell therapy product in vivo. Using this technique, minimal residual disease (MRD) can be monitored in patients to assess treatment response and disease progression. In a study on patients with acute lymphoblastic leukemia, immunoSEQ was validated by analyzing 36 patients with a clonal TCR gamma rearrangement [30]. Wu et al. compared the frequency of cancer clones by high-throughput sequencing versus multiparameter flow cytometry. No false negatives were found for MRD detection through immunoSEQ, and 10 additional patients showed MRD detection through the PCR assay alone, proving higher sensitivity than multiparameter flow cytometry. These data demonstrate an accurate and precise method of monitoring disease through immune cell sequencing, which can be widely applied.

Immune monitoring plays a crucial role in addressing key questions, such as how cell product quantity or composition correlates with clinical responses. In a study by Fraietta et al., patients with chronic lymphocytic leukemia (CLL) were treated with a CD19-targeting CAR, and the polyclonality and expansion of the infused CAR T cells were tracked using immunoSEQ [31]. The use of immunoSEQ to monitor the TCR V betas resulted in a novel and unexpected finding in "Patient 10," who underwent two adoptive transfers of autologous CTL019 cells spaced approximately two months apart. Two months following the second infusion, the expansion of CAR T cells peaked in the peripheral blood, followed by contraction, and the product (CTL019) cell outgrowth occurred in the CD8+ T-cell compartment. High-grade CRS and elevated circulating levels of IFNγ, granulocyte-colony stimulating factor (G-CSF), IL-6, IL-8, and IL-10 were also observed. Alongside the onset of high fevers, rapid clearance of CLL was observed, persisting up to five years post-treatment. Deep sequencing of the TCR beta repertoire (immunoSEQ) showed that pre-infused CD8+ CTL019 CAR T cells and the patient's own peripheral blood CD8+ T-cell compartment one-month following infusion were polyclonal. Multiple different TCR V beta sequences were identified in both compartments. However, two months after the second infusion, there was greater than 50% clonal dominance of TCRVβ5.1 family usage within the CAR T-cell product. More in-depth analysis showed that a single clone made up 94% of the CD8+ CAR T-cell repertoire. Therefore, it was concluded that in the case of this individual, leukemia was largely eliminated by the expansion of a single CAR T-cell clone, which the authors approximated to be roughly 29 population doublings in the peripheral blood. These findings revealed a significant expansion of a single dominant CAR T-cell clone that correlated with high therapeutic efficacy and long-term remission. Similarly, MacLean S. Hall et al. (2023) demonstrated how changes in TCR clonality within tumor-infiltrating lymphocytes (TILs) could be linked to improved patient responses, offering valuable insights into the dynamics of T-cell expansion and persistence post-therapy [32].

Using immunoSEQ and next-generation sequencing (NGS) analysis has proven essential in understanding clinical outcomes. In a study by Shah et al. (2019), immunoSEQ identified the clonal expansion of CAR T cells that harbored lentivector integration in the CBL gene, leading to severe adverse events [33]. This insight allowed for a deeper understanding of the mechanisms behind these adverse events, guiding improvements in CAR T-cell therapies to mitigate such risks.

Advanced techniques like immunoSEQ provide critical insights into the safety, behavior, and efficacy of cell therapy products. These tools enable more precise monitoring and optimization of therapeutic strategies, ultimately contributing to better patient outcomes.

Conclusions

The field of cell therapy has witnessed incredible advancements over the past decade, driven by innovations in immunology, genetic engineering, and sophisticated analytical methodologies. T-cell-based therapies, particularly CAR T cells, have transformed the treatment landscape for various malignancies and immune disorders, offering new hope for patients. Throughout the lifecycle of these therapies, from preclinical development to post-treatment monitoring, analytical methods play a critical role. Techniques such as flow cytometry, gene expression analysis, functional assays, and immune receptor sequencing are indispensable for characterizing therapeutic cells, ensuring their safety and efficacy, and understanding their interactions with the host immune system.

As we look to the future, addressing the inherent challenges associated with the variability and complexity of living cell products remains essential. Continuous advancements in analytical technologies and methodologies are necessary to overcome these challenges, enhance the precision and effectiveness of cell therapies, and ultimately improve patient outcomes. By deepening our understanding of T-cell biology and the mechanisms driving therapeutic efficacy, we can develop more targeted and personalized treatment strategies.

Acknowledgment

Images were created using Biorender.

References

1 Sadelain, M. (2015). CAR therapy: the CD19 paradigm. *J Clin Invest* 125 (9): 3392–3400. https://doi.org/10.1172/jci80010.

2 Kochenderfer, J.N. and Rosenberg, S.A. (2013). Treating B-cell cancer with T cells expressing anti-CD19 chimeric antigen receptors. *Nat. Rev. Clin. Oncol.* 10 (5): 267–276. https://doi.org/10.1038/nrclinonc.2013.46.

3 Fabian, M., Taubmann, J., Bucci, L. et al. (2024). CD19 CAR T-Cell therapy in autoimmune disease — A case

series with follow-up. *N. Engl. J. Med.* 390 (8): 687–700. https://doi.org/10.1056/NEJMoa2308917.

4 Monberg, T.J., Borch, T.H., Svane, I.M., and Donia, M. (2023). TIL therapy: facts and hopes. *Clin. Cancer Res.* 29 (17): 3275–3283. https://doi.org/10.1158/1078-0432.CCR-22-2428.

5 Rosenberg, S.A. and Restifo, N.P. (2015). Adoptive cell transfer as personalized immunotherapy for human cancer. *Science* 348 (6230): 62–68. https://doi.org/10.1126/science.aaa4967.

6 Matsui, K., Boniface, J.J., Reay, P.A. et al. (1991). Low affinity interaction of peptide-MHC complexes with T cell receptors. *Science* 254 (5039): 1788–1791. https://doi.org/10.1126/science.1763329.

7 Altman, J.D., Moss, P.A., Goulder, P.J. et al. (1996). Phenotypic analysis of antigen-specific T lymphocytes. *Science* 274 (5284): 94–96. https://doi.org/10.1126/science.274.5284.94.

8 Schmidt, J., Chiffelle, J., Perez, M.A. et al. (2023). Neoantigen-specific CD8 T cells with high structural avidity preferentially reside in and eliminate tumors. *Nat. Commun.* 14 (1): 3188. https://doi.org/10.1038/s41467-023-38946-z.

9 Bodinier, M., Peyrat, M.A., Tournay, C. et al. (2000). Efficient detection and immunomagnetic sorting of specific T cells using multimers of MHC class I and peptide with reduced CD8 binding. *Nat. Med.* 6 (6): 707–710. https://doi.org/10.1038/76292.

10 Zheng, Z., Chinnasamy, N., and Morgan, R.A. (2012). Protein L: a novel reagent for the detection of Chimeric Antigen Receptor (CAR) expression by flow cytometry. *J. Transl. Med.* 10 (1): 29. https://doi.org/10.1186/1479-5876-10-29.

11 Jena, B., Maiti, S., Huls, H. et al. (2013). Chimeric antigen receptor (CAR)-specific monoclonal antibody to detect CD19-specific T cells in clinical trials. *PLOS ONE* 8 (3): e57838. https://doi.org/10.1371/journal.pone.0057838.

12 Greenman, R., Pizem, Y., Haus-Cohen, M. et al. (2021). Shaping functional avidity of CAR T cells: affinity, avidity, and antigen density that regulate response. *Mol. Cancer Ther.* 20 (5): 872–884. https://doi.org/10.1158/1535-7163.MCT-19-1109.

13 Kiesgen, S., Messinger, J.C., Chintala, N.K. et al. (2021). Comparative analysis of assays to measure CAR T-cell-mediated cytotoxicity. *Nat. Protoc.* 16 (3): 1331–1342. https://doi.org/10.1038/s41596-020-00467-0.

14 Li, F., Zhang, H., Wang, W. et al. (2022). T cell receptor β-chain-targeting chimeric antigen receptor T cells against T cell malignancies. *Nat. Commun.* 13 (1): 4334. https://doi.org/10.1038/s41467-022-32092-8.

15 Croom-Perez, T.J., Robles-Carillo, L.D., Oyer, J.L. et al. (2023). Chapter 5 - Kinetic, imaging based assay to measure NK cell cytotoxicity against adherent cells. In: *Methods in Cell Biology*, vol. 178 (ed. C. Thomas and L. Galluzzi), 63–91. Academic Press. https://doi.org/10.1016/bs.mcb.2022.07.012.

16 Lisby, A.N., Carlson, R.D., Baybutt, T.R. et al. (2022). Chapter 5 - Evaluation of CAR-T cell cytotoxicity: Real-time impedance-based analysis. In: *Methods in Cell Biology*, vol. 167 (ed. S. Spada and L. Galluzzi), 81–98. Academic Press. https://doi.org/10.1016/bs.mcb.2021.08.002.

17 Martinez, E.M. et al. (2018). High-throughput flow cytometric method for the simultaneous measurement of CAR-T cell characterization and cytotoxicity against solid tumor cell lines. *SLAS Discov.* 23 (7): https://doi.org/10.1177/2472555218768745.

18 Li, X., Feng, Y., Shang, F. et al. (2021). Characterization of the therapeutic effects of novel chimeric antigen receptor T cells targeting CD38 on multiple myeloma. *Front. Oncol.* 11: Online. Available: https://www.frontiersin.org/journals/oncology/articles/10.3389/fonc.2021.703087.

19 Tario, J.D., Muirhead, K.A., Pan, D. et al. (2011). Tracking Immune Cell Proliferation and Cytotoxic Potential Using Flow Cytometry. In: *Flow Cytometry Protocols* (ed. T.S. Hawley and R.G. Hawley), 119–164. Totowa, NJ: Humana Press. https://doi.org/10.1007/978-1-61737-950-5_7.

20 Jin, B.Y., Campbell, T.E., Draper, L.M. et al. (2018). Engineered T cells targeting E7 mediate regression of human papillomavirus cancers in a murine model. *JCI Insight* 3 (8). https://doi.org/10.1172/jci.insight.99488.

21 Fürst, D., Neuchel, C., Neagoie, A. et al. (2022). Monitoring the in-Vivo expansion and persistence of CAR-T cells as a tool to help decision making in patients with aggressive B-Cell lymphoma. *Blood* 140 (Supplement 1): 7527–7528. https://doi.org/10.1182/blood-2022-169212.

22 Hamieh, M., Dobrin, A., Cabriolu, A. et al. (2019). CAR T cell trogocytosis and cooperative killing regulate tumour antigen escape. *Nature* 568 (7750): 112–116. https://doi.org/10.1038/s41586-019-1054-1.

23 Fraietta, J.A., Lacey, S.F., Orlando, E.J. et al. (2018). Determinants of response and resistance to CD19 chimeric antigen receptor (CAR) T cell therapy of chronic lymphocytic leukemia. *Nat. Med.* 24 (5): 563–571. https://doi.org/10.1038/s41591-018-0010-1.

24 Tantalo, D.G., Oliver, A.J., von Scheidt, B. et al. (2021). Understanding T cell phenotype for the design of effective chimeric antigen receptor T cell therapies. *J. Immunother. Cancer* 9 (5): e002555. https://doi.org/10.1136/jitc-2021-002555.

25 Good, Z., Spiegel, J.Y., Sahaf, B. et al. (2022). Post-infusion CAR TReg cells identify patients resistant to CD19-CAR therapy. *Nat. Med.* 28 (9): 1860–1871. https://doi.org/10.1038/s41591-022-01960-7.

26 Mousset, C.M., Hobo, W., Woestenenk, R. et al. (2019). Comprehensive phenotyping of T cells using flow cytometry. *Cytometry A* 95 (6): 647–654. https://doi.org/10.1002/cyto.a.23724.

27 Haradhvala, N.J., Leick, M.B., Maurer, K. et al. (2022). Distinct cellular dynamics associated with response to CAR-T therapy for refractory B cell lymphoma. *Nat. Med.* 28 (9): 1848–1859. https://doi.org/10.1038/s41591-022-01959-0.

28 Melenhorst, J.J., Chen, G.M., Wang, M. et al. (2022). Decade-long leukaemia remissions with persistence of CD4+ CAR T cells. *Nature* 602 (7897): 503–509. https://doi.org/10.1038/s41586-021-04390-6.

29 Carlson, C.S., Emerson, R.O., Sherwood, A.M. et al. (2013). Using synthetic templates to design an unbiased multiplex PCR assay. *Nat. Commun.* 4 (1): 2680. https://doi.org/10.1038/ncomms3680.

30 Wu, D., Sherwood, A., Fromm, J.R. et al. (2012). High-throughput sequencing detects minimal residual disease in acute T lymphoblastic leukemia. *Sci. Transl. Med.* 4: 134ra63–134ra63. https://doi.org/10.1126/scitranslmed.3003656.

31 Fraietta, J.A., Nobles, C.L., Sammons, M.A. et al. (2018). Disruption of TET2 promotes the therapeutic efficacy of CD19-targeted T cells. *Nature* 558 (7709): 307–312. https://doi.org/10.1038/s41586-018-0178-z.

32 Hall, M.L.S., Teer, J.K., Yu, X. et al. (2023). Neoantigen-specific $CD4^+$ tumor-infiltrating lymphocytes are potent effectors identified within adoptive cell therapy products for metastatic melanoma patients. *J. Immunother. Cancer* 11 (10): e007288. https://doi.org/10.1136/jitc-2023-007288.

33 Shah, N.N., Qin, H., Yates, B. et al. (2019). Clonal expansion of CAR T cells harboring lentivector integration in the CBL gene following anti-CD22 CAR T-cell therapy. *Blood Adv.* 3 (15): 2317–2322. https://doi.org/10.1182/bloodadvances.2019000219.

15

Characterizing Nucleic Acid-, Gene-, and Cell-Based Therapeutics with Sequencing Technologies

Hayley M. Bennett, Ashley Byrne, William Stephenson, and Zora Modrusan

Department of Proteomic and Genomic Technologies, Genentech Research and Early Development, South San Francisco, CA, USA

Introduction

Nucleic acid sequencing at scale has revolutionized biology and medicine and it is almost impossible to imagine the development of new therapeutics without the involvement of sequencing technologies. Comprehensive genome annotation is required for the identification of actionable gene targets and pathways. Measurement of constitutive and differential gene expression (DGE), along with epigenetic profiling, can facilitate the assessment of biotherapeutic outcomes. Additionally, DNA and RNA themselves are increasingly used as therapeutic modalities, or in the process of generating gene and cell therapies, and sequencing is essential for quality control of the therapeutic agents and confirmation of targeted gene edits. Many of these therapies are evolving at a fast pace and are highly complex treatments that necessitate robust quality control guidelines to ensure patient safety. Sequencing technologies can rapidly perform identity, biosafety, and stability testing for therapeutic characterization. Here we introduce a variety of common sequencing technologies and their application in the characterization of therapeutic agents (Figure 15.1).

DNA Sequencing

Sanger Sequencing

Sanger sequencing is a well-established and widely used technique for characterizing a targeted region of DNA [1]. Initially developed in the 1970s, it utilizes chain terminating dideoxynucleotides (ddNTPs) mixed in at low concentration with dNTPs during polymerase chain reaction (PCR) amplification (Figure 15.1a). The reaction is initiated with primers from a known sequence of 18–24 base pairs (bp) on the template DNA. Modern reactions use fluorescently modified ddNTPs so that all four bases are distinguishable in the same sequencing reaction. The ddNTP bases terminate the extension, and the products are separated by size in a capillary gel to create a chromatogram of position and the incorporated bases. Typically sequences of up to 1,000 bp can be generated, with the balance of ddNTP concentration in the reaction resulting in a sequence of base calls, determined by well-defined chromatogram peaks in the 40–700 bp range. Often a primer on each side of the targeted region will be used to generate a complementary sequence, and the two reads will be collapsed together *in silico* for greater accuracy. In the case of single nucleotide variant calling, the relative peak sizes for each base in the chromatogram can be used to determine the relative abundance in the sample. Sanger sequencing can also utilize primer walking, where primers are iteratively designed for follow up reactions, to keep extending the length of the known sequence, for example, to fully sequence a 4 Kb plasmid. Sanger sequencing has very low throughput by modern standards – the very first draft of the human genome (3 Gb) was an international effort that cost 300 million USD and took 13 years [2]. Nowadays, with next-generation sequencing (NGS), a human genome can be sequenced in 24 hours on one instrument for around 600 USD [2]. Another major drawback of Sanger sequencing is that due to lower sensitivity, substantial PCR amplification is often required to generate enough DNA copies for the assay which can potentially introduce mutations or biases. Despite subsequent leaps in sequencing throughput and sensitivity, Sanger sequencing still remains a useful analytical tool for validating gene targets with predefined flanking regions and constructs for gene delivery.

Next-Generation Sequencing (NGS)

The term NGS was coined in response to the high-throughput parallel sequencing capabilities introduced by technologies like sequencing by synthesis (SBS) [3, 4]. SBS relies on detecting incorporated nucleotides as they are

Characterizing Biotherapeutics: Analytical Methods for Diverse Modalities, First Edition. Edited by Jennie R. Lill and Wendy Sandoval.

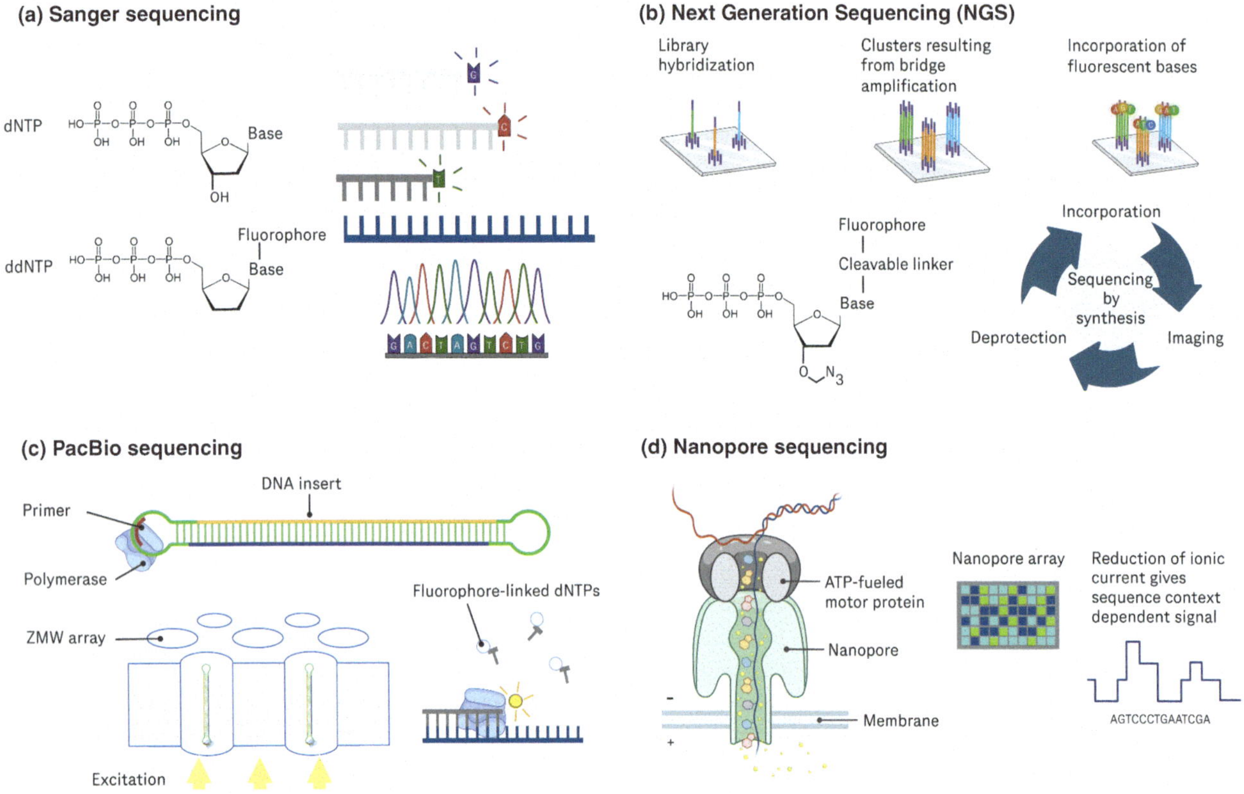

Short-read NGS instruments

Instrument (Illumina)	Max read length	Instrument sequencing time (hrs)	Max number of reads* per flow cell *single end sequencing	Max data per flow cell
iSeq100	2x 150bp	9.5-19	4 million	1.2 Gb
MiniSeq	2x 150bp	5-24	25 million	7.5 Gb
MiSeq	2x 300bp	5-56	25 million	15 Gb
NextSeq550	2x 150bp	11-29	400 million	120 Gb
NextSeq1000	2x 300bp	8-42	400 million	240 Gb
NextSeq2000	2x 300bp	8-44	1.8 billion	540 Gb
NovaSeq6000	2x 250bp	13-44	10 billion	3 Tb
NovaSeqX/NovaSeqXPlus	2x 150bp	13-48	26 billion	8 Tb

Long-read instruments

Instrument (PacBio)	Max read length	Instrument sequencing time (hrs)	Max number of reads per flow cell	Max data per flow cell
Revio	15-20kb	24	25 million	90 Gb
Onso	2x 150bp	48	1.8 billion	150 Gb
Sequel II	15-20kb	30	8 million	20 Gb

Instrument (Nanopore)	Instrument sequencing time (hrs)	Max data per flow cell
MinION/GridION	0.5-72	48 Gb
PromethION	0.5-72	200 Gb

Figure 15.1 Sequencing technologies. (a) Sanger sequencing uses a mix of regular dNTPs and fluorescent terminator ddNTPs in an extension reaction to generate products terminating at different positions. These are separated in capillary electrophoresis, and the signal at each base position generates a chromatogram. (b) Sequencing by synthesis using the Illumina workflow. First sequencing library molecules are hybridized to the flow cell. Amplification on the flow cell generates clusters containing many copies of the same library molecule. Extension on the flow cell sequentially incorporates fluorescent terminator nucleotides. Sequencing progresses in cycles where each nucleotide base is incorporated, imaged and then fluorophore cleaved and the terminal nucleotide deprotected, in preparation for the next cycle. A fleet of instruments cater to different sequencing specifications and throughputs (c) Pacific Biosciences (PacBio) sequencing utilizes ZMWs for single molecule sequencing. Hairpin adaptor molecules are added to DNA inserts to generate SMRTbell library molecules. Signal is detected from fluorophore nucleotides as they are incorporated during extension by a polymerase immobilized in the ZMW. PacBio provides commercially available long read and short read workflows. (d) Nanopore sequencing measures current change during passage of nucleotides through pore proteins situated in a membrane with a set voltage potential. An array of pores generates data from library molecules as they are captured and drawn through the pore (nanopores generating data in green, inactive nanopores blue). The pattern of ionic current is parsed by software to generate sequence data. Oxford Nanopore Technologies (ONT) has commercialized systems for different throughputs with modularized flow cells. Read length is highly dependent on preparation of DNA and application. Tables describing the available instruments are included for Illumina short-read NGS technology, and PacBio and Oxford Nanopore long-read sequencing technology.

used to build a complementary strand on template DNA (Figure 15.1b). Demand for access to high-throughput sequencing has driven innovations in chemistry, microfluidics, optics, and image processing, culminating in sophisticated instrumentation. In the last decade, Illumina has dominated the market for commercialized SBS technology [5].

In the Illumina workflow, DNA adaptors are added to template DNA to create a library which is essentially a collection of sequencing-compatible molecules. Libraries are loaded onto a flow cell which is populated with complementary DNA molecules to the library adaptors which capture individual molecules in a spatially distributed array. Once the molecules are captured, reactions can be performed by sequentially flowing reagents onto the flow cell. Each captured DNA strand is amplified locally using bridge primers to generate multiple copies, in a process referred to as bridge amplification or clustering. The double-stranded DNA is then denatured to produce single strands attached to the flow cell. In the subsequent sequencing reaction, bases attached to a fluorophore are incorporated into the complementary strands, and the copies generate a synchronized signal that can be detected with sensitive optics. Reversible chemistry on the incorporated bases enables the fluorophore to be cleaved and the reaction to progress to the next base, and onward in cycles. In paired-end sequencing, once the first strand has been sequenced for a sufficient number of cycles, a sequencing primer for the complementary strand, generated in the clustering step, can be used to generate a second read originating from the opposite end of the DNA molecule. Due to the high quality of data, and the parallel data produced by multiple copies of a molecule in the sequencing library, it is not necessary to generate a consensus call from both reads. Therefore, paired reads generate either contiguous or linked data from the same molecular origin.

The number of sequencing reads generated in parallel in one sequencing run can now exceed 20 billion – sufficient to sequence 64 whole human genomes (at 1X coverage). However, the technology is also utilized extensively for smaller-scale experiments, and the robust quality of the data enables barcoding (typically 6–10 bp of random sequence added to the library) for multiplexing many samples in one sequencing run, enabling users to benefit from the economy of scale. Due to the widespread adoption of Illumina technologies, there are more recent competing technologies which deliver similar throughput, quality, and sequencing read lengths, but at a reduced cost [6, 7].

Despite the powerful throughput afforded by Illumina sequencing, one of the remaining limitations is read length – the maximum supported read length is 300 bp, and the limit for library size is around 800 bp. These lengths are not sufficient for some applications or require bioinformatic assembly of reads to simulate a longer contiguous sequence [8]. Therefore third-generation sequencing technologies that focus on obtaining longer reads have emerged.

Long-Read Sequencing

Long sequencing reads can elucidate aspects of the transcriptome, including splice variants, and genome structure more effectively. Increased read length improves not only genome alignment accuracy but also genome assembly. Once thought to be difficult to handle, regions that have multiple close matches to a reference, such as repetitive elements or paralogous genes can be assigned with greater ease using long reads. Advances in generating high-throughput long reads were thus critical in improving the reference human genome and completing the first human telomere-to-telomere X chromosome assembly [9]. Two major platforms cater to the market for long reads.

The first high-throughput long-read sequencing technology was developed by Pacific Biosciences (PacBio) [10]. In this technology, a double-stranded library is first created by ligating sequencing adaptors that link the two strands with a hairpin (Figure 15.1c). This effectively generates a circular molecule, containing both the sense and antisense of the DNA fragment. This molecule is complexed with a polymerase and immobilized in zero-mode waveguide (ZMW) structures patterned on a surface. ZMWs act to confine the light field generated when a nucleotide with a phospho-linked fluorophore is incorporated as the polymerase progresses along the molecule, allowing detection of single fluorophore release events [11]. As the molecule is circular, the polymerase can continue to loop repetitively generating signals from nucleotide incorporations on both the sense and antisense strand of the molecule. From this high-quality circular consensus sequence is generated. PacBio also handles base-calling particularly well in high GC/AT regions, repetitive regions, and around centromeres compared to other methods [12, 13]. Further, with sufficient consensus, the inter pulse duration dynamics for the signal from each base can be used to infer the presence of modified deoxyribonucleotides, namely 5-methylcytosine [14], 5-hydroxymethylcytosine, and N6-methyladenine [15].

The highest throughput PacBio instrument, Revio, utilizes four flow cells. Each of these contains 25 million ZMWs, capable of generating 90 Gb of 15–30 Kb – sufficient for sequencing a human genome at an average 30x coverage [16, 17]. Most often long reads generated by PacBio will be used for determining genome structure, in combination with higher coverage afforded by high-throughput short-read data [18–21].

Nanopore sequencing is a long-read sequencing technology rapidly rising in popularity [22]. In nanopore sequencing, single strands of DNA or RNA are passed through a membrane via a pore-forming protein, using an ATP-fueled motor protein (Figure 15.1d). A voltage potential is held across the membrane, and as the nucleic acids pass through, each sequence reduces ionic current in a characteristic and context-dependent manner. Advanced software tools that implement machine learning are used to interpret the signal and call nucleotide bases [23–25]. Oxford Nanopore Technology (ONT) produces flow cells containing either hundreds or thousands of individually addressable nanopores. Each functional nanopore in the array is capable of reading multiple distinct library molecules in a sequential manner throughout the sequencing run. Nanopores can be utilized as long as there is sufficient ATP to fuel the motor proteins and library molecules are available. Optionally, the flow cells can further be washed and reused multiple times, albeit with degradation in performance.

Since early iterations of the technology by ONT [26], substantial advances in both the chemistry and analysis tools have driven improvements that put quality on par with other sequencing technologies for some applications [27, 28], though several comparison studies report that PacBio is still more accurate for variant-calling, insertion and deletion (indel) detection, and structural variants [29]. By 2019, whole genome studies using ONT were demonstrating an average read length of 23.8 Kb [30]. Generating ultra-long reads (over 100 Kb) from ONT is also possible with careful DNA extraction protocols and has been used to phase and reconstruct alleles in the human genome [31].

In the past few years, nanopore sequencing has been applied in a growing suite of new methods. One example is direct RNA sequencing [32], which omits conversions to cDNA, a step in which some biological information can be lost – for example, the length of polyadenosine tails on RNA transcripts, or base modifications. In another example, adaptive sampling can recognize sequences from user-supplied input in real time during the sequencing run and reject molecules from the nanopore to resample the sequencing library effectively enabling enrichment and depletion applications [33]. Further, direct calling of modified bases in native DNA and RNA is accessible through analysis software.

The smallest ONT instrument is the pocket-sized MinION, which takes a single flow cell that can generate a theoretical max of 48 Gb. This instrument's portability has been famously exploited by running sequencing on the International Space Station [34], though more commonly has been utilized in genomic field studies – particularly for genomic surveillance of disease outbreaks [22, 35]. The highest throughput ONT instrument is the PromethION with a theoretical max output of 275 Gb, achieved with 48 independently controlled flow cells. Around three flow cells on the PromethION instrument are sufficient to sequence a human genome [36].

Nucleic-Acid-Based Therapeutics

Anti-sense Oligonucleotides (ASOs) and Small Interfering RNAs (siRNA)

Anti-sense oligonucleotides (ASOs) and small interfering RNAs (siRNA) are short, synthetic nucleic acid-based therapeutics that specifically target RNA transcripts via hybridization to modulate gene expression, protein synthesis, and ultimately, therapeutic outcomes. ASOs bind a complementary sequence in the target mRNA, initiating RNase H-mediated degradation, translation inhibition, miRNA inhibition, or splicing modulation [37, 38]. siRNA therapeutics leverage the RNA interference (RNAi) pathway which comprises a natural defense mechanism against the invasion of exogenous genes [39]. Therapeutic siRNAs are delivered directly and engage with the RNA-induced silencing complex (RISC). The RISC/siRNA complex then guides the cleavage and subsequent degradation of target mRNAs [40].

Sequencing technology is implemented at various stages in ASO/siRNA quality control and development. Verification of the sequence of an ASO can often be achieved through either Sanger sequencing or NGS. Sequencing siRNA requires conversion to cDNA, which can be achieved through small RNA library preparation methods, enabling compatibility with NGS platforms. Sequencing of the ASO/siRNA ensures the purity and integrity of the therapeutic by detecting synthesis-related errors. Whole transcriptome sequencing using RNA sequencing (RNA-seq) is also commonly used to assess biodistribution, pharmacokinetics, and verification of the mechanism of action [41]. RNA-seq enables DGE measurements across ASO/siRNA treated and untreated conditions which can be used to verify modulation of gene expression in the case of RNase-H-mediated degradation (ASO) and siRNA-mediated knockdown (Figure 15.2a). Additionally, RNA-seq DGE can identify undesired off-target effects which can inform further design changes of the ASO/siRNA sequence, modification, or delivery strategy to improve target selectivity. Performing RNA-seq across multiple tissues or single-cell RNA-seq (scRNA-seq) on samples of interest following ASO/siRNA administration allows for the activity, stability, and biodistribution of the therapeutic to be ascertained [42, 43].

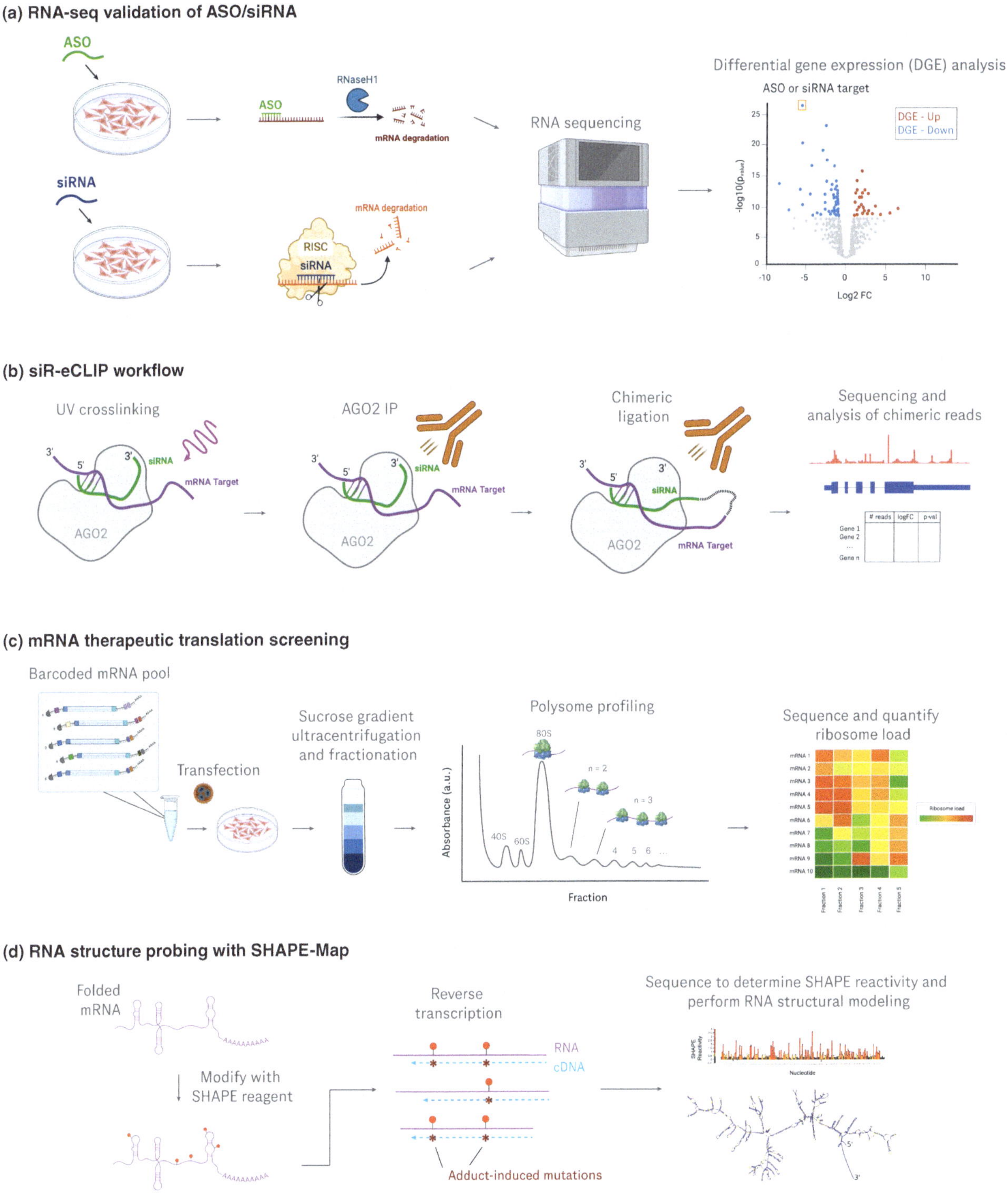

Figure 15.2 Sequencing methods for nucleic acid-based therapeutic development and characterization. (a) Workflow for assessing target knockdown with either an ASO or siRNA. (b) Generalized siRNA–mRNA interaction analysis via UV crosslinking, immunoprecipitation and chimeric ligation. (c) Pooled translation efficiency screen of mRNA constructs with polysome profiling followed by sequencing. (d) RNA chemical probing and secondary structure modeling with SHAPE-MaP.

Off-target interactions are a primary concern for oligonucleotide-based therapeutics, especially siRNAs which are mismatch tolerant. Emerging sequencing methods may provide insight into these unwanted events. While RNA-seq is typically used to assess off-targets for siRNAs, the nature of the interaction is still undetermined (e.g. direct siRNA binding or as an indirect consequence of the siRNA target knockdown). Recently, sequencing approaches that apply cross-linking and immunoprecipitation to microRNAs (miRNA) such as CLEAR-CLIP [44] and miR-eCLIP [45, 46], may provide utility for detecting direct on- and off-target interactions of siRNAs with mRNAs. miR-eCLIP works via crosslinking of either a miRNA or siRNA (e.g. siR-eCLIP) to the effector of miRNA–mRNA binding and cleavage, the AGO2 component of the RISC. Following RNase treatment and immunoprecipitation against AGO2, the miRNA or siRNA can be ligated with associated mRNAs to form chimeric molecules. When sequenced, these chimeric molecules can be used to detect associations of the miRNA or siRNA with off and on-target mRNAs (Figure 15.2b).

DNA Vaccines

DNA vaccines comprise a plasmid containing the genetic information encoding the production of antigen to elicit an immune response. Plasmids typically range from 3 to 10 Kb, as larger plasmids may be difficult to deliver and may suffer from reduced expression efficiency. Sanger sequencing of DNA vaccine plasmids using primers to regions flanking the construct is used to confirm the correct antigen insert(s) and regulatory elements without unwanted mutations. To provide comprehensive coverage over entire plasmids, and detect any aberrant structural variations, or mutations in primer regions, long-read sequencing may also be applied [47]. Currently, no DNA vaccines have been approved for human use in the United States, however DNA vaccines are licensed for use in veterinary settings including for dogs, horses, pigs, and salmon [48].

mRNA Therapeutics

The conception of mRNA as a therapeutic can be traced to the early 1990s; however, multiple advancements in nucleic acid synthesis, delivery, and manufacturing were subsequently required prior to broad adoption of advanced mRNA-based vaccines to combat the SARS-Cov-2 virus during the COVID-19 pandemic [49]. Additionally, mRNA technology is being explored for a variety of protein replacement therapies however this has proven difficult due to challenges associated with delivery to the relevant tissue and the transient nature of mRNA therapeutic expression [50]. mRNA vaccines are designed to elicit an immune response, usually by encoding a viral antigen. After mRNA vaccine administration, the encoded protein is then produced by antigen-presenting cells prompting the immune system to recognize the antigen. mRNA therapeutic design is a complex endeavor that is determined by three primary levers, namely the mRNA sequence itself, RNA modifications and the underlying secondary structure of the RNA [51]. Sequencing methods currently exist to examine each of these aspects of RNA and are poised to contribute to the further development of mRNA as a therapeutic modality.

Of primary concern regarding the characterization of mRNA therapeutics is validation of the non-coding and coding sequences associated with the therapeutic. This can be achieved via reverse transcription (RT) for conversion into cDNA and sequencing with Sanger sequencing or NGS. Recently, long-read sequencing methods such as VAX-seq [52] have been developed for characterization during various stages of mRNA therapeutic production. VAX-seq utilizes DNA nanopore sequencing to verify the plasmid vector harboring the mRNA for in-vitro transcription, followed by both PCR–cDNA nanopore sequencing and direct RNA nanopore sequencing to validate the mRNA therapeutic produced. Using a software tool called *Mana* which was developed as part of VAX-seq, the plasmid and mRNA sequence identity, length, and purity are reported. Nanopore sequencing also provides estimates of poly(A) tail lengths, a critical feature that contributes to the stability and effective translation of linear mRNA therapeutics [53]. Selection of regulatory elements such as the 5′- and 3′-UTR can also have significant effects on mRNA therapeutic stability and translation. High-throughput sequencing-based pooled screens can be implemented to characterize the translation and stability performance of combinatorial UTR selection [54].

RNA modifications, in particular the substitution of uridine for 1-methyl-pseudouridine ($m^1\psi$), are key contributors to mRNA therapeutic immune modulation and stability as demonstrated in the mRNA-based COVID-19 vaccines [55, 56]. RNA modifications can be discerned with NGS, however this requires a chemical conversion specific to the RNA modification in question prior to RT [57, 58]. Direct RNA nanopore sequencing of native mRNA therapeutics offers the attractive capability to investigate the presence and stoichiometries of multiple distinct modifications simultaneously without chemical conversion and is limited only by the training sets used to develop the base-calling models.

Optimization and verification of mRNA translation are paramount to ensure therapeutic efficacy. Sequencing-based polysome profiling screens are an effective means for assessing translational output of an array of mRNA designs that have been delivered to cells in a pooled fashion (Figure 15.2c) [54, 59]. Polysome profiling relies on halting translation (often with a small molecule such as cyclohexamide) and processing a cell lysate through a sucrose gradient for ultracentrifugation, after which the gradient can be fractionated. Each fraction contains RNA with varying numbers of bound ribosomes which is a proxy for translational efficiency. RNA from the different fractions can be converted to cDNA by RT, and an NGS DNA library generated and sequenced to determine translation-optimized mRNA designs. Verification of in-frame translation is also paramount. A recent study observed that $m^1\psi$-containing mRNA therapeutics can lead to +1 ribosomal frameshifting [60], producing unintended proteins. Many assays exist to assess the production of an mRNA-encoded antigen or therapeutic including western blotting, ELISA, and mass spectrometry. However, sequencing technologies such as Ribo-seq (also known as Ribosome profiling) are poised to provide improved resolution of translational characterization. Through enrichment and sequencing of ribosome-protected mRNA fragments, Ribo-seq enables the examination of active translation and open reading frame assignment [61].

RNA structure is also a key feature of mRNA therapeutic design. Emerging evidence suggests that extending the functional mRNA half-life increases protein production [62]. Mechanistically this could be due to reduced accessibility for endonuclease induced cleavage, increased propensity for RNA binding protein interactions that facilitate translation and mitigation of ribosome collisions which can activate mRNA degradation via the no-go decay pathway. Additional investigations indicate that mRNA therapeutic half-lives can be enhanced via structural design by up to 2.1-fold [48, 63]. In order to measure RNA structure, chemical probing techniques such as SHAPE (selective 2′-hydroxyl acylation analyzed by primer extension) or DMS (dimethylsulfate) probing can provide structural information on ribonucleotides in both *in vitro* and *in vivo* contexts. A variant of SHAPE with mutational profiling, SHAPE-MaP, is commonly used for assessing RNA structure in a high-throughput manner using NGS. Briefly, SHAPE-MaP utilizes 2′-hydroxyl-selective reagents which form chemical adducts at conformationally flexible RNA nucleotides [64]. Under modified buffer conditions, the reverse transcriptase records SHAPE adduct sites as mutations in the corresponding cDNA. These cDNAs can then be prepared for NGS which enables counting and localization of mutations consistent with the flexible nucleotide sites in the RNA. Finally, mutation frequencies can be used to derive per-nucleotide reactivities that can be used as an energy parameter constraint in RNA secondary structure modeling (Figure 15.2d).

Gene-Based Therapies

Since the discovery of genes as the fundamental unit of heredity, the field of medicine has strived to manipulate the genome as a therapeutic strategy. As we have learned the necessary techniques, therapeutic manipulation of genes has fallen into two categories: cell therapy and gene therapy. In contrast to cell therapies, gene therapy takes place *in vivo*, editing, supplementing, or regulating genes directly within the patient. Emerging from this concept, gene therapies hold promise in providing precision-based medicine to treat a variety of diseases [65]. In following either of these therapeutic avenues, safety is paramount, as they involve interventions that are long-term and not easily reversible. Sequencing technology provides a comprehensive toolkit for developing gene-based therapies. It allows for the characterization of the genome, transcriptome, and epigenome, which can be performed at both the bulk and single cell levels. By leveraging the power of sequencing technology, we can ensure that gene therapies are not only effective but also safe, paving the way for groundbreaking therapies that can transform patient care.

CRISPR-Mediated Gene Therapy

CRISPR-based genome editing allows for precise modifications to the DNA of living organisms [66], making it particularly suitable for adjusting the genome of patient cells. The CRISPR system utilizes a guide RNA to direct a nuclease, typically Cas9, to a target location in the genome where it introduces a double-stranded break. This break can be repaired by the cell's natural repair mechanisms, either through non-homologous end joining (NHEJ), which often results in small insertions or deletions, or homology-directed repair (HDR), which can be monopolized to introduce specific changes from an accompanying template DNA (Figure 15.3).

CRISPR-based gene therapy treats genetic disorders by directly correcting or modifying the underlying genetic defects. For diseases caused by single-gene mutations, CRISPR can repair the faulty gene or disrupt its harmful effects, typically editing extracted cells *ex vivo* before reintroducing them into the patient, or by *in vivo* delivery of CRISPR components directly to affected tissues. Promising results have been observed in preclinical and early clinical trials for conditions such as sickle cell anemia [67, 68], muscular dystrophy [69], and certain types of blindness [70]. However, challenges such as delivery efficiency, immune responses, and off-target effects need to be addressed to ensure the safety and efficacy of CRISPR-based therapies.

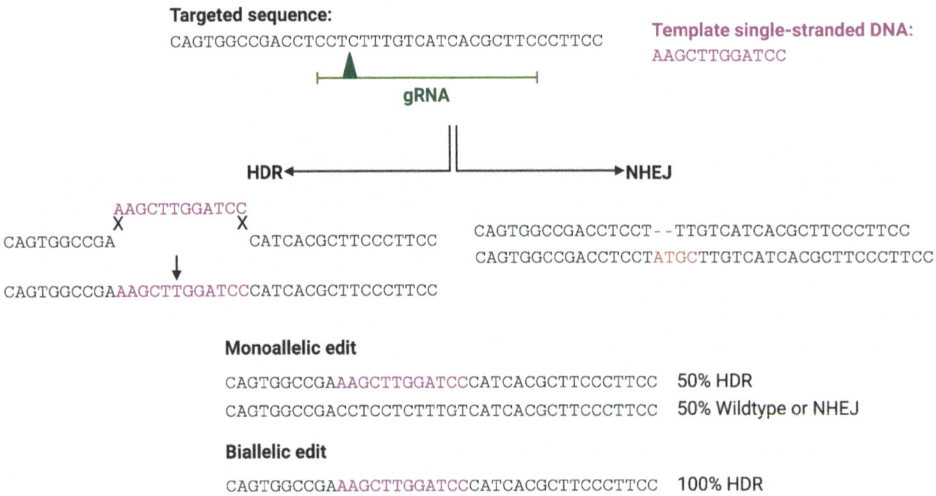

Figure 15.3 CRISPR edits with Cas9. A guide RNA (gRNA) is directed against a target sequence in the genome and directs the Cas9 to cut at a specific site (green arrow) creating a double-stranded break. Template DNA can be incorporated into the cut site by homology-directed repair (HDR). Unintended edits can be introduced by non-homologous end-joining (NHEJ), which can introduce indel mutations. Gene edits can be monoallelic, where only one allele is edited, with the other remaining as wildtype or carrying an unintended edit, or biallelic where both alleles are edited. In clonal populations of edited cells, the genotype can be determined by the percentage of sequencing reads assigned to each category.

Sequencing plays a key role in enabling precise identification of genomic changes and monitoring the outcomes of CRISPR interventions. It offers a comprehensive analysis of both on-target and off-target edits across the genome, enabling detection of small insertions, deletions, and substitutions at the targeted site, as well as unintended changes occurring within the genome. The depth and breadth of sequencing data ensures that the desired genome edits have been achieved with minimal off-target effects, thereby supporting the development of safe and effective gene therapies.

Verification of on-target edits typically involves several techniques, including PCR and Sanger sequencing which are used to amplify and sequence the targeted genomic regions. Comparing the sequenced data to the reference genome, NGS can confirm the presence and exact nature of the CRISPR-induced changes and assess editing efficiency by determining the proportion of successfully edited cells. Techniques like the T7 endonuclease I (T7E1) assay [71] or Surveyor assay [72] can identify mismatches in DNA that indicate successful editing, while functional assays can determine whether the genetic changes result in the desired phenotypic outcomes.

CRISPR off-target effects analysis is essential for ensuring the specificity and safety of gene editing by identifying unintended genetic modifications that may occur at sites other than the intended target. Several empirical and *in silico* methods have been developed to identify these unintended modifications, each with its strengths and limitations [73]. *In silico* tools, such as CasOT [74] and DeepCRISPR [75], predict potential off-target sites by analyzing sequence similarities and scoring systems based on guide RNA sequences. These computational algorithms are useful for a broad survey of potential sites but always require experimental validation. Empirical methods involve experimentation to identify off-target cleavage sites and vary in their sensitivity and throughput. PCR-based assays, such as the T7E1 mismatch cleavage assay, provide a quick and qualitative overview of on-target editing but may underestimate editing efficiency due to their inability to detect single base pair indels [71]. Sanger sequencing offers detailed analysis at a single locus, allowing for the identification of specific insertions and deletions, however, less scalable for multiple loci. NGS is considered the gold standard for comprehensive analysis and provides unbiased genome-wide characterization of off-target editing events. Techniques like High-Throughput Genome-Wide Translocation Sequencing and Digenome-seq sequence the entire genome to identify off-target cleavage, with Digenome-seq being particularly cost-effective and sensitive [76, 77]. CIRCLE-seq enriches circular DNA fragments resulting from CRISPR-induced breaks to detect off-target activity, while DISCOVER-seq uses chromatin immunoprecipitation to enhance detection sensitivity [78, 79]. GUIDE-seq tags double-stranded breaks with oligonucleotides and thus can empirically nominate potential off-target sites [80], which can then be validated using targeted sequencing approaches such as rhAmpSeq [80, 81]. Combining *in silico* predictions with empirical validation provides a robust framework for assessing the specificity and efficiency of CRISPR genome editing.

rAAV Sequencing for Gene Therapy

Adeno-associated viruses (AAVs) are small DNA viruses that are among the most effective vehicles for delivering genetic material to patients' cells. They are well-suited for this purpose due to their nonpathogenic nature, low cytotoxicity, and low immunogenicity. Recombinant AAVs (rAAVs) are made by removing viral DNA from the AAV and replacing it with a transgene tailored to a specific genetic disease. The rAAVs cannot replicate independently, relying instead on the host cell and helper proteins for replication and encapsidation of the rAAV genome. The rAAV provides a template from which mRNA is transcribed and translated to functional proteins in the host cell, compensating for malfunctioning or mutated genes in the patient. Different rAAV serotypes efficiently transduce different parts of the body, such as the eye, central nervous system, and pancreas. However, developing efficient rAAV therapies presents several challenges, including optimal viral capsid design, robust process and formulation conditions, comprehensive quality control, and characterization of rAAV samples.

The development of gene therapies using rAAV requires robust methods of ensuring biosafety, identity of products, and ensuring safety through proper QC analyses. rAAV vectors are manufactured using host cell lines that are either human or insect derived. If not properly purified, these host cells can contribute to lingering DNA contaminants, potentially leading to serious adverse events in the treated patient, specifically causing immunogenic and genotoxic effects [82–84]. As such, health agencies have strict guidelines on the DNA residual limits for recombinant rAAV products [82]. rAAV genome sequencing provides a rapid path to validation data, determining whether the gene of interest has been properly packaged into the rAAV vector, and at the same time flagging whether residual viral genes or other genomic contaminants are present [85].

Traditional rAAV sequencing methods have utilized Sanger or Illumina short-read sequencing, however both have read length limitations and primer dependencies which make identifying full-length products difficult especially in GC-rich regions. More recently, users have adopted long-read sequencing to analyze full-length rAAV products [86, 87]. This method profiles all possible rAAV products helping infer truncations, mutations, and structural alterations. Getting full-length products (i.e. 5′-ITR to 3′-ITR) is not only critical when determining full versus empty capsids during the manufacturing process but it is all pertinent for biosafety testing ensuring that there is no nucleic acid contamination [88].

Cell-Based Therapies

In 2006, Shinya Yamanaka demonstrated the ability to take adult fibroblast cells and reprogram them into an embryonic-like state which we now call induced pluripotent stem cells (iPSCs) [89]. These cells can be injected or grafted into a host to regenerate and repair damaged tissues. This idea of *ex-vivo* manipulation of cells prior to transplantation into a patient has opened the doors to regenerative medicine that could treat diseases once thought to be untargetable or incurable.

Cell therapies can localize and undergo proliferation in specific tissues or compartments. They encompass a wide range of treatments and can be differentiated into allogeneic (i.e. healthy donors provide cells for patients) or autologous (i.e. patient-derived cells). These therapies can be further characterized into subtypes such as stem cells (e.g. iPSCs, ESCs) and gene-edited therapies (e.g. CAR-T).

Most recently, advancements in CRISPR-Cas9 gene editing have led to a genetic renaissance unlocking cell therapies to their full potential. Despite the current successes of using cell therapies, hurdles remain limiting widespread adoption. These include ensuring cell purity, generating potent and viable products that meet both the patient- and disease- specific needs. While allogeneic cells offer advantages in scalability, challenges with regards to immunogenicity and toxicity can hinder therapeutic development. Sequencing technology can characterize these products by determining cell-type identity, monitoring cell dynamics within the patient, ensuring genetic and epigenetic stability, and measuring the host immune response. These insights are essential not only for drug optimization and development but also for monitoring these relatively new therapies.

CAR-T Therapies

Chimeric antigen receptor T-cell (CAR-T) therapy is an advanced cancer treatment that harnesses a patient's own immune cells to combat cancer [90]. T cells are extracted from the patient's blood, then genetically modified to express chimeric antigen receptors (CARs). These synthetic receptors enable the edited T cells to specifically target and bind to antigens on cancer cells. Once modified, the CAR-T cells are cultured to increase their numbers and are subsequently infused back into the patient. FDA-approved CAR-T cell therapies demonstrate extraordinary efficacy in patients with B-cell malignancies and multiple myeloma [91]. However, adverse immunological events such as cytokine release syndrome and immune effector cell-associated neurotoxicity still pose significant risks and challenges to CAR-T cell therapy.

The success of CAR-T therapies relies on multiple factors such as CARs design, T-cell quality, phenotypic characteristics of the infused product, the tumor microenvironment, and the manufacturing process. Recent advances in single cell sequencing technologies make it ideally positioned to address many of these questions. Methods such as scRNA-seq can be combined with other multimodal approaches to evaluate cell surface proteins, chromatin accessibility, and immune repertoire information at a high resolution. In CAR-T cell therapy, scRNA-seq data can reveal cell type composition and gene expression changes in the tumor microenvironment occur pre- and post-treatment in, highlighting possible mechanisms of resistance [92, 93]. For example, one group identified impaired death receptor signaling within the tumor microenvironment as a key contribution to CAR-T dysfunction in patients with Acute Lymphoblastic Leukemia [94]. Single cell sequencing methods combined with immune repertoire can not only help infer clonotypes but also determine phenotypic characteristics such as T cell exhaustion, activation, and memory. Diagnosing T-cell phenotypes is crucial because T-cell exhaustion, characterized by higher expression of inhibitory immune checkpoint receptors, has been associated with poor outcomes in clinical trials [95]. Methods such as scATAC-seq, which analyzes accessible chromatin, can be useful in understanding issues regarding CAR-T cell dysfunction and exhaustion [96, 97]. Additionally, optimization of CAR-T therapies relies heavily on immune repertoire sequencing to ensure proper expansion and enrichment of the CAR-T cell population which is responsible for clinical outcomes [98]. Determining CAR-T cell dynamics, kinetics, and persistence of the exogenous cell product within a patient are also important factors to evaluate when investigating durability. Overall, by integrating advanced single cell technologies, we can gain insights into the complexities of CAR-T cell therapies, ultimately enhancing their design, efficacy, and clinical outcomes.

Induced Pluripotent Stem Cell (iPSC) Derived Therapies

iPSCs are somatic cells from human donors reprogrammed to be pluripotent [89]. Since their conception iPSCs have been extensively useful in drug discovery research to increase the relevance of *in vitro* models to simulate human biology. They provide an avenue for generating human cell types that are otherwise difficult to obtain, and an opportunity to observe differentiation trajectories, or assess the response of specific cell types from different genetic backgrounds (e.g. in a "normal" versus disease paradigm) in *in vitro* assays. There is considerable ongoing interest in using iPSCs for regenerative medicine, primarily to generate specific cell types for therapy [99]. Although there are no FDA-approved iPSC-derived cell therapies, there are currently over 30 clinical trials listed utilizing iPSCs as an intervention [100] (accessed September 2024), with many more trials being enrolled to collect patient cells from specific diagnoses to generate iPSCs for biobanks or further study. These iPSC-derived therapies target a wide range of diseases, including Parkinson's disease and age-related macular degeneration [101, 102]. Single-cell gene expression or epigenetic landscape profiling will help to assess the robustness of protocols for generating particular cell types from iPSCs. Large-scale single-cell atlas projects profiling human tissues *ex vivo* provide a rich source of comparative data [103].

Generating iPSCs can include gene editing, prolonged culturing and expansion, which increases the propensity for the cells to accumulate mutations and karyotypic abnormalities [104]. Furthermore, there is evidence that iPSCs can retain an epigenetic memory from the somatic cells from which they were derived which could potentially hinder the durability or efficacy of the therapy [105]. For an iPSC product, NGS could be employed at multiple stages to ensure consistency and the absence of any deleterious or potentially immunogenic or tumorigenic changes [106, 107].

In allogeneic therapies, where iPSCs would be derived from a universal donor, genotyping the human leukocyte antigen (HLA) subtype of the patient, and matching the haplotype with a donor will likely be key to preventing immune rejection [108]. NGS is now being implemented over PCR with specific primers or Sanger sequencing for high-throughput, low-cost HLA subtyping [109, 110]. Long-read sequencing also holds promise for fast and accurate HLA genotyping and addition to an expanding catalog of known HLA haplotypes [110–112].

Immune Repertoire Sequencing Facilitates Therapeutic Advances

Characterizing the immune system is important for many therapeutics, especially immunotherapy drugs, to assess efficacy and patient safety. An underwhelming immune response to a vaccine could indicate poor efficacy while an overactive immune response could indicate underlying safety issues. One of the most reliable methods to evaluate the state of the immune system is by performing immune repertoire sequencing (Figure 15.4).

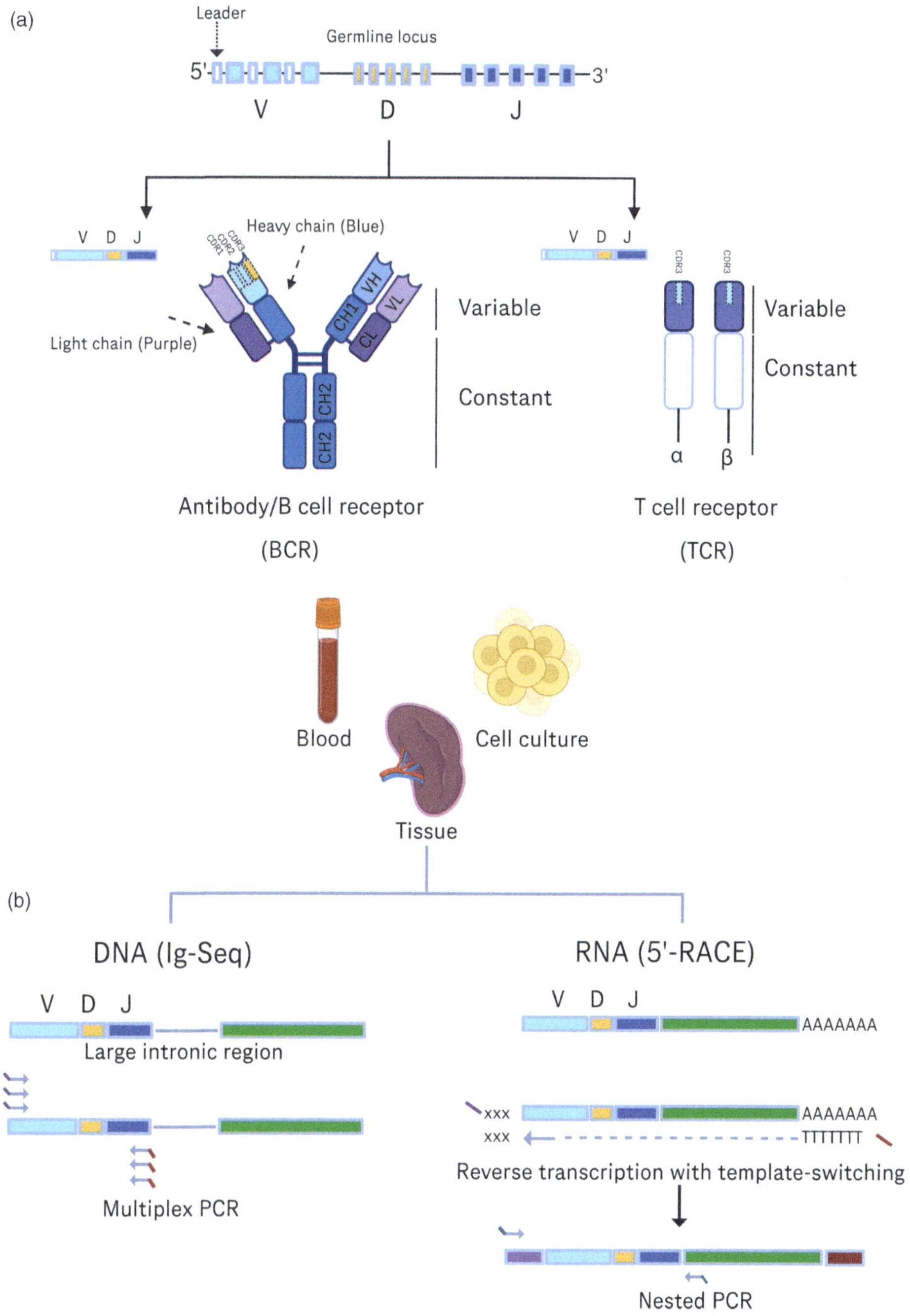

Figure 15.4 Immune repertoire sequencing. (a) Schematic of VDJ recombination and protein structure of BCR and TCR. (b), Typical immune repertoire experiments showing two comparative workflows depending upon starting material. Immunoglobulin-seq uses DNA as input and utilizes a multiplex PCR step whereas 5′-RACE uses RNA as input and relies on template-switching RT step followed by a nested PCR to isolate VDJ regions.

Immune repertoire sequencing involves the sequencing of antigen receptors, derived from a collection of lymphocytes – B and T cells, key players of our adaptive immune system. These cells are unique in that they have the ability to adapt to any foreign pathogen through somatic recombination by shuffling three gene segments: variable (V), diversity (D), and joining (J). Once rearranged, these three gene segments comprise the complementary determining regions (CDR1–CDR3), with CDR3 containing most of the variability and primarily responsible for antigen recognition and specificity. CDR1–CDR3 lie within B and T-cell receptors (BCR/TCR) each comprise a variable heavy chain (VH) and a variable light chain (VL) that dimerize to form antigen binding domains (Figure 15.4a). It is estimated that through recombination alone, 10^{13}–10^{18} possible combinations can be generated within a given repertoire [113]. Upon stimulation, BCRs have an added complexity by undergoing affinity maturation to further improve specificity through somatic hypermutation, which introduces single nucleotide substitutions. High-throughput sequencing is useful in capturing a diversity of combinations of CDR3 rearrangements.

Immune repertoire sequencing can be performed on DNA using a multiplexing strategy with a set of primers targeting V and J segments (Figure 15.4b) [114]. In the 5′ RACE method the 5′ end can be amplified with specific primers using template-switching, followed by nested PCR (Figure 15.4b) and short-read sequencing [115]. To address PCR bias, unique molecular identifiers can be used to collapse sequences originating from the same molecule [114, 116–118]. Due to the transcript length, capturing full-length BCR and TCR sequences requires long-read sequencing approaches [119, 120]. This can be useful to determine the combination of all three CDR3 regions [121], or variants outside of the CDR3 region that may have functional significance [122–124].

It is challenging to deduce true clonal frequencies and VH/VL chain pairing in bulk immune repertoire sequencing. Here, single-cell immune repertoire sequencing can be advantageous, retaining both the VH and VL chain pairing information [125, 126]. Co-incident scRNA-seq can additionally provide information on cell states or phenotypic characteristics of T- or B-cell populations (e.g. Tregs, memory T cells, Plasmablasts, and T-cell activation).

The immune system provides a valuable resource for mining potential therapeutics. TCR sequences from high-avidity, high-specificity tumor infiltrating lymphocytes (TILs) can be engineered into T-cell-based therapies directed against cancer. The low frequency of TILs in tumor tissue often necessitates further selection of the cells via sorting prior to sequencing. Single-cell gene expression can further identify TCRs from cells that have tumor-reactive qualities versus bystander cells [127, 128].

In antibody discovery campaigns, isolating and sequencing plasma B cells that have expanded in response to a target antigen can provide a source of antibody sequences for the development of new therapeutics. Resulting derived therapies include monoclonal antibodies [129–131], bispecific antibodies [132], or antibody–drug conjugates that are targeted to a particular cell type [133].

Conclusions and Future Perspectives

Since the dawn of sequencing technology, advances in molecular biology, chemistry, imaging, microfluidics, engineering, and software have enabled the development of high-throughput, diverse, and accessible sequencing instruments. Sequencing has played, and continues to play, a crucial role in biomedical research enabling deeper understanding of diseases and supporting development of new therapeutics. Increasingly sequencing is also required to characterize the content or efficacy of the therapeutics themselves, particularly with the advent of new modalities such as gene editing or DNA/RNA vaccines.

RNA and DNA therapeutics including ASOs, siRNAs, and vaccines elicit a response by directly impacting gene expression in patient cells. RNA sequencing can be employed to understand the biological effects, which, because of the complexity and interconnectivity of gene networks in different cell types, are hard to comprehensively predict without empirical data. Additionally, sequencing these therapeutics directly identifies unwanted mutations or structural variants, and establishes purity of the therapeutic agents. Advances in sequencing technology – in particular long-read sequencing, allow us to thoroughly assess full-length therapeutic constructs, including therapeutically important modifications.

For cell and gene therapies, understanding the safety and efficacy of a treatment must encompass comprehensive analysis of live biological components. Specifically, for gene-edited cells, the editing efficiency, accuracy, and presence of any off-target edits must be determined as acceptable prior to delivery to the patient. Additionally, confirming the correct cell identity and purity of the product is important. Moreover, by leveraging high-throughput sequencing to capture data on gene expression and cellular composition in cell products and their target tissues, both can be understood at a more nuanced level. Thus, sequencing will play an important role in refining new and emerging therapeutic modalities.

Abbreviations

ASO	antisense oligonucleotides
BCR	B-cell antigen receptor
CAR-T	chimeric antigen receptor T-cell therapy
CDR3	complementary-determining region 3
CRISPR	clustered regularly interspersed short palindromic repeats
dNTPs	deoxynucleotides
ddNTPs	dideoxynucleotides
DGE	differential gene expression
ESC	embryonic stem cell
Gb	gigabyte
gRNA	guide RNA
HDR	homology-directed repair
HLA	human leukocyte antigen

indel	insertions and deletions
iPSC	induced pluripotent stem cell
ITR	inverted terminal repeat
Kb	kilobase
LNP	lipid nanoparticles
$m^1\psi$	1-methyl-pseudouridine
miRNAs	microRNAs
NGS	next-generation sequencing
NHEJ	non-homologous DNA end-joining
ONT	Oxford Nanopore Technologies
PacBio	Pacific Biosciences
PCR	polymerase chain reaction
QC	quality control
rAAV	recombinant adeno-associated virus
RNA-Seq	RNA sequencing
RISC	RNA-induced silencing complex
RT	reverse transcription
scRNA-seq	single-cell RNA sequencing
SHAPE	selective 2′-hydroxyl acylation analyzed by primer extension
siRNA	small interfering RNAs
SBS	sequencing by synthesis
T7E1	T7 endonuclease I
TCR	T-cell receptor
TIL	tumor infiltrating lymphocytes
VDJ	variable, diversity, and joining
VH/VL	variable heavy/variable light chain
UTR	untranslated regions
ZMW	zero-mode waveguide

References

1 Sanger, F., Nicklen, S., and Coulson, A.R. (1992). DNA sequencing with chain-terminating inhibitors. 1977. *Biotechnology* 24: 104–108.

2 Kris A. Wetterstrand, (2019) The Cost of Sequencing a Human Genome. http://Genome.gov.

3 Balasubramanian, S. (2015). Solexa sequencing: decoding genomes on a population scale. *Clin. Chem.* 61 (1): 21–24.

4 Nyrén, P., Pettersson, B., and Uhlén, M. (1993). Solid phase DNA minisequencing by an enzymatic luminometric inorganic pyrophosphate detection assay. *Anal. Biochem.* 208 (1): 171–175.

5 Illumina. https://www.illuminateed.com/

6 Arslan, S., Garcia, F.J., Guo, M. et al. (2024). Sequencing by avidity enables high accuracy with low reagent consumption. *Nat. Biotechnol.* 42 (1): 132–138.

7 Simmons, S.K., Lithwick-Yanai, G., Adiconis, X. et al. (2023). Mostly natural sequencing-by-synthesis for scRNA-seq using ultima sequencing. *Nat. Biotechnol.* 41 (2): 204–211.

8 Redmond, D., Poran, A., and Elemento, O. (2016). Single-cell TCRseq: paired recovery of entire T-cell alpha and beta chain transcripts in T-cell receptors from single-cell RNAseq. *Genome Med.* 8 (1): 80.

9 Miga, K.H., Koren, S., Rhie, A. et al. (2020). Telomere-to-telomere assembly of a complete human X chromosome. *Nature* 585 (7823): 79–84.

10 Eid, J., Fehr, A., Gray, J. et al. (2009). Real-time DNA sequencing from single polymerase molecules. *Science* 323 (5910): 133–138.

11 Iizuka, R., Yamazaki, H., and Uemura, S. (2022). Zero-mode waveguides and nanopore-based sequencing technologies accelerate single-molecule studies. *BPPB* 19: e190032.

12 Zhao, Y., Fang, L.T., Shen, T.-W. et al. (2021). Whole genome and exome sequencing reference datasets from a multi-center and cross-platform benchmark study. *Sci. Data* 8 (1): 296.

13 Foox, J., Tighe, S.W., Nicolet, C.M. et al. (2021). Performance assessment of DNA sequencing platforms in the ABRF next-generation sequencing study. *Nat. Biotechnol.* 39 (9): 1129–1140.

14 Ni, P., Nie, F., Zhong, Z. et al. (2023). DNA 5-methylcytosine detection and methylation phasing using PacBio circular consensus sequencing. *Nat. Commun.* 14 (1): 4054.

15 Dennis, A.F., Xu, Z., and Clark, D.J. (2024). Examining chromatin heterogeneity through PacBio long-read sequencing of M.EcoGII methylated genomes: an m6A detection efficiency and calling bias correcting pipeline. *Nucleic Acids Res.* 52 (9): e45.

16 PacBio. (2022). https://www.pacb.com/.

17 Mahmoud, M., Huang, Y., Garimella, K. et al. (2024). Utility of long-read sequencing for all of us. *Nat. Commun.* 15 (1): 837.

18 Gong, B., Li, D., Łabaj, P.P. et al. (2024). Targeted DNA-seq and RNA-seq of reference samples with short-read and long-read sequencing. *Sci. Data* 11 (1): 892.

19 International Helminth Genomes Consortium (2019). Comparative genomics of the major parasitic worms. *Nat. Genet.* 51 (1): 163–174.

20 Mahmoud, M., Zywicki, M., Twardowski, T., and Karlowski, W.M. (2019). Efficiency of PacBio long read

correction by 2nd generation Illumina sequencing. *Genomics* 111 (1): 43–49.

21 Derakhshani, H., Bernier, S.P., Marko, V.A., and Surette, M.G. (2020). Completion of draft bacterial genomes by long-read sequencing of synthetic genomic pools. *BMC Genomics* 21 (1): 519.

22 Wang, Y., Zhao, Y., Bollas, A. et al. (2021). Nanopore sequencing technology, bioinformatics and applications. *Nat. Biotechnol.* 39 (11): 1348–1365.

23 McIntyre, A.B.R., Alexander, N., Grigorev, K. et al. (2019). Single-molecule sequencing detection of N6-methyladenine in microbial reference materials. *Nat. Commun.* 10 (1): 579.

24 Xu, L. and Seki, M. (2020). Recent advances in the detection of base modifications using the Nanopore sequencer. *J. Hum. Genet.* 65 (1): 25–33.

25 Hendra, C., Pratanwanich, P.N., Wan, Y.K. et al. (2022). Detection of m6A from direct RNA sequencing using a multiple instance learning framework. *Nat. Methods* 19 (12): 1590–1598.

26 Clarke, J., Wu, H.-C., Jayasinghe, L. et al. (2009). Continuous base identification for single-molecule nanopore DNA sequencing. *Nat. Nanotechnol.* 4 (4): 265–270.

27 Soto-Serrano, A., Li, W., Panah, F.M. et al. (2024). Matching excellence: ONT's rise to parity with PacBio in genome reconstruction of non-model bacterium with high GC content. *bioRxiv* 2024.02.26.582104.

28 You, Y., Prawer, Y.D.J., De Paoli-Iseppi, R. et al. (2023). Identification of cell barcodes from long-read single-cell RNA-seq with BLAZE. *Genome Biol.* 24 (1): 66.

29 Harvey, W.T., Ebert, P., Ebler, J. et al. (2023). Whole-genome long-read sequencing downsampling and its effect on variant-calling precision and recall. *Genome Res.* 33 (12): 2029–2040.

30 Deschamps, S., Zhang, Y., Llaca, V. et al. (2018). A chromosome-scale assembly of the sorghum genome using nanopore sequencing and optical mapping. *Nat. Commun.* 9 (1): 4844.

31 Jain, M., Koren, S., Miga, K.H. et al. (2018). Nanopore sequencing and assembly of a human genome with ultra-long reads. *Nat. Biotechnol.* 36 (4): 338–345.

32 Jain, M., Abu-Shumays, R., Olsen, H.E., and Akeson, M. (2022). Advances in nanopore direct RNA sequencing. *Nat. Methods* 19 (10): 1160–1164.

33 Martin, S., Heavens, D., Lan, Y. et al. (2022). Nanopore adaptive sampling: a tool for enrichment of low abundance species in metagenomic samples. *Genome Biol.* 23 (1): 11.

34 Castro-Wallace, S.L., Chiu, C.Y., John, K.K. et al. (2017). Nanopore DNA sequencing and genome assembly on the international Space Station. *Sci. Rep.* 7 (1): 18022.

35 Goenka, S.D., Gorzynski, J.E., Shafin, K. et al. (2022). Accelerated identification of disease-causing variants with ultra-rapid nanopore genome sequencing. *Nat. Biotechnol.* 40 (7): 1035–1041.

36 Shafin, K., Pesout, T., Lorig-Roach, R. et al. (2020). Nanopore sequencing and the Shasta toolkit enable efficient de novo assembly of eleven human genomes. *Nat. Biotechnol.* 38 (9): 1044–1053.

37 Collotta, D., Bertocchi, I., Chiapello, E., and Collino, M. (2023). Antisense oligonucleotides: a novel frontier in pharmacological strategy. *Front. Pharmacol.* 14: 1304342.

38 Kulkarni, J.A., Witzigmann, D., Thomson, S.B. et al. (2021). The current landscape of nucleic acid therapeutics. *Nat. Nanotechnol.* 16 (6): 630–643.

39 Hu, B., Zhong, L., Weng, Y. et al. (2020). Therapeutic siRNA: state of the art. *Signal Transduct. Target. Ther.* 5 (1): 1–25.

40 Watts, J.K. and Corey, D.R. (2012). Silencing disease genes in the laboratory and the clinic. *J. Pathol.* 226 (2): 365–379.

41 Michel, S., Klar, R., and Jaschinski, F. (2021). Investigation of the activity of antisense oligonucleotides targeting multiple genes by RNA-sequencing. *Nucleic Acid Ther.* 31 (6): 427–435.

42 Holgersen, E.M., Gandhi, S., Zhou, Y. et al. (2021). Transcriptome-wide off-target effects of steric-blocking oligonucleotides. *Nucleic Acid Ther.* 31 (6): 392–403.

43 Mortberg, M.A., Gentile, J.E., Nadaf, N.M. et al. (2023). A single-cell map of antisense oligonucleotide activity in the brain. *Nucleic Acids Res.* 51 (14): 7109–7124.

44 Moore, M.J., Scheel, T.K.H., Luna, J.M. et al. (2015). miRNA–target chimeras reveal miRNA 3′-end pairing as a major determinant of Argonaute target specificity. *Nat. Commun.* 6 (1): 1–17.

45 Manakov, S.A., Shishkin, A.A., Yee, B.A. et al. (2022). Scalable and deep profiling of mRNA targets for individual microRNAs with chimeric eCLIP. *bioRxiv* 2022.02.13.480296.

46 Eclipsebio. (2024). https://eclipsebio.com/

47 Mumm, C., Drexel, M.L., McDonald, T.L. et al. (2023). Multiplexed long-read plasmid validation and analysis using OnRamp. *Genome Res.* 33 (5): 741.

48 Redding, L. and Weiner, D.B. (2009). DNA vaccines in veterinary use. *Expert Rev. Vaccines* 8 (9): 1251–1276.

49 Dolgin, E. (2021). The tangled history of mRNA vaccines. *Nature* 597 (7876): 318–324.

50 Qin, S., Tang, X., Chen, Y. et al. (2022). mRNA-based therapeutics: powerful and versatile tools to combat diseases. *Signal Transduct. Target. Ther.* 7 (1): 166.

51 Metkar, M., Pepin, C.S., and Moore, M.J. (2024). Tailor made: the art of therapeutic mRNA design. *Nat. Rev. Drug Discov.* 23 (1): 67–83.

52 Gunter, H.M., Idrisoglu, S., Singh, S. et al. (2023). mRNA vaccine quality analysis using RNA sequencing. *Nat. Commun.* 14 (1): 5663.

53 Liu, A. and Wang, X. (2022). The pivotal role of chemical modifications in mRNA therapeutics. *Front. Cell Dev. Biol.* 10: 901510.

54 Leppek, K., Byeon, G.W., Kladwang, W. et al. (2022). Combinatorial optimization of mRNA structure, stability, and translation for RNA-based therapeutics. *Nat. Commun.* 13 (1): 1536.

55 Nance, K.D. and Meier, J.L. (2021). Modifications in an emergency: the role of N1-Methylpseudouridine in COVID-19 vaccines. *ACS Cent. Sci.* 7 (5): 748–756.

56 Morais, P., Adachi, H., and Yu, Y.-T. (2021). The critical contribution of Pseudouridine to mRNA COVID-19 vaccines. *Front. Cell Dev. Biol.* 9: 789427.

57 Schwartz, S., Bernstein, D.A., Mumbach, M.R. et al. (2014). Transcriptome-wide mapping reveals widespread dynamic-regulated pseudouridylation of ncRNA and mRNA. *Cell* 159 (1): 148–162.

58 Shen, W., Sun, H., Liu, C. et al. (2024). GLORI for absolute quantification of transcriptome-wide mA at single-base resolution. *Nat. Protoc.* 19 (4): 1252–1287.

59 Bicknell, A.A., Reid, D.W., Licata, M.C. et al. (2024). Attenuating ribosome load improves protein output from mRNA by limiting translation-dependent mRNA decay. *Cell Rep.* 43 (4): 114098.

60 Mulroney, T.E., Pöyry, T., Yam-Puc, J.C. et al. (2024). N-methylpseudouridylation of mRNA causes +1 ribosomal frameshifting. *Nature* 625 (7993): 189–194.

61 Brar, G.A. and Weissman, J.S. (2015). Ribosome profiling reveals the what, when, where and how of protein synthesis. *Nat. Rev. Mol. Cell Biol.* 16 (11): 651–664.

62 Mauger, D.M., Cabral, B.J., Presnyak, V. et al. (2019). mRNA structure regulates protein expression through changes in functional half-life. *Proc. Natl. Acad. Sci. USA* 116 (48): 24075–24083.

63 Wayment-Steele, H.K., Kim, D.S., Choe, C.A. et al. (2021). Theoretical basis for stabilizing messenger RNA through secondary structure design. *Nucleic Acids Res.* 49 (18): 10604–10617.

64 Siegfried, N.A., Busan, S., Rice, G.M. et al. (2014). RNA motif discovery by SHAPE and mutational profiling (SHAPE-MaP). *Nat. Methods* 11 (9): 959–965.

65 Wang, J.-H., Gessler, D.J., Zhan, W. et al. (2024). Adeno-associated virus as a delivery vector for gene therapy of human diseases. *Signal Transduct. Target. Ther.* 9 (1): 78.

66 Wang, J.Y. and Doudna, J.A. (2023). CRISPR technology: a decade of genome editing is only the beginning. *Science* 379 (6629): eadd8643.

67 Frangoul, H., Altshuler, D., Cappellini, M.D. et al. (2021). CRISPR-Cas9 gene editing for sickle cell disease and β-thalassemia. *N. Engl. J. Med.* 384 (3): 252–260.

68 Office of the Commissioner (2024). FDA Approves First Gene Therapies to Treat Patients with Sickle Cell Disease. U.S. Food and Drug Administration.

69 Erkut, E. and Yokota, T. (2022). CRISPR therapeutics for Duchenne muscular dystrophy. *Int. J. Mol. Sci.* 23 (3).

70 Pierce, E.A., Aleman, T.S., Jayasundera, K.T. et al. (2024). Gene editing for -associated retinal degeneration. *N. Engl. J. Med.* 390 (21): 1972–1984.

71 Kim, H.J., Lee, H.J., Kim, H. et al. (2009). Targeted genome editing in human cells with zinc finger nucleases constructed via modular assembly. *Genome Res.* 19 (7): 1279–1288.

72 Guschin, D.Y., Waite, A.J., Katibah, G.E. et al. (2010). A rapid and general assay for monitoring endogenous gene modification. In: *Engineered Zinc Finger Proteins*, 247–256.

73 Guo, C., Ma, X., Gao, F., and Guo, Y. (2023). Off-target effects in CRISPR/Cas9 gene editing. *Front. Bioeng. Biotechnol.* 11: 1143157.

74 Xiao, A., Cheng, Z., Kong, L. et al. (2014). CasOT: a genome-wide Cas9/gRNA off-target searching tool. *Bioinformatics* 30 (8): 1180–1182.

75 Chuai, G., Ma, H., Yan, J. et al. (2018). DeepCRISPR: optimized CRISPR guide RNA design by deep learning. *Genome Biol.* 19 (1): 1–18.

76 Chiarle, R., Zhang, Y., Frock, R.L. et al. (2011). Genome-wide translocation sequencing reveals mechanisms of chromosome breaks and rearrangements in B cells. *Cell* 147 (1): 107–119.

77 Kim, D., Bae, S., Park, J. et al. (2015). Digenome-seq: genome-wide profiling of CRISPR-Cas9 off-target effects in human cells. *Nat. Methods* 12 (3): 237–243.

78 Tsai, S.Q., Nguyen, N.T., Malagon-Lopez, J. et al. (2017). CIRCLE-seq: a highly sensitive in vitro screen for genome-wide CRISPR–Cas9 nuclease off-targets. *Nat. Methods* 14 (6): 607–614.

79 Wienert, B., Wyman, S.K., Richardson, C.D. et al. (2019). Unbiased detection of CRISPR off-targets in vivo using DISCOVER-Seq. *Science* 364 (6437): 286–289.

80 Tsai, S.Q., Zheng, Z., Nguyen, N.T. et al. (2014). GUIDE-seq enables genome-wide profiling of off-target

cleavage by CRISPR-Cas nucleases. *Nat. Biotechnol.* 33 (2): 187–197.

81 IDT. https://www.idtdna.com/page.

82 Wright, J.F. (2014). Product-related impurities in clinical-grade recombinant AAV vectors: characterization and risk assessment. *Biomedicine* 2 (1): 80–97.

83 Yang, T.-Y., Braun, M., Lembke, W. et al. (2022). Immunogenicity assessment of AAV-based gene therapies: an IQ consortium industry white paper. *Mol. Ther. Methods Clin. Dev.* 26: 471–494.

84 Dalwadi, D.A., Calabria, A., Tiyaboonchai, A. et al. (2021). AAV integration in human hepatocytes. *Mol. Ther.* 29 (10): 2898–2909.

85 Penaud-Budloo, M., Lecomte, E., Guy-Duché, A. et al. (2017). Accurate identification and quantification of DNA species by next-generation sequencing in adeno-associated viral vectors produced in insect cells. *Hum. Gene Ther. Methods* 28 (3): 148–162.

86 Namkung, S., Tran, N.T., Manokaran, S. et al. (2022). Direct ITR-to-ITR Nanopore sequencing of AAV vector genomes. *Hum. Gene Ther.* 33 (21–22): 1187–1196.

87 Radukic, M.T., Brandt, D., Haak, M. et al. (2020). Nanopore sequencing of native adeno-associated virus (AAV) single-stranded DNA using a transposase-based rapid protocol. *NAR Genom. Bioinform.* 2 (4): lqaa074.

88 Brimble, M.A., Winston, S.M., and Davidoff, A.M. (2023). Stowaways in the cargo: contaminating nucleic acids in rAAV preparations for gene therapy. *Mol. Ther.* 31 (10): 2826–2838.

89 Takahashi, K. and Yamanaka, S. (2006). Induction of pluripotent stem cells from mouse embryonic and adult fibroblast cultures by defined factors. *Cell* 126 (4): 663–676.

90 Sterner, R.C. and Sterner, R.M. (2021). CAR-T cell therapy: current limitations and potential strategies. *Blood Cancer J.* 11 (4): 69.

91 UPMC Hillman Cancer Center. FDA-approved CAR T-cell Therapies.

92 Dhodapkar, K.M., Cohen, A.D., Kaushal, A. et al. (2022). Changes in bone marrow tumor and immune cells correlate with durability of remissions following BCMA CAR T therapy in myeloma. *Blood Cancer Discov* 3 (6): 490–501.

93 Jiang, V.C., Hao, D., Jain, P. et al. (2022). TIGIT is the central player in T-cell suppression associated with CAR T-cell relapse in mantle cell lymphoma. *Mol. Cancer* 21 (1): 185.

94 Singh, N., Lee, Y.G., Shestova, O. et al. (2020). Impaired death receptor Signaling in Leukemia causes antigen-independent resistance by inducing CAR T-cell dysfunction. *Cancer Discov.* 10 (4): 552–567.

95 Gumber, D. and Wang, L.D. (2022). Improving CAR-T immunotherapy: overcoming the challenges of T cell exhaustion. *EBioMedicine* 77: 103941.

96 Jiang, P., Zhang, Z., Hu, Y. et al. (2022). Single-cell ATAC-seq maps the comprehensive and dynamic chromatin accessibility landscape of CAR-T cell dysfunction. *Leukemia* 36 (11): 2656–2668.

97 Gennert, D.G., Lynn, R.C., Granja, J.M. et al. (2021). Dynamic chromatin regulatory landscape of human CAR T cell exhaustion. *Proc. Natl. Acad. Sci. USA* 118 (30).

98 Tian, Y., Li, Y., Shao, Y., and Zhang, Y. (2020). Gene modification strategies for next-generation CAR T cells against solid cancers. *J. Hematol. Oncol.* 13 (1): 54.

99 Cerneckis, J., Cai, H., and Shi, Y. (2024). Induced pluripotent stem cells (iPSCs): molecular mechanisms of induction and applications. *Signal Transduct. Target. Ther.* 9 (1): 1–26.

100 National Library of Medicine. ClinicalTrials http://ClinicalTrials.gov.

101 Schwartz, S.D., Regillo, C.D., Lam, B.L. et al. (2015). Human embryonic stem cell-derived retinal pigment epithelium in patients with age-related macular degeneration and Stargardt's macular dystrophy: follow-up of two open-label phase 1/2 studies. *Lancet* 385 (9967): 509–516.

102 Takahashi, J. (2020). iPS cell-based therapy for Parkinson's disease: a Kyoto trial. *Regen Ther* 13: 18–22.

103 Rozenblatt-Rosen, O., Stubbington, M.J.T., Regev, A., and Teichmann, S.A. (2017). The human cell atlas: from vision to reality. *Nature* 550 (7677): 451–453.

104 Lund, R.J., Närvä, E., and Lahesmaa, R. (2012). Genetic and epigenetic stability of human pluripotent stem cells. *Nat. Rev. Genet.* 13 (10): 732–744.

105 Kim, K., Doi, A., Wen, B. et al. (2010). Epigenetic memory in induced pluripotent stem cells. *Nature* 467 (7313): 285–290.

106 Zhao, T., Zhang, Z.-N., Rong, Z., and Xu, Y. (2011). Immunogenicity of induced pluripotent stem cells. *Nature* 474 (7350): 212–215.

107 Lee, A.S., Tang, C., Rao, M.S. et al. (2013). Tumorigenicity as a clinical hurdle for pluripotent stem cell therapies. *Nat. Med.* 19 (8): 998–1004.

108 Sullivan, S., Fairchild, P.J., Marsh, S.G. et al. (2020). Haplobanking induced pluripotent stem cells for clinical use. *Stem Cell Res.* 49: 102035.

109 Schöfl, G., Lang, K., Quenzel, P. et al. (2017). 2.7 million samples genotyped for HLA by next generation sequencing: lessons learned. *BMC Genomics* 18 (1): 161.

110 Mosbruger, T.L., Dinou, A., Duke, J.L. et al. (2020). Utilizing nanopore sequencing technology for the rapid

and comprehensive characterization of eleven HLA loci; addressing the need for deceased donor expedited HLA typing. *Hum. Immunol.* 81 (8): 413–422.

111 Anukul, N., Jenjaroenpun, P., Sirikul, C. et al. (2023). Ultrarapid and high-resolution HLA class I typing using transposase-based nanopore sequencing applied in pharmacogenetic testing. *Front. Genet.* 14: 1213457.

112 Hurley, C.K., Kempenich, J., Wadsworth, K. et al. (2020). Common, intermediate and well-documented HLA alleles in world populations: CIWD version 3.0.0. *Hla* 95 (6): 516–531.

113 Murphy, K.M. and Weaver, C. (2016). *Janeway's Immunobiology: Ninth International Student Edition.* W.W. Norton & Company.

114 Vollmers, C., Sit, R.V., Weinstein, J.A. et al. (2013). Genetic measurement of memory B-cell recall using antibody repertoire sequencing. *Proc. Natl. Acad. Sci. USA* 110 (33): 13463–13468.

115 Friedensohn, S., Khan, T.A., and Reddy, S.T. (2017). Advanced methodologies in high-throughput sequencing of immune repertoires. *Trends Biotechnol.* 35 (3): 203–214.

116 Shugay, M., Britanova, O.V., Merzlyak, E.M. et al. (2014). Towards error-free profiling of immune repertoires. *Nat. Methods* 11 (6): 653–655.

117 Rosati, E., Dowds, C.M., Liaskou, E. et al. (2017). Overview of methodologies for T-cell receptor repertoire analysis. *BMC Biotechnol.* 17 (1): 61.

118 Barennes, P., Quiniou, V., Shugay, M. et al. (2021). Benchmarking of T cell receptor repertoire profiling methods reveals large systematic biases. *Nat. Biotechnol.* 39 (2): 236–245.

119 Singh, M., Al-Eryani, G., Carswell, S. et al. (2019). High-throughput targeted long-read single cell sequencing reveals the clonal and transcriptional landscape of lymphocytes. *Nat. Commun.* 10 (1): 3120.

120 Byrne, A., Beaudin, A.E., Olsen, H.E. et al. (2017). Nanopore long-read RNAseq reveals widespread transcriptional variation among the surface receptors of individual B cells. *Nat. Commun.* 8 (1): 16027.

121 Kunik, V. and Ofran, Y. (2013). The indistinguishability of epitopes from protein surface is explained by the distinct binding preferences of each of the six antigen-binding loops. *Protein Eng. Des. Sel.* 26 (10): 599–609.

122 Zhou, J.O., Zaidi, H.A., Ton, T., and Fera, D. (2020). The effects of framework mutations at the variable domain interface on antibody affinity maturation in an HIV-1 broadly neutralizing antibody lineage. *Front. Immunol.* 11: 1529.

123 Kerwin, B.A., Bennett, C., Brodsky, Y. et al. (2020). Framework mutations of the 10-1074 bnAb increase conformational stability, manufacturability, and stability while preserving full neutralization activity. *J. Pharm. Sci.* 109 (1): 233–246.

124 Pettit, D.K., Rogers, R.S., Arthur, K. et al. (2016). CHO cell production and sequence improvement in the 13C6FR1 anti-Ebola antibody. *MAbs* 8 (2): 347–357.

125 DeKosky, B.J., Kojima, T., Rodin, A. et al. (2015). In-depth determination and analysis of the human paired heavy- and light-chain antibody repertoire. *Nat. Med.* 21 (1): 86–91.

126 Cole, C., Byrne, A., Beaudin, A.E. et al. (2018). Tn5Prime, a Tn5 based 5′ capture method for single cell RNA-seq. *Nucleic Acids Res.* 46 (10): e62.

127 Pétremand, R., Chiffelle, J., Bobisse, S. et al. (2024). Identification of clinically relevant T cell receptors for personalized T cell therapy using combinatorial algorithms. *Nat. Biotechnol.* 1–6.

128 Tan, C.L., Lindner, K., Boschert, T. et al. (2024). Prediction of tumor-reactive T cell receptors from scRNA-seq data for personalized T cell therapy. *Nat. Biotechnol.* 1–9.

129 Pantaleo, G., Correia, B., Fenwick, C. et al. (2022). Antibodies to combat viral infections: development strategies and progress. *Nat. Rev. Drug Discov.* 21 (9): 676–696.

130 Weiner, L.M., Surana, R., and Wang, S. (2010). Monoclonal antibodies: versatile platforms for cancer immunotherapy. *Nat. Rev. Immunol.* 10 (5): 317–327.

131 Sabahi, Z., Daei Sorkhabi, A., Sarkesh, A. et al. (2023). A systematic review of the safety and efficacy of monoclonal antibodies for progressive multiple sclerosis. *Int. Immunopharmacol.* 120: 110266.

132 Klein, C., Brinkmann, U., Reichert, J.M., and Kontermann, R.E. (2024). The present and future of bispecific antibodies for cancer therapy. *Nat. Rev. Drug Discov.* 23 (4): 301–319.

133 Dumontet, C., Reichert, J.M., Senter, P.D. et al. (2023). Antibody-drug conjugates come of age in oncology. *Nat. Rev. Drug Discov.* 22 (8): 641–661.

16

Mass Spectrometry-Based Methods to Investigate Protein–Protein Interactions in Cells and Their Application in Drug Discovery

Tess Branon

Proteomic & Genomic Technologies Department, Genentech Inc, South San Francisco, CA, USA

Introduction

In 1838, Swedish chemist Jöns Jacob Berzelius first coined the term "protein." Berzelius derived this word from the Greek *proteios,* meaning "of the first rank," to emphasize the fundamental importance of proteins to all living organisms [1]. Even now nearly 200 years later, the central role of proteins in nearly every biological process is still recognized and they remain the subject of large swaths of biological research. Proteins are the skeletal structures that give our cells shape, they allow us to sense and interact with the world, they are messengers and receivers that allow our organs and cells to communicate with each other, they are transducers that translate these stimuli into actionable change, and, of course, they are the agents that catalyze the energetically unfavorable chemical reactions that allow life to exist.

Given the extensive role of proteins in biology, it is no coincidence that their dysfunction is also central to many human diseases. This critical role in disease is reflected in the fact that drug discovery has historically focused on modulating the function of proteins. In 1894, German chemist Emil Fisher introduced the "lock and key" model which proposed that the specificity of enzymes toward their substrates is based on their respectively complementary geometric structures, where the substrate "key" is perfectly shaped to fit in the enzyme "lock" keyhole [2]. This concept has led to the development of small molecule ligands that fit into these keyhole pockets to activate or inhibit the activity of proteins, giving rise to many of the drugs that have improved the quality and length of human lives over the past hundred years.

According to genomic and protein structural analysis, however, only approximately 10–15% of proteins encoded in the human genome possess these hydrophobic keyhole pockets that are suitable for binding small drug-like molecules [3]. While the emergence of larger molecule drugs such as biologics has allowed the targeting of larger surfaces on proteins, these protein therapeutics are unable to traverse cellular membranes and are therefore restricted to extracellular targets, which are predicted to comprise less than 10% of all human proteins [4]. Therefore, as many of these lower-hanging fruit have been successfully drugged, focus in drug discovery has had to shift toward targeting what have been classically considered to be "undruggable" proteins. Toward this end, rather than targeting a pocket or surface on a single protein, there has been increasing interest in targeting the composite interface between two or more interacting proteins. With this new vision in mind, it is critical that the methods and assays used for drug discovery shift toward using model systems that more accurately reflect the cellular environment and preserve these protein–protein interactions (PPIs) of interest.

While there is much to be learned from studying the biochemical and biophysical properties of a protein *in vitro*, understanding the true nature of its biological properties can only come from studying it in the protein-dense, complex native habitat of the cell. Each human cell is predicted to harbor billions of protein molecules that represent over ten thousand different types of proteins, and these proteins are capable of forming over five hundred thousand different PPIs with each other [5, 6]. In addition to this dense and diverse protein-rich environment, cells are highly spatially compartmentalized which gives rise to many unique biochemical environments that in turn facilitate distinct cellular functions. For example, the oxidizing chemical environment of the endoplasmic reticulum (ER) facilitates efficient folding of newly synthesized proteins [7], while differences in pH between compartments on either side of the mitochondrial inner membrane drive the synthesis of adenosine triphosphate (ATP) [7, 8]. Cellular compartmentalization also allows specific molecules to be concentrated together or to be segregated from others. For example, enzymes involved in a biosynthetic pathway

Characterizing Biotherapeutics: Analytical Methods for Diverse Modalities, First Edition. Edited by Jennie R. Lill and Wendy Sandoval.

may be concentrated together in a cellular compartment to increase the rate of synthesis; or signaling proteins may be segregated from their downstream targets to control the timing of signal transduction. Therefore, the ability to characterize the localization, trafficking, and interaction dynamics of proteins within the native context of the living cell is critical to advancing our understanding of these systems, and consequently, to successfully drug PPIs.

In this chapter, we will focus on mass spectrometry (MS)-based techniques for investigating PPIs. MS is a versatile and powerful analytical technique that has significantly advanced our ability to study proteins in complex biological samples, like those from tissues or cells [9]. It enables the identification of thousands of endogenous proteins from a single sample, as well as their quantification and characterization of their post-translational modifications. However, identification of proteins from biological samples by MS requires cell lysis, which disrupts membranes and macromolecular structures, and results in the loss of spatial information associated with the proteins identified. By integrating MS with the biochemical techniques outlined below, researchers can recover some of this spatial and contextual information to gain insight into the interactions between proteins, shedding a brighter light on their biological functions and roles in disease. The biochemical techniques we will discuss in this chapter are affinity purification, cross-linking, and proximity labeling. When combined with MS, these assays can be leveraged to not only identify therapeutically relevant PPIs from biological samples but also to characterize the dynamics of these interactions in cells and assess how they can be modulated by different environmental conditions or drug treatments.

In our closing discussion, we will examine several examples of current PPI modulation approaches in drug discovery. We will also share a forward-looking perspective on efforts to increase the automation and throughput of these assays that aim to improve their tractability at the scale required for drug discovery research. We will also close with a brief discussion on other technologies designed to probe and manipulate specific PPIs, such as two-hybrid assays, protein complementation assays, and resonance energy transfer assays. These methodologies are invaluable for screening chemical libraries to identify compounds that can either induce or inhibit specific PPIs of interest within cells. By leveraging these advanced screening technologies, researchers can uncover novel therapeutic agents that modulate critical biological pathways. By the end of this chapter, readers will have a robust understanding of the current landscape of PPI interrogation methods, their applications in drug discovery, and the future directions of this exciting field.

Affinity Purification Mass Spectrometry (AP-MS)

AP-MS is one of the oldest and most widely used approaches to studying protein complexes from biological samples. While the agent employed for affinity purification can vary, most often it is based on immunoprecipitation (IP) where antibodies coupled to a precipitation apparatus, such as beads or columns, are used to bind and extract specific target proteins of interest [10]. The methods used to disrupt cells for protein extraction and the steps employed to wash away nonspecific proteins from the precipitation apparatus are often gentle, allowing proteins that interact with the target protein to be retained for identification (Figure 16.1). The immunoprecipitated proteins and their interactors are subsequently digested into peptides that can then be identified by MS-based analysis, often preceded by a high-performance liquid chromatography (HPLC) fractionation step which separates the peptide mixtures to reduce the peptide mixture complexity, thus increasing the sensitivity of the method.

One of the greatest advantages of AP-MS is that it is possible to apply this technique to entirely endogenous systems (Figure 16.1a), which can lead to more physiologically relevant results and allows employment of this technique in biological systems that are intractable to genetic manipulation [11]. However, because antibodies suitable for efficient immunoprecipitation of target proteins are often limited, AP-MS is most commonly performed using affinity tags that are fused to the target protein of interest (Figure 16.1b), such as the FLAG-tag or HA-tag [12]. While recombinant tagging and overexpression of the target protein bait can perturb the trafficking, conformation, or formation of relevant interactions, focus and effort can be applied to optimize the tag insertion sites and the expression levels to achieve more physiologically relevant results. Furthermore, only the target protein bait needs to be tagged, the interacting partner prey that is captured and identified is endogenous proteins.

There has been a plethora of studies published in the past several decades that employ AP-MS to map PPIs in a number of different biological models. While earlier studies may have mostly consisted of lower-throughput, focused studies looking at a small handful of specific PPIs, larger-scale studies in more recent years have provided a more comprehensive view of cellular interactomes which can yield new biological insight. In 2007, for example, Ewing et al. reported the first large-scale study of PPIs in human cells using AP-MS. In this study, the authors mapped the interactomes of 338 FLAG-tagged bait proteins with known or suspected disease associations to produce a high-confidence dataset of 6,463 interactions between

2,235 distinct proteins that provided a valuable source of novel PPIs relevant to human disease [13]. More recently, there have been even larger-scale efforts, notably Bioplex 3.0, where AP-MS has been used to map the interactomes of 10,128 human proteins – half of the human proteome – to report 118,162 interactions among 14,586 proteins [14]. By comparing this dataset collected in 293T cells to a second dataset from 5,522 proteins from the human cancer cell line HCT116, the authors were also able to elucidate cell-line-specific interactions and link these complexes to rewired subnetworks that could be relevant to disease [14].

There have also been a number of targeted AP-MS studies focusing on important therapeutic targets. For example, the proteasome has emerged as an important therapeutic target for multiple myeloma (MM) – several marketed proteasome inhibitors have improved survival and quality of

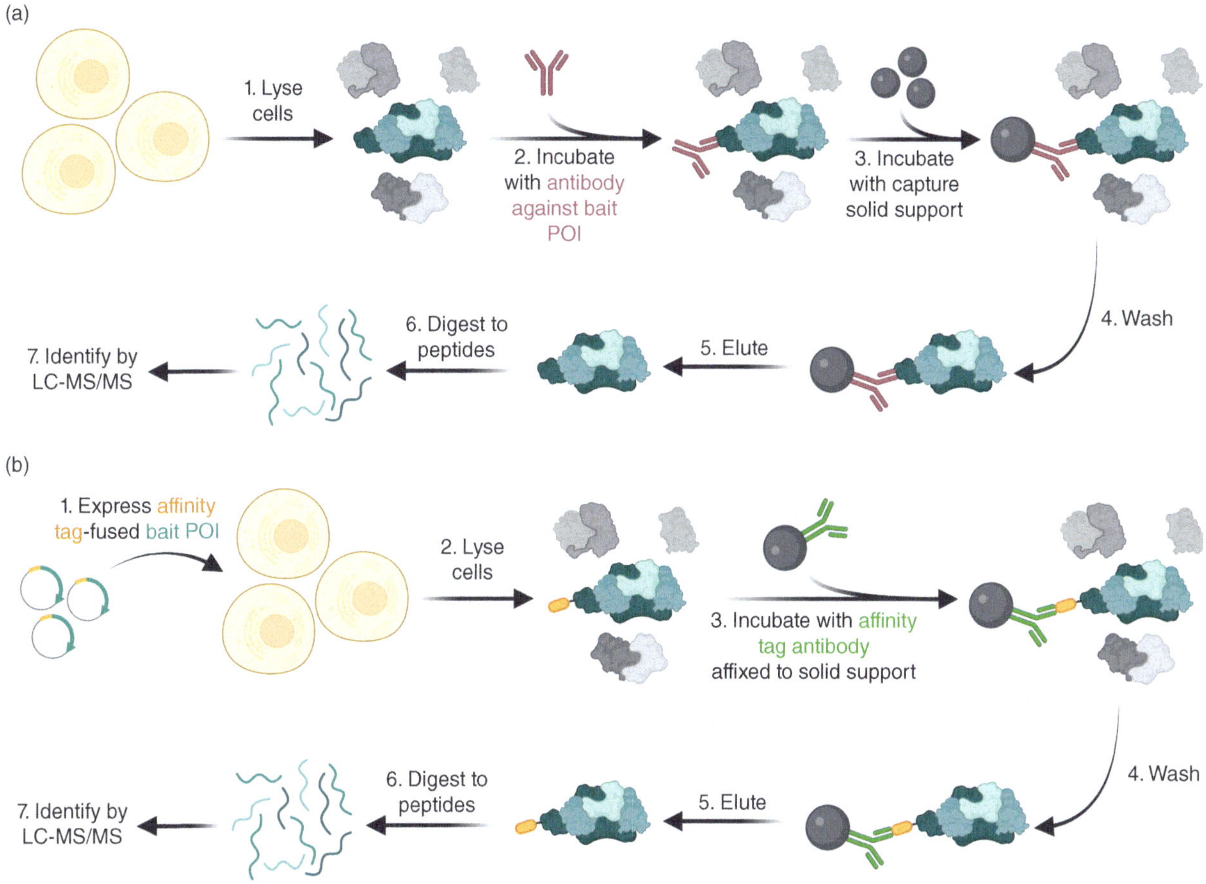

Figure 16.1 Affinity purification methods for AP-MS. (a) Common immunoprecipitation method for endogenous proteins. After cell lysis, lysate mixtures are incubated with an antibody against the bait protein of interest. If the bait is a membrane protein and an appropriate antibody is available, this antibody incubation step can be performed before lysis. After the antibody is incubated for a sufficient amount of time to bind the bait protein of interest, the antibodies can be conjugated to a solid support; for example, through Protein A/G-conjugated beads. Once the protein–antibody complex is bound to the solid support, gentle washing can be performed to remove unbound proteins and nonspecific binders. The bait protein of interest and its interactors can then be eluted, digested, and analyzed by MS to identify the interacting proteins. (b) Common affinity purification method utilizing affinity tags. The bait protein of interest is genetically fused with an affinity tag and expressed in cells. Cells can then be lysed, and the lysate is incubated with a high-affinity antibody against the affinity tag that is affixed to a solid support. Gentle washing can then be performed to remove unbound proteins and nonspecific binders. The bait protein of interest and its interactors can then be eluted, digested, and analyzed by MS to identify the interacting proteins. (c) Common tandem affinity purification (TAP) method. The bait protein of interest is genetically fused with two different affinity tags in tandem and expressed in cells. Cells can then be lysed, and the lysate is incubated with a high-affinity antibody against the first affinity tag that is affixed to a solid support. Gentle washing can then be performed to remove unbound proteins and nonspecific binders. The bait protein of interest and its interactors can then be eluted, or in some cases proteolytically cleaved depending on the tandem affinity tag design. The eluted protein mixture is then incubated with a high-affinity antibody against the second affinity tag that is affixed to a solid support. Gentle washing can then be performed to further remove unbound proteins and nonspecific binders. The bait protein of interest and its interactors can then be eluted, digested, and analyzed by MS to identify the interacting proteins. Source: BioRender.com

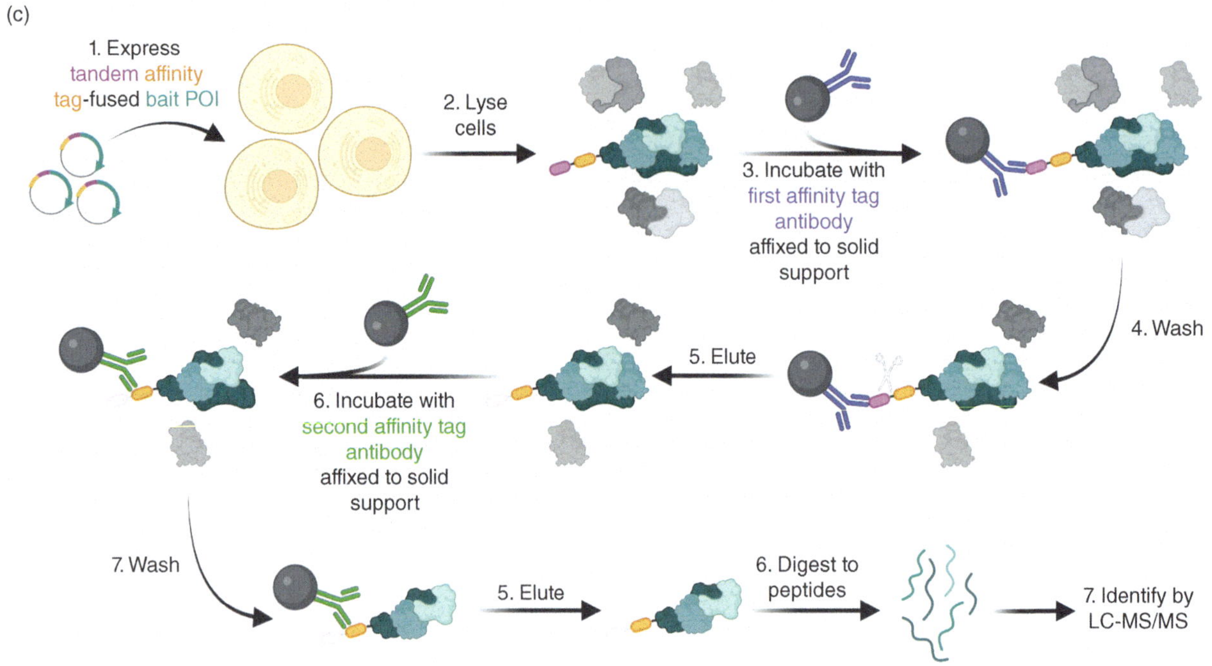

Figure 16.1 (*Continued*)

life for MM patients and have grossed over three billion US dollars in sales annually [15]. A number of AP-MS studies have focused on identifying important proteasome–protein interactions, utilizing both endogenous proteasome IPs [16] and affinity tag AP-MS [17–19]. These studies have yielded many disease-relevant insights, such as the differential regulators of standard- and immuno-proteasomes [20]. Other important therapeutic targets that have been investigated using AP-MS include differential interactors of cancer-driving RAS mutants [21, 22], host-viral interacting proteins for viruses such as SARS-CoV-2 [23–26] and HIV-1 [27], metabolic and autophagy-related complexes [28], and neurodegenerative disease targets such as Amyloid beta precursor protein (APP) and Presenilin-1 (PSEN1) for Alzheimer's disease, Huntingtin (HTT) for Huntington's disease, Parkin (PARK2) for Parkinson's disease, and Ataxin-1 (ATXN1) for spinocerebellar ataxia type 1 [29].

Beyond mapping disease-relevant PPI networks, AP-MS can also be applied to characterize the effects of small molecules in drug discovery. For example, in combination with global proteome analysis, AP-MS can be used to investigate mechanisms of molecules inducing targeted protein degradation like proteolysis targeting chimeras (PROTACs) and molecular glues by monitoring the induction of substrate interactions with E3 ligases like Cereblon (CRBN) [30–32]. Other innovations, such as protein interaction kinetics and estimation of stoichiometries (PIKES), have allowed AP-MS to be used in combination with chemical modulation to investigate the dynamics of interchangeable multiprotein complexes like Cullin RING ubiquitin ligases (CRLs) [33]. AP-MS has also been used to investigate other targeted protein degradation agents with therapeutic potential, such as proteolysis-targeting antibodies (PROTABs), where AP-MS was used in combination with mass spectrometric analysis of post-translational modifications to characterize ubiquitination of targets following PROTAB treatment [34]. Furthermore, beyond affinity purification using affinity tags or antibodies against endogenous proteins, AP-MS can also be performed using small molecules themselves, such as was reported with small molecule kinase inhibitors and immunophilin binders to provide an unbiased way to identify small-molecule targets [35, 36].

As evidenced by the vast number of studies and impactful biological discoveries made over the past several decades, AP-MS is undoubtedly an effective technique for discovering PPIs and mapping protein complexes. However, there are some inherent limitations to this technique that are worth noting. From a practical perspective, the use of antibodies for IP can lead to high reagent costs and limit the number of samples that can be generated. Furthermore, there is potential that an antibody's binding epitope is masked by an endogenous binding partner of the target, leading to false negatives. Beyond caveats related to antibodies, several studies have shown that some of the interaction partners detected in AP-MS experiments can result from nonspecific interactions that occur after cell lysis. A study investigating the interactions of ribonucleoprotein complexes definitively showed that some of the enriched interactors had resulted from rearrangement post-lysis and did not recapitulate the *in vivo* state [37]. Upon cell lysis, biologically irrelevant interactions can form when proteins, which could be partially unfolded by the detergents used for lysis, encounter proteins with which they would not interact under physiological conditions. This can be particularly troublesome for proteins that are intrinsically unstructured or unfolded, such as some microproteins [38]. Furthermore, these nonspecific interactions and those forming with the affinity solid-support can be difficult to dissociate due to the low stringency of washes that are necessary to ensure true-positive interactions are not lost. Steps such as biochemical fractionation can be employed prior to AP-MS to isolate specific cell compartments and limit PPIs that form during affinity purification to more physiologically relevant interactions that may naturally occur in a particular organelle [39]. Tandem affinity purification (TAP)-MS can also reduce false positives from the experiment due to the two different consecutive purification steps (Figure 16.1c); however, this can also lead to considerable losses of true-positive interactors [40]. Steps can also be taken during data analysis, such as utilizing background proteins commonly detected in AP-MS datasets to filter out false-positive hits and achieve higher selectivity and specificity in determining confident interaction partners [41]. In addition to high rates of false-positive detections, AP-MS often has a high false-negative rate because only stable interactions that remain intact throughout the experimental protocol are detected; therefore, many of the weaker and more transient PPIs are lost [42]. This false-negative rate can be reduced by employing cross-linking techniques, which stabilizes weak interactions between proteins, but this can also further stabilize nonspecific interactions [16, 20, 43]. This technique as well as others that can improve the detection of weak and transient interactions such as proximity labeling are discussed in more detail in the subsequent sections.

Cross-Linking Mass Spectrometry (XL-MS)

Cross-linking is a technique that covalently links two proximal protein residues, often using soluble chemical cross-linking reagents that target protein surface residues. After the cross-linking reaction, protein samples can be extracted and digested for downstream MS analysis to identify the cross-linked peptides, thus providing information on PPIs at the protein level as well as structural information on the PPI interface [44] (Figure 16.2a). These chemical cross-linkers often contain two reactive end groups that can react with a range of protein residue groups, including primary amines on lysines and N-termini [45–47], carboxyl groups on C-termini and aspartic and glutamic residues [48–50], and sulfhydryl groups on cysteine residues [51, 52]. Cross-linker reactive end groups can also consist of photo-reactive moieties that are light-activated, often with UV-light, allowing for much greater temporal control of the cross-linking reaction [53–57]. More recently, instead of the addition of a cross-linking reagent, photoactivatable amino acids have been incorporated directly into proteins globally or site-specifically during translation which also allows interior and hydrophobic patches to be probed [57–62] (Figure 16.2b). Similarly, metabolic incorporation of photoactivatable sugars has also been used to map the interactomes of glycosylated proteins [63]. The physicochemical properties, and importantly the specificity, of cross-linking reagents are largely determined by their reactive moieties. Therefore, there should be careful consideration in the selection of which reactive moieties are used. Increasing specificity will likely result in fewer linked peptides and thus lower detectability of interacting proteins; however, lower specificity will make subsequent analysis more difficult.

Another critical component of cross-linking reagents is the spacer elements that connect the two reactive moieties, which can vary in both length and functionality. The length of the cross-linking reagent spacer provides spatial constraints on which residues have the potential to be cross-linked, affecting the spatial resolution and the density of data obtained in the XL-MS experiment. Shorter cross-linking reagents can provide stronger evidence of a direct interaction between cross-linked proteins, but will likely yield fewer intermolecular cross-links due to more limited availability of reactive residues near the PPI interface. Some cross-linking reagents lack a spacer element altogether, like the carbomiimides that allow "zero-length" amide cross-linking [64]. A longer cross-linker spacer enables more reactive protein residues to fall within the cross-linker's range and thus may increase the detection of PPIs; however, the potential to gain structural insight into the PPI interface decreases because a cross-link produces only an upper-bound distance restraint. Longer cross-linkers also generally have reduced cellular permeability which can reduce their efficacy for in-cell interactome mapping.

Detecting cross-linked peptides by MS involves several inherent challenges, including the complexity and heterogeneity of cross-linked peptides, low abundance, complex fragmentation patterns, and the need for specialized data analysis algorithms. To address some of these challenges, many modern cross-linkers also incorporate functional components in their spacer elements. For example, cleavable connectors are often used to facilitate MS analysis. Disuccinimidylsulfoxide (DSSO) is a popular cross-linker with two symmetrical bonds that are labile with certain types of MS dissociation mechanisms which allows separation of functional cross-links between two peptides from nonfunctional reactions where only one end of the cross-linker has reacted with a protein residue [65]. Furthermore, separating the cross-linked peptides during MS fragmentation allows identification using standard bioinformatic tools, considerably simplifying subsequent data analysis and making identification of the cross-linked peptides and parent proteins less ambiguous. Another common element that is included in cross-linker spacers is affinity handles which facilitate enrichment of the cross-linked peptides to enhance sensitivity [66, 67], although there are also other strategies

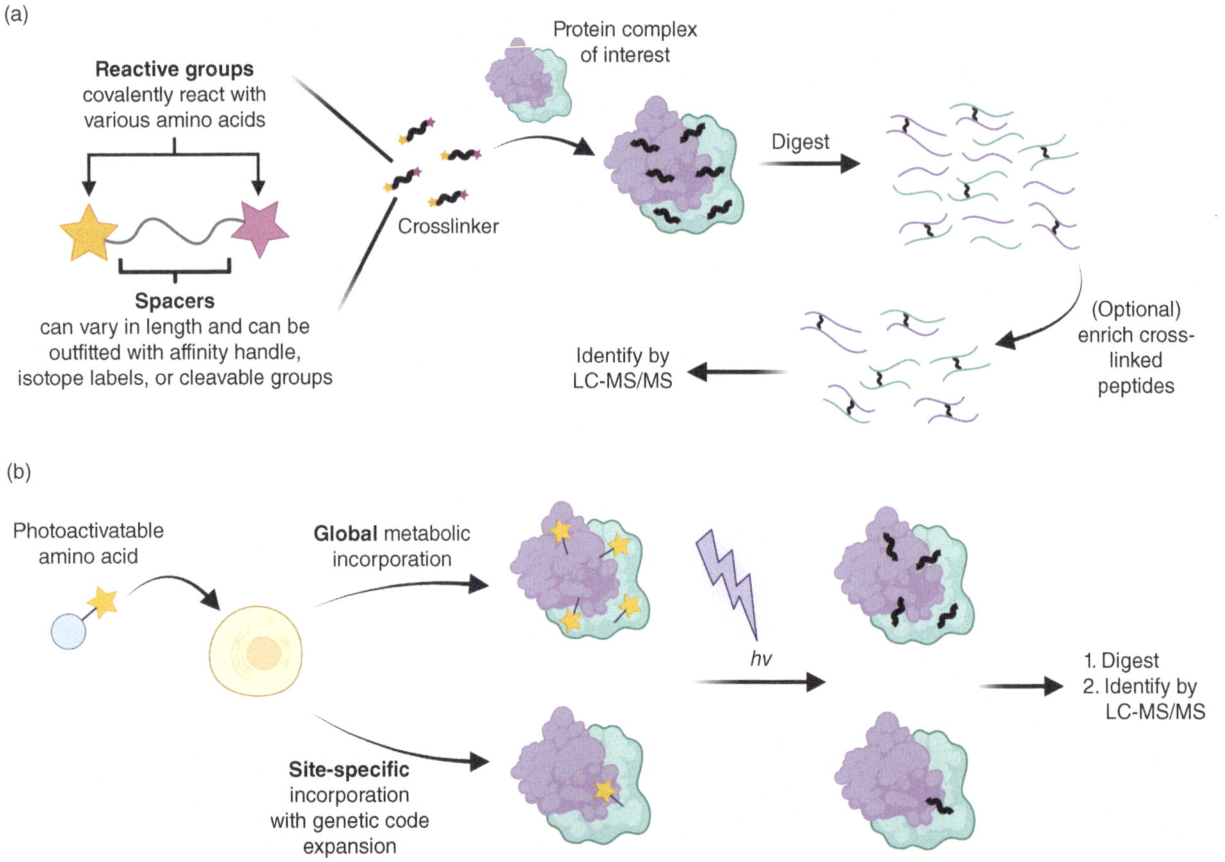

Figure 16.2 Cross-linking methods for XL-MS. (a) Cross-linking reagents, generally composed of two reactive groups that can covalently react with various amino acids separated by a spacer that can vary in length and properties to aid downstream workflows, are incubated with protein-containing samples (e.g. purified protein complexes, cell lysates, or cells). The cross-linker reagent reacts with proteins and cross-links different amino acids, either within the same protein (intramolecular cross-link) or between different proteins (intermolecular cross-link). After the cross-linking reaction, proteins can be digested. If the cross-linking reagent used contains an affinity handle, cross-linked peptides can be enriched to improve sensitivity. The cross-linked peptides can then be analyzed by MS to identify interacting proteins and gain information on the interaction interface. (b) Photoactivatable amino acids can be supplied to cells and incorporated into proteins either globally through endogenous metabolic incorporation or site-specifically through genetic code expansion methods. Cells or cell lysate is then exposed to ultraviolet light to covalently react the photoactivatable side chains to other protein residues, forming cross-links. Cross-linked protein samples can then be processed as above for MS-mediated identification of cross-links between interacting proteins. (c) Common method to combine cross-linking with AP-MS to improve identification of weak and transient interactors. The bait protein of interest is genetically fused with a strep affinity tag and expressed in cells. Cells can then be lysed, and proteins are cross-linked using formalin (40% formaldehyde solution) to stabilize protein interactions. The cross-linked lysate is then incubated with streptavidin affixed to a solid support. Due to the cross-linked interactors and the strong interaction between streptavidin and the strep-tag, stringent washing can be performed to thoroughly remove unbound proteins and nonspecific binders. Samples are then boiled to reverse the formalin cross-linking, releasing the interacting proteins which are then digested and identified by MS. Source: BioRender.com.

(c)

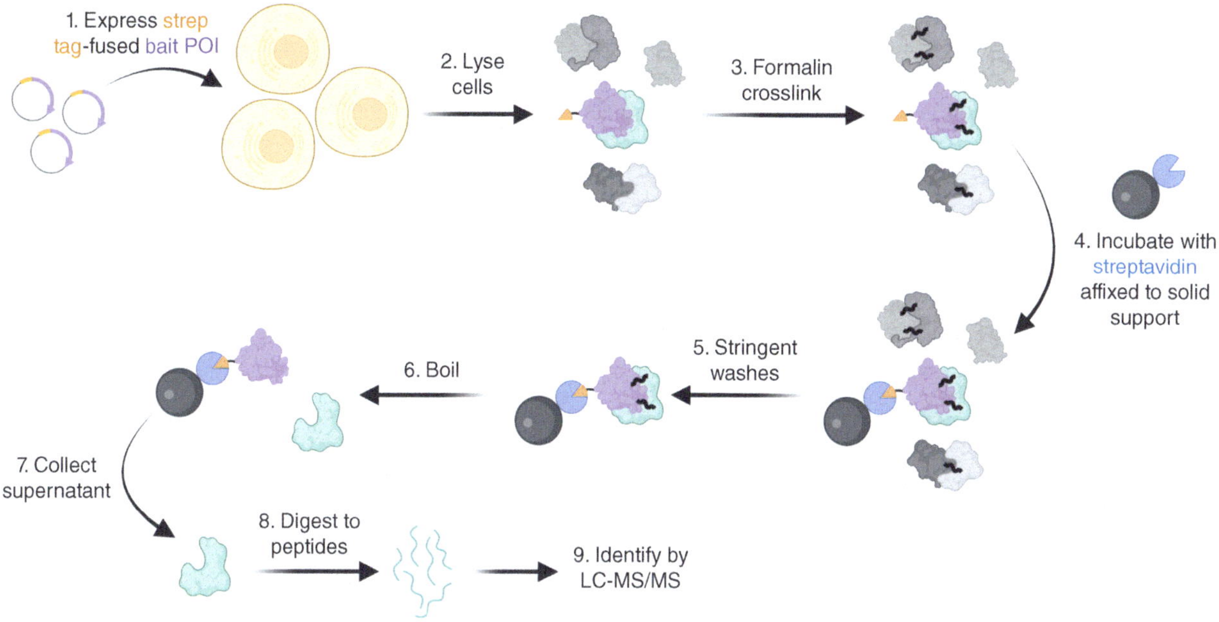

Figure 16.2 (*Continued*)

that allow enrichment of cross-linked peptides such as strong cation exchange [68], size-exclusion chromatography [69], isotopic tags [70], or fluorescent labels [71]. Some cross-linking reagents include both cleavable and affinity components in their spacers, such as protein interaction reporter (PIR) probes [72, 73], which allow extensive tuning to optimize sensitivity, specificity, and facile bioinformatic analysis.

As discussed in the previous section, cross-linking techniques can complement other interactomic techniques like AP-MS to enhance the detection of weak and transient interactions [16, 20]. For example, strep-protein interaction experiments (SPINE) utilize formaldehyde, one of the oldest and most popular cross-linkers that cross-link proteins at very short distances (2 Å), is cell-permeable and reversible by boiling [74] (Figure 16.2c). In SPINE, a strep-tag affinity tag is fused to the bait protein of interest which allows for affinity purification using a streptavidin or strep-tactin purification matrix after the cell samples are cross-linked with formaldehyde [75]. The strong binding of the affinity tag and the formaldehyde cross-linking allow for more stringent washing steps to remove the background as much as possible while maintaining capture of the cross-linked interacting protein preys. The complexes are eluted using biotin [76] and the formaldehyde cross-links can be reversed by boiling prior to digestion and subsequent MS analysis. Another variant, Membrane-SPINE, combines this technique with membrane-specific purification protocols to target membrane PPIs [77]. Another example is reversible cross-link immuno-precipitation (ReCLIP) which uses cell-permeable thiol-cleavable cross-linkers that facilitate stringent IP conditions and selective elution of the interaction partners for MS identification, thus reducing background and conserving specificity [78].

Beyond complex stabilization for AP-MS, cross-linking can also be used to map global interactomes [79–81] and to gain proteome-wide structural information on complex and membrane topologies [79, 82–84]. There have been many XL-MS experiments applied to whole proteomes in cell lysates from many different human cell lines [81, 85–88]; however, most of these datasets are sparse relative to other lysate-based interactomic experiments like AP-MS, with most cross-links found within abundant and well-characterized complexes like the nucleosome, ribosome, and proteasomes. Others have made attempts at the even more ambitious goal of performing XL-MS *in situ* in cells [84, 89–91] or organelles [79, 92, 93] to preserve the important contribution of molecular crowding and spatial organization. Most of these studies utilize enrichable cross-linking reagents containing cleavable spacers to address the high sample complexity and have detected hundreds and sometimes thousands of cross-links. Despite the challenges of in-cell XL-MS, there have been novel biological insights gained from these experiments that are relevant to disease and drug discovery. For example, XL-MS in cells and organisms have revealed insight into interaction network dynamics of protein chaperones like HSP70 [94], bacterial resistance mechanisms [91], host–pathogen interactions [95], virion structural information [96], synaptic transmission [79, 97, 98], and identification of new cancer cell surface targets for chimeric antigen receptor (CAR) T-cell therapies [84].

Although XL-MS has been in use for many decades, its widespread use to map whole-cell interactomes *in situ* is still considered an emerging technology due to challenges in sensitivity, information density, and data analysis. In concept, XL-MS poses many notable advantages over other technologies; for example, these technologies can often be performed in completely endogenous biological systems, which not only allows the bypassing of any genetic manipulation, overexpression, or protein tagging that could perturb the physiological state of the cell but also allows the interactomes of genetically intractable systems like primary cells and tissues or clinical samples to be investigated. In addition, because of the spatial constraints imposed by the cross-linker, XL-MS is capable of yielding some of the strongest evidence for direct, physiologically relevant PPIs. Lastly and importantly, XL-MS is also capable of providing structural information on the PPI interface, which is invaluable information for drug discovery campaigns. However, at this current time, due to the complexity of the cellular proteome and the wide dynamic range of abundances for its proteins and their proteoforms, detection of cross-links is difficult for all but the most abundant proteins. Even with enrichment protocols to maximize sensitivity, data acquisition and analysis methods to confidently identify cross-linked peptides remain challenging due to difficulties with accurate error estimation and thresholding of the results [99]. Nevertheless, rapid technological progress in the XL-MS field is sure to continue making strides toward comprehensive structural interactome mapping in cells to yield important contributions to drug discovery.

Proximity Labeling Mass Spectrometry (PL-MS)

PL-MS is a method that has emerged in more recent years to map macromolecular complexes, organelles, and PPI networks in living cells [100, 101]. In PL-MS, a catalyst is targeted to a specific protein of interest or subcellular organelle and is used to locally activate a small molecule substrate that bears an affinity handle, most often biotin. This activated biotin substrate molecule is then able to covalently biotinylate specific residues on the nearby proteins but will quench before reacting with more distal proteins (Figure 16.3a). Because tagging of these endogenous proteins is based on proximity and not affinity, weak and transient PPIs are able to be captured. After the labeling reaction is complete, the cells are lysed and the proximal biotinylated proteins are harvested using streptavidin-coated beads, digested, and identified by MS. Unlike AP-MS, PL-MS enables the recording of interactomic information while the cell is intact, allowing preservation of natural crowding and spatial organization. Furthermore, due to the high-affinity interaction between biotin and streptavidin, more stringent washes can be used to dissociate nonspecific binders while retaining tagged proteins, resulting in decreased false-positive rates and higher signal-to-noise ratios.

While there is a new generation of small-molecule photocatalysts that have emerged only recently which will be discussed later in this section, PL-MS has been traditionally performed using enzymatic methods that generally fall into two categories: peroxidase-based methods and biotin ligase-based methods.

Peroxidase-based PL-MS Methods

Peroxidase-based methods utilize enzymes like horseradish peroxidase (HRP) [102–104] and engineered ascorbate peroxidase 2 (APEX2) [105, 106], both of which are derived from naturally occurring peroxidases in plants. Due to the four disulfide bonds required for proper folding and function of HRP, it is only active in oxidizing environments like the secretory pathway and extracellular spaces. Therefore, HRP-mediated PL-MS has proved useful for mapping PPIs between membrane proteins, either through genetic fusion to proteins of interest [102, 104] or through antibody targeting as is used in selective proteomic proximity labeling assay using tyramide (SPPLAT) [103]. In contrast, APEX2 is derived from a cytosolic hydrogen peroxide scavenger from soybean and retains activity in all cellular compartments [107]. In PL-MS, these peroxidases are most often used to oxidize biotin phenol (also known as biotin tyramide) to generate biotin phenoxyl radicals which can then diffuse away from the enzyme active site and covalently react with electron-rich side chains on proteins, such as tyrosine [105] (Figure 16.3b). Because these radicals are very reactive and have a half-life <1 ms, the labeling radius of peroxidase-mediated labeling is small (<20 nm) and only proteins proximal to the enzymes will be sufficiently tagged [105]. A major advantage of peroxidase-mediated PL-MS is its speed: with a kcat = 299 s^{-1}, APEX2 can quickly generate sufficient quantities of biotin phenoxyl radicals to tag proximal proteins in 1 minute or less, allowing for probing of dynamic biological process that occur on very short timescales such as signaling events [108, 109]. Although only active in the secretory pathway or extracellularly, HRP is even more catalytically active in these spaces, allowing for even shorter labeling windows [102].

Peroxidase chemistry is also versatile: the biotin phenoxyl radicals react with not only proteins but also nucleic acids, enabling mapping of the spatial distribution of endogenous RNAs throughout the cell [110, 111]. The versatility of peroxidases also lies in their promiscuous substrate specificity, for example, HRP can be used to activate biotin or fluorescein aryl azides in enzyme-mediated activation of radical sources (EMARS) [104, 112]. This substrate versatility also allows applications of these peroxidases outside of proteomics. For example, APEX2 can oxidize the fluorogenic dye Amplex red in the presence of hydrogen peroxide, allowing it to be used as a genetically encoded hydrogen peroxide sensor with a colorimetric readout [113]. Peroxidases can also catalyze the hydrogen peroxide-dependent polymerization of 3,3′-diaminobenzidine (DAB) into a localized precipitate that gives contrast in electron microscopy (EM) after treatment with OsO4, allowing it to be used as a genetically encoded reporter for EM [114].

Peroxidases have been used in cell culture to map various subcellular structures, such as mitochondrial sub-compartments [105, 115–117], the ER membrane [117], and the neuronal synaptic cleft [102], as well as protein interaction networks [108, 109]. While these peroxidases have been implemented successfully for PL-MS in cultured cells, the substrates required for peroxidase-catalyzed proximity labeling pose challenges for its use *in vivo*. Peroxidase labeling requires the use of hydrogen peroxide, which is toxic to cells and difficult to

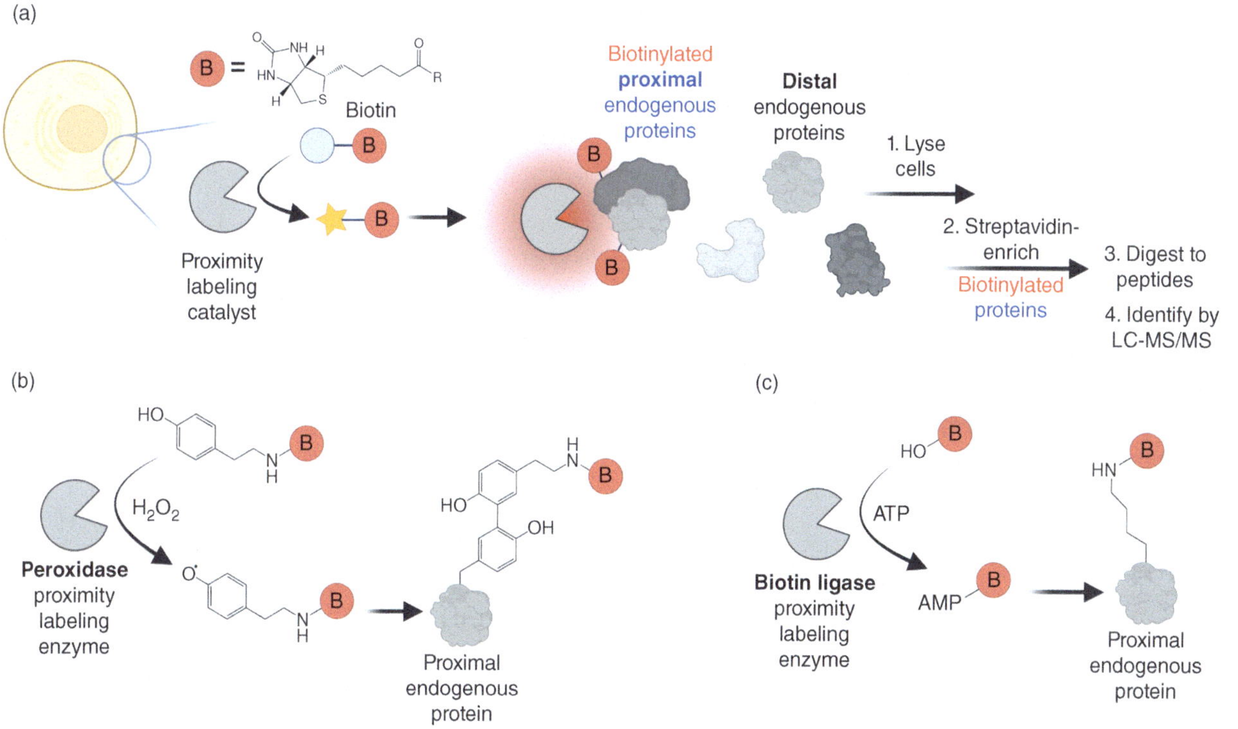

Figure 16.3 Proximity labeling methods for PL-MS. (a) Proximity-labeling reagents are introduced into living cells, either through genetic fusion, antibodies, or ligands. The proximity-labeling catalysts can then activate a probe containing an affinity handle, most commonly biotin. The activated probe can then go on to covalently react with proximal proteins but will quench before reaching distal proteins. After the labeling reaction is complete, cells are lysed and the biotinylated proximal proteins are enriched using streptavidin beads. After stringent washes, proteins are often digested on-bead to peptides which are then identified by MS. (b) Labeling chemistry for peroxidase proximity-labeling enzymes, which commonly use hydrogen peroxide to oxidize biotin phenol probes to generate biotin phenoxyl radicals that can then go on to react with electron-rich side-chains, such as tyrosine, on proximal proteins. (c) Labeling chemistry for biotin ligase proximity-labeling enzymes, which adenylate biotin using cellular ATP to generate biotin-AMP that can then go on to react with primary amines, such as those on the N-termini or lysine, on proximal proteins. (d) Small-molecule photocatalysts can be used for proximity labeling using a wide variety of chemistry. Upon irradiation, photocatalysts can activate a wide variety of probes outfitted with affinity handles, most commonly biotin. Some probes, such as diazirines, can be converted to highly reactive species that will covalently react with very proximal proteins. Other probes, such as aryl azides, can be converted to moderately reactive species that will covalently react with proteins in a larger radius compared to short-range probes. Longer-range probes, such as phenols, can also be used to covalently react with proteins in even larger radii.

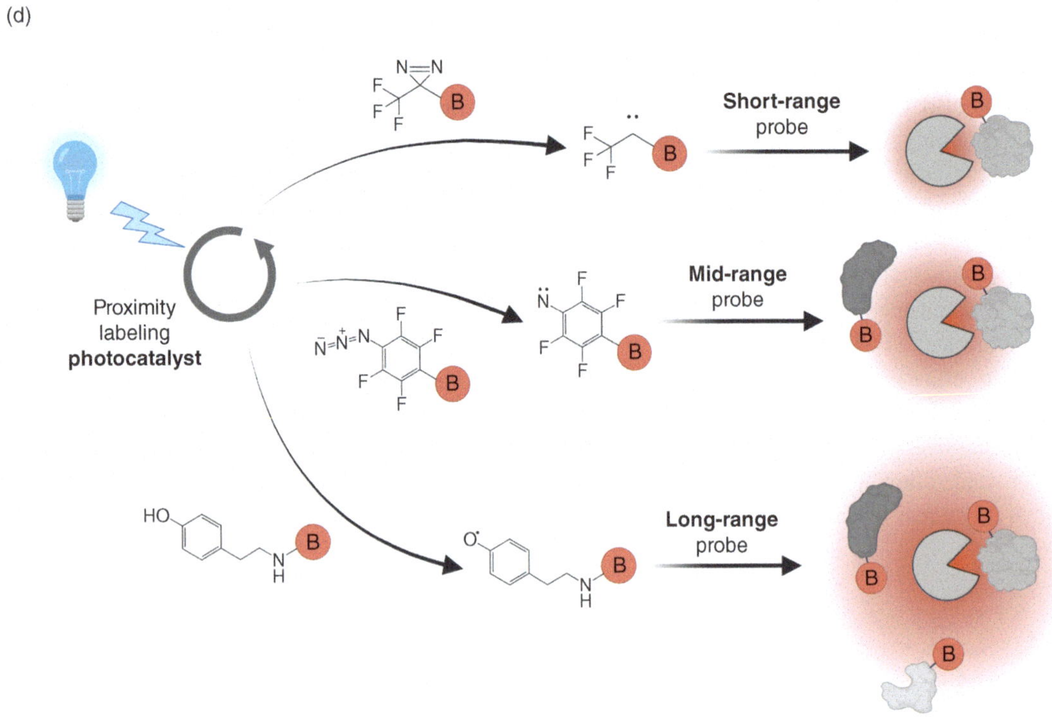

Figure 16.3 (*Continued*)

deliver into live organisms without causing severe tissue damage. Furthermore, biotin phenol has limited cell permeability and requires long incubation times to diffuse through tissues. These limitations have resulted in APEX peroxidase being used in only a small handful of *in vivo* studies, and in each case required either genetic modification to compromise cuticle integrity [118, 119] or manual dissection of tissue [120, 121] to deliver APEX2 chemical substrates to the relevant cells.

Biotin ligase-based PL-MS Methods

The second class of PL-MS enzymes is biotin ligases. The first generation of these methods, BioID [122, 123], is derived from the *Escherichia coli* biotin ligase (BirA) which covalently attaches the cofactor biotin to the active site of enzymes whose chemistry is biotin dependent, such as carboxylases involved in fatty acid metabolism [124]. To carry out biotinylation of biotin-dependent enzymes, BirA adenylates biotin to create a reactive intermediate, biotin adenosine monophosphate (biotin-5′-AMP), which is held tightly in its active site until it contacts a target protein. Upon contact, BirA transfers the biotin to a specific lysine side-chain on the target protein. In the entire *E. coli* proteome, BirA recognizes and biotinylates only a single lysine residue – lysine 122 of the biotin carboxyl carrier protein subunit of acetyl-coenzyme A (CoA) carboxylase [125, 126]. In 1993, a 15 amino acid peptide was found using a library screening approach that can be efficiently and specifically biotinylated by BirA, thus named the biotinylation acceptor peptide (BAP). This peptide has allowed researchers to use wild-type BirA to detect interactions between two proteins of interest by fusing BirA to one protein and BAP to the other and using biotinylation to readout interaction events [127].

Over a decade later, however, Cronan et al. reported that the high sequence-specificity of BirA biotinylation could be abolished by a single point mutation at arginine 118 to glycine [122], which was incorporated into BirA by Roux et al. to develop BioID [123]. The R118G mutation is predicted to weaken the affinity of the enzyme for the reactive biotin-5′-AMP species, allowing it to react promiscuously with nearby nucleophiles, such as deprotonated lysine side-chains or N-termini (Figure 16.3c). The rate of biotin-5′-AMP release from BioID increases approximately 440-fold from that of wild-type; however, the rate of biotin-5′-AMP formation by BioID also decreases approximately fivefold relative to wild-type [128, 129]. The half-life of adenylates such as biotin-5′-AMP has been measured to be on the order of several minutes *in vitro* [130], giving it a potential reactive radius that is on the order of millimeters. Although the reactive biotin-5′-AMP intermediate is likely hydrolyzed slowly in solution, the high concentrations of nucleophilic protein side chains and small molecules inside cells are likely to shorten its lifetime. Roux et al. experimentally approximated the reactive radius of BioID-generated biotin-5′-AMP in living cells to be roughly 10 nm [131].

In contrast to peroxidases, BioID labeling is simple and nontoxic: only biotin needs to be supplemented to initiate labeling. This feature has resulted in BioID being applied to a wide variety of biological questions since its introduction 12 years ago, including studies in cultured mammalian cells [123, 131–134], plants [135–137], parasites [138–146], slime molds [147, 148], mice [149], and yeast [150]. BioID has been used, for example, to map the protein composition of the centrosome–cilium interface [132] and the inhibitory postsynaptic region [149], each with nanometer spatial specificity. Because biotin is nontoxic and easily deliverable, BioID is more amenable to proximity labeling *in vivo*. Despite this, however, there have been only a few *in vivo* demonstrations [149, 151]. This is likely related to BioID's low catalytic activity, which makes BioID difficult or impossible to apply in some contexts - such as in worms, flies, or the ER lumen [152]. Furthermore, its slow kinetics necessitates labeling with biotin for 18–24 hours, and sometimes much longer [149], to produce sufficient biotinylated material for proteomic analysis. This precludes the use of BioID for studying dynamic processes that occur on the timescale of minutes or even a few hours.

Since the introduction of BioID, there have been several new promiscuous biotin ligase variants introduced for PL-MS. BioID2, for example, was derived from *Aquifex aeolicus* biotin ligase and made promiscuous by introducing the point mutation R36G, which is analogous to the R118G mutation in BioID2 [153]. The *A. aeolicus* biotin ligase is approximately 30% smaller than *E. coli BirA*, which makes it a more attractive option as it could result in less perturbation of protein folding or trafficking when fused to a bait protein. Additionally, because *A. aeolicus* is a thermophilic bacterium, BioID2 can be used at temperatures higher than 37°C, which may be useful for some applications. However, BioID2 still requires long labeling times of over 16 hours [153–156], therefore the scope of dynamic processes that can be probed using BioID2 is still narrow. Furthermore, BioID2 has a higher affinity for biotin than BioID and can carry out promiscuous biotinylation with lower concentrations of biotin [153]. While this may be advantageous when performing proximity labeling in organisms with poor biotin uptake mechanisms, it can be disadvantageous in contexts such as cell culture where free biotin concentrations in the media and biotin uptake are high enough to induce promiscuous biotinylation by BioID2 constitutively. The use of endogenous biotin present in media by BioID2 means that it can carry out biotinylation of endogenous proteins the entire time it is being expressed, which removes any temporal control of labeling by the user (through the addition of exogenous biotin to initiate labeling). This lack of temporal control further prevents the use of BioID2 to probe dynamic proteomes, limiting its application in certain contexts. Users can regain temporal control of the labeling window by depleting biotin from the media; however, this starves cells of biotin and can lead to poor cell health and proteome perturbation.

BASU is another promiscuous biotin ligase derived from the *Bacillus subtilis* biotin ligase [157]. To develop BASU, the N-terminal domain was removed, and four mutations were introduced: R124G, which is analogous to the *E. coli* BirA R118G mutation to impart promiscuity, and three additional mutations in the C-terminal domain that are analogous to residues in other biotin ligases and are thought to improve ATP binding [157]. BASU has been used in the context of RNA–protein interaction detection (RaPID), which is a method that allows for the identification of RNA-binding proteins [157, 158]. In RaPID, BASU is fused to the protein λN, which can bind to the RNA stem-loop BoxB. This RaPID-BASU construct is co-transfected with an RNA component that contains the BoxB stem-loop flanking an RNA motif of interest. λN binding to BoxB brings BASU into proximity with the RNA motif of interest, allowing biotinylation of RNA-binding proteins that interact with the RNA motif of interest.

TurboID is an *E. coli*-derived promiscuous biotin ligase that was heavily engineered using yeast display directed evolution to greatly increase its activity and reduce the required labeling time over 100-fold compared to BioID, BioID2, and BASU, requiring only 10 minutes or less in most cultured cell systems [152]. This increased activity has also allowed TurboID to be widely used *in vivo* in many organisms, extending widespread use of PL-MS in plants [159–165], worms [152, 166–169], flies [152, 170, 171], zebrafish [172, 173], and mice [174–176]. Like BioID2, however, TurboID is able to utilize biotin much more efficiently which can result in labeling activity in the absence of exogenous biotin addition in some situations where endogenous biotin concentrations are high [152]. In these situations, researchers may have less control over the labeling window which can make probing dynamic processes more difficult. Furthermore, under extreme conditions of high expression or longer than prescribed labeling times, toxicity can be observed, potentially due to biotin sequestration or over-biotinylation of the proteome [152]. For example, toxicity has been observed in flies when TurboID is expressed constitutively and ubiquitously in every tissue; however, this does not occur when its expression is inducible or tissue specific [152]. To address this, the inventors of TurboID also created a variant called miniTurbo, which is ~20% smaller than TurboID and exhibits lower biotinylation activity in the absence of exogenous biotin [152]. This allows miniTurbo to exhibit lower toxicity and also imparts tighter user control of labeling windows for probing dynamic processes; however, there have been some reported instances of poor stability [177].

AirID is another biotin ligase-based PL-MS enzyme that was generated using a combination of ancestral reconstruction and site-directed mutagenesis [178]. AirID shares 82% sequence identity with *E. coli* BirA and is approximately the same size as BioID and TurboID. Proximal biotinylation activity is improved compared to BioID but still substantially lower than TurboID, thus AirID requires labeling times on the order of 6–24 hours [178–180]. While this longer labeling time reduces temporal resolution, it does not exhibit background biotinylation prior to exogenous biotin addition or cellular toxicity during extended labeling periods. Therefore, like BioID and BioID2, it may be better suited than TurboID if the goal is to integrate biotinylation signal over long periods of time.

UltraID is the most recently reported promiscuous biotin ligase for PL-MS, which was found somewhat serendipitously during a campaign to create a split version of BioID2 [181]. The authors found that after splitting the BioID2 enzyme between residues K171 and S172, the larger N-terminal fragment not only retained complete activity on its own but even showed more efficient biotinylation at shorter labeling times. From this truncation of BioID2, the authors thus created what they went on to name microID. To further improve the activity of microID, the authors used a yeast-display-directed evolution approach similar to what was used to engineer TurboID [181]. This engineering effort resulted in two additional point mutations near the active site that increased the activity of microID approximately twofold – this engineered mutant was selected as their final enzyme which they named UltraID. UltraID is the smallest of the proximity labeling enzymes with a size of only 20 kDa, 43% smaller than BioID or TurboID. While its activity is slightly lower than that of TurboID, UltraID is able to produce robust biotinylation during short labeling windows (10 minutes) and has lower background biotinylation in the absence of exogenous biotin. While this method is new and has not yet been applied to any biological investigations beyond the original reporting study, it is a promising new development for enzymatic PL-MS.

Light-Activated Enzymatic PL-MS Methods

There have also been several light-activated enzymatic PL-MS methods developed that enable additional temporal and spatial control [182–185]. Many of these methods utilize singlet-oxygen generating (SOG) enzymes like miniSOG which are flavin-containing enzymes that generate singlet oxygen species upon irradiation. These singlet oxygen species can then diffuse away and activate a wide range of molecules. In some of these methods, the singlet oxygen is used to oxidize thiol-conjugated biotin probes that can react with cysteine residues on nearby proteins [182], in others the singlet oxygen is used to oxidize residues like histidine directly on nearby proteins which can then be reacted with chemical handles like nucleophilic alkyne probes [183]. Due to the longer half-life and diffusion radius of singlet oxygen (~70 nm), these methods may be more amenable for mapping longer-range interactomes [183]. Other methods like light-induced interactome tagging (LITag) utilize different flavin-containing enzymes, such as light-oxygen-voltage (LOV) domains, to generate the singlet oxygen species. LOV domains have also been used in other ways as well, for example, Lee et al. used structure-guided screening and directed evolution to install the LOV domain into TurboID to impart rapid and reversible light control of its labeling activity [185]. These methods collectively add additional spatial and temporal control over PL-MS experiments; however, because most of these methods rely on blue light, their use *in vivo* may be limited.

Split Enzyme-based PL-MS Methods

In addition to the enzymatic PL-MS technologies discussed above, there have also been split-enzyme versions developed from nearly each reported PL-MS enzyme [175, 179, 186–192]. These split-technologies are typically composed of two inactive protein fragments that can come together to reconstitute an active enzyme for PL-MS. Because these protein fragments have low affinity for each other, they must be brought in proximity by other means to form the active enzyme. This is often done by fusing each fragment into two proteins that interact with each other, thus bringing the split-enzyme fragments together upon a PPI [188]. However, this split-enzyme approach has also been used to map organelle contact sites and cell contacts like synapses by localizing each fragment to opposing membranes [175, 186, 190, 192]. These split-PL-MS approaches allow an additional layer of selectivity to be achieved, as proximity labeling is only carried out in specific complexes or subcellular locales of interest.

Photocatalyst-based PL-MS Methods

As mentioned above, there has recently been a new generation of PL-MS reagents reported that are not based on enzymes, but rather light-activated small molecules

[193–200]. Many of these photocatalysts function through electron exchange mechanisms, like Dexter energy transfer [193, 199] or single-electron transfer [196, 197, 201]. Upon irradiation, the photocatalysts are excited into a state where they can exchange electrons with nearby probe molecules through a radiationless mechanism. These locally activated probes, outfitted with chemical handles like biotin or alkynes, can then go on to covalently react with nearby proteins (Figure 16.3d). One advantage of these small-molecule photocatalysts is that they bypass the need for genetic fusion of a protein and thus any genetic manipulation, which allows PL-MS in genetically intractable systems and under more physiological conditions without tag fusions or overexpression. Because targeting of the PL-MS reagent is not done genetically, however, targeting must be achieved through other means. Typically this has been done using antibodies [193, 194, 199], which has restricted most of their applications to extracellular spaces; however, there are also instances where specific photocatalysts that are naturally attracted to particular subcellular environments like the mitochondria [198] or nucleus [202] have been employed. There are also examples where the photocatalysts are conjugated directly to ligands or drugs, allowing probing of drug–protein interactions [203, 204]. The limitations of these targeting strategies when compared to the genetic-fusion strategies of most enzymatic PL-MS methods is that complete or near-complete targeting can only be achieved with highly specific, high-affinity targeting agents. With decreasing selectivity and affinity, the off-target and unbound populations of the photocatalyst will increase, thus decreasing the signal-to-noise ratio. Although some of this background noise can be mitigated during data analysis by comparing it to appropriate controls (as discussed further below), the necessary affinity and selectivity criteria for targeting strategies in these emerging photocatalyst PL-MS methods have yet to be established. Furthermore, using larger targeting agents such as antibodies can place the photocatalyst too far from the interactions of interest, preventing their detection [205].

Another advantage of the photocatalysts is their diverse chemistry – there have been many reported small-molecule photocatalysts that can be activated with different wavelengths of visible light, including blue [193, 196–198], yellow [199], and red light [194, 195], allowing for applications that may require more or less penetrance into cells and tissues. Different photocatalyst chemistries also allow activations of a variety of probes that have varying labeling radii to map different ranges of interactomes [199], ranging from longer-range mapping using singlet oxygen mechanisms [196, 201] to ranges similar to enzyme PL-MS methods [199] to even shorter ranges that may provide stronger evidence of direct interactions or even structural information on binding interfaces [203]. The diverse probe chemistry also allows tagging of different protein residues, which could allow higher coverage of different protein populations or richer structural information. This versatility could allow photocatalyst-based PL-MS experiments to be carefully tuned for the biological question at hand; however, it should be noted that the efficiency of tagging by different probes varies widely, with the shorter-range probes that could provide the most informative structural information typically displaying the lowest tagging efficiency [199]. Thus, while photocatalysts represent an exciting advancement for PL-MS, they remain an emerging technology with much to be understood regarding their limitations and requirements.

Most PL-MS methods utilize biotin as an enrichment handle due to its strong interaction with streptavidin. However, this can be problematic at times when the target proteome of interest is small due to low signal-to-noise with endogenously biotinylated proteins such as carboxylases, particularly *in vivo*; for example, when the bait protein is in low abundance or present only in rare cell types. Some researchers overcome this by depleting the endogenously biotinylated carboxylases prior to streptavidin enrichment of proteins biotinylated in the PL-MS experiment [206]. An alternative is to use a non-biotin-based probe for PL-MS, like an alkyne handle [183], or other labeling systems like PUP-IT which utilize the prokaryotic enzyme PafA to ligate a small peptide (Pup) to proteins for affinity purification [207]. These alternatives are not possible when using biotin ligase-based PL-MS methods, as their obligate substrate is biotin, which as discussed above, can also cause issues of toxicity *in vivo* due to biotin sequestration [152].

One of the unique advantages of PL-MS is that because labeling is dependent on proximity and not affinity, detection of transient or weak interactions is more likely. However, this, in turn, means that any interactions that are detected cannot be confidently assigned as direct. Furthermore, because any protein in proximity to the PL catalyst that has sterically available reactive residues could in theory be labeled, a protein can be labeled and detected even if it is not considered physiologically relevant to the protein being studied. The enrichment of these "bystander" proteins can be reduced during analysis by applying a ratiometric method, where the enrichment of proteins tagged by the bait protein of interest-targeted PL catalyst is compared to a control condition where the PL catalyst is free but targeted generally to the same subcellular location [102, 115, 117, 203, 205, 208].

Despite these limitations, PL-MS has been applied in hundreds if not thousands of studies over the past decade, many of which have made novel and significant discoveries that are relevant to human diseases and drug discovery. For example, PL-MS has been used to gain a better understanding of the complex and dynamic signaling processes that occur downstream of G-protein coupled receptors (GPCRs), which are the targets of over 30% of all marketed drugs [209], after treatment with different agonists or antagonist drugs [108, 109, 210–213]. PL-MS has also been used to characterize the interactomes of E3 ligases like Cereblon which have been leveraged for targeted protein degradation therapies, and to assay how these interactomes change in response to treatments with different degradation-inducing molecules [180, 214–219]. Many of these PL-MS strategies, therefore, provide new avenues to screen chemical matter to discover novel degraders that could be developed into new therapies. Other notable examples of PL-MS applications for disease biology include protein homeostasis and its role in neurodegeneration [134, 220], pathological protein aggregates in neurological disease [221, 222], stress granules and their role in neurodegeneration [223], cancer signaling pathways [205, 219, 224–232], chromatin state changes during oncogenic mutation or epigenetic drug treatment [233–235], host–pathogen interactions [138–146, 236–244], immune signaling [193, 196, 201], and interorgan signaling in disease [171, 245]. PL-MS can also be used to identify the proteins engaged by a small-molecule drug, by conjugating or recruiting PL catalysts to small-molecule ligands and using the drugs themselves as the bait [203, 204, 246, 247]. Therefore, PL-MS has established itself as a powerful technology to map PPI networks in living biological systems to gain invaluable insights into human disease that can fuel impactful drug discovery research.

Concluding Remarks and Other Interactomic Technologies

One drug, one target – this has been the prevailing paradigm in drug discovery. While many drugs in this category do function by modulating protein interaction networks, a promising approach has emerged in recent years where drugs are specifically designed to target an interaction interface between multiple proteins in order to drug disease-relevant proteins that do not have well-defined pockets to target alone. One such approach is induced proximity, where therapeutic bioactives strengthen or create new protein interactions that can modulate biological processes with high selectivity [248]. Examples of this mechanism of action (MOA) can be seen as early as the 1990s, where immunosuppressive drugs for organ transplant patients like rapamycin were retrospectively discovered to inhibit important proteins like mTOR by recruiting another protein binder [249]. Similarly, thalidomide, a sedative that was tragically used to treat morning sickness in pregnant women which led to thousands of birth defects in the 1950s and early 1960s, and its derivatives lenalidomide and pomalidomide were later discovered to recruit new protein substrates to the E3 ligase Cereblon (CRBN) [30, 31, 250]. By acting as a molecular glue, these immunomodulatory drugs recruit substrates like CK1α and IKZF1/3 to CRBN, mediating their ubiquitination and ultimate degradation. Today, these drugs are used in combination with other drugs to engage the immune system to treat diseases such as MM and leprosy.

There are a number of newer generation molecular glue degraders under clinical or preclinical investigation [251–253]; however, most of these molecules have been found serendipitously through phenotypic screens rather than rational design. In contrast, rationally designed bivalent molecules like PROTACs have shown much promise with over 20 molecules in clinical trials as of 2023 [254]. PROTACs tether a target-binding ligand with an E3-binding ligand through a linker. Because they are modular by nature, PROTACs in theory could be made to recruit any target of interest to an E3 ligase to induce its degradation. In practice, however, PROTAC design has remained largely empirical and has been limited to a small number E3 ligases such as CRBN and von Hippel–Lindau (VHL) [255, 256]. Furthermore, due to their larger size, it is often challenging and time consuming to achieve acceptable drug-like properties compared to smaller molecular glue-like degraders.

In addition to molecular glues and PROTACs, many other promising targeted protein degradation strategies utilize induced proximity. For example, antibody-based PROTACs(AbTACs) and proteolysis-targeting antibodies (PROTABs) target membrane proteins for proteolytic degradation by recruiting them to plasma membrane-localized E3 ligases through their extracellular domains using antibodies [34, 257]. Similarly, lysosome-targeting chimeras (LYTACs) degrade targets of interest using antibodies against them that are linked to a ligand of the lysosomal trafficking shuttle (e.g. asialoglycoprotein receptor (ASGPR) [258]) and that pull the LYTAC–target complex into lysosomes for degradation. Other modalities like autophagy targeting chimeras (AUTACs and AUTOTACs) and cytokine receptor-targeting chimeras (KineTACs) also leverage lysosome-based targeted degradation [259–262]. Beyond degradation, there are also induced proximity strategies that have emerged to stabilize proteins such as deubiquitinase-targeting chimeras (DUBTACs) [263], modulate post-translational modifications such as acetylation [264] or phosphorylation [265–267], induce interactions that sterically block downstream signaling interactions

[268–270], or strategies to incorporate chemical genetic control into cell therapies [271].

With such encouraging therapeutic promise in modulating PPIs by targeting their interactions directly, the MS-based technologies outlined in this chapter, which are used to study PPI networks and how they can be modulated with bioactive agents, are becoming increasingly critical to drug discovery research. We chose to focus on three such technologies here, AP-MS, XL-MS, and PL-MS, but there are many other creative strategies beyond the scope of this chapter that have been employed to characterize PPIs. One such example is Virotrap, where a target protein of interest is fused to the HIV-1 GAG protein to trap interaction partners in virus-like particles that bud from the cell [272] (Figure 16.4a). This strategy therefore does not require cell homogenization and preserves the protein complexes during purification for subsequent identification by MS. Virotrap has been used to make important discoveries for drug discovery; for example, it was used to identify LZTR1 as a novel E3 ligase regulator of oncogenic RAS proteins [273]. Other interaction-based MS technologies have proved useful in studying the intrinsic properties of proteins that could be leveraged for PPI drug discovery. For example, a recent study by Poirson et al. leverages molecular glue-induced PPIs to artificially induce proximity between ubiquitin–proteasome pathway components and a reporter substrate to assess the degradation or stabilization capability of different E3 ligases and deubiquitinases [274]. This study will surely be an important resource that infuses targeted protein degradation and stabilization drug design with new effector options to explore.

This chapter has focused primarily on the biochemical techniques that can be employed to gain structural and spatiotemporally resolved PPI information from cells when coupled with MS; however, there have also been many advancements on the MS side to improve sensitivity, throughput, and data quality. Traditionally, label-free MS methods have been commonly employed for AP-MS and PL-MS experiments, typically followed by the application of statistical frameworks like SAINT (Significance Analysis of INTeractome) [133, 275] or CompPASS [14, 276] to distinguish true PPIs from background noise. This workflow is straightforward and does not require costly reagents, making it very accessible to the scientific community at large. However, these workflows are based on peptide spectral counts which can only provide semiquantitative information due to sampling bias, dynamic range limitations, and experimental variability [277]. To obtain more quantitative interactomic data, researchers with the necessary resources and expertise can utilize reagents like stable isotope labeling by amino acids in cell culture (SILAC) [208, 278] or tandem mass tags (TMTs) [279, 280] which leverage different isotopic compositions to "barcode" samples so they can be mixed and analyzed together, allowing accurate quantification of proteins.

The MS workflows described above have traditionally utilized data-dependent acquisition (DDA) MS methods, where the instrument first performs a survey scan to detect all the ionized peptides present in the sample fraction being analyzed and then selects the most abundant ions for further fragmentation and analysis. This selectivity can allow for more accurate spectral matching to identify the peptides and therefore the proteins in the samples, but it can also miss low-abundance peptides and introduce more variability from stochastic selection when there are too many co-eluting peptides [281]. To address these limitations, researchers have been increasingly utilizing data-independent acquisition (DIA) methods for their interactomic experiments, where the instrument captures all of the fragment ions within a predetermined m/z range in an unbiased fashion [134, 213]. While this can make the data analysis more complicated, it can improve detection of low abundance peptides, as well as allow more accurate and reproducible measurements. These properties of DIA allow quantification of peptides without the need for isotopic labels like TMT, which means the workflows can be scaled up to process larger numbers of samples in shorter amounts of time. Faster MS data acquisition with DIA, especially when combined with automated sample preparation workflows [213, 282], presents the potential to greatly improve the throughput and reproducibility of these interactomic experiments to the scale required for drug discovery research.

Recent developments in MS workflows to improve sensitivity and lower required sample inputs also hold promise to advance PPI drug discovery research. While there have been great advances in transcriptomics that have made gene expression analysis at the single-cell level rather routine, efficiently measuring protein levels at the single-cell level remains challenging [283]. There have been several methods developed to automate and miniaturize sample preparation to maximize the number of proteins from small sample inputs that reach the MS for detection [284–286]. For example, nanodroplet processing in one pot for trace samples (nanoPOTS), a microfluidic sample-preparation platform, has been used to robustly quantify ~1,500 proteins from single-cell samples [286]. Other methods, like Single-Cell ProtEomics by Mass Spectrometry (SCoPE-MS), have also miniaturized sample preparation as well as optimized MS detection methods to quantify over 3,000 proteins in single monocytes and macrophages and demonstrated the ability discern single cells by cell type [285]. There have also been methods that implement DIA workflows for single-cell proteomic

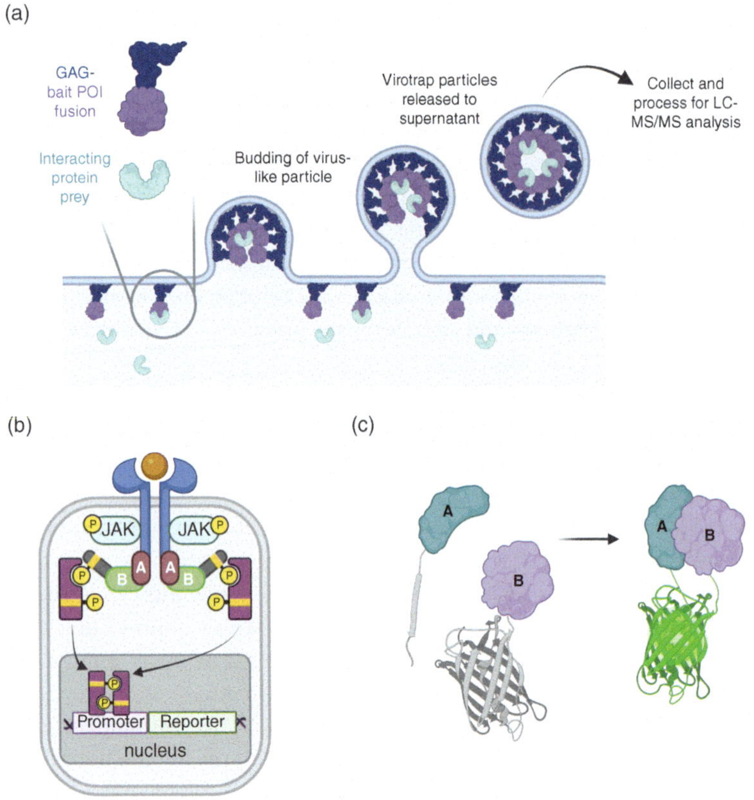

Figure 16.4 Other interactomic methods. (a) In Virotrap, a bait protein of interest is fused with the HIV-1 GAG protein and expressed in cells, where it interacts with its natural binding partner preys. The GAG protein facilitates the budding of virus-like particles (VLPs) from the cell membrane, encapsulating the bait protein of interest and its interacting partners within these particles. The VLPs are then collected from the culture medium and processed for analysis by MS to identify interacting proteins. (b) Common mammalian two-hybrid method, mammalian protein–protein interaction trap (MAPPIT). Protein of interest A (bait) is fused to a modified, signaling-deficient cytokine receptor. The potential interacting partner, protein B (prey), is fused to a signaling-competent domain, such as a STAT (signal transducer and activator of transcription) recruitment domain. Cells are stimulated with a cytokine that normally activates the signaling pathway of the modified receptor, therefore if the bait and prey proteins interact, the signaling-competent domain of the prey recruits STAT proteins to the receptor complex, leading to the activation of a reporter gene. (c) Example of bifunctional complementation. Two proteins of interest, protein A and protein B, are each fused to complementary, nonfluorescent fragments of a fluorescent protein or a luminescent protein. These fusion proteins are co-expressed in living cells, and if they interact, the complementary fragments of the fluorescent or luminescent protein are brought into close proximity, allowing them to reassemble into a functional, fluorescent, or luminescent protein. The reconstituted fluorescent or luminescent signal can be detected and quantified using fluorescence microscopy, flow cytometry, or a luminometer, providing a direct readout of the PPI. (d) Proximity ligation assay. Two primary antibodies are incubated with cells, each specific to two proteins of interest, protein A and protein B. Secondary antibodies, conjugated to unique DNA oligonucleotides (probes), are then added so that each primary antibody is recognized by a specific secondary antibody with a unique DNA probe. If the two proteins are in close proximity, the DNA probes on the secondary antibodies can hybridize with a connector oligonucleotide, forming a circular DNA template. The circular DNA template is ligated and then amplified using rolling circle amplification. This amplification step generates a long single-stranded DNA molecule, which can be detected using fluorescently labeled complementary oligonucleotides. The fluorescence signal can be visualized and quantified using fluorescence microscopy, providing a direct readout of the PPI. (e) Resonance energy transfer assay. The proteins of interest, protein A and protein B, are genetically fused to the donor and acceptor fluorophores, respectively. The fluorophores are selected so that the donor fluorophore absorbs light at a specific wavelength, while the acceptor fluorophore absorbs the energy emitted by the donor. These fusion proteins are co-expressed in cells, and when the donor fluorophore is excited by an external light source, it can transfer energy to the acceptor fluorophore if they are in close proximity. This energy transfer results in a decrease in donor fluorescence and an increase in acceptor fluorescence. The changes in fluorescence are detected and measured using fluorescence microscopy or spectroscopy, with increasing acceptor fluorescence indicating an interaction.

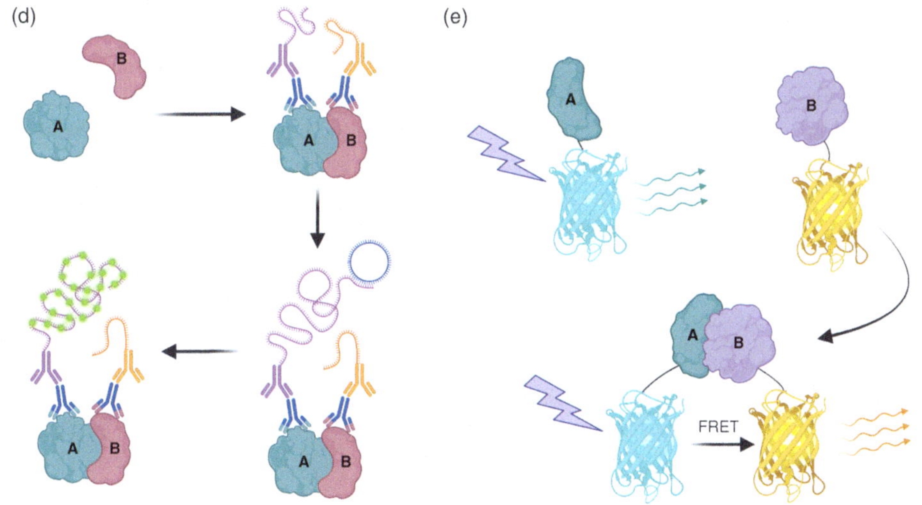

Figure 16.4 (*Continued*)

experiments, such as the study presented by Brunner et al. where the authors characterized the proteomes of single cells arrested in various stages of the cell cycle, quantifying up to 2,000 proteins per cell at different cell cycle stages [287]. While most of these demonstrated applications of ultra-low input proteomics have been used to map global proteomes, there have also been some demonstrations with low-input interactomic experiments [288]. Beyond MS, there is also increasing attention on emerging single-molecule protein sequencing technologies [289], such as nanopore sequencing [290] and fluorosequencing [291], which could greatly improve the sensitivity for identification and quantification of proteins from complex samples with limited or single-cell input once these technologies are fully realized. Advancements in these areas will hold great promise for future clinical applications of interactomics where sample input is limited.

The interactomic methods described in this chapter allow for the identification of PPIs at the protein level, but there is also an opportunity in each of the technologies to gain structural information on the interaction interfaces. For example, XL-MS, whether used alone for global interactome mapping or in combination with AP-MS, provides spatial information on proximal residues present near PPI interfaces when detecting cross-linked peptides [44]. The vast majority of PL-MS experiments function to identify interactors at the protein level alone; however, there are examples where the modified sites are specifically sought to gain information on protein topology. For example, photocatalyst-mediated PL utilizing short-range probes like diazirine-based probes has been used to reveal residues proximal to drug-binding sites [203, 204]. There have also been several advancements in sample preparation protocols that have increased the recovery of PL-labeled residues on interacting proteins, such as cleavable or modified probes [292, 293], antibody-based biotinylated peptide enrichments [294, 295], or modified streptavidin elution protocols [296, 297]. These tools collectively allow an additional layer of structural information to be gleaned from interactomic experiments, which can be used to elucidate drug MOAs and improve the efficiency of drug optimizations.

For all of the MS-based interactomic technologies and applications discussed above, a last but critical step is the data analysis and interpretation. After the MS data is acquired and processed, different analytical methods can be applied to draw out biological insights that can inspire follow-up investigation and validation experiments, which in turn can eventually lead to meaningful biological discoveries – it is important to note that these interactomic experiments cannot exist in a vacuum, they are often the "first step" that can generate hypotheses, but these hypotheses must be followed up on with (often many) additional experiments before any conclusions can be made. Data interpretation for interactomics experiments begins with the experimental design, which is most meaningful when they are designed around a specific question so that the appropriate controls and comparisons can be included. For example, if AP-MS or PL-MS is being used to gain more information about the interactome of a protein of interest, a negative control or spatial control must be employed to remove background proteins as discussed above. If the question is how a particular mutation affects the interactome of a protein of interest, a wild-type control must be used for comparison. If the question is how a particular drug treatment affects the interactome of a protein of interest, a no-treatment control must be used for comparison.

As described above, it is best to employ a quantitative proteomics approach to accurately measure interactome changes between samples. Significant changes, often reported in fold-changes between compared samples, can be found using statistical methods like receiver operating characteristic (ROC) analysis [208, 280] or packages like MSstats [298, 299]. However, the interpretation of these changes must consider the nature of the system used for the interactomic experiment. For example, in AP-MS and PL-MS experiments where the bait protein of interest is overexpressed, effects of this overexpression on the global proteome must be considered. Overexpressing a biologically active protein can result in downstream signaling, remodeling of the proteome, and changing of the cell state. Therefore, if one is comparing the interactome from one overexpressed protein sample to a different overexpressed protein sample, changes in the abundance of potential interactor protein prey should be considered to improve confidence and reduce the rate of false positives. To reduce the proteome remodeling and perturbation of cell state, experimenters can consider using inducible expression of their baits to minimize the effects of downstream signaling and transcriptional changes. Similarly, drug treatments can be kept to shorter time points when possible.

In addition to the MS-based interactomic technologies discussed in this chapter, there are several cell-based pair-wise screening assays that are widely used in drug discovery research to identify PPIs and assess their modulation with bioactive compounds. For example, luminescence-based mammalian interactome mapping (LUMIER) is an affinity purification-based method that pulls down on a FLAG-tagged bait and uses luminescence instead of MS to readout interactions with luciferase-fused prey interactor proteins [300]. Another class of widely used binary assays is two-hybrid assays, which are some of the oldest and widely used assays to screen for PPIs [40]. The original two-hybrid assay was created in yeast, where two proteins of interest are fused to two halves of a split transcription factor. When the proteins of interest interact, the function of the transcription factor is restored, which enables the expression of the reporter protein [301]. Over the years, this assay has evolved more complex designs which have allowed its adoption into mammalian cells [302–304] (Figure 16.4b). Beyond the ability to identify PPIs, the relatively low cost and ease of two-hybrid assays allow them to be used in high-throughput efforts to screen chemical matter for PPI modulators [305–308].

Another popular class of binary PPI assays is protein complementation assays (PCA). In comparison to two hybrid methods which can have rigid and artificial restrictions on the subcellular localization of PPIs due to their indirect reporter-based detection, PCAs read the PPI directly through reconstitution of a fluorescent or luminescent protein [40]. For example, in bimolecular fluorescence complementation (BiFC), two halves of a split fluorescent protein are fused to different proteins that can interact to reconstitute the fluorescence protein for detection [309] (Figure 16.4c). Other methods, such as bimolecular luminescence complementation (BiLC), utilize split enzymes like luciferase to amplify the signal and boost sensitivity [310]. In addition to using these technologies to monitor PPIs, these technologies are widely used in drug discovery to screen chemical matter. For example, the split-nanoluciferase system HiBit is widely used to monitor protein dynamics to screen libraries of potential PPI modulators [311, 312]. Proximity ligation assays (PLA) are similar but use oligonucleotide-conjugated antibodies to detect proteins of interest, which if in close proximity, can facilitate rolling-circle amplification of the oligonucleotides that can then recruit fluorophores to create measurable signal (Figure 16.4d) [40]. The signal amplification in PLA allows greater sensitivity, but this method, of course, is dependent on antibodies against the targets of interest and specialized reagents which can limit their application. Nonetheless, PLA has been used in some high-content assays to screen for PPI modulators, including in difficult biological models like primary cells [313].

Another similar class of binary assays is energy transfer assays (Figure 16.4e), such as Förster resonance energy transfer (FRET) and bioluminescence resonance energy transfer (BRET) [40]. In FRET assays, proteins of interest are fused to two different fluorophores, a donor and an acceptor. When the proteins are in near proximity with each other, irradiation is applied to excite the donor fluorophore and through a non-radiative process, this energy is transferred to the acceptor fluorophore to induce fluorescent emission in the acceptor fluorophore's emission wavelength. Because the energy process is non-radiative, FRET reads out very proximal interactions and provides stronger evidence for direct interactions [314]. FRET also has a high temporal resolution, allowing probing of dynamic processes in real time. However, FRET generally has a lower sensitivity than many of the other PPI techniques as it has no signal amplification, thus lowering signal-to-noise. BRET is similar, but instead uses luciferase-induced chemiluminescence to provide the donor energy that is transferred to an acceptor fluorophore, which generally allows greater sensitivity and more straightforward workflows [40]. Variations of these assays, such as homogeneous time-resolved fluorescence (HTRF), are commonly used for drug target studies in high-throughput screening including for PPI modulators [315].

PPI modulation is an exciting and growing area for drug discovery research, presenting a promising opportunity to better access the elusive "undruggable" proteome. As reviewed in this chapter, there are many different approaches to study PPIs and their modulation, and from these approaches, we as a scientific community have generated a massive amount of data. Traditionally, this data has been analyzed manually by researchers leveraging their own expertise and bespoke computational tools; however, the rapidly developing field of artificial intelligence has already begun to revolutionize how interactomic data is being used for drug discovery [316]. AI algorithms like machine learning models can be trained on experimental PPI datasets to predict new interactions or infer the functions of uncharacterized proteins [317]. Deep learning models like those used by AlphaFold also allow us to incorporate structural prediction into our analysis of PPI data [318, 319], while others can facilitate protein pocket-ligand prediction and PPI site prediction [320]. Deep learning can also be used in conjunction with virtual screening to integrate molecular and interactome data for protein–compound interaction prediction [321, 322]. With these exciting advances in both experimental and computational capabilities to map and modulate PPI networks, the future for protein interaction-focused drug discovery is bright and presents a hopeful future for the next generation of life-changing therapies.

Abbreviations

PPI	protein–protein interaction
ATP	adenosine triphosphate
ER	endoplasmic reticulum
MS	mass spectrometry
AP-MS	affinity purification-mass spectrometry
IP	immunoprecipitation
HPLC	high-performance liquid chromatography
MM	multiple myeloma
APP	amyloid beta precursor protein
PSEN1	Presenilin-1
HTT	Huntingtin
PARK2	Parkin
ATXN1	Ataxin-1
PROTAC	proteolysis targeting chimera
CRBN	Cereblon
PIKES	protein interaction kinetics and estimation of stoichiometries
CRL	Cullin RING ubiquitin ligases
PROTAB	proteolysis targeting antibodies
TAP	tandem affinity purification
XL-MS	cross-linking mass spectrometry
SPINE	strep-protein interaction experiment
ReCLIP	reversible cross-link immuno-precipitation
CAR	chimeric antigen receptor
PL-MS	proximity labeling-mass spectrometry
HRP	horseradish peroxidase
APEX2	engineered ascorbate peroxidase 2
SPPLAT	selective proteomic proximity labeling assay using tyramide
EMARS	enyme-mediated activation of radical sources
DAB	3,3′-diaminobenzidine
EM	electron microscopy
BirA	*Escherichia coli* biotin ligase
AMP	adenosine monophosphate
CoA	acetyl-coenzyme A
BAP	biotinylation acceptor peptide
RaPID	RNA–protein interaction detection
SOG	singlet-oxygen generating
LITag	light-induced interactome tagging
LOV	light-oxygen-voltage
GPCR	G-protein-coupled receptor
MOA	mechanism of action
VHL	von Hippel–Lindau
AbTAC	antibody-based proteolysis targeting chimera
LYTAC	lysosome-targeting chimera
ASGPR	asialoglycoprotein receptor
AUTAC and AUTOTAC	autophagy targeting chimeras
KineTACs	cytokine receptor-targeting chimeras
DUBTACs	deubiquitinase-targeting chimera
SAINT	Significance Analysis of INTeractome
SILAC	stable isotope labeling by amino acids in cell culture
TMT	tandem mass tag
DDA	data-dependent acquisition
DIA	data-independent acquisition
nanoPOTS	nanodroplet processing in one pot for trace samples
SCoPE-MS	Single-Cell ProtEomics by Mass Spectrometry
ROC	receiver operating characteristic
LUMIER	luminescence-based mammalian interactome mapping
PCA	protein complementation assays

BiFC	bimolecular fluorescence complementation
BiLC	bimolecular luminescence complementation
PLA	proximity ligation assays
FRET	Förster resonance energy transfer
BRET	bioluminescence resonance energy transfer
HTRF	homogeneous time resolved fluorescence
VLPs	virus-like particles
MAPPIT	mammalian protein–protein interaction trap
STAT	signal transducer and activator of transcription

References

1 Hartley, H. (1951). Origin of the Word 'Protein'. *Nature* 168: 244–244.

2 Fischer, E. (1894). Einfluss der Configuration auf die Wirkung der Enzyme. *Ber. Dtsch. Chem. Ges.* 27: 2985–2993.

3 Russ, A.P. and Lampel, S. (2005). The druggable genome: an update. *Drug Discov. Today* 10: 1607–1610.

4 Verdine, G.L. and Walensky, L.D. (2007). The challenge of drugging Undruggable targets in Cancer: lessons learned from targeting BCL-2 family members. *Clin. Cancer Res.* 13: 7264–7270.

5 Milo, R. (2013). What is the total number of protein molecules per cell volume? A call to rethink some published values. *BioEssays* 35: 1050–1055.

6 Shrestha, R., Fajardo, J.E., and Fiser, A. (2021). Residue-based pharmacophore approaches to study protein–protein interactions. *Curr. Opin. Struct. Biol.* 67: 205–211.

7 Braakman, I. and Hebert, D.N. (2013). Protein folding in the endoplasmic reticulum. *Cold Spring Harb. Perspect. Biol.* 5: a013201.

8 Alberts, B., Johnson, A., Lewis, J. et al. (2014). *Molecular Biology of the Cell*, 597–640. https://doi.org/10.1201/9781315735368-11.

9 Aebersold, R. and Mann, M. (2003). Mass spectrometry-based proteomics. *Nature* 422: 198–207.

10 Gingras, A.-C., Gstaiger, M., Raught, B., and Aebersold, R. (2007). Analysis of protein complexes using mass spectrometry. *Nat. Rev. Mol. Cell Biol.* 8: 645–654.

11 Malovannaya, A., Lanz, R.B., Jung, S.Y. et al. (2011). Analysis of the human endogenous Coregulator Complexome. *Cell* 145: 787–799.

12 Dunham, W.H., Mullin, M., and Gingras, A. (2012). Affinity-purification coupled to mass spectrometry: basic principles and strategies. *Proteomics* 12: 1576–1590.

13 Ewing, R.M., Chu, P., Elisma, F. et al. (2007). Large-scale mapping of human protein–protein interactions by mass spectrometry. *Mol. Syst. Biol.* 3: 89–89.

14 Huttlin, E.L., Bruckner, R.J., Navarrete-Perea, J. et al. (2021). Dual proteome-scale networks reveal cell-specific remodeling of the human interactome. *Cell* 184: 3022–3040.e28.

15 Park, J.E., Miller, Z., Jun, Y. et al. (2018). Next-generation proteasome inhibitors for cancer therapy. *Transl. Res.* 198: 1–16.

16 Bousquet-Dubouch, M.-P., Baudelet, E., Guerin, F. et al. (2009). Affinity purification strategy to capture human endogenous proteasome complexes diversity and to identify proteasome-interacting proteins* S. *Mol. Cell. Proteomics* 8: 1150–1164.

17 Hamazaki, J., Iemura, S.I., Natsume, T. et al. (2006). A novel proteasome interacting protein recruits the deubiquitinating enzyme UCH37 to 26S proteasomes. *EMBO J.* 25: 4524–4536.

18 Qiu, X., Ouyang, S.Y., Li, C.J. et al. (2006). hRpn13/ADRM1/GP110 is a novel proteasome subunit that binds the deubiquitinating enzyme, UCH37. *EMBO J.* 25: 5742–5753.

19 Wang, X. and Huang, L. (2008). Identifying dynamic interactors of protein complexes by quantitative mass spectrometry*. *Mol. Cell. Proteomics* 7: 46–57.

20 Fabre, B., Lambour, T., Garrigues, L. et al. (2015). Deciphering preferential interactions within supramolecular protein complexes: the proteasome case. *Mol. Syst. Biol.* 11: 771.

21 Kelly, M.R., Kostyrko, K., Han, K. et al. (2020). Combined proteomic and genetic interaction mapping reveals new RAS effector pathways and susceptibilities. *Cancer Discov.* 10: 1950–1967.

22 Nolan, A., Raso, C., Kolch, W. et al. (2023). Proteomic mapping of the Interactome of KRAS mutants identifies new features of RAS signalling networks and the mechanism of action of Sotorasib. *Cancers* 15: 4141.

23 Stukalov, A., Girault, V., Grass, V. et al. (2021). Multilevel proteomics reveals host perturbations by SARS-CoV-2 and SARS-CoV. *Nature* 594: 246–252.

24 Liu, X., Huuskonen, S., Laitinen, T. et al. (2021). SARS-CoV-2–host proteome interactions for antiviral drug discovery. *Mol. Syst. Biol.* 17: e10396.

25 Gordon, D.E., Jang, G.M., Bouhaddou, M. et al. (2020). A SARS-CoV-2 protein interaction map reveals targets for drug repurposing. *Nature* 583: 459–468.

26 Bouhaddou, M., Reuschl, A.K., Polacco, B.J. et al. (2023). SARS-CoV-2 variants evolve convergent strategies to remodel the host response. *Cell* 186: 4597–4614.

27 Schynkel, T., Snippenberg, W.V., Verniers, K. et al. (2024). Interactome of the HIV-1 proteome and human host *RNA*. *EMBO Rep.* 25: 4078–4090. https://doi.org/10.1038/s44319-024-00222-6.

28 Ostacolo, K., López García de Lomana, A., Larat, C. et al. (2024). ATG7(2) interacts with metabolic proteins and regulates central energy metabolism. *Traffic* 25: e12933.

29 Hosp, F., Vossfeldt, H., Heinig, M. et al. (2015). Quantitative interaction proteomics of neurodegenerative disease proteins. *Cell Rep.* 11: 1134–1146.

30 Krönke, J., Udeshi, N.D., Narla, A. et al. (2014). Lenalidomide causes selective degradation of IKZF1 and IKZF3 in multiple myeloma cells. *Science* 343: 301–305.

31 Lu, G., Middleton, R.E., Sun, H. et al. (2014). The myeloma drug Lenalidomide promotes the Cereblon-dependent destruction of Ikaros proteins. *Science* 343: 305–309.

32 Horn-Ghetko, D., Hopf, L.V., Tripathi-Giesgen, I. et al. (2024). Noncanonical assembly, neddylation and chimeric cullin–RING/RBR ubiquitylation by the 1.8 MDa CUL9 E3 ligase complex. *Nat. Struct. Mol. Biol.* 31: 1083–1094.

33 Reichermeier, K.M., Straube, R., Reitsma, J.M. et al. (2020). PIKES analysis reveals response to degraders and key regulatory mechanisms of the CRL4 network. *Mol. Cell* 77: 1092–1106.

34 Marei, H., Tsai, W.T., Kee, Y.S. et al. (2022). Antibody targeting of E3 ubiquitin ligases for receptor degradation. *Nature* 610: 182–189.

35 Ong, S.-E., Schenone, M., Margolin, A.A. et al. (2009). Identifying the proteins to which small-molecule probes and drugs bind in cells. *Proc. Natl. Acad. Sci.* 106: 4617–4622.

36 Bantscheff, M., Eberhard, D., Abraham, Y. et al. (2007). Quantitative chemical proteomics reveals mechanisms of action of clinical ABL kinase inhibitors. *Nat. Biotechnol.* 25: 1035–1044.

37 Mili, S. and Steitz, J.A. (2004). Evidence for reassociation of RNA-binding proteins after cell lysis: implications for the interpretation of immunoprecipitation analyses. *RNA* 10: 1692–1694.

38 Chu, Q., Rathore, A., Diedrich, J.K. et al. (2017). Identification of microprotein-protein interactions via APEX tagging. *Biochemistry* 56: 3299–3306.

39 Lavallée-Adam, M., Rousseau, J., Domecq, C. et al. (2013). Discovery of cell compartment specific protein-protein interactions using affinity purification combined with tandem mass spectrometry. *J. Proteome Res.* 12: 272–281.

40 Titeca, K., Lemmens, I., Tavernier, J., and Eyckerman, S. (2019). Discovering cellular protein-protein interactions: technological strategies and opportunities. *Mass Spectrom. Rev.* 38: 79–111.

41 Mellacheruvu, D., Wright, Z., Couzens, A.L. et al. (2013). The CRAPome: a contaminant repository for affinity purification–mass spectrometry data. *Nat. Methods* 10: 730–736.

42 Bludau, I. and Aebersold, R. (2020). Proteomic and interactomic insights into the molecular basis of cell functional diversity. *Nat. Rev. Mol. Cell Biol.* 21: 327–340.

43 Rappsilber, J. (2011). The beginning of a beautiful friendship: cross-linking/mass spectrometry and modelling of proteins and multi-protein complexes. *J. Struct. Biol.* 173: 530–540.

44 O'Reilly, F.J. and Rappsilber, J. (2018). Cross-linking mass spectrometry: methods and applications in structural, molecular and systems biology. *Nat. Struct. Mol. Biol.* 25: 1000–1008.

45 Dutton, A., Adams, M., and Singer, S.J. (1966). Bifunctional imidoesters as cross-linking reagents. *Biochem. Biophys. Res. Commun.* 23: 730–739.

46 Kalkhof, S. and Sinz, A. (2008). Chances and pitfalls of chemical cross-linking with amine-reactive N-hydroxysuccinimide esters. *Anal. Bioanal. Chem.* 392: 305–312.

47 Nguyen-Huynh, N., Sharov, G., Potel, C. et al. (2015). Chemical cross-linking and mass spectrometry to determine the subunit interaction network in a recombinant human SAGA HAT subcomplex. *Protein Sci.* 24: 1232–1246.

48 Back, J.W., de Jong, L., Muijsers, A.O., and de Koster, C.G. (2003). Chemical cross-linking and mass spectrometry for protein structural Modeling. *J. Mol. Biol.* 331: 303–313.

49 Grabarek, Z. and Gergely, J. (1990). Zero-length crosslinking procedure with the use of active esters. *Anal. Biochem.* 185: 131–135.

50 Leitner, A., Joachimiak, L.A., Unverdorben, P. et al. (2014). Chemical cross-linking/mass spectrometry targeting acidic residues in proteins and protein complexes. *Proc. Natl. Acad. Sci.* 111: 9455–9460.

51 Ishmael, F.T., Alley, S.C., and Benkovic, S.J. (2001). Identification and mapping of protein-protein interactions between gp32 and gp59 by cross-linking*. *J. Biol. Chem.* 276: 25236–25242.

52 Smyth, D., Blumenfeld, O., and Konigsberg, W. (1964). Reactions of N-ethylmaleimide with peptides and amino acids. *Biochem. J.* 91: 589–595.

53 Brunner, J. (1993). New photolabeling and crosslinking methods. *Annu. Rev. Biochem.* 62: 483–514.

54 Dorman, G. and Prestwich, G.D. (1994). Benzophenone photophores in biochemistry. *Biochemistry* 33: 5661–5673.

55 Dubinsky, L., Krom, B.P., and Meijler, M.M. (2012). Diazirine based photoaffinity labeling. *Bioorg. Med. Chem.* 20: 554–570.

56 Gomes, A.F. and Gozzo, F.C. (2010). Chemical cross-linking with a diazirine photoactivatable cross-linker investigated by MALDI- and ESI-MS/MS. *J. Mass Spectrom.* 45: 892–899.

57 Pham, N.D., Parker, R.B., and Kohler, J.J. (2013). Photocrosslinking approaches to interactome mapping. *Curr. Opin. Chem. Biol.* 17: 90–101.

58 Beck-Sickinger, A.G. and Budisa, N. (2012). Genetically encoded Photocrosslinkers as molecular probes to study G-protein-coupled receptors (GPCRs). *Angew. Chem. Int. Ed.* 51: 310–312.

59 Grunbeck, A., Huber, T., Sachdev, P., and Sakmar, T.P. (2011). Mapping the ligand-binding site on a G protein-coupled receptor (GPCR) using genetically encoded Photocrosslinkers. *Biochemistry* 50: 3411–3413.

60 Suchanek, M., Radzikowska, A., and Thiele, C. (2005). Photo-leucine and photo-methionine allow identification of protein-protein interactions in living cells. *Nat. Methods* 2: 261–268.

61 Yanagisawa, T., Hino, N., Iraha, F. et al. (2012). Wide-range protein photo-crosslinking achieved by a genetically encoded N ε -(benzyloxycarbonyl)lysine derivative with a diazirinyl moiety. *Mol. BioSyst.* 8: 1131–1135.

62 Liu, J., Yang, B., and Wang, L. (2023). Residue selective crosslinking of proteins through photoactivatable or proximity-enabled reactivity. *Curr. Opin. Chem. Biol.* 74: 102285.

63 Bond, M.R., Zhang, H., Vu, P.D., and Kohler, J.J. (2009). Photocrosslinking of glycoconjugates using metabolically incorporated diazirine-containing sugars. *Nat. Protoc.* 4: 1044–1063.

64 Rivera-Santiago, R.F., Sriswasdi, S., Harper, S.L., and Speicher, D.W. (2015). Probing structures of large protein complexes using zero-length cross-linking. *Methods* 89: 99–111.

65 Kao, A., Chiu, C.L., Vellucci, D. et al. (2011). Development of a novel cross-linking strategy for fast and accurate identification of cross-linked peptides of protein complexes*. *Mol. Cell. Proteomics* 10: M110.002170.

66 Buncherd, H., Nessen, M.A., Nouse, N. et al. (2012). Selective enrichment and identification of cross-linked peptides to study 3-D structures of protein complexes by mass spectrometry. *J. Proteome* 75: 2205–2215.

67 Petrotchenko, E.V., Serpa, J.J., and Borchers, C.H. (2011). An isotopically coded CID-cleavable biotinylated cross-linker for structural proteomics*. *Mol. Cell. Proteomics* 10: S1–S8.

68 Fritzsche, R., Ihling, C.H., Götze, M., and Sinz, A. (2012). Optimizing the enrichment of cross-linked products for mass spectrometric protein analysis. *Rapid Commun. Mass Spectrom.* 26: 653–658.

69 Leitner, A., Reischl, R., Walzthoeni, T. et al. (2012). Expanding the chemical cross-linking toolbox by the use of multiple proteases and enrichment by size exclusion chromatography*. *Mol. Cell. Proteomics* 11: M111.014126.

70 Müller, D.R., Schindler, P., Towbin, H. et al. (2001). Isotope-tagged cross-linking reagents. A new tool in mass spectrometric protein interaction analysis. *Anal. Chem.* 73: 1927–1934.

71 Petrotchenko, E.V., Xiao, K., Cable, J. et al. (2009). BiPS, a Photocleavable, isotopically coded, fluorescent cross-linker for structural proteomics. *Mol. Cell. Proteomics* 8: 273–286.

72 Tang, X. and Bruce, J.E. (2010). A new cross-linking strategy: protein interaction reporter (PIR) technology for protein – protein interaction studies. *Mol. BioSyst.* 6: 939–947.

73 Yang, L., Tang, X., Weisbrod, C.R. et al. (2010). A Photocleavable and mass spectrometry identifiable cross-linker for protein interaction studies. *Anal. Chem.* 82: 3556–3566.

74 Sutherland, B.W., Toews, J., and Kast, J. (2008). Utility of formaldehyde cross-linking and mass spectrometry in the study of protein–protein interactions. *J. Mass Spectrom.* 43: 699–715.

75 Herzberg, C., Weidinger, L.A., Dörrbecker, B. et al. (2007). SPINE: a method for the rapid detection and analysis of protein–protein interactions in vivo. *Proteomics* 7: 4032–4035.

76 Terpe, K. (2003). Overview of tag protein fusions: from molecular and biochemical fundamentals to

commercial systems. *Appl. Microbiol. Biotechnol.* 60: 523–533.

77 Müller, V.S., Jungblut, P.R., Meyer, T.F., and Hunke, S. (2011). Membrane-SPINE: an improved method to identify protein–protein interaction partners of membrane proteins in vivo. *Proteomics* 11: 2124–2128.

78 Smith, A.L., Friedman, D.B., Yu, H. et al. (2011). ReCLIP (reversible cross-link Immuno-precipitation): an efficient method for interrogation of labile protein complexes. *PLoS One* 6: e16206.

79 Zhu, Y., Akkaya, K.C., Ruta, J. et al. (2024). Cross-link assisted spatial proteomics to map sub-organelle proteomes and membrane protein topologies. *Nat. Commun.* 15: 3290.

80 Liu, F. and Heck, A.J. (2015). Interrogating the architecture of protein assemblies and protein interaction networks by cross-linking mass spectrometry. *Curr. Opin. Struct. Biol.* 35: 100–108.

81 Mendes, M.L., Fischer, L., Chen, Z.A. et al. (2019). An integrated workflow for crosslinking mass spectrometry. *Mol. Syst. Biol.* 15: e8994.

82 Petrotchenko, E.V. and Borchers, C.H. (2010). Crosslinking combined with mass spectrometry for structural proteomics. *Mass Spectrom. Rev.* 29: 862–876.

83 Yu, C. and Huang, L. (2018). Cross-linking mass spectrometry: an emerging Technology for Interactomics and Structural Biology. *Anal. Chem.* 90: 144–165.

84 Mandal, K., Wicaksono, G., Yu, C. et al. (2023). Structural surfaceomics reveals an AML-specific conformation of integrin β2 as a CAR T cellular therapy target. *Nat. Cancer.* 4: 1592–1609.

85 Liu, F., Rijkers, D.T.S., Post, H., and Heck, A.J.R. (2015). Proteome-wide profiling of protein assemblies by cross-linking mass spectrometry. *Nat. Methods* 12: 1179–1184.

86 Tan, D., Li, Q., Zhang, M.J. et al. (2016). Trifunctional cross-linker for mapping protein-protein interaction networks and comparing protein conformational states. *elife* 5: e12509.

87 Yang, B., Wu, Y.J., Zhu, M. et al. (2012). Identification of cross-linked peptides from complex samples. *Nat. Methods* 9: 904–906.

88 Liu, F., Lössl, P., Scheltema, R. et al. (2017). Optimized fragmentation schemes and data analysis strategies for proteome-wide cross-link identification. *Nat. Commun.* 8: 15473.

89 Kaake, R.M., Wang, X., Burke, A. et al. (2014). A new in vivo cross-linking mass spectrometry platform to define protein-protein interactions in living cells*. *Mol. Cell. Proteomics* 13: 3533–3543.

90 Chavez, J.D., Weisbrod, C.R., Zheng, C. et al. (2013). Protein interactions, Post-translational modifications and topologies in human cells*. *Mol. Cell. Proteomics* 12: 1451–1467.

91 Wu, X., Chavez, J.D., Schweppe, D.K. et al. (2016). In vivo protein interaction network analysis reveals porin-localized antibiotic inactivation in Acinetobacter baumannii strain AB5075. *Nat. Commun.* 7: 13414.

92 Schweppe, D.K., Chavez, J.D., Lee, C.F. et al. (2017). Mitochondrial protein interactome elucidated by chemical cross-linking mass spectrometry. *Proc. Natl. Acad. Sci.* 114: 1732–1737.

93 Liu, F., Lössl, P., Rabbitts, B.M. et al. (2018). The interactome of intact mitochondria by cross-linking mass spectrometry provides evidence for coexisting respiratory supercomplexes*. *Mol. Cell. Proteomics* 17: 216–232.

94 Chavez, J.D., Schweppe, D.K., Eng, J.K., and Bruce, J.E. (2016). In vivo conformational dynamics of Hsp90 and its interactors. *Cell Chem. Biol.* 23: 716–726.

95 Schweppe, D.K., Harding, C., Chavez, J.D. et al. (2015). Host-microbe protein interactions during bacterial infection. *Chem. Biol.* 22: 1521–1530.

96 Bogdanow, B., Gruska, I., Mühlberg, L. et al. (2023). Spatially resolved protein map of intact human cytomegalovirus virions. *Nat. Microbiol.* 8: 1732–1747.

97 Gonzalez-Lozano, M.A., Koopmans, F., Sullivan, P.F. et al. (2020). Stitching the synapse: cross-linking mass spectrometry into resolving synaptic protein interactions. *Sci. Adv.* 6: eaax5783.

98 Wittig, S., Ganzella, M., Barth, M. et al. (2021). Cross-linking mass spectrometry uncovers protein interactions and functional assemblies in synaptic vesicle membranes. *Nat. Commun.* 12: 858.

99 Lee, K. and O'Reilly, F.J. (2023). Cross-linking mass spectrometry for mapping protein complex topologies in situ. *Essays Biochem.* 67: 215–228.

100 Kim, D.I. and Roux, K.J. (2016). Filling the void: proximity-based Labeling of proteins in living cells. *Trends Cell Biol.* 26: 804–817.

101 Qin, W., Cho, K.F., Cavanagh, P.E., and Ting, A.Y. (2021). Deciphering molecular interactions by proximity labeling. *Nat. Methods* 18: 133–143.

102 Loh, K.H., Stawski, P.S., Draycott, A.S. et al. (2016). Proteomic analysis of unbounded cellular compartments: synaptic clefts. *Cell* 166: 1295–1307.

103 Rees, J.S., Li, X., Perrett, S. et al. (2017). Selective proteomic proximity Labeling assay using Tyramide (SPPLAT): a quantitative method for the proteomic analysis of localized membrane-bound protein clusters. *Curr. Protoc. Protein Sci.* 88: 19.27.1–19.27.18.

104 Jiang, S., Kotani, N., Ohnishi, T. et al. (2012). A proteomics approach to the cell-surface interactome using the enzyme-mediated activation of radical sources reaction. *Proteomics* 12: 54–62.

105 Rhee, H.-W., Zou, P., Udeshi, N.D. et al. (2013). Proteomic mapping of mitochondria in living cells via spatially restricted enzymatic tagging. *Science* 339: 1328–1331.

106 Lam, S.S., Martell, J.D., Kamer, K.J. et al. (2015). Directed evolution of APEX2 for electron microscopy and proximity labeling. *Nat. Methods* 12: 51–54.

107 Raven, E.L. (2003). Understanding functional diversity and substrate specificity in haem peroxidases: what can we learn from ascorbate peroxidase? *Nat. Prod. Rep.* 20: 367–381.

108 Paek, J., Kalocsay, M., Staus, D.P. et al. (2017). Multidimensional tracking of GPCR Signaling via peroxidase-catalyzed proximity Labeling. *Cell* 169: 338–349.

109 Lobingier, B.T., Hüttenhain, R., Eichel, K. et al. (2017). An approach to spatiotemporally resolve protein interaction networks in living cells. *Cell* 169: 350–360.

110 Kaewsapsak, P., Shechner, D.M., Mallard, W. et al. (2017). Live-cell mapping of organelle-associated RNAs via proximity biotinylation combined with protein-RNA crosslinking. *elife* 6: e29224.

111 Fazal, F.M., Han, S., Parker, K.R. et al. (2019). Atlas of subcellular RNA localization revealed by APEX-Seq. *Cell* 178: 473–490.

112 Honke, K. and Kotani, N. (2012). Identification of cell-surface molecular interactions under living conditions by using the enzyme-mediated activation of radical sources (EMARS) method. *Sensors* 12: 16037–16045.

113 Dwyer, D.J., Belenky, P.A., Yang, J.H. et al. (2014). Antibiotics induce redox-related physiological alterations as part of their lethality. *Proc. Natl. Acad. Sci.* 111: E2100–E2109.

114 Martell, J.D., Deerinck, T.J., Lam, S.S. et al. (2017). Electron microscopy using the genetically encoded APEX2 tag in cultured mammalian cells. *Nat. Protoc.* 12: 1792–1816.

115 Hung, V., Zou, P., Rhee, H.W. et al. (2014). Proteomic mapping of the human mitochondrial intermembrane space in live cells via Ratiometric APEX tagging. *Mol. Cell* 55: 332–341.

116 Han, S., Udeshi, N.D., Deerinck, T.J. et al. (2017). Proximity Biotinylation as a method for mapping proteins associated with mtDNA in living cells. *Cell Chem. Biol.* 24: 404–414.

117 Hung, V., Lam, S.S., Udeshi, N.D. et al. (2017). Proteomic mapping of cytosol-facing outer mitochondrial and ER membranes in living human cells by proximity biotinylation. *elife* 6: e24463.

118 Reinke, A.W., Balla, K.M., Bennett, E.J., and Troemel, E.R. (2017). Identification of microsporidia host-exposed proteins reveals a repertoire of rapidly evolving proteins. *Nat. Commun.* 8: 14023.

119 Reinke, A.W., Mak, R., Troemel, E.R., and Bennett, E.J. (2017). In vivo mapping of tissue- and subcellular-specific proteomes in Caenorhabditis elegans. *Sci. Adv.* 3: e1602426.

120 Chen, C.-L., Hu, Y., Udeshi, N.D. et al. (2015). Proteomic mapping in live Drosophila tissues using an engineered ascorbate peroxidase. *Proc. Natl. Acad. Sci.* 112: 12093–12098.

121 Li, J., Han, S., Li, H. et al. (2020). Cell-surface proteomic profiling in the Fly brain uncovers wiring regulators. *Cell* 180: 373–386.e15.

122 Choi-Rhee, E., Schulman, H., and Cronan, J.E. (2004). Promiscuous protein biotinylation by Escherichia coli biotin protein ligase. *Protein Sci.* 13: 3043–3050.

123 Roux, K.J., Kim, D.I., Raida, M., and Burke, B. (2012). A promiscuous biotin ligase fusion protein identifies proximal and interacting proteins in mammalian cells. *J. Cell Biol.* 196: 801–810.

124 Tong, L. (2013). Structure and function of biotin-dependent carboxylases. *Cell. Mol. Life Sci.* 70: 863–891.

125 Chapman-Smith, A. and Cronan, J.E. (1999). The enzymatic biotinylation of proteins: a post-translational modification of exceptional specificity. *Trends Biochem. Sci.* 24: 359–363.

126 Cronan, J.E. and Waldrop, G.L. (2002). Multi-subunit acetyl-CoA carboxylases. *Prog. Lipid Res.* 41: 407–435.

127 Kulyyassov, A., Shoaib, M., Pichugin, A. et al. (2011). PUB-MS: a mass spectrometry-based method to monitor protein–protein proximity in vivo. *J. Proteome Res.* 10: 4416–4427.

128 Kwon, K. and Beckett, D. (2000). Function of a conserved sequence motif in biotin holoenzyme synthetases. *Protein Sci.* 9: 1530–1539.

129 Xu, Y. and Beckett, D. (1994). Kinetics of biotinyl-5′-adenylate synthesis catalyzed by the Escherichia coli repressor of biotin biosynthesis and the stability of the enzyme-product complex. *Biochemistry* 33: 7354–7360.

130 Berg, P. (1958). Studies on the enzymatic utilization of amino acyladenylates; the formation of adenosine triphosphate. *J. Biol. Chem.* 233: 601–607.

131 Kim, D.I., KC, B., Zhu, W. et al. (2014). Probing nuclear pore complex architecture with proximity-dependent biotinylation. *Proc. Natl. Acad. Sci.* 111: E2453–E2461.

132 Gupta, G.D., Coyaud, É., Gonçalves, J. et al. (2015). A dynamic protein interaction landscape of the human centrosome-cilium Interface. *Cell* 163: 1484–1499.

133 Go, C.D., Knight, J.D., Rajasekharan, A. et al. (2021). A proximity-dependent biotinylation map of a human cell. *Nature* 595: 120–124.

134 Bartolome, A., Heiby, J.C., Fraia, D.D. et al. (2024). ProteasomeID: quantitative mapping of proteasome interactomes and substrates for in vitro and in vivo studies. *elife* 13: https://doi.org/10.7554/elife.93256.2.

135 Lin, Q., Zhou, Z., Luo, W. et al. (2017). Screening of proximal and interacting proteins in Rice protoplasts by proximity-dependent Biotinylation. *Front. Plant Sci.* 8: 749.

136 Khan, M., Youn, J.-Y., Gingras, A.-C. et al. (2018). In planta proximity dependent biotin identification (BioID). *Sci. Rep.* 8: 9212.

137 Conlan, B., Stoll, T., Gorman, J.J. et al. (2018). Development of a rapid in planta BioID system as a probe for plasma membrane-associated immunity proteins. *Front. Plant Sci.* 9: 1882.

138 Morriswood, B., Havlicek, K., Demmel, L. et al. (2013). Novel Bilobe components in Trypanosoma brucei identified using proximity-dependent Biotinylation. *Eukaryot. Cell* 12: 356–367.

139 Chen, A.L., Moon, A.S., Bell, H.N. et al. (2016). Novel insights into the composition and function of the toxoplasma IMC sutures: IMC sutures composition and function. *Cell. Microbiol.* 19: e12678.

140 Nadipuram, S.M., Kim, E.W., Vashisht, A.A. et al. (2016). In vivo biotinylation of the Toxoplasma parasitophorous vacuole reveals novel dense granule proteins important for parasite growth and pathogenesis. *MBio* 7, https://doi.org/10.1128/mbio.00808-16.

141 Chen, A.L., Kim, E.W., Toh, J.Y. et al. (2015). Novel components of the toxoplasma inner membrane complex revealed by BioID. *MBio* 6: e02357-14.

142 Long, S., Brown, K.M., Drewry, L.L. et al. (2017). Calmodulin-like proteins localized to the conoid regulate motility and cell invasion by toxoplasma gondii. *PLoS Pathog.* 13: e1006379.

143 Zhou, Q., Hu, H., and Li, Z. (2016). An EF-hand-containing protein in Trypanosoma brucei regulates cytokinesis initiation by maintaining the stability of the cytokinesis initiation factor CIF1*. *J. Biol. Chem.* 291: 14395–14409.

144 Dang, H.Q., Zhou, Q., Rowlett, V.W. et al. (2017). Proximity interactions among basal body components in Trypanosoma brucei identify novel regulators of basal body biogenesis and inheritance. *MBio* 8: e02120-16.

145 Kehrer, J., Frischknecht, F., and Mair, G.R. (2016). Proteomic analysis of the plasmodium berghei gametocyte Egressome and vesicular bioID of Osmiophilic body proteins identifies Merozoite TRAP-like protein (MTRAP) as an essential factor for parasite transmission*. *Mol. Cell. Proteomics* 15: 2852–2862.

146 Gaji, R.Y., Johnson, D.E., Treeck, M. et al. (2015). Phosphorylation of a myosin motor by TgCDPK3 facilitates rapid initiation of motility during toxoplasma gondii egress. *PLoS Pathog.* 11: e1005268.

147 Batsios, P., Ren, X., Baumann, O. et al. (2016). Src1 is a protein of the inner nuclear membrane interacting with the Dictyostelium Lamin NE81. *Cells* 5: 13.

148 Meyer, I., Peter, T., Batsios, P. et al. (2017). CP39, CP75 and CP91 are major structural components of the Dictyostelium centrosome's core structure. *Eur. J. Cell Biol.* 96: 119–130.

149 Uezu, A., Kanak, D.J., Bradshaw, T.W. et al. (2016). Identification of an elaborate complex mediating postsynaptic inhibition. *Science* 353: 1123–1129.

150 Opitz, N., Schmitt, K., Hofer-Pretz, V. et al. (2017). Capturing the Asc1p/receptor for activated C kinase 1 (RACK1) microenvironment at the head region of the 40S ribosome with quantitative BioID in yeast*. *Mol. Cell. Proteomics* 16: 2199–2218.

151 Dingar, D., Kalkat, M., Chan, P.K. et al. (2015). BioID identifies novel c-MYC interacting partners in cultured cells and xenograft tumors. *J. Proteome* 118: 95–111.

152 Branon, T.C., Bosch, J.A., Sanchez, A.D. et al. (2018). Efficient proximity labeling in living cells and organisms with TurboID. *Nat. Biotechnol.* 36: 880–887.

153 Kim, D.I., Jensen, S.C., Noble, K.A. et al. (2016). An improved smaller biotin ligase for BioID proximity labeling. *Mol. Biol. Cell* 27: 1188–1196.

154 KC, B., May, D.G., Benson, B.V. et al. (2017). VRK2A is an a-type Lamin-dependent nuclear envelope kinase that phosphorylates BAF. *Mol. Biol. Cell* 28: 2241–2250.

155 Redwine, W.B., ME, D.S., Hollyer, I. et al. (2017). The human cytoplasmic dynein interactome reveals novel activators of motility. *elife* 6: e28257.

156 Jung, E.-M., Moffat, J.J., Liu, J. et al. (2017). Arid1b haploinsufficiency disrupts cortical interneuron development and mouse behavior. *Nat. Neurosci.* 20: 1694–1707.

157 Ramanathan, M., Majzoub, K., Rao, D.S. et al. (2018). RNA–protein interaction detection in living cells. *Nat. Methods* 15: 207–212.

158 Lu, M. and Wei, W. (2019). Proximity labeling to detect RNA–protein interactions in live cells. *FEBS Open Bio* 9: 1860–1868.

159 Feng, C., Roitinger, E., Hudecz, O. et al. (2023). TurboID-based proteomic profiling of meiotic chromosome axes in Arabidopsis thaliana. *Nat. Plants* 9: 616–630.

160 Mair, A., Xu, S.-L., Branon, T.C. et al. (2019). Proximity labeling of protein complexes and cell-type-specific organellar proteomes in Arabidopsis enabled by TurboID. *elife* 8: e47864.

161 Zhang, Y., Song, G., Lal, N.K. et al. (2019). TurboID-based proximity labeling reveals that UBR7 is a regulator of N NLR immune receptor-mediated immunity. *Nat. Commun.* 10: 3252.

162 Wallner, E.-S., Mair, A., Handler, D. et al. (2024). Spatially resolved proteomics of the Arabidopsis stomatal lineage identifies polarity complexes for cell divisions and stomatal pores. *Dev. Cell* 59: 1096–1109.

163 Cho, H., Seo, D., Kim, M. et al. (2024). SERKs serve as co-receptors for SYR1 to trigger systemin-mediated defense responses in tomato. *J. Integr. Plant Biol.* https://doi.org/10.1111/jipb.13747.

164 Sun, F., Hamada, N., Montes, C. et al. (2024). TurboID-based proteomic profiling reveals proxitome of ASK1 and CUL1 of the SCF ubiquitin ligase in plants. *New Phytol.* https://doi.org/10.1111/nph.20014.

165 Wen, Z., Hu, R., Pi, Q. et al. (2024). DEAD-box RNA helicase RH20 positively regulates RNAi-based antiviral immunity in plants by associating with SGS3/RDR6 bodies. *Plant Biotechnol. J.* https://doi.org/10.1111/pbi.14448.

166 Sanchez, A.D., Branon, T.C., Cote, L.E. et al. (2021). Proximity labeling reveals non-centrosomal microtubule-organizing center components required for microtubule growth and localization. *Curr. Biol.* 31: 3586–3600.

167 Artan, M., Barratt, S., Flynn, S.M. et al. (2021). Interactome analysis of Caenorhabditis elegans synapses by TurboID-based proximity labeling. *J. Biol. Chem.* 297: 101094.

168 Profes, M.S., Tiroumalechetty, A., Patel, N. et al. (2024). Characterization of the intracellular neurexin interactome by in vivo proximity ligation suggests its involvement in presynaptic actin assembly. *PLoS Biol.* 22: e3002466.

169 Kutzner, C.E., Bauer, K.C., Lackmann, J.W. et al. (2024). Optogenetic induction of mechanical muscle stress identifies myosin regulatory ubiquitin ligase NHL-1 in C. Elegans. *Nat. Commun.* 15: 6879.

170 Uçkun, E., Wolfstetter, G., Anthonydhason, V. et al. (2021). In vivo profiling of the Alk Proximitome in the developing Drosophila brain. *J. Mol. Biol.* 433: 167282.

171 Droujinine, I.A., Meyer, A.S., Wang, D. et al. (2021). Proteomics of protein trafficking by in vivo tissue-specific labeling. *Nat. Commun.* 12: 2382.

172 Rosenthal, S.M., Misra, T., Abdouni, H. et al. (2021). A toolbox for efficient proximity-dependent Biotinylation in zebrafish embryos. *Mol. Cell. Proteomics* 20: 100128.

173 Xiong, Z., Lo, H.P., KA, M.M. et al. (2021). In vivo proteomic mapping through GFP-directed proximity-dependent biotin labelling in zebrafish. *elife* 10: e64631.

174 Kim, K., Park, I., Kim, J. et al. (2021). Dynamic tracking and identification of tissue-specific secretory proteins in the circulation of live mice. *Nat. Commun.* 12: 5204.

175 Takano, T., Wallace, J.T., Baldwin, K.T. et al. (2020). Chemico-genetic discovery of astrocytic control of inhibition in vivo. *Nature* 588: 296–302.

176 Liu, X., Yam, P.T., Schlienger, S. et al. (2024). Numb positively regulates hedgehog signaling at the ciliary pocket. *Nat. Commun.* 15: 3365.

177 May, D.G., Scott, K.L., Campos, A.R., and Roux, K.J. (2020). Comparative application of BioID and TurboID for protein-proximity Biotinylation. *Cells* 9: 1070.

178 Kido, K., Yamanaka, S., Nakano, S. et al. (2020). AirID, a novel proximity biotinylation enzyme, for analysis of protein–protein interactions. *elife* 9: e54983.

179 Schaack, G.A., Sullivan, O.M., and Mehle, A. (2023). Identifying protein-protein interactions by proximity Biotinylation with AirID and splitAirID. *Curr. Protoc.* 3: e702.

180 Yamanaka, S., Horiuchi, Y., Matsuoka, S. et al. (2022). A proximity biotinylation-based approach to identify protein-E3 ligase interactions induced by PROTACs and molecular glues. *Nat. Commun.* 13: 183.

181 Kubitz, L., Bitsch, S., Zhao, X. et al. (2022). Engineering of ultraID, a compact and hyperactive enzyme for proximity-dependent biotinylation in living cells. *Commun. Biol.* 5: 657.

182 To, T.-L., Medzihradszky, K.F., Burlingame, A.L. et al. (2016). Photoactivatable protein labeling by singlet oxygen mediated reactions. *Bioorg. Med. Chem. Lett.* 26: 3359–3363.

183 Zheng, F., Yu, C., Zhou, X., and Zou, P. (2023). Genetically encoded photocatalytic protein labeling enables spatially-resolved profiling of intracellular proteome. *Nat. Commun.* 14: 2978.

184 Hananya, N., Ye, X., Koren, S., and Muir, T.W. (2023). A genetically encoded photoproximity labeling approach for mapping protein territories. *Proc. Natl. Acad. Sci.* 120: e2219339120.

185 Lee, S.-Y., Cheah, J.S., Zhao, B. et al. (2023). Engineered allostery in light-regulated LOV-Turbo enables precise spatiotemporal control of proximity labeling in living cells. *Nat. Methods* 20: 908–917.

186 Martell, J.D., Yamagata, M., Deerinck, T.J. et al. (2016). A split horseradish peroxidase for the detection of intercellular protein–protein interactions and sensitive visualization of synapses. *Nat. Biotechnol.* 34: 774–780.

187 Han, Y., Branon, T.C., Martell, J.D. et al. (2019). Directed evolution of Split APEX2 peroxidase. *ACS Chem. Biol.* 14: 619–635.

188 Schopp, I.M., Amaya Ramirez, C.C., Debeljak, J. et al. (2017). Split-BioID a conditional proteomics approach to monitor the composition of spatiotemporally defined protein complexes. *Nat. Commun.* 8: 15690.

189 Munter, S.D., Görnemann, J., Derua, R. et al. (2017). Split-BioID: a proximity biotinylation assay for dimerization-dependent protein interactions. *FEBS Lett.* 591: 415–424.

190 Cho, K.F., Branon, T.C., Rajeev, S. et al. (2020). Split-TurboID enables contact-dependent proximity labeling in cells. *Proc. Natl. Acad. Sci.* 117: 12143–12154.

191 Boassa, D., Lemieux, S.P., Lev-Ram, V. et al. (2019). Split-miniSOG for spatially detecting intracellular protein-protein interactions by correlated light and Electron microscopy. *Cell Chem. Biol.* 26: 1407–1416.e5.

192 Kwak, C., Shin, S., Park, J.S. et al. (2020). Contact-ID, a tool for profiling organelle contact sites, reveals regulatory proteins of mitochondrial-associated membrane formation. *Proc. Natl. Acad. Sci.* 117: 12109–12120.

193 Geri, J.B., Oakley, J.V., Reyes-Robles, T. et al. (2020). Microenvironment mapping via Dexter energy transfer on immune cells. *Science* 367: 1091–1097.

194 Buksh, B.F., Knutson, S.D., Oakley, J.V. et al. (2022). μMap-red: proximity Labeling by red light Photocatalysis. *J. Am. Chem. Soc.* 144: 6154–6162.

195 Tay, N.E.S., Ryu, K.A., Weber, J.L. et al. (2023). Targeted activation in localized protein environments via deep red photoredox catalysis. *Nat. Chem.* 15: 101–109.

196 Bechtel, T.J., Bertoch, J.M., Olow, A.K. et al. (2022). Proteomic mapping of intercellular synaptic environments via flavin-dependent photoredox catalysis. *Org. Biomol. Chem.* 21: 98–106.

197 Müller, M., Gräbnitz, F., Barandun, N. et al. (2021). Light-mediated discovery of surfaceome nanoscale organization and intercellular receptor interaction networks. *Nat. Commun.* 12: 7036.

198 Liu, Z., Guo, F., Zhu, Y. et al. (2024). Bioorthogonal photocatalytic proximity labeling in primary living samples. *Nat. Commun.* 15: 2712.

199 Lin, Z., Schaefer, K., Lui, I. et al. (2024). Multiscale photocatalytic proximity labeling reveals cell surface neighbors on and between cells. *Science* 385: eadl5763.

200 Takato, M., Sakamoto, S., Nonaka, H. et al. (2024). Photoproximity labeling of endogenous receptors in the live mouse brain in minutes. *Nat. Chem. Biol.* 1–11. https://doi.org/10.1038/s41589-024-01692-4.

201 Oslund, R.C., Reyes-Robles, T., White, C.H. et al. (2022). Detection of cell–cell interactions via photocatalytic cell tagging. *Nat. Chem. Biol.* 18: 850–858.

202 Tamura, T., Takato, M., Shiono, K., and Hamachi, I. (2019). Development of a photoactivatable proximity Labeling method for the identification of nuclear proteins. *Chem. Lett.* 49: 145–148.

203 Huth, S.W. et al. (2023). μMap Photoproximity Labeling enables small molecule binding site mapping. *J. Am. Chem. Soc.* 145: 16289–16296.

204 Trowbridge, A.D., Seath, C.P., Rodriguez-Rivera, F.P. et al. (2022). Small molecule photocatalysis enables drug target identification via energy transfer. *Proc. Natl. Acad. Sci.* 119: e2208077119.

205 Bartholow, T.G., Burroughs, P.W.W., Elledge, S.K. et al. (2023). Photoproximity Labeling from single catalyst sites allows calibration and increased resolution for Carbene Labeling of protein partners in vitro and on cells. *ACS Cent. Sci.* 10: 199–208. https://app.readcube.com/library/?style=Nature+%7B%22language%22:%22en-US%22%7D(2023) https://doi.org/10.1021/acscentsci.3c01473.

206 Artan, M., Hartl, M., Chen, W., and de Bono, M. (2022). Depletion of endogenously biotinylated carboxylases enhances the sensitivity of TurboID-mediated proximity labeling in Caenorhabditis elegans. *J. Biol. Chem.* 298: 102343.

207 Yue, S., Xu, P., Cao, Z., and Zhuang, M. (2022). PUP-IT2 as an alternative strategy for PUP-IT proximity labeling. *Front. Mol. Biosci.* 9: 1007720.

208 Hung, V., Udeshi, N.D., Lam, S.S. et al. (2016). Spatially resolved proteomic mapping in living cells with the engineered peroxidase APEX2. *Nat. Protoc.* 11: 456–475.

209 Addis, P., Bali, U., Baron, F. et al. (2024). Key aspects of modern GPCR drug discovery. *SLAS Discov.* 29: 1–22.

210 Chandan, N.R., Abraham, S., SenGupta, S. et al. (2022). A network of Gαi signaling partners is revealed by proximity labeling proteomics analysis and includes PDZ-RhoGEF. *Sci. Signal.* 15: eabi9869.

211 Zhuo, Y., Robleto, V.L., and Marchese, A. (2023). Proximity Labeling to identify β-Arrestin1 binding partners downstream of ligand-activated G protein-coupled receptors. *Int. J. Mol. Sci.* 24: 3285.

212 Polacco, B.J., Lobingier, B.T., Blythe, E.E. et al. (2024). Profiling the proximal proteome of the activated μ-opioid receptor. *Nat. Chem. Biol.* 1–11. https://doi.org/10.1038/s41589-024-01588-3.

213 Zhong, X., Li, Q., Polacco, B.J. et al. (2024). A proximity proteomics pipeline with improved reproducibility and throughput. *Mol. Syst. Biol.* 20: 952–971.

214 Barroso-Gomila, O., Merino-Cacho, L., Muratore, V. et al. (2023). BioE3 identifies specific substrates of ubiquitin E3 ligases. *Nat. Commun.* 14: 7656.

215 Huang, H.-T., Lumpkin, R.J., Tsai, R.W. et al. (2024). Ubiquitin-specific proximity labeling for the identification of E3 ligase substrates. *Nat. Chem. Biol.* 1–10. https://doi.org/10.1038/s41589-024-01590-9.

216 Coyaud, E., Mis, M., Laurent, E.M. et al. (2015). BioID-based identification of Skp Cullin F-box (SCF)β-TrCP1/2 E3 ligase substrates*[S]. *Mol. Cell. Proteomics* 14: 1781–1795.

217 Chen, Y.F., Rahman, A., Sax, J.L. et al. (2024). C646 degrades Exportin-1 to modulate p300 chromatin occupancy and function. *Cell Chem. Biol.* 31: 1363–1372.e8.

218 Mukhopadhyay, U., Levantovsky, S., Carusone, T.M. et al. (2024). A ubiquitin-specific, proximity-based labeling approach for the identification of ubiquitin ligase substrates. *Sci. Adv.* 10: eadp3000.

219 Bansod, S., Dodhiawala, P.B., Geng, Y. et al. (2024). The TRIM4 E3 ubiquitin ligase degrades TPL2 and is modulated by oncogenic KRAS. *Cell Rep.* 43: 114667.

220 Zhai, Y., Huang, X., Zhang, K. et al. (2022). Spatiotemporal-resolved protein networks profiling with photoactivation dependent proximity labeling. *Nat. Commun.* 13: 4906.

221 Chung, C.Y., Khurana, V., Yi, S. et al. (2017). In situ peroxidase Labeling and mass-spectrometry connects alpha-Synuclein directly to endocytic trafficking and mRNA metabolism in neurons. *Cell Syst.* 4: 242–250.

222 Chou, C.-C., Zhang, Y.I., Umoh, M.E. et al. (2018). TDP-43 pathology disrupts nuclear pore complexes and nucleocytoplasmic transport in ALS/FTD. *Nat. Neurosci.* 21: 228–239.

223 Markmiller, S., Soltanieh, S., Server, K.L. et al. (2018). Context-dependent and disease-specific diversity in protein interactions within stress granules. *Cell* 172: 590–604.

224 Couzens, A.L., Knight, J.D., Kean, M.J. et al. (2013). Protein interaction network of the mammalian hippo pathway reveals mechanisms of kinase-phosphatase interactions. *Sci. Signal.* 6: rs15.

225 Couzens, A.L., Xiong, S., Knight, J.D. et al. (2017). MOB1 mediated Phospho-recognition in the Core mammalian hippo pathway*. *Mol. Cell. Proteomics* 16: 1098–1110.

226 Lempiäinen, J.K., Niskanen, E.A., Vuoti, K.M. et al. (2017). Agonist-specific protein Interactomes of glucocorticoid and androgen receptor as revealed by proximity mapping*. *Mol. Cell. Proteomics* 16: 1462–1474.

227 Lu, J., Jiang, H., Li, D. et al. (2021). Proximity Labeling, quantitative proteomics, and biochemical studies revealed the molecular mechanism for the inhibitory effect of Indisulam on the proliferation of gastric Cancer cells. *J. Proteome Res.* 20: 4462–4474.

228 Phelan, J.D., Young, R.M., Webster, D.E. et al. (2018). A multiprotein supercomplex controlling oncogenic signalling in lymphoma. *Nature* 560: 387–391.

229 Uçkun, E., Siaw, J.T., Guan, J. et al. (2021). BioID-screening identifies PEAK1 and SHP2 as components of the ALK Proximitome in neuroblastoma cells. *J. Mol. Biol.* 433: 167158.

230 Hsu, E., Hutchison, K., Liu, Y. et al. (2024). Reduction of ZFX levels decreases histone H4 acetylation and increases Pol2 pausing at target promoters. *Nucleic Acids Res.* 52: 6850–6865.

231 Ahmed, N., Preisinger, C., Wilhelm, T., and Huber, M. (2024). TurboID-based IRE1 Interactome reveals participants of the endoplasmic reticulum-associated protein degradation machinery in the human mast cell Leukemia cell line HMC-1.2. *Cells* 13: 747.

232 Yu, C., Shen, Q., Holmes, A.B. et al. (2024). MEF2B C-terminal mutations enhance transcriptional activity and stability to drive B cell lymphomagenesis. *Nat. Commun.* 15: 7195.

233 Seath, C.P., Burton, A.J., Sun, X. et al. (2023). Tracking chromatin state changes using nanoscale photo-proximity labelling. *Nature* 616: 574–580.

234 Myers, S.A., Wright, J., Peckner, R. et al. (2018). Discovery of proteins associated with a predefined genomic locus via dCas9–APEX-mediated proximity labeling. *Nat. Methods* 15: 437–439.

235 Villaseñor, R., Pfaendler, R., Ambrosi, C. et al. (2020). ChromID identifies the protein interactome at chromatin marks. *Nat. Biotechnol.* 38: 728–736.

236 Zhang, Y., Shang, L., Zhang, J. et al. (2022). An antibody-based proximity labeling map reveals mechanisms of SARS-CoV-2 inhibition of antiviral immunity. *Cell Chem. Biol.* 29: 5–18.

237 Cygan, A.M., Jean Beltran, P.M., Mendoza, A.G. et al. (2021). Proximity-Labeling reveals novel host and parasite proteins at the toxoplasma Parasitophorous vacuole membrane. *MBio* 12: e00260-21.

238 Kumar, A., Salemi, M., Bhullar, R. et al. (2021). Proximity biotin Labeling reveals Kaposi's sarcoma-associated herpesvirus interferon regulatory factor networks. *J. Virol.* 95, https://doi.org/10.1128/jvi.02049-20.

239 Quan, J., Fan, Q., Simons, L.M. et al. (2024). Leveraging biotin-based proximity labeling to identify cellular factors governing early alphaherpesvirus infection. *MBio* 15: e0144524. https://doi.org/10.1128/mbio.01445-24.

240 Fu, J., Li, S., Guan, H. et al. (2024). Legionella maintains host cell ubiquitin homeostasis by effectors with unique catalytic mechanisms. *Nat. Commun.* 15: 5953.

241 Chen, P., Chen, Y., Xia, N. et al. (2024). A pyruvate transporter in the apicoplast of apicomplexan parasites. *Proc. Natl. Acad. Sci.* 121: e2314314121.

242 Herneisen, A.L., Peters, M.L., Smith, T.A. et al. (2024). SPARK regulates AGC kinases central to the Toxoplasma gondii asexual cycle. *elife* 13: https://doi.org/10.7554/elife.93877.2.

243 Chaabene, R.B., Martinez, M., Bonavoglia, A. et al. (2024). Toxoplasma gondii rhoptry discharge factor 3 is essential for invasion and microtubule-associated vesicle biogenesis. *PLoS Biol.* 22: e3002745.

244 Wang, C., Sun, P., Jia, Y. et al. (2024). Protein disulfide isomerase PDI8 is indispensable for parasite growth and associated with secretory protein processing in toxoplasma gondii. *MBio* e0205124. https://doi.org/10.1128/mbio.02051-24.

245 Yang, R., Meyer, A.S., Droujinine, I.A. et al. (2022). A genetic model for in vivo proximity labelling of the mammalian secretome. *Open Biol.* 12: 220149.

246 Kwak, C., Park, C., Ko, M. et al. (2022). Identification of proteomic landscape of drug-binding proteins in live cells by proximity-dependent target ID. *Cell Chem. Biol.* 29: 1739–1753.

247 Tao, A.J., Jiang, J., Gadbois, G.E. et al. (2023). A biotin targeting chimera (BioTAC) system to map small molecule interactomes in situ. *Nat. Commun.* 14: 8016.

248 Liu, X. and Ciulli, A. (2023). Proximity-based modalities for biology and medicine. *ACS Cent. Sci.* 9: 1269–1284.

249 Banaszynski, L.A., Liu, C.W., and Wandless, T.J. (2005). Characterization of the FKBP·rapamycin·FRB ternary complex. *J. Am. Chem. Soc.* 127: 4715–4721.

250 Ito, T., Ando, H., Suzuki, T. et al. (2010). Identification of a primary target of thalidomide teratogenicity. *Science* 327: 1345–1350.

251 Słabicki, M., Kozicka, Z., Petzold, G. et al. (2020). The CDK inhibitor CR8 acts as a molecular glue degrader that depletes cyclin K. *Nature* 585: 293–297.

252 Simonetta, K.R., Taygerly, J., Boyle, K. et al. (2019). Prospective discovery of small molecule enhancers of an E3 ligase-substrate interaction. *Nat. Commun.* 10: 1402.

253 Słabicki, M., Yoon, H., Koeppel, J. et al. (2020). Small-molecule-induced polymerization triggers degradation of BCL6. *Nature* 588: 164–168.

254 Wang, X., Qin, Z.L., Li, N. et al. (2024). Annual review of PROTAC degraders as anticancer agents in 2022. *Eur. J. Med. Chem.* 267: 116166.

255 Diehl, C.J. and Ciulli, A. (2022). Discovery of small molecule ligands for the von Hippel-Lindau (VHL) E3 ligase and their use as inhibitors and PROTAC degraders. *Chem. Soc. Rev.* 51: 8216–8257.

256 Bricelj, A., Steinebach, C., Kuchta, R. et al. (2021). E3 ligase ligands in successful PROTACs: an overview of syntheses and linker attachment points. *Front. Chem.* 9: 707317.

257 Cotton, A.D., Nguyen, D.P., Gramespacher, J.A. et al. (2021). Development of antibody-based PROTACs for the degradation of the cell-surface immune checkpoint protein PD-L1. *J. Am. Chem. Soc.* 143: 593–598.

258 Ahn, G., Banik, S.M., Miller, C.L. et al. (2021). LYTACs that engage the asialoglycoprotein receptor for targeted protein degradation. *Nat. Chem. Biol.* 17: 937–946.

259 Takahashi, D., Moriyama, J., Nakamura, T. et al. (2019). AUTACs: cargo-specific degraders using selective autophagy. *Mol. Cell* 76: 797–810.e10.

260 Pei, J., Pan, X., Wang, A. et al. (2021). Developing potent LC3-targeting AUTAC tools for protein degradation with selective autophagy. *Chem. Commun.* 57: 13194–13197.

261 Ji, C.H., Kim, H.Y., Lee, M.J. et al. (2022). The AUTOTAC chemical biology platform for targeted protein degradation via the autophagy-lysosome system. *Nat. Commun.* 13: 904.

262 Pance, K., Gramespacher, J.A., Byrnes, J.R. et al. (2023). Modular cytokine receptor-targeting chimeras for targeted degradation of cell surface and extracellular proteins. *Nat. Biotechnol.* 41: 273–281.

263 Liu, J., Yu, X., Chen, H. et al. (2022). TF-DUBTACs stabilize tumor suppressor transcription factors. *J. Am. Chem. Soc.* 144: 12934–12941.

264 Wang, W.W., Chen, L.Y., Wozniak, J.M. et al. (2021). Targeted protein acetylation in cells using heterobifunctional molecules. *J. Am. Chem. Soc.* 143: 16700–16708.

265 Hu, Z., Chen, P.H., Li, W. et al. (2023). Targeted Dephosphorylation of tau by phosphorylation targeting chimeras (PhosTACs) as a therapeutic modality. *J. Am. Chem. Soc.* 145: 4045–4055.

266 Yamazoe, S., Tom, J., Fu, Y. et al. (2020). Heterobifunctional molecules induce Dephosphorylation of kinases – a proof of concept study. *J. Med. Chem.* 63: 2807–2813.

267 Siriwardena, S.U., Munkanatta Godage, D.N., Shoba, V.M. et al. (2020). Phosphorylation-inducing chimeric small molecules. *J. Am. Chem. Soc.* 142: 14052–14057.

268 Wasko, U.N., Jiang, J., Dalton, T.C. et al. (2024). Tumour-selective activity of RAS-GTP inhibition in pancreatic cancer. *Nature* 629: 927–936.

269 Holderfield, M., Lee, B.J., Jiang, J. et al. (2024). Concurrent inhibition of oncogenic and wild-type RAS-GTP for cancer therapy. *Nature* 629: 919–926.

270 Raina, K., Forbes, C.D., Stronk, R. et al. (2024). Regulated induced proximity targeting chimeras—RIPTACs—A heterobifunctional small molecule strategy for cancer selective therapies. *Cell Chem. Biol.* 31: 1490–1502.e42.

271 Jan, M., Scarfò, I., Larson, R.C. et al. (2021). Reversible ON- and OFF-switch chimeric antigen receptors controlled by lenalidomide. *Sci. Transl. Med.* 13.

272 Eyckerman, S., Titeca, K., Van Quickelberghe, E. et al. (2016). Trapping mammalian protein complexes in viral particles. *Nat. Commun.* 7: 11416.

273 Steklov, M., Pandolfi, S., Baietti, M.F. et al. (2018). Mutations in LZTR1 drive human disease by dysregulating RAS ubiquitination. *Science* 362: 1177–1182.

274 Poirson, J., Cho, H., Dhillon, A. et al. (2024). Proteome-scale discovery of protein degradation and stabilization effectors. *Nature* 628: 878–886. https://doi.org/10.1038/s41586-024-07224-3.

275 Choi, H., Larsen, B., Lin, Z.Y. et al. (2011). SAINT: probabilistic scoring of affinity purification–mass spectrometry data. *Nat. Methods* 8: 70–73.

276 Sowa, M.E., Bennett, E.J., Gygi, S.P., and Harper, J.W. (2009). Defining the human deubiquitinating enzyme interaction landscape. *Cell* 138: 389–403.

277 Lundgren, D.H., Hwang, S.-I., Wu, L., and Han, D.K. (2010). Role of spectral counting in quantitative proteomics. *Expert Rev Proteomics* 7: 39–53.

278 Ong, S.-E., Blagoev, B., Kratchmarova, I. et al. (2002). Stable isotope Labeling by amino acids in cell culture, SILAC, as a simple and accurate approach to expression proteomics*. *Mol. Cell. Proteomics* 1: 376–386.

279 Thompson, A., Schäfer, J., Kuhn, K. et al. (2003). Tandem mass tags: a novel quantification strategy for comparative analysis of complex protein mixtures by MS/MS. *Anal. Chem.* 75: 1895–1904.

280 Cho, K.F., Branon, T.C., Udeshi, N.D. et al. (2020). Proximity labeling in mammalian cells with TurboID and split-TurboID. *Nat. Protoc.* 15: 3971–3999.

281 Hu, A., Noble, W. S. and Wolf-Yadlin, A. (2016) Technical advances in proteomics: new developments in data-independent acquisition. *F1000Research* 5, F1000 Faculty Rev-419.

282 Cirri, E., Knaudt, H., Fraia, D.D. et al. (2023). Automated workflow for BioID improves reproducibility and identification of protein-protein interactions. *bioRxiv* 2023.09.08.556804. https://doi.org/10.1101/2023.09.08.556804.

283 Bennett, H.M., Stephenson, W., Rose, C.M., and Darmanis, S. (2023). Single-cell proteomics enabled by next-generation sequencing or mass spectrometry. *Nat. Methods* 20: 363–374.

284 Schoof, E.M., Furtwängler, B., Üresin, N. et al. (2021). Quantitative single-cell proteomics as a tool to characterize cellular hierarchies. *Nat. Commun.* 12: 3341.

285 Specht, H., Emmott, E., Petelski, A.A. et al. (2021). Single-cell proteomic and transcriptomic analysis of macrophage heterogeneity using SCoPE2. *Genome Biol.* 22: 50.

286 Woo, J., Williams, S.M., Markillie, L.M. et al. (2021). High-throughput and high-efficiency sample preparation for single-cell proteomics using a nested nanowell chip. *Nat. Commun.* 12: 6246.

287 Brunner, A., Thielert, M., Vasilopoulou, C. et al. (2022). Ultra-high sensitivity mass spectrometry quantifies single-cell proteome changes upon perturbation. *Mol. Syst. Biol.* 18: e10798.

288 Furlan, C., Dirks, R.A., Thomas, P.C. et al. (2019). Miniaturised interaction proteomics on a microfluidic platform with ultra-low input requirements. *Nat. Commun.* 10: 1525.

289 Alfaro, J.A., Bohländer, P., Dai, M. et al. (2021). The emerging landscape of single-molecule protein sequencing technologies. *Nat. Methods* 18: 604–617.

290 Motone, K. and Nivala, J. (2023). Not if but when nanopore protein sequencing meets single-cell proteomics. *Nat. Methods* 20: 336–338.

291 Reed, B.D., Meyer, M.J., Abramzon, V. et al. (2022). Real-time dynamic single-molecule protein sequencing on an integrated semiconductor device. *Science* 378: 186–192.

292 Lee, S.-Y., Kang, M.G., Shin, S. et al. (2017). Architecture mapping of the inner mitochondrial membrane proteome by chemical tools in live cells. *J. Am. Chem. Soc.* 139: 3651–3662.

293 Li, H., Frankenfield, A.M., Houston, R. et al. (2021). Thiol-cleavable biotin for chemical and enzymatic Biotinylation and its application to mitochondrial

TurboID proteomics. *J. Am. Soc. Mass Spectrom.* 32: 2358–2365.

294 Udeshi, N.D., Pedram, K., Svinkina, T. et al. (2017). Antibodies to biotin enable large-scale detection of biotinylation sites on proteins. *Nat. Methods* 14: 1167–1170.

295 Kim, D.I., Cutler, J.A., Na, C.H. et al. (2018). BioSITe: a method for direct detection and quantitation of site-specific Biotinylation. *J. Proteome Res.* 17: 759–769.

296 Lee, S.-Y., Lee, H., Lee, H.K. et al. (2016). Proximity-directed Labeling reveals a new rapamycin-induced heterodimer of FKBP25 and FRB in live cells. *ACS Cent. Sci.* 2: 506–516.

297 Shin, S., Lee, S.Y., Kang, M.G. et al. (2024). Super-resolution proximity labeling with enhanced direct identification of biotinylation sites. *Commun. Biol.* 7: 554.

298 Kohler, D., Staniak, M., Yu, F. et al. (2024). An MSstats workflow for detecting differentially abundant proteins in large-scale data-independent acquisition mass spectrometry experiments with FragPipe processing. *Nat. Protoc.* 19: 2915–2938. https://doi.org/10.1038/s41596-024-01000-3.

299 Huang, T., Choi, M., Tzouros, M. et al. (2020). MSstatsTMT: statistical detection of differentially abundant proteins in experiments with isobaric Labeling and multiple mixtures. *Mol. Cell. Proteomics* 19: 1706–1723.

300 Barrios-Rodiles, M., Brown, K.R., Ozdamar, B. et al. (2005). High-throughput mapping of a dynamic Signaling network in mammalian cells. *Science* 307: 1621–1625.

301 Fields, S. and Song, O. (1989). A novel genetic system to detect protein–protein interactions. *Nature* 340: 245–246.

302 Eyckerman, S., Verhee, A., der Heyden, J.V. et al. (2001). Design and application of a cytokine-receptor-based interaction trap. *Nat. Cell Biol.* 3: 1114–1119.

303 Petschnigg, J., Groisman, B., Kotlyar, M. et al. (2014). The mammalian-membrane two-hybrid assay (MaMTH) for probing membrane-protein interactions in human cells. *Nat. Methods* 11: 585–592.

304 Masschaele, D., Gerlo, S., Lemmens, I. et al. (2018). Two-hybrid systems, methods and protocols. *Methods Mol. Biol.* 1794: 269–278.

305 Li, J., Zhang, S., Gao, L. et al. (2011). A cell-based high-throughput assay for the screening of small-molecule inhibitors of p53–MDM2 interaction. *SLAS Discov.* 16: 450–456.

306 Bhatia, N., Yang, B., Xiao, T.Z. et al. (2011). Identification of novel small molecules that inhibit protein–protein interactions between MAGE and KAP-1. *Arch. Biochem. Biophys.* 508: 217–221.

307 Cherian, M.T., Lin, W., Wu, J., and Chen, T. (2015). CINPA1 is an inhibitor of constitutive Androstane receptor that does not activate Pregnane X receptor. *Mol. Pharmacol.* 87: 878–889.

308 Yasui, T., Yamamoto, T., Sakai, N. et al. (2017). Discovery of a novel B-cell lymphoma 6 (BCL6)–corepressor interaction inhibitor by utilizing structure-based drug design. *Bioorg. Med. Chem.* 25: 4876–4886.

309 Kerppola, T.K., Brown, K.R., and Ozdamar, B. (2008). Bimolecular fluorescence complementation (BiFC) analysis as a probe of protein interactions in living cells. *Biophysics* 37: 465–487.

310 Verhoef, L.G.G.C. and Wade, M. (2017). Visualization of protein interactions in living cells using bimolecular luminescence complementation (BiLC). *Curr. Protoc. Protein Sci.* 90: 30.5.1–30.5.14.

311 Schwinn, M.K., Steffen, L.S., Zimmerman, K. et al. (2020). A simple and scalable strategy for analysis of endogenous protein dynamics. *Sci. Rep.* 10: 8953.

312 Reyes-Alcaraz, A., Lucero Garcia-Rojas, E.Y., Merlinsky, E.A. et al. (2022). A NanoBiT assay to monitor membrane proteins trafficking for drug discovery and drug development. *Commun. Biol.* 5: 212.

313 Leuchowius, K.-J., Jarvius, M., Wickström, M. et al. (2010). High content screening for inhibitors of protein interactions and Post-translational modifications in primary cells by proximity ligation*. *Mol. Cell. Proteomics* 9: 178–183.

314 Okamoto, K. and Sako, Y. (2017). Recent advances in FRET for the study of protein interactions and dynamics. *Curr. Opin. Struct. Biol.* 46: 16–23.

315 Degorce, F., Card, A., Soh, S. et al. (2009). HTRF: a technology tailored for drug discovery – a review of theoretical aspects and recent applications. *Curr. Chem. Genomics* 3: 22–32.

316 Lee, M. (2023). Recent advances in deep learning for protein-protein interaction analysis: a comprehensive review. *Molecules* 28: 5169.

317 Wang, L., Wang, H.-F., Liu, S.-R. et al. (2019). Predicting protein-protein interactions from matrix-based protein sequence using convolution neural network and feature-selective rotation Forest. *Sci. Rep.* 9: 9848.

318 Jumper, J., Evans, R., Pritzel, A. et al. (2021). Highly accurate protein structure prediction with AlphaFold. *Nature* 596: 583–589.

319 Lee, C.Y., Hubrich, D., Varga, J.K. et al. (2024). Systematic discovery of protein interaction interfaces

using AlphaFold and experimental validation. *Mol. Syst. Biol.* 20: 75–97.

320 Gainza, P., Sverrisson, F., Monti, F. et al. (2020). Deciphering interaction fingerprints from protein molecular surfaces using geometric deep learning. *Nat. Methods* 17: 184–192.

321 Watanabe, N., Ohnuki, Y., and Sakakibara, Y. (2021). Deep learning integration of molecular and interactome data for protein–compound interaction prediction. *J. Cheminformatics* 13: 36.

322 Ghosh, N., Saha, I., and Gambin, A. (2023). Interactome-based machine learning predicts potential therapeutics for COVID-19. *ACS Omega* 8: 13840–13854.

17

Receptomics and Extracellular Interactomics for Biotherapeutic Characterization

Naincy R. Chandan[2], Abel Ferrel[3], and Shengya Cao[1]

[1]Development, Inceptive, Palo Alto, CA, USA
[2]Discovery Technology, Nurix Therapeutics, Inc, San Francisco, CA, USA
[3]Independent Researcher, San Francisco, CA, USA

Introduction

Biotherapeutics have demonstrated a range of success in addressing pressing patient needs. Recent years saw the US Food and Drug Administration (FDA) approval of the first adeno-associated virus (AAV)-based gene therapy [1] and chimeric antigen receptor (CAR)-T-cell therapy [2] in 2017, for Leber congenital amaurosis type 2 and B-cell acute lymphoblastic leukemia respectively. The approval of biotherapeutics has only accelerated since then with the approval of 9 antibody-drug conjugates (ADCs), 16 enzymes and proteins, 20 natural products, 26 peptide therapeutics, and 60 monoclonal antibodies (mAbs) between 2016 and 2022 [3]. 2023 saw the approval of the first Clustered regularly interspaced short palindromic repeats (CRISPR)-Cas9 gene editing therapy, for sickle cell disease [4]. With the growing list of approved biotherapeutic modalities, many researchers have found that the greater complexity of biotherapeutics is offset by their therapeutic potential.

That is not to say that biotherapeutics do not pose significant biological and financial risks. Even one of the best-established modalities, the mAb, which was first approved by the FDA in 1986 for the prevention of transplant rejection [5], still experiences many late-stage failures [6]. This is despite the extensive research into the biology governing antibody target engagement, effector function, immunogenicity, pharmacokinetics (PK)/pharmacodynamics (PD) and other variables that contribute to efficacy and safety. For newer, less established modalities, many characterization challenges remain, though the particular challenge varies by modality. For example, while Fc receptors that bind to the constant regions of antibodies were first identified by Boyden and Sorkin in 1960 [7], the universal receptor for AAVs was first reported in 2016 and is a gene with relatively unknown function [8]. These recent entries into patients can also bring dangerous surprises. While AAV treatments have generally shown good safety profiles, the safety record was recently scarred by an attempt to cure life-threatening X-linked myotubular myopathy which led to the unexpected death of four children due to gene therapy-related liver failure [9]. Explorations into the cause of liver toxicity are still ongoing.

With the acknowledgment that all clinical trials must balance risk and reward, the biotherapeutic community has continually advanced technologies to tip the scale toward overall benefit. One subsection of these efforts has been focused on identifying and characterizing the extracellular protein interactions of biotherapeutics. Due to their biophysical properties, biotherapeutics typically have limited inherent cell permeability. Therefore, both desired and undesired effects are often mediated by interactions with secreted or cell – surface proteins. Some interactions are the basis for biotherapeutic function, being foundational to therapeutic mechanisms of action (MOA) such as immune checkpoint blockade [10] or engineered targeting of the cell surface marker CD19 by the first approved CAR-T-cell therapy [2] to promote immune cell killing of cancer cells. Other interactions are absorption, distribution, metabolism, excretion, and toxicity (ADMET), design considerations, regulating cellular and tissue uptake and clearance, biodistribution, drug catabolism, drug–drug interactions and drug disease interactions [11–13]. This includes modality-dependent desired effects like antibody-dependent cell-mediated cytotoxicity (ADCC) mediated by Fc receptors for mAbs [14], receptor-mediated internalization of ADCs and lipid nanoparticle (LNPs) necessary for payload release [15] or receptor-mediated

Characterizing Biotherapeutics: Analytical Methods for Diverse Modalities, First Edition. Edited by Jennie R. Lill and Wendy Sandoval.

viral uptake for AAVs-dependent gene transduction [8]. Unanticipated interactions can also cause patient adverse events. A case study for biotherapeutics-specific challenges is cetuximab, an anti-EGFR mAb that was approved for colorectal cancer and head and neck squamous cell carcinoma. The therapeutic mAb was produced in mouse SP2/0 cells which resulted in a non-primate galactose-α-1,3-galactose post-translational modification. In a subset of the patient, exposure to cetuximab resulted in severe adverse effects like anaphylaxis. This was later determined to be the result of pre-existing IgE antibodies for this sugar modification present in a subset of the human population [16]. While this would have been almost impossible to predict at the time, with this knowledge gained, we can use the latest technologies to identify putative interactions and their downstream consequences before a biotherapeutic enters a patient's body.

Although the field of using broad, "omics"-level techniques for biotherapeutic characterization is nascent, extracellular protein-biotherapeutic interactions characterization has been an essential part of biotherapeutic development. Going forward, the ability to detect interactions in a broad and unbiased manner can therefore complement and accelerate biotherapeutic design. In this chapter, we will provide perspective on the usage of receptomic or more broadly extracellular interactomic technologies to increase both safety and efficacy through better characterization and design of biotherapeutics for MOA and ADMET considerations. Rather than providing an exhaustive list of techniques, which have been covered elsewhere [17–20], in this chapter, we'will provide an overview of the types of approaches followed by specific use cases broken down by biotherapeutic modality (Figure 17.1).

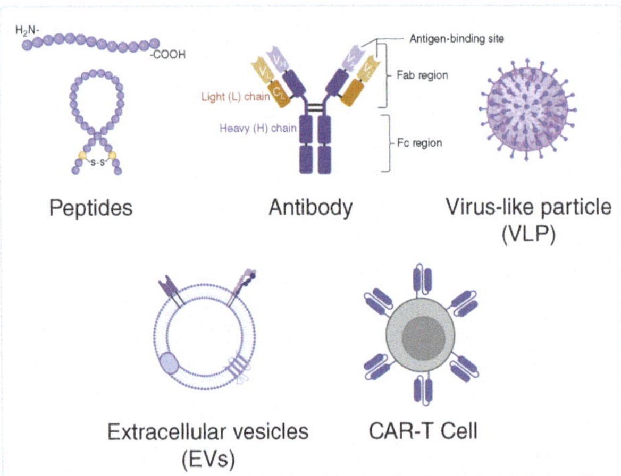

Figure 17.1 Schematic of biotherapeutics.

Receptomic and Extracellular Interactomics Technology Overview

Cell surface receptomics and extracellular interactomics are particularly challenging subcategories of the general problem of interactomics. While the goal remains to identify protein–protein interactions in an unbiased manner, additional difficulty results from the involvement of membrane proteins and their unique biophysical properties, high rates of heterogeneous post-translational modifications, and more variable solution environments with different salts, reductive potential, and pH ranges than the intracellular environment [21–23]. As a result, extracellular protein interactions, including those with biotherapeutics, can often be missed by more general interactomics techniques. In this chapter, we will address some of the approaches specifically designed for extracellular protein–protein interactions and their application to biotherapeutics.

Most interactomics technologies fall into roughly five classes: biochemical fractionation, affinity purification (AP), proximity-based covalent modification, protein-fragment complementation, and interaction screening [20]. Each of these approaches has different advantages and disadvantages such as differences in throughput, material requirements, data interpretability, and tradeoffs between complexity and physiological relevance. When applied to biotherapeutics, the particular concerns for specific modalities and the biological question being asked typically govern which approaches would be most suitable.

Biochemical Fractionation and Affinity Purification

Biochemical fractionation is an approach where complex lysates or biological fluids are broken down into simpler mixtures (or fractions) while retaining specific biological activities. While the activity can be any quantifiable behavior, this chapter will only look at identifying cell-surface or extracellular protein binding for biotherapeutics, meaning that the relevant activity is the presence of the therapeutic molecule. Modern renditions of this approach couple fractionation methods to mass spectrometry (MS) to identify proteins that co-elute with the biotherapeutic (Figure 17.2, left), typically termed co-fractionation MS with thermal solubility MS being a special variant [19].

For this activity, AP can also be considered a special case of biochemical fractionation. Rather than simplifying mixtures based on biophysical properties agnostic to the biotherapeutic or its interacting proteins (i.e. charge, size, or hydrophobicity), AP uses either the biotherapeutic or a suspected interacting protein coupled to a solid substrate (i.e. a dense or magnetic bead) that can simplify solutions

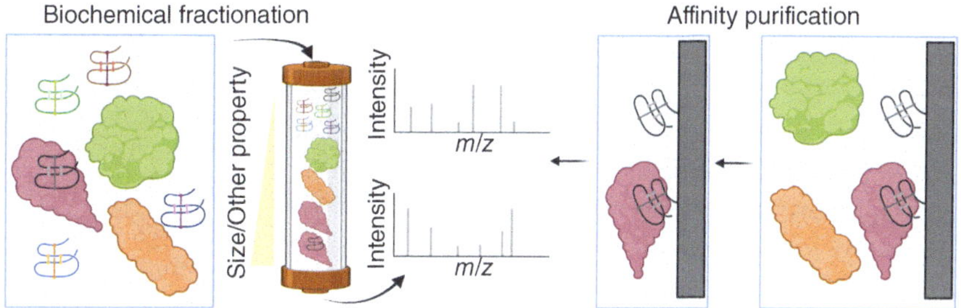

Figure 17.2 Biochemical fractionation and affinity purification both start with complex mixtures incubated with biotherapeutics. The resulting interaction is most commonly readout using mass spectrometry.

by "pulling out" direct and indirect binding partners from lysates or biofluids. The "pulled out" proteins can then be identified using MS (Figure 17.2, right) though simpler readouts are sometimes sufficient [24].

Historically, these workhorse techniques have identified several therapeutic interactions and continue to excel at detecting simple, high-affinity interactions. However, these approaches can easily miss interactions that are low affinity, membrane-dependent, or with low-abundance proteins. For low affinity interactions, such as many glycan–protein interactions, dissociation can occur on the order of seconds making it unlikely to survive all processing steps [25]. Processing also includes solubilization of cellular membranes, perturbing interactions dependent on the physical or geometric properties of the membrane. Lastly, these approaches do not include signal amplification and struggle with rare or context-dependent interactions. Since membrane and secreted proteins are categorically less abundant, they are often missed without additional processing [26]. Abundances can also vary based on the choice of the starting lysate. Important interactions not represented by that lysate (e.g. proteins expressed in other tissues) will be absent.

While there are some limitations to keep in mind, this class of approaches can be some of the most accommodating to native untagged molecules. These methods can also detect multiple interactions simultaneously making it fairly unbiased beyond the starting lysate selection. Therefore, these approaches are well suited for confirmation of on-target engagement for high-affinity specific biotherapeutics that can be coupled to solid supports but can be less capable for identifying weak off-target effects or for complex, delivery-based modalities such as LNPs, EVs, and AAVs.

Proximity-Based Covalent Modification

Proximity-based covalent modification (Figure 17.3, top) is a group of approaches that covalently modifies proteins in a distance-dependent manner. Since proteins need to be close to one another to interact, this class of approaches uses proximity as a proxy for interactions. Specific methodologies include chemical crosslinking and proximity labeling. Given that the modifications are covalent, they are stable to downstream processing, making this class of approaches suitable for weak or transient interactions and overcoming disadvantages typically associated with membrane and extracellular protein interactions [22, 23]. By using modifications that are enrichable (e.g. contains a biotin group), interactions can be isolated and identified by MS.

The other advantage of this class of approaches is its broad applicability. These approaches have been used on everything from purified recombinant proteins to complex living organisms [27, 28]. Recent developments also added tunable parameters such as labeling distance [29, 30], extracellular selectivity [31–34], or timing [35]. The main

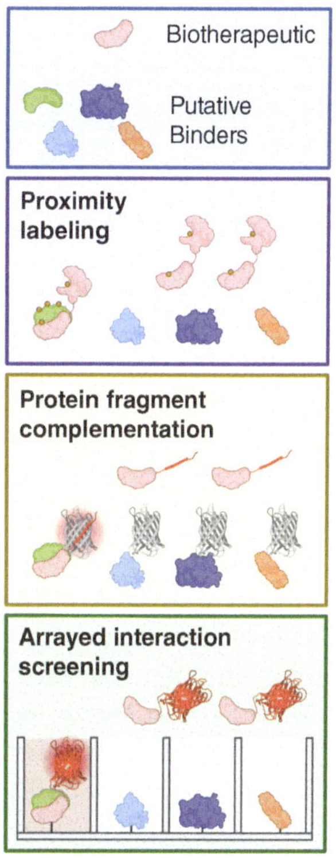

Figure 17.3 Tagging and format requirements differentiates the different approaches.

downside, however, is data interpretability. Since these approaches read out on proximity, high-abundance proteins that are simply present can overwhelm lower-abundance bona fide binders. Data interpretability for cross-linking-based approaches can also be confounded by challenges processing and identifying cross-linked proteins by MS. Though proximity labeling overcomes the processing challenge, the approach requires a peroxidase, promiscuous biotin ligase or photocatalyst to be coupled to the target whereas crosslinking does not require tagging. These bulky additions can occlude interactions or disrupt target function [19, 27]. However, for the extracellular space and the growing availability of proximity-labeling compatible binding reagents for many biotherapeutics (i.e. peroxidase or photocatalyst-tagged antibodies), these challenges are often surmountable.

Protein-Fragment Complementation

Protein-fragment complementation (Figure 17.3, middle) describes a range of approaches where a reporter activity is split between two protein fragments. The activity is only reconstituted when the halves are physically brought together. The archetypal technique of this class is yeast two-hybrid where a transcription factor that drives the expression of a reporter gene is broken up into two individually inactive parts [36]. When those fragments are fused to proteins that interact, the transcriptional activity is reconstituted, resulting in detectable reporter expression. This original conception actually cannot be used for membrane anchored or extracellular proteins because of the inability of the transcription factor to enter the nucleus. However, other variants of the approach exist, including split reporters such as split fluorescent proteins [37] or split luciferase [38]. In particular, the NanoBiT system is a popular split NanoLuc luciferase which separates the enzyme into an 11 amino acid fragment and a large fragment consisting of the remainder of the protein. Since 11 amino acids is a small tag, it is a relatively innocuous tag for most proteins [39].

The main disadvantage to this approach is that both proteins of interest (POI), in this context the biotherapeutic and the potential binder, must be tagged. Aside from concerns of tags altering the function of the proteins, tags also limit the scale and versatility of the approach. Since few groups have comprehensive libraries with the extracellular proteome tagged with one fragment that can be complemented by biotherapeutics tagged with the other fragment, this approach is rarely used for unbiased, omics-level discovery. However, it can be a sensitive method for hypothesis driven approaches, specific targeted assays, or confirmatory studies.

Interaction Screening

Interaction screening (Figure 17.3, bottom) is a broad class of approaches that involves screening a target protein, likely the biotherapeutic, against a library of receptors or extracellular proteins. Both the target protein and the library can vary in format, making these approaches extremely tweakable to achieve the desired balance between interpretability, ease of deployment, and physiological relevance.

Interaction screening can further be divided into two subclasses, pooled and arrayed. In pooled approaches, the entire library is presented in the same solution. The physical act of binding can then be used to isolate just the binding-competent species. While this is like traditional AP, the distinction is that the library composition is one that is specifically engineered rather than relying on natural abundances. Pooled approaches are generally scalable and adding one additional library member either does not or only marginally alters the workflow. The downside is that they are often harder to design and interpret. Since all the library species are present simultaneously, every species needs to be distinguishable from each other. In some cases, MS can be used for untagged proteins. However, to be reliably detectable by MS, library species will likely need to be present at high copy numbers, resulting in trade-offs in library diversity. MS can also be unsuitable for some formats such as cell-based libraries since irrelevant proteins would dominate the mixture. Alternatively, a unique tag can be coupled with each library species. Practically, this is typically implemented using nucleic acid barcodes which can be readout using sequencing. While this is feasible, it does require more effort to be placed in library design. The other downside to pooled approaches is that there could be interference between members of the library. This could mean competition for binding sites or secondary signaling effects due to specific responses in cells. Despite these challenges, pooled libraries have been successful in a variety of scenarios, and recent effects to express proteins from relevant cell types using endogenous genomic loci have helped reproduce more *in vivo*-like binding conditions [40, 41].

In addition to pooled approaches, interaction screening can take on arrayed formats. For arrayed screens, the identity of the library species is encoded in a spatial location (i.e. a well on a plate or a position on a slide). This eliminates the need for barcoding and results in simpler handling and interpretation since the same readout can be applied across the entire library. For example, one library format is the protein microarray. There, hundreds of different proteins are adsorbed onto a slide, with each protein getting a small but defined spot on a grid. To identify binders for a mAb, the mAb is flowed over the library, any unbound mAb is washed away, and the bound mAb can be detected with a fluorescent secondary antibody. Any fluorescent spot can be identified by looking up what protein was placed at that location.

For both pooled and arrayed approaches, generating a library takes careful consideration. While large, comprehensive libraries allow for more interactions to be captured, they come at significant monetary costs. This cost can vary depending on the specific format of the library. Some of the easiest to work with are libraries of recombinant proteins but purifying and quality controlling hundreds of proteins can be onerous. This is especially true for membrane and secreted proteins where it is not uncommon to get little to no expression, non-native modifications or disulfide bonding patterns, and poor responses to any tags that might be needed for characterization, immobilization, or readout (i.e. barcoding). Alternative library formats include non-purified proteins overexpressed from DNA libraries using *in vitro* transcription/translation systems [42] or transfected into cells and secreted into the cell media to be directly captured into a plate [43]. Lastly, libraries can be directly expressed on cells using transfected cDNA libraries or virally induced protein expression. While this can get the most native behaviors, control over expression levels or secondary effects resulting from protein overexpression is hard to achieve.

Other than the contents of the library, the format can also be decided by the amount of screening material available. For many of these approaches, specifically pooled assays, low input can be accommodated using signal amplification through enzymatic readouts or PCR amplifying nucleic acid barcodes.

Interaction screening can often give the most unbiased detection of interactions (for a given library), returning a list of what *can* bind rather than what binds given a specific system, making it one of the preferred approaches for identifying unexpected off-target effects.

Additional Approaches

Several of these approaches (i.e. biochemical fractionation and interaction screening) do not have to be limited to binding. While beyond the scope of this chapter, interactions can often be inferred from functional readouts. This might involve fractionating based on pathway activity, using reporter cell lines for interaction screening or using RNA-sequencing to detect broad cellular changes. Especially with recent developments in functional genomics (i.e. Perturb-seq and related approaches [41, 44, 45]), binding and functional readouts can be achieved simultaneously.

Natural Products and Peptides

This class of biotherapeutics is the smallest in terms of molecular weight, with many natural products weighing in at around 500 Da [46] to peptides reaching up to 50 amino-acids or 10 kilodaltons [47]. In part due to the small size, this class has some unique challenges in terms of PK/PD and have, on average, the shortest circulating half-lives among biotherapeutics. Even clinically approved cyclic peptides, such as romidepsin, can have circulating half-lives of merely 3.5 hours [48]. In contrast, human IgG1 mAbs like daclizumab and omalizumab weigh around 150 kDa and have circulating half-lives of 20 days [11]. The short half-lives of natural products and peptide therapeutics is both due to the size-based filtration by the kidneys and susceptibility to drug-metabolizing enzymes, peptidases, and proteases [13, 47]. Both elimination paths can and have been addressed through altering extracellular protein interactions. Other methods of addressing elimination include adding covalent modifications and fusing therapeutics to other proteins, natural products, or peptides. Both modifications and fusions can generate unexpected protein interactions that can impact both drug efficacy and safety.

For natural products, receptomics or extracellular interatomics is relevant for both the identification of the active compound and optimization or derivatization of the compounds once found. While the latter challenge is similar for natural products and peptides, natural product discovery is still a place where biochemical fractionation can shine. Even though a natural product is often synthetically or semisynthetically made once identified and characterized, the initial discovery starts with complex natural extracts [49]. These extracts can be mixed with biofluids, cells, cell lysates, or cellular membrane preparations [50] and then divided into many fractions by biophysical properties like size or charge. In the past, fractionation had to be extensive but with modern metabolomics and proteomics methods, interactions can be identified in fairly complex mixtures reducing the number of fractionation steps required [51]. Especially with recent attention on the contributions of the human microbiome to disease through natural products and secondary metabolites and an improved ability to grow microbiome organisms, microbiome extracts can be an important source of active natural products with therapeutic potential [51, 52].

For peptide therapeutics, there are additional challenges including biochemical and physical instability, tendency to aggregate, immunogenicity, low membrane permeability, typically low orally availability [53]. Many of these challenges can be addressed by designing for or around extracellular protein interactions with the drug.

This is exemplified by the development history of GLP-1 receptor agonists (RAs). Various GLP-1 RAs have been approved for type 2 diabetes and, most recently, have gained significant attention for dominating the $80 billion weight loss market [54, 55]. Like many approved peptide therapeutics, GLP-1 RAs are based on an endogenously active peptide, bind and act through cell surface receptors, specifically GPCRs [53], and are subject to short half-lives both through renal filtration and breakdown [48]. In the last two decades, approved GLP-1 RAs went from having circulating half-lives spanning 2.6 hours for lixisenatide to almost a week for semaglutide. These gains came from either creating an extracellular interaction with an abundant plasma protein, albumen, or preventing or inhibiting inactivation by DPP4, a membrane-bound peptidase that inactivates endogenous GLP-1 peptide [54]. Additional development paths for GLP-1 RAs are also following a common path for peptides, including hybridization with a targeting arm. In particular, fusion with a NMDA receptor agonist has been shown to achieve more potent weight-loss effects [56]. While the development of GLP-1 RAs occurred over the last two decades and did not rely on modern receptormics approaches [57], GLP-1 RA development and design has been dominated by altering extracellular interaction partners and presents a roadmap for development of similar therapeutics. Using modern approaches, development can be accelerated by identifying interactions with proteases and peptidases, on- and off-target receptors and potential carrier proteins in the same experiment.

In general, mutation, cyclization, and other conformational alterations (typically through disulfide bonding) [48], chemical modifications [53], and peptide fusions are commonly used to improve the properties of peptide therapeutics. However, these changes can have unpredictable effects. One example is peginesatide, a peptide PEGylated for stability, was first approved but later withdrawn from the market for safety considerations surrounding high rates of anaphylaxis [48]. One hypothesis for this adverse effect is the presence of anti-PEG and, therefore, anti-peginesatide antibodies detected in patients using a bead-based extracellular interactomics approach [24]. Like cetuximab, these methods picked up patient derive IgE that existed in a subset of the population that could drive adverse effects. An additional covalent modification, lipidation has also had somewhat unpredictable effects on target receptor selectivity. Though not a therapeutic peptide, the length of an acyl chain modification on the N-terminal region of melanocortin neuropeptides has been shown to change the receptor selectivity in a highly nonlinear manner between melanocortin receptors 1 and 4 [58]. Lastly, hybridization is a common strategy to either increase potency or have both an effector and targeting arm. In some cases, the fusion of multiple peptides has brought both pleasant and unpleasant surprises. In a study of several chimeric cyclic peptides as antibiotics, fusion between lipopolysaccharide binder and a lipopolysaccharide transporter protein, LptD resulted in several molecules that gained plasma protein binding that prevented their use for unfavorable PK/PD characteristics. However, one chimeric molecule gained an BamA interaction that helped it counter resistance mechanisms [59]. This shows that peptide fusions can often result in unknown interactions that need to be addressed to improve both safety and efficacy of putative therapeutics.

While most peptide therapeutics are aimed at cell-surface or extracellular targets, there has been some success and strong desire for intracellular targets. Though two cyclic peptides with intracellular targets have been approved for therapeutic use, cell penetrance remains a challenge [48]. Few peptides can rely on passive diffusion to cross the cellular membrane. Therefore, many strategies rely on receptor-mediated endocytosis. There are many cell-surface receptors and cell-surface molecules like heparan sulfates that can be used to trigger internalization of biologics [60]. For example, the LYTAC approach conjugates a dedicated small molecule or antibody protein binder to a glycopeptide substrate for the cation-independent mannose-6-phosphate receptor which results in internalization and degradation of extracellular target proteins [61]. However, in the small peptide context, engineering target binding on top of internalization receptor binding is challenging. One strategy that has worked well for molecular glues is to strengthen a weak interaction with an existing interaction partner rather than engineering an interaction de novo [62]. In recent years, the use of different degrees of multimerization has allowed for the tuning of detectable affinity ranges which has been crucial for detecting weak or transient extracellular interaction through direct screening approaches [21, 22]. This can identify promising extracellular interactions that can be worked off of using libraries of cell-surface receptor proteins [63].

Antibodies and Related Proteins

Antibodies and related proteins are widely used in therapeutic applications, ranging from small 10 kDa protein fragments to 750 kDa macromolecular assemblies such as IgM. A key advantage of antibody therapeutics is their extended circulation half-life, typically ranging from 2 to 3 weeks for full-length IgG antibodies. Although engineered proteins share similar size ranges and functional capabilities with antibodies, they differ significantly in their interaction profiles, as many are designed without interactions with Fc receptors or other antibody-specific features [64].

Antibodies are large, Y-shaped proteins made up of two heavy chains and two light chains arranged in the Fab (fragment antigen-binding) and the Fc (fragment crystallizable) domains. The antigen-binding site consists of the Fv (fragment variable) domain, the part of the Fab region located at the top of Y. These regions are extensively engineered to enhance therapeutic properties [65]. Antibodies products are typically categorized into two main types: monoclonal and polyclonal. mAbs are produced from clones derived from a single cell resulting in all antibodies in the mixture having identical sequences. In contrast, polyclonal antibody products contain antibodies generated from multiple clones, resulting in a diverse mixture of antibodies with varying sequences though all typically selected to bind to the same antigen.

Antibodies mediate therapeutic effects through various mechanisms. Antibodies can bind and block antigens like toxins or viruses from exerting pathogenic effects, a process called neutralization. Antibodies can coat pathogens leading to the uptake and destruction of the pathogens by immune cells, a process called opsonization. Additionally, antibodies can alter cell signaling by crosslinking cell surface receptors, blocking receptor–ligand interactions or stimulating receptor activity. They can also serve as drug delivery vehicles as ADCs, which link potent small molecule drugs to antibodies for precise targeting. These mechanisms highlight the versatility of antibodies and how they can be used in treating a wide range of diseases.

With over 100 approved products, antibody-based therapeutics have become highly successful biotherapeutics [66]. While most products are IgG-based mAbs, formats like single-chain variable fragments (scFvs), antigen-binding fragments (Fabs), and nanobodies (VHHs) which lack a Fc region can offer reduced immunogenicity, smaller size, and improved tissue penetration [67]. Additionally, bispecific antibodies (bsAbs), which can bind to two different antigens, and other formats like ADCs have broadened the therapeutic potential of mAbs by providing design flexibility in size, targeting, valency, and pharmacokinetics. This also results in new functions as seen in targeted degradation approaches like degrader antibody conjugate (DAC), lysosome-targeting chimeras (LYTACs) and Ab-based PROTAC (AbTAC) [61, 68, 69]. These innovations significantly improve therapeutic precision but come with their own challenges.

Advances in molecular engineering have refined mAbs to enhance their therapeutic actions and reduce immunogenicity, thereby improving their overall safety and efficacy profiles. Depending on the disease indication, the approval rates for mAbs range from 14% to 32%, with higher rates observed with antibodies developed for non-cancer indications [70]. Despite advancements, developing therapeutic antibodies remains complex and costly, with relatively low clinical approval rates. Successful development relies on optimizing key properties such as antigen binding affinity and specificity, minimizing off-target effects, biological efficacy, PK/PD, immunogenicity, tissue penetration, and stability profile [71]. Before delving into antibody format-specific factors, we will first examine antibody – receptor interactions, classify major engineering strategies, challenges, and suggest how receptomics and interactomics approaches can address some of these challenges to enhance therapeutic outcomes.

Antibodies consist of two main regions: the Fab domain, which binds to specific antigens, and the Fc domain, which mediates immune responses. While antigen binding is crucial for an antibody's function, many of the unique advantages of antibodies come from the interactions between the Fc region and the numerous Fc receptors. The Fc region of IgG interacts with Fc gamma receptors (FcγRs) and the neonatal Fc receptor (FcRn). It mediates antibody-dependent cell-mediated cytotoxicity (ADCC) and phagocytosis (ADCP) through interactions with FcγRs on immune cells such as macrophages, NK cells, and neutrophils. Among the five immunoglobulin isotypes, IgG is the most prevalent and commonly used in antibody therapeutics due to its Fc region interactions, which impact properties like efficacy and pharmacokinetics [28]. IgG subclasses differ in their effector functions: IgG1, the most potent complement activator, and Fcγ receptor binder is often used in cancer therapies due to its strong effector functions. IgG2 offers limited effector functions and is chosen for intermediate activity or high-density antigen targets. IgG3, while like IgG1 in function, is more prone to proteolysis. IgG4, with the weakest effector functions, is preferred in scenarios requiring minimal immune activation, such as in autoimmune or inflammatory conditions [72]. On the receptor side, the human FcγR family includes five activating and one inhibitory receptor with distinct tissue-specific distributions and regulatory mechanisms: hFcγ RI, hFcγ RIIA, hFcγ RIIB (inhibitory), hFcγ RIIC, hFcγ RIIIA, and hFcγ RIIIB [73]. Other Fc receptors, such as Fc epsilon Receptor (FcεR), predominantly found on mast cells and basophils, play a pivotal role in the immune response by binding to IgE antibodies [74]. FcRn, expressed on endothelial cells, binds to IgGs in a pH dependent manner, extending the half-life of antibodies by binding to IgGs at acidic pH in endosomes and recycling it to the blood and releasing the antibody at neutral pH [75, 76]. FcRn binds all IgG isotypes and albumin but does not interact with other immunoglobulins. FcRn blockers (Abdegs), like Efgartigimod (IgG1-Fc), are utilized to reduce pathogenic

IgG autoantibodies in autoimmune diseases. Efgartigimod is the first FDA-approved FcRn antagonist for treating myasthenia gravis with anti-acetylcholine receptor antibodies [77].

Since much of an antibody's function acts through its binding to cell surface receptors, several strategies have been employed to alter these interactions to improve therapeutic efficacy [65, 66, 78, 79]. Many changes made to the Fc region of therapeutic antibodies alter the binding of antibodies with their receptors to address specific therapeutic needs. Enhancing binding to activating receptors like FcγRIIIa improves ADCC and ADCP for treating cancer and infectious diseases, or boosts complement-dependent cytotoxicity (CDC) via increased C1q [80–82]. For example, margetuximab is an anti-HER2 antibody with five Fc mutations to enhance ADCC [83]. Afucosylation can further enhance ADCC by boosting FcγRIIIa binding affinity [79]. Conversely, in autoimmune diseases where reduced effector functions are desired, mutations can diminish FcγR and complement binding as seen with mutations like LALA or increase interaction with the inhibitory receptor FcγRIIb [80–82]. Mutations can also optimize FcRn binding to extend an antibody's half-life by improving its recycling, and modifications can induce receptor clustering to boost agonistic activity [72]. Additionally, the Fc region can be engineered to create antibody transport vehicles (ATVs) that cross the blood–brain barrier (BBB) by targeting receptors like transferrin receptor (TfR) or CD98hc on brain endothelial cells to enhance therapeutic cargo delivery to the brain [84–86], often carrying therapeutic payloads like toxins or radioactive molecules linked to the Fc region. Even a single amino acid change can form or lose off-target interactions as demonstrated with CLDN6 antibodies [87]. This emphasizes the importance of rigorous evaluation of the final candidates following modifications like humanization and affinity maturation.

Monoclonal antibodies are typically developed and isolated using a target antigen, leading to the expectation that they bind exclusively to that target. However, off-target binding, whether resulting from nonspecific interactions or polyspecificity of the CDRs – an ability to recognize multiple structurally related but distinct epitopes, is often overlooked in antibody discovery. A study investigating off-target binding of the lead candidates using membrane proteome array found that about 33% of evaluated samples (83 out of 254) exhibited polyspecific off-target binding [88]. This type of cross-reactivity can lead to unintended toxicity, adverse events, significantly contributing to clinical trial failures [88, 89]. For example, the anti-PD1 antibody camrelizumab caused severe, capillary hemangioma in Phase 1 clinical trials [90]. Receptor proteome screening with ~4,000 human membrane receptors revealed that camrelizumab binds to VEGFR2, FZD5, and ULBP2 [91]. Rescue therapy with a VEGFR2 antagonist alleviated hemangioma, confirming the role of off-target in adverse effects [92]. This underscores the importance of using receptomics to identify polyspecificity in therapeutic antibodies, and how it can guide the mitigation of unpredicted toxicities.

Unfortunately, camrelizumab is not an isolated example. Antibody therapies have encountered serious and even fatal adverse effects due to off-target or unexpected binding liabilities not caught by preclinical screening. For instance, theralizumab, an IgG4 superagonistic anti-CD28 antibody, caused severe toxicity in a first-in-human study due to differences between human and monkey CD28 amino acid sequences that preclinical tests missed [93, 94]. Similarly, humanized IgG1 antibody ABT-736, which targets amyloid β (Aβ) oligomers, caused significant toxicity in cynomolgus monkeys due to off-target binding to platelet factor-4 (PF-4), which is implicated in heparin-induced thrombocytopenia (HIT) in humans [95]. Humanized anti-FGFR4 antibody hLD1.vB exhibited rapid clearance in rodents due to off-target binding with complement C3, a rodent-specific interaction [96]. Similarly, the humanized antibody Ab2, targeting Aβ, had a significantly shorter half-life in cynomolgus monkeys due to binding with monkey fibrinogen, resulting in reduced serum recovery [97]. These examples highlight the necessity of using receptomics to identify the range of protein interactions beyond just binding to the primary target. All these interactions can impact safety, toxicology, and pharmacokinetics. Identifying protein interaction liabilities early on can improve risk–benefit analysis and potentially reduce late-stage failures, leading to financial benefits.

Due to the complexity of antibodies and the proteome, off-target interactions are difficult to predict. Many proteins can contain segments that resemble target epitope features. There is binding-site plasticity due to the conformational flexibility of the antibody paratope, and allosteric conformational changes in different regions of the antibody upon antigen binding to consider as well [88, 89, 98–101]. Understanding the mechanisms behind unintended interactions and toxicities can help mitigate risks while leveraging the benefits of polyspecific antibodies. Recent guidelines from the US FDA require evaluation of unintended antigen targeting by biotherapeutics for biotherapeutics investigational new drug (IND) filing [102]. Several commercially available services, including Integral Molecule's Membrane Proteome Array™, Charles River Labs' Retrogenix® Human Cell Microarray, and Creative Biolabs' human membrane protein array, offer antibody specificity profiling and off-target identification services.

A range of assays have been developed to characterize engineered antibodies for their interactions and off-target effects. Multiplexed Fc array is employed to investigate the interaction of antigen-specific antibodies with various Fc receptors, providing a comprehensive profile of their effector functions [103]. In addition, bead-based high-throughput assays, such as alpha screens, assess the binding affinities of mutant IgGs for FcγRs [104]. Improvements on traditional protein microarrays have also been developed to better replicate the natural environment of receptor proteins. Human cell-based microarray technology express membrane proteins directly on cells to present native-appearing proteins to test for on- and off-target interactions [105]. For example, a study assessed the specificity of the mAb developed against the extracellular epitope of glucose transporter 4 (GLUT4) expressed on VLPs by screening it against ~4,500 human membrane proteins expressed on the cell surface [106]. They observed that one of the antibodies showed low-level cross-reactivity with Notch1 due to molecular mimicry, where a shared "LGXXGP" motif was present in both GLUT4 and Notch1. Despite their structural and sequence dissimilarity, this conserved loop motif allowed the antibody to bind to both proteins. However, cell-based screening approaches need to be approached with care. For example, fixation of the cells can obscure off-target interactions, as demonstrated by a CD19 antibody showing off-target binding only under unfixed conditions [88].

Screening antibodies using relevant conditions can often reveal unpredicted interactions that might affect efficacy or safety and point to pharmacokinetic and mechanistic issues *in vivo*. For example, post-translational modifications such as glycosylations can change with disease. Glycan-dependent interactions can be identified by screening receptor library with glycosidase treated library and provide comprehensive information on both protein and glycan-dependent off-target binding. Additionally, pH-dependent antibody binding can be assessed by screening at different pH buffer conditions to mimic endosomal binding of FcRn or the acidic tumor microenvironment.

Other than screening approaches, antibody – protein complexes can also be identified using immunoprecipitation (IP) followed by MS to reveal both intended target and off-target interactions. Quantitative proteomics approaches in MS analysis can offer insights into the relative strengths of off-target interactions. Emerging computational prediction methods can complement experimental data by predicting potential off-target interactions based on structural and sequence similarities [107].

Other than traditional antibodies that only target one antigen, bsAbs are engineered to simultaneously bind to two different epitopes. This can allow for the simultaneous stimulation or blocking of multiple interactions and signaling pathways. This can also bring two typically spatially separated things into proximity. In the case of bispecific T-cell engagers (BiTEs), this could be redirecting immune cells to target tumor cells by binding to CD3, a T cell marker, and a tumor-associated antigen. This has been shown to stimulate immune killing of cancer cells. One example of a clinically approved BiTE is blinatumomab, a bsAb targeting CD19 and CD3 which redirects T cells to attack B cell malignancies [108]. Other than bringing together cells, bsAbs can also bring together proteins that don't normally interact on the same cells. This is demonstrated by proteolysis-targeting antibodies (PROTABs) or antibody-based proteolysis targeting chimeras (AbTACs). These are designed to induce proximity between membrane-localized protein-of-interest and membrane-bound E3-ligase leading to the degradation of the protein-of-interest [69, 109].

For cases where the bsAb is designed to bring proteins together, proximity labeling can be an excellent method to identify desired and undesired induced proximity. Here, the labeling enzyme can be attached to the antibody, which identifies the surface interactome of the bsAb, or to one of the bsAb antigens, which identifies changes in the target's interactome. With two different binding arms, bsAbs have twice the potential for off-target interactions. Understanding these interactions is crucial for assessing bsAb safety and guiding the design of strategies to mitigate unintended adverse effects, thereby enhancing therapeutic efficacy and safety.

Due to the ability of antibodies to precisely locate their antigens, antibodies have been used to deliver other molecules. ADCs represent a class of biotherapeutics that combine the high affinity and selectivity of monoclonal antibodies to deliver cytotoxic payloads directly to target cells, minimizing systemic toxicity while enabling the use of agents that are 100–1,000-fold more potent than conventional chemotherapy [110]. Therefore, precise delivery of cytotoxic payloads in ADCs demands rigorous characterization to guarantee that the therapeutics target only the intended cells, minimizing serious off-target effects. ADC mechanism involves antigen binding, receptor-mediated endocytosis, lysosomal trafficking, linker cleavage, payload release, and subsequent cell death induction. This targeted approach significantly enhances the therapeutic index by concentrating the cytotoxic effect on target cells while sparing other tissues, potentially revolutionizing cancer treatment strategies. Advanced biotherapeutics like multi-payload ADCs are designed to deliver multiple cytotoxic agents uniformly, offering enhanced therapeutic efficacy and overcoming resistance mechanisms, unlike single-payload ADCs that deliver only one type of cytotoxic agent [111].

ADCs consist of a mAb, a cytotoxic agent, and a linker moiety. ADCs can have cleavable and non-cleavable linkers. Cleavable linkers are designed to release the cytotoxic drug in response to changes in environmental conditions (pH, redox potential, glutathione concentration, or enzymatic activity) that occur between the bloodstream, the tumor cells, and the tumor microenvironment (TME). They are stable under physiological conditions and, following the internalization of the ADC into the tumor cell, they are rapidly cleaved to ensure selective release of the cytotoxic agent. Non-cleavable linkers require complete lysosomal degradation of the antibody for payload release, offering enhanced stability in circulation and potentially reduced systemic toxicity.

ADCs are complex biotherapeutics that undergo biotransformation *in vivo*, impacting their efficacy and safety [112]. This biotransformation, which includes payload metabolism, linker cleavage, or deconjugation, can result in heterogeneous mixtures that may lose their therapeutic activity partially or entirely. Additionally, factors related to mAbs, linker-specific and payload-specific factors, and their interaction with the tumor and its microenvironment all influence the efficacy and toxicity of ADCs. Payload conjugation often increases the hydrophobicity of mAbs, resulting in reduced stability and increased aggregation, which can lead to non-specific endocytosis in cells not expressing the target antigen [113]. For example, FcγRs mediated uptake or mannose receptor-mediated endocytosis, especially in the liver, can lead to hepatotoxicity and other adverse effects [110, 113]. Furthermore, conjugating payload and linker may prevent antigen binding of the antibody component of ADC or could functionalize it for off-target interactions and toxicity. Understanding off-target toxicities and their mechanisms is essential for developing safer and more effective ADCs. Receptomics methods, including MS-based interactome identification and comparative, unbiased interactome screening of mAbs and mAb-based ADCs, can reveal changes in interactions due to linker and payload conjugation. This information helps design ADCs with reduced off-target effects while preserving therapeutic efficacy. Additionally, the glycosylation patterns of ADCs can impact pharmacokinetics; for example, elevated G0F oligosaccharide content may lead to the uptake of ADC by mannose receptors, particularly in the liver [114]. Strategies used for antibodies can also be applied to ADCs to identify glycan-mediated interactions.

Similarly, immunocytokines are antibody–cytokine fusion molecules that deliver cytokines specifically to target sites, enhancing localized immune stimulation. This targeted approach maximizes the therapeutic efficacy of cytokines while reducing the dose-dependent toxicity often seen with systemic cytokine therapies [115]. These constructs face challenges similar to ADCs in optimizing pharmacokinetics, reducing immunogenicity, as well as managing payload-related adverse effects. Despite several candidates advancing through clinical trials, no immunocytokines have yet received regulatory approval, highlighting the ongoing refinement needed for this promising therapeutic modality.

First-generation lysosome-targeting chimeras (LYTACs) represent another advanced modality. These engineered bifunctional molecules target extracellular or membrane-associated POI for degradation. They consist of an antibody that binds to the target protein and a ligand that binds to lysosomal surface receptors. This dual-targeting approach facilitates the internalization and degradation of the target protein in lysosomes, allowing for precise modulation without inducing cell death unlike ADCs [61, 68].

To address on-target off-tissue toxicity, prodrugs such as probodies are engineered that remain inactive until activated by disease tissues. These molecules feature a mAb with a masking peptide linked to the light chain's N-terminus via a protease-cleavable linker, enabling selective activation within the tumor microenvironment [116]. This strategy enhances safety and allows for higher dosing compared to conventional antibodies. For instance, a probody version of cetuximab aims to reduce skin toxicity while preserving anti-tumor efficacy in head-and-neck and colorectal cancers [117, 118].

Advanced antibody-based modalities such as ADCs, LYTACs, and probodies can benefit from receptomics approaches used for analyzing the entire molecule for off-target interactions. Characterizing the extracellular interactome for target proteins such as lysosomal targeting receptors, membrane E3 ligases and proteases can inform the rational design of more specific antibody binding domains. For instance, selecting binding pockets distinct from orthogonal binding sites on the target can help prevent competition with endogenous proteins or inform about the need for engineering antibodies with higher affinity for orthogonal binding sites. Additionally, unbiased interactomics can provide insights into the coincident degradation of endogenous interactors because of protein-of-interest degradation and the potential range of off-target effects. This comprehensive approach can enhance efficacy, reduce off-target effects, and minimize immunogenicity while predicting potential synergistic or antagonistic interactions.

Lipid Nanoparticles (LNPs), Extracellular Vesicles (EVs), Virus-Like Particles (VLPs) and Viruses

Other than cell therapies, this class of therapeutics is some of the largest and is typically hundreds of nanometers in size [119, 120]. They also range from having a few defined components with LNPs to having surface complexities resembling a cell with EVs and enveloped viruses. In exchange for the increased size and complexity (and associated challenges in characterization, manufacturing, and pricing), this class of therapeutics gains unique and flexible capabilities. While we have grouped these approaches because of their similar interactomic concerns, these approaches are typically used to overcome shortcomings of other therapeutic cargo. Often considered packaging, these modalities can shield cargo therapeutics from degradation and modification, shield patients from systemic exposure to toxic compounds and alter cargo targeting and uptake [119]. This enables applications like gene expression and editing *in vivo*. However, between their large size, multivalent surfaces, and complex nature, these complexes often bind to a variety of proteins that exist in the extracellular environment. Receptomics and extracellular interactomics, especially direct interaction screens, can and have helped ensure that these assemblies are delivered to the right place and trigger the right responses.

LNPs are lipid covered synthetic particles used to protect unstable drugs, solubilized hydrophobic compounds, increase drug safety, provide orthogonal targeting, and promote drug uptake by cells. LNPs are typically manufactured with a few well-defined components and therefore are better understood and have fewer manufacturing challenges than EVs, VLPs, and viruses. Due to the formation of a micelle-like structure, the LNP interior can contain a variety of hydrophobic or hydrophilic environments designed to best contain and release cargo depending on therapeutic needs. Since the LNP interior is inaccessible, the interaction between LNPs and patients is heavily governed by the outer surface. For example, though most LNPs are taken up by the liver and spleen, different lipid composition can affect the distribution between the tissues [119]. The outer surface can also be deliberately modified to enhance more specific targeting. ApoE-conjugated LNPs can shuttle LNPs across the blood brain barrier [121]. EGF antibody-conjugated LNPs can target LNPs to EGF expressing cells [122]. However, with each modification comes different surface properties and the potential for unknown bioactive interactions. ApoE, antibodies, and lipids all have cell-surface receptor interactions and a receptomics approach may identify which interactions matter before they occur in a patient [123]. Several receptomics methods can be used but given the manufacturability of LNPs and the central question of what can bind in any context, a direct screening approach like that used for EVs may be ideal [124].

EVs and VLPs are vesicles or viral capsid-like structures that either bud off cells organically or are induced through the expression of budding factors like viral proteins [120, 125]. Since both the term EV and the term VLP encompass a variety of nanoparticles, the exact delineation between the two varies by publication. Regardless, for EVs and membrane enclosed VLPs, the exterior is often as complicated as the membranes of cells and contain a variety of macromolecules [126]. As a result, these particles can bind hundreds of receptors and secreted factors that recognize protein, lipid and glycan components. Due to the large fluid surfaces, these particles can also have emergent interactions that do not occur for the individual proteins on the surfaces in isolation. This means that the extracellular interactome of these structures cannot be predicted from composition alone and have to be determined empirically [124]. Though this can be a characterization challenge, these particles also have pre-existing interactions that can be exploited. For example, EVs can modulate uptake and circulating half-life through the "Eat Me" and "Don't Eat Me" pathways by naturally displaying the pro-phagocytic phosphatidylserine phospholipid and the anti-phagocytic CD47 protein [120, 127]. EVs can also induce signaling and particular EVs have been shown to display immunomodulatory receptors which can distally repress immune responses [120, 128]. As with LNPs, many approaches can be used for interactomic characterization. However, because of both the expression and binding of EVs to a heterogeneous mix of glycans, this class may present a challenge for MS based approaches. Instead, direct interaction screens have successfully characterized glycan-dependent EV uptake through lectin microarray [129] and identified hundreds of EV binders using an arrayed protein library [124].

Like EVs which have evolved for intercellular communication in the host, viruses have evolved to hijack host cells to express viral proteins making them suited for applications like gene delivery. However, they also carry risks of being mutagenic and immunogenic and have been costly not just monetarily [119]. In the late twentieth century, several patients died due to severe immune responses and induced leukemia from viral therapies [130]. Given these safety concerns, it was critical to better understand virus–patient interactions in order for viruses to be used as therapies. One aspect known to govern viral behavior is interactions between the virus and the host's extracellular proteins. These interactions influence tissue and cell type targeting, immunoregulation, internalization and gene transduction [131]. As a result, there is a long history of using receptomics approaches to study virus-cell surface receptor interactions. For example, the identification of the poliovirus receptor was done by overexpressing a cDNA library derived from a poliovirus sensitive cell in insensitive cells in 1989 [132]. The universal AAV receptor necessary for AAV gene transduction was identified through an unbiased genetic knockout screen in 2016 [8]. Novel entry receptors were identified for cytomegalovirus using arrayed receptor protein libraries in 2018 [63]. All these approaches were also rapidly deployed to find additional receptors and regulators of SARS-CoV-2 virus during the COVID-19 pandemic [133, 134].

While most of those viruses are not a part of the therapeutic repertoire, there are a variety of viral vectors that have been used especially for delivering gene therapies. These primarily include adenoviruses, adeno-associated viruses (AAVs) and herpes simplex viruses (HSV) though other vectors are also being pursued in clinic. Due to the low immunogenicity, low pathogenicity and high capacity for stably transducing both dividing and nondividing cells, AAVs are the most popular choice [130]. Despite a fairly good safety profile, AAVs have low specificity which necessitates high dosages that drive manufacturing challenges, safety concerns, and long-term efficacy loss [1]. One way around these challenges is to engineer the AAV capsids for more specific targeting or tissue accumulation. Insertional mutagenesis engineered AAVs capable of crossing the blood brain barrier. However, the effort also resulted in unexpected, unrelated protein interactions with even a few minor changes to the AAV capsid, leading the recommendations for a receptomics profiling to predict off-target effects of engineered viruses before they enter patients [135].

Aside from gene delivery, therapeutic viruses have also been used for their lytic capability in oncology applications. The first FDA approved viral therapy was an oncolytic HSV, talimogene laherparepvec. While this was still engineered to deliver a transgene, specifically GM-CSF to help call in dendritic cells to induce immune responses, the virus can also directly replicate and kill tumor cells [136]. Therefore, to increase efficacy and lower toxicity, further improvements are often focused on increasing the specificity for cancer cells. This is happening by exploring a range of viral vectors that bind to different cell-surface receptors or by engineering viruses that have higher affinities to tumor associated antigens [137]. As with AAVs, these less explored or engineered viruses interact with host proteins to differing effects. For example, many viruses can induce immunosuppression by binding to immunomodulatory host receptors [131]. Especially as oncolytic viruses are being considered in combination with immunotherapies, a greater understanding of both the pro- and anti-inflammatory interactions of these viruses may increase potency of the dual therapeutic approach.

Cell Therapies

Cell therapies are categorized into stem cell-based and non-stem cell-based approaches. Stem cell therapies utilize adult stem cells, such as mesenchymal stem cells (MSCs), hematopoietic stem cells (HSCs), and neural stem cells (NSCs), for conditions ranging from blood cancers to neurodegenerative diseases [138]. These cells can differentiate into various types and secrete factors that promote healing and modulate the immune system. Non-stem cell-based therapies involve differentiated somatic cells, including T cells, Natural Killer (NK) cells, dendritic cells, fibroblasts, macrophages, and pancreatic islet cells, used for therapeutic purposes. Adoptive cell transfer (ACT) is a type of non-stem cell-based immunotherapy that manipulates or expands immune cells to target cancer [139]. Key ACT modalities include CAR-T cells, CAR-NK, CAR-macrophage (M), tumor-infiltrating lymphocytes, viral-specific cytotoxic T lymphocytes (v-CTL), and regulatory T cells (Treg) [140, 141].

Unlike conventional biotherapeutics, cell therapies have the potential to directly modify or replace damaged tissues or cells. These cells, ranging from 1 to 25 μm, are significantly larger than other modalities. The complexity of cell therapy extends beyond mere size to include diverse surface compositions, and extensive cellular machinery. Additionally, the composition of the cells can dynamically change by sensing a variety of extrinsic clues such as interaction with other cells or extracellular matrix, physical force, and microenvironment factors like gradient of cytokines, nutrient, and oxygen availability [142]. These dynamic interactions lead to heterogeneous alterations in proteins, lipids, glycans, and secretome expression. Such variabilities pose significant challenges in maintaining the consistent potency and safety of therapies, and ultimately securing regulatory approval [143, 144].

Many of these responses are driven by receptor–ligand changes on the surface of cells. Therefore, understanding the complete receptor binding profiling of both therapeutic cells and their target tissues can reveal molecular mechanisms driving or undermining efficacy. This is especially true for immune cell therapies like CAR-T cells, where receptor-mediated interactions drive therapeutic effects. These interactions are dependent not only on the proteins but can also be modulated by post-translational modifications (PTMs) of the receptors. Glycosylation has been shown to strongly affect cell therapy and regulates NK cell function [145]. Additionally, interactions with heparan sulfate proteoglycans (HSPGs) can stabilize the receptor on the cell surface, enhancing engagement with ligands and leading to a stronger cellular response [146]. MS can identify the differences in the post-translational modifications of surface protein in the therapeutic cells, which can impact their function and stability [147]. In stem cell therapies, characterizing protein composition changes during differentiation can ensure the desired cell type is achieved.

Advanced techniques such as single-cell proteomics and flow cytometry are used for characterization of therapeutic cells [148, 149]. For example, SCoPE2, a single-cell proteomics method, revealed protein expression heterogeneity in macrophages derived from the same type of monocytes [149]. Similarly, single-cell microarray method has been used to compare intracellular calcium release before and after ligand stimulation in B cells, offering insights into cellular signaling pathways [150]. These methods can be extended to study various cell-type therapies and cellular signaling, providing valuable information about protein binders, activation mechanisms, and cellular responses.

To identify novel regulators of therapeutic cells, unbiased reporter-based, membrane receptor screening method is employed. For example, screening of ~6,500 membrane receptors identified Siglec-15 as a novel immune suppressor of T cells, which is upregulated on human cancer and tumor-infiltrating myeloid cells. Targeting Siglec-15 could be a novel approach enhancing T-cell function to improve cell therapies [151]. Extending this method to other types of cell therapies can be advantageous in multiple ways. For example, identification of critical ligand–receptor pairs that regulate cell fate, differentiation, and function of therapeutic cells could be leveraged to optimize therapeutic cells. Phenotypic screening with high-content imaging (HCI) or signaling-based receptor library screening can be performed under relevant conditions, such as co-culture with tumor cells or in presence of immunosuppressive factors. Studying interactions between stem cells and their microenvironment can provide insights into the complex signaling networks that regulate their behavior. For instance, a study used a reporter-based Jurkat expressing CAR and NF-κB and NFAT dual reporter for high-throughput screening of CAR constructs. By co-culturing these reporter cells with stimulator cells, high-throughput screening of CAR constructs successfully identified a rare functional CAR variant present at a frequency of only 6 in around 1 million [152].

Identifying receptors or secreted factors that regulate cell fate can be valuable for optimizing cell therapy manufacturing protocols. For example, IFNγ regulates the immunomodulatory activity of human bone marrow mesenchymal stem cells [153]. Similar insights into factors influencing cell expansion and survival can enhance the effectiveness and consistency of cell-based therapies. Receptomics can also reveal how cryopreservation affects cell surface receptors, enabling the development of improved preservation methods that maintain cellular function. Understanding the receptor profile changes during cell expansion and manipulation can inform the development of optimized manufacturing protocols that maintain desired cellular characteristics. Additionally, screening with clinically relevant samples, for example, cells from multiple donors can be performed to identify common and donor-specific factors, potentially leading to more personalized approaches.

Label-free, cell-based BLI interferometry (c-BLI) can be used to rapidly screen cell therapy candidates, like different CAR constructs, for desired binding and functional properties. It can also measure the expression levels and binding capabilities of key receptors on therapeutic cells, such as CAR on CAR-T cells, which can help in quality control and optimization. The interaction between therapeutic and target cells, for example, CAR-T cells and tumor cells, can provide insights into the kinetics and strength of cell–cell binding. The high-throughput nature of BLI enables efficient screening and characterization of cell therapy candidates [154]. Additionally, identification and characterization of the secretome of the therapeutic cell by MS-based methods can also aid in understanding the mechanism of effector function [153, 155].

The PK/PD of cell therapies differs significantly from other biotherapeutics, as they do not adhere to standard dose–response relationships due to their unique behaviors. Cells in these therapies can proliferate, differentiate, and migrate within the body, complicating the prediction of their pharmacological effects [156]. Personalized dosing strategies are often necessary to account for these dynamic processes [157]. Besides understanding the mechanism of effector function, comprehensive receptor screening of therapeutic cells can also help in pharmacological modeling [157]. Adverse side effects are significant concerns for cell therapies. For example, CAR-T cells have been associated with cytokine release syndrome (CRS) and neurotoxicity, on target-off tumor recognition [158–160]. Identification of the extracellular interactome of the engineered therapeutic cells can identify potential immunogenic epitopes, allowing for engineering of cells with reduced risk of immune rejection, crucial for improving the safety, and efficacy of allogeneic cell therapies. In addition, reporter-based screening can enhance selectivity, and safety by reducing on-target-off-tissue effects. For example, IL-2 signaling reporters were used to select CARs that activate only at high antigen levels on tumor cells, minimizing activation on healthy cells with lower antigen levels. This strategy enhanced CAR selectivity and safety, leading to more effective CAR-T cell therapies [161].

Genetic modifications, such as the insertion of CARs or the knockout of inhibitory receptors (e.g. PD-1), can direct the cells to specific tissues or enhance their activity. Additionally, surface modifications using ligands or antibodies can further refine their targeting, as complex as chemically conjugating drug loaded nanoparticles onto the plasma membrane of the cells [162]. However, these modifications also introduce the risk of unexpected interactions. For instance, a study revealed that T-cell receptor (TCR) targeting the MAGE A3 cancer antigen showed no off-target effects in preclinical tests but caused fatal cardiac toxicity in patients. Further analysis revealed that TCR targeted a peptide from the Titin muscle protein, leading to toxicity [163]. This highlights the need for rigorous and comprehensive preclinical testing with receptor and secreted factor libraries to avoid off-target effects.

In summary, cell therapies are on the cutting edge of therapeutic innovation. However, their complexity, variability, and safety concerns necessitate a deep understanding of the interactions between therapeutic cells with the host environment. By leveraging advances in receptomics, interactomics, and genetic engineering offer the potential for developing highly targeted and effective treatments, tailored to the specific needs of individual patients.

Summary

The extracellular interactions of biotherapeutics govern important properties like drug targeting, distribution, clearance, uptake, and variability in patient responses. Having a comprehensive list of extracellular interactions for biotherapeutics informs rational improvements to biotherapeutic design. A range of different approaches can be deployed to meet the different challenges posed by the different biotherapeutic modalities, whether it be the fast clearance and degradation for peptide therapeutics, Fc receptor effects for antibodies, targeting and uptake for LNPs, EV, VLPs, and viruses or survival and signaling for cell therapies. The coupling of the continually increasing coverage and sensitivity of extracellular interactomics approaches with biotherapeutic design and discovery will accelerate the drive toward better safety and efficacy for better patient outcomes.

Abbreviations

AAV	adeno-associated virus
AbTAC	Ab-based PROTAC
ADC	antibody–drug conjugate
ADCC	antibody-dependent cell-mediated cytotoxicity
ADCP	antibody-dependent cellular phagocytosis
ADMET	absorption, distribution, metabolism, excretion, and toxicity
AP	affinity purification
BiTE	bispecific T-cell engager
BLI	biolayer interferometry
bsAb	bispecific antibody
CAR-T	chimeric antigen receptor T-cell
CDC	complement-dependent cytotoxicity
CDR	complementarity-determining region
CRISPR	clustered regularly interspaced short palindromic repeats
DAC	degrader antibody conjugate
DCP	disulfide-constrained peptide
EV	extracellular vesicle
Fab	fragment antigen binding
Fc	fragment crystallizable
FcγR	Fc gamma receptor
FcRn	neonatal Fc receptor
FcεR	Fc epsilon receptor
FDA	Food and Drug Administration
Fv	fragment variable
GLP-1	glucagon-like peptide-1
HSV	Herpes simplex virus
Kd	dissociation constant
LNP	lipid nanoparticle
LYTAC	lysosome-targeting chimera
mAb	monoclonal antibody
MOA	mechanism of action
MS	mass spectrometry
PROTAB	proteolysis-targeting antibodies
PCR	polymerase chain reaction
PK/PD	pharmacokinetics/pharmacodynamics
PTM	post-translational modifications
VLP	virus-like particle

References

1 Wang, J.H., Gessler, D.J., Zhan, W. et al. (2024). Adeno-associated virus as a delivery vector for gene therapy of human diseases. *Signal Transduct. Target. Ther.* 9 (1): 1–33.

2 Mullard, A. (2017). FDA approves first CAR T therapy. *Nat. Rev. Drug Discov.* 16 (10): 669–669.

3 Al Musaimi, O., Al Shaer, D., Albericio, F., and de la Torre, B.G. (2023). 2022 FDA TIDES (peptides and oligonucleotides) harvest. *Pharm.* 16 (3): 336.

4 Sheridan, C. (2023). The world's first CRISPR therapy is approved: who will receive it? *Nat. Biotechnol.* 42 (1): 3–4.

5 Barnes, K. (2018). The first monoclonal antibody therapy. *Nat. Resour. Res.* (cited 2024 Jul 3); Available from: https://www.nature.com/articles/d42859-018-00024-6.

6 Sun, A. and Benet, L.Z. (2020). Late-stage failures of monoclonal antibody drugs: a retrospective case study analysis. *Pharmacology* 105 (3–4): 145–163.

7 Grey, H.M., Anderson, C.L., Heusser, C.H. et al. (1977). Structural and functional heterogeneity of Fc receptors. *Cold Spring Harb. Symp. Quant. Biol.* 41: 315–321.

8 Pillay, S., Meyer, N.L., Puschnik, A.S. et al. (2016). An essential receptor for adeno-associated virus infection. *Nature* 530 (7588): 108–112.

9 Shieh, P.B., Kuntz, N.L., Dowling, J.J. et al. (2023). Safety and efficacy of gene replacement therapy for X-linked myotubular myopathy (ASPIRO): a multinational, open-label, dose-escalation trial. *Lancet Neurol.* 22 (12): 1125–1139.

10 Chen, D.S. and Mellman, I. (2013). Oncology meets immunology: the Cancer-immunity cycle. *Immunity* 39 (1): 1–10.

11 Lobo, E.D., Hansen, R.J., and Balthasar, J.P. (2004). Antibody pharmacokinetics and pharmacodynamics. *J. Pharm. Sci.* 93 (11): 2645–2668.

12 van den Anker, J., Reed, M.D., Allegaert, K., and Kearns, G.L. (2018). Developmental changes in pharmacokinetics and pharmacodynamics. *J. Clin. Pharmacol.* 58 (S10): S10–S25.

13 Li, Y., Meng, Q., Yang, M. et al. (2019). Current trends in drug metabolism and pharmacokinetics. *Acta Pharm. Sin. B* 9 (6): 1113–1144.

14 Yu, J., Song, Y., and Tian, W. (2020). How to select IgG subclasses in developing anti-tumor therapeutic antibodies. *J Hematol Oncol.* 13 (1): 45.

15 Jain, N., Smith, S.W., Ghone, S., and Tomczuk, B. (2015). Current ADC linker chemistry. *Pharm. Res.* 32 (11): 3526–3540.

16 Chung, C.H., Mirakhur, B., Chan, E. et al. (2008). Cetuximab-induced anaphylaxis and IgE specific for galactose-α-1,3-galactose. *N. Engl. J. Med.* 358 (11): 1109–1117.

17 Gonzalez, L.C. (2012). Protein microarrays, biosensors, and cell-based methods for secretome-wide extracellular protein–protein interaction mapping. *Methods* 57 (4): 448–458.

18 Wood, L. and Wright, G.J. (2019). Approaches to identify extracellular receptor–ligand interactions. *Curr. Opin. Struct. Biol.* 56: 28–36.

19 Low, T.Y., Syafruddin, S.E., Mohtar, M.A. et al. (2021). Recent progress in mass spectrometry-based strategies for elucidating protein–protein interactions. *Cell. Mol. Life Sci.* 78 (13): 5325–5339.

20 Cao, S., Martinez-Martin, N., Cao, S., and Martinez-Martin, N. (2021). Unbiased identification of extracellular protein–protein interactions for drug target and biologic drug discovery. In: *High-Throughput Screening for Drug Discovery [Internet].* IntechOpen (cited 2024 Jul 5). Available from: https://www.intechopen.com/chapters/76914.

21 Wright, J. and Signal, G. (2009). Initiation in biological systems: the properties and detection of transient extracellular protein interactions. *Mol. BioSyst.* 5 (12): 1405–1412.

22 Wright, G.J., Martin, S., Bushell, K.M., and Söllner, C. (2010). High-throughput identification of transient extracellular protein interactions. *Biochem. Soc. Trans.* 38 (4): 919–922.

23 Martinez-Martin, N. (2017). Technologies for Proteome-Wide Discovery of extracellular host-pathogen interactions. *J Immunol Res* 2017 (1): 2197615.

24 Zhou, Z.H., Stone, C.A., Jakubovic, B. et al. (2021). Anti-PEG IgE in anaphylaxis associated with polyethylene glycol. *J. Allergy Clin. Immunol Pract.* 9 (4): 1731, e3–1733.

25 Cohen, M. (2015). Notable aspects of glycan-protein interactions. *Biomol. Ther.* 5 (3): 2056–2072.

26 Wang, M., Heo, G.Y., Omarova, S. et al. (2012). Sample Prefractionation for mass spectrometry quantification of Low-abundance membrane proteins. *Anal. Chem.* 84 (12): 5186–5191.

27 Qin, W., Cho, K.F., Cavanagh, P.E., and Ting, A.Y. (2021). Deciphering molecular interactions by proximity labeling. *Nat. Methods* 18 (2): 133–143.

28 Sanchez, A.D., Branon, T.C., Cote, L.E. et al. (2021). Proximity labeling reveals non-centrosomal

microtubule-organizing center components required for microtubule growth and localization. *Curr. Biol.* 31 (16): 3586–3600.e11.

29 Oakley, J.V., Buksh, B.F., Fernández, D.F. et al. (2022). Radius measurement via super-resolution microscopy enables the development of a variable radii proximity labeling platform. *Proc. Natl. Acad. Sci.* 119 (32): e2203027119.

30 Lin, Z., Schaefer, K., Lui, I. et al. (2023). Multi-scale photocatalytic proximity labeling reveals cell surface neighbors on and between cells [Internet]. *bioRxiv* 2023.10.28.564055. (cited 2024 Jul 9) Available from: https://www.biorxiv.org/content/10.1101/2023.10.28.564055v1.

31 Frei, A.P., Jeon, O.Y., Kilcher, S. et al. (2012). Direct identification of ligand-receptor interactions on living cells and tissues. *Nat. Biotechnol.* 30 (10): 997–1001.

32 Loh, K.H., Stawski, P.S., Draycott, A.S. et al. (2016). Proteomic analysis of unbounded cellular compartments: synaptic clefts. *Cell* 166 (5): 1295–1307.e21.

33 Tremblay, T.L. and Hill, J.J. (2017). Biotin-transfer from a trifunctional crosslinker for identification of cell surface receptors of soluble protein ligands. *Sci. Rep.* 7 (1): 46574.

34 Sobotzki, N., Schafroth, M.A., Rudnicka, A. et al. (2018). HATRIC-based identification of receptors for orphan ligands. *Nat. Commun.* 9 (1): 1519.

35 Lee, S.Y., Cheah, J.S., Zhao, B. et al. (2023). Engineered allostery in light-regulated LOV-Turbo enables precise spatiotemporal control of proximity labeling in living cells. *Nat. Methods* 20 (6): 908–917.

36 Fields, S., Song, O., and kyu. (1989). A novel genetic system to detect protein–protein interactions. *Nature* 340 (6230): 245–246.

37 Kerppola, T.K. (2008). Bimolecular fluorescence complementation (BiFC) analysis as a probe of protein interactions in living cells. *Annu. Rev. Biophys.* 37: 465–487.

38 Azad, T., Tashakor, A., and Hosseinkhani, S. (2014). Split-luciferase complementary assay: applications, recent developments, and future perspectives. *Anal. Bioanal. Chem.* 406 (23): 5541–5560.

39 Dixon, A.S., Schwinn, M.K., Hall, M.P. et al. (2016). NanoLuc complementation reporter optimized for accurate measurement of protein interactions in cells. *ACS Chem. Biol.* 11 (2): 400–408.

40 Chong, Z.S., Ohnishi, S., Yusa, K., and Wright, G.J. (2018). Pooled extracellular receptor-ligand interaction screening using CRISPR activation. *Genome Biol.* 19: 205.

41 Hart, T., Tong, A.H.Y., Chan, K. et al. (2017). Evaluation and Design of Genome-Wide CRISPR/SpCas9 knockout screens. *G3* 7 (8): 2719–2727.

42 Glick, Y., Ben-Ari, Y., Drayman, N. et al. (2016). Pathogen receptor discovery with a microfluidic human membrane protein array. *Proc. Natl. Acad. Sci.* 113 (16): 4344–4349.

43 Bushell, K.M., Söllner, C., Schuster-Boeckler, B. et al. (2008). Large-scale screening for novel low-affinity extracellular protein interactions. *Genome Res.* 18 (4): 622–630.

44 Adamson, B., Norman, T.M., Jost, M. et al. (2016). A multiplexed single-cell CRISPR screening platform enables systematic dissection of the unfolded protein response. *Cell* 167 (7): 1867–1882.e21.

45 Dixit, A., Parnas, O., Li, B. et al. (2016). Perturb-Seq: dissecting molecular circuits with scalable single-cell RNA profiling of pooled genetic screens. *Cell* 167 (7): 1853, e17–1866.

46 Bade, R., Chan, H.F., and Reynisson, J. (2010). Characteristics of known drug space. Natural products, their derivatives and synthetic drugs. *Eur. J. Med. Chem.* 45 (12): 5646–5652.

47 Diao, L. and Meibohm, B. (2013). Pharmacokinetics and pharmacokinetic–pharmacodynamic correlations of therapeutic peptides. *Clin. Pharmacokinet.* 52 (10): 855–868.

48 Zhang, H. and Chen, S. (2022). Cyclic peptide drugs approved in the last two decades (2001–2021). *RSC Chem. Biol.* 3 (1): 18–31.

49 Butler, M.S. (2004). The role of natural product chemistry in drug discovery. *J. Nat. Prod.* 67 (12): 2141–2153.

50 DePierre, J.W. and Karnovsky, M.L. (1973). Plasma membranes of mammalian cells. *J. Cell Biol.* 56 (2): 275–303.

51 Atanasov, A.G., Zotchev, S.B., Dirsch, V.M., and Supuran, C.T. (2021). Natural products in drug discovery: advances and opportunities. *Nat. Rev. Drug Discov.* 20 (3): 200–216.

52 Weersma, R.K., Zhernakova, A., and Fu, J. (2020). Interaction between drugs and the gut microbiome. *Gut* 69 (8): 1510.

53 Fosgerau, K. and Hoffmann, T. (2015). Peptide therapeutics: current status and future directions. *Drug Discov. Today* 20 (1): 122–128.

54 Nauck, M.A., Quast, D.R., Wefers, J., and Meier, J.J. (2021). GLP-1 receptor agonists in the treatment of type 2 diabetes – state-of-the-art. *Mol. Metab.* 46: 101102.

55 Bloomberg. The Pharmaceutical Rivalry Behind the Weight-Loss Drug Boom. http://Bloomberg.com

[Internet]. 2024 (cited 2024 Jul 25); Available from: https://www.bloomberg.com/news/articles/2024-01-25/zepbound-vs-wegovy-the-race-to-corner-the-weight-loss-drug-market

56 Petersen, J., Ludwig, M.Q., Juozaityte, V. et al. (2024). GLP-1-directed NMDA receptor antagonism for obesity treatment. *Nature* 629 (8014): 1133–1141.

57 Mentlein, R., Gallwitz, B., and Schmidt, W.E. (1993). Dipeptidyl-peptidase IV hydrolyses gastric inhibitory polypeptide, glucagon-like peptide-1(7–36)amide, peptide histidine methionine and is responsible for their degradation in human serum. *Eur. J. Biochem.* 214 (3): 829–835.

58 Menacho-Melgar, R., Decker, J.S., Hennigan, J.N., and Lynch, M.D. (2019). A review of lipidation in the development of advanced protein and peptide therapeutics. *J. Control. Release* 295: 1–12.

59 Luther, A., Urfer, M., Zahn, M. et al. (2019). Chimeric peptidomimetic antibiotics against gram-negative bacteria. *Nature* 576 (7787): 452–458.

60 (2014). Heparan sulfate proteoglycan as a cell-surface endocytosis receptor. *Matrix Biol.* 35: 51–55.

61 Banik, S.M., Pedram, K., Wisnovsky, S. et al. (2020). Lysosome-targeting chimaeras for degradation of extracellular proteins. *Nature* 584 (7820): 291–297.

62 Soini, L., Leysen, S., Davis, J., and Ottmann, C. (2022). Molecular glues to stabilise protein–protein interactions. *Curr. Opin. Chem. Biol.* 69: 102169.

63 Martinez-Martin, N., Marcandalli, J., Huang, C.S. et al. (2018). An unbiased screen for human cytomegalovirus identifies Neuropilin-2 as a central viral receptor. *Cell* 174 (5): 1158, e19–1171.

64 Trier, N., Hansen, P., and Houen, G. (2019). Peptides, Antibodies, Peptide Antibodies and More. *Int. J. Mol. Sci.* 20 (24): 6289.

65 Goulet, D.R. and Atkins, W.M. (2020). Considerations for the Design of Antibody-Based Therapeutics. *J. Pharm. Sci.* 109 (1): 74–103.

66 Carter, P.J. and Rajpal, A. (2022). Designing antibodies as therapeutics. *Cell* 185 (15): 2789–2805.

67 Weisser, N.E. and Hall, J.C. (2009). Applications of single-chain variable fragment antibodies in therapeutics and diagnostics. *Biotechnol. Adv.* 27 (4): 502–520.

68 Ahn, G., Banik, S.M., Miller, C.L. et al. (2021). LYTACs that engage the asialoglycoprotein receptor for targeted protein degradation. *Nat. Chem. Biol.* 17 (9): 937–946.

69 Cotton, A.D., Nguyen, D.P., Gramespacher, J.A. et al. (2021). Development of antibody-based PROTACs for the degradation of the cell-surface immune checkpoint protein PD-L1. *J. Am. Chem. Soc.* 143 (2): 593–598.

70 Crescioli, S., Kaplon, H., Chenoweth, A. et al. (2024). Antibodies to watch in 2024. *MAbs* 16 (1): 2297450.

71 Keri, D., Walker, M., Singh, I. et al. (2024). Next generation of multispecific antibody engineering. *Antib. Ther.* 7 (1): 37–52.

72 Delidakis, G., Kim, J.E., George, K., and Georgiou, G. (2022). Improving antibody therapeutics by manipulating the Fc domain: immunological and structural considerations. *Annu. Rev. Biomed. Eng.* 24: 249–274.

73 Li, F., Smith, P., and Ravetch, J.V. (2014 Apr 1). Inhibitory Fcγ receptor is required for the maintenance of tolerance through distinct mechanisms. *J. Immun., Balt.* 192 (7): 3021–3028.

74 Kang, T.H. and Jung, S.T. (2019). Boosting therapeutic potency of antibodies by taming Fc domain functions. *Exp. Mol. Med.* 51 (11): 1–9.

75 Martin, W.L. and Bjorkman, P.J. (1999). Characterization of the 2:1 complex between the class I MHC-related Fc receptor and its Fc ligand in solution. *Biochemistry* 38 (39): 12639–12647.

76 Popov, S., Hubbard, J.G., Kim, J. et al. (1996). The stoichiometry and affinity of the interaction of murine Fc fragments with the MHC class I-related receptor, FcRn. *Mol. Immunol.* 33 (6): 521–530.

77 Heo, Y.A. (2022 Feb). Efgartigimod: first approval. *Drugs* 82 (3): 341–348.

78 Damelang, T., Brinkhaus, M., van Osch, T.L.J. et al. (2023). Impact of structural modifications of IgG antibodies on effector functions. *Front. Immunol.* 14: 1304365.

79 Chiu, M.L., Goulet, D.R., Teplyakov, A., and Gilliland, G.L. (2019). Antibody structure and function: the basis for engineering therapeutics. *Antibodies (Basel)* 8 (4): 55.

80 Lund, J., Pound, J.D., Jones, P.T. et al. (1992). Multiple binding sites on the CH_2 domain of IgG for mouse Fc gamma R11. *Mol. Immunol.* 29 (1): 53–59.

81 Wang, X., Mathieu, M., and Brezski, R.J. (2018). IgG Fc engineering to modulate antibody effector functions. *Protein Cell* 9 (1): 63–73.

82 Richards, J.O., Karki, S., Lazar, G.A. et al. (2008). Optimization of antibody binding to FcgammaRIIa enhances macrophage phagocytosis of tumor cells. *Mol. Cancer Ther.* 7 (8): 2517–2527.

83 Bang, Y.J., Giaccone, G., Im, S.A. et al. (2017). First-in-human phase 1 study of margetuximab (MGAH22), an Fc-modified chimeric monoclonal antibody, in patients with HER2-positive advanced solid tumors. *Ann. Oncol. Off J. Eur. Soc Med. Oncol.* 28 (4): 855–861.

84 van Lengerich, B., Zhan, L., Xia, D. et al. (2023). A TREM2-activating antibody with a blood-brain barrier transport vehicle enhances microglial metabolism

in Alzheimer's disease models. *Nat. Neurosci.* 26 (3): 416–429.

85 Chew, K.S., Wells, R.C., Moshkforoush, A. et al. (2023). CD98hc is a target for brain delivery of biotherapeutics. *Nat. Commun.* 14 (1): 5053.

86 Yu, Y.J., Zhang, Y., Kenrick, M. et al. (2011). Boosting brain uptake of a therapeutic antibody by reducing its affinity for a transcytosis target. *Sci. Transl. Med.* 3 (84): 84ra44.

87 Screnci, B., Stafford, L.J., Barnes, T. et al. (2022). Antibody specificity against highly conserved membrane protein Claudin 6 driven by single atomic contact point. *iScience* 25 (12): 105665.

88 Norden, D.M., Navia, C.T., Sullivan, J.T., and Doranz, B.J. (2024). The emergence of cell-based protein arrays to test for polyspecific off-target binding of antibody therapeutics. *MAbs* 16 (1): 2393785.

89 Ausserwöger, H., Schneider, M.M., Herling, T.W. et al. (2022). Non-specificity as the sticky problem in therapeutic antibody development. *Nat. Rev. Chem.* 6 (12): 844–861.

90 Mo, H., Huang, J., Xu, J. et al. (2018). Safety, anti-tumour activity, and pharmacokinetics of fixed-dose SHR-1210, an anti-PD-1 antibody in advanced solid tumours: a dose-escalation, phase 1 study. *Br. J. Cancer* 119 (5): 538–545.

91 Finlay, W.J.J., Coleman, J.E., Edwards, J.S., and Johnson, K.S. (2019). Anti-PD1 "SHR-1210" aberrantly targets pro-angiogenic receptors and this polyspecificity can be ablated by paratope refinement. *MAbs* 11 (1): 26–44.

92 Li, W., Wei, Z., Yang, X. et al. (2019). Salvage therapy of reactive capillary hemangiomas: Apatinib alleviates the unique adverse events induced by camrelizumab in non-small cell lung cancer. *J. Cancer Res. Ther.* 15 (7): 1624–1628.

93 Ohresser, M., Olive, D., Vanhove, B., and Watier, H. (2006). Risk in drug trials. *Lancet (London, Engand)* 368 (9554): 2205–2206.

94 Waibler, Z., Sender, L.Y., Merten, C. et al. (2008). Signaling signatures and functional properties of anti-human CD28 superagonistic antibodies. *PLoS One* 3 (3): e1708.

95 Loberg, L.I., Chhaya, M., Ibraghimov, A. et al. (2021). Off-target binding of an anti-amyloid beta monoclonal antibody to platelet factor 4 causes acute and chronic toxicity in cynomolgus monkeys. *MAbs* 13 (1): 1887628.

96 Bumbaca, D., Wong, A., Drake, E. et al. (2011). Highly specific off-target binding identified and eliminated during the humanization of an antibody against FGF receptor 4. *MAbs* 3 (4): 376–386.

97 Vugmeyster, Y., Szklut, P., Wensel, D. et al. (2011 Jul). Complex pharmacokinetics of a humanized antibody against human amyloid beta peptide, anti-abeta Ab2, in nonclinical species. *Pharm. Res.* 28 (7): 1696–1706.

98 Finton, K.A.K., Larimore, K., Larman, H.B. et al. (2013). Autoreactivity and exceptional CDR plasticity (but not unusual polyspecificity) hinder elicitation of the anti-HIV antibody 4E10. *PLoS Pathog.* 9 (9): e1003639.

99 Bostrom, J., Yu, S.F., Kan, D. et al. (2009). Variants of the antibody herceptin that interact with HER2 and VEGF at the antigen binding site. *Science* 323 (5921): 1610–1614.

100 Yang, D., Kroe-Barrett, R., Singh, S. et al. (2017). IgG cooperativity - is there allostery? Implications for antibody functions and therapeutic antibody development. *MAbs* 9 (8): 1231–1252.

101 Cornwell, O., Bond, N.J., Radford, S.E., and Ashcroft, A.E. (2019). Long-range conformational changes in monoclonal antibodies revealed using FPOP-LC-MS/MS. *Anal. Chem.* 91 (23): 15163–15170.

102 FDA. (2024) Considerations for the Development of Chimeric Antigen Receptor (CAR) T Cell Products [Internet]. U.S. Department of Health and Human Services. Available from: https://www.google.com/url?q=https://www.fda.gov/regulatory-information/search-fda-guidance-documents/considerations-development-chimeric-antigen-receptor-car-t-cell-products&sa=D&source=docs&ust=1726365406444079&usg=AOvVaw0n-PSE9pyfMnH3Mqn7kXyR

103 Brown, E.P., Dowell, K.G., Boesch, A.W. et al. (2017). Multiplexed Fc array for evaluation of antigen-specific antibody effector profiles. *J. Immunol. Methods* 443: 33–44.

104 Lazar, G.A., Dang, W., Karki, S. et al. (2006). Engineered antibody Fc variants with enhanced effector function. *Proc. Natl. Acad. Sci. USA* 103 (11): 4005–4010.

105 Freeth, J. and Soden, J. (2020). New advances in cell microarray technology to expand applications in target deconvolution and off-target screening. *SLAS Discov Adv. Life Sci. R D.* 25 (2): 223–230.

106 Tucker, D.F., Sullivan, J.T., Mattia, K.A. et al. (2018). Isolation of state-dependent monoclonal antibodies against the 12-transmembrane domain glucose transporter 4 using virus-like particles. *Proc. Natl. Acad. Sci. USA* 115 (22): E4990–E4999.

107 Saksena, S.D., Liu, G., Banholzer, C. et al. (2022). Computational counterselection identifies nonspecific therapeutic biologic candidates. *Cell Rep. Methods* 2 (7): 100254.

108 Kauer, J., Vogt, F., Hagelstein, I. et al. (2021). CD18 antibody application blocks unwanted off-target T cell activation caused by bispecific antibodies. *Cancer* 13 (18): 4596.

109 Marei, H., Tsai, W.T.K., Kee, Y.S. et al. (2022). Antibody targeting of E3 ubiquitin ligases for receptor degradation. *Nature* 610 (7930): 182–189.

110 Nguyen, T.D., Bordeau, B.M., and Balthasar, J.P. (2023). Mechanisms of ADC toxicity and strategies to increase ADC tolerability. *Cancer* 15 (3): 713.

111 Journeaux, T. and Bernardes, G.J.L. (2024). Homogeneous multi-payload antibody-drug conjugates. *Nat. Chem.* 16 (6): 854–870.

112 Tumey, L.N., Rago, B., and Han, X. (2015). In vivo biotransformations of antibody-drug conjugates. *Bioanalysis* 7 (13): 1649–1664.

113 Aoyama, M., Tada, M., Yokoo, H. et al. (2022). Fcγ receptor-dependent internalization and off-target cytotoxicity of antibody-drug conjugate aggregates. *Pharm. Res.* 39 (1): 89–103.

114 Gorovits, B. and Krinos-Fiorotti, C. (2013). Proposed mechanism of off-target toxicity for antibody-drug conjugates driven by mannose receptor uptake. *CII.* 62 (2): 217–223.

115 Neri, D. and Sondel, P.M. (2016). Immunocytokines for cancer treatment: past, present and future. *Curr. Opin. Immunol.* 40: 96–102.

116 Polu, K.R. and Lowman, H.B. (2014). Probody therapeutics for targeting antibodies to diseased tissue. *Expert. Opin. Biol. Ther.* 14 (8): 1049–1053.

117 Boone, S.L., Rademaker, A., Liu, D. et al. (2007). Impact and management of skin toxicity associated with anti-epidermal growth factor receptor therapy: survey results. *Oncology* 72 (3–4): 152–159.

118 Desnoyers, L.R., Vasiljeva, O., Richardson, J.H. et al. (2013). Tumor-specific activation of an EGFR-targeting probody enhances therapeutic index. *Sci. Transl. Med.* 5 (207): 207ra144.

119 Xu, L., Wang, X., Liu, Y. et al. (2022). Lipid nanoparticles for drug delivery. *Adv NanoBiomed Res.* 2 (2): 2100109.

120 van der Meel, R., Fens, M.H.A.M., Vader, P. et al. (2014). Extracellular vesicles as drug delivery systems: lessons from the liposome field. *J. Control. Release* 195: 72–85.

121 Topal, G.R., Mészáros, M., Porkoláb, G. et al. (2021). ApoE-Targeting Increases the Transfer of Solid Lipid Nanoparticles with Donepezil Cargo across a Culture Model of the Blood–Brain Barrier. *Pharm.* 13 (1): 38.

122 Zhai, J., Luwor, R.B., Ahmed, N. et al. (2018). Paclitaxel-loaded self-assembled lipid nanoparticles as targeted drug delivery Systems for the Treatment of aggressive ovarian Cancer. *ACS Appl. Mater. Interfaces* 10 (30): 25174–25185.

123 Freeman, G.J., Casasnovas, J.M., Umetsu, D.T., and DeKruyff, R.H. (2010). TIM genes: a family of cell surface phosphatidylserine receptors that regulate innate and adaptive immunity. *Immunol. Rev.* 235 (1): 172–189.

124 Cao, S., Peterson, S.M., Müller, S. et al. (2021). A membrane protein display platform for receptor interactome discovery. *Proc. Natl. Acad. Sci.* 118 (39): e2025451118.

125 Nooraei, S., Bahrulolum, H., Hoseini, Z.S. et al. (2021). Virus-like particles: preparation, immunogenicity and their roles as nanovaccines and drug nanocarriers. *J. Nanobiotechnology* 19 (1): 59.

126 Keerthikumar, S., Chisanga, D., Ariyaratne, D. et al. (2016). ExoCarta: a web-based compendium of Exosomal cargo. *J. Mol. Biol.* 428 (4): 688–692.

127 Kim, H.S., Choi, D.Y., Yun, S.J. et al. (2012). Proteomic analysis of microvesicles derived from human mesenchymal stem cells. *J. Proteome Res.* 11 (2): 839–849.

128 Poggio, M., Hu, T., Pai, C.C. et al. (2019). Suppression of Exosomal PD-L1 induces systemic anti-tumor immunity and memory. *Cell* 177 (2): 414–427.e13.

129 Williams, C., Pazos, R., Royo, F. et al. (2019). Assessing the role of surface glycans of extracellular vesicles on cellular uptake. *Sci. Rep.* 9 (1): 11920.

130 Zhao, Z., Anselmo, A.C., and Mitragotri, S. (2022). Viral vector-based gene therapies in the clinic. *BioTM.* 7 (1): e10258.

131 Schneider-Schaulies, J. (2000). Cellular receptors for viruses: links to tropism and pathogenesis. *J. Gen. Virol.* 81 (6): 1413–1429.

132 Mendelsohn, C.L., Wimmer, E., and Racaniello, V.R. (1989). Cellular receptor for poliovirus: molecular cloning, nucleotide sequence, and expression of a new member of the immunoglobulin superfamily. *Cell* 56 (5): 855–865.

133 Husain, B., Yuen, K., Sun, D. et al. (2022). Cell-based receptor discovery identifies host factors specifically targeted by the SARS CoV-2 spike. *Commun. Biol.* 5 (1): 788.

134 Daniloski, Z., Jordan, T.X., Wessels, H.H. et al. (2021). Identification of required host factors for SARS-CoV-2 infection in human cells. *Cell* 184 (1): 92–105.e16.

135 Shay, T.F., Jang, S., Chen, X. et al. (2024). Human cell surface-AAV interactomes identify LRP6 as blood-brain-barrier transcytosis receptor and immune cytokine IL3 as AAV9 binder [Internet]. *bioRxiv* 2024.01.05.574399. (cited 2024 Jul 13); Available from: https://www.biorxiv.org/content/10.1101/2024.01.05.574399v1.

136 Ferrucci, P.F., Pala, L., Conforti, F., and Cocorocchio, E. (2021). Talimogene Laherparepvec (T-VEC): an Intralesional Cancer immunotherapy for advanced melanoma. *Cancer* 13 (6): 1383.

137 Kohlhapp, F.J. and Kaufman, H.L. (2016). Molecular pathways: mechanism of action for Talimogene Laherparepvec, a new oncolytic virus immunotherapy. *Clin. Cancer Res.* 22 (5): 1048–1054.

138 Hoang, D.M., Pham, P.T., Bach, T.Q. et al. (2022). Stem cell-based therapy for human diseases. *Signal Transduct. Target. Ther.* 7 (1): 272.

139 Weber, E.W., Maus, M.V., and Mackall, C.L. (2020). The emerging landscape of immune cell therapies. *Cell* 181 (1): 46–62.

140 June, C.H., Riddell, S.R., and Schumacher, T.N. (2015). Adoptive cellular therapy: a race to the finish line. *Sci. Transl. Med.* 7 (280): 280ps7.

141 June, C.H., O'Connor, R.S., Kawalekar, O.U. et al. (2018). CAR T cell immunotherapy for human cancer. *Science* 359 (6382): 1361–1365.

142 Bashor, C.J., Hilton, I.B., Bandukwala, H. et al. (2022). Engineering the next generation of cell-based therapeutics. *Nat. Rev. Drug Discov.* 21 (9): 655–675.

143 Morotti, M., Albukhari, A., Alsaadi, A. et al. (2021). Promises and challenges of adoptive T-cell therapies for solid tumours. *Br. J. Cancer* 124 (11): 1759–1776.

144 KD, S.S., Joga, R., Srivastava, S. et al. (2024). Regulatory landscape and challenges in CAR-T cell therapy development in the US, EU, Japan, and India. *Eur J Pharm Biopharm Off J Arbeitsgemeinschaft Pharm Verfahrenstechnik EV.* 201: 114361.

145 Margraf-Schönfeld, S., Böhm, C., and Watzl, C. (2011). Glycosylation affects ligand binding and function of the activating natural killer cell receptor 2B4 (CD244) protein. *J. Biol. Chem.* 286 (27): 24142–24149.

146 Sarrazin, S., Lamanna, W.C., and Esko, J.D. (2011). Heparan sulfate proteoglycans. *Cold Spring Harb. Perspect. Biol.* 3 (7): a004952.

147 Lombard-Banek, C. and Schiel, J.E. (2020). Mass spectrometry advances and perspectives for the characterization of emerging adoptive cell therapies. *Mol Basel Switz.* 25 (6): 1396.

148 Budnik, B., Levy, E., Harmange, G., and Slavov, N. (2018). SCoPE-MS: mass spectrometry of single mammalian cells quantifies proteome heterogeneity during cell differentiation. *Genome Biol.* 19 (1): 161.

149 Specht, H., Emmott, E., Petelski, A.A. et al. (2021). Single-cell proteomic and transcriptomic analysis of macrophage heterogeneity using SCoPE2. *Genome Biol.* 22 (1): 50.

150 Yamamura, S., Kishi, H., Tokimitsu, Y. et al. (2005). Single-cell microarray for analyzing cellular response. *Anal. Chem.* 77 (24): 8050–8056.

151 Wang, J., Sun, J., Liu, L.N. et al. (2019). Siglec-15 as an immune suppressor and potential target for normalization cancer immunotherapy. *Nat. Med.* 25 (4): 656–666.

152 Rydzek, J., Nerreter, T., Peng, H. et al. (2019). Chimeric antigen receptor library screening using a novel NF-κB/NFAT reporter cell platform. *Mol Ther J Am Soc Gene Ther.* 27 (2): 287–299.

153 Krampera, M., Cosmi, L., Angeli, R. et al. (2006). Role for interferon-gamma in the immunomodulatory activity of human bone marrow mesenchymal stem cells. *Stem Cells (Dayton, Ohio)* 24 (2): 386–398.

154 Verzijl, D., Riedl, T., Parren, P.W.H.I., and Gerritsen, A.F. (2017). A novel label-free cell-based assay technology using biolayer interferometry. *Biosens. Bioelectron.* 87: 388–395.

155 Ren, G., Zhang, L., Zhao, X. et al. (2008). Mesenchymal stem cell-mediated immunosuppression occurs via concerted action of chemokines and nitric oxide. *Cell Stem Cell* 2 (2): 141–150.

156 Aalipour, A., Chuang, H.Y., Murty, S. et al. (2019). Engineered immune cells as highly sensitive cancer diagnostics. *Nat. Biotechnol.* 37 (5): 531–539.

157 Aijaz, A., Vaninov, N., Allen, A. et al. (2019). Convergence of cell pharmacology and drug delivery. *Stem Cells Transl. Med.* 8 (9): 874–879.

158 Bonifant, C.L., Jackson, H.J., Brentjens, R.J., and Curran, K.J. (2016). Toxicity and management in CAR T-cell therapy. *Mol. Ther. Oncolytics.* 3: 16011.

159 Curran, K.J., Pegram, H.J., and Brentjens, R.J. (2012). Chimeric antigen receptors for T cell immunotherapy: current understanding and future directions. *J. Gene Med.* 14 (6): 405–415.

160 Lee, D.W., Gardner, R., Porter, D.L. et al. (2014). Current concepts in the diagnosis and management of cytokine release syndrome. *Blood* 124 (2): 188–195.

161 Di Roberto, R.B., Castellanos-Rueda, R., Frey, S. et al. (2020). A functional screening strategy for engineering chimeric antigen receptors with reduced on-target, off-tumor activation. *Mol. Ther. J. Am. Soc. Gene. Ther.* 28 (12): 2564–2576.

162 Stephan, M.T., Moon, J.J., Um, S.H. et al. (2010). Therapeutic cell engineering with surface-conjugated synthetic nanoparticles. *Nat. Med.* 16 (9): 1035–1041.

163 Cameron, B.J., Gerry, A.B., Dukes, J. et al. (2013). Identification of a Titin-derived HLA-A1-presented peptide as a cross-reactive target for engineered MAGE A3-directed T cells. *Sci. Transl. Med.* 5 (197): 197ra103.

18

Membrane Proteins: Challenging Biotherapeutic Targets

Hiruni S. Jayasekera, Farhana Afrin Mohona, and Michael T. Marty

Department of Chemistry and Biochemistry, The University of Arizona, Tucson, AZ, USA

Membrane Proteins: Challenging Biotherapeutic Targets

Families of Membrane Proteins

Nearly all physiological processes intersect the membrane. As the primary conduits of material and information across the lipid bilayer, membrane proteins play critical biochemical roles. They account for about 30% of the proteome and roughly half of drug targets [1–5]. However, membrane proteins are challenging targets for biotherapeutic development for several reasons [6–8]. First, membrane proteins tend to have low expression levels, which creates an obstacle to gaining enough purified protein for antibody development [9, 10]. Second, they have low stability outside membranes, so it is challenging to deliver these proteins as immunogens [11]. For example, using detergents to solubilize membrane proteins can interfere with the exposure of extracellular epitopes and may be incompatible with *in vivo* antibody generation [11]. Finally, post-translational modification, especially glycosylation, can lead to antigenic heterogeneity and is difficult to correctly reproduce in non-human expression systems [12, 13]. In this section, we will begin our discussion by highlighting the most significant membrane protein families targeted by biotherapeutics and consider throughout the chapter how these challenges are addressed [6, 14].

Receptor Tyrosine Kinases

Receptor tyrosine kinases (RTKs) play roles in diverse cellular processes, including growth, differentiation, cell proliferation, metabolism, migration, and vascular function. They are also involved in oncogenesis, making them important drug targets [15–17]. The extracellular ligand binding domain of RTKs generally interacts with protein ligands (Figure 18.1a), including growth factors, hormones, and cytokines. When a ligand binds to the extracellular domain, RTKs dimerize to promote downstream signaling through the intracellular kinase domain [15, 18].

Humans have 58 known RTKs, and they are classified based on their interactions and the structure of their extracellular domains into 20 subfamilies, including epidermal growth factor receptors (EGFRs, class I), insulin receptors (IRs, class II), platelet-derived growth factor receptors (PDGFRs, class III), vascular endothelial growth factor receptors (VEGFR, class IV), fibroblast growth factor receptors (FGFRs, class V), cholecystokinin receptor family (CCK, class VI), and others [18–20].

Because they are easily accessible on the surface of the cell, targeting the extracellular ligand binding domain of RTKs has been an effective strategy for biotherapeutics, especially for cancer. The advantage of this approach is that a truncated, soluble ectodomain can be used for antibody generation, which avoids issues of membrane protein expression and solubilization [21]. With this approach, monoclonal antibodies (mAbs) can be developed to inhibit ligand binding to receptors or to inhibit receptor dimerization/activation after ligand binding (Table 18.1a) [15]. For example, trastuzumab, pertuzumab, cetuximab, panitumumab, and nimotuzumab target EGFRs to inhibit cell proliferation, angiogenesis, and cell survival [15, 24, 25]. Initially approved as Herceptin, trastuzumab was the first antibody to target an RTK and was a breakthrough in targeted cancer therapeutics [15].

Beyond the EGFR family, the PDGFR family has also shown growing interest as a target for several cancer types. For example, olaratumab and tovetumab are directed against PDGFRs for treating solid tumors [15]. The VGFR family has also been a target in cancer therapy due to their role in angiogenesis. For example, icrucumab inhibits tumor activation and angiogenesis, ramucirumab blocks neo-angiogenesis, and tanibirumab shows antitumor activity against lung, breast, colorectal, and glioblastoma cancer models [15, 19, 26–28].

Characterizing Biotherapeutics: Analytical Methods for Diverse Modalities, First Edition. Edited by Jennie R. Lill and Wendy Sandoval.

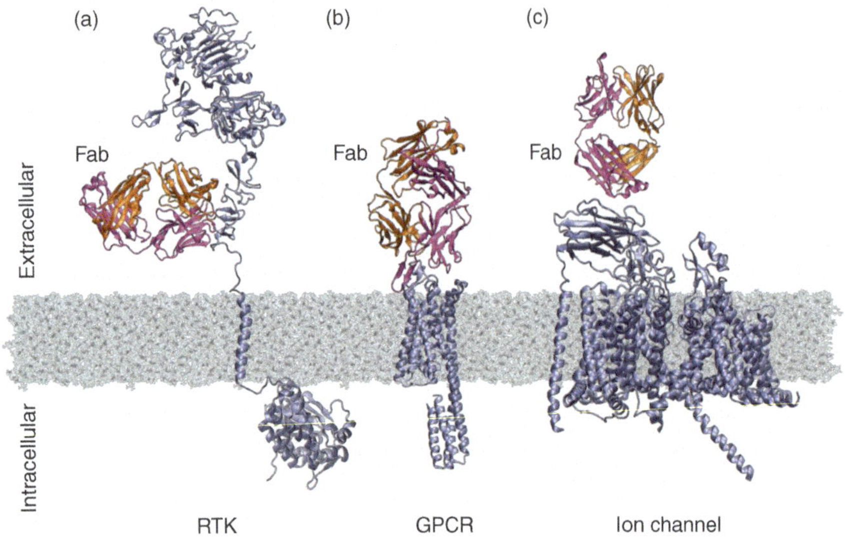

Figure 18.1 Examples of Fab binding to different membrane protein types: (a) an example RTK, HER2 (PDB: 1N87 for trastuzumab bound extracellular region, 2KS1 for transmembrane domain, 3PPO for intracellular kinase domain); (b) an example GPCR, a serotonin receptor (PDB: 5TUD); (c) an example ion channel, $Na_v1.4$ (PDB: 5XSY).

Table 18.1 List of FDA-approved mAbs for membrane proteins.

	Name of drug	Company	Year of approval	Target protein
Receptor tyrosine kinases (RTK)	Trastuzumab	Genentech	September 1998	Human epidermal growth factor receptor 2 (HER2)
	Pertuzumab	Genentech	June 2012	HER2
	Cetuximab	ImClone	February 2004	Epidermal growth factor receptor (EGFR)
	Panitumumab	Amgen Inc.	September 2006	EGFR
	Necitumumab	Eli Lilly Co.	November 2015	EGFR
	Margetuximab	Macrogenics Inc.	December 2020	HER2
	Teprotumumab	Horizon Therapeutics Ireland	January 2020	Insulin-like growth factor type I receptor (IGF-IR)
	Ramucirumab	Eli Lilly Co	April 2014	Vascular endothelial growth factor (VEGF)
	Olaratumab	Eli Lilly Co	October 2016	Platelet-derived growth factor receptor alpha (PDGFRα)
	Amivantamab	Janssen Biotech	May 2021	Bispecific EGFR and mesenchymal-epithelial transition (MET) receptor
	Bevacizumab	Genentech	February 2004	VEGF-A
	Brolucizumab	Novartis Pharms Corp	October 2019	VEGF-A
	Burosumab	Kyowa Kirin	April 2018	Fibroblast growth factor 23 (FGF23)
G protein-coupled receptors (GPCR)	Mogamulizumab	Kyowa Hakko Kirin	August 2018	C-C chemokine receptor type 4 (CCR4)
	Erenumab	Amgen Inc	May 2018	Calcitonin gene-related peptide (CGRP)
	Eptinezumab	Lundbeck Seattle BioPharmaceuticals, Inc.	February 2020	CGRP

Source: Hutchings et al., Fauvel and Yasri, Hutchings, FDA. [11, 15, 22, 23].

Although many mAbs targeting RTKs have been successful, several drug candidates have failed in clinical trials. For example, onartuzumab, which targeted hepatocyte growth factor receptor, and cixutumumab, which targeted insulin receptors, failed due to lack of efficacy in cancer trials [15, 29, 30]. Other mAb therapies were abandoned because tumors developed resistance. For antibodies targeting EGFR and HER2, developing resistance within a year is common through different types of mechanisms including mutations, upregulation or downregulation of signaling, and activation of alternative pathways [15, 31].

To address challenges with resistance and efficacy, more advanced biotherapeutic strategies can be used. To improve antitumor efficacy and limit resistance, bispecific antibodies can be used to target two receptors simultaneously [15, 32–34]. Similarly, combination therapies, such as trastuzumab and pertuzumab, can enhance efficacy through synergistic effects [15, 35]. Finally, antibody–drug conjugates can be used to deliver cytotoxic agents to tumor cells, enhancing treatment efficacy and simultaneously minimizing side effects [36]. For example, Ado-trastuzumab emtansine is a notable antibody-drug conjugate for HER2-positive breast cancer [15]. As these advanced biotherapeutic modalities mature, RTKs will continue to be an important membrane protein target for biotherapeutics.

G Protein-Coupled Receptors

G protein-coupled receptors (GPCRs) are the largest family of membrane proteins in number and one of the most important classes of proteins for drug discovery [11]. They are involved in diverse cellular processes, including cell growth, metabolism, sensory perception, immune response, and neurotransmission [9, 37, 38]. These diverse functions come from binding diverse ligands, ranging from small molecules to larger peptides and proteins [22, 39].

There are around 800 GPCRs, but more than half are olfactory or sensory receptors, leaving roughly 370 GPCRs that are considered potential drug targets. GPCRs are grouped into four families: family A (the rhodopsin family), family B (the secretin and adhesion subfamilies), family C (the metabotropic glutamate family), and family D (the frizzled family) [9, 11]. The rhodopsin family represents the largest and most diverse group of receptors. They are a major focus for drug development with receptors for histamine, dopamine, glycoprotein hormones, adrenergic agents, neuropeptides, and chemokines [11, 22]. Family B GPCRs are split into two subfamilies where secretin binds large peptide ligands, but so far lacks small-molecule drug targets. Most of the adhesion subfamily ligands have not been identified yet. Family B of GPCRs is well known for their metabolic function, regulating homeostasis, endocrine functions, and neuronal activity [22]. The metabotropic glutamate family binds small metabolites like glutamate and plays important roles in neuronal and calcium homeostasis. The frizzled family is a promising target because they are involved in developmental biology and tissue homeostasis [11, 40, 41].

Although GPCRs are important therapeutic targets, they have largely been limited to small molecule or peptide therapeutics [6, 42]. Developing antibodies for GPCRs has been especially challenging for several reasons. First, unlike RTKs, GPCRs generally do not have large soluble domains that can be detached for antibody development (Figure 18.1b). Thus, the immunogenic regions are often limited to the N-terminal domain and extracellular loops. Second, GPCRs are conformationally dynamic and difficult to stabilize outside the membrane [22]. Combined with their low expression levels, these challenges mean that it is difficult to express, purify, and stabilize GPCRs for antibody development. Finally, the development of GPCR biotherapeutics is also limited by biased signaling, where different ligands binding to the same receptor can induce different downstream signaling responses [22, 43, 44]. It can be challenging to develop antibodies that target these biases.

Due to these challenges, only three FDA-approved antibodies are directed toward GPCRs (Table 18.1). Mogamulizumab is a humanized IgG1 antibody directed against C-C chemokine receptor type 4 (CCR4) to induce antibody-dependent cellular cytotoxicity for the treatment of blood cancers [45]. Erenumab and eptinezumab are antagonists for calcitonin gene-related peptide (CGRP) and are used to treat migraine [22].

There are several GPCR-targeted antibodies currently in clinical trials, but several recent attempts have failed. For example, plozalizumab failed in the phase 2 trial due to a lack of efficacy for treating rheumatoid arthritis [11]. Here, one challenge was the low homology between human chemokine receptor type 2 (CCR2) and other species, which limited preclinical studies in nonhuman models [22].

Despite these challenges, innovative new technologies are being employed for GPCR biotherapeutics. For example, protein mutations are widely used to improve the expression and stability of GPCRs for large-scale purification [11, 43, 46]. Similar engineering approaches can also be used to bias the conformation. Another approach to stabilize the protein for antibody development is to use different membrane mimetics (discussed below), and a range of these different platforms have been used for antigen presentation [22, 47]. Using these novel approaches, it is likely that future biotherapeutics for GPCR targets will help to treat a variety of diseases.

Ion Channels

Ion channels are the second most abundant class of membrane proteins after GPCRs, comprising almost 400 members. Ion channels transport ions across membranes and play crucial roles in maintaining ion homeostasis

[48, 49]. Based on their gating mechanisms, ion channels are broadly classified into three main groups: voltage-gated ion channels, ligand-gated ion channels, and mechanosensitive ion channels [48].

Despite their importance in diseases like cancer, glaucoma, infectious disease, inflammation, and migraine, ion channels are challenging for the development of biotherapeutics for many of the same reasons as GPCRs. First, they have smaller extracellular loops that limit the potential immunogenic regions (Figure 18.1c). They can also be poorly expressing and challenging to stabilize and purify. Finally, ion channels often have highly conserved sequences between species, so they are not immunogenic enough to elicit strong antibody responses in mammals used for antibody development [48, 50].

Due to these challenges, ion channels are less explored for biotherapeutics than GPCRs, and there are currently no approved biotherapeutic drugs for these targets [51]. Several companies have attempted antibody-drug development for ion channels. For example, Visterra Inc. developed mAbs that targeted voltage-gated sodium channels (Na_V) for the treatment of pain. Here, they used several of the strategies described above for GPCRs. For example, to improve stability, they engineered a chimeric construct fused with the more stable prokaryotic form and reconstituted the protein in nanodisc for efficient immunization [48]. Amgen also explored developing antibodies toward transient receptor potential ankyrin 1 (TRPA1) by using DNA transfection rather than protein immunization [48, 52]. Unfortunately, none of these mAbs appear to be under ongoing development.

Although there has been considerable focus on Na_V channels, other ion channels have been explored as well. For example, the ligand-gated purinergic channel (P2X) family shows larger extracellular regions compared to other ion channels, so it is easier to target using biotherapeutics. A polyclonal antibody that targets nfP2X7, a non-functional form of P2X7, is in phase 1 clinical trials to treat basal cell carcinoma [48, 53]. Also, a mAb targeting the potassium channel, $K_V1.3$, is in preclinical trials for treatment of autoimmune and inflammatory diseases [54, 55].

Overall, ion channels present challenging but promising targets for biotherapeutic development. For ion channels and GPCRs, continued advances in molecular biology, protein engineering, and membrane mimetics will help improve platforms for drug development [11, 15, 22, 48]. For the more mature field of RTK therapeutics, new modalities to combat resistance will lead to more effective therapies. The next section will discuss advances in detergents and membrane mimetics to solve challenges in purification and presentation of membrane proteins. The final section will then discuss the unique considerations and opportunities for MS analysis of membrane proteins in a biotherapeutic context.

Membrane Protein Solubilization Methods

Many applications of membrane proteins in biotherapeutics require the extraction of membrane proteins from their natural environments (Figure 18.2a) into membrane mimetics [57, 58]. However, different membrane mimetics vary in their properties and require different considerations for analysis [59–61]. Here, we will give an overview of different membrane protein solubilization techniques and their applications for membrane protein biotherapeutics.

Detergent Micelles

Detergents are the most widely used agents to solubilize membrane proteins (Figure 18.2b) [62–65]. Detergents are amphiphilic small molecules with a hydrophilic head group and a hydrophobic tail. The amphiphilic nature of detergents enables them to protect the hydrophobic regions of membrane proteins inside micelles. There are a variety of different types of detergents, and it is important to choose the right detergent for the target protein and analysis performed [64, 66, 67]. In particular, there are trade-offs between effective solubilization and preservation of protein structure.

Detergents can be broadly classified based on the charge they carry on their hydrophilic head groups [60, 61, 66, 67]. Ionic detergents carry a net charge, such as sodium dodecyl sulfate (SDS) [61, 68]. These detergents are generally very effective at solubilization but can be too harsh, causing denaturing of proteins. Therefore, SDS is typically used for solubilization under denaturing conditions.

In contrast, nonionic detergents contain neutral hydrophilic headgroups and offer relatively mild solubilizing properties. Examples include Triton X-100, *n*-dodecyl-ß-D-maltoside (DDM), and *n*-octyl-ß-D-glucoside (OG). Because these nonionic detergents are less likely to disrupt membrane protein structure and function during the purification process, they are commonly used for membrane protein extraction and analysis under non-denaturing conditions [65, 68].

Due to their importance in structural biology applications, recent research has developed novel nonionic detergents with modifications that improve solubilization efficacy, protein stability, and analytical compatibility [69, 70]. Examples of these newer classes of detergents include oligo-glycerol detergents [71, 72], neopentyl glycols

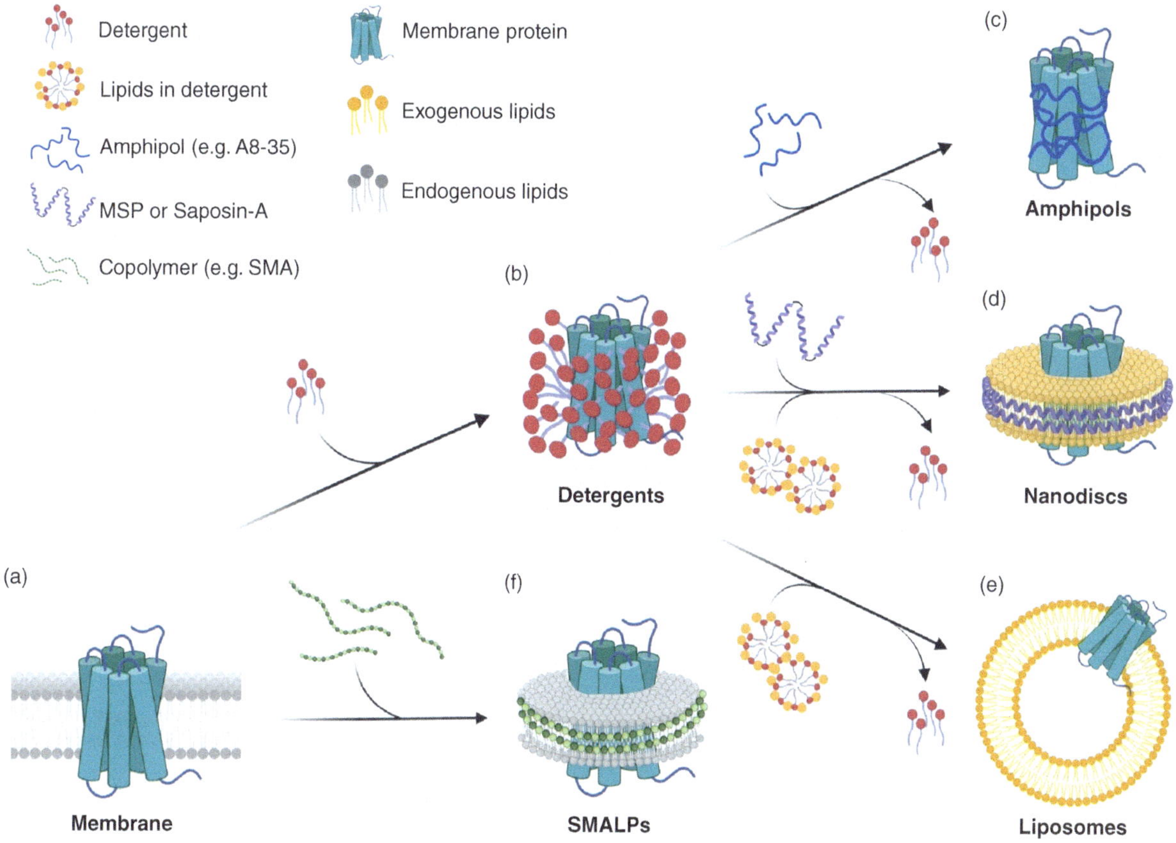

Figure 18.2 Illustration of different approaches used to solubilize membrane proteins. Membrane proteins are naturally present in (a) lipid bilayer membranes. Proteins can be extracted from the membrane into (b) detergent micelles. From here, it can be reconstituted into (c) amphipols, (d) nanodiscs, or (e) liposomes. Alternatively, the protein can be extracted from the membrane directly into (f) SMALPs. Source: Woubshete et al. [56]/John Wiley & Sons/CC BY 4.0.

(such as lauryl maltose neopentyl glycol or LMNG) [73, 74], and glyco-diosgenin (GDN) [75]. An important feature of newer detergents such as LMNG and GDN is their low critical micelle concentration (CMC) values, which allow them to be used in dilute solutions like those used in cryo-EM [69].

Zwitterionic detergents have both positively and negatively charged groups and fall in between ionic and nonionic detergents [61, 76]. Some zwitterionic detergents, such as fos-cholines, can be harsher and behave more like ionic detergents [77]. Others, like 3-[(3-cholamidopropyl) dimethylammonio]-1-propanesulfonate (CHAPS) and *n*-dodecyl-*N,N*-dimethylamine-*N*-oxide (LDAO), tend to be gentler and have been broadly used in structural biology applications [76, 78, 79].

There are several important properties to consider in detergents, including the CMC, aggregation number, and hydrophilic–lipophilic balance (HLB) [64, 70, 80]. At lower concentrations, detergents remain as individual molecules in solution. As the concentration increases, they assemble into micellar structures. The minimum concentration required for detergent molecules to form micelles is called the CMC, and the average number of detergent molecules per micelle is the aggregation number [61, 68, 76]. Both properties are influenced by the length of the detergent's alkyl chain, where longer chains tend to decrease the CMC and increase the aggregation number [81]. HLB describes the surface activity based on the hydrophilic and hydrophobic properties of a detergent. Typically, detergents with HLB values from 12 to 15 at concentrations above the CMC are recommended for membrane protein extraction [72, 81].

In a recent example [82], Urner and colleagues created a library of detergents with diverse properties, such as HLB and packing parameter (a structural property that relates the head group area with the tail volume [83]). They then solubilized proteins from bacterial inner membranes and used native MS to measure how each protein co-purified with phospholipids and lipopolysaccharide (LPS) in each unique detergent condition (Figure 18.3). By adjusting detergent HLB and packing density (Figure 18.3a), they enhanced phospholipid retention during extraction

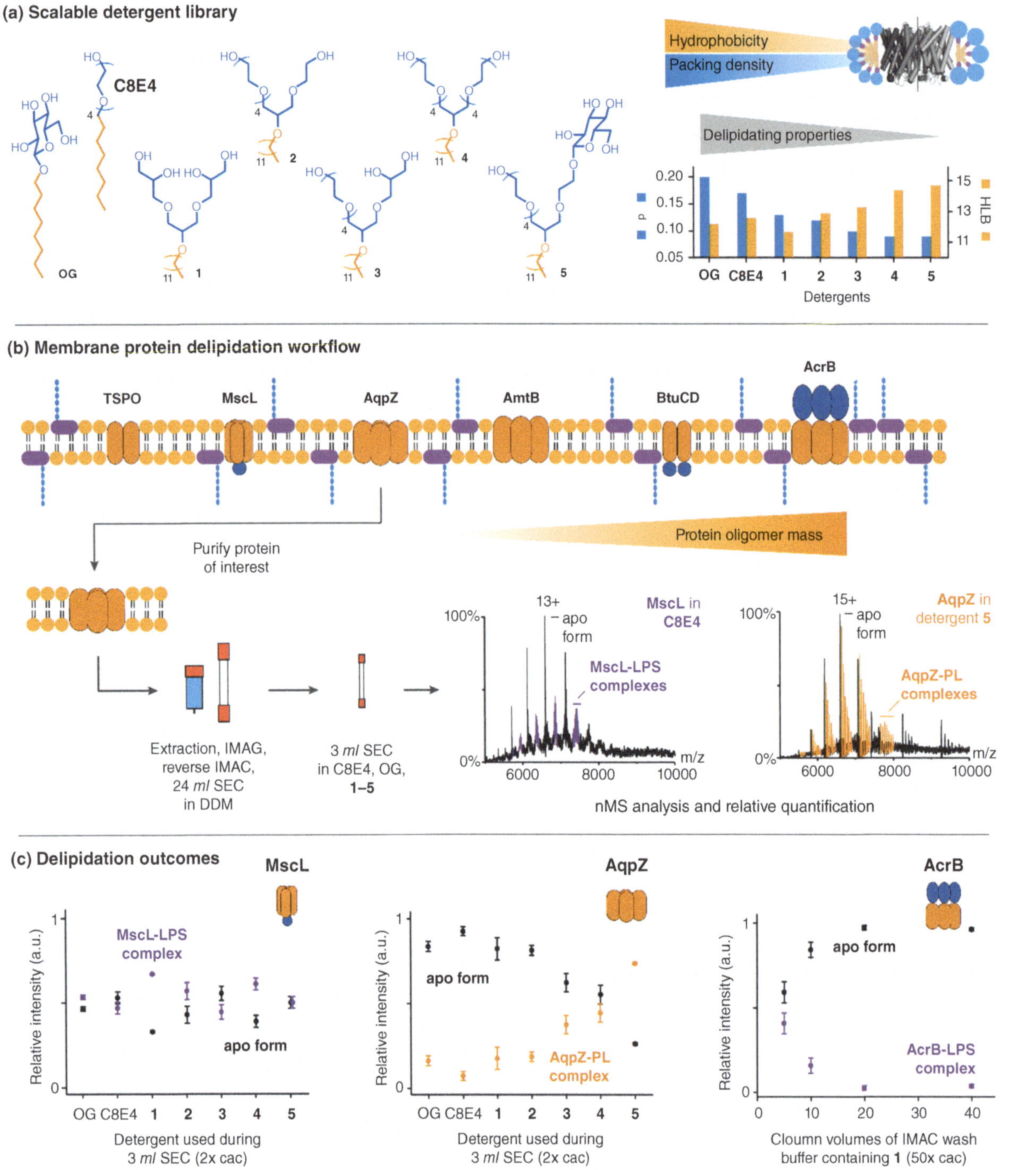

Figure 18.3 (a) Structures of detergents with the different HLB and packing parameters. (b) Schematic of solubilizing different inner membrane proteins from bacteria and using native MS to quantify delipidation outcomes. (c) Relative intensities of the apo state, protein–phospholipid complexes, and protein–LPS complexes acquired using native MS. Source: Urner et al. [82]/American Chemical Society/CC BY-NC-ND 4.0.

and purification of membrane proteins (Figure 18.3b, c). This study demonstrated how detergents can be designed to optimize membrane protein solubilization and interactions.

Although detergents are useful for solubilizing membrane proteins, they are not natural lipid bilayers and may not be effective at preserving membrane protein activity and interactions [64, 84–86]. In membrane protein biotherapeutics, identifying optimal detergent conditions to stabilize membrane proteins is a critical step. Often, a range of detergents need to be screened to find the optimal type. For example, Kotov and coworkers developed a high-throughput screen that tested 94 detergents to find which best stabilized different membrane proteins [87]. Although each protein had a unique profile of optimal detergents, LMNG and DDM tended to be stabilizing, and fos-choline tended to be destabilizing. Overall, careful selection of the detergent is essential for studying membrane proteins.

Alternative Membrane Mimetics

To address challenges in using detergents to solubilize membrane proteins, alternative membrane mimetics have been developed. Although detergents may be required to initially extract proteins from the membrane, they can be reconstituted into various membrane mimetics for characterization in defined lipid bilayers (Figure 18.2). Other mimetics enable direct reconstitution without intermediate detergent, such as styrene maleic acid-lipid particles (SMALPs) or some forms of liposomes. In this section, we will provide an overview of some of the key membrane mimetic systems and their applications in biotherapeutic development.

Amphipols

Amphipols are amphipathic polymers used to solubilize membrane proteins (Figure 18.2c) [88–91]. Different structures of amphipols have been developed [92, 93], with A8-35 being the most extensively used [93–95]. To incorporate membrane proteins into amphipols, detergent-extracted membrane proteins are mixed with the amphipol at a predefined ratio and incubated. The amphipol molecules interact with the membrane proteins, and the detergent is then removed using polystyrene beads, detergent removal columns, or dialysis [95, 96].

Unlike detergent micelles, which must be above their CMC, the primary advantage of amphipols is that they bind tightly to membrane proteins and can work at dilute concentrations. However, like detergents, they do not provide a lipid bilayer environment and thus may not preserve the natural structure and interactions found in lipid membranes [84]. Overall, although amphipols have been effectively used for the structural and biophysical characterization of membrane proteins [56, 97–99] and have been used to study antibody binding to membrane proteins [100, 101], their application in biotherapeutics has been relatively limited.

Nanodiscs

Nanodiscs are nanoscale discoidal lipid bilayers encircled by two amphipathic membrane scaffold proteins (MSPs), which are derived from human ApoA1 (Figure 18.2d) [57, 102–105]. To reconstitute membrane proteins into nanodiscs, MSP and lipids dissolved in detergent are mixed with the membrane protein. Then, porous polystyrene beads are added to the mixture to remove the detergent and drive self-assembly.

In nanodiscs, the hydrophilic regions of the MSP are exposed to the surrounding solution, and the hydrophobic regions interact with the lipids in the core [105–107]. The size of the nanodisc can be adjusted by altering the length of the MSP, and the thickness can be customized by modifying the lipid composition [57, 85, 107, 108]. Recently, circularized nanodiscs have been engineered by covalently linking the N-terminus to the C-terminus, which has further improved their homogeneity, stability, and size range [109, 110].

Due to their stability and biocompatibility, nanodiscs have become a useful platform for a range of biotherapeutic applications [105, 111, 112]. To develop therapeutic antibodies targeting membrane proteins, nanodiscs have been used to present antigens in more natural conformations. One example developed a single-domain antibody targeting the human apelin receptor (APJ), which is a GPCR and a target for treating chronic heart failure [113]. Here, Zhang and coworkers reconstituted APJ into nanodiscs that were then injected into camels as immunogens. Other approaches, such as by Kossiakoff and coworkers, have used nanodiscs with phage display technology to generate antibodies against membrane proteins [114]. Finally, it is possible to use nanodiscs for vaccines. In one study, injecting nanodiscs with embedded influenza hemagglutinin generated a robust and protective antibody response [115]. However, due to cost and stability limitations, nanodiscs are more likely to be used in antibody development rather than direct vaccine delivery.

In addition to presenting membrane proteins for vaccines or antibody development, nanodiscs are also useful as drug-delivery vehicles, especially in delivery of hydrophobic drugs [108, 111]. For example, one study used nanodisc to deliver the anticancer drug, cabazitaxel, where they found a sustained release profile and higher drug efficacy [116]. Moreover, both the MSP belt and the lipids can be engineered to attach chemical functionality and enable targeted delivery. For example, another study attached an antibody that targeted carcinoembryonic antigen to a PEGylated lipid in nanodiscs [117]. Further conjugation of a radioactive Cu chelator allowed PET imaging of these targeted nanodiscs and demonstrated localization to the tumor. Future research using nanodiscs as a drug-delivery vehicle holds the potential to develop unique biotherapeutics capable of targeted delivery of lipids and hydrophobic drugs.

Peptide Nanodiscs

Alongside nanodiscs encircled by protein belts, there are peptide-based nanodiscs that also serve as membrane mimetics. For example, saposin-lipid nanoparticles (SapNPs) use a small protein, saposin A, as the belt [118, 119]. Unlike MSP, SapA covers both leaflets of the bilayer with a single monomer without encircling the particle. This modular assembly enables the incorporation of varying numbers of SapA molecules per complex, making SapNPs adaptable in size and composition [120]. SapNPs have been used in a variety of structural and biophysical studies [121] and have been used to present membrane proteins for antibody development [112].

Another novel peptide nanodisc system is the peptidisc [122]. In peptidisc assembly, hydrophobic regions of membrane proteins are protected by an amphipathic bi-helical peptide. Unlike nanodiscs, peptidiscs eliminate the need for additional lipids, incorporating only the co-purified annular lipids with the membrane protein [122, 123]. The peptidisc system has been useful in determining structures of membrane proteins and profiling membrane proteomes [123–127].

Several recent studies have showcased the use of peptide nanodiscs as vaccines (Figure 18.4) [128]. Here, Moon and coworkers used a synthetic peptide from ApoA1 called 22A to create peptide nanodiscs and delivered antigens and adjuvants to lymphoid organs as a vaccine. Administration of these peptide nanodiscs resulted in more efficient and prolonged antigen presentation on major histocompatibility complex molecules, leading to stronger T-cell responses. These studies have since been extended to use 22A nanodiscs in vaccination for a variety of different types of cancer [129, 130]. These applications of peptide nanodiscs demonstrate a range of potential uses in biotherapeutic development.

SMALPs

Although the mimetics discussed above provide lipid environments for studying membrane proteins, most of

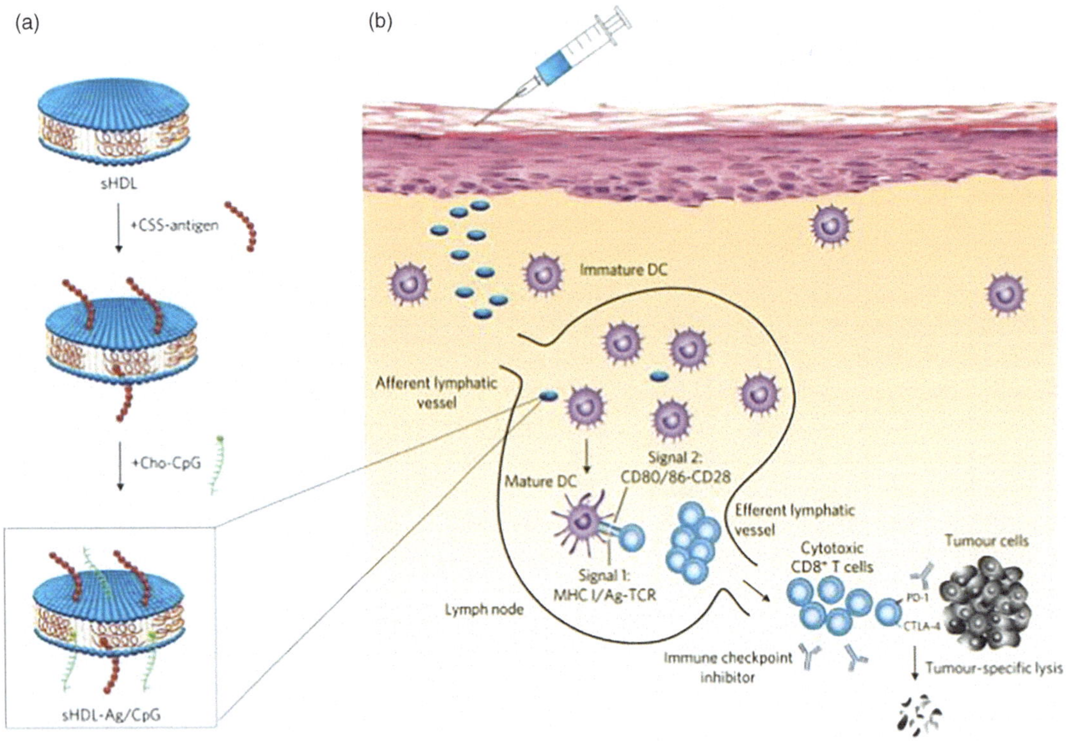

Figure 18.4 Synthetic high-density lipoprotein (sHDL) nanodiscs made using 22A peptides and embedded antigens as a platform for vaccines. (a) sHDL nanodiscs co-deliver antigen peptides and adjuvants. (b) Following application, sHDL nanodiscs deliver antigen and adjuvants (CpG) to lymph nodes, promoting antigen presentation by dendritic cells (DCs). Source: Kuai et al. [128]/with permission of Springer Nature.

those mimetics still require the use of detergents for the initial solubilization of the proteins. In contrast, SMALPs are a detergent-free membrane mimetic that allows the extraction of membrane proteins directly from their natural lipid bilayers (Figure 18.2f) [63, 131]. SMALPs are created from lipids and poly-styrene-co-maleic acid (SMA), which is an amphipathic copolymer comprising hydrophobic styrene and hydrophilic maleic acid moieties [132, 133]. When added to membranes, SMA inserts into the lipid bilayer [132, 134], allowing the SMA to extract the protein from lipid membranes [135, 136]. SMALPs are compatible with a wide range of targets, including integral and peripheral membrane proteins [85, 136].

Early SMALP technology had several limitations, most notably they required a higher pH and were sensitive to divalent cations, making them challenging to use in systems that require calcium and magnesium [136]. They were also limited in size and could not accommodate very large membrane protein complexes [136–138]. To address these challenges, researchers have developed a variety of new polymers, including diisobutylene maleic acid (DIBMA), styrene maleic imide (SMI), and styrene maleimide quaternary ammonium (SMA-QA) [137, 139]. These new polymers have different optimal pH levels, sizes, and tolerances to divalent cations, enhancing their applicability in diverse applications [137, 139].

The unique ability of SMALPs to disrupt membranes directly to extract membrane proteins facilitates higher yields of membrane proteins from low-expressing cell lines, which enables drug development with more challenging membrane protein targets [140]. For example, SMALPs can be used to study binding of ligands and antibodies to GPCRs [138, 141]. Moreover, SMALPs can be used in antibody development, vaccines, and drug delivery in similar ways to the peptide and protein nanodiscs described above [112]. Although each system has unique properties and distinct advantages, nanoscale lipid bilayer particles enable diverse applications in biotherapeutics for membrane proteins.

Liposomes and Bicelles

Finally, we will introduce membrane mimetics composed primarily of lipids, which can take different structural forms. Among the different mimetic systems discussed above, liposomes can be the simplest membrane mimetic in composition, requiring only a phospholipid bilayer vesicle (Figure 18.2e) [84, 91]. However, liposomes can also be customized for different target membrane systems by changing the lipid composition, size, and protein-to-lipid ratio [142, 143]. In conventional assembly of liposomes, purified membrane proteins in detergents are reconstituted into preformed liposomes, with the detergent being removed during the assembly process [144–146]. However, it is also possible to create liposomes directly from cell membranes, such as with extracellular vesicles [147, 148].

Liposomes have numerous biotherapeutic applications [149–151]. In one example, Lenormand and coworkers used liposomes to deliver two proapoptotic membrane proteins, voltage-dependent anion channel (VDAC) and Bcl-2 homologous antagonist/killer (Bak), to cells [149]. This study demonstrated that functional VDAC and Bak proteins were successfully delivered into cancer cells, where they triggered apoptosis. Thus, liposomes can be used to deliver membrane proteins as therapeutic agents.

Bicelles are similar to liposomes but have a mixture of two different lipid-like molecules [152]. Typically, one lipid forms a bilayer while the other has more detergent-like properties and caps the edges of the bilayer. Bicelles can form a variety of structures, including discoidal lipid nanoparticles like nanodiscs. Bicelles have been broadly used for membrane protein analysis, especially by NMR, and have been applied for drug delivery [153].

In summary, mimetics do not always eliminate the need for detergents, but they provide novel strategies to study membrane proteins within bilayer environments. They provide versatile platforms for membrane protein biotherapeutic development and drug delivery. Continued development and optimization of these approaches will enhance our understanding of membrane proteins and advance biotherapeutic development.

MS Analysis of Membrane Proteins

Mass spectrometry has increasingly become an indispensable tool in biotherapeutics. In this section, we will explore how various structural MS techniques are used, both independently and in combination, to address critical biological questions related to membrane proteins in biotherapeutic applications. Drawing on the prior section, we will consider how different membrane mimetics are used.

Proteoform Characterization

One of the main questions in MS of biotherapeutics is proteoform characterization. Proteins synthesized from a single gene can have a high level of heterogeneity due to post-translational modifications (PTMs), genetic variations, and splice variants [154–156]. Each unique variant of the protein is called a proteoform. For membrane proteins, the proteoform landscape is particularly complex due to frequent glycosylation of proteins on the cell surface [157, 158]. Characterizing this proteoform landscape is important to ensure consistent biotherapeutic manufacturing and generation of antibodies that hit the correct target [139].

Recent advancements in MS have enhanced the ability to quantitatively analyze PTMs at a variety of different levels. Intact mass analysis directly profiles the mass distribution, revealing the ratios of each proteoform that has a distinct mass. Often, intact mass analysis is performed on denatured proteins to reveal the mass distribution of the protein alone [159–161]. However, it is also possible to perform native MS by using non-denaturing ionization conditions, which help maintain non-covalent interactions [57, 162]. Native MS typically requires careful consideration of the membrane mimetic system to keep the protein soluble during analysis, as discussed below.

Complementing these intact MS measurements, more detailed sequence information can be gained from bottom-up and top-down proteomics approaches. Bottom-up analysis uses enzymes to digest proteins to peptides, which are then sequenced to identify sites of PTMs [163]. Top-down approaches perform intact mass analysis first and then use fragmentation within the mass spectrometer to sequence the protein. Top-down proteomics enables detailed characterization of the position of the PTMs and disulfide bridge patterns [164–166] Increasingly, proteomic approaches are being combined with native MS to characterize PTMs in protein complexes [167–169]. For bottom-up and denatured intact analysis methods, there are fewer special considerations for membrane proteins because the protein structure does not need to be preserved.

MS analysis of all different flavors has been applied to characterize proteoforms in a range of applications. Native MS has been used to characterize membrane protein proteoforms in detergents [170] and nanodiscs [171]. As discussed in prior reviews, detergents have dominated native MS analysis, but new membrane mimetics have enabled unique new types of experiments [57, 84, 172, 173]. However, there are not many applications characterizing PTMs of membrane proteins for biotherapeutic applications.

To address this gap, advanced native MS methods have recently been developed for highly heterogeneous biotherapeutics with multiple glycosylation sites [174]. Building on these studies, future research will dive deeper into characterizing proteoforms for biotherapeutic development of membrane proteins. For example, developing therapeutics that target particular glycosylation sites could be a powerful application where MS methods would be essential in defining the composition of the target.

Complex Architecture

Another key question in MS of biotherapeutics is how proteins assemble into complexes. Many membrane proteins either oligomerize or form macromolecular assemblies to perform their biological roles [175, 176]. Understanding the architecture of membrane protein complexes is important to develop effective biotherapeutics to target the correct form of the complex.

Native MS (introduced above) is commonly used to study complex architecture [162]. Although the direct applications to membrane protein biotherapeutics are limited, a number of studies have explored the architecture of therapeutically important membrane protein targets [177]. Early examples used native MS to study protein complex architecture of membrane proteins in detergents, with a striking example being ATPase complexes [178]. These studies primarily relied on common nonionic detergents, such as DDM. However, more challenging targets have required advances in detergent selection [71, 72, 85]. For example, Robinson and coworkers have advanced native MS to study GPCR targets, which have relied on careful detergent selection and design [179–181].

Alternative membrane mimetics have also been used to study membrane protein complex architecture [57, 84, 172, 173]. In one example, Townsend et al. embedded the influenza A M2 protein in lipid nanodiscs with different lipids [182]. They discovered different stoichiometries of complex formation in different lipid environments and observed drug binding to the tetramer species. To capture interactions in more natural environments, Robinson and coworkers used sonicated lipid vesicles prepared directly from cell membranes [183–185]. Together these methods demonstrate the potential for native MS to study membrane protein complex architecture in complex environments.

Another technique to characterize complex architecture is cross-linking (XL)-MS. XL-MS measures the proximity of amino acid residues through covalent cross-linking with bifunctional linkers [186, 187]. These linkers typically target specific amino acids, such as the primary amine groups on lysine side chains, with the spacer arm defining the distance constraints between the cross-linked residues. After digesting the cross-linked protein, analyzing the resulting peptides using bottom-up proteomics reveals the spatial relationships between residues.

Complementing native MS studies, XL-MS provides unique information on which subunits are close together in the complex. For example, Schmidt and coworkers used XL-MS to further characterize the ATPase complex, first in detergents [188] and recently in synaptic vesicles [189]. This example highlights the ability of XL-MS to study protein complex architecture *in situ*, which we will also discuss below. Overall, these MS methods help define how membrane protein complexes are assembled.

Interface Mapping

Interface mapping is important for biotherapeutic applications to identify binding epitopes for antigen–antibody complexes. Mapping membrane protein interfaces can be achieved using various structural MS techniques. Each technique provides complementary information on binding surfaces by measuring unique properties such as solvent accessibility, distant restraints, and structural dynamics.

Hydrogen-deuterium exchange (HDX)-MS is widely used in biotherapeutic applications to map protein–protein binding interfaces [190–193]. In HDX-MS, labile hydrogens on the protein backbone exchange with deuterium from the solvent. This exchange is possible when hydrogen atoms are solvent-accessible and not involved in intra- or inter-molecular bonding [191, 194]. After quenching the exchange, the protein is digested, and the resulting peptides are analyzed using MS.

HDX-MS has been used for a range of biotherapeutic applications, especially epitope mapping [192, 195]. In one example with membrane proteins, Kim et al. [196] used HDX-MS to perform epitope mapping on broadly neutralizing antibodies for human immunodeficiency virus (HIV) that targeted the membrane-proximal external region (MPER) of the HIV glycoprotein 41 (GP41) envelope protein. Comparing the MPER peptide presented in free solution and bound to liposomes, they discovered that significantly less exchange occurred when the antibody bound the MPER in liposomes, revealing a different presentation when the antigen is bound to a lipid bilayer. This study demonstrated the use of HDX to perform epitope mapping on membrane proteins presented in different environments.

Due to the complexity of natural membranes, HDX-MS requires purified proteins solubilized in membrane mimetics. With detergents, no additional sample preparation may be needed. For example, Chung et al. investigated the formation of a complex between the β2 adrenergic receptor (β_2AR), a GPCR, and Gs, a stimulatory G protein, in a neopentyl glycol detergent without any special cleanup steps [197]. However, when membrane mimetics with lipids are used, additional sample preparation is often needed to protect the reverse phase column used to separate peptides. Because the labeling is reversible, these sample cleanup steps must be performed quickly. For example, in the liposome study of MPER described above, a second guard column was used to trap lipids [196]. Recent approaches have used zirconium oxide beads to remove lipids [198–201]. Sequence coverage can also be improved by using urea to denature membrane proteins prior to digestion [202]. Using these techniques, HDX has been applied to a variety of different membrane mimetic systems.

Covalent labeling (CL)-MS is another bottom-up method used to characterize protein structures by covalently modifying solvent-accessible amino acids [203, 204]. These labeling reagents can either target specific amino acids (such as using *N*-ethylmaleimide to target cystine) or can non-specifically label multiple residue types on the protein surface (such as with hydroxyl radicals and carbenes) [205–207]. The type of labeling reagent used in CL-MS determines the locations on the protein structure that can be probed. For example, nonspecific hydroxyl radicals provide broader structural coverage compared to labels that target specific amino acid residue types [204]. A few factors also influence how the protein structure gets labeled using covalent labeling, such as solvent accessibility, the inherent reactivity of amino acids, and the changes in microenvironment introduced by the binding of ligands [208]. After the protein is labeled, the labeled protein is digested and analyzed by liquid chromatography (LC)-MS/MS to identify the relative amount of labeling at specific amino acid residues [208].

CL-MS has also been used in biotherapeutics to characterize binding interfaces of membrane proteins, mostly on monoclonal antibodies to determine antigen interactions [209]. With purified samples, fast photochemical oxidation of proteins (FPOP) with hydroxyl radicals has been used for epitope mapping on protein ectodomains from RTKs [210, 211]. FPOP has also been applied in detergents, nanodiscs [212], liposomes [213], and cells [214]. Unlike HDX, the irreversible labeling in CL-MS enables a range of downstream sample preparation strategies, such as protein precipitation [212], to remove lipids and clean up the sample prior to MS analysis.

In one example of using CL-MS to map membrane protein interfaces with biotherapeutics, Vachet and coworkers used CL-MS to explore the binding interactions of three monoclonal antibodies with membrane-bound tumor necrosis factor α (mTNFα), in living HEK293T cells (Figure 18.5) [207]. They first transfected HEK293T cells with a plasmid encoding mTNFα with a C-terminal purification tag. Next, they mixed the cells with the monoclonal antibodies for binding and performed labeling using diethylpyrocarbonate (DEPC) as the covalent labeling reagent to modify solvent-accessible amino acid residues on proteins while still *in situ*. After labeling, they quenched the reaction and isolated the membrane protein using MS-compatible buffers with DDM detergent. Finally, they digested the proteins and conducted LC-MS/MS analysis. The authors observed decreased DEPC labeling in residues buried in the epitope upon antibody binding. Additionally, they observed changes in labeling away from the epitope, suggesting alterations in mTNFα homotrimer packing or conformational changes. Importantly, because CL-MS methods can be performed in cells (unlike HDX), they do not require any membrane mimetics during the actual experiment, allowing the membrane proteins to be probed

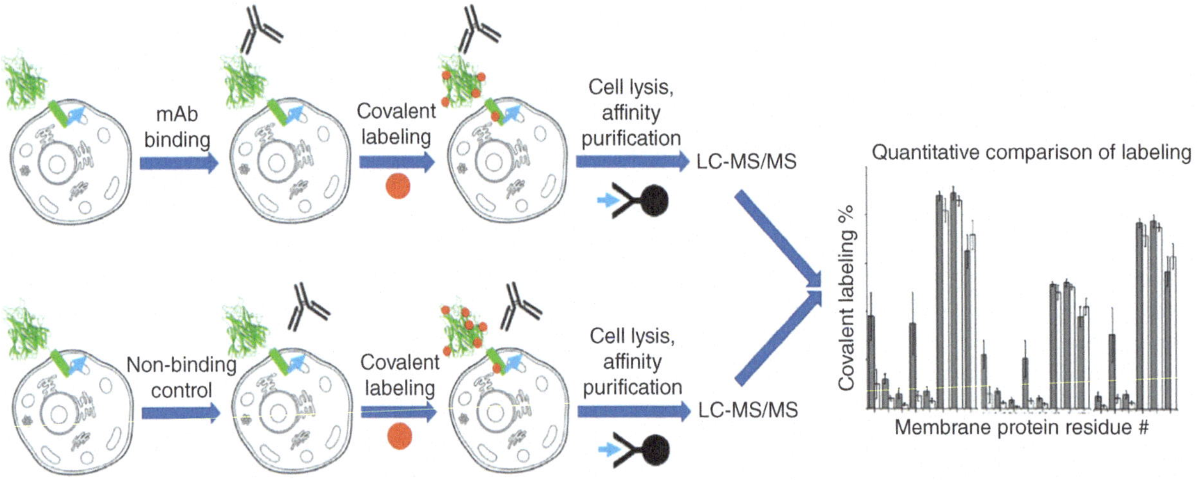

Figure 18.5 Workflow for in-cell labeling of mTNFα. mTNFα expressed in HEK293T cells was incubated with a binding or nonbinding mAb and then labeled by DEPC. After the DEPC reaction, the cells were lysed, and the protein was purified from the cell lysate using a C-terminal EPEA affinity tag. LC-MS/MS was then used to compare the labeling extents between the two conditions. Source: Kirsch et al. [207]/with permission of American Chemical Society.

in their native environment. Thus, CL-MS provides a powerful method for detecting binding interfaces of membrane proteins in living cells.

XL-MS can also be performed in cells or organelles, providing the most natural environment for studying membrane protein interaction interfaces [189, 215, 216]. For example, Bruce and coworkers developed an innovative XL-MS workflow to study a peptide interacting with mitochondrial proteins (Figure 18.6) [217]. Here, intact mitochondria were isolated and treated with a peptide therapeutic, SS-31. After cross-linking SS-31 with mitochondrial proteins, the cross-links were enriched through a biotin tag on the SS-31. Twelve inner mitochondrial membrane proteins were identified as interacting

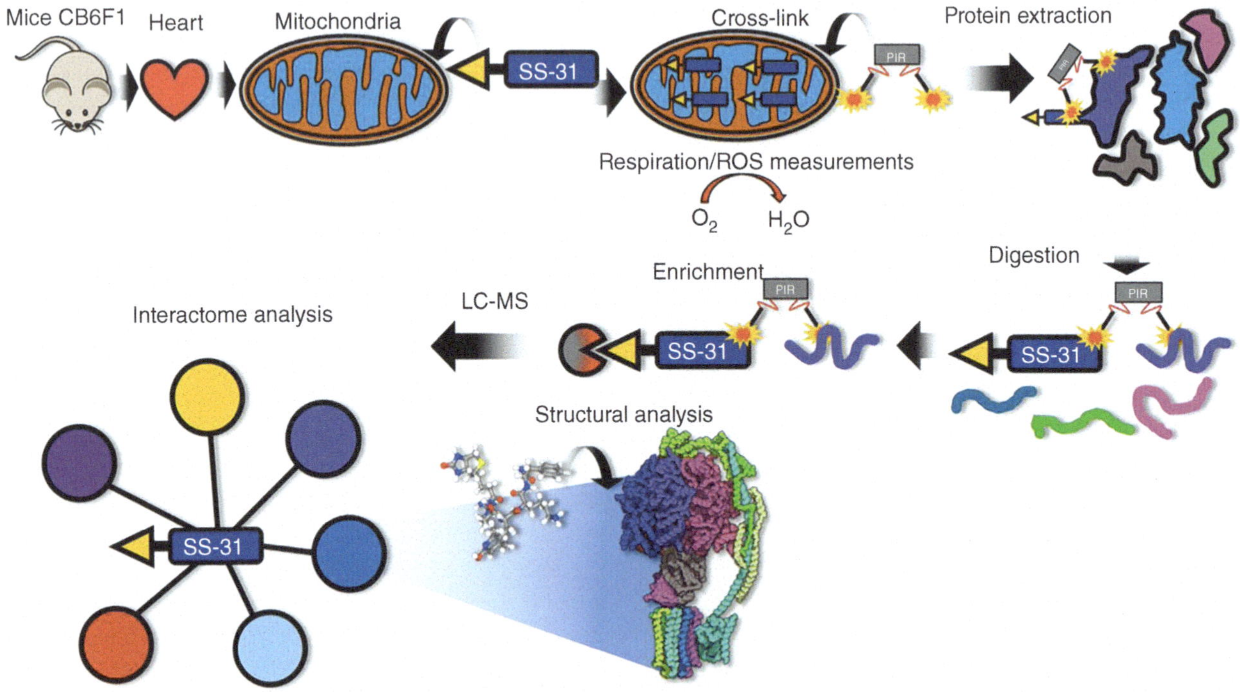

Figure 18.6 Mitochondria isolated from mouse heart tissue were treated with the SS-31 peptide labeled with a biotin tag (indicated by a triangle). Addition of a protein interaction reporter (PIR) cross-linker cross-linked the SS-31 peptide to interacting proteins. The peptide was then extracted with cross-linked proteins. After digestion and enrichment of cross-linked peptides, LC-MS/MS analysis revealed the interacting proteins and interaction interfaces. Source: Chavez et al. [217]/with permission of National Academy of Sciences.

with the protein, helping to reveal the targets of the therapeutic and interfaces on these proteins where SS-31 bound. The ability to expand these measurements to proteome scales enables membrane protein interaction networks and interfaces to be uncovered [189].

Finally, these structural MS methods can be used in combination for more complex characterization of interaction interfaces. For example, Li et al. used an integrative MS-based approach for epitope mapping and structural characterization [218]. Here, the authors used a combination of HDX, FPOP, and site-specific carboxyl group footprinting to investigate the binding of the soluble ectodomain of human interleukin-6 receptor (IL-6R) to adnectin protein therapeutics. They found that the conserved epitope for both adnectins is a flexible loop connecting two β-strands in the cytokine-binding domain of IL-6R. These findings revealed the value of combining information from multiple techniques to characterize interfaces of protein–protein complexes. Overall, MS methods can provide unique insights that address key biological questions surrounding membrane protein biotherapeutics.

Conclusion

In this chapter, we have discussed various families of membrane proteins that are of biotherapeutic interest, including RTKs, GPCRs, and ion channels. Next, we explored the challenges associated with the solubilization of membrane proteins from their native lipid bilayers into aqueous solutions. Finally, we discussed various structural MS techniques used to address challenging biological questions in the field of biopharmaceuticals for membrane protein targets.

In conclusion, although significant progress has been made in the characterization and therapeutic targeting of membrane proteins, ongoing research and innovative approaches are important to overcome existing challenges. The integration of advanced solubilization techniques and structural MS approaches holds promise for the future of membrane protein biotherapeutics, potentially leading to the development of new treatments for a variety of diseases.

Acknowledgments

This work was supported by the National Institutes of Health/National Institute of General Medical Sciences under award number R35 GM128624 to MTM. The content is solely the responsibility of the authors and does not necessarily represent the official views of the National Institutes of Health.

References

1 Vinothkumar, K.R. and Henderson, R. (2010). Structures of membrane proteins. *Q. Rev. Biophys.* 43: 65–158. https://doi.org/10.1017/S0033583510000041.

2 Yin, H. and Flynn, A.D. (2016). Drugging Membrane Protein Interactions. *Annu. Rev. Biomed. Eng.* 18: 51–76. https://doi.org/10.1146/annurev-bioeng-092115-025322.

3 Lacapère, J.-J., Pebay-Peyroula, E., Neumann, J.-M., and Etchebest, C. (2007). Determining membrane protein structures: still a challenge! *Trends Biochem. Sci.* 32: 259–270. https://doi.org/10.1016/j.tibs.2007.04.001.

4 Lindahl, E. and Sansom, M.S. (2008). Membrane proteins: molecular dynamics simulations. *Curr. Opin. Struct. Biol.* 18: 425–431. https://doi.org/10.1016/j.sbi.2008.02.003.

5 Cournia, Z., Allen, T.W., Andricioaei, I. et al. (2015). Membrane Protein Structure, Function, and Dynamics: a Perspective from Experiments and Theory. *J. Membr. Biol.* 248: 611–640. https://doi.org/10.1007/s00232-015-9802-0.

6 Gulezian, E., Crivello, C., Bednenko, J. et al. (2021). Membrane protein production and formulation for drug discovery. *Trends Pharmacol. Sci.* 42: 657–674. https://doi.org/10.1016/j.tips.2021.05.006.

7 Li, F., Egea, P.F., Vecchio, A.J. et al. (2021). Highlighting membrane protein structure and function: A celebration of the Protein Data Bank. *J. Biol. Chem.* 296: https://doi.org/10.1016/j.jbc.2021.100557.

8 Jelokhani-Niaraki, M. (2023). Membrane Proteins: Structure, Function and Motion. *Int. J. Mol. Sci.* 24: 468. https://doi.org/10.3390/ijms24010468.

9 McCusker, E.C., Bane, S.E., O'Malley, M.A., and Robinson, A.S. (2007). Heterologous GPCR Expression: A Bottleneck to Obtaining Crystal Structures. *Biotechnol. Prog.* 23: 540–547. https://doi.org/10.1021/bp060349b.

10 Grisshammer, R. (2009). Purification of recombinant G-protein-coupled receptors. *Methods Enzymol.* 463: 631–645. https://doi.org/10.1016/S0076-6879(09)63036-6.

11 Hutchings, C.J., Koglin, M., and Marshall, F.H. (2010). Therapeutic antibodies directed at G protein-coupled receptors. *MAbs* 2: 594–606. https://doi.org/10.4161/mabs.2.6.13420.

12 Carnec, X., Quan, L., Olson, W.C. et al. (2005). Anti-CXCR4 monoclonal antibodies recognizing overlapping epitopes differ significantly in their ability to inhibit entry of human immunodeficiency virus type 1. *J. Virol.* 79: 1930–1933. https://doi.org/10.1128/JVI.79.3.1930-1933.2005.

13 Yasuoka, H., Tsujimoto, M., Yoshidome, K. et al. (2008). Cytoplasmic CXCR4 expression in breast cancer: induction by nitric oxide and correlation with lymph node metastasis and poor prognosis. *BMC Cancer* 8: 340. https://doi.org/10.1186/1471-2407-8-340.

14 Santos, R., Ursu, O., Gaulton, A. et al. (2017). A comprehensive map of molecular drug targets. *Nat. Rev. Drug Discov.* 16: 19–34. https://doi.org/10.1038/nrd.2016.230.

15 Fauvel, B. and Yasri, A. (2014). Antibodies directed against receptor tyrosine kinases: current and future strategies to fight cancer. *MAbs* 6: 838–851. https://doi.org/10.4161/mabs.29089.

16 Bajinting, A. and Ng, H.L. (2021). Chapter Nine - Structural studies of full-length receptor tyrosine kinases and their implications for drug design. In: *Advances in Protein Chemistry and Structural Biology* (ed. R. Donev), 311–336. Academic Press https://doi.org/10.1016/bs.apcsb.2020.10.007.

17 Lemmon, M.A. and Schlessinger, J. (2010). Cell signaling by receptor tyrosine kinases. *Cell* 141: 1117–1134. https://doi.org/10.1016/j.cell.2010.06.011.

18 Lemmon, M.A., Schlessinger, J., and Ferguson, K.M. (2014). The EGFR family: Not so prototypical receptor tyrosine kinases. *Cold Spring Harb. Perspect. Biol.* 6: a020768–a020768. https://doi.org/10.1101/cshperspect.a020768.

19 Schwartz, J.D., Rowinsky, E.K., Youssoufian, H. et al. (2010). Vascular endothelial growth factor receptor-1 in human cancer. *Cancer* 116: 1027–1032. https://doi.org/10.1002/cncr.24789.

20 Batool, Z., Azfal, A., Liaquat, L. et al. (2023). Chapter 4 - Receptor tyrosine kinases (RTKs): from biology to pathophysiology. In: *Receptor Tyrosine Kinases in Neurodegenerative and Psychiatric Disorders* (ed. H.M. Mansour, M.M. Khattab, and A.S. El-khatib), 117–185. Academic Press https://doi.org/10.1016/B978-0-443-18677-6.00012-9.

21 Ségaliny, A.I., Tellez-Gabriel, M., Heymann, M.-F., and Heymann, D. (2015). Receptor tyrosine kinases: Characterisation, mechanism of action and therapeutic interests for bone cancers. *J. Bone Oncol.* 4: 1–12. https://doi.org/10.1016/j.jbo.2015.01.001.

22 Hutchings, C.J. (2020). A review of antibody-based therapeutics targeting G protein-coupled receptors: an update. *Expert. Opin. Biol. Ther.* 20: 925–935. https://doi.org/10.1080/14712598.2020.1745770.

23 FDA, Drugs@FDA: FDA-Approved Drugs, (n.d.). https://www.accessdata.fda.gov/scripts/cder/daf/index.cfm (accessed August 27, 2024).

24 Scheuer, W., Friess, T., Burtscher, H. et al. (2009). Strongly enhanced antitumor activity of trastuzumab and pertuzumab combination treatment on HER2-positive human xenograft tumor models. *Cancer Res.* 69: 9330–9336. https://doi.org/10.1158/0008-5472.CAN-08-4597.

25 Patel, D., Bassi, R., Hooper, A. et al. (2009). Anti-epidermal growth factor receptor monoclonal antibody cetuximab inhibits EGFR/HER-2 heterodimerization and activation. *Int. J. Oncol.* 34: 25–32.

26 Lee, S.H. (2011). Tanibirumab (TTAC-0001): a fully human monoclonal antibody targets vascular endothelial growth factor receptor 2 (VEGFR-2). *Arch. Pharm. Res.* 34: 1223–1226. https://doi.org/10.1007/s12272-011-0821-9.

27 Garcia, J.A., Hudes, G.R., Choueiri, T.K. et al. (2014). A phase 2, single-arm study of ramucirumab in patients with metastatic renal cell carcinoma with disease progression on or intolerance to tyrosine kinase inhibitor therapy. *Cancer* 120: 1647–1655. https://doi.org/10.1002/cncr.28634.

28 Hsu, J.Y. and Wakelee, H.A. (2009). Monoclonal antibodies targeting vascular endothelial growth factor. *BioDrugs* 23: 289–304. https://doi.org/10.2165/11317600-000000000-00000.

29 Solomon, V.R., Alizadeh, E., Bernhard, W. et al. (2020). Development and preclinical evaluation of cixutumumab drug conjugates in a model of insulin growth factor receptor I (IGF-1R) positive cancer. *Sci. Rep.* 10: 18549. https://doi.org/10.1038/s41598-020-75279-z.

30 Roche (n.d.). Roche provides update on phase III study of onartuzumab in people with specific type of lung cancer. *Eur. Pharm. Rev.* https://www.europeanpharmaceuticalreview.com/news/24343/roche-provides-update-phase-iii-study-onartuzumab-people-specific-type-lung-cancer/ (accessed July 1, 2024).

31 Passaro, A., Jänne, P.A., Mok, T., and Peters, S. (2021). Overcoming therapy resistance in EGFR-mutant lung cancer. *Nat. Cancer* 2: 377–391. https://doi.org/10.1038/s43018-021-00195-8.

32 Tebbutt, N., Pedersen, M.W., and Johns, T.G. (2013). Targeting the ERBB family in cancer: couples therapy. *Nat. Rev. Cancer* 13: 663–673. https://doi.org/10.1038/nrc3559.

33 Emde, A., Pradeep, C.-R., Ferraro, D.A. et al. (2011). Combining epitope-distinct antibodies to HER2: cooperative inhibitory effects on invasive growth. *Oncogene* 30: 1631–1642. https://doi.org/10.1038/onc.2010.547.

34 Schaefer, G., Haber, L., Crocker, L.M. et al. (2011). A two-in-one antibody against HER3 and EGFR has superior inhibitory activity compared with monospecific antibodies. *Cancer Cell* 20: 472–486. https://doi.org/10.1016/j.ccr.2011.09.003.

35 Tsao, L.-C., Crosby, E.J., Trotter, T.N. et al. (2022). Trastuzumab/pertuzumab combination therapy stimulates antitumor responses through complement-dependent cytotoxicity and phagocytosis. *JCI Insight* 7: e155636. https://doi.org/10.1172/jci.insight.155636.

36 Drachman, J.G. and Senter, P.D. (2013). Antibody-drug conjugates: the chemistry behind empowering antibodies to fight cancer. *Hematol. Am. Soc. Hematol. Educ. Program* 2013: 306–310. https://doi.org/10.1182/asheducation-2013.1.306.

37 Zvonok, N., Xu, W., Williams, J. et al. (2010). Mass spectrometry-based GPCR proteomics: Comprehensive characterization of the human cannabinoid 1 receptor. *J. Proteome Res.* 9: 1746–1753. https://doi.org/10.1021/pr900870p.

38 Rosenbaum, D.M., Rasmussen, S.G.F., and Kobilka, B.K. (2009). The structure and function of G-protein-coupled receptors. *Nature* 459: 356–363. https://doi.org/10.1038/nature08144.

39 Di Liberto, V., Mudò, G., and Belluardo, N. (2019). Crosstalk between receptor tyrosine kinases (RTKs) and G protein-coupled receptors (GPCR) in the brain: Focus on heteroreceptor complexes and related functional neurotrophic effects. *Neuropharmacology* 152: 67–77. https://doi.org/10.1016/j.neuropharm.2018.11.018.

40 Tyndall, J.D.A. and Sandilya, R. (2005). GPCR agonists and antagonists in the clinic. *Med. Chem.* 1: 405–421. https://doi.org/10.2174/1573406054368675.

41 Pin, J.-P., Kniazeff, J., Goudet, C. et al. (2004). The activation mechanism of class-C G-protein coupled receptors. *Biol. Cell.* 96: 335–342. https://doi.org/10.1016/j.biolcel.2004.03.005.

42 Jacobson, K.A. and Costanzi, S. (2012). New insights for drug design from the X-ray crystallographic structures of G-protein-coupled receptors. *Mol. Pharmacol.* 82: 361–371. https://doi.org/10.1124/mol.112.079335.

43 Davenport, A.P., Scully, C.C.G., de Graaf, C. et al. (2020). Advances in therapeutic peptides targeting G protein-coupled receptors. *Nat. Rev. Drug Discov.* 19: 389–413. https://doi.org/10.1038/s41573-020-0062-z.

44 Hughes, C.E. and Nibbs, R.J.B. (2018). A guide to chemokines and their receptors. *FEBS J.* 285 (16): 2944–2971. https://www.ncbi.nlm. nih.gov/pmc/articles/PMC6120486/ (accessed August 23, 2024).

45 Ollila, T.A., Sahin, I., and Olszewski, A.J. (2019). Mogamulizumab: a new tool for management of cutaneous T-cell lymphoma. *Onco. Targets Ther.* 12: 1085–1094. https://doi.org/10.2147/OTT.S165615.

46 Serrano-Vega, M.J., Magnani, F., Shibata, Y., and Tate, C.G. (2008). Conformational thermostabilization of the beta1-adrenergic receptor in a detergent-resistant form. *Proc. Natl. Acad. Sci. USA* 105: 877–882. https://doi.org/10.1073/pnas.0711253105.

47 Van Hout, A., Klarenbeek, A., Bobkov, V. et al. (2018). CXCR4-targeting nanobodies differentially inhibit CXCR4 function and HIV entry. *Biochem. Pharmacol.* 158: 402–412. https://doi.org/10.1016/j.bcp.2018.10.015.

48 Hutchings, C.J., Colussi, P., and Clark, T.G. (2019). Ion channels as therapeutic antibody targets. *MAbs* 11: 265–296. https://doi.org/10.1080/19420862.2018.1548232.

49 Picci, G., Marchesan, S., and Caltagirone, C. (2022). Ion channels and transporters as therapeutic agents: From biomolecules to supramolecular medicinal chemistry. *Biomedicine* 10: 885. https://doi.org/10.3390/biomedicines10040885.

50 Wilkinson, T.C.I. (2016). Discovery of functional monoclonal antibodies targeting G-protein-coupled receptors and ion channels. *Biochem. Soc. Trans.* 44: 831–837. https://doi.org/10.1042/BST20160028.

51 Haustrate, A., Hantute-Ghesquier, A., Prevarskaya, N., and Lehen'kyi, V. (2019). Monoclonal antibodies targeting ion channels and their therapeutic potential. *Front. Pharmacol.* 10: https://doi.org/10.3389/fphar.2019.00606.

52 Lee, K.J., Wang, W., Padaki, R. et al. (2014). Mouse monoclonal antibodies to transient receptor potential ankyrin 1 act as antagonists of multiple modes of channel activation. *J. Pharmacol. Exp. Ther.* 350: 223–231. https://doi.org/10.1124/jpet.114.215574.

53 Gilbert, S.M., Gidley Baird, A., Glazer, S. et al. (2017). A phase I clinical trial demonstrates that nfP2X7-targeted antibodies provide a novel, safe and tolerable topical therapy for basal cell carcinoma. *Br. J. Dermatol.* 177: 117–124. https://doi.org/10.1111/bjd.15364.

54 Selvakumar, P., Fernández-Mariño, A.I., Khanra, N. et al. (2022). Structures of the T cell potassium channel Kv1.3 with immunoglobulin modulators. *Nat. Commun.* 13: 3854. https://doi.org/10.1038/s41467-022-31285-5.

55 Minotaur Therapeutics MNT 002 - AdisInsight, (n.d.). https://adisinsight.springer.com/drugs/800078077 (accessed August 26, 2024).

56 Woubshete, M., Cioccolo, S., and Byrne, B. (2024). Advances in membrane mimetic systems for manipulation and analysis of membrane proteins: detergents, polymers, lipids and scaffolds. *ChemPlusChem* e202300678. https://doi.org/10.1002/cplu.202300678.

57 Keener, J.E., Zhang, G., and Marty, M.T. (2020). Native mass spectrometry of membrane proteins. *Anal. Chem.* https://doi.org/10.1021/acs.analchem.0c04342.

58 Mehmood, S., Marcoux, J., Hopper, J.T.S. et al. (2014). Charge reduction stabilizes intact membrane protein complexes for mass spectrometry. *J. Am. Chem. Soc.* 136: 17010–17012. https://doi.org/10.1021/ja510283g.

59 Tanford, C. and Reynolds, J.A. (1976). Characterization of membrane proteins in detergent solutions. *Biochim. Biophys. Acta Rev. Biomembr.* 457: 133–170. https://doi.org/10.1016/0304-4157(76)90009-5.

60 Kalipatnapu, S. and Chattopadhyay, A. (2005). Membrane protein solubilization: recent advances and challenges in solubilization of serotonin1A receptors. *IUBMB Life* 57: 505–512. https://doi.org/10.1080/15216540500167237.

61 Seddon, A.M., Curnow, P., and Booth, P.J. (2004). Membrane proteins, lipids and detergents: not just a soap opera. *Biochim. Biophys. Acta Biomembr.* 1666: 105–117. https://doi.org/10.1016/j.bbamem.2004.04.011.

62 Neugebauer, J.M. (1990). [18] Detergents: an overview. In: *Methods in Enzymology* (ed. M.P. Deutscher), 239–253. Academic Press https://doi.org/10.1016/0076-6879(90)82020-3.

63 Young, J.W. (2023). Recent advances in membrane mimetics for membrane protein research. *Biochem. Soc. Trans.* 51: 1405–1416. https://doi.org/10.1042/BST20230164.

64 Ratkeviciute, G., Cooper, B.F., and Knowles, T.J. (2021). Methods for the solubilisation of membrane proteins: the micelleaneous world of membrane protein solubilisation. *Biochem. Soc. Trans.* 49: 1763–1777. https://doi.org/10.1042/BST20210181.

65 Stetsenko, A. and Guskov, A. (2017). An overview of the top ten detergents used for membrane protein crystallization. *Cryst.* 7: 197. https://doi.org/10.3390/cryst7070197.

66 Hjelmeland, L.M. (1990). [19] Solubilization of native membrane proteins. In: *Methods in Enzymology* (ed. M.P. Deutscher), 253–264. Academic Press. https://doi.org/10.1016/0076-6879(90)82021-S.

67 Schimerlik, M.I. (1998). Overview of membrane protein solubilization. *Curr. Protoc. Neurosci.* 2: 5–9. https://doi.org/10.1002/0471142301.ns0509s02.

68 le Maire, M., Champeil, P., and Møller, J.V. (2000). Interaction of membrane proteins and lipids with solubilizing detergents. *Biochim. Biophys. Acta Biomembr.* 1508: 86–111. https://doi.org/10.1016/S0304-4157(00)00010-1.

69 Lee, H.J., Lee, H.S., Youn, T. et al. (2022). Impact of novel detergents on membrane protein studies. *Chem* 8: 980–1013. https://doi.org/10.1016/j.chempr.2022.02.007.

70 Wycisk, V., Wagner, M.-C., and Urner, L.H. (2024). Trends in the diversification of the detergentome. *ChemPlusChem* 89: e202300386. https://doi.org/10.1002/cplu.202300386.

71 Urner, L.H., Liko, I., Yen, H.-Y. et al. (2020). Modular detergents tailor the purification and structural analysis of membrane proteins including G-protein coupled receptors. *Nat. Commun.* 11: 564. https://doi.org/10.1038/s41467-020-14424-8.

72 Urner, L.H., Junge, F., Fiorentino, F. et al. (2023). Rationalizing the Optimization of Detergents for Membrane Protein Purification. *Chem. Eur. J.* 29: e202300159. https://doi.org/10.1002/chem.202300159.

73 Chae, P.S., Rasmussen, S.G.F., Rana, R.R. et al. (2010). Maltose–neopentyl glycol (MNG) amphiphiles for solubilization, stabilization and crystallization of membrane proteins. *Nat. Methods* 7: 1003–1008. https://doi.org/10.1038/nmeth.1526.

74 Chae, P.S., Rana, R.R., Gotfryd, K. et al. (2013). Glucose-Neopentyl Glycol (GNG) amphiphiles for membrane protein study. *Chem. Commun.* 49: 2287–2289. https://doi.org/10.1039/C2CC36844G.

75 Chae, P.S., Rasmussen, S.G.F., Rana, R.R. et al. (2012). A new class of amphiphiles bearing rigid hydrophobic groups for solubilization and stabilization of membrane proteins. *Chem. Eur. J.* 18: 9485–9490. https://doi.org/10.1002/chem.201200069.

76 Orwick-Rydmark, M., Arnold, T., and Linke, D. (2016). The use of detergents to purify membrane proteins. *Curr. Protoc. Protein Sci.* 84: 4–8. https://doi.org/10.1002/0471140864.ps0408s84.

77 Tulumello, D.V. and Deber, C.M. (1818). Efficiency of detergents at maintaining membrane protein structures in their biologically relevant forms. *Biochim. Biophys. Acta Biomembr.* 2012: 1351–1358. https://doi.org/10.1016/j.bbamem.2012.01.013.

78 Patrick, J.W. and Laganowsky, A. (2019). Generation of charge-reduced ions of membrane protein complexes for native ion mobility mass spectrometry studies. *J. Am. Soc. Mass Spectrom.* 30: 886–892. https://doi.org/10.1007/s13361-019-02187-6.

79 Privé, G.G. (2007). Detergents for the stabilization and crystallization of membrane proteins. *Methods* 41: 388–397. https://doi.org/10.1016/j.ymeth.2007.01.007.

80 Helenius, A. and Simons, K. (1975). Solubilization of membranes by detergents. *Biochim. Biophys. Acta Rev. Biomembr.* 415: 29–79. https://doi.org/10.1016/0304-4157(75)90016-7.

81 Anandan, A. and Vrielink, A. (2016). Detergents in membrane protein purification and crystallisation. In: *The Next Generation in Membrane Protein Structure Determination* (ed. I. Moraes), 13–28. Cham: Springer International Publishing. https://doi.org/10.1007/978-3-319-35072-1_2.

82 Urner, L.H., Fiorentino, F., Shutin, D. et al. (2024). Detergents with scalable properties identify non-canonical lipopolysaccharide binding to bacterial inner membrane proteins. *J. Am. Chem. Soc.* 146: 11025–11030. https://doi.org/10.1021/jacs.3c14358.

83 Nagarajan, R. (2002). Molecular packing parameter and surfactant self-assembly: the neglected role of the surfactant tail. *Langmuir* 18: 31–38. https://doi.org/10.1021/la010831y.

84 Marty, M.T., Hoi, K.K., and Robinson, C.V. (2016). Interfacing membrane mimetics with mass spectrometry. *Acc. Chem. Res.* 49: 2459–2467. https://doi.org/10.1021/acs.accounts.6b00379.

85 Urner, L.H. (2022). Advances in membrane mimetics and mass spectrometry for understanding membrane structure and function. *Curr. Opin. Chem. Biol.* 69: 102157. https://doi.org/10.1016/j.cbpa.2022.102157.

86 Oluwole, A., Shutin, D., and Bolla, J.R. (2023). Mass spectrometry of intact membrane proteins: shifting towards a more native-like context. *Essays Biochem.* 67: 201–213. https://doi.org/10.1042/EBC20220169.

87 Kotov, V., Bartels, K., Veith, K. et al. (2019). High-throughput stability screening for detergent-solubilized membrane proteins. *Sci. Rep.* 9: 10379. https://doi.org/10.1038/s41598-019-46686-8.

88 Tribet, C., Audebert, R., and Popot, J.-L. (1996). Amphipols: polymers that keep membrane proteins soluble in aqueous solutions. *Proc. Natl. Acad. Sci.* 93: 15047–15050. https://doi.org/10.1073/pnas.93.26.15047.

89 Österlund, N., Luo, J., Wärmländer, S.K.T.S., and Gräslund, A. (2019). Membrane-mimetic systems for biophysical studies of the amyloid-β peptide. *Biochim. Biophys. Acta BBA - Proteins Proteomics* 1867: 492–501. https://doi.org/10.1016/j.bbapap.2018.11.005.

90 Klöpfer, K. and Hagn, F. (2019). Beyond detergent micelles: the advantages and applications of non-micellar and lipid-based membrane mimetics for solution-state NMR. *Prog. Nucl. Magn. Reson. Spectrosc.* 114–115: 271–283. https://doi.org/10.1016/j.pnmrs.2019.08.001.

91 Majeed, S., Ahmad, A.B., Sehar, U., and Georgieva, E.R. (2021). Lipid membrane mimetics in functional and structural studies of integral membrane proteins. *Membranes* 11: 685. https://doi.org/10.3390/membranes11090685.

92 Popot, J.-L., Althoff, T., Bagnard, D. et al. (2011). Amphipols from A to Z*. *Annu. Rev. Biophys.* 40: 379–408. https://doi.org/10.1146/annurev-biophys-042910-155219.

93 Popot, J.-L. (2010). Amphipols, nanodiscs, and fluorinated surfactants:three nonconventional approaches to studying membrane proteins in aqueous solutions. *Annu. Rev. Biochem.* 79: 737–775. https://doi.org/10.1146/annurev.biochem.052208.114057.

94 Zoonens, M. and Popot, J.-L. (2014). Amphipols for each season. *J. Membr. Biol.* 247: 759–796. https://doi.org/10.1007/s00232-014-9666-8.

95 Le Bon, C., Marconnet, A., Masscheleyn, S. et al. (2018). Folding and stabilizing membrane proteins in amphipol A8-35. *Methods* 147: 95–105. https://doi.org/10.1016/j.ymeth.2018.04.012.

96 Schmidpeter, P.A.M., Sukomon, N., and Nimigean, C.M. (2020). Reconstitution of membrane proteins into platforms suitable for biophysical and structural analyses. In: *Expression, Purification, and Structural Biology of Membrane Proteins* (ed. C. Perez and T. Maier), 191–205. New York, NY: Springer https://doi.org/10.1007/978-1-0716-0373-4_14.

97 Bosco, M., Damian, M., Chauhan, V. et al. (2020). Biotinylated non-ionic amphipols for GPCR ligands screening. *Methods* 180: 69–78. https://doi.org/10.1016/j.ymeth.2020.06.001.

98 Jia, X., Long, Y., Yu, X. et al. (2023). Construction of stable membranal CMTM6-PD-L1 full-length complex to evaluate the PD-1/PD-L1-CMTM6 interaction and develop anti-tumor anti-CMTM6 nanobody. *Acta Pharmacol. Sin.* 44: 1095–1104. https://doi.org/10.1038/s41401-022-01020-3.

99 Fuss, M.F., Wieferig, J.-P., Corey, R.A. et al. (2023). Cyclic di-AMP traps proton-coupled K+ transporters of the KUP family in an inward-occluded conformation. *Nat. Commun.* 14: 3683. https://doi.org/10.1038/s41467-023-38944-1.

100 Charvolin, D., Perez, J.-B., Rouvière, F. et al. (2009). The use of amphipols as universal molecular adapters to immobilize membrane proteins onto solid supports. *Proc. Natl. Acad. Sci.* 106: 405–410. https://doi.org/10.1073/pnas.0807132106.

101 Clénet, D., Clavier, L., Strobbe, B. et al. (2021). Full-length G glycoprotein directly extracted from

rabies virus with detergent and then stabilized by amphipols in liquid and freeze-dried forms. *Biotechnol. Bioeng.* 118: 4317–4330. https://doi.org/10.1002/bit.27900.

102 McLean, M.A., Gregory, M.C., and Sligar, S.G. (2018). Nanodiscs: a Controlled Bilayer Surface for the Study of Membrane Proteins. *Annu. Rev. Biophys.* 47: 107–124. https://doi.org/10.1146/annurev-biophys-070816-033620.

103 Bayburt, T.H., Grinkova, Y.V., and Sligar, S.G. (2002). Self-assembly of discoidal phospholipid bilayer nanoparticles with membrane scaffold proteins. *Nano Lett.* 2: 853–856. https://doi.org/10.1021/nl025623k.

104 Brough, Z., Zhao, Z., and van Hoa, F.D. (2024). From bottom-up to cell surface proteomics: detergents or no detergents, that is the question. *Biochem. Soc. Trans.* BST20231020. https://doi.org/10.1042/BST20231020.

105 Elzoghby, A.O., Samir, O., Soliman, A. et al. (2023). Nanodiscs: game changer nano-therapeutics and structural biology tools. *Nano Today* 53: 102026. https://doi.org/10.1016/j.nantod.2023.102026.

106 Bayburt, T.H., Grinkova, Y.V., and Sligar, S.G. (2006). Assembly of single bacteriorhodopsin trimers in bilayer nanodiscs. *Arch. Biochem. Biophys.* 450: 215–222. https://doi.org/10.1016/j.abb.2006.03.013.

107 Puthenveetil, R. and Vinogradova, O. (2013). Optimization of the design and preparation of nanoscale phospholipid bilayers for its application to solution NMR. *Proteins: Struct., Funct., Bioinf* 81: 1222–1231. https://doi.org/10.1002/prot.24271.

108 Mu, Q., Deng, H., An, X. et al. (2024). Designing nanodiscs as versatile platforms for on-demand therapy. *Nanoscale* 16: 2220–2234. https://doi.org/10.1039/D3NR05457H.

109 Nasr, M.L. and Wagner, G. (2018). Covalently circularized nanodiscs; challenges and applications. *Curr. Opin. Struct. Biol.* 51: 129–134. https://doi.org/10.1016/j.sbi.2018.03.014.

110 Zhang, S., Ren, Q., Novick, S.J. et al. (2021). One-step construction of circularized nanodiscs using SpyCatcher-SpyTag. *Nat. Commun.* 12: 5451. https://doi.org/10.1038/s41467-021-25737-7.

111 Bariwal, J., Ma, H., Altenberg, G.A., and Liang, H. (2022). Nanodiscs: a versatile nanocarrier platform for cancer diagnosis and treatment. *Chem. Soc. Rev.* 51: 1702–1728. https://doi.org/10.1039/D1CS01074C.

112 Baskakova, K.O., Kuzmichev, P.K., and Karbyshev, M.S. (2024). Advanced applications of Nanodiscs-based platforms for antibodies discovery. *Biophys. Chem.* 313: 107290. https://doi.org/10.1016/j.bpc.2024.107290.

113 Ma, Y., Ding, Y., Song, X. et al. (2020). Structure-guided discovery of a single-domain antibody agonist against human apelin receptor. *Sci. Adv.* 6: eaax7379. https://doi.org/10.1126/sciadv.aax7379.

114 Dominik, P.K., Borowska, M.T., Dalmas, O. et al. (2016). Conformational chaperones for structural studies of membrane proteins using antibody phage display with nanodiscs. *Structure* 24: 300–309. https://doi.org/10.1016/j.str.2015.11.014.

115 Bhattacharya, P., Grimme, S., Ganesh, B. et al. (2010). Nanodisc-incorporated hemagglutinin provides protective immunity against influenza virus infection. *J. Virol.* 84: 361–371. https://doi.org/10.1128/JVI.01355-09.

116 Pandey, A.K., Piplani, N., Mondal, T. et al. (2021). Efficient delivery of hydrophobic drug, Cabazitaxel, using Nanodisc: a nano sized free standing planar lipid bilayer. *J. Mol. Liq.* 339: 116690. https://doi.org/10.1016/j.molliq.2021.116690.

117 Wong, P., Li, L., Chea, J. et al. (2020). Antibody targeted PET imaging of 64Cu-DOTA-Anti-CEA PEGylated lipid nanodiscs in CEA positive tumors. *Bioconjug. Chem.* 31: 743–753. https://doi.org/10.1021/acs.bioconjchem.9b00854.

118 Popovic, K., Holyoake, J., Pomès, R., and Privé, G.G. (2012). Structure of saposin A lipoprotein discs. *Proc. Natl. Acad. Sci.* 109: 2908–2912. https://doi.org/10.1073/pnas.1115743109.

119 Leney, A.C., Rezaei Darestani, R., Li, J. et al. (2015). Picodiscs for facile protein-glycolipid interaction analysis. *Anal. Chem.* 87: 4402–4408. https://doi.org/10.1021/acs.analchem.5b00170.

120 Kostelic, M.M., Ryan, A.M., Reid, D.J. et al. (2019). Expanding the types of lipids amenable to native mass spectrometry of lipoprotein complexes. *J. Am. Soc. Mass Spectrom.* 30: 1416–1425. https://doi.org/10.1007/s13361-019-02174-x.

121 Li, J., Han, L., Li, J. et al. (2018). Detecting protein–glycolipid interactions using CaR-ESI-MS and Model membranes: comparison of pre-loaded and passively loaded picodiscs. *J. Am. Soc. Mass Spectrom.* 29: 1493–1504. https://doi.org/10.1007/s13361-018-1936-8.

122 Carlson, M.L., Young, J.W., Zhao, Z. et al. (2018). The Peptidisc, a simple method for stabilizing membrane proteins in detergent-free solution. *Elife* 7: e34085. https://doi.org/10.7554/eLife.34085.

123 Angiulli, G., Dhupar, H.S., Suzuki, H. et al. (2020). New approach for membrane protein reconstitution into peptidiscs and basis for their adaptability to different proteins. *Elife* 9: e53530. https://doi.org/10.7554/eLife.53530.

124 Morizumi, T., Kim, K., Li, H. et al. (2023). Structures of channel rhodopsin paralogs in peptidiscs explain their contrasting K+ and Na+ selectivities. *Nat. Commun.* 14: 4365. https://doi.org/10.1038/s41467-023-40041-2.

125 Young, J.W., Wason, I.S., Zhao, Z. et al. (2022). Development of a method combining peptidiscs and proteomics to identify, stabilize, and purify a detergent-sensitive membrane protein assembly. *J. Proteome Res.* 21: 1748–1758. https://doi.org/10.1021/acs.jproteome.2c00129.

126 Antony, F., Brough, Z., Zhao, Z., and Duong van Hoa, F. (2024). Capture of the mouse organ membrane proteome specificity in peptidisc libraries. *J. Proteome Res.* 23: 857–867. https://doi.org/10.1021/acs.jproteome.3c00825.

127 Rantalainen, K., Berndsen, Z.T., Antanasijevic, A. et al. (2020). HIV-1 envelope and MPER antibody structures in lipid assemblies. *Cell Rep.* 31: https://doi.org/10.1016/j.celrep.2020.107583.

128 Kuai, R., Ochyl, L.J., Bahjat, K.S. et al. (2017). Designer vaccine nanodiscs for personalized cancer immunotherapy. *Nat. Mater.* 16: 489–496. https://doi.org/10.1038/nmat4822.

129 Najafabadi, A.H., Abadi, Z.I.N., Aikins, M.E. et al. (2021). Vaccine nanodiscs plus polyICLC elicit robust CD8+ T cell responses in mice and non-human primates. *J. Control. Release* 337: 168–178. https://doi.org/10.1016/j.jconrel.2021.07.026.

130 Scheetz, L., Kadiyala, P., Sun, X. et al. (2020). Synthetic high-density lipoprotein nanodiscs for personalized immunotherapy against gliomas. *Clin. Cancer Res.* 26: 4369–4380. https://doi.org/10.1158/1078-0432.CCR-20-0341.

131 Ravula, T., Hardin, N.Z., and Ramamoorthy, A. (2019). Polymer nanodiscs: advantages and limitations. *Chem. Phys. Lipids* 219: 45–49. https://doi.org/10.1016/j.chemphyslip.2019.01.010.

132 Unger, L., Ronco-Campaña, A., Kitchen, P. et al. (2021). Biological insights from SMA-extracted proteins. *Biochem. Soc. Trans.* 49: 1349–1359. https://doi.org/10.1042/BST20201067.

133 Nestorow, S.A., Dafforn, T.R., and Frasca, V. (2021). Biophysical characterisation of SMALPs. *Biochem. Soc. Trans.* 49: 2037–2050. https://doi.org/10.1042/BST20201088.

134 Hoi, K.K., Bada Juarez, J.F., Judge, P.J. et al. (2021). Detergent-free lipodisq nanoparticles facilitate high-resolution mass spectrometry of folded integral membrane proteins. *Nano Lett.* 21: 2824–2831. https://doi.org/10.1021/acs.nanolett.0c04911.

135 Gulamhussein, A.A., Meah, D., Soja, D.D. et al. (2019). Examining the stability of membrane proteins within SMALPs. *Eur. Polym. J.* 112: 120–125. https://doi.org/10.1016/j.eurpolymj.2018.12.008.

136 Lee, S.C., Knowles, T.J., Postis, V.L.G. et al. (2016). A method for detergent-free isolation of membrane proteins in their local lipid environment. *Nat. Protoc.* 11: 1149–1162. https://doi.org/10.1038/nprot.2016.070.

137 Simon, K.S., Pollock, N.L., and Lee, S.C. (2018). Membrane protein nanoparticles: the shape of things to come. *Biochem. Soc. Trans.* 46: 1495–1504. https://doi.org/10.1042/BST20180139.

138 Routledge, S.J., Jamshad, M., Little, H.A. et al. (2020). Ligand-induced conformational changes in a SMALP-encapsulated GPCR. *Biochim. Biophys. Acta Biomembr.* 1862: 183235. https://doi.org/10.1016/j.bbamem.2020.183235.

139 Dodd, R., Schofield, D.J., Wilkinson, T., and Britton, Z.T. (2020). Generating therapeutic monoclonal antibodies to complex multi-spanning membrane targets: overcoming the antigen challenge and enabling discovery strategies. *Methods* 180: 111–126. https://doi.org/10.1016/j.ymeth.2020.05.006.

140 Tedesco, D., Maj, M., Malarczyk, P. et al. (2021). Application of the SMALP technology to the isolation of GPCRs from low-yielding cell lines. *Biochim. Biophys. Acta Biomembr.* 1863: 183641. https://doi.org/10.1016/j.bbamem.2021.183641.

141 Sharma, P., Plant, M., Lam, S.K., and Chen, Q. (2021). Kinetic analysis of antibody binding to integral membrane proteins stabilized in SMALPs. *BBA Adv.* 1: 100022. https://doi.org/10.1016/j.bbadva.2021.100022.

142 Panda, A., Giska, F., Duncan, A.L. et al. (2023). Direct determination of oligomeric organization of integral membrane proteins and lipids from intact customizable bilayer. *Nat. Methods* 20: 891–897. https://doi.org/10.1038/s41592-023-01864-5.

143 Frick, M., Schwieger, C., and Schmidt, C. (2021). Liposomes as Carriers of Membrane-Associated Proteins and Peptides for Mass Spectrometric Analysis. *Angew. Chem. Int. Ed.* 60: 11523–11530. https://doi.org/10.1002/anie.202101242.

144 Goddard, A.D., Dijkman, P.M., Adamson, R.J., and Watts, A. (2013). Chapter 18 - Lipid-dependent GPCR dimerization. In: *Methods in Cell Biology* (ed. P.M. Conn), 341–357. Academic Press https://doi.org/10.1016/B978-0-12-408143-7.00018-9.

145 Rigaud, J.-L. and Lévy, D. (2003). Reconstitution of membrane proteins into liposomes. In: *Methods in Enzymology*, 65–86. Academic Press https://doi.org/10.1016/S0076-6879(03)72004-7.

146 Nsairat, H., Khater, D., Sayed, U. et al. (2022). Liposomes: structure, composition, types, and clinical applications. *Heliyon* 8: e09394. https://doi.org/10.1016/j.heliyon.2022.e09394.

147 Buzas, E.I. (2023). The roles of extracellular vesicles in the immune system. *Nat. Rev. Immunol.* 23: 236–250. https://doi.org/10.1038/s41577-022-00763-8.

148 Zeev-Ben-Mordehai, T., Vasishtan, D., Siebert, C.A. et al. (2014). Extracellular Vesicles: a Platform for the Structure Determination of Membrane Proteins by Cryo-EM. *Structure* 22: 1687–1692. https://doi.org/10.1016/j.str.2014.09.005.

149 Liguori, L., Marques, B., Villegas-Mendez, A. et al. (2008). Liposomes-mediated delivery of pro-apoptotic therapeutic membrane proteins. *J. Control. Release* 126: 217–227. https://doi.org/10.1016/j.jconrel.2007.12.004.

150 Alves, N.J., Turner, K.B., Medintz, I.L., and Walper, S.A. (2015). Emerging therapeutic delivery capabilities and challenges utilizing enzyme/protein packaged bacterial vesicles. *Ther. Deliv.* https://doi.org/10.4155/tde.15.40.

151 Schwendener, R.A. (2014). Liposomes as vaccine delivery systems: a review of the recent advances. *Ther. Adv. Vaccines* 2: 159–182. https://doi.org/10.1177/2051013614541440.

152 Dürr, U.H.N., Gildenberg, M., and Ramamoorthy, A. (2012). The magic of bicelles lights up membrane protein structure. *Chem. Rev.* 112: 6054–6074. https://doi.org/10.1021/cr300061w.

153 Amengual, J., Notaro-Roberts, L., and Nieh, M.-P. (2023). Morphological control and modern applications of bicelles. *Biophys. Chem.* 302: 107094. https://doi.org/10.1016/j.bpc.2023.107094.

154 Smith, L.M. and Kelleher, N.L. (2013). Proteoform: a single term describing protein complexity. *Nat. Methods* 10: 186–187. https://doi.org/10.1038/nmeth.2369.

155 Kaltashov, I.A., Ivanov, D.G., and Yang, Y. (2024). Mass spectrometry-based methods to characterize highly heterogeneous biopharmaceuticals, vaccines, and nonbiological complex drugs at the intact-mass level. *Mass Spectrom. Rev.* 43: 139–165. https://doi.org/10.1002/mas.21829.

156 Liu, H., Gaza-Bulseco, G., Faldu, D. et al. (2008). Heterogeneity of monoclonal antibodies. *J. Pharm. Sci.* 97: 2426–2447. https://doi.org/10.1002/jps.21180.

157 Wu, D. and Robinson, C.V. (2022). Understanding glycoprotein structural heterogeneity and interactions: insights from native mass spectrometry. *Curr. Opin. Struct. Biol.* 74: 102351. https://doi.org/10.1016/j.sbi.2022.102351.

158 Chandler, K.B. and Costello, C.E. (2016). Glycomics and glycoproteomics of membrane proteins and cell-surface receptors: present trends and future opportunities. *Electrophoresis* 37: 1407–1419. https://doi.org/10.1002/elps.201500552.

159 Lippens, J.L., Egea, P.F., Spahr, C. et al. (2018). Rapid LC–MS method for accurate molecular weight determination of membrane and hydrophobic proteins. *Anal. Chem.* 90: 13616–13623. https://doi.org/10.1021/acs.analchem.8b03843.

160 Campuzano, I.D.G., Robinson, J.H., Hui, J.O. et al. (2019). Native and denaturing MS protein deconvolution for biopharma: monoclonal antibodies and antibody–drug conjugates to polydisperse membrane proteins and beyond. *Anal. Chem.* 91: 9472–9480. https://doi.org/10.1021/acs.analchem.9b00062.

161 Campuzano, I.D.G. and Sandoval, W. (2021). Denaturing and native mass spectrometric analytics for biotherapeutic drug discovery research: historical, current, and future personal perspectives. *J. Am. Soc. Mass Spectrom.* 32: 1861–1885. https://doi.org/10.1021/jasms.1c00036.

162 Tamara, S., den Boer, M.A., and Heck, A.J.R. (2022). High-resolution native mass spectrometry. *Chem. Rev.* 122: 7269–7326. https://doi.org/10.1021/acs.chemrev.1c00212.

163 Lössl, P., van de Waterbeemd, M., and Heck, A.J. (2016). The diverse and expanding role of mass spectrometry in structural and molecular biology. *EMBO J.* 35: 2634–2657. https://doi.org/10.15252/embj.201694818.

164 Kelleher, N.L., Lin, H.Y., Valaskovic, G.A. et al. (1999). Top down versus bottom up protein characterization by tandem high-resolution mass spectrometry. *J. Am. Chem. Soc.* 121: 806–812. https://doi.org/10.1021/ja973655h.

165 Cui, W., Rohrs, H.W., and Gross, M.L. (2011). Top-down mass spectrometry: recent developments, applications and perspectives. *Analyst* 136: 3854–3864. https://doi.org/10.1039/C1AN15286F.

166 Toby, T.K., Fornelli, L., and Kelleher, N.L. (2016). Progress in top-down proteomics and the analysis of proteoforms. *Annu. Rev. Anal. Chem.* 9: 499–519. https://doi.org/10.1146/annurev-anchem-071015-041550.

167 Habeck, T., Brown, K.A., Des Soye, B. et al. (2024). Top-down mass spectrometry of native proteoforms and their complexes: a community study. *Nat. Methods* https://doi.org/10.1038/s41592-024-02279-6.

168 Habeck, T. and Lermyte, F. (2023). Seeing the complete picture: proteins in top-down mass spectrometry. *Essays Biochem.* 67: 283–300. https://doi.org/10.1042/EBC20220098.

169 Zhou, M., Lantz, C., Brown, K.A. et al. (2020). Higher-order structural characterisation of native proteins and complexes by top-down mass spectrometry. *Chem. Sci.* 11: 12918–12936. https://doi.org/10.1039/D0SC04392C.

170 Harvey, S.R., O'Neale, C., Schey, K.L., and Wysocki, V.H. (2022). Native mass spectrometry and surface induced dissociation provide insight into the post-translational modifications of tetrameric AQP0 isolated from bovine eye lens. *Anal. Chem.* 94: 1515–1519. https://doi.org/10.1021/acs.analchem.1c04322.

171 Ro, S.Y., Schachner, L.F., Koo, C.W. et al. (2019). Native top-down mass spectrometry provides insights into the copper centers of membrane-bound methane monooxygenase. *Nat. Commun.* 10: 2675. https://doi.org/10.1038/s41467-019-10590-6.

172 Marty, M.T. (2020). Nanodiscs and mass spectrometry: making membranes fly. *Int. J. Mass Spectrom.* 458: 116436. https://doi.org/10.1016/j.ijms.2020.116436.

173 Townsend, J.A. and Marty, M.T. (2023). What's the defect? Using mass defects to study oligomerization of membrane proteins and peptides in nanodiscs with native mass spectrometry. *Methods* 218: 1–13. https://doi.org/10.1016/j.ymeth.2023.07.004.

174 Schachner, L.F., Mullen, C., Phung, W. et al. (2024). Exposing the molecular heterogeneity of glycosylated biotherapeutics. *Nat. Commun.* 15: 3259. https://doi.org/10.1038/s41467-024-47693-8.

175 Zilkenat, S., Grin, I., and Wagner, S. (2017). Stoichiometry determination of macromolecular membrane protein complexes. *Biol. Chem.* 398: 155–164. https://doi.org/10.1515/hsz-2016-0251.

176 Ahnert, S.E., Marsh, J.A., Hernández, H. et al. (2015). Principles of assembly reveal a periodic table of protein complexes. *Science* 350: aaa2245. https://doi.org/10.1126/science.aaa2245.

177 Fiorentino, F., Rotili, D., and Mai, A. (2023). Native mass spectrometry-directed drug discovery: recent advances in investigating protein function and modulation. *Drug Discov. Today* 28: 103548. https://doi.org/10.1016/j.drudis.2023.103548.

178 Zhou, M., Morgner, N., Barrera, N.P. et al. (2011). Mass spectrometry of intact V-Type ATPases reveals bound lipids and the effects of nucleotide binding. *Science* 334: 380–385. https://doi.org/10.1126/science.1210148.

179 Yen, H.-Y., Hoi, K.K., Liko, I. et al. (2018). PtdIns(4,5)P2 stabilizes active states of GPCRs and enhances selectivity of G-protein coupling. *Nature* 559: 423–427. https://doi.org/10.1038/s41586-018-0325-6.

180 Yen, H.-Y., Liko, I., Song, W. et al. (2022). Mass spectrometry captures biased signalling and allosteric modulation of a G-protein-coupled receptor. *Nat. Chem.* 14: 1375–1382. https://doi.org/10.1038/s41557-022-01041-9.

181 Yen, H.-Y., Jazayeri, A., and Robinson, C.V. (2023). G protein-coupled receptor pharmacology—insights from mass spectrometry. *Pharmacol. Rev.* 75: 397. https://doi.org/10.1124/pharmrev.120.000237.

182 Townsend, J.A., Sanders, H.M., Rolland, A.D. et al. (2021). Influenza AM2 channel oligomerization is sensitive to its chemical environment. *Anal. Chem.* 93: 16273–16281. https://doi.org/10.1021/acs.analchem.1c04660.

183 Chorev, D.S., Tang, H., Rouse, S.L. et al. (2020). The use of sonicated lipid vesicles for mass spectrometry of membrane protein complexes. *Nat. Protoc.* 15: 1690–1706. https://doi.org/10.1038/s41596-020-0303-y.

184 Chorev, D.S. and Robinson, C.V. (2020). The importance of the membrane for biophysical measurements. *Nat. Chem. Biol.* 16: 1285–1292. https://doi.org/10.1038/s41589-020-0574-1.

185 Zhu, Y., Yun, S.D., Zhang, T. et al. (2023). Native mass spectrometry of proteoliposomes containing integral and peripheral membrane proteins. *Chem. Sci.* 14: 14243–14255. https://doi.org/10.1039/D3SC04938H.

186 Liu, F., Rijkers, D.T.S., Post, H., and Heck, A.J.R. (2015). Proteome-wide profiling of protein assemblies by cross-linking mass spectrometry. *Nat. Methods* 12: 1179–1184. https://doi.org/10.1038/nmeth.3603.

187 O'Reilly, F.J. and Rappsilber, J. (2018). Cross-linking mass spectrometry: methods and applications in structural, molecular and systems biology. *Nat. Struct. Mol. Biol.* 25: 1000–1008. https://doi.org/10.1038/s41594-018-0147-0.

188 Schmidt, C., Zhou, M., Marriott, H. et al. (2013). Comparative cross-linking and mass spectrometry of an intact F-type ATPase suggest a role for phosphorylation. *Nat. Commun.* 4: 1985. https://doi.org/10.1038/ncomms2985.

189 Wittig, S., Ganzella, M., Barth, M. et al. (2021). Cross-linking mass spectrometry uncovers protein interactions and functional assemblies in synaptic vesicle membranes. *Nat. Commun.* 12: 858. https://doi.org/10.1038/s41467-021-21102-w.

190 Yan, X. and Maier, C.S. (2009). Hydrogen/deuterium exchange mass spectrometry. In: *Mass Spectrometry of Proteins and Peptides: methods and Protocols* (ed. M.S. Lipton and L. Paša-Tolic), 255–271. Totowa, NJ: Humana Press https://doi.org/10.1007/978-1-59745-493-3_15.

191 Narang, D., Lento, C., and Wilson, D.J. (2020). HDX-MS: an analytical tool to capture protein motion in action. *Biomedicine* 8: 224. https://doi.org/10.3390/biomedicines8070224.

192 Wei, H., Mo, J., Tao, L. et al. (2014). Hydrogen/deuterium exchange mass spectrometry for probing higher order structure of protein therapeutics: methodology and applications. *Drug Discov. Today* 19: 95–102. https://doi.org/10.1016/j.drudis.2013.07.019.

193 James, E.I., Murphree, T.A., Vorauer, C. et al. (2022). Advances in hydrogen/deuterium exchange mass spectrometry and the pursuit of challenging biological systems. *Chem. Rev.* 122: 7562–7623. https://doi.org/10.1021/acs.chemrev.1c00279.

194 Katta, V., Chait, B.T., and Carr, S. (1991). Conformational changes in proteins probed by hydrogen-exchange electrospray-ionization mass spectrometry. *Rapid Commun. Mass Spectrom.* 5: 214–217. https://doi.org/10.1002/rcm.1290050415.

195 Zhu, S., Liuni, P., Chen, T. et al. (2022). Epitope screening using hydrogen/deuterium exchange mass spectrometry (HDX-MS): an accelerated workflow for evaluation of lead monoclonal antibodies. *Biotechnol. J.* 17: 2100358. https://doi.org/10.1002/biot.202100358.

196 Kim, M., Sun, Z.-Y.J., Rand, K.D. et al. (2011). Antibody mechanics on a membrane-bound HIV segment essential for GP41-targeted viral neutralization. *Nat. Struct. Mol. Biol.* 18: 1235–1243. https://doi.org/10.1038/nsmb.2154.

197 Chung, K.Y., Rasmussen, S.G.F., Liu, T. et al. (2011). Conformational changes in the G protein Gs induced by the β2 adrenergic receptor. *Nature* 477: 611–615. https://doi.org/10.1038/nature10488.

198 Clouser, A.F., Alam, Y.H., and Atkins, W.M. (2020). Cholesterol asymmetrically modulates the conformational ensemble of the nucleotide-binding domains of P-glycoprotein in lipid nanodiscs. *Biochemistry*. https://doi.org/10.1021/acs.biochem.0c00824.

199 Hebling, C.M., Morgan, C.R., Stafford, D.W. et al. (2010). Conformational analysis of membrane proteins in phospholipid bilayer nanodiscs by hydrogen exchange mass spectrometry. *Anal. Chem.* 82: 5415–5419. https://doi.org/10.1021/ac100962c.

200 Anderson, K.W., Gallagher, E.S., and Hudgens, J.W. (2018). Automated removal of phospholipids from membrane proteins for H/D exchange mass spectrometry workflows. *Anal. Chem.* 90: 6409–6412. https://doi.org/10.1021/acs.analchem.8b00429.

201 Reading, E., Hall, Z., Martens, C. et al. (2017). Interrogating membrane protein conformational dynamics within native lipid compositions. *Angew. Chem. Int. Ed.* 56: 15654–15657. https://doi.org/10.1002/anie.201709657.

202 Möller, I.R., Slivacka, M., Hausner, J. et al. (2019). Improving the sequence coverage of integral membrane proteins during hydrogen/deuterium exchange mass spectrometry experiments. *Anal. Chem.* 91: 10970–10978. https://doi.org/10.1021/acs.analchem.9b00973.

203 Mendoza, V.L. and Vachet, R.W. (2009). Probing protein structure by amino acid-specific covalent labeling and mass spectrometry. *Mass Spectrom. Rev.* 28: 785–815. https://doi.org/10.1002/mas.20203.

204 Zhou, Y. and Vachet, R.W. (2013). Covalent labeling with isotopically encoded reagents for faster structural analysis of proteins by mass spectrometry. *Anal. Chem.* 85: 9664–9670. https://doi.org/10.1021/ac401978w.

205 Kirsch, Z.J. and Vachet, R.W. (2024). Proteins can withstand more extensive labeling while providing accurate structural information in covalent labeling-mass spectrometry. *J. Am. Soc. Mass Spectrom.* 35: 1030–1039. https://doi.org/10.1021/jasms.4c00043.

206 Pan, X. and Vachet, R.W. (2022). Membrane protein structures and interactions from covalent labeling coupled with mass spectrometry. *Mass Spectrom. Rev.* 41: 51–69. https://doi.org/10.1002/mas.21667.

207 Kirsch, Z.J., Blake, J.M., Huynh, U. et al. (2023). Membrane protein binding interactions studied in live cells via diethylpyrocarbonate covalent labeling mass spectrometry. *Anal. Chem.* 95: 7178–7185. https://doi.org/10.1021/acs.analchem.2c05616.

208 Limpikirati, P., Liu, T., and Vachet, R.W. (2018). Covalent labeling-mass spectrometry with non-specific reagents for studying protein structure and interactions. *Methods* 144: 79–93. https://doi.org/10.1016/j.ymeth.2018.04.002.

209 Tremblay, C.Y., Kirsch, Z.J., and Vachet, R.W. (2022). Epitope mapping with diethylpyrocarbonate covalent labeling-mass spectrometry. *Anal. Chem.* 94: 1052–1059. https://doi.org/10.1021/acs.analchem.1c04038.

210 Yan, Y., Chen, G., Wei, H. et al. (2014). Fast photochemical oxidation of proteins (FPOP) maps the epitope of EGFR binding to adnectin. *J. Am. Soc. Mass Spectrom.* 25: 2084–2092. https://doi.org/10.1007/s13361-014-0993-x.

211 Zhang, Y., Wecksler, A.T., Molina, P. et al. (2017). Mapping the binding interface of VEGF and a monoclonal antibody Fab-1 fragment with fast photochemical oxidation of proteins (FPOP) and mass spectrometry. *J. Am. Soc. Mass Spectrom.* 28: 850–858. https://doi.org/10.1007/s13361-017-1601-7.

212 Lu, Y., Zhang, H., Niedzwiedzki, D.M. et al. (2016). Fast photochemical oxidation of proteins maps the topology of intrinsic membrane proteins: light-harvesting complex 2 in a nanodisc. *Anal. Chem.* 88: 8827–8834. https://doi.org/10.1021/acs.analchem.6b01945.

213 Sun, J., Liu, X.R., Li, S. et al. (2021). Nanoparticles and photochemistry for native-like transmembrane protein footprinting. *Nat. Commun.* 12: 7270. https://doi.org/10.1038/s41467-021-27588-8.

214 Shortt, R.L., Wang, Y., Hummon, A.B., and Jones, L.M. (2023). Development of spheroid-FPOP: an in-cell protein footprinting method for 3D tumor spheroids. *J. Am. Soc. Mass Spectrom.* 34: 417–425. https://doi.org/10.1021/jasms.2c00307.

215 Chavez, J.D. and Bruce, J.E. (2019). Chemical cross-linking with mass spectrometry: a tool for systems structural biology. *Curr. Opin. Chem. Biol.* 48: 8–18. https://doi.org/10.1016/j.cbpa.2018.08.006.

216 Piersimoni, L., Kastritis, P.L., Arlt, C., and Sinz, A. (2022). Cross-linking mass spectrometry for investigating protein conformations and protein–protein interactions—A method for all seasons. *Chem. Rev.* 122: 7500–7531. https://doi.org/10.1021/acs.chemrev.1c00786.

217 Chavez, J.D., Tang, X., Campbell, M.D. et al. (2020). Mitochondrial protein interaction landscape of SS-31. *Proc. Natl. Acad. Sci.* 117: 15363–15373. https://doi.org/10.1073/pnas.2002250117.

218 Li, K.S., Chen, G., Mo, J. et al. (2017). Orthogonal mass spectrometry-based footprinting for epitope mapping and structural characterization: the IL-6 receptor upon binding of protein therapeutics. *Anal. Chem.* 89: 7742–7749. https://doi.org/10.1021/acs.analchem.7b01748.

19

Chemoproteomics Approaches for Diverse Modalities

H. Christian Eberl[1], Sascha Knecht[1], Markus A. Queisser[2], and Marcus Bantscheff[1,3]

[1] *Target Discovery, GSK, Heidelberg, Germany*
[2] *Modality Platform Technologies, GSK, Medicines Research Centre, Stevenage, UK*
[3] *Research and Early Development, Roche Pharma, Basel, Switzerland*

Introduction

Recent innovations in Omics technologies and Chemical Biology have led to a revolution in the analytical toolbox available to support drug discovery [1]. Chemical Biology employs chemistry and chemical tools to investigate biological processes, offering insights into disease and drug mechanisms and their modulation by bioactive molecules. Chemical proteomics, a subdiscipline of chemical biology, examines the impact of chemicals and bioactive molecules in general on the proteome. Numerous chemoproteomics approaches have emerged, including affinity enrichment of sub-proteomes, proximity labeling, and assessing the impact of small-molecule binding on protein physicochemical properties (e.g. thermal stability). Together, Omics-based profiling and chemoproteomics approaches enable the study of target proteins and downstream processes modulated by bioactive molecules in the context of disease and models thereof.

In recent years, drug discovery and development have undergone an impressive transformation driven by the emergence of new chemical and therapeutic modalities including targeted covalent inhibitors, protein degraders, and molecular glues, RNA therapeutics including small interfering RNA, recombinant peptides and proteins, monoclonal antibodies, antibody–drug conjugates, and gene therapy (Figure 19.1a). These modalities are part of a broader strategy to enable previously "undruggable" targets and address unmet medical needs, including rare diseases and cancers [2, 3]. This diversification of modalities necessitated an evolution of the analytical procedures applied to characterize the effects of candidate molecules on the proteome leading to the development of new chemoproteomics workflows and adaptations of protocols to reflect particular mechanistic features. Scientists can now choose from an ever-expanding arsenal of (chemo-)proteomics approaches supporting drug discovery efforts addressing critical questions such as ligandability of targets and target engagement, selectivity, and mechanism of action of drug candidates and thus providing key insights for modality selection, prioritization of lead molecules, drug differentiation, and understanding of adverse events.

In order to give guidance which methods are most suitable for individual modalities, it is useful to distinguish molecules acting via an occupancy-driven mechanism such as non-covalent small-molecule inhibitors from molecular modalities eliciting event-driven pharmacology where the restoration of function requires resynthesis (Figure 19.1b). This distinction has been prominently discussed in the context of Proteolysis Targeting Chimera [4, 5] but can easily be expanded to other modalities. Molecules – being these small or large, chemical compounds, peptides, proteins, or antibodies – that modulate target activity reversibly by blocking a specific site might need to outcompete endogenous ligands and may bind in a dose-dependent way to other proteins having similar functions and binding sites. Understanding occupancy under equilibrium conditions and off-rates provides a reasonable orientation on the expected activities. In contrast, low-affinity binding of ProTACs can lead to comprehensive depletion of the protein target of interest with effect duration being coupled to the protein synthesis rate, hence understanding binding affinities across the proteome provides an insufficient proxy of the expected biological effects. In this chapter, we introduce important chemoproteomics concepts and workflows and provide an overview of suitable approaches for a set of emerging chemical modalities.

Characterizing Biotherapeutics: Analytical Methods for Diverse Modalities, First Edition. Edited by Jennie R. Lill and Wendy Sandoval.

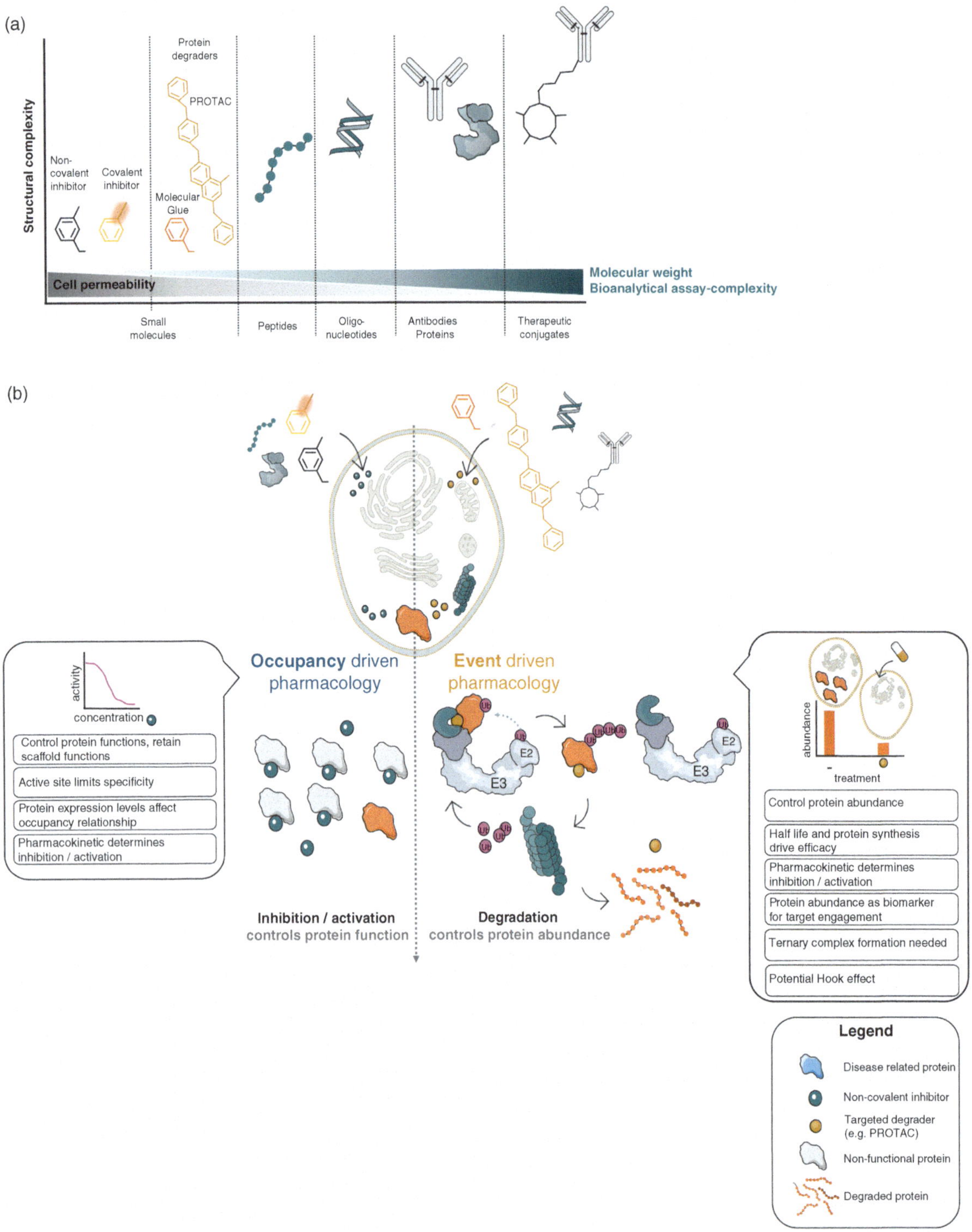

Figure 19.1 | Emerging therapeutic modalities and pharmacological concepts in drug discovery. (a) Overview of emerging modalities in drug discovery. New modalities, including targeted covalent inhibitors, protein degraders, RNA therapeutics, peptides, antibodies and antibody–drug conjugates, have collectively transformed drug discovery efforts and are designed to target previously "undruggable" targets. The diversification of therapeutic modalities, accompanied by increased structural complexity, has driven the evolution of chemoproteomics workflows and newly developed protocols, enabling precise characterization of candidate molecules' effects on the proteome. (b) Occupancy-driven and event-driven pharmacology are two distinct concepts used to describe how drugs interact with their targets and produce therapeutic effects. Differentiating between occupancy-driven pharmacology (e.g. non-covalent small-molecule inhibitors) and event-driven pharmacology (e.g. proteolysis targeting chimeras, or PROTACs) is essential for choosing the right analytical methods to enable accurate evaluation of drug activity and effect duration. Occupancy-driven molecules modulate target activity by reversible binding, often requiring outcompeting endogenous ligands, while event-driven modalities, like covalent inhibitors and PROTACs, induce effects that depend on target protein resynthesis.

Chemoproteomics Approaches

Experimental methods for identifying cellular targets of bioactive compounds can be grouped into three major strategies (Figure 19.2). The first strategy focuses on identifying direct interactions between a bioactive compound and its protein target(s) using the compound or chemically derived probes in cell extracts or live cell systems (Figure 19.2a,b). The compound or probe can act as a "bait" to directly capture the protein target(s) or covalently modify them [1, 6].

The second strategy examines changes in protein structure or biophysical properties, such as thermal stability or resistance to proteolysis (Figure 19.2c,d), upon compound binding [7–12] and, finally, the third strategy uses proximity-labeling techniques such as *in situ* biotinylation or alkylation to mark the targets of bioactive molecules with the respective labels [13–15].

These compound-centric chemoproteomics [16] strategies are complemented by indirect, mechanism-centric proteomics approaches (Figure 19.2e,f,g). While these indirect methods do not directly identify targets, they provide functional context by characterizing downstream effects of target modulation, including proteome-wide changes in protein abundance, turnover, subcellular localization, post-translational modifications, and protein–protein interactions [6].

Probe-Based (Direct) Target Deconvolution

Probe-based direct target identification methods classically aim at the enrichment of drug targets by utilizing the compound under investigation as "bait" [17, 18] (Figure 19.2a,b).

Affinity enrichment using tagged bioactive small molecules combined with quantitative MS-based proteomics provides a sensitive and specific tool for comprehensive profiling of drug interactions within cellular proteomes. Compounds can hereby interact and capture the protein target(s) by covalent (activity-based proteomics) or non-covalent interactions (affinity-based proteomics) [1, 6]. In both cases, this requires a derivatization of the bioactive molecule with an affinity tag or a reactive group, to enable the isolation of captured target proteins from cellular extracts and subsequent identification by mass spectrometry (MS). To retain bioactivity and the desired target binding profile of the modified compound used as probe, detailed information on the structure–activity relationship (SAR) of a bioactive molecule is required [1, 19, 20]. The bioactive molecule is derivatized via a spacer and immobilized to a bead matrix (e.g. sepharose beads), prior incubation with a cell extract. Alternatively, the bioactive molecule can be derivatized with an affinity tag, such as biotin and added to a cell extract before capturing the protein target(s) on, for example, streptavidin beads. Further, modification of the bioactive molecule with bioorthogonal-chemistry compatible tags such as alkyne-, azide-, or trans-cyclooctene tags enables probing for target engagement in live cells instead of only in cell extracts, as the small size of the tag retains cell permeability of the probe [21, 22]. Captured proteins can then be enriched through covalent coupling to dibenzylcyclooctyne (DBCO) activated agarose resins or in a two-step procedure by reaction with biotin-labeling reagents to facilitate enrichment via streptavidin agarose or similar tools [22].

In cases where compound-specific probes are not available or have too low affinity to the target to enable enrichment, activity-based probe profiling (ABPP) provides a suitable alternative [23, 24] (Figure 19.2b). Activity-based probes (ABPs) are designed to covalently bind to the active site of a certain protein target or target class and typically consist of two elements: a reactive group (also called a "warhead") and a tag are both connected via a linker [25]. The warhead is usually an electrophilic group that binds covalently to nucleophilic residues within active sites while the reporter group enables the enrichment of proteins and either consists of a biotinylated moiety or a group for biorthogonal reactions. The linker provides a spacer between the reporter tag and the warhead and modulates the reactivity of the probe as well as selectivity [25]. Depending on the reporter tag, ABPs can in many cases be adapted for *in situ* and *in vivo* labeling.

Protein-binding profiles obtained by affinity- or activity-based probes are only qualitative and do not assess target potency. A conceptually simple way to further classify small-molecule candidates is competition-binding experiments in which the cell extract is first incubated with increasing concentrations of the original, underivatized bioactive compound followed by incubation with the affinity- or activity-based probe [1, 26]. The preincubation step with the underivatized bioactive compound leads to a dose-dependent increase in target protein occupancy, hence, target proteins already bound by the free drug cannot be pulled down by the affinity- or activity-based probe. Subsequent analysis of the pulled-down proteome by LC-MS/MS and plotting the fraction of protein intensity in each sample relative to the vehicle control allows fitting of sigmoidal curves and determination of half-maximal binding concentrations (IC_{50} values) for all captured proteins. If additional experiments to study the influence of the probe molecule on the equilibrium between compound of interest and targets are performed, even apparent dissociation

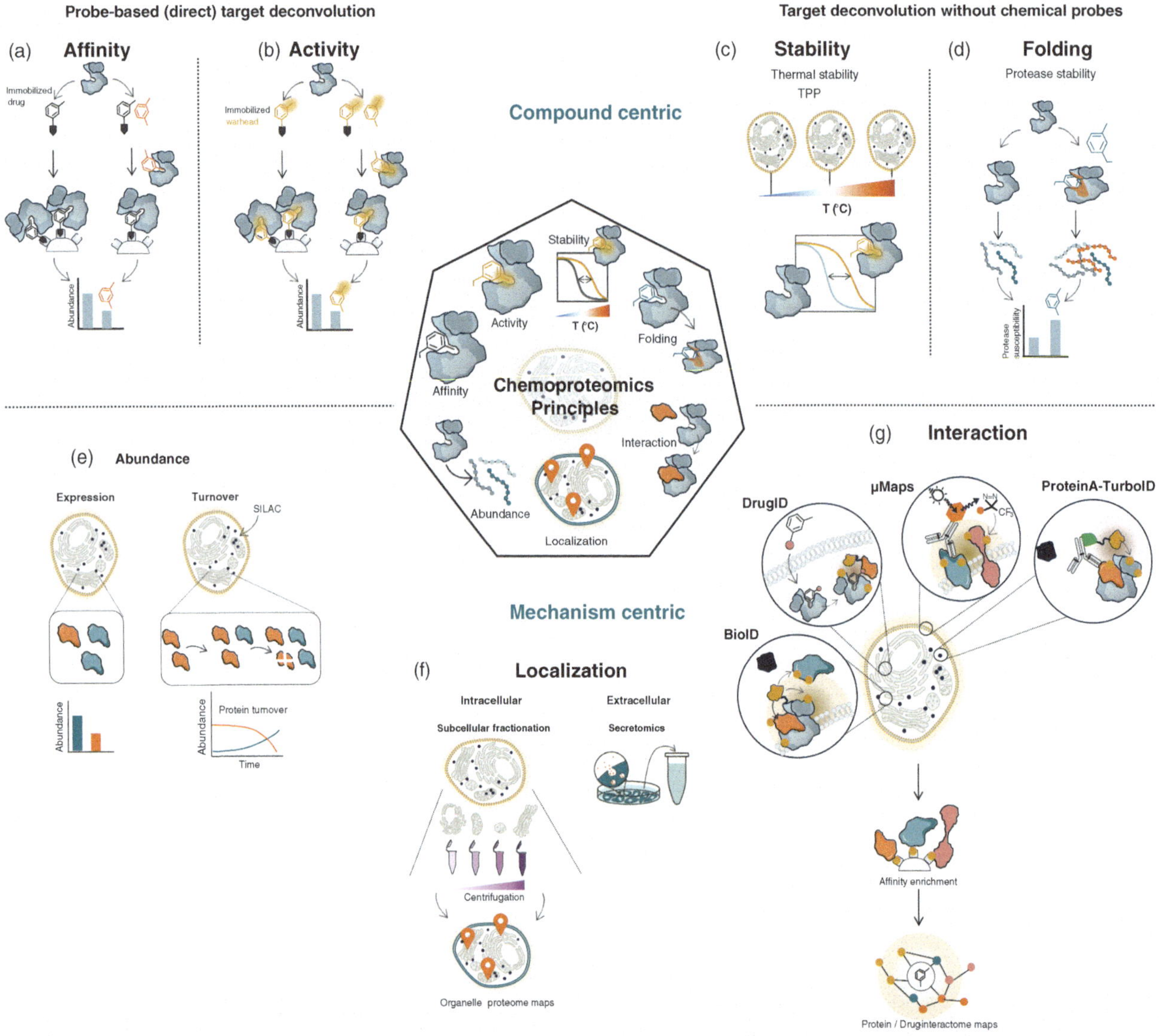

Figure 19.2 Schematic overview of chemoproteomics principles to characterize (off-)targets by compound centric chemoproteomics strategies such as the analysis of protein affinity, activity, stability and folding and indirect mechanism-centric proteomic strategies by the analysis of protein abundance, localization and protein–drug interactions. Probe-based (direct) target deconvolution methods utilize the compound under investigation as "bait" to enrich off-targets by either non-covalent interactions (a, affinity-based proteomics) or by covalent interactions (b, activity-based proteomics) with a chemical probe. Both methods require the derivatization of the bioactive molecule with an affinity tag or a reactive group, to enable the isolation of captured target proteins from cellular extracts and subsequent identification by mass spectrometry. In cases where compound-specific probes are not available or have insufficient affinity to the target to enable enrichment activity, activity-based probes can be designed to contain an (electrophilic) warhead that binds covalently to target proteins. To determine IC50 values for all captured proteins, competition-binding assays can be performed where cell extracts are incubated with increasing concentrations of the bioactive compound, followed by incubation with an affinity-based probe. Target deconvolution without chemical probes relies on measuring ligand-induced changes in biophysical properties of target proteins, such as thermal stability (c) or interrogating protein stability and conformation changes in response to drug binding (d). In addition to directly measuring drug–target engagement, potential targets or pathways can be inferred by indirect mechanism-centric proteomics methods. Assessment of global protein abundance in cells or tissues (e) can indicate drug effects and side effects. For investigating a drug's impact on global protein turnover, newly synthesized proteins can be dynamically labeled with SILAC, which allows the determination of protein synthesis and degradation rates. Protein localizations (f) within cells, tissues, or the extracellular space can be obtained by fractionation-based methods or by investigation of the secretome and cell surface proteome. Proximity ligation-based protein–protein interaction methods (g) rely on the fusion of individual proteins with engineered ascorbate and biotin ligases that generate reactive affinity handles – such as biotin – to modify proteins covalently in a close spatial radius and are useful for studying interactions under different conditions or perturbations. Small-molecule–drug or oligonucleotide–drug–protein interactions inside living cells can be identified with Drug-ID which uses a Halo-tag-biotin ligase fusion construct that binds to chloroalkane-modified drugs, allowing biotinylation of the drug's protein targets and nearby proteins. Protein interactions on the cell surface can be studied with μ-Maps, that uses a target specific antibody, and an iridium catalyst activated by light to generate short-lived biotin intermediates. By using a ProteinA-TurboID fusion enzyme together with a target protein-specific antibody bait proximal proteins can be biotinylated without prior genetic manipulation of the target cell.

constants can be determined for non-covalent interactions [27, 28]. Hence, competition-binding experiments can provide selectivity data of the test compound against all probe-captured proteins or protein target classes.

Target Deconvolution Without Chemical Probes

While the probe-based methods mentioned above are effective, they require the creation of functionalized small molecules, a process that can be complex and time consuming.

In recent years, proteomics techniques have emerged that detect small-molecule–protein interactions without the need for chemical probes (Figure 19.2c,d). These methods rely on measuring ligand-induced changes in the biophysical properties of target proteins, such as thermal stability, chemical denaturation by solvent or acidic stresses, proteolysis resistance, or the surface exposure of amino acid side chains in the presence of chaotropes. When combined with quantitative mass spectrometry, these techniques enable proteome-wide characterizations of small-molecule–protein interactions [29]).

Thermal Stability Profiling

Thermal proteome profiling (TPP), also known as cellular thermal shift assay (CETSA)–MS, can reveal a drug's mechanism of action by assessing drug-induced changes in protein thermal stability across the proteome [7, 29] (Figure 19.2c). In TPP, living cells, cell extracts or tissues are treated with either the drug or a vehicle and then subjected to a gradient of increasing temperatures, typically ranging from 37°C to 70°C or higher.

As the temperature increases, proteins begin to unfold and lose their tertiary structure, leading to denaturation and aggregation. The samples are then centrifuged at high speed to separate the aggregated proteins from the soluble proteins that remain in the supernatant. Finally, the soluble fraction from each temperature point is then analyzed using quantitative mass spectrometry. The relative amounts of a given protein remaining in the supernatant across the different heating conditions are used to infer the protein's thermal stability and to generate melting curves.

The inflection point of the denaturation curve at which 50% of the protein remains in solution is typically referred to as melting temperature (Tm). It is important to note that not all proteins will precipitate easily, especially at temperatures that are compatible with the structural integrity of cells and, therefore, aggregation as a consequence of unfolding may be incomplete or not happen at all for small and hydrophilic proteins [30]. The dose-dependent format of this approach (isothermal titration [31], 2D-TPP) provides a quantitative view on compound effects allowing the determination of EC_{50}s across the proteome. Several compressed formats of the overall approach have been reported (PISA, STPP-UP) that minimize the required LC-MS analysis time, thus providing more throughput for larger compound characterization efforts while limiting insights to a qualitative assessment of compound effects. In cell-based experiments, indirect compound effects can also affect the thermal stability of proteins. While this makes data analysis more complex, it offers the opportunity to use thermal stability as a descriptor of protein state complementing other (prote-)omics approaches and to infer protein complex formation [32]. Specialized workflows also assess subcellular localization, for example, effects on plasma membrane proteins [10].

Recently, similar to heat-induced denaturation, solvent-induced denaturation has been described as a suitable approach to probe the influence of compound binding on protein stability across the proteome [33, 34].

Protease Stability

Effects on protein stability and conformation in response to drug binding can also be interrogated with proteolytic methods such as drug affinity responsive target stability (DARTS) [12, 35] and limited proteolysis-mass spectrometry (LIP-MS) [9, 36] (Figure 19.2d).

DARTS is based on the concept that when a small molecule binds to a protein, it stabilizes the protein, making it less susceptible to proteolytic degradation. Proteins that are stabilized by the ligand are identified by comparing the degree of proteolysis in the presence and absence of the ligand.

Similarly, LIP-MS uses controlled, limited proteolysis to selectively cleave accessible and flexible regions of proteins. The accessibility of proteolytic cleavage sites can be influenced by the protein's conformation and interactions. If a protein binds to a ligand, undergoes a conformational change, or interacts with other proteins, the accessibility of cleavage sites might change. Resulting peptide patterns from the treated and untreated samples act as reporters for protein folding and differences in the proteolytic patterns indicate changes in protein conformation or binding sites.

Similar to thermal stability, proteolytic resistance can be influenced by multiple factors and proteome-wide analysis can be a descriptor of proteome state complementing abundance-based methods, for example, for biomarker analysis [37].

Proximity-Labeling Approaches

Proximity-labeling techniques (Figure 19.2g) like APEX (ascorbate peroxidase) and BIO-ID (proximity-dependent biotin identification) use engineered ascorbate- and biotin ligases to generate reactive affinity handles – such as biotin – to modify proteins covalently in a close spatial radius and are useful for studying interactions under different conditions or perturbations. Biotinylated proteins can then be enriched from crude lysates using streptavidin beads following their identification by quantitative LC-MS analysis.

Recently, an "off-the-shelf" proximity biotinylation approach [38] was introduced that uses a ProteinA-coupled TurboID enzyme that can be targeted to baits of interest using antibodies against endogenous proteins or their post-translational modifications. Upon permeabilization of live cells, addition of a bait-specific antibody and subsequent incubation with the ProteinA-TurboID enzyme, bait proximal proteins are biotinylated and analyzed by quantitative LC-MS analysis. This approach in principle can also be used to identify (off-)targets of therapeutic antibodies and other small or large-molecule drug candidates [39–44]. For example, in the recently described Drug-ID [13] a Halo-tag is genetically fused to a biotin ligase. The Halo-tag can covalently bind to a chloroalkane (CA)-modified small or large-molecule drug and upon the addition of biotin, protein targets of the drug as well as interaction partners in close proximity are then biotinylated.

Another variation of antibody-based, but ligase-free, proximity-labeling techniques is μMap [14] (microenvironment mapping). This approach was designed to study protein interactions and localization within specific microenvironments, for example, on the cell surface of cells or tissues. Similar to ProteinA-TurboID, an antibody targeted to the bait of interest is used. Activation of an iridium catalyst by light, which is attached to a secondary antibody generates carbene reactive biotin intermediates with a short half-life that allow nearby proximity labeling of proteins in the extracellular space.

Inspired by this unique activation mode, the principle of photocatalysis was further leveraged for small molecule target ID through the incorporation of an iridium photocatalyst onto a bioactive small molecule [41].

Mechanism-Centric (Indirect) Proteomics Methods

In addition to directly measuring drug–target engagement, potential targets or pathways can be inferred indirectly from phenotypic observations and molecular signatures following drug treatment and genetic perturbations.

Changes in protein abundances or posttranslational modifications across the proteome over time can indicate the intended effects of the compound, as well as potential adverse effects, side effects, or compensatory mechanisms aimed at restoring homeostasis, which may lead to resistance [45–47] (Figure 19.2e).

To assess whether drug treatment impacts global protein turnover, newly synthesized proteins can be labeled metabolically in dynamic or pulse–chase experiments using stable isotope labeling with amino acids in cell culture (SILAC) [48, 49]. This technique allows for the determination of protein synthesis and degradation rates, providing insights into protein stability and turnover, which is especially beneficial for analyzing changes in protein abundance and dynamics in response to PROTAC treatment (see Degrader chapter) [50–53].

Beyond protein abundance, proteomics can provide additional functional information including protein localization within cells or tissues (Figure 19.2f). Fractionation-based methods such as LOPIT [54, 55], SubCellBarCode [56], and dynamic organellar maps (DOMs, [57]) or size-exclusion chromatography-based strategies [58, 59] as well as targeted approaches such as cell surface biotinylation identify drug effects on protein localization and differentiate drug action across cellular compartments [10] or in the extracellular space [14, 60–63].

For mapping spatial protein dynamics across multiple subcellular compartments, proteomics analyses of fractionated organelles and proximity ligation-based protein–protein interaction (PPI) networks have likewise become powerful tools [64, 65].

Covalent Drugs

Many of the methods described above have been predominantly used in the context of studying drugs and probe molecules that act as non-covalent, reversible inhibitors displaying prototypic occupancy-driven pharmacology. Recently, there has been renewed interest in covalent inhibitors as an attractive opportunity to unlock challenging targets with significant therapeutic potential which are not druggable with conventional inhibitors, for example, kRAS [66–70]. Moreover, covalent inhibition in some cases can increase selectivity within an enzyme class by targeting non-conserved nucleophilic residues and by displaying different pharmacodynamics and pharmacokinetics [71, 72].

However, the irreversible nature of covalent inhibitors demands an adjusted method portfolio to characterize their activity and selectivity. While for non-covalent inhibitors, selectivity and pharmacokinetics are mainly off-rate driven due to systemic clearance, for covalent inhibitors it is important to assess the rate of covalent bond formation on top of affinity, ideally in cells. In addition, due to the event-driven pharmacology, protein synthesis rates are of interest to understand effect duration [1, 73].

Consequently, drug binding to its target cannot be described as a first order equilibrium, but instead is characterized by an initial non-covalent recognition event (Ki) which is directly followed by target inactivation via a covalent reaction (Kinact). Thus, the covalent reaction is described by the second-order rate constant Kinact / Ki. Due to complex kinetics, data at a single time point cannot comprehensively describe activity and selectivity of a covalent drug [74]. Rather, multiple time points, kinetics, and a critical assessment of protein turnover for both targets and off-targets are necessary.

During early drug discovery, recombinant target proteins are often used to identify first probes and optimize molecules [75] and for proteome-wide assessment of covalent inhibitors.

Cell-based chemoproteomics methods are preferred over lysate-based work as binding events are strongly dependent on factors such as temperature, cofactor concentrations, and protein levels that are difficult to replicate in cell extracts [76].

The proteome-wide assessment of drug-modified amino acids is conceptually simple, as modified peptides can be directly identified by deep LC-MS-based proteomics profiling of drug-treated cells only requiring to add the compound adduct as a variable modification into the database search algorithm (Figure 19.3a). This approach does not need any additional sample preparation steps and enables the identification of substoichiometric modification of abundant proteins. However, the lack of an enrichment step greatly affects sensitivity and standard proteomics analyses using trypsin as sole protease frequently achieve low sequence coverage resulting in a very incomplete assessment of modified peptides. In addition,

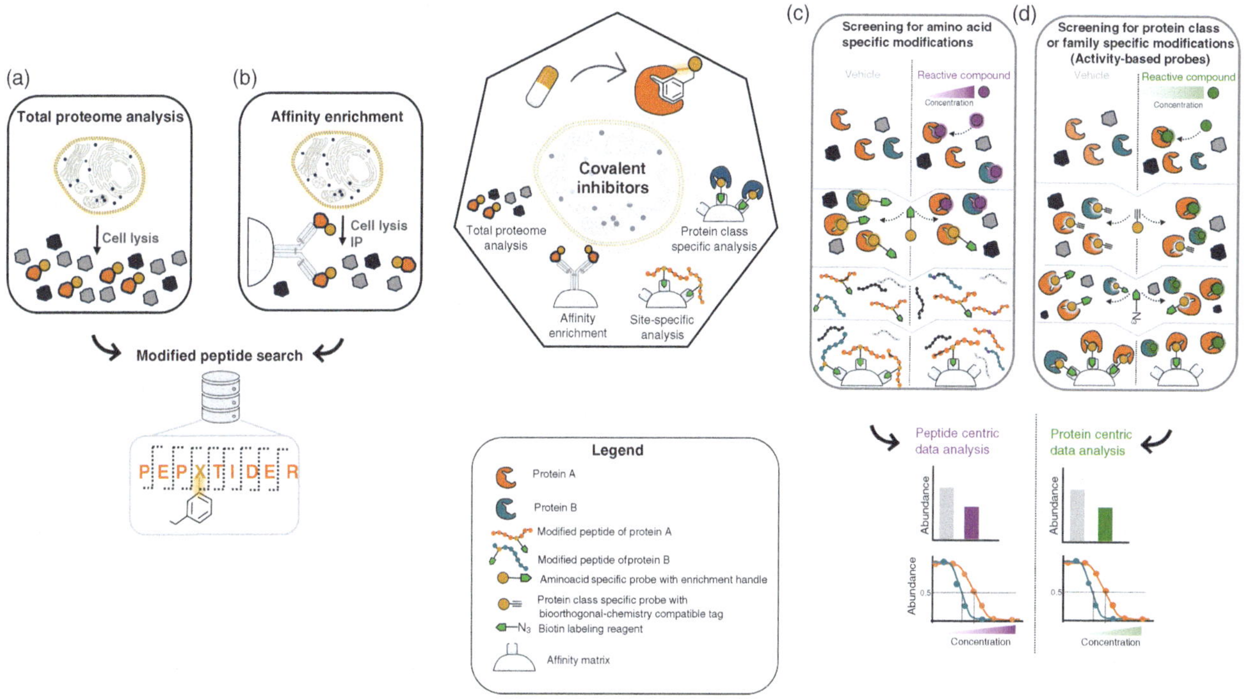

Figure 19.3 Chemoproteomics method portfolio to characterize activity and selectivity of covalent drugs. (a) Proteome-wide assessment of drug-modified amino acids via total proteome analysis allows for the direct identification of modified peptides in drug-treated cells by incorporating the drug adduct as a variable modification in the database search algorithm. This streamlined approach does not require additional sample preparation steps and can identify substoichiometric modifications in abundant proteins but comes at the costs of reduced sensitivity. (b) Direct monitoring of covalent target occupancy in complex systems by affinity enrichment step, for example, by immunoprecipitation to increase sensitivity and to ensure sufficient sequence coverage. Similar to a, identification of modified enriched peptides requires the incorporation of the drug adduct as variable modification in the database search algorithm. (c) Amino acid-specific enrichment utilizing amino acid-specific bifunctional covalent probes, that consist of an amino acid specific functional group (e.g. iodoacetamide that reacts with cysteine moieties) and an enrichment handle (e.g. biotin). (d) Protein-class or family-specific enrichment utilizing covalent mechanisms conserved within a protein family (activity-based probes (ABPs)).

determining target occupancy with such an approach is not straightforward and hence results tend to be qualitative. To increase sensitivity and to directly monitor covalent target occupancy in complex systems, an enrichment step, for example, by immunoprecipitation, ensures sufficient sequence coverage (Figure 19.3b). Reproducibility can be even further improved by targeted data acquisition methods. Immunoprecipitation of human KRAS from a mouse xenograft study enabled monitoring time-dependent KRAS G12C covalent modification by AMG 510 and comparison to the downstream pERK signal [77]. This approach is of course limited to one or few targets of interest, mainly addressing on-target occupancy but not covering proteome-wide selectivity.

Similar to investigating naturally occurring post-translational modifications like phosphorylation, dedicated enrichment steps enable a broader coverage of specific residues or modification sites in a single experiment (Figure 19.3c,d). These enrichment steps can be broadly grouped into three categories: (i) amino acid-specific enrichments utilizing widely unselective covalent probes; (ii) protein-class or family-specific enrichments utilizing covalent mechanisms conserved within a protein family – often referred to as activity-based probes (ABPs); and (iii) compound-specific enrichment by utilizing functionalized probes derived from an inhibitor of interest to cover the target protein and the specific off-target space of a series (see also earlier).

To probe the occupancy of specific amino acids in a proteome-wide scale, enrichment tools have been developed (Figure 19.3c). The most frequently targeted amino acids for targeted covalent strategies are cysteine residues for which several highly efficient profiling methods are available. These methods make use of bifunctional probes containing a cysteine-reactive moiety, for example, iodoacetamide and a moiety for enrichment, for example, a biotin. The first probe that was employed for proteome-scale experiments was isoTOP-ABPP [78] which utilizes an alkyne for Cu-click mediated enrichment and an isotopically labeled and proteolytically cleavable linker for relative quantification. This workflow enabled coverage of around 1,000 cysteine residues from cell lines and was initially utilized to pinpoint hyperreactive cysteines. By monitoring the reduction of the cysteine signal after reaction with covalent fragments, this assay was transformed into a powerful selectivity profiling assay for cysteine-reactive fragments [79]. This assay was used to screen a reactive fragment library which identified covalent ligands for more than 700 cysteines. Recently, this workflow was further refined by introducing the desthiobiotin moiety for enrichment. Desthiobiotin can be efficiently eluted from neutravidin beads with biotin and in combination with an isobaric mass tagging approach for quantification enabled a scalable and sensitive cysteine profiling workflow covering more than 20,000 cysteine residues [80]. With these refinements, profiling of larger reactive fragment libraries has become feasible to understand ligandability across the proteome. This enabled an expansion of chemogenomic space and can identify a starting point for drug discovery projects. This type of workflow has been broadly adopted in industry and was for example recently employed as a key method to identify and profile selective covalent inhibitors targeting a conserved and functionally relevant cysteine residue in Werner helicase [72, 81] a synthetic lethal target for microsatellite instable cancers.

In addition to proteome-wide profiling of reactive cysteine residues, methods and probes have been described to enrich further amino acids. Examples include iodoacetamide-alkyne under low pH conditions to profile selenocystines, SuTex probes to probe tyrosines, amine-reactive pentynoic acid sulfotetrafluorophenyl ester to probe lysines [82], aryl halides to profile cysteines and lysines [83] or NHS-esters to profile nucleophilic amino acids in general [84]. Application of these approaches in the context of drug discovery however is lacking behind the cysteine-centric efforts.

In summary, amino acids-specific enrichment techniques are generic and scalable and provide a reasonable albeit not complete assessment of the reactivity of reactive drugs across the proteome. It should be noted that these approaches only monitor the reduction of signal for the unmodified amino acid and not the direct labeling event. Especially studies in live cells carry the risk of false positive observations due to treatment-induced changes in protein abundance or labeling independent changes in accessibility.

Activity-based proteome profiling (ABPP) utilizes probes that react selectively and covalently with the active sites of proteins, thus capturing a protein class or family for interrogation of activity or for probing occupancy after treatment with a covalent inhibitor (Figure 19.3d). Activity-based probes have been described for a large variety of different protein classes which either contain a reactive amino acid in their active site, for example, serin hydrolases [85], cysteine proteases [86] or glycosidases [87] or for which the probe is sufficiently well positioned in the active site to enable a covalent reaction with a nearby reactive amino acid, for example, activity-based probes for kinases – XO44 [88] or acyl phosphates [89] – or probes for cytochrome P450 enzymes [90].

Activity-based probes are bifunctional molecules containing a reactive warhead which specifically engages and covalently reacts with the target proteins, a linker and a functional group for enrichment. Cleavable linkers, for example, protease cleavable linkers like the TOP-ABPP [91], redox-cleavable linkers, $Na_2S_2O_4$-cleavable azobenzene-based linkers [92], or linkers containing vicinal diols that can be cleaved with sodium periodate [93] enable the specific elution of captured proteins or modified peptides for reduced complexity and for localization of the modified amino acid residues. Alternatively, when no cleavable linkers are available, the preferred option is to perform an on-bead proteolytic digest leaving the modified peptide on the beads or eluting these in a second step [94, 95].

The enrichment function of the probe needs to either bind directly to beads, for example, biotin to streptavidin beads or a chloroalkane to Halo-tag-coated beads [96] or a biorthogonal reaction is utilized to react to either biotin or directly onto pre-coupled beads. Most commonly, probes are functionalized with alkynes which enable a Cu-click reaction with biotin-azide. Alkynes are synthetically easily accessible; probes are stable and the tag only introduces minimal sterical hindrance. However, due to its slow kinetics, the Cu-click reaction occurs in the excess of azide-biotin, necessitating the removal of the unreacted azide-biotin through protein precipitation. These additional handling steps can result in increased variability and protein loss. Strain-promoted azide–alkine cycloaddition provides an alternative faster and more efficient two-component biorthogonal reaction. The inverse-electron demand Diels-Alder reaction between trans cyclooctene (TCO) and tetrazine has been demonstrated to be a highly versatile addition to the chemoproteomics toolbox [21]. The high efficiency and simplified click reaction directly on beads generates a simpler and more robust workflow which compensates for the risk of losing activity due to the more bulky TCO group.

ABPs can be a key tool in the selectivity assessment of covalent inhibitors. For instance, the phase I trial of FAAH inhibitor BIA 10-2474 was stopped in January 2016 after serious adverse events including one fatality. A comprehensive – unfortunately retrospective – chemoproteomics analysis of BIA 10-2474 and its metabolite BIA 10-2639 using a fluorophosphonate probe for serine hydrolases identified off-target activity on multiple lipases. A follow-up lipidomics analysis confirmed effects on multiple lipids beyond the expected modulation of the target FAAH [97]. Although this study did not directly provide a causal link between the observed off-target activity and the observed toxicity, the identified off-targets demonstrated an unanticipated lack of selectivity of the drug and provided valuable starting points for further investigations.

The feasibility to synthesize probes from covalent molecules has been demonstrated for multiple proteins from very different protein families: the E3 ligase HOIP [98], multiple kinases [99–101], glutathione peroxidase GPX4 [102], or the Ubiquitin carboxy-terminal hydrolase L1 (UCHL1) [103]. A study on the E3 ligase HOIP (RNF3) illustrates the utility of such probes. Here, fragment-based screening identified a covalent inhibitor of the RBR E3 ligase HOIP. A TCO probe derived from this hit was used as ABP to demonstrate cellular target engagement in two different cell lines. Further, several off-targets including the kinase IRAK4, several aldehyde dehydrogenases, or the transcription factor DIDO1 were identified.

In addition to methods that specifically probe the covalent reaction, almost all chemoproteomics methods established for non-covalent targets can also be utilized for covalent inhibitors; however, the experimental design needs to be adjusted to not assume an equilibrium but for an irreversible reaction. For example, bead-based affinity enrichments can be applied to lysates after live cell treatment to probe for irreversibly modified targets. Kinobeads (a mixed inhibitor matrix with promiscuous kinase inhibitors) have been utilized to profile ibrutinib and to distinguish between covalent and non-covalent targets [76]. In addition, proximity-based approaches as well as probe-free chemoproteomics methods probing protein stability/conformation such as TPP, DARTS, or LiP-MS can also be applied to study covalent inhibitors. For example, in the initial CETSA paper, a very pronounced thermal stabilization of MetAP2 was observed with the covalent inhibitor TNP-470 [8].

Protein Degraders

For decades, traditional drug development has concentrated on the creation of inhibitors with high binding potency. These efforts typically focus on optimizing *in vitro* characteristics to achieve cellular efficacy. However, nontraditional modalities with event-driven pharmacology require a range of analytical techniques to address their specific consequences on the cell, in particular, to measure effects on protein stability and protein homeostasis. MS-based proteomics became essential to identify these changes and to characterize small molecules (Figure 19.2) [1]. Recent studies have revealed ligand-induced changes in protein expression, uncovering the principles by which compounds regulate their protein targets. Quantitative proteomics profiled hundreds of compounds in a human cancer cell line with significant effects on protein expression by preclinical and clinical drugs exemplifying the scale of these events [45].

One significant example of the benefits of targeting protein stability is seen in selective estrogen receptor downregulators (SERDs), which reduce ERα protein levels and block ER activity. Traditional SERDs are thought to induce misfolding of the ER protein, leading to proteasome-dependent degradation of ERα. Despite their effectiveness, the most potent SERD molecules induce degradation of the ER protein in ER+ breast cancer cells but do not completely degrade the ER protein. These cases highlight the potential and challenges in analyzing and characterizing degraders [104].

Targeted protein degradation (TPD) is an emerging field which describes the mechanism of action of small-molecule degraders which induce the degradation of target proteins, affecting the protein abundance and subsequently the homeostasis of the cell [105, 106]. Most prominent are bifunctional degraders also known as proteolysis-targeting chimeras (PROTACs) which are composed of an E3 ligase recruiting moiety and a target warhead, recruiting the E3 ligase in proximity to the protein of interest which leads to ubiquitination and degradation of the target [107]. These molecules function catalytically and display event-driven pharmacology, akin to covalent inhibitors, rather than being driven by occupancy. Measuring protein degradation and monitoring protein abundance are essential and have become an active field of the development of new technologies. Protein degradation is a dynamic process which is characterized by several parameters such as DC_{50} (half-maximal degradation concentration) and D_{max} (maximum level of degradation achievable) values at a defined time point. Optimization comprises the fine-tuning of multiple kinetic parameters such as the rate, the extent, and the duration of the degradation event [108]. *In vitro* binding can be used to characterize the strength and specificity of the inhibition or activation of proteins. However, *in vitro* binding studies are not sufficient to fully characterize PROTACs. In fact, binding and target engagement are not sufficient to predict degradation, as demonstrated by promiscuous kinase experiments and PROTAC optimization studies [109, 110]. This is because the degrader's function depends not only on binding to the target but also on the formation of a productive ternary complex between E3 ligase and target which allows degradation [111] (Figure 19.1b). Therefore, additional assays that can measure these processes are typically needed to fully understand and predict the behavior of PROTACs in a biological system. Ternary complexes can be measured by FRET, aLISA, or SPR *in vitro* and in a cellular context by NanoBRET, co-IP, or BioID.

Proteomic technologies are suitable to be applied to characterize degraders and have become important analytical tools in the TPD field [112–114]. Occupancy-driven chemoproteomics can describe the interactions between small molecules and proteins in equilibrium which provides information about their engagement with cellular targets and gives valuable information about drug–target relationships. Global proteomics profiling (Figure 19.4) is frequently employed to characterize degraders, showing their selectivity and degradation potency on the proteome level [109, 115]. Profiling of one of the early RIPK2 PROTACs identified MAPKAPK3 as the single off-target which was affected on the protein level. Interestingly, neither the RIPK2 binder nor the VHL ligase ligand is binding or engaging with MAPKAPK3, highlighting the fact that degraders have unexpected effects on homeostasis and need to be profiled in unbiased proteomic analysis. These effects can differ vastly at different concentrations and time points and need to be considered in the characterization. A functional experiment to monitor degradation events composes of testing the degrader at high and low concentrations as well short and longer time points to capture potential off-targets and cell homeostasis response. The combination of chemoproteomics and global proteomics can be a very powerful combination to fully characterize protein degraders and to understand their mechanism of action [116].

The continuous synthesis and degradation to maintain protein homeostasis is referred to as protein turnover. Both, protein synthesis and degradation jointly determine the responsiveness of the cellular proteome to a treatment. Probing the rates at which proteins are synthesized and degraded (kinetic) and the mechanisms of protein synthesis and degradation (dynamic) of protein turnover lends insight into how cells regulate essential processes such as growth, differentiation, and stress response and are major factors to be considerate in TPD [117].

The application of biochemical tools, such as small-molecule inhibitors of synthesis (e.g. cycloheximide, puromycin) or degradation (e.g. MG-132, bortezomib), enabled insights into proteome stability and turnover [52]. However, the inhibition of protein synthesis or degradation may lead to compensatory and off-target effects that affect physiological turnover rates. For example, NF-κB is located in the cytoplasm bound to an inhibitory protein called IκB. Upon receiving a signal like a stress signal or a pro-inflammatory signal, IκB is phosphorylated and then ubiquitinated, marking it for degradation. NF-κB remains bound to IκB in the cytoplasm and is not able to translocate to the nucleus to activate gene transcription, that is, BTK promoter which leads to decreased BTK expression and thus is not suitable to block PROTAC-mediated BTK degradation [110, 118]. Equally problematic, cycloheximide can lead to phosphorylation/activation of E3 ligases such as MDM2 [119].

Stable Isotope Labeling by Amino acids in Cell culture (SILAC) is a powerful technique to study differential protein expression in cells. It involves growing cells in a medium containing a "light" or "heavy" form of a

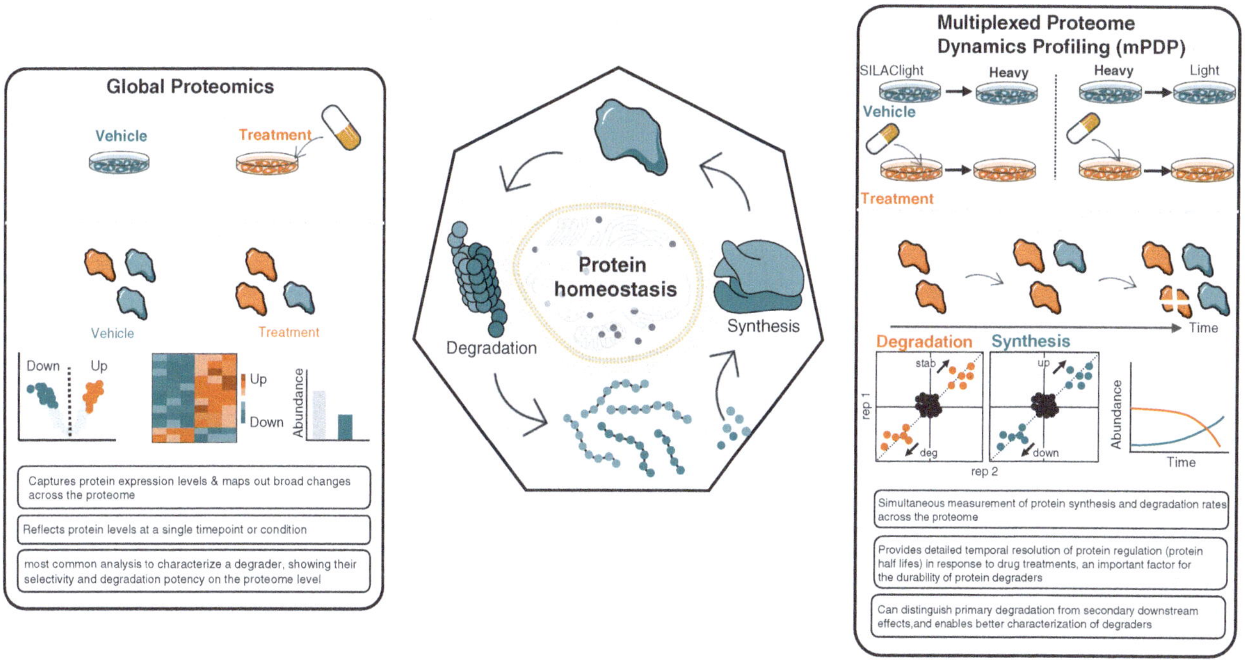

Figure 19.4 Distinguishing on-target, off-target and downstream effects of protein degraders by global proteomics and multiplexed proteome dynamics profiling (mPDP). Total proteome and protein turnover analyses provide a comprehensive view of protein dynamics within a cell to characterize degraders, their mechanism of action, and potential off-target effects. While global proteome data (left panel) provide a snapshot on the proteome and give insight about the selectivity and potential off-targets, a more detailed temporal resolution of protein regulation (protein half lifes) in response to drug treatments can be obtained with multiplexed proteome dynamics profiling" (mPDP, right panel). mPDP combines dynamic-SILAC labeling with isobaric mass tagging for multiplexed analysis of protein degradation and synthesis and allows monitoring and discrimination of transcriptional and degradational changes on the proteome level.

particular amino acid. The "heavy" form of the amino acid is labeled with stable isotopes, typically carbon-13 (^{13}C) or nitrogen-15 (^{15}N). Monitoring the incorporation of these 'heavy' amino acids into proteins over time by mass spectrometry allows for the precise and accurate measurement of protein turnover rates in the cell.

Refinements of this approach aim at distinguishing primary degradation from secondary downstream effects (Figure 19.4). "Multiplexed Protein Dynamics Profiling" (mPDP) combines dynamic-SILAC labeling with isobaric mass tagging for multiplexed analysis of protein degradation and synthesis. mPDP specifically allows monitoring and discriminating drug effects on protein synthesis versus degradation by quantifying the different labeled protein populations [50]. The complexity of degradation events was demonstrated by mPDP profiling of FAK PROTACs. The FAK PROTACs used a selective FAK/PTK2 warhead but showed active degradation of the whole CAK complex, consisting of Cyclin K, CDK7, and FAK, indicated by the protein downregulation in both SILAC-labeled protein populations [120]. Moreover, mPDP can be used to characterize Hsp90 client relationship, discriminating between constitutive and clients. When the effect of a selective BRD4 degrader was investigated by mPDP BRD4 was significantly downregulated in a time-dependent manner. However, transcriptional regulation was induced which led to upregulation of BRD2 and BRD3 suggesting that selective degradation can be tried to be compensated and can have bystander effects. Recently, the effect of the molecular glue degrader Indisulam on the proteome was investigated. Indisulam selectively degraded RBM39 (Caper-alpha) within several hours. Interestingly, the degradation of the protein RBM39 by Indisulam led to a feedback mechanism in the RBM39 pathway which led to an increase in protein synthesis rate as measured by SILAC incorporation [121]. This is an example of a cellular compensation mechanism upon induced targeted protein degradation and highlights the importance of half-life and protein synthesis rate for this modality but is highly relevant and has similar consequences for irreversible covalent modifications. Proteome analysis is an established cost-effective and fast analytical tool to quantify reliable changes in protein levels. mPDP requires SILAC incorporation and cell line labeling which can be challenging for some cell models.

The importance of profiling for off-target drug degradation became particularly evident with the discovery of the mechanism of action of Thalidomide, which tragically became infamous for its teratogenicity [122]. Thalidomide

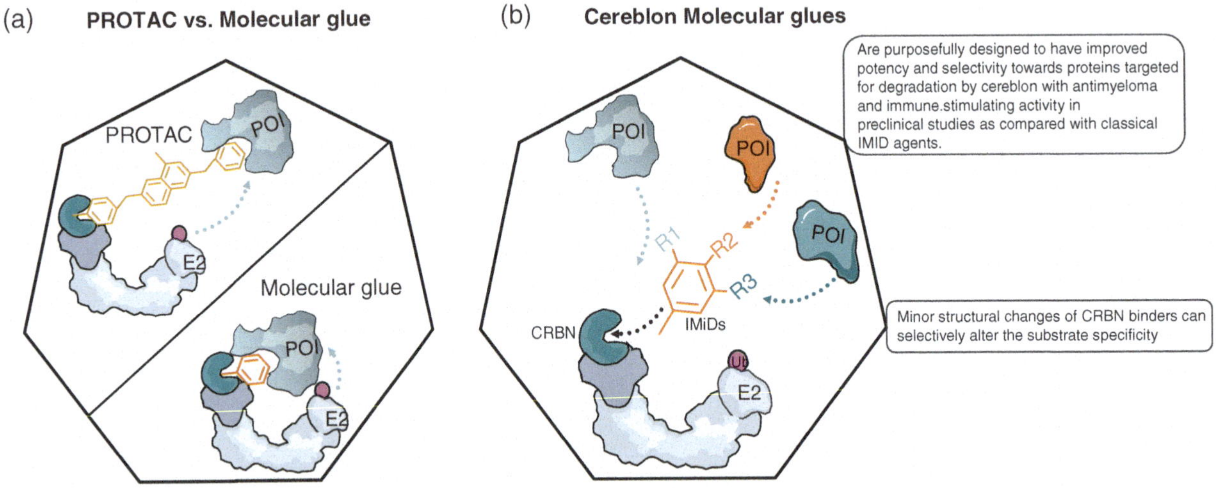

Figure 19.5 Types of targeted protein degradation technologies. (a) Differences in degrader modalities. PROTACs (proteolysis targeting chimeras) are bifunctional molecules connected by a linker, bringing an E3 ligase in proximity to the target protein which allows the ubiquitin transfer and subsequent degradation of the target. Unlike PROTACs, which require two small-molecule binders, molecular glues simplify targeted protein degradation by stabilizing and enhancing the natural affinity with E3 ligases by single molecules. Due to their size, molecular glues typically offer advantages such as better solubility, membrane permeability and oral bioavailability, owing to their alignment with traditional small-molecule drug design principles. (b) Structure-targeted degradation relationships of cereblon E3 ligase modulators. Thalidomide derivatives, called immunomodulatory drugs (IMiDs), are a class of drugs that function as a molecular glue and induce proteasomal degradation of neosubstrates by recruiting the E3 ligase cereblon (CRBN) via interactions between IMiDs and CRBN. Many neosubstrates involved in the molecular action of IMiDs have been reported. Minor structural changes of CRBN binders can selectively alter the substrate specificity, in particular the 5,6-position modifications of the glutaramide ring are essential for controlling neosubstrate selectivity.

belongs to a class of immuno-modulating drugs (IMiDs) which interact with the E3 ligase cereblon (CRBN) and form a new surface which attracts new substrates such as transcription factors Ikaros (IKZF), SALL4, and GSTP1 which undergo subsequently ubiquitination and degradation. This mechanism is now well known as molecular glue degraders (Figure 19.5a). These small molecules enhance protein–protein interactions to promote proteasomal degradation by increasing the natural affinity between target proteins and E3 ligases. Unlike PROTACs, which require two small-molecule binders, molecular glue degraders simplify targeted protein degradation by enhancing the natural affinity with E3 ligases. There are numerous other putative IMiD neosubstrates and it was demonstrated that the degradation of SALL4 and p63 via IMiD-CRBN leads to teratogenicity. However, no consensus exists on all potential neosubstrates driven by cereblon engagers [123, 124]. Neosubstrate degradation is cell-type dependent, and the profiling can be an analytical challenge, some new substrates have low abundance or lower ternary complex formation rate, which limits degradation. This might be the reason that new substrates are still being used today [125]. Cereblon recruitment as mechanism of action not only occurs in several approved drugs which function as molecular glues but is also most utilized in clinical stage PROTACs. This has important consequences from a safety perspective: using these drugs in oncology and beyond, neosubstrate degradation will depend on tissue distribution, substrate specificity, biological function, selectivity of the warhead, linker and linking position in bifunctional degraders (Figure 19.5b). In particular, the CRBN linkage position from the C5,6 position of the glutarimide ring is notoriously liable for GSPT1 degradation activity which should be monitored as guidance for all CRBN-C5,6-linked bifunctional PROTACs. GSPT1 functions as protein synthesis regulator, and degradation of GSPT1 impairs protein synthesis overall which can result in decreased protein levels and complicates the interpretation of PROTAC activity [126].

With the development of clinical-stage degraders, the described off-targets beg the challenging question of what degree off-target proteome coverage is necessary for safety assessment. Liu et al. described comprehensive off-target profiling by utilizing four different cell lines which covers 80% of known neosubstrates as a general approach to cover the majority of meaningful proteomes [127]. This approach represents a good starting point to cover the majority of neosubstrates; in an ideal scenario, the same neosubstrate off-targets are identified in multiple cell models. Interestingly, profiling of many cereblon degraders can lead to analytical results which demonstrate degradation of different single off-targets in each cell line which can complicate the interpretation and would require further profiling in other cell models [109]. Primary cells (hepatocytes, PBMCs, splenocytes, etc.) closely mimic the *in vivo* state and maintain physiological relevance and are preferred over cell line, but cell availability, donor variability, assay drift, and other technical issues present challenges for this approach. A potential gold standard for neosubstrate profiling for global proteomics represents iPSCs which express high levels of transcription factors and many of the known neosubstrates. Moreover, human iPSCs can be used to assess the teratogenicity risk by mesendoderm differentiation [128].

Profiling degraders in higher species models is essential to explore off-targets in more complex systems and de-risking degraders in preclinical studies [129]. In general, the goal is to demonstrate complete target degradation ($D_{max} > 90\%$) in at least one nonclinical species and exposure multiples above that in general toxicology studies. The challenge is to identify species early on (e.g. during early lead-optimization stages) using predominantly *in vitro* data. Ideally, proteomic analyses should demonstrate comparable degradation *in vitro* across species that supports the potential to achieve complete degradation *in vivo*. Whole embryo culture (WEC) can detect changes in embryogenesis which makes it an attractive model for IMiDs and their teratogenicity. The rat WEC model is well established in the area of drug toxicity. Unfortunately, this is not suitable to investigate IMiD pharmacology, nonresponsive organisms like mouse and rat have two critical amino acid substitutions in this domain, E377V and V388I, that prevents stable associations with SALL4 and IKZF1/3 in the presence of IMiDs [130]. Instead, the rabbit WEC is a mainstay of regulatory developmental toxicity testing; Degradation of one or more of these CRBN neosubstrates is believed to play an important role in thalidomide-related developmental toxicity observed in rabbits and primates. Federspiel et al. developed a targeted assay for these proteins combining peptide immunoaffinity enrichment and high-resolution mass spectrometry and successfully applied this assay to rabbit embryo samples from pregnant rabbits dosed with IMiDs [131]. They identified a set of 25 proteins of interest associated with CRBN-related protein homeostasis and/or embryo/fetal development. Further, they confirmed previously reported *in vivo* decreases in neosubstrates with known developmental biology implications in the rabbit embryos following IMiD exposure. While more research is needed to use the existing information for risk assessment purposes, this data may be useful to screen for compounds without these neosubstrate changes such as novel non-glutaramide-containing cereblon binders. As TPD moves toward clinical applications, the challenge of balancing on-target efficacy with off-target safety remains. Rigorous proteomic analysis across multiple models and species will be crucial to mitigate risks and optimize the therapeutic potential of protein degraders.

Antibody–Drug Conjugates and Oligonucleotides

Antibody–drug conjugates (ADCs) and oligonucleotides (Oligos) represent other event-driven modalities which are promising and effective approaches to drug discovery to overcome limitations of typical occupancy-driven approaches. Traditionally, ADCs combine the specificity of antibodies with the cancer-killing ability of cytotoxic drugs, offering a targeted method to deliver these potent agents directly to tumor cells [132]. However, the effectiveness of ADCs can be very limited by factors such as insufficient payload delivery, acquired resistance to the payload, and adverse effects related to non-specific release. To enhance both the safety and efficacy of ADCs [133], significant advancements are being made in the various components that make up these conjugates and lead to a revival of the ADC as a drug modality [1]. This includes the antibody itself, the warhead and linker that connect the antibody to the payload, and the chemistry involved in conjugating all of these components together. In response to the challenges faced, several innovative ADC formats are being developed. These include bispecific ADCs, which can target two different antigens simultaneously, and probody–drug conjugates, which are designed to be activated specifically in the tumor microenvironment, thereby improving tumor specificity [134]. Dual-payload ADCs are another novel format that holds the potential in combating drug resistance and addressing tumor heterogeneity by delivering two different cytotoxic agents [135].

The identification of antigens presented by targeted cells is crucial for designing effective ADCs. Thus, MS-based technique became essential to quantify antigen expression levels in an antibody-independent way [45, 136, 137]. Cell surface receptors, which are frequently bound by various drugs and endogenous ligands, often lead to adaptations in the plasma membrane proteome. Conducting an in-depth analysis of dynamic processes at the cell surface presents significant challenges due to the biochemical properties and low abundances of plasma membrane proteins. Cell surface thermal proteome profiling allows for a comprehensive characterization of ligand-induced changes in both protein abundances and thermal stabilities at the plasma membrane [10]. To provide detailed information about the microenvironment of cell surface target proteins, micro-environment mapping (μMap) [14] enables selective labeling of antibody binding targets and their protein neighbors on live cells. This spatial information can help in selecting best targets by enabling a judgment on internalization efficiency and indirect effects on the cell.

While ADCs employ broadly cytotoxic payloads, degrader–antibody conjugates (DACs) incorporate degraders as payloads with more targeted biological activities, often associated with specific cancers or other tissues/cell types [138]. The construction of DACs, however, presents additional challenges that must be overcome. The antigen selected for DAC generation must not only meet the ADC internalization and trafficking criteria but should also be highly expressed on tumors, tissues, or other cells that are sensitive to the desired biological pathway targeted by the degrader. Low antigen expression in additional tissues and cells could be desirable, but this need may be mitigated by the biological selectivity of the degrader payload itself. Traditional ADCs require high potencies for their cytotoxic payloads to ensure sufficient intracellular concentration to induce cell death which can complicate the creation of the desired conjugates due to aggregation issues of high drug-to-antibody ratio (DAR), which may negatively influence their *in vivo* pharmacokinetic attributes and developability. Using DACs might be an attractive way to lower DAR and the efficacious DAC concentration due to the catalytic nature of degraders. This necessitates careful consideration regarding whether to incorporate novel functional handles into the degrader structure or to exploit existing degrader functionality using novel ADC linking approaches. Other challenges in degrader-conjugate designs include ensuring good stability of the chimeric degrader payload in the lysosomal environment, the payload's ability to efficiently escape the lysosome compartment, and the propensity of the payload to generate bystander effects.

Endosomal escape is also a significant challenge for oligonucleotide therapeutics. Oligos are short DNA or RNA molecules that can be designed to interact with specific genetic sequences, thereby modulating the expression of targeted genes. Oligos enter cells via endocytosis and are enclosed in an endosome, which limits their activity. Only a small percentage of Oligos escape endosomes, with estimates as low as 1% [139]. The endosomal–lysosomal pathway is complex and not fully understood, making it difficult to enhance endosomal escape. Current research is focused on developing strategies to enhance endosomal escape without causing unwanted side effects [140]. Understanding and optimizing these characteristics is important for the optimization of both ADCs and Oligos. Combining multiplexed proteome dynamics with subcellular proteomics could provide a powerful approach to investigate endosome escape. Multiplexed proteome dynamics could be used to track the movement and interactions of proteins involved in endosomal escape, providing insights into the temporal aspects of this process. On the other hand, subcellular proteomics focuses on the analysis of proteins within specific cellular compartments, such as endosomes. By isolating and analyzing proteins from endosomes, it could provide detailed information about the proteins involved in endosome escape and their functions. By combining these two approaches, it would be possible to track the dynamic behavior of proteins within endosomes over time. This could provide valuable insights into the mechanisms of endosome escape, potentially leading to the development and prioritization of strategies to enhance this process for therapeutic applications [141]. Another promising approach to monitor mechanism, distribution, and interactions of oligonucleotide drugs within the cell is Drug-ID [13]. In this specific case, the oligo of interest is functionalized either with a benzylguanine moiety or a chloralkane, recruiting either a SNAP-tag or Halo-tag biotin ligase fusion protein to the oligo which leads to biotin labeling of interacting and proximal proteins thus revealing localization and interactions of the oligo.

Concluding Remarks

Recent advancements in Omics technologies and Chemical Biology have significantly transformed the landscape of drug discovery and development. The emergence of new chemical and therapeutic modalities, including targeted covalent inhibitors, protein degraders, molecular glues, RNA therapeutics, recombinant peptides and proteins, monoclonal antibodies, antibody–drug conjugates, and gene therapy, have broadened the scope of potential drug targets and addressed previously unmet medical needs. Chemoproteomics has played a crucial role in understanding the impact of these bioactive molecules on the proteome. It has provided key insights into the ligandability of targets, target engagement, selectivity, and mechanism of action of drug candidates, thereby aiding in modality selection, prioritization of lead molecules, drug differentiation, and understanding of adverse events.

The complexity of these new modalities, particularly those with event-driven pharmacology such as covalent inhibitors and protein degraders, necessitates continuous innovation in analytical tools and methodologies. As we continue to explore the potential of these new modalities, it is crucial that we also advance our understanding of their mechanisms of action and their impact on the proteome. This will not only aid in the development of more effective and safer drugs but also contribute to our broader understanding of disease mechanisms and drug action.

Acknowledgments

We would like to thank all our GSK colleagues for their valuable scientific input and support throughout this work.

Abbreviations

ADCs	antibody–drug conjugates
DACs	degrader–antibody conjugates
DC_{50}	half-maximal degradation concentration
D_{max}	maximum level of degradation achievable
DARTS	drug affinity responsive target stability
IMIDs	immuno-modulating drugs
LIP-MS	limited proteolysis-mass spectrometry
mPDP	multiplexed proteome dynamics profiling
MS	mass spectrometry
PROTACs	proteolysis-targeting chimeras
SERDs	selective estrogen receptor downregulators
SILAC	stable isotope labeling by amino acids in cell culture
TPD	targeted protein degradation
TPP	thermal proteome profiling
WEC	whole embryo culture

References

1 Meissner, F., Geddes-McAlister, J., Mann, M., and Bantscheff, M. (2022). The emerging role of mass spectrometry-based proteomics in drug discovery. *Nat. Rev. Drug Discov.* 21 (9): 637–654.

2 Blanco, M.J., Gardinier, K.M., and Namchuk, M.N. (2022). Advancing new chemical modalities into clinical studies. *ACS Med. Chem. Lett.* 13 (11): 1691–1698.

3 Blanco, M.J. and Gardinier, K.M. (2020). New chemical modalities and strategic thinking in early drug discovery. *ACS Med. Chem. Lett.* 11 (3): 228–231.

4 Lai, A.C. and Crews, C.M. (2017). Induced protein degradation: an emerging drug discovery paradigm. *Nat. Rev. Drug Discov.* 16 (2): 101–114.

5 Toure, M. and Crews, C.M. (2016). Small-molecule PROTACS: new approaches to protein degradation. *Angew. Chem. Int. Ed. Engl.* 55 (6): 1966–1973.

6 Drewes, G. and Knapp, S. (2018). Chemoproteomics and chemical probes for target discovery. *Trends Biotechnol.* 36 (12): 1275–1286.

7 Savitski, M.M., Reinhard, F.B., Franken, H. et al. (2014). Tracking cancer drugs in living cells by thermal profiling of the proteome. *Science* 346 (6205): 1255784.

8 Martinez Molina, D., Jafari, R., Ignatushchenko, M. et al. (2013). Monitoring drug target engagement in cells and tissues using the cellular thermal shift assay. *Science* 341 (6141): 84–87.

9 Feng, Y., De Franceschi, G., Kahraman, A. et al. (2014). Global analysis of protein structural changes in complex proteomes. *Nat. Biotechnol.* 32 (10): 1036–1044.

10 Kalxdorf, M., Gunthner, I., Becher, I. et al. (2021). Cell surface thermal proteome profiling tracks perturbations and drug targets on the plasma membrane. *Nat. Methods* 18 (1): 84–91.

11 West, G.M., Tang, L., and Fitzgerald, M.C. (2008). Thermodynamic analysis of protein stability and ligand binding using a chemical modification- and mass spectrometry-based strategy. *Anal. Chem.* 80 (11): 4175–4185.

12 Pai, M.Y., Lomenick, B., Hwang, H. et al. (2015). Drug affinity responsive target stability (DARTS) for small-molecule target identification. *Methods Mol. Biol.* 1263: 287–298.

13 Hanswillemenke, A., Hofacker, D.T., Sorgenfrei, M. et al. (2024). Profiling the interactome of oligonucleotide drugs by proximity biotinylation. *Nat. Chem. Biol.*

14 Geri, J.B., Oakley, J.V., Reyes-Robles, T. et al. (2020). Microenvironment mapping via Dexter energy transfer on immune cells. *Science* 367 (6482): 1091–1097.

15 Qin, W., Cho, K.F., Cavanagh, P.E., and Ting, A.Y. (2021). Deciphering molecular interactions by proximity labeling. *Nat. Methods* 18 (2): 133–143.

16 Wright, M.H. and Sieber, S.A. (2016). Chemical proteomics approaches for identifying the cellular targets of natural products. *Nat. Prod. Rep.* 33 (5): 681–708.

17 Moellering, R.E. and Cravatt, B.F. (2012). How chemoproteomics can enable drug discovery and development. *Chem. Biol.* 19 (1): 11–22.

18 Bantscheff, M. and Drewes, G. (2012). Chemoproteomic approaches to drug target identification and drug profiling. *Bioorg. Med. Chem.* 20 (6): 1973–1978.

19 Schirle, M., Bantscheff, M., and Kuster, B. (2012). Mass spectrometry-based proteomics in preclinical drug discovery. *Chem. Biol.* 19 (1): 72–84.

20 Conway, L.P., Li, W., and Parker, C.G. (2021). Chemoproteomic-enabled phenotypic screening. *Cell Chem. Biol.* 28 (3): 371–393.

21 Rutkowska, A., Thomson, D.W., Vappiani, J. et al. (2016). A modular probe strategy for drug localization, target identification and target occupancy

measurement on single cell level. *ACS Chem. Biol.* 11 (9): 2541–2550.

22 Parker, C.G. and Pratt, M.R. (2020). Click chemistry in proteomic investigations. *Cell* 180 (4): 605–632.

23 Niphakis, M.J. and Cravatt, B.F. (2014). Enzyme inhibitor discovery by activity-based protein profiling. *Annu. Rev. Biochem.* 83: 341–377.

24 Hagenstein, M.C., Mussgnug, J.H., Lotte, K. et al. (2003). Affinity-based tagging of protein families with reversible inhibitors: a concept for functional proteomics. *Angew. Chem. Int. Ed. Engl.* 42 (45): 5635–5638.

25 Porta, E.O.J. and Steel, P.G. (2023). Activity-based protein profiling: a graphical review. *Curr Res Pharmacol Drug Discov.* 5: 100164.

26 Bantscheff, M., Eberhard, D., Abraham, Y. et al. (2007). Quantitative chemical proteomics reveals mechanisms of action of clinical ABL kinase inhibitors. *Nat. Biotechnol.* 25 (9): 1035–1044.

27 Sharma, K., Weber, C., Bairlein, M. et al. (2009). Proteomics strategy for quantitative protein interaction profiling in cell extracts. *Nat. Methods* 6 (10): 741–744.

28 Eberl, H.C., Werner, T., Reinhard, F.B. et al. (2019). Chemical proteomics reveals target selectivity of clinical Jak inhibitors in human primary cells. *Sci. Rep.* 9 (1): 14159.

29 George, A.L., Duenas, M.E., Marin-Rubio, J.L., and Trost, M. (2024). Stability-based approaches in chemoproteomics. *Expert Rev. Mol. Med.* 26: e6.

30 Leuenberger, P., Ganscha, S., Kahraman, A. et al. (2017). Cell-wide analysis of protein thermal unfolding reveals determinants of thermostability. *Science* 355 (6327): eaai7825.

31 Ball, K.A., Webb, K.J., Coleman, S.J. et al. (2020). An isothermal shift assay for proteome scale drug-target identification. *Commun. Biol.* 3 (1): 75.

32 Mateus, A., Kurzawa, N., Becher, I. et al. (2020). Thermal proteome profiling for interrogating protein interactions. *Mol. Syst. Biol.* 16 (3): e9232.

33 Van Vranken, J.G., Li, J., Mitchell, D.C. et al. (2021). Assessing target engagement using proteome-wide solvent shift assays. *elife* 10: e70784.

34 Zhang, X., Wang, Q., Li, Y. et al. (2020). Solvent-induced protein precipitation for drug target discovery on the proteomic scale. *Anal. Chem.* 92 (1): 1363–1371.

35 Lomenick, B., Hao, R., Jonai, N. et al. (2009). Target identification using drug affinity responsive target stability (DARTS). *PNAS USA* 106 (51): 21984–21989.

36 Schopper, S., Kahraman, A., Leuenberger, P. et al. (2017). Measuring protein structural changes on a proteome-wide scale using limited proteolysis-coupled mass spectrometry. *Nat. Protoc.* 12 (11): 2391–2410.

37 Mackmull, M.T., Nagel, L., Sesterhenn, F. et al. (2022). Global, in situ analysis of the structural proteome in individuals with Parkinson's disease to identify a new class of biomarker. *Nat. Struct. Mol. Biol.* 29 (10): 978–989.

38 Santos-Barriopedro, I., van Mierlo, G., and Vermeulen, M. (2021). Off-the-shelf proximity biotinylation for interaction proteomics. *Nat. Commun.* 12 (1): 5015.

39 Benhalevy, D., Anastasakis, D.G., and Hafner, M. (2018). Proximity-CLIP provides a snapshot of protein-occupied RNA elements in subcellular compartments. *Nat. Methods* 15 (12): 1074–1082.

40 Kaewsapsak, P., Shechner, D.M., Mallard, W. et al. (2017). Live-cell mapping of organelle-associated RNAs via proximity biotinylation combined with protein-RNA crosslinking. *Elife* 6: e29224.

41 Trowbridge, A.D., Seath, C.P., Rodriguez-Rivera, F.P. et al. (2022). Small molecule photocatalysis enables drug target identification via energy transfer. *PNAS USA* 119 (34): e2208077119.

42 Kwak, C., Park, C., Ko, M. et al. (2022). Identification of proteomic landscape of drug-binding proteins in live cells by proximity-dependent target ID. Cell. *Chem. Biol.* 29 (12): 1739–53 e6.

43 Yamanaka, S., Horiuchi, Y., Matsuoka, S. et al. (2022). A proximity biotinylation-based approach to identify protein-E3 ligase interactions induced by PROTACs and molecular glues. *Nat. Commun.* 13 (1): 183.

44 Milione, R.R., Schell, B.B., Douglas, C.J., and Seath, C.P. (2024). Creative approaches using proximity labeling to gain new biological insights. *Trends Biochem. Sci.* 49 (3): 224–235.

45 Mitchell, D.C., Kuljanin, M., Li, J. et al. (2023). A proteome-wide atlas of drug mechanism of action. *Nat. Biotechnol.* 41 (6): 845–857.

46 Eckert, S., Berner, N., Kramer, K. et al. (2024). Decrypting the molecular basis of cellular drug phenotypes by dose-resolved expression proteomics. *Nat. Biotechnol.* 1–10.

47 Zecha, J., Bayer, F.P., Wiechmann, S. et al. (2023). Decrypting drug actions and protein modifications by dose- and time-resolved proteomics. *Science* eade3925.

48 Ong, S.E., Blagoev, B., Kratchmarova, I. et al. (2002). Stable isotope labeling by amino acids in cell culture, SILAC, as a simple and accurate approach to expression proteomics. *Mol. Cell. Proteomics* 1 (5): 376–386.

49 Ong, S.E. and Mann, M. (2006). A practical recipe for stable isotope labeling by amino acids in cell culture (SILAC). *Nat. Protoc.* 1 (6): 2650–2660.

50 Savitski, M.M., Zinn, N., Faelth-Savitski, M. et al. (2018). Multiplexed proteome dynamics profiling reveals mechanisms controlling protein homeostasis. *Cell* 173 (1): 260–74 e25.

51 Bortecen, T., Muller, T., and Krijgsveld, J. (2023). An integrated workflow for quantitative analysis of the newly synthesized proteome. *Nat. Commun.* 14 (1): 8237.

52 Ross, A.B., Langer, J.D., and Jovanovic, M. (2021). Proteome turnover in the spotlight: approaches, applications, and perspectives. *Mol. Cell. Proteomics* 20: 100016.

53 Zecha, J., Meng, C., Zolg, D.P. et al. (2018). Peptide level turnover measurements enable the study of proteoform dynamics. *Mol. Cell. Proteomics* 17 (5): 974–992.

54 Dunkley, T.P., Watson, R., Griffin, J.L. et al. (2004). Localization of organelle proteins by isotope tagging (LOPIT). *Mol. Cell. Proteomics* 3 (11): 1128–1134.

55 Geladaki, A., Kocevar Britovsek, N., Breckels, L.M. et al. (2019). Combining LOPIT with differential ultracentrifugation for high-resolution spatial proteomics. *Nat. Commun.* 10 (1): 331.

56 Orre, L.M., Vesterlund, M., Pan, Y. et al. (2019). SubCellBarCode: proteome-wide mapping of protein localization and relocalization. *Mol. Cell* 73 (1): 166–182.

57 Itzhak, D.N., Tyanova, S., Cox, J., and Borner, G.H. (2016). Global, quantitative and dynamic mapping of protein subcellular localization. *Elife* 5: e16950.

58 Samant, R.S., Batista, S., Larance, M. et al. (2023). Native size-exclusion chromatography-based mass spectrometry reveals new components of the early heat shock protein 90 inhibition response among limited global changes. *Mol. Cell. Proteomics* 22 (2): 100485.

59 Heusel, M., Frank, M., Kohler, M. et al. (2020). A global screen for assembly state changes of the mitotic proteome by SEC-SWATH-MS. *Cell Syst.* 10 (2): 133–155.

60 Swietlik, J.J., Sinha, A., and Meissner, F. (2020). Dissecting intercellular signaling with mass spectrometry-based proteomics. *Curr. Opin. Cell Biol.* 63: 20–30.

61 Knecht, S., Eberl, H.C., Kreisz, N. et al. (2023). An introduction to analytical challenges, approaches and applications in mass-spectrometry-based secretomics. *Mol. Cell. Proteomics* 22 (9): 100636.

62 Buksh, B.F., Knutson, S.D., Oakley, J.V. et al. (2022). muMap-red: proximity labeling by red light photocatalysis. *J. Am. Chem. Soc.* 144 (14): 6154–6162.

63 Muller, M., Grabnitz, F., Barandun, N. et al. (2021). Light-mediated discovery of surfaceome nanoscale organization and intercellular receptor interaction networks. *Nat. Commun.* 12 (1): 7036.

64 Lundberg, E. and Borner, G.H.H. (2019). Spatial proteomics: a powerful discovery tool for cell biology. *Nat. Rev. Mol. Cell Biol.* 20 (5): 285–302.

65 Christopher, J.A., Stadler, C., Martin, C.E. et al. (2021). Subcellular proteomics. *Nat Rev Methods Primers.* 1 (1): 32.

66 Lanman, B.A., Allen, J.R., Allen, J.G. et al. (2020). Discovery of a covalent inhibitor of KRAS(G12C) (AMG 510) for the treatment of solid Tumors. *J. Med. Chem.* 63 (1): 52–65.

67 Rathod, L.S., Dabhade, P.S., and Mokale, S.N. (2023). Recent progress in targeting KRAS mutant cancers with covalent G12C-specific inhibitors. *Drug Discov. Today* 28 (5): 103557.

68 Li, H.Y., Qi, W.L., Wang, Y.X., and Meng, L.H. (2023). Covalent inhibitor targets KRas(G12C): a new paradigm for drugging the undruggable and challenges ahead. *Genes Dis.* 10 (2): 403–414.

69 Liu, J., Kang, R., and Tang, D. (2022). The KRAS-G12C inhibitor: activity and resistance. *Cancer Gene Ther.* 29 (7): 875–878.

70 Huang, L., Guo, Z., Wang, F., and Fu, L. (2021). KRAS mutation: from undruggable to druggable in cancer. *Signal Transduct. Target. Ther.* 6 (1): 386.

71 Zhang, T., Kwiatkowski, N., Olson, C.M. et al. (2016). Covalent targeting of remote cysteine residues to develop CDK12 and CDK13 inhibitors. *Nat. Chem. Biol.* 12 (10): 876–884.

72 Baltgalvis, K.A., Lamb, K.N., Symons, K.T. et al. (2024). Chemoproteomic discovery of a covalent allosteric inhibitor of WRN helicase. *Nature* 629 (8011): 435–442.

73 Singh, J., Petter, R.C., Baillie, T.A., and Whitty, A. (2011). The resurgence of covalent drugs. *Nat. Rev. Drug Discov.* 10 (4): 307–317.

74 De Cesco, S., Kurian, J., Dufresne, C. et al. (2017). Covalent inhibitors design and discovery. *Eur. J. Med. Chem.* 138: 96–114.

75 Chan, W.C., Sharifzadeh, S., Buhrlage, S.J., and Marto, J.A. (2021). Chemoproteomic methods for covalent drug discovery. *Chem. Soc. Rev.* 50 (15): 8361–8381.

76 Dittus, L., Werner, T., Muelbaier, M., and Bantscheff, M. (2017). Differential Kinobeads profiling for target identification of irreversible kinase inhibitors. *ACS Chem. Biol.* 12 (10): 2515–2521.

77 Canon, J., Rex, K., Saiki, A.Y. et al. (2019). The clinical KRAS(G12C) inhibitor AMG 510 drives anti-tumour immunity. *Nature* 575 (7781): 217–223.

78 Weerapana, E., Wang, C., Simon, G.M. et al. (2010). Quantitative reactivity profiling predicts functional cysteines in proteomes. *Nature* 468 (7325): 790–795.

79 Backus, K.M., Correia, B.E., Lum, K.M. et al. (2016). Proteome-wide covalent ligand discovery in native biological systems. *Nature* 534 (7608): 570–574.

80 Kuljanin, M., Mitchell, D.C., Schweppe, D.K. et al. (2021). Reimagining high-throughput profiling of reactive cysteines for cell-based screening of large electrophile libraries. *Nat. Biotechnol.* 39 (5): 630–641.

81 Picco, G., Rao, Y., Al Saedi, A. et al. (2024). Novel WRN helicase inhibitors selectively target microsatellite-unstable Cancer cells. *Cancer Discov.* 14 (8): 1457–1475.

82 Hacker, S.M., Backus, K.M., Lazear, M.R. et al. (2017). Global profiling of lysine reactivity and ligandability in the human proteome. *Nat. Chem.* 9 (12): 1181–1190.

83 Shannon, D.A., Banerjee, R., Webster, E.R. et al. (2014). Investigating the proteome reactivity and selectivity of aryl halides. *J. Am. Chem. Soc.* 136 (9): 3330–3333.

84 Ward, C.C., Kleinman, J.I., and Nomura, D.K. (2017). NHS-esters as versatile reactivity-based probes for mapping proteome-wide Ligandable hotspots. *ACS Chem. Biol.* 12 (6): 1478–1483.

85 Liu, Y., Patricelli, M.P., and Cravatt, B.F. (1999). Activity-based protein profiling: the serine hydrolases. *PNAS USA* 96 (26): 14694–14699.

86 Greenbaum, D., Medzihradszky, K.F., Burlingame, A., and Bogyo, M. (2000). Epoxide electrophiles as activity-dependent cysteine protease profiling and discovery tools. *Chem. Biol.* 7 (8): 569–581.

87 Vocadlo, D.J. and Bertozzi, C.R. (2004). A strategy for functional proteomic analysis of glycosidase activity from cell lysates. *Angew. Chem. Int. Ed. Engl.* 43 (40): 5338–5342.

88 Zhao, Q., Ouyang, X., Wan, X. et al. (2017). Broad-Spectrum kinase profiling in live cells with lysine-targeted sulfonyl fluoride probes. *J. Am. Chem. Soc.* 139 (2): 680–685.

89 Patricelli, M.P., Szardenings, A.K., Liyanage, M. et al. (2007). Functional interrogation of the kinome using nucleotide acyl phosphates. *Biochemistry* 46 (2): 350–358.

90 Wright, A.T. and Cravatt, B.F. (2007). Chemical proteomic probes for profiling cytochrome p450 activities and drug interactions in vivo. *Chem. Biol.* 14 (9): 1043–1051.

91 Weerapana, E., Speers, A.E., and Cravatt, B.F. (2007). Tandem orthogonal proteolysis-activity-based protein profiling (TOP-ABPP)--a general method for mapping sites of probe modification in proteomes. *Nat. Protoc.* 2 (6): 1414–1425.

92 Yang, Y.Y., Grammel, M., Raghavan, A.S. et al. (2010). Comparative analysis of cleavable azobenzene-based affinity tags for bioorthogonal chemical proteomics. *Chem. Biol.* 17 (11): 1212–1222.

93 Yang, Y., Hahne, H., Kuster, B., and Verhelst, S.H. (2013). A simple and effective cleavable linker for chemical proteomics applications. *Mol. Cell. Proteomics* 12 (1): 237–244.

94 Speers, A.E. and Cravatt, B.F. (2009). Activity-based protein profiling (ABPP) and click chemistry (CC)-ABPP by MudPIT mass spectrometry. *Curr Protoc Chem Biol* 1: 29–41.

95 Cravatt, B.F., Wright, A.T., and Kozarich, J.W. (2008). Activity-based protein profiling: from enzyme chemistry to proteomic chemistry. *Annu. Rev. Biochem.* 77: 383–414.

96 Friedman Ohana, R., Levin, S., Hurst, R. et al. (2021). Streamlined target deconvolution approach utilizing a single Photoreactive Chloroalkane capture tag. *ACS Chem. Biol.* 16 (2): 404–413.

97 van Esbroeck, A.C.M., Janssen, A.P.A., Cognetta, A.B. 3rd, et al. (2017). Activity-based protein profiling reveals off-target proteins of the FAAH inhibitor BIA 10-2474. *Science* 356 (6342): 1084–1087.

98 Johansson, H., Isabella Tsai, Y.C., Fantom, K. et al. (2019). Fragment-based covalent ligand screening enables rapid discovery of inhibitors for the RBR E3 ubiquitin ligase HOIP. *J. Am. Chem. Soc.* 141 (6): 2703–2712.

99 Lanning, B.R., Whitby, L.R., Dix, M.M. et al. (2014). A road map to evaluate the proteome-wide selectivity of covalent kinase inhibitors. *Nat. Chem. Biol.* 10 (9): 760–767.

100 Niessen, S., Dix, M.M., Barbas, S. et al. (2017). Proteome-wide map of targets of T790M-EGFR-directed covalent inhibitors. Cell. *Chem. Biol.* 24 (11): 1388–400 e7.

101 Dalton, S.E., Dittus, L., Thomas, D.A. et al. (2018). Selectively targeting the Kinome-conserved lysine of PI3Kdelta as a general approach to covalent kinase inhibition. *J. Am. Chem. Soc.* 140 (3): 932–939.

102 Eaton, J.K., Furst, L., Ruberto, R.A. et al. (2020). Selective covalent targeting of GPX4 using masked nitrile-oxide electrophiles. *Nat. Chem. Biol.* 16 (5): 497–506.

103 Panyain, N., Godinat, A., Lanyon-Hogg, T. et al. (2020). Discovery of a potent and selective covalent inhibitor and activity-based probe for the Deubiquitylating enzyme UCHL1, with Antifibrotic activity. *J. Am. Chem. Soc.* 142 (28): 12020–12026.

104 Kieser, K.J., Kim, D.W., Carlson, K.E. et al. (2010). Characterization of the pharmacophore properties of novel selective estrogen receptor downregulators (SERDs). *J. Med. Chem.* 53 (8): 3320–3329.

105 Wu, T., Yoon, H., Xiong, Y. et al. (2020). Targeted protein degradation as a powerful research tool in basic biology and drug target discovery. *Nat. Struct. Mol. Biol.* 27 (7): 605–614.

106 Chamberlain, P.P. and Hamann, L.G. (2019). Development of targeted protein degradation therapeutics. *Nat. Chem. Biol.* 15 (10): 937–944.

107 Bond, M.J. and Crews, C.M. (2021). Proteolysis targeting chimeras (PROTACs) come of age: entering the third decade of targeted protein degradation. *RSC Chem. Biol.* 2 (3): 725–742.

108 Riching, K.M., Caine, E.A., Urh, M., and Daniels, D.L. (2022). The importance of cellular degradation kinetics for understanding mechanisms in targeted protein degradation. *Chem. Soc. Rev.* 51 (14): 6210–6221.

109 Donovan, K.A., Ferguson, F.M., Bushman, J.W. et al. (2020). Mapping the degradable Kinome provides a resource for expedited degrader development. *Cell* 183 (6): 1714–31 e10.

110 Tinworth, C.P., Lithgow, H., Dittus, L. et al. (2019). PROTAC-mediated degradation of Bruton's tyrosine kinase is inhibited by covalent binding. *ACS Chem. Biol.* 14 (3): 342–347.

111 Gadd, M.S., Testa, A., Lucas, X. et al. (2017). Structural basis of PROTAC cooperative recognition for selective protein degradation. *Nat. Chem. Biol.* 13 (5): 514–521.

112 Schwalm, M.P., Saxena, K., Muller, S., and Knapp, S. (2024). Luciferase- and HaloTag-based reporter assays to measure small-molecule-induced degradation pathway in living cells. *Nat. Protoc.*

113 Sathe, G. and Sapkota, G.P. (2023). Proteomic approaches advancing targeted protein degradation. *Trends Pharmacol. Sci.*

114 Zhang, A.X., Cassidy, K., Dahl, G. et al. (2021). The vital role of proteomics in characterizing novel protein degraders. *SLAS Discov.* 26 (4): 518, 2472555220985776–523.

115 Bondeson, D.P., Mares, A., Smith, I.E. et al. (2015). Catalytic in vivo protein knockdown by small-molecule PROTACs. *Nat. Chem. Biol.* 11 (8): 611–617.

116 Grandi, P. and Bantscheff, M. (2019). Advanced proteomics approaches to unravel protein homeostasis. *Drug Discov. Today Technol.* 31: 99–108.

117 Schwanhausser, B., Busse, D., Li, N. et al. (2011). Global quantification of mammalian gene expression control. *Nature* 473 (7347): 337–342.

118 Guo, Q., Jin, Y., Chen, X. et al. (2024). NF-kappaB in biology and targeted therapy: new insights and translational implications. *Signal Transduct. Target. Ther.* 9 (1): 53.

119 Dai, C.L., Shi, J., Chen, Y. et al. (2013). Inhibition of protein synthesis alters protein degradation through activation of protein kinase B (AKT). *J. Biol. Chem.* 288 (33): 23875–23883.

120 Law, R.P., Nunes, J., Chung, C.W. et al. (2021). Discovery and characterisation of highly cooperative FAK-degrading PROTACs. *Angew. Chem. Int. Ed. Engl.* 60 (43): 23327–23334.

121 Zinn, N., Werner, T., Doce, C. et al. (2021). Improved proteomics-based drug mechanism-of-action studies using 16-Plex isobaric mass tags. *J. Proteome Res.* 20 (3): 1792–1801.

122 Ito, T., Ando, H., and Handa, H. (2011). Teratogenic effects of thalidomide: molecular mechanisms. *Cell. Mol. Life Sci.* 68 (9): 1569–1579.

123 Matyskiela, M.E., Couto, S., Zheng, X. et al. (2018). SALL4 mediates teratogenicity as a thalidomide-dependent cereblon substrate. *Nat. Chem. Biol.* 14 (10): 981–987.

124 Asatsuma-Okumura, T., Ando, H., De Simone, M. et al. (2019). p63 is a cereblon substrate involved in thalidomide teratogenicity. *Nat. Chem. Biol.* 15 (11): 1077–1084.

125 Chamberlain, P.P., Lopez-Girona, A., Miller, K. et al. (2014). Structure of the human Cereblon-DDB1-lenalidomide complex reveals basis for responsiveness to thalidomide analogs. *Nat. Struct. Mol. Biol.* 21 (9): 803–809.

126 Vetma, V., Perez, L.C., Elias, J. et al. (2024). Confounding factors in targeted degradation of short-lived proteins. *ACS Chem. Biol.* 19 (7): 1484–1494.

127 Liu, X., Zhang, Y., Ward, L.D. et al. (2021). A proteomic platform to identify off-target proteins associated with therapeutic modalities that induce protein degradation or gene silencing. *Sci. Rep.* 11 (1): 15856.

128 Belair, D.G., Lu, G., Waller, L.E. et al. (2020). Thalidomide inhibits human iPSC Mesendoderm differentiation by modulating CRBN-dependent degradation of SALL4. *Sci. Rep.* 10 (1): 2864.

129 Ball, J.S., Stedman, D.B., Hillegass, J.M. et al. (2014). Fishing for teratogens: a consortium effort for a harmonized zebrafish developmental toxicology assay. *Toxicol. Sci.* 139 (1): 210–219.

130 Asatsuma-Okumura, T., Ito, T., and Handa, H. (2020). Molecular mechanisms of the teratogenic effects of thalidomide. *Pharm.* 13 (5): 95.

131 Federspiel, J.D., Catlin, N.R., Nowland, W.S. et al. (2024). Differential analysis of Cereblon Neosubstrates in rabbit embryos using targeted proteomics. *Mol. Cell. Proteomics* 23 (7): 100797.

132 Dean, A.Q., Luo, S., Twomey, J.D., and Zhang, B. (2021). Targeting cancer with antibody-drug conjugates: promises and challenges. *MAbs* 13 (1): 1951427.

133 Gu, Y., Wang, Z., and Wang, Y. (2024). Bispecific antibody drug conjugates: making 1+1>2. *Acta Pharm. Sin. B* 14 (5): 1965–1986.

134 Autio, K.A., Boni, V., Humphrey, R.W., and Naing, A. (2020). Probody therapeutics: an emerging class of therapies designed to enhance on-target effects with reduced off-tumor toxicity for use in Immuno-oncology. *Clin. Cancer Res.* 26 (5): 984–989.

135 Yamazaki, C.M., Yamaguchi, A., Anami, Y. et al. (2021). Antibody-drug conjugates with dual payloads for combating breast tumor heterogeneity and drug resistance. *Nat. Commun.* 12 (1): 3528.

136 Bausch-Fluck, D., Hofmann, A., Bock, T. et al. (2015). A mass spectrometric-derived cell surface protein atlas. *PLoS One* 10 (3): e0121314.

137 Kalxdorf, M., Gade, S., Eberl, H.C., and Bantscheff, M. (2017). Monitoring cell-surface N-Glycoproteome dynamics by quantitative proteomics reveals mechanistic insights into macrophage differentiation. *Mol. Cell. Proteomics* 16 (5): 770–785.

138 Dragovich, P.S. (2022). Degrader-antibody conjugates. *Chem. Soc. Rev.* 51 (10): 3886–3897.

139 Dowdy, S.F., Setten, R.L., Cui, X.S., and Jadhav, S.G. (2022). Delivery of RNA therapeutics: the great endosomal escape! *Nucleic Acid Ther.* 32 (5): 361–368.

140 Mangla, P., Vicentini, Q., and Biscans, A. (2023). Therapeutic oligonucleotides: an outlook on chemical strategies to improve endosomal trafficking. *Cells* 12 (18): 2253.

141 Tsuchikama, K., Anami, Y., Ha, S.Y.Y., and Yamazaki, C.M. (2024). Exploring the next generation of antibody-drug conjugates. *Nat. Rev. Clin. Oncol.* 21 (3): 203–223.

20

Drug Delivery Modalities and Mechanisms

Yu Tong Tam[1], Madyson Migliozzi[1], Devin B. Tesar[1], Amin Famili[2], Aaron T. Wecksler[3], Sara Wichner[4], Alavattam Sreedhara[1], and Whitney Shatz-Binder[1]

[1]Department of Pharmaceutical Development, Genentech, South San Francisco, CA, USA
[2]Department of Synthetic Molecule Analytical Chemistry, Genentech, South San Francisco, CA, USA
[3]Department of Protein Analytical Chemistry, Genentech, South San Francisco, CA, USA
[4]Department of Protein Chemistry, Genentech, South San Francisco, CA, USA

Introduction

The development of non-small-molecule therapeutics is an important sector of the pharmaceutical industry and has recently seen an explosion in novel modalities and accompanying drug-delivery systems (DDSs). Since the first approval of recombinant human insulin by the Food and Drug Administration (FDA) in 1982, various formats of peptides, monoclonal antibodies (mAbs), fusion proteins, antibody fragments and bispecific antibodies have been developed, achieving significant breakthroughs in disease treatment [1–4]. A major milestone in drug delivery technologies arose with the development of insulin glargine, the first genetically engineered, long-acting insulin analog to be introduced into clinical practice. The key advancement was the ability to deliver a once daily treatment by modulating the peptide solubility at physiological pH. By making alterations to its sequence, the peptide precipitated into a local depot at the injection site, ensuring delivery of consistent basal insulin levels for around 24 hours [5]. More recently, glucagon-like peptide-1 (GLP-1) agonists have accounted for an estimated 35 billion dollars in sales in 2023, and with more FDA approvals on the horizon, those numbers are expected to skyrocket [6]. Moreover, the total market share of protein therapeutics has approached 400 billion dollars with hundreds of protein-based products on the market, whereby 6 out of the top 10 selling drugs in 2023 were protein-based therapeutics [7].

Although oral administration is the most commonly used administration route for small molecule (and some peptide) therapeutics, due to the susceptibility to degradation by digestive enzymes, limited bioavailability, and large molecular size of proteins, oral administration of many therapeutic modalities is largely restricted [8]. As such, the preferred administration routes for protein and peptide therapeutics have mainly been via parenteral routes (Figure 20.1), that is, delivered to tissues and the circulatory system by injection through intravenous (IV), subcutaneous (SC), and intramuscular (IM) routes. Furthermore, direct administration of protein therapeutics via intravitreal (IVT) injection has become one of the gold standards for many intraocular diseases. Although these therapeutics have proven to be highly effective in treating a variety of diseases, patient safety and compliance with parenterally administered medicines remain considerable hurdles to their most effective use. For example, potent protein therapeutics have the potential to elicit an inflammatory response upon systemic administration [9]. Specifically, acute cytokine release syndromes are often associated with the initial administration of some therapeutic mAbs, causing clinical symptoms ranging from nausea, fever, to hypotension, and in some cases can potentially be lethal [10]. Moreover, although studies have shown that poor adherence to monthly IVT injections can lead to substantial reduction in vision acuity [11, 12], according to a recent survey about half of patients with retinal diseases faced substantial challenges pertaining to time commitment and travel cost associated with their doctor's visits [13]. Thus, the incorporation of novel DDSs can facilitate a patient-centric approach, by improving safety, adherence, and efficacy to better serve patients with chronic diseases.

A cornerstone of the field of drug delivery is the modulation of the pharmacokinetic (PK) properties of therapeutic molecules. PK is defined as combined biological processes and events that determine the absorption, distribution, metabolism, and excretion (ADME) of a drug within the target organism [14]. The factors governing

Characterizing Biotherapeutics: Analytical Methods for Diverse Modalities, First Edition. Edited by Jennie R. Lill and Wendy Sandoval.

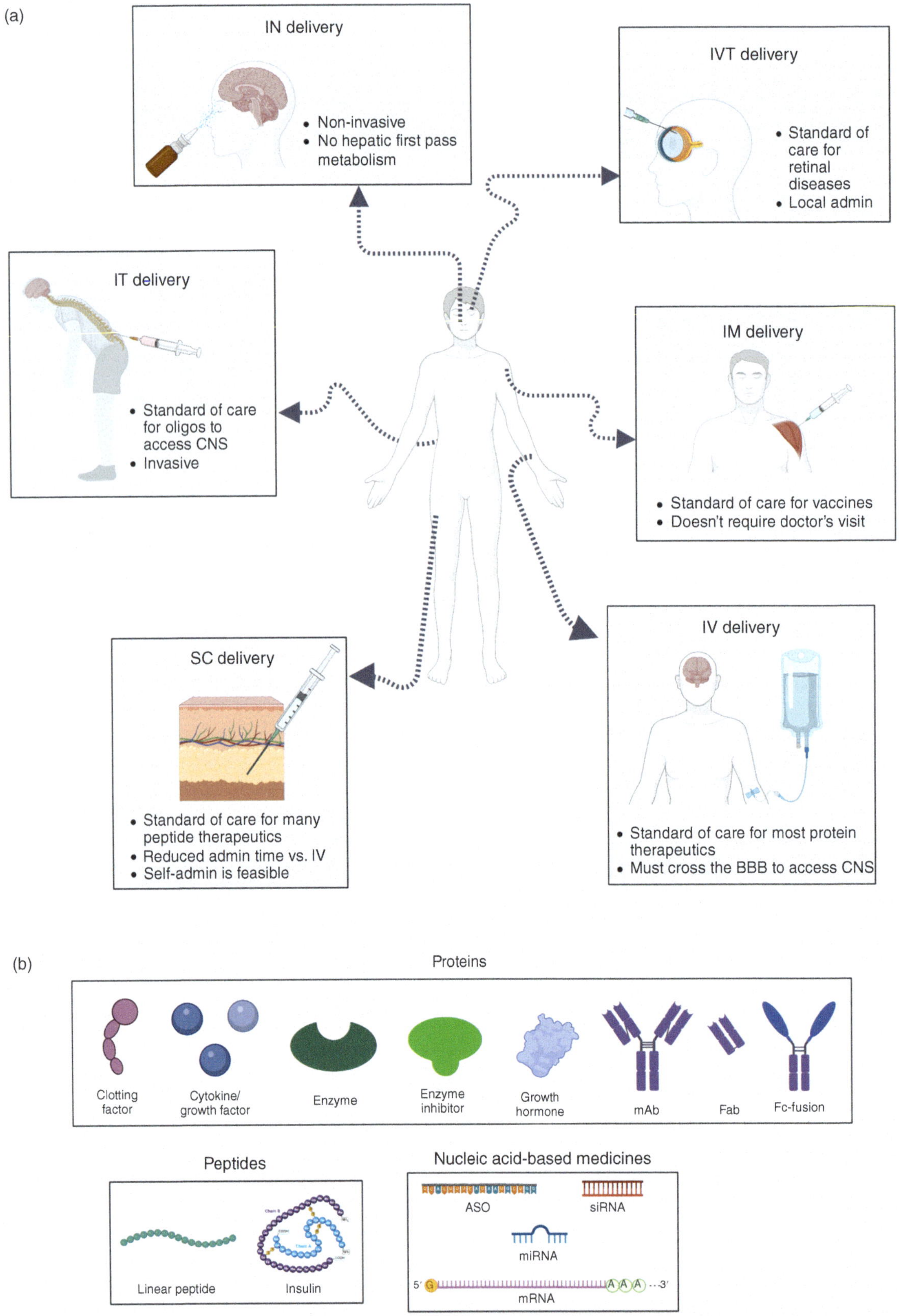

Figure 20.1 (a) Common parenteral routes of administration for proteins, peptides and nucleic acid-based medicines (admin: administration; BBB: blood–brain barrier; CNS: central nervous system; IM: intramuscular; IN: intranasal; IT: intrathecal; IV: intravenous; IVT: intravitreal; SC: subcutaneous). (b) Classes of therapeutics discussed in this chapter that have been successfully paired with drug-delivery systems (ASO: antisense oligonucleotide; Fc: fragment crystallizable; mAb: monoclonal antibody; miRNA: microRNA; messenger RNA; siRNA: small interfering RNA). Source: Adapted from BioRender.

these biological processes have been well documented in literature and will only be briefly discussed in this section [14–22]. In general, the absorption and distribution of a biotherapeutic is related in part to the route of administration (RoA). For delivery of proteins, the most direct RoA is IV administration which relies on direct infusion into the bloodstream, leading to maximum exposure (C_{max}) in the shortest amount of time. Alternatively, SC and IM administration of proteins rely on indirect delivery and typically result in both a lower C_{max} and delay in the maximum exposure. On the other hand, for peptides, although SC administration may result in a delay in C_{max} relative to IV administration, the achieved exposure can ultimately be equivalent to IV. Drug metabolism of biotherapeutics is also a complicated process that may include events such as on- and off-target degradation, receptor-mediated degradation, and clearance by anti-drug antibodies (ADA). The excretion of biotherapeutics can also be complicated as the route of excretion is linked to the physicochemical properties of the drug such as molecular size, hydrodynamic radius (R_H), surface charge, etc.

More recently, DDSs have enjoyed an expansion in applications, well beyond traditional enhancement of PK or safety profiles. A wave of novel modalities has been ushered in such as nucleic acid-based medicines (NABM) [23] and cell-based therapies [24] among others, that have already established their potential to provide access to targets that were previously deemed to be intractable or inaccessible [25, 26]. However, these new modalities have brought a host of new challenges for the field of drug delivery to address, such as crossing membrane barriers that were historically considered impermeable, enabling exquisite target selectivity, reaching difficult to access sites of action, or enabling higher-order oligomerization for receptor activation. While the objective of this chapter is to provide a broad overview of the current state of drug delivery, this is a field with continuous innovation. Thus, the authors have selected the following therapeutic modalities for in-depth exploration: biologics (mAbs, proteins, and select peptides) and nucleic-acid-based medicines (siRNA, ASO, and mRNA) (Figure 20.1b). The chapter will focus on drug delivery modalities and mechanisms, both from a historical perspective, in addition to examining newer developments in the field. The authors have paid particular attention to areas that have either already demonstrated a big impact on the standard of care or a potential to have a dramatic impact on pushing the boundaries of therapeutic delivery in the future. Of note, since Part 2 of this book is exclusively dedicated to discussing analytical methods for characterization of biotherapeutics, the section of this chapter entitled 'An Analytical Perspective to Drug-Delivery Technology Development' will only address this topic insofar as it relates to DDSs and the therapeutics that they are accompanying.

Delivery of Proteins

Systems That Covalently Modify the Protein API

Most protein and peptide therapeutics that have been FDA approved in combination with a DDS have leveraged the same strategy: covalent modification. This clinically proven strategy is typically utilized to improve their PK profile; however, there are other reasons to covalently modify a protein or peptide, such as to improve solubility or decrease immunogenicity among others. Although the use of polyethylene glycol (PEG) is by far the most common approach for proteins, lipidation has gained in popularity for peptides and in some cases may confer more desirable properties relative to PEGylation.

PEGylation

PEG polymers (linear or branched) are water soluble and biocompatible and have demonstrated limited toxicity when administered in conjunction with therapeutic agents [27]. PEGylation of proteins, in which PEG is covalently attached to a bioactive molecule, has been studied extensively over the past few decades. Due to the hydrophilic nature of the PEG polymer, it can provide a shielding corona that makes conjugated proteins less recognizable by the immune system, thus improving safety and extending circulation time [28]. Since the first publication of the PEGylation strategy, which demonstrated extended circulation half-life and reduced immunogenicity of bovine liver catalase [29, 30], there have been more than a dozen approved PEGylated protein products in the clinical setting, dating back to the 1990s and spanning nearly all routes of administration RoA (Table 20.1). The cross-linking strategies, along with the design of PEG molecules and the versatility in applications have been well reviewed in the past and readers can find more information in recent review articles on various PEGylation strategies [28, 31–34].

FcRn

Another strategy for protein half-life extension through covalent modification relies on the neonatal fragment crystallizable receptor (FcRn). FcRn is a member of the Fc-receptors and is the primary driver for the long endogenous half-life observed for antibodies [35]. The mechanism involves the pinocytosis of antibodies into endothelial cells and the subsequent pH-dependent binding to FcRn during endosome acidification. This interaction prevents the degradation of antibodies by

Table 20.1 Various clinically approved PEGylated protein therapeutics.

Drug product	Company	Indication	Launch year	Delivery route	Protein format	PEG format	Elimination half-life[a)]
Adagen™	Enzon	Adenosine deaminase (ADA) deficiency	1990	IM	Enzyme	single chain	3–6 d
Oncaspar™	Enzon	Acute lymphoblastic leukemia	1994	IM / IV	Enzyme	single chain	3–6 d
Pegintron™	Schering	Hepatitis C	2001	SC	Cytokine	single chain	40–60 h
Pegasys™	Roche	Chronic hepatitis B, chronic hepatitis C	2002	SC	Cytokine	more than one chain	50–140 h
Neulasta®	Amgen	Infection during chemotherapy	2002	SC	Cytokine	single chain	15–80 h
Somavert™	Pfizer	Acromegaly	2003	SC	Growth hormone	branched chain	6 d
Mircera™	Roche	Anemia associated with chronic kidney disease	2007	SC	Cytokine	single chain	130 h
Cimzia™	UCB	Rheumatoid arthritis	2008	SC	mAb	single chain	14 d
Krystexxa®	Horizon Pharma	Chronic gout	2010	IV	Enzyme	more than one chain	10–12 d
Sylatron™	Merck	Melanoma	2011	SC	Cytokine	single chain	43 h
Plegridy™	Biogen	Multiple sclerosis	2014	SC	Cytokine	single chain	78 h
Adynovate®	Baxalta	Hemophilia A	2015	IV	Clotting factor	one or more chains	11–15 h
Rebinyn®	Novo Nordisk	Hemophilia B	2017	IV	Clotting factor	more than one chain	93 h
Udenyca™	Coherus Biosciences	Infection during chemotherapy	2018	SC	Cytokine	single chain	15–80 h
Fulphila™	Mylan GmbH	Infection during chemotherapy	2018	SC	Cytokine	single chain	15–80 h
Asparlas™	Servier Pharma	Acute lymphoblastic leukemia	2018	IV	Enzyme	single chain	16 d
Revcovi™	Leadiant Bioscience	ADA-SCID	2018	IM	Enzyme	single chain	24–72 h
Palynziq™	BioMarin Pharmaceutical	Phenylketonuria	2018	SC	Enzyme	single chain	47–60 h
Jivi™	Bayer Healthcare	Hemophilia A	2018	IV	Clotting factor	branched chain	17–22 h
Ziextenzo™	Sandoz	Infection during chemotherapy	2019	SC	Cytokine	single chain	15–80 h
Esperoct®	Novo Nordisk	Hemophilia A	2019	IV	Clotting factor	single chain	14–22 h
Nyvepria™	Pfizer, Inc.	Infection during chemotherapy	2020	SC	Cytokine	single chain	15–80 h
Besremi™	PharmaEssentia Corp	Polycythemia vera	2021	SC	Cytokine	single chain	7 d
Skytrofa™	Ascendis	Growth hormone deficiency	2021	SC	Growth hormone	single chain	30 h
Rolvedon™	Spectrum Pharmaceuticals	Infection during chemotherapy	2022	SC	Cytokine	single chain	36 h
Stimufend®	Fresenius Kabi	Infection during chemotherapy	2022	SC	Cytokine	single chain	15–80 h
Fylnetra™	Amneal Pharmaceuticals LLC	Infection during chemotherapy	2022	SC	Cytokine	single chain	15–80 h
Elfabrio®	Chiesi Farmaceutici S.p.A	Fabry disease	2023	IV	Enzyme	more than one chain	80 h

ADA-SCID, adenosine deaminase-deficient severe combined immune deficiency; IM, intramuscular; IV, intravenous; mAb, monoclonal antibody; PEG, polyethylene glycol; SC, subcutaneous.

a) Information obtained from www.accessdata.fda.gov.

Source: [28]/John Wiley & Sons/CC BY 4.0.

the lysosome, enabling their recycling back to the cell surface and reintroduction into the circulatory system after dissociation from FcRn at physiological pH [36]. Likewise, human serum albumin (HSA) maintains a similar long half-life compared to that of antibodies due to its interactions with FcRn [37]. X-ray crystallography has been performed on the HSA–FcRn–antibody Fc complex, revealing a detailed understanding of the amino acids involved in the pH-dependent binding interaction for both proteins [36]. In turn, this information has been harnessed to develop novel therapeutic DDSs.

In an effort to increase the circulation time of both small-molecule and protein-based therapeutics, FcRn has been leveraged by developing both HSA-fusion proteins and Fc-fusion proteins [38, 39]. By linking the therapeutic, also known as active pharmaceutical ingredient (API) to a protein that interacts with FcRn, half-life extension has successfully been achieved. Moreover, the detailed mapping of the binding interactions has led to the ability to modulate interactions to FcRn through rational mutagenesis. For example, introduction of the triple mutation, YTE (M252Y/S254T/T256E), in the Fc region has demonstrated significant increase in affinity to FcRn, with a commensurate increase in mAb half-life [40]. Conversely, introduction of the HAHQ (H310A/H435Q) mutation abolishes its affinity towards FcRn. While it may seem counterintuitive to desire faster clearance of biotherapeutics through eliminating FcRn activity, this strategy can become useful if the site of administration is also localized to the target site of action. For example, for a therapeutic that was administered IVT, it is typically desirable to enable slow clearance from the vitreous, but fast clearance once it reaches systemic circulation. Such is the case with the anti-VEGF molecules Lucentis and Avastin. While both display similar IVT half-life, the lack of FcRn binding capabilities for Lucentis leads to a faster systemic clearance, thus diminishing the potential for off-target interactions relative to Avastin, which binds FcRn [41–43].

Lipidation

Lipidation is the coupling of a fatty chain to a protein or peptide molecule. In nature, attachment of lipids to proteins is found in all organisms. Cells leverage lipidation to control the function and intracellular localization of a selection of proteins [44, 45]. In drug delivery, lipidation is commonly used to either extend peptide half-life or improve immunogenicity and/or cell membrane permeability [46]. One of the primary ways by which fatty acid addition enables half-life extension is through its ability to interact with albumin, which has been demonstrated to bind nonesterified fatty acids with high affinity at specific sites on the albumin molecule [47]. Several lipidated peptide drugs have been approved by the FDA starting as far back as 1964 (Table 20.2). More recently, peptide treatments such as semaglutide and tirzepatide treatments, both GLP-1 agonists, have gained notoriety for their blockbuster status, with Mounjaro (tirzepatide) having an estimated ~$2 billion in sales for Q1 2024 alone. And with the recent FDA approval of tirzepatide in November for weight loss, and likely more indications on the way, those numbers are expected to skyrocket. Both semaglutide and tirzepatide take advantage of the half-life

Table 20.2 Various clinically approved lipidated peptide therapeutics.

Drug product	Peptide name	Company	Type	Indication	Launch year	Delivery route	Elimination half-life (h)
Cubicin®	Daptomycin	Merck	N/A	Bacterial infection	2003	IV	8.5
Poly-Rx	Polymyxin B	Multiple	N/A	Bacterial infection	1964	Topical/IM, IV, IT	4–6
Egrifta SV®	Tesamorelin	Thera-technologies	GHRF	Excess abdominal fat in HIV patients with lipodystrophy	2010	SC	0.4–0.6
Levemir®	Detemir	Novo Nordisk	Insulin	Diabetes	2005	SC	5–7
Tresiba®	Degludec				2015	SC	25
Victoza®	Liraglutide		GLP-1	T2D	2010	SC	13
Saxenda®				Obesity	2014	SC	
Ozempic®	Semaglutide		GLP-1	T2D	2017	SC	168
Rybelsus®					2019	SC	
Wegovy®		Lilly		Obesity	2021	SC	
Mounjaro®	Tirzepatide		GLP-1/GIP	T2D	2022	SC	120
Zepbound®				Obesity	2023	SC	

GHRF, growth hormone-release factor; GIP, gastric inhibitory polypeptide; GLP-1, glucagon-like peptide-1; IM, intramuscular; IT, intrathecal; IV, intravenous; N/A, not applicable; SC, subcutaneous; T2D, type-2 diabetes.

extension provided by lipid addition, which has enabled dosing to be a once weekly SC administration. While there are challenges with the lipidation of peptides, such as lack of a mechanistic understanding of the effect of lipidation on cell uptake, cell processing, and immunogenicity, this strategy has already established itself as one of the go-to delivery approaches for modulating peptide PK.

Long Acting Delivery Systems That Do Not Modify the Protein API

The development of long acting delivery (LAD) systems for protein therapeutics has gained significant attention over the past few decades. LAD offers distinct advantages in sustaining release duration of therapeutics upon single administration, including (i) extending drug exposure while maintaining the desired effective therapeutic level, (ii) reducing the frequency of injection to improve patient compliance, and (iii) improving the safety profile by lowering the risk of high initial drug concentration in blood [48, 49]. With recent advances in DDSs, a diverse set of polymers, inorganic materials, and fatty acid components are now being adopted to demonstrate sustained release of proteins and peptides over extended duration. However, in order to define the possibility of technical success when evaluating these novel LAD technologies in combination with protein therapeutics, considerations such as effective protein loading within the drug delivery matrix, protein stability in the matrix, drug release, as well as control over product quality within a manufacturing setting are of the utmost importance. The target product profile of a specific molecule needs to be well-defined in order to assess suitability of a LAD system.

There is vast clinical precedence for using polymer-based DDSs to extend the release duration of therapeutic agents, such as *in situ* forming depot, polymeric implants, and polymeric microparticles (Figure 20.2). In recent decades, most of the clinically approved products for LAD are based on the use of biodegradable polymers such as poly(lactic-*co*-glycolic acid) (PLGA) or nonbiodegradable polymers such as ethylene vinyl acetate (EVA). Depending on the polymer compositions and the desired target product profile, such polymer-based delivery systems can be tuned to achieve sustained drug release ranging from as short as one week to a period of several years [50–52]. However, due to the unique physicochemical properties of protein therapeutics, developing a polymer-based DDS that is tailored for the sustained release of protein is immensely challenging. Early success was found in 1999 with the FDA approval of a PLGA depot that enabled monthly injection of human growth hormone (~22 kDa) (Nutropin). Nevertheless, it was withdrawn from the market in 2004 as a result of a complicated production process and undesired burst drug release [53]. Below, we will review more recent clinical development of polymer-based DDSs based on PLGA and EVA polymers.

Biodegradable Polymers

Biodegradable polymers have shown great promise in the field of LAD. These polymers can be safely degraded at the site of injection and metabolized by the body without causing harmful complications or an undesired immune response. Thus, one of the prominent benefits of using biodegradable polymers for the design of LAD carriers is that surgical removal is not necessary after administration in patients. Polymers including poly(lactic acid) (PLA), PLGA, and poly(caprolactone) (PCL) (Figure 20.3) have been widely studied and have shown significant potential for tissue engineering [54–58]. Among them, PLGA is one of the most established polymer-based DDSs with various clinically approved products for sustained drug release (Table 20.3). One particular merit of using PLGA as a drug delivery carrier is its ability to generate biodegradable byproducts following hydrolytic degradation, including lactic and glycolic acid [59, 60]. Furthermore, the relative ratio of the two monomeric units within PLGA copolymer can be tuned to manipulate its physicochemical properties, such as crystallinity, glass transition temperature (T_g), hydrophilicity/hydrophobicity, and can therefore influence the release rate of the encapsulated molecules [59].

In general, the possible mechanisms driving release of a therapeutic protein or peptide from a PLGA-based DDS are diffusion, osmotic pumping, and polymer erosion (Figure 20.4) [61–63]. Nonetheless, it is often observed

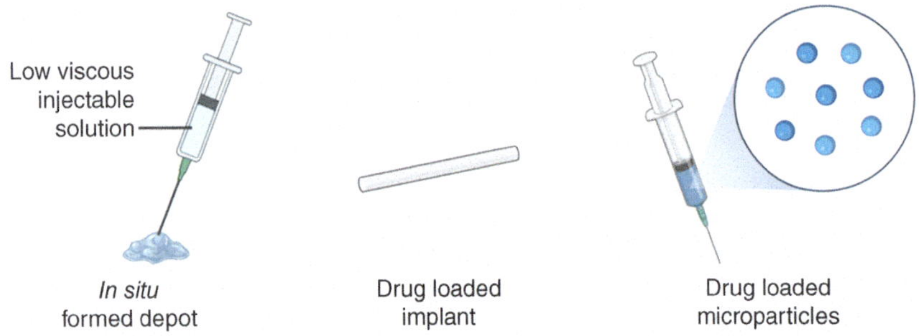

Figure 20.2 Various examples of clinically approved nonmodifying polymeric LAD delivery systems. Source: Adapted from BioRender.

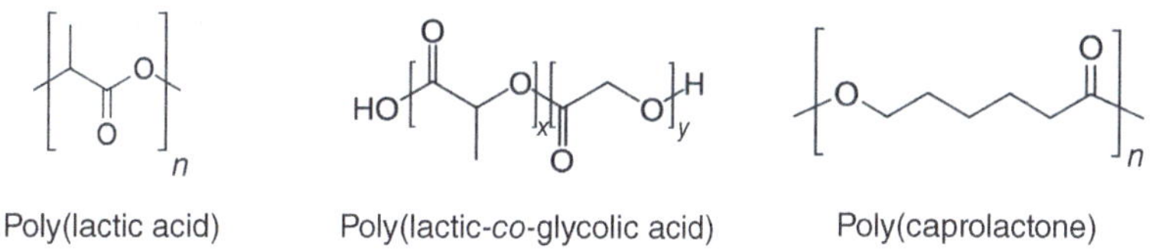

Figure 20.3 Chemical structures of poly(lactic acid) (PLA), poly(lactic-co-glycolic acid) (PLGA), and poly(caprolactone) (PCL).

Table 20.3 Various clinically approved products of PLGA-based LAD systems for peptide therapeutics.

Drug product	Indication	LAD system	Drug MW (kg/mol)	Durability (months)	Delivery route	Company	Launch year
Lupron®	Prostate cancer	PLGA microparticles	1.2	1–6	IM	Abbvie	1989
Zoladex®	Prostate cancer	PLGA rod	1.3	1, 3, or 6	SC	AstraZeneca	1989
Suprefact®	Prostate cancer	PLGA rod	1.2	2–3	SC	Xediton	1996
Sandostatin®	Acromegaly	PLGA microparticles	1.0	1	IM	Novartis	1998
Eligard®	Prostate cancer	PLGA *in situ* forming implant	1.2	1–6	SC	Astellas	2002
Trelstar®	Prostate cancer	PLGA microparticles	1.7	1–6	IM	Verity	2010
Bydureon®	T2D	PLGA microparticles	4.2	1 week	SC	AstraZeneca	2012
Signifor®	Acromegaly	PLGA microparticles	1.3	1	SC	Novartis	2014
Scenesse®	Protoporphyria	PLGA rod	1.7	6	SC	Clinuvel	2016
Camcevi®	Prostate cancer	PLGA *in situ* forming implant	1.3	6	SC	Accord BioPharma	2021

IM, intramuscular; LAD, long-acting delivery; PLGA, poly(lactic-*co*-glycolic acid); SC, subcutaneous; T2D, type 2 diabetes.
Source: [50, 51].

that upon initial exposure of PLGA carriers to the aqueous physiological environment, an undesired burst release of the therapeutic occurs within the first few hours to several days after administration [64]. This has been mainly attributed to nonencapsulated molecules that are dispersed near the surface of the PLGA matrix and are easily accessible by hydration [62]. Beyond this phenomenon, diffusion-based release is the most common mechanism governing therapeutic release (Figure 20.4a,b). This process lasts several days to weeks, where the encapsulated therapeutic molecules need to first be dissolved in aqueous media, followed by diffusion through water-filled pores in the PLGA matrix [62, 64]. The rate of release can be influenced by many different factors, including the physicochemical properties of the therapeutic, type and concentration of the polymer and excipients, and morphology and shape of the drug delivery carrier [62]. The osmotic pressure may be increased by the influx of water, driving the transport of the drug (Figure 20.4c). However, osmotic pumping is more common for a non-swelling DDS [61]. Another mechanism governing therapeutic release is polymer erosion (Figure 20.4d), eventually leading to complete dissolution of encapsulated molecules [65]. The erosion phase can happen contemporaneously or consecutively with the diffusion phase [66]. Degradation of the polymer backbone can create more pores and channels within the polymer matrix, which accelerate molecule release [61]. In short, therapeutic release from PLGA-based DDSs is a complex process governed by both the diffusion of the therapeutic and the degradation of the polymer matrix. A more detailed discussion about the release mechanism from PLGA-based DDSs can be found in other extensive review articles [61, 62].

In order to develop a PLGA-based LAD carrier for proteins, several important parameters need to be taken into account to ensure that the product quality is maintained or improved. As alluded to earlier, PLGA carriers can be degraded to form acidic by-products. The microclimate pH

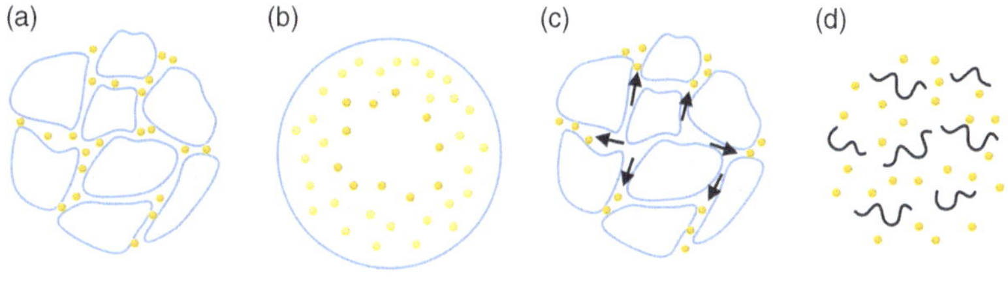

Figure 20.4 Therapeutic release mechanisms of PLGA by (a) diffusion through water-filled pores, (b) diffusion through the polymer, (c) osmotic pumping and (d) erosion. Yellow denotes the therapeutic, light blue is the PLGA depot. Source: [61, 62].

distribution within the PLGA matrix can drop to a range of low pH values during incubation as a result of polymer degradation and accumulation of acidic by-products. The range of microclimate pH can be influenced by the molecular weight and the composition of the PLGA polymer [67]. Ding and Schwendeman showed that the most acidic microclimate pH value was recorded to be an average of ~2.9 after 1 day of incubation of PLGA microspheres in aqueous buffer [67]. Although the average pH value shifted to a less acidic range (pH ~3–5) after extended incubation due to the release of acids from the polymer and influx of buffered media, this still has the potential to hugely impact protein stability [67]. Kang et al. evaluated the stability of bovine serum albumin (BSA, ~66 kDa) when encapsulated in PLGA microspheres [68] and attributed the observed insoluble aggregates of BSA during *in vitro* release to the low microclimate pH within the PLGA matrix [68]. Although incorporation of basic excipients such as magnesium oxide could suppress formation of the insoluble aggregates, the recovery of soluble fraction of protein was low. On the other hand, in Chang et al. a model therapeutic antibody fragment (Fab, ~50 kDa) was successfully loaded within an *in situ* forming PLGA depot as an injectable formulation and protein release was tuned by changing the molecular weight of the polymer, achieving sustained protein release for up to 80 days [69]. Interestingly, although the Fab fragment exhibited a pH-dependent stability profile, the rate of main peak loss of extracted Fab samples from the depot was similar to the rate of main peak loss when incubated in aqueous buffer at pH 7.4. This was attributed to the improved stability of the protein, either in spray-dried form or within the porous interiors of the PLGA depot, which led to rapid exchange of the buffer media thereby neutralizing the pH within the depot. Furthermore, the percolation network within the polymer matrix can dictate the release behavior of the encapsulated molecules. When loading is high within the polymer matrix, it can create a good porous network, which can improve diffusion of the therapeutic and contribute to the initial phase of drug release. Conversely, when loading is low, it cannot form a porous network, thus reducing the initial burst of the encapsulated molecules. Zhang et al. demonstrated that greater than or equal to 20% drug loading by weight caused an initial burst release of 65%–80% in the first 7 days from a PLGA rod [70]. By contrast, lower drug loading content at 15% by weight only released 4.4% of encapsulated molecules in 1 day [70]. Thus, in order to enable the development of PLGA-based carriers for LAD of proteins, it is crucial to identify suitable formulation strategies with optimized attributes that maintain protein stability as well as suitable polymer characteristics and drug loading for better control of polymer degradation and release kinetics.

Non-biodegradable Polymers

Non-biodegradable polymers have been investigated heavily over the past decades due to their versatility and strong durability for biomedical applications. The main classes of non-biodegradable synthetic polymers include poly(ethylene-vinyl acetate), poly(urethanes), poly (siloxanes), and other types of polymers (Figure 20.5) [71]. These materials are biocompatible, meaning they are compatible with living tissues with minimal toxicity. However, the majority of non-biodegradable devices or implants require surgical implantation and removal procedures by trained medical professionals. Thus, depending on the therapeutic application, it is ideal to design a non-biodegradable carrier that can last for a longer release duration to minimize the frequency of implantation and explantation procedures. Compared to a biodegradable DDS, non-biodegradable polymers have higher mechanical strength and abrasion resistance, and the release kinetics can therefore be better controlled to achieve a constant release rate. Thus, a non-biodegradable polymer is considered more durable as a drug reservoir for sustained drug release in a biological environment [71].

EVA is a thermoplastic polymer, who's stiffness, crystallinity, and melting temperature can be tuned by varying the ratio between ethylene and vinyl acetate monomers [52]. Pharmaceutical processing of EVA generally involves hot-melt extrusion of the polymer blend and injection molding. Other techniques such as solvent casting or spray coating can also be employed [52]. Since the melt temperature of EVA polymer can be varied with increasing vinyl acetate content (approximately 45–55°C at 40% by weight of vinyl acetate), it is more compatible with thermosensitive molecules [52]. Given the non-biodegradable nature of EVA, the proposed mechanism for drug release from an EVA-based drug delivery system is mainly driven by drug diffusion [52]. As a reservoir-type drug release system,

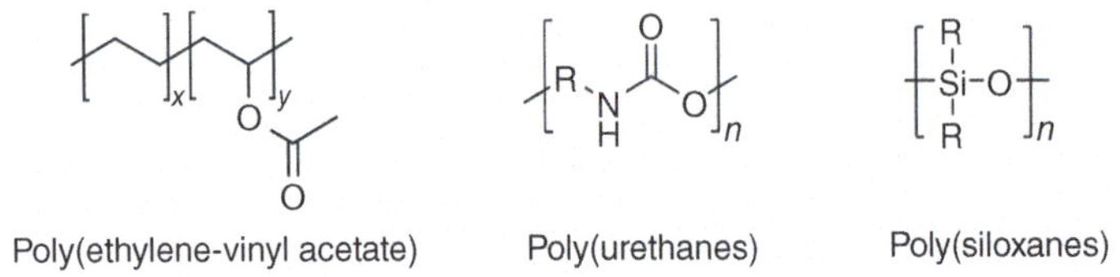

Figure 20.5 Chemical structures of poly(ethylene-vinyl acetate), poly(urethanes), and poly(siloxanes).

after coating with a rate-controlling membrane, drug diffusion of the loaded therapeutic agent is rate limiting and constant. EVA offers a wide range of release duration based on polymer characteristics and membrane geometries. This is of particular interest to achieve a zero-order release kinetics for a certain therapeutic application, which can maintain a constant release rate from the drug delivery carrier until complete exhaustion of the encapsulated molecules [72]. In addition, one distinct advantage of EVA-based carriers is the ability to achieve high drug loading levels over 40% by weight while maintaining the desired mechanical properties [73].

EVA has been utilized for the development of various controlled drug release systems since the 1970s. Early studies by Langer and Folkman demonstrated the use of EVA to achieve sustained protein release for up to 100 days using catalase (~250 kDa) as the model protein [74]. In addition, Saltzman et al. investigated the use of an EVA disk for intracranial delivery of recombinant human nerve growth factor (~86 kDa) in rats [75]. The EVA implants showed a sustained release profile for over 2 months in buffered saline (*in vitro*) which was linearly correlated to protein concentration in the brain [75]. Despite significant advancement in the use of EVA for sustained drug release, thus far most of the clinically approved products have been in combination with small molecule drugs (<500 Da) [52, 76, 77]. This may be attributed to considerations such as protein stability during processing at a high temperature, protein–polymer interaction, and achieving high protein concentration which are crucial in order to enable an EVA-based LAD system.

Protein Formulation Within a Polymer LAD System

The manufacturing process to encapsulate protein molecules within a LAD system is not trivial and can induce a number of external stresses on proteins. Thus, identifying optimized formulations that can maintain or improve protein stability during the loading process is of paramount importance. Since water plays such an important role in chemical modifications that ultimately lead to protein degradation, developing solid formulations of proteins has the potential to greatly improve long-term protein stability and extend product shelf life [78, 79]. Moreover, handling protein solutions in the presence of additional stresses such as mechanical stress, heat stress, and exposure to organic solvents can be detrimental in maintaining the structural integrity and stability of proteins [78]. Currently, the most established drying technologies to produce solid formulations of proteins include freeze drying (or lyophilization) and spray drying [80]. For freeze drying, a typical process involves a freezing step, where the protein solution is first frozen at a controlled rate. Subsequently, the frozen water can be removed under vacuum via sublimation during the primary drying process [80]. This is then followed by a secondary drying step to remove additional moisture to reach the desired moisture level [80]. However, the produced solid after freeze drying usually forms a cake, which can raise product quality concerns and potentially impact protein stability [81]. Additional processing steps, such as milling, may also be required to produce particles for different applications. In addition, proteins also experience extra stresses such as cold denaturation, exposure to the interfaces between ice and water, and the freeze concentration, which can exacerbate protein stability [82]. In order to define a successful lyophilized formulation for proteins, Carpenter et al. described a list of minimum criteria needed for the proper design of process parameters. These include: (i) inhibition of protein unfolding during the process of freeze drying; (ii) T_g of the produced solids is higher than the storage temperature; (iii) the final moisture content is ideally less than 1% by mass; (iv) formation of a uniform and elegant lyophilized cake appearance; and (v) additional steps to minimize protein chemical degradation [83]. Detailed discussions on how to optimize the freeze drying process for proteins can be found in other in depth review articles [83–85].

Spray drying typically involves an atomization step where the protein solution is first atomized in a drying chamber, followed by a subsequent drying step in the presence of a high temperature gaseous medium to produce dried powders [80]. It is a single step continuous process, which provides better control over powder quality and can be used as a viable alternative to lyophilization. Various processing parameters such as the feeding rate of the stock solution, inlet and outlet temperature, and different types of nozzles or atomizers can influence physical properties such as size, final moisture content, and morphology of the dried particles [86]. Although it avoids many of the stressors of freeze drying, thermal instability of proteins is a major hurdle to the successful development of spray dried formulations. Since most protein therapeutics are formulated in a buffered solution, the spray drying process generally requires a relatively high temperature for solvent evaporation [80]. Furthermore, the presence of air–liquid interfacial stress and shear stress during spray drying can also affect protein stability [86]. Ameri and Maa gave a detailed description of the dehydration mechanisms of spray drying and the influence on protein stability in response to stress events [87]. Thus, identifying suitable spray drying conditions and formulation strategy is paramount for obtaining solid powders with a controlled particle size range and desired physical properties.

One way of preserving the structural integrity of proteins upon drying is by using excipients such as sugars [88].

The mechanism of how sugars protect the structure of proteins upon drying is still in constant debate. Nonetheless, a few proposed mechanisms have been widely discussed in the literature including water replacement, vitrification, and preferential exclusion [89–91]. In the water replacement theory, the hydroxyl groups of sugars can play a substitute role in replacing the hydrogen bond interactions between water and protein, thereby maintaining protein conformation [89]. This stabilization mechanism was evidenced by the ability of sugars to reverse dehydration-induced shifts of amide bands via infrared spectroscopic studies [92]. The vitrification theory is based on the formation of a rigid, amorphous sugar glassy matrix to immobilize proteins during drying, thereby limiting protein mobility and slowing down degradation [89]. Such an immobilization mechanism is largely dictated by the T_g of the sugar matrix, above which the kinetic immobilization of sugars is greatly lost [89]. On the other hand, the preferential exclusion theory suggests that sugars can form a protective layer to trap water around the surface of proteins, thus protecting proteins during the drying process [88]. Among different sugar excipients, trehalose and sucrose have been widely used to preserve protein stability upon drying [89, 93]. In particular, both sugars are classified as nonreducing sugars, which can minimize the risk of covalent reactions with proteins via the Maillard reaction [93]. In most cases, trehalose is preferred as a protectant for drying of proteins when compared to sucrose due to its high resistance to chemical degradation under acidic conditions, high T_g, and greater flexibility to form hydrogen bonds with proteins [90, 93]. However, some studies have suggested that sucrose outperforms trehalose in maintaining protein stability in the solid state [94, 95]. The protection effect of sugars can be molecule and process dependent. Additional formulation screening effort to identify the optimal protein to excipient ratio, protein concentration, and choice of excipients is required for specific applications using solid formulations of proteins.

To incorporate solid proteins in an LAD implant typically requires additional processing steps such as a hot-melt extrusion process. This process generally operates at a high temperature, which can be detrimental to protein stability. Rajagopol et al. demonstrated that the use of trehalose can limit the aggregation of BSA at high temperature [96]. BSA was first spray dried to form solid powders in the presence of trehalose, where the high T_g of trehalose protected against aggregation at 100°C with less than 3% monomer loss after 1 hour [96]. Spray-dried BSA powders were then successfully loaded within PLGA via hot-melt extrusion at elevated temperature without significant aggregation [96]. In another study, a similar protective effect of trehalose was found for a spray-dried formulation of Fab, maintaining protein stability upon loading and release in a PLGA depot [69]. However, the formation of subvisible particles during processing of solid protein formulations could impose additional risk for safety and efficacy of the product. Gikanga et al. showed that solid formulations of model mAbs can induce the formation of subvisible particles upon stress [97]. Although there was minimal change in soluble aggregates in the mixer comparing both spray-dried and lyophilized protein formulation, spray-dried solids generated a significant amount of subvisible particles relative to the lyophilized solids after stress [97]. Overall, while solid formulations of proteins can provide a different means of protein loading in a drug-delivery carrier, considerations such as subvisible particle formation and protein stability in response to external stresses need to be taken into account for formulation development.

Device-Based Drug-Delivery Systems

Patient compliance with treatment plans involving parenteral dosing has remained a concern due to frequent doctor visit requirements and uncomfortable procedures. Thus, device-based DDS′ have offered an alternative for a wide range of administration routes for nearly 50 years. Beginning with the insulin infusion pump in the late 1970s [98], commercially available reservoir-type devices now include nebulizers, which deliver aerosolized liquid mAb formulations; autoinjectors, and prefilled syringes (PFS), and more recently transdermal microneedle patches [99–102] (Figure 20.6). By developing means for easy, at-home administration or longer redosing periods, it

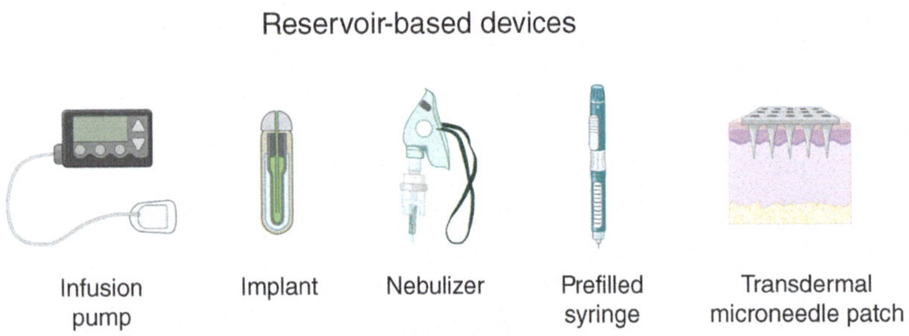

Figure 20.6 Examples of reservoir-based devices. Source: Adapted from BioRender.

has been possible to greatly improve clinical outcomes [103, 104]. In addition to improving patient compliance, dosing inaccuracy at home and in the clinic can be reduced through highly tunable or precise delivery systems. For instance, in contrast to standard IVT injections, PFS have shown utility in reducing overall preparation times in busy clinics and have enabled delivery of consistent doses [105]. In countries where clinical practices are less stringent, PFS use has greatly decreased the potential for contamination and mis-dosing [106]. This section will explore two active and passive controlled release devices, their release kinetics, and development considerations for these DDSs.

Reservoir-based controlled release devices are intended for therapeutic administration over extended periods. These devices utilize an internal reservoir to house enough material for multiday dosing. Insulin pumps are broadly recognized reservoir-based drug release devices. Insulin is delivered SC via a wearable pump with connected infusion tubing, or through a remote-controlled adhesive patch pump connected to an infusion cannula [107]. The pump aids in the delivery of basal insulin, as well as allows the user to self-administer bolus insulin during meals or for correcting spikes in blood glucose. Basal insulin release requires physician- or patient-input blood glucose parameters, based on patient lifestyle and predicted insulin needs. Pump users can also manually input parameters such as carbohydrate intake and blood glucose level to calculate bolus dosing [107]. Alternatively, insulin needs can be monitored through app-integrated continuous glucose monitors (CGMs), allowing diabetics to respond to changes in blood glucose in near real time and dose with greater precision [108]. More recently, medical researchers have evaluated the efficacy of the artificial pancreas, which uses machine learning to enable automated, closed-loop insulin delivery [109–111].

A more recently approved drug-delivery device for an indication other than diabetes is the Port Delivery System (PDS), or Susvimo™, a refillable, implantable LAD device, for the treatment of wet age-related macular degeneration. Unlike IVT injections, which require patients to visit the clinic monthly, Susvimo allows for 6 month continuous release of ranibizumab, an anti-VEGF Fab, after which the implant is refilled with fresh drug product [112]. The system consists of two parts: initial fill and implantation, and the refill exchange. First, the implant is filled with ranibizumab drug product using a co-packaged initial fill needle (Figure 20.7a–c). The filled implant is then surgically implanted, where it will continuously deliver the drug to the vitreous humor, using a sintered titanium drug release controlling element (RCE) that acts as a membrane to modulate protein diffusion through micrometer-sized pores [113]. After 6 months of release, the concentration of drug inside the implant is sufficiently low that redosing is required to maintain a therapeutic concentration of ranibizumab. During the refill exchange procedure, a specialized vented needle attached to a collection reservoir (Figure 20.7d) facilitates redosing in a single step. Different views of what the PDS implant looks like within the eye are demonstrated in Figure 20.7e–g [114]. In 2022, one year after commercial launch, Roche issued a voluntary recall of the PDS implant due to loss of septum integrity after repeated needle punctures. Modifications made to the refill needle and redesign of the implant septum were successful in resolving the issue, and Susvimo was FDA approved for relaunch in July 2024 [115].

Drug release and elimination kinetics go hand-in-hand to achieve what is known as the steady-state drug concentration, that is, the equilibrium state during which the amount of drug eliminated is equal to the amount entering the system [116]. In circulation, drug elimination can be concentration independent (zero order) or dependent (first order) (Figure 20.8) [116]. Salicyluric acid and salicyl phenolic glucuronide, two metabolites of aspirin, are among few drugs eliminated from the body by zero-order elimination when the dose is sufficiently high, while most other medications, including protein therapeutics, are first order [18, 117]. Moreover, zero- and first-order kinetics can also refer to drug release from a DDS. Though similar, the distinction between rate of systemic elimination and kinetic drug release from a delivery system is important. While elimination describes the rate of excretion of an API from a physiological system, drug release kinetics refer to the rate at which the API is administered from the delivery device to the target. Drug release kinetics are tunable through modifications to the mode of delivery, such as through API encapsulation in a non-degradable system.

The advantage of zero-order drug release systems over conventional dosing regimens is the longer therapeutic window, where zero-order drug release allows for a constant release of drug to the patient, thereby minimizing spikes in plasma drug concentration (Figure 20.8) [118]. Of note, layer-by-layer (LBL) polymeric films have been successful in achieving zero-order delivery of peptides and small molecules. Hydrogen-bonded films of tannic acid (TA) and PEGylated salmon calcitonin (PEG-sCT) are alternately layered on a quartz substrate. Film disintegration by hydrolysis gradually frees the PEG-sCT from the TA for delivery [119]. In contrast to zero-order release, first-order drug release is proportional to the concentration of drug remaining in the system (Figure 20.8). On-body devices like insulin pumps which deliver insulin based on plasma concentration of glucose, or implantable

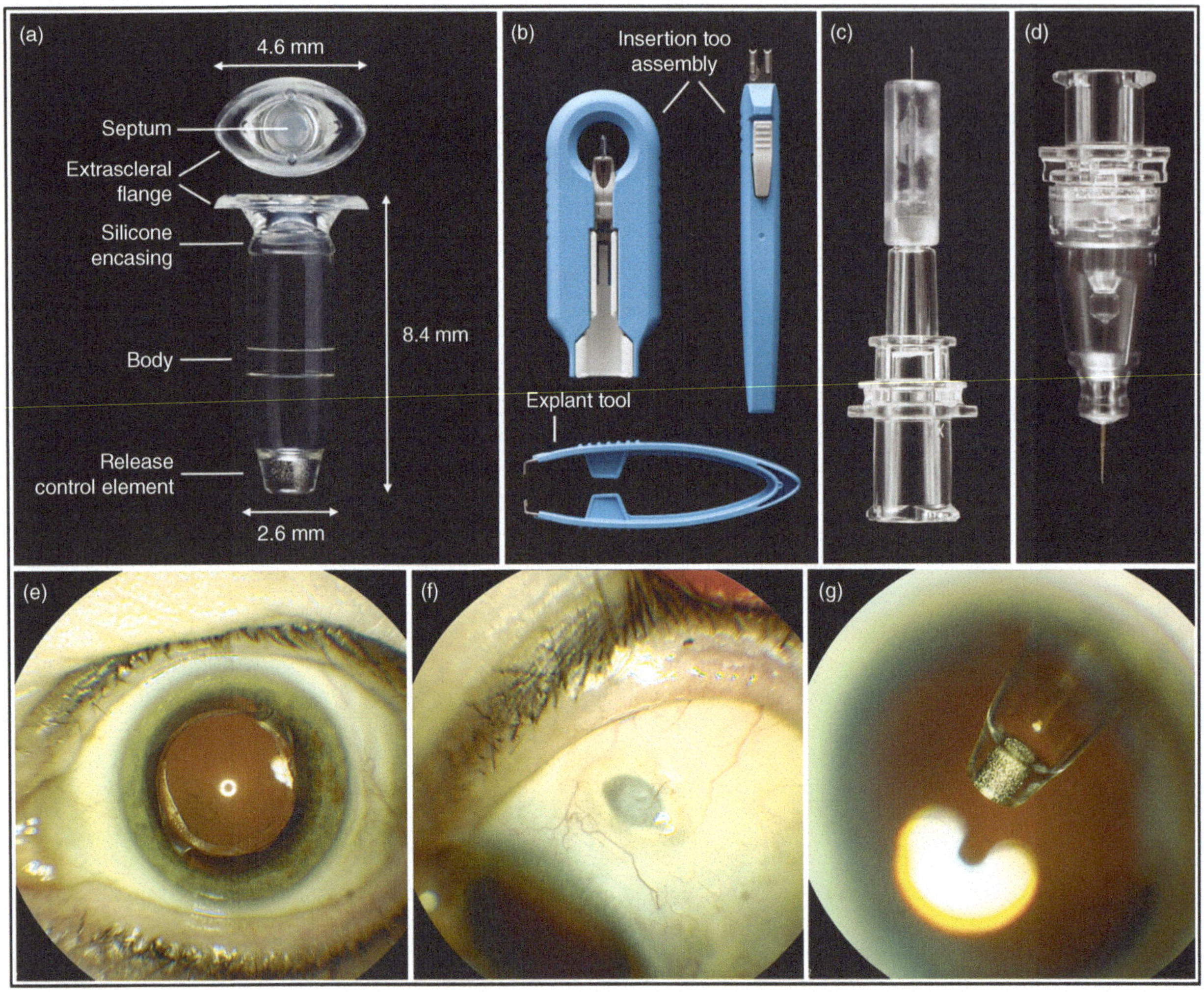

Figure 20.7 The PDS ocular implant, ancillary devices, and position *in situ*. (a) PDS implant showing four key features: an extra scleral flange that anchors the implant in the sclera, a self-sealing septum that allows for implant refills, an implant body that contains the drug reservoir for the customized formulation of ranibizumab, and the porous titanium RCE that controls the rate of ranibizumab diffusion into the vitreous. (b) Insertion tool assembly to facilitate handling of the implant during the initial filling and insertion procedures, and explant tool (lower image) to grasp and securely hold the implant flange during the explant procedure (if explanation is clinically indicated). (c) Initial fill needle to fill the implant with ranibizumab before insertion. (d) Refill needle to refill-exchange, *in situ*, the implant with ranibizumab. (e–g) The PDS implant *in situ*. Images from a PDS-implanted patient showing (e) eye in primary position (implant not visible), (f) eye looking down with the PDS implant septum visible, and (g) eye looking up with implant visible through dilated pupil. Abbreviations: PDS: Port Delivery System with ranibizumab; RCE: release control element. Source: [113] / Taylor & Francis Group.

devices such as the Susvimo ocular implant are both examples of reservoir-based controlled release devices which follow first-order release kinetics [112, 120]. In Figure 20.9a, *in vitro* drug release from the PDS is easily seen as first order, since different starting protein concentrations directly influence the amount of drug released. Consequently, the serum API concentration is also first order. Compared to ranibizumab IVT, sustained dosing from the PDS eliminates the need for frequent dosing by preventing large fluctuations in serum concentration of ranibizumab for up to 12 months when dosed at 100 mg/ml (Figure 20.9b) [112, 121].

Protein Formulation Within a Reservoir or Controlled Release Delivery System

Although solid formulations achieve high loading capacity by circumventing solubility issues, many LAD systems look to use high concentration versions of already marketed liquid therapeutics, as a means of improving ease of use of those products. Techniques for increasing protein solubility at high concentrations in liquid formulation include the addition of excipients such as salts, nonionic surfactants like polysorbates, or by protein re-engineering or addition of peptide tags to eliminate or mask protein charge patches [122–124]. Furthermore, increased protein

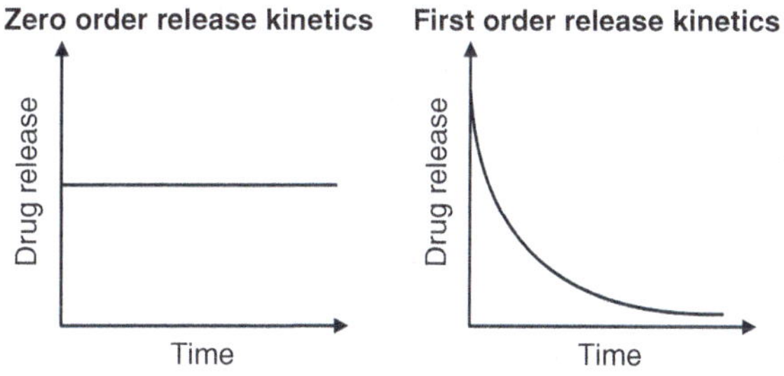

Figure 20.8 Zero- and first-order drug release kinetics as a function of time. Zero-order drug release allows for constant release of API regardless of the concentration remaining in the reservoir, while first-order drug release is concentration-dependent, and exponentially decreases over time. Source: Adapted from BioRender.

concentration can lead to higher formulation viscosity. High-viscosity drug products can lead to clogging of filter membranes during production and require greater injection forces, making their administration challenging.

While liquid formulations are typically stored at 2–8°C, in reservoir-based implants such as the PDS, the filled drug must maintain stability at a vitreous temperature of 34°C over the entire dosing period [125, 126]. For insulin pumps, reservoir vials of insulin are exposed to changing ambient conditions when worn on the body, with FDA guidance that "insulin contained in the infusion set of a pump device and exposed to temperature exceeding 98.6°F (37°C) should be discarded [127]." Additionally, this poses a unique challenge in proper self-administration and at-home handling. In a study of 255 patients receiving mAb treatment for rheumatic disease indications, Vlieland et al. observed that 45% of the medication storage time was spent outside the recommended temperature range of 2–8°C [128]. More research is needed for success in room temperature storage of biologics; however, approaches to increase thermal stability, such as through the use of novel PEG hydrogels, are currently being developed [129–131].

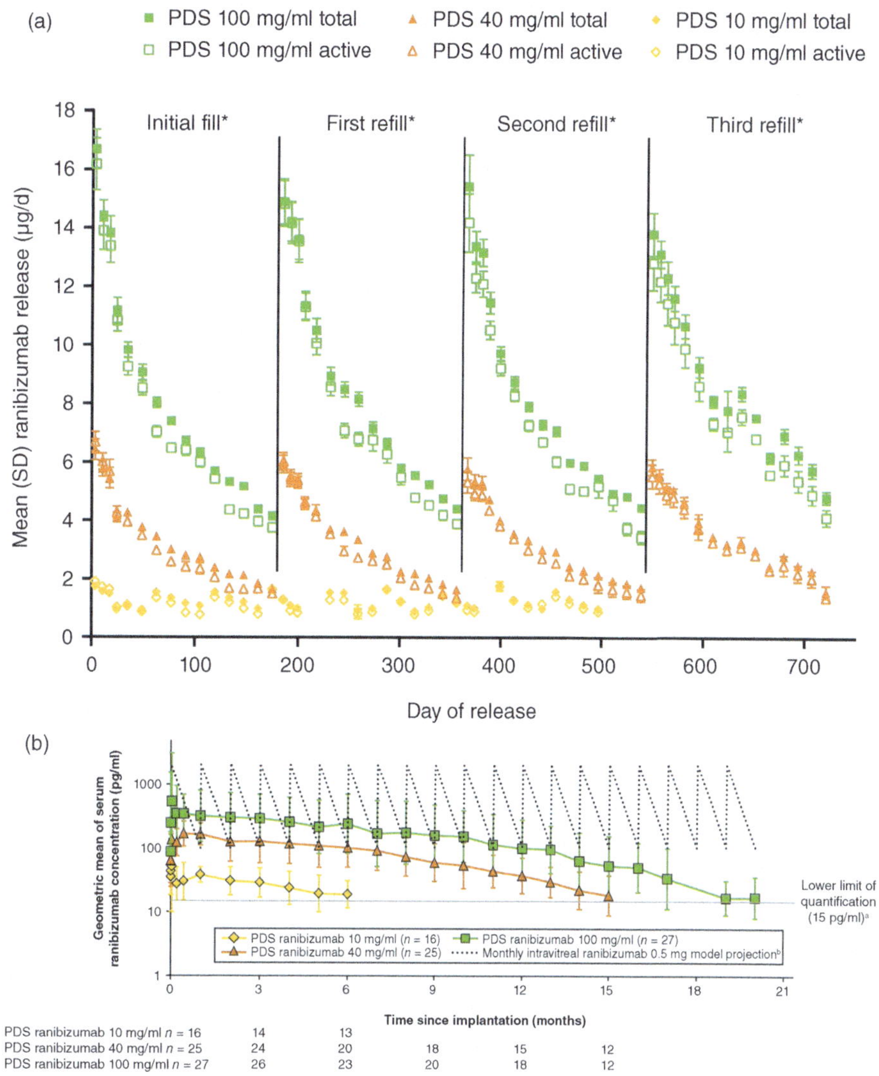

Figure 20.9 *In vitro* drug release of ranibizumab from the Port Delivery System (PDS) compared to *in vivo* serum concentration from the phase II Ladder clinical trial. (a) *In vitro* studies before the phase II Ladder trial: average ranibizumab active and total release rates with PDS 10 mg/ml (N = 10 for initial fill and N = 6 for each refill), 40 mg/ml (N = 37), or 100 mg/ml (N = 37). *Refill cycles were every 6 months for PDS 40 mg/ml and 100 mg/ml, and (due to the lower concentration) every 8 weeks for PDS 10 mg/ml; hence, there were a total of 3 refills for PDS 40 mg/ml and 100 mg/ml, and 8 refills for PDS 10 mg/ml [112]. (b) Serum pharmacokinetic (PK) profiles among PK-evaluable patients in the PDS with ranibizumab arms without refill. Patient numbers decreased over time as patients received refills. Vertical bars represent the geometric standard deviation. Y-axis is shown on a logarithmic scale. [a]Lowe J, et al. J Immunol Methods. 2018;461:44–52. [b]Xu L, et al. Invest Ophthalmol Vis Sci. 2013;54:1616–1624. Figure generated based on the updated population PK analysis (including data from the HARBOR clinical trial [NCT00891735]). Abbreviation: PDS: Port Delivery System with ranibizumab. Source: [121]/Springer Nature/CC BY-NC 4.0.

During development of sustained release drug–device combination products, it is critical to evaluate biocompatibility as well as drug compatibility with the device materials of construction (MOC). Leachable and extractable studies are performed to determine if the MOC contains any toxic volatile or nonvolatile compounds. Potential contaminants from the drug or device manufacturing process must also be examined. Jenke and Carlson compiled an extensive list of extractables common to polymers used in drug and device packaging and development, creating a database based on risk index and available toxicological information [132]. Of the 550 materials identified as common extractables, approximately 55% were considered potentially mutagenic by *in silico* predictions, or had little *in vitro* information available in order to determine safety [132]. Since the materials of construction are often in contact with the aqueous drug product, leachable and extractable studies are often performed by incubation of the MOC in an appropriate aqueous media or through exhaustive extractions with water. Extractions with organic solvents are also done as a worst case [133]. The FDA has issued guidance on industry best practices for evaluating leachables and extractables [134].

Emerging Drug-Delivery Technologies

Inorganic Materials

In addition to polymer-based materials, inorganic materials have become increasingly popular in recent years for use in pharmaceutical development. Unique properties of inorganic materials (e.g. silica, gold, and carbon nanotube), can provide potential benefits such as protein stabilization, as well as the ability to be manipulated for different applications [135]. In particular, the development of mesoporous silica nanoparticles (NP) has demonstrated recent success with its high loading capacity of proteins [136]. Due to the ease of manufacturing and designable morphology, physical properties such as particle stiffness, pore size, and surface chemistry can be tuned [136]. In addition, silica can degrade in water to form silicic acid and be renally excreted from the body [137]. Although a large pore size is required to achieve high drug loading, manipulation of core chemistry and co-formulation with excipients can tailor the interaction between encapsulated molecules and the silica framework. Lee et al. demonstrated that a proteasome subunit (~29 kDa) can form a complex with functionalized mesoporous silica NPs with nickel moieties via electrostatic interaction, therefore controlling protein association and protein release [138]. Moreover, Xu et al. designed a mesoporous silica particle carrier with a large pore size of 45 nm capable of achieving high loading capacity of a model mAb (~150 kDa) for intracellular delivery [139]. Recently, Tyagi et al. developed a silica microparticle hydrogel depot technology to achieve sustained release of a model antibody *in vivo* [140]. In their system, the mAb was first loaded into silica microparticles via spray drying, followed by dispersion in a silica matrix to form a hydrogel depot. Upon SC administration of the depot in mice, the authors demonstrated that the encapsulated protein was released from the depot for up to 62 days [140]. Although there are no silica-based LAD systems on the market yet, the safety profile of silica-based technologies has been proven by multiple clinical trials in humans [141]. Thus, further clinical investigation of the use of inorganic silica as a LAD approach for different therapeutic modalities is warranted as there exists considerable market potential.

Lipid-Based Materials

The use of lipids as a non-API-modifying DDS has gained increasing interest due to their biocompatibility, biodegradability, the ease of scale-up, and low toxicity. The most commonly used lipid-based drug-delivery carriers include liposomes, solid lipid nanoparticles (LNPs), and nanostructured lipid carriers [142–144]. Generally, lipid carriers are manufactured via homogenization, emulsification, or solvent evaporation [144, 145]. Due to the hydrophilic nature of protein therapeutics, proteins can be encapsulated within the aqueous phase of the lipid carriers [142]. Upon encapsulation within the lipid matrix, the release kinetics of the encapsulated molecules can be manipulated by changing the compositions of lipid components. Although lipid-based DDSs are considered efficient drug delivery carriers for various modalities, the loading capacity of proteins tends to be low within lipid carriers, thus limiting their potential use for LAD [143]. Recently, Vollrath et al. demonstrated a solid lipid extrusion strategy to encapsulate a model mAb (~150 kDa) and Fab (~50 kDa) in a lipid implant [146, 147]. A stabilized protein lyophilizate was co-mixed with the lipid components using a hot-melt extruder. The proposed drug release mechanism was driven by drug diffusion via the micro channels and the interconnected pore network within the lipid implant [146]. Both model proteins were reported to achieve a sustained release profile for 120 days *in vitro*. In terms of protein stability, the monomer content showed minimal decreases, about 7% and 5% for the model mAb and ranibizumab respectively, over a period of 126 days [146]. In another study, Wang and Friess developed a particle coating strategy to tune the release rate of protein microparticles [148]. A model mAb was successfully loaded on mannitol beads in a fluid bed system, followed by coating with a lipid layer on top of the protein-loaded mannitol carrier microparticles as a controlled-release membrane layer [148]. Increasing the apparent amount of lipid coating resulted in a slower release rate, thereby extending protein release from 14 days to 42 days [148]. Overall, despite challenges in low drug loading, lipid-based DDSs hold great promise for sustained release of proteins.

More recently, structurally complex polar lipids have gained increased attention as drug delivery vehicles due to their ability to form different liquid crystalline phases [149, 150]. As amphiphilic molecules, polar lipids can self-assemble in water to form different structures, generating long-range orientational order in the presence of a solvent (lyotropic liquid crystalline phases) [150, 151]. The main classes of lyotropic liquid crystalline phases include lamellar, hexagonal, and cubic phases [150]. Specifically, the high viscosity of the cubic phase upon contact with aqueous media has huge potential as a slow release matrix [151, 152]. Rizwan et al. demonstrated that *in vitro* release of a fluorescently labeled ovalbumin (~45 kDa) from the lipid carrier could be extended to reach approximately 30% after 3 weeks [153]. The proposed drug release mechanism is mainly driven by drug diffusion through a tortuous network of channels to the external media, or through the destruction events of micellar structure [153]. Moreover, the FDA has recently approved a lipid liquid crystalline platform for extended release of buprenorphine (Brixadi) [154]. Developed by Camurus, it is a weekly or monthly SC injection for patients with moderate to severe opioid use disorder based on their FluidCrystal injectable depot technology [155]. Since its approval, a lot of attention has turned toward developing lipid liquid crystalline injectable formulations for various drugs, mainly in the therapeutic peptides space. Given the low toxicity of the lipid components, along with multiple ongoing clinical trials based on the lipid liquid crystalline platform, more clinical approvals are surely imminent.

Delivering Genes to the Immune System: Enabling Cell-Specific Targeting

In the past decade, LNPs have established themselves as suitable carriers for gene delivery to the immune system. Specifically, LNP encapsulation has proven useful for delivering mRNA payloads while evading enzymatic degradation. These NPs typically include cholesterol, an ionizable lipid, a PEG lipid, and a structural lipid. They are assembled in a single step, which mixes the mRNA in aqueous buffer, and the lipids in a mild organic solvent such as ethanol or methanol. LNPs on the market such as the mRNA vaccines against coronavirus disease 2019 (COVID-19) rely on passive targeting through surface charge modulation using IM injection [156, 157]. They exit muscle tissue through draining lymph nodes, eventually leading to the generation of antiviral antibodies via B cells [158, 159]. Successful delivery to the spleen or lung has also been demonstrated using passive targeting by formulating LNPs with excess negative charge or excess positive charge, respectively [160]. Nonetheless, there are challenges with this strategy that still need to be addressed. For example, it has also been shown that NPs formulated to target the spleen are mostly taken up by macrophages, which predominantly release cytokines upon uptake, thereby impacting their therapeutic index (TI) [156, 161]. To fully realize the potential of gene delivery, newer strategies have been developed that are moving beyond charge modulation as a means to influence biodistribution (Figure 20.10). For example, selective targeting has also been demonstrated by either modulating lipid ratio [162] or lipid design [163]. However, these approaches are labor intensive, requiring in-depth new chemical matter discovery and/or formulation development for each new target or tissue. Thus, another approach that is gaining momentum is cell-specific gene delivery using targeted LNPs.

Targeted LNPs are similar to traditional LNPs but include a ligand on their surface designed to target a specific receptor. Full-length mAb, cleaved antibody fragment ($F[ab']_2$) [164], Fab [165], and scFv [166] have all been successfully used as targeting ligands (Figure 20.10b). Typically, production of targeted LNPs is a two-step process, whereby the NABM cargo is encapsulated in the first step, followed by conjugation of the ligand in the second step. To enable surface modification, a lipid with a conjugation handle, such as PEG-maleimide or PEG-azide, is included in the first step [164, 165, 167–171]. Another approach with demonstrated success leverages formation of the LNP first, followed by addition of micelles with a conjugation ligand, for example, an scFv targeting the C terminus of a mAb [167, 170, 172–174]. In this paradigm, the two types of particles fuse to grant functionality to the LNPs and in a third step the targeting ligand is chemically linked, either through formation of a covalent bond via conjugation to interchain cysteines [168, 175], random conjugation to lysines [167, 170, 171], or click chemistry [176].

The roles of immune cells in vaccination and immunotherapies make them a sought after target for gene delivery. Recently, targeted LNP delivery to immune cells has gained significant momentum. This can be attributed of mRNA LNP vaccines for the COVID-19 pandemic, the development of chimeric antigen receptor T-cell (CAR-T) therapies for cancer, and the possibility of personalized

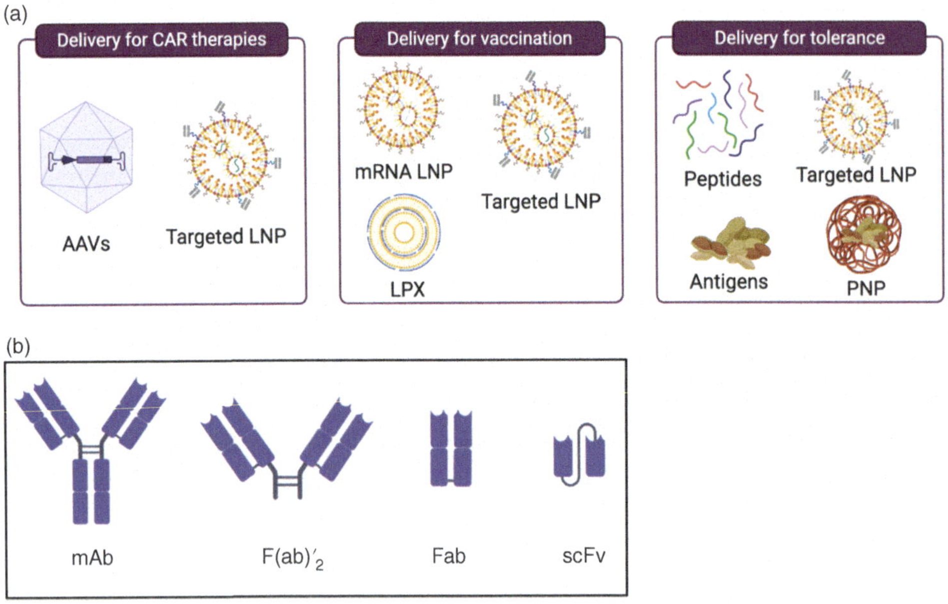

Figure 20.10 (a) Targeted LNPs can be leveraged for immune-cell-specific delivery within three paradigms and (b) some of the more commonly leveraged antibody-based targeting ligand structures. Source: Adapted from BioRender.

cancer vaccines. For CAR-T therapies, utilizing targeted delivery has the potential to greatly reduce toxicity and manufacturing costs, both of which are current hurdles to effectively treat autoimmune disorders. Moreover, better targeting of dendritic cells (DCs) opens up the possibility for next-generation cancer immunotherapy treatments, as well as vaccinations against infectious diseases.

Although T-cell-specific *in vivo* transfection can be challenging, the many receptors on their surface offer the possibility for entry using receptor-mediated targeting. Pan T cells have been targeted using CD2 [177], CD3 [165, 174], CD5 [165, 167, 170], CD7 [165], and CD28 [178]. CD4 has been used as a targeting ligand for T helper cells [169, 171, 174, 178], and CD8 has also been targeted [179]. PD1 can be used to target non-naive T cells and Tregs [164]. The choice of targeting ligand, as well as the type of antibody employed, must be carefully considered within the context of the intended mechanism of action. For example, since CD5 is a negative regulator of T cell activation [180], care must be taken to choose an antibody that does not activate CD5, unless tolerance is the desired outcome. Similar considerations must be applied for anti-PD1, which can exacerbate pre-existing autoimmune diseases [181]. In this section, an overview of activities using next-generation strategy for selective delivery of genes to T cells is described. Because this field is still in its infancy, this section will review research mostly in preclinical stages. Nonetheless, these opportunities in cell-specific delivery are paving the way for improved CAR therapies, personalized cancer vaccines, and immunotherapy.

Delivery of CAR-T Therapies

The premise of CAR-T therapies is based on the idea that T cells can be programmed to enable them to recognize and kill specific cancer cells [182]. Currently approved CAR-T therapies rely on a workflow whereby T cells are removed from the patient, genetically engineered *ex vivo* to express antigen-binding and T cell-activating proteins (typically using a retroviral vector delivering plasmid DNA), expanded in culture for two weeks, then readministered to the patient (Figure 20.11). Although this method has led to high response rates [183], the strategy is time-consuming, expensive, and often requires multi-day inpatient care or treatment with immunosuppressive therapies. The long manufacturing time for these therapies (2–3 weeks) is necessitated by the fact that each therapy must be custom-made for the patient, to enable requirements such as major histocompatibility complex (MHC) matching. While there are strategies being evaluated to generate "universal" CAR-T cells that can be made for many patients that share the same targeting antigen, such as by knockout of factors like T cell receptors (TCRs) and MHCs, gene editing is often imperfect or incomplete, leading to additional side effects [184]. Currently, many patients are unable to take advantage of CAR-T treatment because of the long lead time for these therapies, which can impact patient eligibility due to complications with their illness. Therefore, developing alternative strategies to streamline delivery of the necessary genes to T cells would greatly increase the likelihood of success.

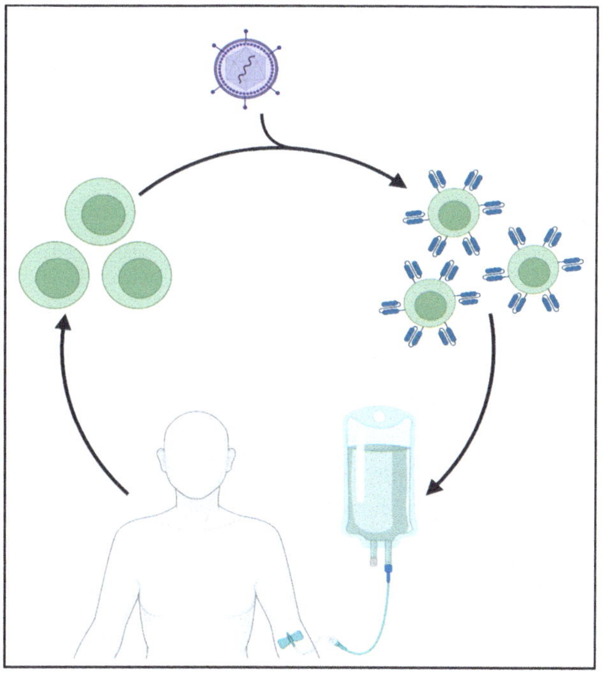

Figure 20.11 General workflow for approved CAR-T therapies. Blood is taken from the patient and T cells are isolated. Following transfection with DNA that encodes for chimeric antigen receptors that bind to cancer cell-surface markers such as CD19 or B-cell maturation antigen, the T cells are then expanded *in vitro*. The T cells that have now been engineered to recognize and kill cancer cells are then administered back to the patient. Source: BioRender/with permission of ELSEVIER.

One approach that may improve the CAR-T therapy workflow is *in situ* transfection of T cells. In this paradigm, T cells could be targeted through IV administration inside LNPs or AAVs, enabling *in vivo* transfection of cells in their native environment. This has the potential to eliminate many steps in the current workflow, including the time and cost to isolate T cells (leukapheresis), engineering and expansion, and readministration to the patient. Moreover, although retroviral transfection of T cells with a DNA plasmid is efficient, delivery of mRNA within NPs offers the advantage of not editing the genome and is short-lived (up to 7 days). While this shorter duration might require multiple dosing, a targeted approach could increase likelihood of delivery to the intended cells, while also minimizing safety issues due to unintended expression of the target antigen in healthy tissue. Several examples of this approach that have demonstrated success in *in vivo* models are listed in Table 20.4. These preclinical studies highlight the potential of *in situ* targeted CAR-T engineering to address the manufacturing cost, and safety liabilities associated with currently available therapies, thereby increasing the potential for CAR-T therapies to be affordable and widely available.

Vaccine Delivery

Targeting DCs for gene delivery also has the potential to significantly enhance current therapeutic treatments for both infectious disease and cancer vaccines. In approved COVID-19 vaccines, mRNA LNPs are used to elicit an antibody response against the surface antigen of the SARS-CoV2 virus, which typically wanes after several months [158, 159]. However, turning naive T cells into antigen-specific T cells by DC priming can lead to a much longer lasting effect, since memory T cells can remain in circulation for up to 317 days [186]. Targeting DCs can also play an important role in cancer immunology. In the cancer immunity cycle, dying cells release antigens, which are taken up by DCs, processed in the cytoplasm and loaded onto MHC class I molecules, which migrate to the surface of the DC [187]. They prime T cells to look for these antigens on the surface of other cells and kill them.

Table 20.4 Preclinical studies with successful *in situ* delivery of CAR-T therapies.

Disease model	Delivery vehicle	Targeting moiety	Conjugation strategy	Results	Reference
Human leukemia, prostate cancer and hepatitis B-induced hepatocellular carcinoma mouse models	PNP	Anti-CD8, anti-CD3 (both effectorless mAbs)	Electrostatic adsorption	Increased splenic targeting and efficacy in liquid and tumor models	[185]
Cardiac disease mouse model	LNP	Anti-CD5 mAb	Covalent conjugation using lipid mal-PEG	Observed CAR-T formation, destruction of fibrosis, restoration of heart health	[167]
GFP mouse model	LNP	Anti-CD3, anti-CD7 (both Fabs)	Covalent conjugation using lipid mal-PEG	Higher specificity of T cell transfection relative to clinical standard, lower toxicity	[165]

Fab, antibody fragment; GFP, green fluorescent protein; LNP, lipid nanoparticle; mal-PEG, maleimide-polyethylene glycol; mAb, monoclonal antibody; PNP, polymeric nanoparticle.

Since in many cancers, immune checkpoint molecules such as PD1 and CTLA4 impede this process, checkpoint inhibitor therapies have been designed to block immune checkpoints and allow the endogenous immunity cycle to continue. Within this paradigm, vaccination against shared or personal cancer antigens can further enable the body's own ability to fight cancer, by targeting DCs to enable T cell priming. In one such example, mRNA lipoplexes were designed to target CD11c+ DCs in the spleen by modulating the negative net charge of the NPs [160]. However, one drawback to this strategy is that it requires exquisite selectivity, as not all immune cells are appropriate targets. Macrophages, for example, do not express mRNA as well as DCs and in addition release cytokines in response to stimuli, which cause side effects, thereby limiting the feasible dose. Therefore, there is a need for additional strategies to enable DC-specific targeting to enhance the effectiveness of these vaccine therapies.

While DCs have fewer options for surface receptor targeting than T cells, some do exist such as conventional DC1 (cDC1) and DC2 (cDC2) cells [188]. While cDC1 cells are responsible for cross-priming of CD8+ T cells, cDC2 cells are responsible for priming CD4+ T cells. Moreover, cDC1s display XCR1 in both mice and humans, as well as CD8αα, DEC-205, and CD103, as do conventional DC2s. In contrast, plasmacytoid DCs (pDCs), primarily produce type 1 interferons and do not present antigens as well as cDC1 and cDC2. DCs also display the mannose receptor CD206 [189], although macrophages also display this receptor, and thus CD206 may not be an appropriate target. There are several instances of enhanced delivery to DCs described in the literature. These studies are summarized in Table 20.5.

Future Direction: Delivery for Immune Tolerance

Autoimmune disorders have the potential to greatly benefit from T cell reprogramming. In this approach, autoantigen-reactive T effector cells – the immune cells responsible for attacking healthy tissue – are converted to regulatory T cells, thus limiting the potential for further damage [193]. This can be accomplished by delivering mRNA that encodes for antigen and anti-inflammatory cytokines. Though many disorders could be targeted with this approach, considerably fewer publications exist relative to the field of cancer vaccines. This discrepancy may be due to the lack of an immune-neutral carrier, as commonly used lipids are known to stimulate the immune system [194]. Experimental research described so far has relied on optimization of NP composition. In one study, Ly6c-targeted LNPs were used to deliver IL10 mRNA [172], while a couple others demonstrated encouraging proof-of-concept in preclinical animal models targeting the spleen [195, 196]. However, these approaches have yet to translate to the clinic. In yet an alternate approach, successful targeting has been demonstrated by modulating NP physicochemical properties to mimic apoptotic cells (negatively charged), instead of mimicking pathogens (positively charged) as has been traditionally done [197]. Such is the case for carboxylated PLGA NPs, which have been shown to prevent onset and progression of type 1 diabetes in mice [198]. A similar strategy was also successfully utilized in a phase 2a clinical trial, where a negatively charged PLGA NP is being used to prevent gluten-induced immune activation in patients with celiac disease [199]. In both of these cases, to induce an antigen-specific T cell response, the antigen proteins themselves were encapsulated within polymeric nanoparticles (PNP).

Tissue-Specific Delivery: Targeting the Brain Parenchyma

Direct Injection

Cerebroventricular Delivery

Delivering drugs to the brain has been an ongoing pursuit for over 60 years, where early efforts relied on more straightforward and invasive methods. As early as the

Table 20.5 Strategies for DC-targeted LNPs.

Disease model	Delivery vehicle	DC-targeting strategy	Targeting ligands	Results	Reference
E.G7-OVA tumor mouse model	LNP	Modulation of LNP size, lipid types and composition	Electrostatic adsorption	Induction of clear therapeutic antitumor effects compared to two clinically relevant LNPs	[190]
B6 mice	LNP	Conjugation to LNP	DEC205 scFv	4X increase in LNP uptake by DCs and specific costimulatory gene KD	[166]
TC-1 mouse model	LPR	Conjugation to lipid	TriMan	T-cell activation and antitumor activity using higher dose than LPX	[191]
BALB/c mice	MLNP	Conjugation to lipid	Man-cholesterol	Faster antibody responses	[192]

KD, knockdown; MLNP, mannosylated LNP; LNP, lipid nanoparticle; LPR, hybrid lipid-shell polymer core mRNA nanoparticle; LPX, lipoplex.

1960s, researchers developed a subcutaneously implanted reservoir under the skin of the skull to feed a catheter into the cerebrospinal fluid (CSF) of one lateral brain ventricle [200]. This type of direct delivery into brain ventricles was aimed to treat diseases originating at the surface of the brain such as cryptococcal meningitis (the first use case), as these affected areas are directly accessible by the CSF flow path [200]. Further work of the time validated the notion that direct intracerebroventricular (ICV) injection resulted in overall greater exposure of the brain surface to a chemotherapeutic agent as compared with IV or lumbar intrathecal (IT) injection [201]. As noted, drugs delivered directly to the CSF are exposed only to the ependymal surface of the brain along the CSF flow path and do not reach deeper regions of the brain parenchyma [202–204]. In addition, CSF-injected drugs rapidly equilibrate into the surrounding blood at such a rate that full turnover of the CSF volume into the blood occurs ~5 times per day [203, 205, 206], making the functional distinction between CSF (ICV or IT) delivery and IV delivery more difficult to ascertain. As a result of these factors, there have been multiple instances in which ICV injection of a drug targeting a disease of the deeper brain tissue failed to demonstrate any clinical effect [207, 208]. In fact, only one FDA-approved treatment (cerliponase α, or TPP1) for a parenchymal brain disease (ceroid cipofuscinosis 2, or CLN2) by ICV injection exists [209]. In this case, it has been argued that the observed efficacy was not due to the drug acting at the level of the brain parenchyma, but due to the redistribution of the drug to the blood, where it was rapidly taken up into peripheral organs through known receptor-mediated interactions and exerted a therapeutic effect [203, 210, 211]. As a result of these experiences in leveraging ICV injection for the treatment of deep brain tissue disease, it has been cautioned that studies exploring such a delivery strategy should include an IV control arm to accurately account for any impact that CSF-to-blood transfer of the therapeutic agent may have on resulting clinical observations and efficacy [203].

Hyperosmolar Agents

Hyperosmolar agents exhibit the ability to act as tight junction modulators and have been used as blood–brain barrier (BBB) modulators in the past [212–214]. The mechanism of this modulation is thought to be dehydration of cerebrovascular endothelial cells, which in turn leads to shrinkage of these cells and opening of tight junction (TJ) gaps, enabling the free paracellular flow of solutes and other molecules [215, 216]. Subsequent injection of a drug of interest in this timeframe can allow the drug access to the brain in amounts that are up to orders of magnitude greater than when injected alone. Early efforts to identify substances that could modulate BBB opening in this way used concentrated solutions of electrolytes as well as non-charged substances such as propionamide, methanol, ethyl carbamate, propylene glycol, arabinose, and others [213, 217]. Further work identified mannitol as an effective choice for BBB permeability enhancement and it has since become the most thoroughly investigated agent [214]. While early investigations into this strategy were performed in rats [217] and rabbits [218], subsequent work has validated the approach for use in other species including dogs [219] and non-human primates (NHP) [220], in addition to humans [203]. In humans, the method typically involves the injection of a moderate volume (200 ml or more) of ~20% aqueous solution in less than a minute [218]. The transient window of increased BBB permeability has been observed to last approximately 10–40 minutes in humans and NHP [203, 220, 221], although some studies have shown that the increased permeability can last up to 2 hours before returning to baseline [222]. Studies using mannitol have been used to deliver various compounds, including small chemotherapeutic agents such as cisplatin, 5-fluorouracil, methotrexate, vincristine, and others [223–225], as well as large-molecule therapeutics such as mAbs [226, 227].

Next-Generation Approaches with Some Clinical Data

Focused Ultrasound

Focused ultrasound (FUS) combined with microbubbles is a promising method for opening the BBB in a targeted and reversible manner. Various studies have explored the use of FUS and microbubbles to enhance drug delivery to the brain [228]. FUS has shown promise in enhancing the delivery of mAbs to the brain for the treatment of neurodegenerative diseases like Alzheimer's disease (AD) and glioblastoma (GBM). Studies have demonstrated that FUS-induced BBB opening can significantly increase the levels of therapeutic antibodies in the brain, such as anti-amyloid and anti-tau mAbs [229, 230]. Furthermore, FUS combined with intranasal (IN) delivery has shown a substantial increase in the delivery of immune checkpoint inhibitors to gliomas, demonstrating the potential of FUS for targeted mAb delivery to brain tumors [231]. Overall, FUS holds significant promise as a noninvasive method to enhance the delivery of mAbs to the brain for therapeutic purposes.

In addition to the delivery of mAbs, FUS has also been successfully used to open the BBB for targeted AAV vector delivery in NHPs and Parkinson's disease (PD) patients, facilitating potential gene therapy interventions for neurodegenerative disorders [232]. Additional studies have demonstrated successful BBB opening in NHPs and patients with PD using low-intensity FUS, allowing for targeted drug delivery into specific brain regions [233]. Real-time cavitation mapping during

FUS procedures has been crucial for monitoring safety and efficacy, with enhanced resolution techniques like coherence-factor-based passive acoustic mapping (PAM), facilitating accurate targeting and monitoring of treatment effects [234]. These advancements in FUS technology hold significant potential for noninvasive and precise delivery of therapeutics to the brain by temporarily opening the BBB.

Cell-Penetrating Peptides

Another peptide-based approach to overcoming cellular barriers for drug delivery is the utilization of the so-called cell-penetrating peptides (CPPs), also known as protein transduction domains (PTD) (Figure 20.12). CPPs are functional carrier peptides capable of internalizing cargo molecules across the cell membrane [235]. Generally, these are relatively short (less than 30 amino acid residues) and can be cationic or amphipathic [236, 237]. The key characteristic of CPPs is that they are able to permeate through the cell membrane in the absence of any target receptor and without significant damage to the cell membrane. Additionally, these peptides can transport covalently bound or non-covalently bound (e.g. electrostatically associated) cargo molecules with high efficiency [238, 239]. A variety of cargos have been used with CPPs including ASO, siRNA, peptides, peptide–nucleic acid conjugates, proteins, and liposomes. The ability of CPPs to mediate delivery of conjugated compounds such as these into the cytosol of cells *in vitro* and *in vivo* has long been known [240–242]. Although not fully understood, CPPs gain entry into the cell by co-opting the natural processes of both caveolin- and clathrin-mediated endocytosis [243–245], followed by endosomal escape [246] (for review, see [240]). CPP sequences have been derived from multiple sources including phage receptors, neurotoxins, neurotrophic peptides, and, perhaps most notably, the TAT protein from HIV-1 [240, 247, 248]. The ability of CPPs to cross cellular membrane barriers has allowed them to be used as a means to gain entry past the BBB [240, 249, 250]. The earliest demonstration of this used the PTD of the TAT protein to mediate

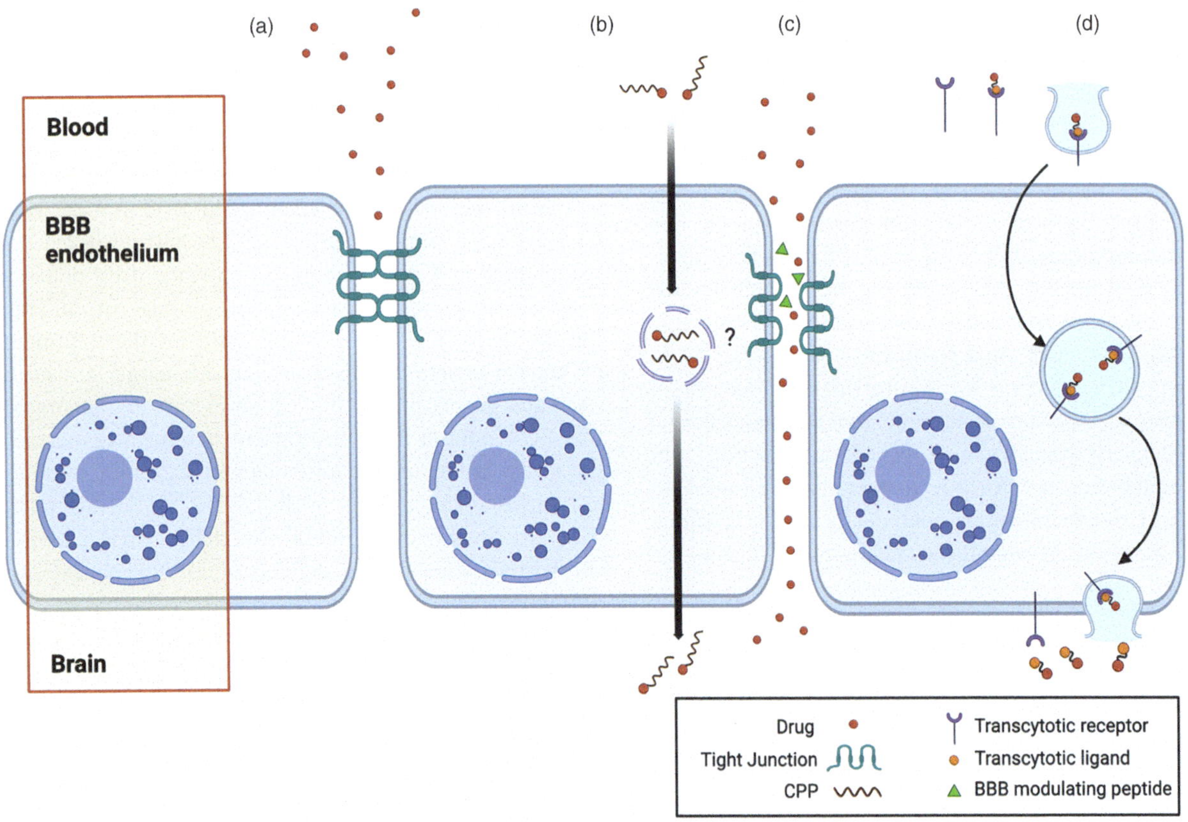

Figure 20.12 Schematic representation of select strategies used for delivery across the BBB. Endothelial cells comprising the BBB are connected to one another via tight junctions. This forms a physical barrier that prevents the free diffusion of most molecules into the CNS (a). Platform methodologies for overcoming this barrier include fusion of drug to cell-penetrating peptides (CPPs) to directly cross cellular membranes (b), transient disruption and opening of the BBB tight junctions to allow free diffusion of large drug molecules across the barrier (c), and co-opting the natural process of transcytosis via fusion of the drug to transcytotic receptor-specific ligands (d). Source: Adapted from BioRender.

transfer of a conjugated reporter protein (β-galactosidase) across the BBB [251]. Some conflicting evidence exists, however, as *in vitro* studies have failed to show transport of TAT peptide across intact endothelial monolayers [252]. While effective CPPs tend to be arginine rich, investigations in the late 2000s and beyond demonstrated that cyclization of arginine-rich peptides generally improves their cell-penetrating delivery properties, even when as few as three arginine residues are present [253–255]. Subsequent work on these cyclic peptides established motifs rich in arginine and hydrophobic residues (e.g. FXRRRR, where X is the L-2-naphthylalanine) which can be embedded into small- or medium-sized cyclic peptides (~7–13 amino acids) [255]. These CPPs were shown to bind to the plasma membrane of mammalian cultured cells and internalized via the endosomal pathway with subsequent translocation to the cytoplasm and nucleus at rates significantly higher (~2–5 fold) than simple Arg_9 CPP sequences [255]. Work over the following decade further refined this, and similar cyclic CPP technologies, as methodologies for the delivery of therapeutic molecules. In one notable case, such cyclic CPPs were used to deliver an ASO for the treatment of Duchenne muscular dystrophy in a murine model of the disease [256]. Subsequent work has further extended this and other approaches of using CPPs fused to cargos of interest and/or incorporated into NPs to deliver small molecule drugs, proteins, and nucleic acids to the brain in animal models [257]. A few of these include human metallothionein 1A and dopamine fusion proteins for PD, as well as Curcumin and BACE1 siRNA as LNP complexes for AD in murine models (see [258] for review).

Peptidomimetic BBB Modulators

An alternative method for gaining access to the brain from systemic circulation, that does not require crossing of cell membranes, is the use of peptide agents. These can selectively disrupt the capacity of proteins within the TJ to maintain their stable *cis* or *trans* interactions [259], thus compromising the integrity of the paracellular barrier and allowing the free movement of molecules into the CNS [223] (Figure 20.12). One well characterized family of such peptides target the interactions of E-cadherin at the junctional interface of the cerebrovascular endothelial cells that comprise the BBB. These are short (6- to 12-mer) peptides derived from the EC1 domains of E-cadherin and associated mutants, where HAV6 and ADTC5 are the two most well-studied members of this group. HAV6 is a linear peptide that binds to the EC1 domain of E-cadherin and blocks the EC1-EC2 *cis*-interaction of E-cadherin molecules on the same membrane face. In contrast, ADTC5 is a cyclic peptide that binds to the domain-swapping region of EC1 and inhibits EC1–EC1 *trans*-interactions of E-cadherin on opposing membrane faces in the TJ. This disruption of the BBB integrity is transient, with the duration of effect dependent on the PK properties of the modulating peptide in serum [259–261]. As a result, investigators have employed design strategies to enhance resistance to endogenous serum peptidases, such as making the modulating peptides cyclic (as in the case of ADTC5) [259]. Both of these peptide designs can enhance the delivery of small molecules such as mannitol, gadopentetic acid, and dextran in addition to larger molecules such as PEG and proteins [260, 262, 263]. However, owing to the improved serum persistence of ADTC5 relative to HAV6, as well as the fact that the opening of the cellular junction is more pronounced when *trans*-cadherin interactions are disrupted, ADTC5 has a significantly greater effect on allowing access of IV-administered large molecules such as lysozyme, albumin, mAb, and fibronectin into the brain [259, 264]. Because cadherin proteins are distributed throughout cells in higher mammals [265], special consideration must be given to the effects of these peptidomimetic modulators on the whole body and organ distribution of the therapeutic molecules delivered with this strategy to avoid off-target effects [259]. While E-cadherin has been the focus of much of the work in this field, similar peptide modulation strategies have been validated for other adhesive proteins such as claudin-5 [266] and occludin [267]. Ongoing and future research will aim to identify adhesive protein targets and peptide sequences that offer specific barrier disruption at the BBB as compared to other tissues ands organs.

Receptor-Mediated Transport Shuttles

As an alternative to passively increasing the overall permeability of the BBB, some strategies have sought to target active transport mechanisms to gain access to the brain. The brain capillary endothelial cells that form the BBB contain specific receptor-mediated transport mechanisms that can be leveraged to transport therapeutic molecules into the brain (Figure 20.12). This strategy utilizes receptors involved in transcytosis within the BBB to selectively transport their target ligands from systemic circulation to the CNS tissue [203, 268]. By chemically or recombinantly fusing a therapeutic molecule to a transcytotic ligand or peptide fragment thereof, an increased distribution of the therapeutic in the brain can be achieved as compared to administration of the therapeutic alone [203]. Much like targeting of the BBB TJ proteins by peptidomimetics, the target receptor should be highly expressed in brain microvasculature, preferably in a fairly specific fashion relative to other tissues. To this end, proteomic studies have sought to identify strong candidates for such a delivery strategy [269, 270]. Candidate receptors for other ligands such as insulin-like growth factor 1 and

2 (IGF-1/IGF-2) [271, 272] and low-density lipoprotein (LDL) [273–275] as well as the LDL receptor-related protein 1 (LRP1) [269] have been investigated but were found not to mediate biologically meaningful transport of ligand-fused cargos. This was due to either the receptor not having biologically relevant expression at the level of the brain microvasculature or saturating competition from endogenous ligand or ligand binding proteins in serum [203]. Some candidates that have demonstrated success for receptor-mediated transport to the brain include insulin receptor [276, 277], CD98 [278, 279] as well as transferrin receptor (TfR) (also known as CD71) [203]. With regards to CD98, preclinical studies have demonstrated that targeting CD98 heavy chain (CD98hc) is a promising strategy [280] with favorable PK, biodistribution, and safety profiles in both mice and NHPs [278]. Nonetheless, unlike TfR-targeting, this strategy has yet to be validated in the clinic.

While TfR has two widely distributed isoforms (TfR1 and TfR2) [281], only TfR1 has been shown to be highly expressed in the BBB [282]. TfR is an attractive candidate as a brain shuttle target due to being a highly studied model receptor for intracellular trafficking research [283] and having a known 3D crystal structure [284]. Studies have shown that TfR-mediated transcytosis of its Tf ligand in the BBB is bidirectional [285], and in a fashion saturable by the high levels of Tf naturally found in serum [286]. Further work has since demonstrated that, in addition to its natural Tf ligand, other proteins that bind to TfR on the luminal side of the brain vasculature can be transported across the BBB and delivered into the brain [287–289]. In one example, transgenic mice were treated with either a mAb designed to target AD-related protein amyloid-β (Aβ) or anti-Aβ mAb with a scFv against TfR fused to each light chain of the mAb by IV injection. The results demonstrated that delivery efficiency of the bivalent anti-TfR therapeutic was 2–3% of the total injected dose per gram of brain tissue, or 80-fold greater than the anti-Aβ mAb without TfR-targeting capabilities [288]. After years of research, therapeutic molecules leveraging this approach eventually entered clinical trials when a study was conducted to test TfR-mediated BBB transfer of a lysosomal enzyme to treat a rare lysosomal storage disease (LSD) [290]. One key challenge of this approach is how to engineer a "release mechanism" for the TfR-bound cargo once it has reached abluminal surface of the brain vasculature. A shuttle carrier with high affinity will be more robustly taken up from the circulation by luminal TfR; however, a shuttle with more modest affinity will have a greater chance of being released into the CNS upon arriving at the abluminal vascular surface [291]. Moreover, as with any strategy that requires direct binding to a biologically relevant receptor, special consideration must be given to the potential impact on normal systemic function of the target in the presence of the targeting molecules. In one case, dosing of a high-affinity anti-TfR mAb in mice resulted in a decrease in reticulocyte count along with associated clinical signs (lethargy, lack of locomotion, and responsiveness, etc.), whereas dosing of a bispecific anti-TfR/BACE1 mAb in which only one Fab arm targets TfR did not show these clinical signs, although a more modest reduction in reticulocytes was still observed at higher doses (>50 mg/kg) [292]. It was further established that reticulocyte reduction could be ameliorated either by reducing anti-TfR affinity (presumably at the cost of transport efficiency) or by eliminating effector function of the Fc region of the IgG [292]. Overall, important lessons have been demonstrated regarding the biological considerations when selecting a target receptor to meditate BBB transport, as well as factors such as affinity for the transporter, potential peripheral effects, and mitigation strategies in the design and engineering of the therapeutic molecule itself.

Intrathecal Administration of ASOs

ASOs are promising drugs for the treatment of CNS disorders due to their robust molecular targeting properties and enhanced pharmacological profiles [293]. ASOs are complementary nucleic acid fragments of 12–25 nucleotides designed to hybridize endogenous mRNAs in a targeted fashion to modulate biological function. ASOs elicit their pharmacological effect by multiple means including RNase-mediated degradation of target mRNAs, targeted modification of mRNA splicing to include or exclude specific exons and introns, or altering the functions of regulatory RNAs [294–296]. Early work with ASOs as biotherapeutic agents demonstrated that they were capable of *in vitro* replication of Rous sarcoma virus, although this effect was not seen *in vivo* due to enzymatic degradation of the phosphodiester backbone and poor tissue distribution [295, 297]. Subsequent improvements in the synthesis and chemical composition of ASO molecules, such as modification of the phosphodiester backbone to be more resistant to degradation by systemic and cellular nucleases, as well as incorporation of modified or unnatural nucleotides, led to next-generation ASOs with greatly enhanced *in vivo* persistence and therapeutic effect [295, 298]. A key challenge in the delivery of ASOs for the treatment of neurological diseases is that getting these drugs to their site of action in the CNS requires direct delivery to the central compartment, as these molecules do not readily cross the BBB [203]. The IT dosing route offers a relatively straightforward solution for bypassing the BBB and delivering drugs directly to the CNS and has been a favored alternative to direct ICV injection for ASOs and other NABMs [299–301].

IT dosing is done by injection directly into the lumbar spinal column, usually with a catheter that allows for the subsequent infusion of drug over a period of time [302]. IT injection can be a single dose but is also used with strategies that incorporate implanted catheters with pumps or ports for prolonged delivery or repeated dosing over the course of a given therapy [302]. Following IT delivery of a therapeutic, a gradient is typically observed with drug concentration being highest at the catheter and lowest at the superior portion of the spinal column and brain [303, 304]. This gradient is established due to a combination of slow diffusion from the site of injection as well as the intrinsic brain-to-lumbar flow of CSF that exists in the spinal column [302]. To ensure that effective drug exposure at the site of action is achieved, IT dosing strategies must account for this gradient when setting variables such as infusion rate, infusion time, and drug concentration. Dosing in the IT space offers some clear advantages over other delivery strategies [305]. Unlike IV (systemic) delivery, IT delivery allows direct access to the CNS without having to contend with the BBB, which is particularly important for the case of large molecule (antibody) therapeutics, where very limited uptake (0.1–0.2%) of an injected dose may reach the brain from the systemic circulation [306]. Furthermore, such direct delivery to the site of action limits systemic exposure to the drug, which could be critical to avoiding undesirable side effects of the therapy. The first successful development and regulatory approval of an IT-delivered ASO-based drug for spinal muscular atrophy (Spinraza) was achieved in 2016 [293, 294]. While IT is the standard route for ASO delivery for CNS indications, the delivery route has also been paired with other therapeutic molecules. Along with ASOs, several other IT-dosed therapeutics, proteins, gene therapy vectors, and stem cells have also entered clinical development for CNS disorders [299, 307, 308]. The risk–benefit ratio of IT administration makes it a valuable therapeutic approach. Although IT delivery is not without risk, when used to administer therapeutic compounds with a strong safety profile by a skilled practitioner, the approach has a robust safety margin [302]. However, IT delivery is not without its own limitations. One key challenge for IT drug delivery is that the distribution kinetics and PKPD properties of dosed therapeutics remain poorly understood due to the unique anatomical and functional characteristics of the IT space [294, 309, 310]. As a result of various clinical failures over the years [23, 311, 312], it has been suggested that the CNS disease targeted by a given therapy must be restricted to the surface of the brain or spinal cord, as these are the sites that the CSF of the spinal column and ventricles has access to [202–204]. Attempts to treat brain diseases originating at other sites in the CNS are likely to be futile with this delivery methodology [203].

Intranasal Administration for Brain Delivery

Since the late 1990s, intranasal (IN) injection has been leveraged as a RoA to access the brain. However, until recently, this RoA had been nearly exclusively paired with small molecules, as is demonstrated by the numerous FDA-approved molecules [313]. Small molecules are the easiest modality to deliver intranasally because they are typically small, lipophilic and can tolerate formulation conditions that are amenable to a powder or fine mist product. However, the explosion of novel therapeutic modalities, along with the expansion of NP strategies over the past 10 years, has illuminated a path forward for therapeutic delivery beyond small molecules [314–319]. The idea of IN administration of biologics to the CNS has been contemplated as far back as 2012 in the seminal review by Lochhead and Thorne [320], where they provided probable mechanisms of uptake and distribution, along with a collection of the most promising studies demonstrating feasibility. However, the advent of novel therapeutic modalities such as NABMs combined with the ongoing commitment to identify mAbs for the treatment of CNS diseases has pushed this RoA to be more seriously considered for proteins, peptides, antibodies, and NABMs. For mAbs, the appeal is in the potential to bypass the BBB, which typically prevents >99% partitioning to the brain [321]. However, one major drawback is the dose volume (100–150 ul per nare) compared with other routes of administration, which limits many mAbs from reaching a therapeutic dose. With respect to administration of NABMs, the desirability lies in the potential to positively impact patient compliance, convenience, and burden to the healthcare system by offering an alternative to IT administration. Convenience is an especially important consideration within the context of neurodegenerative diseases, where patient populations may be especially adverse to repeat doctor's office or hospital visits. Thus, IN administration has the potential to have an immeasurable impact on patient well-being.

To date, there has yet to be FDA approval of an intranasally administered protein therapeutic candidate for treatment of CNS disorders. However, there are several ongoing trials to evaluate peptides and one mAb via this route which have been summarized in Table 20.6. Peptides have advantages over proteins because they are smaller, can be synthetically made, and often have more straightforward CMC than proteins [326, 327]. To date, the most extensive studies delivering a peptide to the CNS via IN administration have been with insulin, to mixed results.

Table 20.6 Clinical trials with peptides and proteins being administered IN for CNS delivery.

Name	Type	Disease/condition	Clinical trial no.	Sponsor	Phase	Reference
Insulin	Peptide	Mild cognitive impairment or AD	NCT01767909	University of Southern California	2/3	[322]
			NCT02503501	HealthPartners Institute	2	[323]
			NCT02462161	Wake Forest University Health Sciences	1	[324]
		Diseased Alzheimer patients	NCT01595646	Wake Forest University Health Sciences	2	https://clinicaltrials.gov/study/NCT01595646?term=NCT01595646&rank=1
			NCT01547169	University of Washington	2	https://clinicaltrials.gov/study/NCT01547169?term=NCT01547169&rank=1
		PD/MSA	NCT02064166	Brigham & Women's Hospital	2	[325]
Oxytocin	peptide	Hyperphagia in Prader-Willi syndrome	NCT03197662	Montefiore Medical Center	2	https://clinicaltrials.gov/study/NCT03197662?term=NCT03197662&rank=1
		Anorexia Nervosa	NCT03414112	University of Minnesota	1	https://clinicaltrials.gov/study/NCT03414112?term=NCT03414112&rank=1
		Premenstrual dysphoric disorder	NCT02508103	University of North Carolina, Chapel Hill	2/3	https://clinicaltrials.gov/study/NCT02508103?term=NCT02508103&rank=1&tab=results
		Schizophrenia	NCT02567032	UCSF	1	https://clinicaltrials.gov/study/NCT02567032?term=NCT02567032&rank=1
Arginine Vasopressin	peptide	Autism	NCT03204786	Stanford University	2/3	https://clinicaltrials.gov/study/NCT03204786?term=NCT03204786&rank=1
		Schizophrenia	NCT04190004	National Institute of Mental Health and Neuro Sciences, India	N/A	https://clinicaltrials.gov/study/NCT04190004?term=NCT04190004&rank=1
		Attention Control	NCT04493554	University of Electronic Science and Technology of China	N/A	https://clinicaltrials.gov/study/NCT04493554?term=NCT04493554&rank=1
Calcitonin	peptide	Acute treatment of migraine	NCT04408794	Pfizer	2/3	https://clinicaltrials.gov/study/NCT04408794?term=NCT04408794&rank=1
Progesterone	peptide	Brain injury, stroke	NCT04143880	Second Affiliated Hospital, School of Medicine, Zhejiang University	4	https://clinicaltrials.gov/study/NCT04143880?term=NCT04143880&rank=1
Foralumab	mAb	MS	NCT06292923	Tiziana Life Sciences LTD	2	https://classic.clinicaltrials.gov/ct2/show/NCT06292923

AD, Alzheimer's disease; mAb, monoclonal antibody; MSA, multiple systems atrophy; MS, multiple sclerosis; PD, Parkinson's disease.

While safety has been demonstrated, robust efficacy has still not been clearly established [322–324].

In addition to the multitude of peptide clinical trials for delivery to the CNS, there is also an ongoing phase 2 trial of an IN administered mAb (Foralumab) to treat multiple sclerosis (MS). A phase 1 clinical trial of Foralumab in healthy volunteers was completed in 2019 where all 3 doses (10, 50, and 250 µg/dose), nasally administered for 5 consecutive days, were well tolerated and demonstrated no drug-related safety issues. One potential reason for the success of this particular anti-CD3 antibody thus far, and which makes it an anomaly compared to most mAb therapeutics, may be related to its high potency which appears to be enabling therapeutically relevant dose volumes amenable to IN administration. Moreover, it is worth noting that in this case the clinical benefit is derived from systemic stimulation of Tregs and does not require delivery of the mAb to the CNS. Nonetheless, these studies are very encouraging for the field and provide a path forward for IN administration of mAbs therapeutics to treat the CNS.

Beyond the clinical activities highlighted above, there has been quite a bit of ongoing preclinical activity over the past decade. Table 20.7 is a selection of peptides, proteins, and NABMs of particular interest within the context of IN delivery to the brain (a more exhaustive list can be found in Patharapankal et al. [318]. It is interesting to note that while many of the studies claim direct nose-to-brain (N2B) delivery, as opposed to IN delivery with nasal absorption followed by passage into systemic circulation, most of these preclinical studies have not been designed such that they can rule out other mechanisms of action, such as by including a comparison with brain activity and/or distribution after IV administration. Therefore, it has been difficult thus far to prove that N2B is the sole mechanism of delivery. In most cases, formulations hae been developed to bypass mucociliary clearance and/or enhance permeation of the nasal epithelia, allowing for a combination of delivery mechanisms to reach the brain (brain-targeting versus nose-to-brain). Nonetheless, the studies compiled in Table 20.7 all demonstrate efficacy in the brain after IN administration of either a peptide, protein, or NABM and thus represent some of the most successful case studies of the moment.

An Analytical Perspective to Drug-Delivery Technology Development

General Considerations for Analytical Characterization

Understanding the MoA for any therapeutic is one of the most important aspects of drug development in order to ensure patient safety. Thus, it is critical to understand how the API and its DDS are driving the desired therapeutic effect, and any potential undesirable or off-target effects. Because the therapeutic, the DDS, and the combination of both together can each display their own form of heterogeneity, thorough characterization is paramount to ensure batch-to-batch manufacturing consistency and therefore, safety and efficacy for the patient. The many different potential modifications (and combination thereof) result in an overwhelming possibility of product quality heterogeneity, which is not feasible to exhaustively characterize. Therefore, deciding which product quality attributes will help determine what analytical techniques to employ (Figure 20.13). The most important attributes are those that have a significant impact on potency, PK, immunogenicity, and/or safety. These are designated as critical quality attributes (CQAs), and require the highest level of analytical characterization and product quality control. Analytical techniques used for CQA characterization typically require a high degree of assay suitability, which may include determination of assay accuracy, sensitivity, repeatability, and stability indicating properties.

A second category of product quality attributes and their associated analytical characterization are those that provide an "image" or "snapshot" of the overall heterogeneity of the system, such as size, polydispersity, charge, and (post translational modification) PTM heterogeneity. These techniques typically rely on the separation and visualization of the product variants using techniques such as chromatography and capillary electrophoresis and are heavily relied upon throughout development. Moreover, because these methods provide a measurement of drug purity, they are typically mandatory as release assays on control systems. Assays used for the latter purpose require validation and the highest level of assay suitability.

Finally, a third category can be defined as relating to such aspects as identification of the molecule, structural characteristics, site-specific attributes, noncritical attributes. These typically fall into the category denoted as "extended characterization," with these analytical techniques having less stringent requirements for assay suitability than those used for CQA assessments and release assays.

As outlined above and described in detail elsewhere within this book, there are many established methods commonly used for analytical characterization (Figure 20.13, top panel). These methods include separation-based techniques for batch heterogeneity characterization (i.e. charge variant, size variant, glycosylation variant analyses, etc.), structural characterization techniques (i.e. secondary structure, ternary structure, etc.), residue-specific characterization (i.e. oxidation, deamidation, glycation, etc.), and matrix/formulation characterization (i.e. pH, cryoprotectant, osmolality, surfactant, etc.). Many of these techniques

Table 20.7 Preclinical studies demonstrated efficacy or distribution to the brain after IN administration of a peptide, protein, or NABM.

Modality	Name	Delivery	Disease application	Animal model	Outcome	Reference, year
Neuropeptide	Cystatin C-peptide	N/A	AD	Mouse	Reduction of amyloid aggregates and improved memory	[328], 2013
	CGRP	1% chitosan	Migraine	Mouse	Reduction in CGRP-induced plasma protein	[329], 2020
	NPS	Self-assembling nanofibers	Anxiety disorder	Mouse	Produced prolonged anxiolytic-like effects	[330], 2021
	GDNF	Cationic liposome	PD, MS	Rat	Neuroprotective improvement	[331], 2014
Protein	bFGF	Gelatin NLCs	PD	Rat	Neuroprotective improvement	[332], 2014
	Activin A and Serpin B2	TDM	AD, HD, ALS, PD, brain damage, stroke	Mouse	Maintains the structural and functional integrity of neurons, slows down the progressive cognitive disease	[333], 2018
	Bevacizumab	PLGA NP	Glioblastoma	Mouse	Reduction in the tumor growth accompanied by a higher anti-angiogenic effect compared to the free bevacizumab	[334], 2019
	anti-TRAIL	PLGA NPs or NLC nanosystems	AD, PD, epilepsy	Mouse	Reduced neuroinflammation	[335], 2022
	anti-Nogo-A	N/A	EAE	Mouse	Widespread penetration of in the CNS, including the thoracolumbar spinal cord, resulted in the improvement of motor symptoms and in the preservation of myelin in the spinal cord	[336], 2023
	anti-CD3	N/A	AD	Mouse	Ameliorated disease in a murine model of AD by targeting microglial activation in the brain independent of amyloid beta deposition	[337], 2023
	TTCM2	Pluronic block copolymer micelles	AD	Mouse	Effectively cleared pathological tau, elevated synaptic proteins, and improved cognitive functions in aged tauopathy mice	[338], 2024

	α-synuclein	IND conjugate	PD	Mouse	mRNA levels in the SNc/VTA, DR, and LC were significantly lower relative to control	[339], 2018
ASO	α-synuclein	IND-conjugate	PD	Cyno	IND-ASO treatment led to decreased levels of endogenous a-synuclein protein in the midbrain monoamine nuclei	[340], 2020
	HuR	N/A	Neuropathic pain	Mouse	ASO showed myelin protection at analgesic doses with multitarget mechanisms	[341, 342], 2020, 2023
miRNA	GATA2, Rb1	ENCP	AD	Mouse	Statistically significant increase of a potentially therapeutic miRNA mimic in the hippocampus area and its further effect on two mRNA targets, following its intranasal administration	[343], 2020
siRNA	SERT	Sertraline conjugate	Major depression	Mouse	All doses significantly decreased mRNA expression 24 h after last administration, with no significant differences between doses	[344], 2016
	TNF-α	Cationic nano emulsion	Prevention of neuroinflammation	Rat	Site-specific downregulation of TNF-α cytokines	[345], 2016
	HTT	Chitosan NP	HD	Mouse	Effective in lowering protein expression but effect required more time to become evident	[346], 2020
	VEGF/PLK1	*In vivo* jet PEI hyaluronan nano micelles	Glioma	Mouse	Successful delivery of siRNA in different cell types of the prefrontal cortex as soon as 4 hours after intranasal admin, and the presence of siRNA was still detected after 24 hours of admin	[347], 2022
	BACE	DGL	AD	Mouse	Rapa@DAK/siRNA is verified to improve the cognition of transgenic AD mice after intranasal administration	[348], 2022
mRNA	GFP	Cationic LNP	N/A	Mouse	Luciferase-mRNA/cationic liposome group showed significantly higher luciferase activity in the cortex, striatum, and midbrain regions as compared with the control groups	[349], 2020

AD, Alzheimer's disease; ALS, amyotrophic lateral sclerosis; ASO, antisense oligonucleotide; BACE, β-site amyloid precursor protein cleaving enzyme; bFGF, basic fibroblast growth factor; CGRP, human calcitonin gene-related peptide; CNS, central nervous system; DGL, dendrigraft poly-l-lysines; DR, dorsal raphe nucleus; EAE, experimental autoimmune encephalomyelitis; ENCP, enveloped nanocomplex; GDNF, glial cell-line-derived neurotrophic factor; GFP, green fluorescent protein; HD, Huntington's disease; HTT, Hereditary hemorrhagic telangiectasia; HuR, human antigen R; IND, indatraline; LC, locus coeruleus; miRNA, microRNA; mRNA; messenger RNA; MS, multiple sclerosis; N/A, not applicable; NLC, nanostructured lipid carrier; NP, nanoparticle; PD, Parkinson's disease; PEI, polyethylenimine; PLK1, polo-like kinase 1; SERT, serotonin transporter; siRNA, silencing RNA; SNc, substantia nigra compacta; TDM, tetradecyl-β-D-maltoside; TNF, tumor necrosis factor; TRAIL, TNF-related apoptosis inducing ligand; TTCM2, toxic tau conformation-specific monoclonal antibody-2; VEGF, vascular endothelial growth factor; VTA, ventral tegmental area.

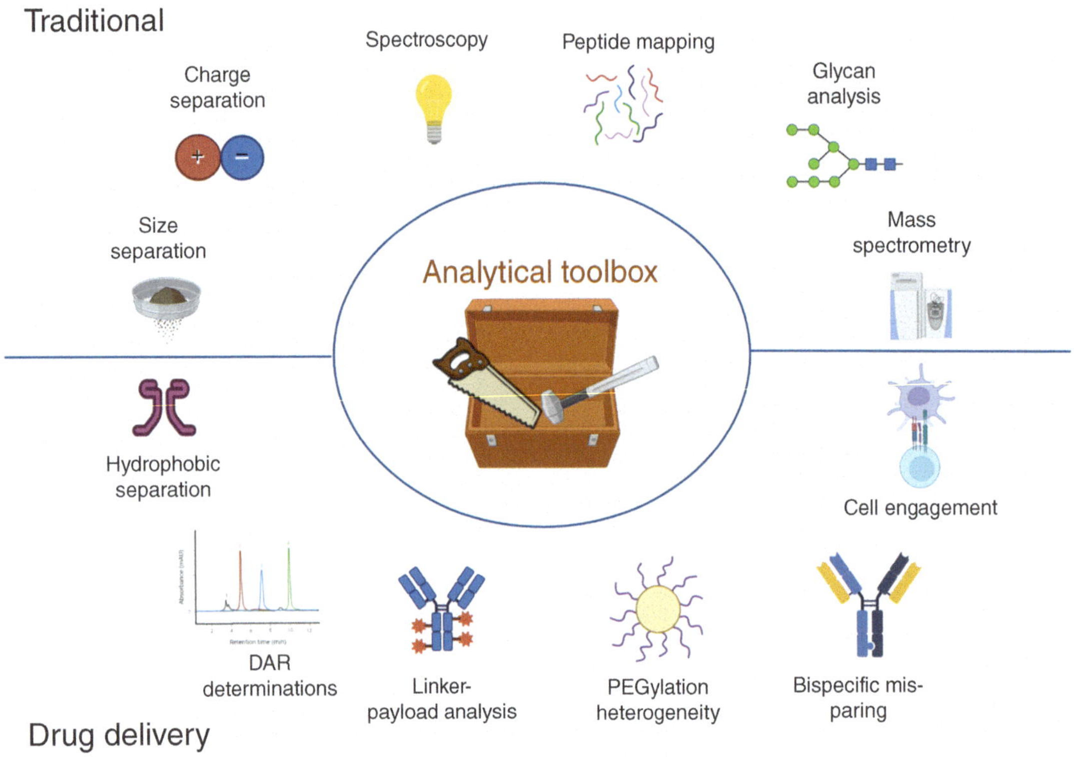

Figure 20.13 Analytical toolbox with traditional and drug-delivery-specific methods. Source: Adapted from BioRender.

may be perfectly suitable to support the characterization of novel DDSs; however, improvements, advancements, and modifications to these methods are typically required to achieve the desired level of characterization. For example, non-covalent/formulation-based and/or device-based strategies used as a DDS may not require significant modifications to existing methods for analytical characterization. The API for these strategies is typically unaltered and thus, characterization of the biotherapeutic may be similar to those administered without a tailored delivery strategy. However, covalent modifications made to the API to improve PK or target selectivity are more likely to impact the physicochemical properties of the biotherapeutic (i.e. hydrophobicity, R_H, molecular size, reactivity, etc.), thus necessitating novel analytical techniques.

With the advent of new and more exotic strategies for therapeutic drug delivery, comes the commensurate need for novel analytical techniques (Figure 20.13, bottom panel). In the following sections, we will examine how traditional techniques are used to support novel DDSs and describe advancements that have been implemented to achieve the desired level of analytical characterization. We will also provide an overview of analytical challenges associated with the evolution of DDSs that influence therapeutic targeting and control strategies for product release. The first section will focus on DDSs that entail covalent modifications to the API and describe considerations for the use of either traditional and/or advanced methods to support analytical characterization. The next section will focus on those that utilize non-covalent/device strategies in which there is typically no direct modification to the API.

Strategies for Systems Using Covalent Modifications

PEGylation

PEGylation is a widely used strategy whereby a PEG chain is covalently conjugated to a hydroxyl group (serine, threonine), thiol (cysteine), or primary amine (lysine); however, the selectivity and resulting heterogeneity can be difficult to control, and therefore it is not always straightforward to fully characterize [350]. Moreover, the extent of PEGylation can impact biological activity and thus, the PEG-to-therapeutic ratio may be of utmost importance to control. Lastly, the efficiency of PEGylation and characterization of unmodified therapeutic and unbound (free) PEG molecules require analytical considerations.

Traditional separation techniques such as high-performance liquid chromatography (HPLC) have shown utility for the analytical characterization of PEGylated molecules. For example, size-exclusion chromatography (SEC) and reverse phase-HPLC (RP-HPLC) have been shown to enable separation and quantification of free PEG versus PEGylated-protein, by taking advantage of differences in R_H and other physicochemical properties [351]. However, the use of HPLC becomes exponentially more powerful when coupled to secondary detection systems [350, 352]. Mass spectrometry (MaS) is by far one of the most powerful detection methods for characterizing PEGylated molecules. MaS has been used to characterize the average molecular weight and heterogeneity of PEGylated proteins at the intact level [353], the efficiency and site-specific PEGylation at the peptide level [354], and to characterize PEGylated proteins under native conditions [355]. SEC-HPLC has been coupled to alternative detection methods including charged aerosol detection (CAD) [356], evaporative light scattering (ELSD) [357], multiangle light scattering (MALS) [358], and quasi electric light scattering (QELS) [358], all of which can provide higher resolution of size heterogeneity for PEG molecules and PEGylated proteins. In addition to HPLC-coupled techniques, there is a myriad of others that have been employed to determine the size and R_H heterogeneity in PEGylated proteins including static light scattering (SLS), dynamic light scattering (DLS), and analytical ultracentrifugation (AUC) [359]. Combining these techniques can provide a clear understanding of the overall heterogeneity and physicochemical properties of PEGylated molecules.

Beyond these extensions of traditional technologies, the use of biological material for PEGylation has been developed. For example, enzyme-linked immunosorbent assay (ELISA), anti-PEG assays, and other activity bioassays have been developed to characterize and quantify PEGylated therapeutics. The details of these assays and the analytical considerations have been previously overviewed [352]. In general, these assays need to ensure the readout is specific for the PEGylated therapeutic, and that there is no interference from native proteins if used to quantify drug exposure in serum samples for the determination of impact on PK.

Lipidation

Another popular approach for half-life extension has been through conjugation of a fatty acid to a protein or peptide. This approach has especially garnered a lot of attention for peptides recently due to the tremendous success of GLP-1 peptide agonists. One of the mechanisms through which lipidation prolongs half-life is by enabling the formation of depots upon SC administration, thereby delaying absorption from the injection site into the bloodstream [360]. The formation of these micellar depots is driven by the fatty acid hydrophobicity, which also makes most lipidated peptides easy to purify and analyze by RP-HPLC, as their increased hydrophobicity allows easy separation from unmodified and/or much smaller precursors. In most cases, this is combined with MaS, akin to analysis of PEGylated molecules [361]. In some cases, hydrophobic interaction chromatography (HIC) has been leveraged as an alternative means of purification and characterization. For this method, a hydrophobic stationary phase (such as agarose with butyl or phenyl ligands) is used and the samples are applied to the column in a high-salt buffer. A decreasing salt gradient is used to elute proteins from the column in the order of increasing hydrophobicity.

Antibody Conjugation

An attractive and successful means for increasing drug selectivity is to conjugate APIs to mAbs, such as is the approach with ADCs, and more recently antibody-oligonucleotide conjugates (AOC). In the case of ADCs, this strategy combines the highly selective targeting of a mAb with the potent cytotoxicity of a small molecule [362]. In the case of AOCs, this strategy has leveraged many of the advances of ADC molecule design to enable half-life extension, as well as increased selectivity and bioavailability of an oligonucleotide [34, 363]. While antibody conjugation provides an elegant drug delivery solution to increase selectivity, there are unique analytical challenges that are associated with their covalent addition (either specific or nonspecific) such as the type of compound they are conjugating to (payload), number of conjugated payloads (drug-to-antibody ratio, DAR), and choice of linker used for conjugation. In addition, characterizing and separating the biological activity from the payload, mAb alone, and/or any free drug within a bioassay, can be challenging.

The payloads for ADCs have historically been focused on molecules with known anticancer properties such as chemotherapeutic agents, tubulin inhibitors, and DNA damaging molecules [364]. The first generation of ADCs were primarily attempting to take advantage of the known molecular actions of small molecules against cancer, while reducing systemic toxicity through target selectivity and delivery. While these ADC molecules mitigate toxicity for patients, the intrinsic hazard for scientists performing analytical characterization persists. The Health Hazard Categories (HHC) and Occupational Exposure Limits (OEL) for the payload may become substantially less when conjugated to the ADC; however, the opposite can be said for the naked antibody versus the conjugated molecule. This means that any assay development and analytical analysis of ADCs becomes substantially more challenging, not just because of the physicochemical modifications to the API but also because of safety concerns for the scientist.

The second generation of ADCs have moved beyond cytotoxic payloads and towards exotic molecule attachments. For example, mAb conjugation to antibiotics has shown promise for eliminating bacteria [365], conjugation to degrader molecules can target the antibody–antigen complex for proteasome degradation [366], and antibody–oligonucleotide conjugates show promise for combining the precision of nucleotide-based therapies with the specificity and deliverability of antibodies [363]. The next generation of antibody-conjugate modalities will include bispecific molecules, dual-drug molecules, increasing DAR, and probody–drug conjugates (conditionally active) [367]. With every iteration that will come to fruition, so will come commensurate challenges with ensuring developability of the actual biotherapeutic.

Payload conjugation to a mAb typically occurs through a chemical reaction with side chains of select amino acids, and, therefore, the type of conjugation has a profound impact on the DAR, typically has a range between 1 and 8 for an ADC and 1–2 for AOCs. For example, ado-trastuzumab emtansine (Kadcyla®) and polatuzumab vedotin (POLIVY®) both have a reported average DAR equal to 3.5. This is attributed to the chemistry used for conjugation, which is through lysines [368] and interchain disulfides [369] respectively, both of which are considered nonspecific conjugation strategies. More recently, efforts to simplify and control the DAR by engineering the antibody to contain free cysteine residues (THIOMAB technology) have enabled selective labeling with DAR equal to 2.0 [370]. Moreover, AOCs have followed suit by taking advantage of site-specific conjugation strategies, thereby enabling greater control over payload addition during molecule design [363].

Differences in the type of conjugation and resulting heterogeneity in DAR significantly impacts analytical characterization. The conjugation of small molecules typically results in a significantly more hydrophobic molecule than that of the naked mAb, leading to physiochemical changes ranging from solubility to stability, to simple absorbance interference impacting concentration determinations. The use of HIC is a proven technique for separation and quantitation of heterogeneity in small molecule payload DAR [371], as well as the antibody intermediate (naked mAb), the full ADC, and any free drug (unbound) [372]. On the other hand, since oligonucleotides are larger compounds, hydrophilic, and negatively charged, ion-exchange chromatography has been much more well suited for species separation and determination of DAR within AOCs.

Linkers are the chemical bridge between the antibody and its payload, and typically come in a cleavable or non-cleavable form. The use of the type of linker will depend on the MoA of the ADC and affects the analytical characterization required. For example, if the mechanism involves receptor-mediated endocytotic uptake of the mAb, and the payload can interact with the target while attached to the ADC, then a non-cleavable linker may be suitable. The analytical characterization of non-cleavable linkers is typically limited to the free drug and DAR characterization, since the molecular chemistry of the linker is unmodified for efficacy.

Cleavable linkers have chemical moieties that are susceptible to the environmental differences between systemic circulation and cellular/tissue uptake. These types of linkers can be further subdivided into either chemical or enzymatic cleavage linkers. The former utilizes changes to the physicochemical environments such as pH or reduction potential, while the latter takes advantage of specific enzymatic mechanisms at the desired site of administration [373]. Cleavable linkers are needed when the payload requires release from the mAb to be efficacious, which makes the cleavability a critical aspect for the ADC. Analytical characterization of payload release may be necessary under conditions that mimic the *in vivo* conditions expected for release. Ligand binding assays (LBA) are analogous to sandwich ELISA format for capturing the ADC, but are tailored to detect either the total mAb or the ADC only, and can be utilized in biomatrices [374]. MaS has been the method of choice for quantitating the release of payloads [375]; however, radiolabeling has been used to study catabolites *in vivo* [376], and fluorescence strategies using microscopy and flow cytometry have to be used to study and quantitate cellular trafficking [377, 378].

Strategies for Drug-Delivery Systems Employing Release Matrices

Controlled-release DDSs are designed to modulate the release of a therapeutic agent in a controlled, reproducible manner over a predetermined target duration of time. From an analytical characterization and control system perspective, the encapsulating matrix may be viewed as a (highly complex) excipient while the API either remains unaltered or is altered by linkage to the matrix. In either case, controlled-release DDSs are typically characterized primarily in drug release experiments, in which the performance of the DDS is assessed as a function of time. These release experiments are performed primarily *in vitro*, especially during development when a large number of experiments are needed to guide their optimization. Due to the controlled environment and facile manipulation of *in vitro* experiments, analytical characterization can be performed more often and more in-depth than during *in vivo* experiments, so most characterization is expected to occur in the *in vitro* context. Nonetheless, there still exists a need for select experiments to be performed using *in vivo* conditions to validate that the *in vitro* conditions are representative of the *in vivo* system. Thus, to provide maximum value during development and to ensure seamless transfer to the *in vivo* setting, *in vitro*

Table 20.8 Biological compositions in humans for different parenteral sites of injection.

ROA	Site of injection	Compositions	pH	Reference
IV	Blood plasma	~91% water, ~7% protein, ~2% nutrients, hormones, electrolytes such as sugar, lipids, sodium, potassium, and calcium excluding red blood cells.	~7.3–7.5	[379]
IVT	Vitreous humor	~98%–99% water. ~1%–2% collagen, hyaluronic acid, chondroitin sulfate, proteins, and other components such as lactate, xanthine, hypoxanthine, creatinine, and urea.	~7.0–7.4	[380, 381]
SC	Subcutaneous	Extracellular matrix formed by 50% dry weight of collagen, ~1% hyaluronic acid, chondroitin sulfate, and interstitial fluid composed of sodium, potassium, magnesium, calcium, carbonate, bicarbonate, chloride, sulfate.	~7.3–7.5	[382]
IT	Spinal canal to reach CSF	Similar to blood plasma but with ~0.59% plasma proteins	~7.4	[383, 384]

CSF, cerebrospinal fluid; IT, intrathecal; IV, intravenous, IVT, intravitreal; ROA, route of administration; SC, subcutaneous.

conditions and overall analytical strategy must be carefully considered.

Although physicochemical properties of the DDS can be tailored to improve the stability of encapsulated proteins and influence the release rate of proteins from the delivery vehicle by manipulation of material chemistries, excipient compositions, as well as protein formulations, due to structural complexity of proteins, product quality attributes of proteins within the desired DDS need to be well-controlled to ensure that the stability of protein is maintained. Moreover, assessing *in vitro* release kinetics and selecting appropriate animal species for *in vivo* PK analysis are crucial in evaluating the possibility of technical success for a particular drug delivery technology. Importantly, additional toxicology evaluation of novel excipients is required in preclinical models to ensure their safety for human use. Despite immense challenges, continued efforts to develop novel DDSs for sustained release proteins remain critical for improving overall well-being and health outcomes for patients.

Selection of Appropriate *In Vitro* Release Conditions

As mentioned above, *in vitro* characterization of DDSs serves to facilitate development by allowing iteration in shorter time frames with less material and without the use of animals, in keeping with the 3Rs principles of replacement, reduction, and refinement. Generally, this also permits evaluation of more conditions and can be performed in experimentally optimized designs such as Design of Experiments (DOE). However, in order for *in vitro* experiments to be valuable to the overall development cycle of a DDS, they must be as representative of the final *in vivo* conditions as possible. The first and most important consideration for designing the most representative *in vitro* conditions is the target RoA and injection/implantation site. As much as is feasible, the *in vitro* conditions should seek to accurately mimic the physiological conditions and the intended site of administration so that the DDS is exposed to the same conditions *in vitro* as in the eventual *in vivo* setting. The biological composition of different parenteral sites of administration are shown in Table 20.8. Although the biological compartments are highly complex and not always straightforward to replicate *in vitro*, assessing stability and release rate of encapsulated API *in vitro* is a necessary first step to provide insight on how to optimize the formulations as well as the design of the DDS. In practice, it is not practical to mimic all conditions of the *in vivo* environment due to the complexity of living systems and active biological processes. Instead, an attempt should be made to capture the most salient physiological conditions that may directly or indirectly impact the performance of the DDS. In general practice, the primary conditions that are mimicked include pH, temperature and salt concentration/composition, where the vast majority of *in vitro* experiments employ PBS at pH 7.4 to achieve this.

PBS is a commonly used buffer medium to evaluate stability of proteins in a physiologically relevant *in vitro* condition. It is an isotonic solution composed of sodium chloride, disodium hydrogen phosphate, potassium chloride, and monopotassium dihydrogen phosphate, and it has a physiological pH of ~7.4 at 37°C [385]. Not only are its compositions similar to those of electrolytes found in serum, it is also nontoxic to cells. Hence, PBS has also been used as a washing step in cell culture and different immunoassays [386]. Moreover, PBS has been shown to be a good mimic of IVT chemical instability as demonstrated by Loyet et al. [387]. Nonetheless, in some cases, PBS may not be most representative of *in vivo* physical stability. In one case study reported by Sreenivasan et al., the amount of soluble aggregates of the model mAb tested in PBS was higher than the amount of soluble aggregates in human serum fluid upon stress over 72 hours [386]. In another case study, it was found that stressed protein samples in PBS generated less particles over 2 μm when compared to those incubated in human serum fluid [386]. This reinforces the idea that although evaluating the stability profile of proteins in PBS at pH 7.4 and 37°C is a good first step, it may not be sufficient in depicting the full picture of how proteins behave in a biological system, especially for systems that exhibit pH- or temperature-dependent properties such as gelation or triggered release.

Ultimately, certain compromises must be made between a perfect replication of the *in vivo* condition at the target site of administration and what is readily achievable in an *in vitro* setup. Although, the presence of certain proteins, enzymes, or metabolites can be replicated by spiking them into the release medium, it is not always possible for them to retain their activity over the entire *in vitro* experimental duration, since in biological systems these may be produced and maintained by active cellular processes that cannot reasonably be mimicked *in vitro*. Another complex attribute of physiological matrices that is experimentally challenging to mimic is the reductive/oxidative environment. Different biological matrices have different redox environments that are actively regulated, and some may be reducing or oxidizing. As drugs (especially proteins or peptides with disulfide bonds) are exposed to these conditions over an extended period of time in the context of a controlled-release DDS, the reducing or oxidizing environment may lead to undesirable changes in drug integrity that need to be carefully monitored. In practice, maintaining a physiologically representative redox environment in an *in vitro* setting is highly complex as it requires specific ratios of reducing and oxidizing agents (e.g. reduced and oxidized glutathione), adding chelating agents, and sparging dissolved oxygen, among other delicate considerations. In practice, this is likely not achievable for all *in vitro* drug release experiments but can be implemented in a targeted manner to understand whether the system under investigation is sensitive to redox conditions.

Other physiologically relevant conditions that are challenging to adequately capture in an *in vitro* setting include species-to-species differences, animal-to-animal variability, and potential differences in biological matrix properties between the healthy state and the disease state. All of these are expected to contribute to the differential performance of a DDS in preclinical and clinical studies, but their complexity is beyond what can rationally be mimicked *in vitro* as of today. Instead, a scientific risk-based assessment is typically adopted, to understand how a specific DDS is expected to perform in different contexts.

Establishing an understanding of how the *in vitro* release performance correlates with the *in vivo* PK performance is a powerful tool when evaluating a DDS. An *in vitro–in vivo* correlation (IVIVC) is a predictive mathematical model which describes the relationship between drug release kinetics *in vitro* and *in vivo* absorption [388]. In general, IVIVC can be categorized into five levels: level A, multiple level C, level B, level C, and level D as listed from the most to the least informative [388]. Developing a meaningful IVIVC can help better define clinically relevant dissolution profiles, and it can act as a surrogate for bioequivalence studies, thus expediting the approval process. A recent study showed that PLGA polymers from different sources have different intrinsic properties such as viscosity and molecular weight, which can significantly impact the physicochemical properties of PLGA microspheres, thereby affecting *in vitro* release and *in vivo* performance [389]. Developing IVIVC can predict the *in vivo* performance of different sources of PLGA microsphere formulations and be able to identify which formulation is out of specification [389]. Nevertheless, evaluating drug release *in vitro* can require a considerable amount of time for testing, ranging from weeks to years. Therefore, the development of accelerated testing methods is often a necessity. As demonstrated by Goel et al., the use of a mixed media of cosolvent (5% tetrahydrofuran) and surfactant (5% Triton X-100) in PBS buffer (pH 7.4) at an elevated temperature (50°C) enabled complete release of a peptide from PLGA within 14 to 21 days as opposed to 3 to 6 months of real-time release [390]. In addition, Jain et al. developed an accelerated *in vitro* release method with the inclusion of 0.75% w/v cremophor surfactant in phosphate buffer using USP apparatus 2 [391]. A biopredictive *in vitro* release model in 2 days was scaled to predict *in vivo* performance of approximately 75 days using a long acting injectable formulation of crystalline suspension [391]. However, there is no clear regulatory guidance on IVIVC for a DDS via the parenteral routes, mainly due to the complex release mechanisms for *in vitro* and *in vivo* evaluations.

Finally, a variety of excipients are typically used to maintain the stability of therapeutic proteins within a DDS. Some examples of protein stabilizers include sugars, salts, and surfactants [392]. Upon injection at the site of administration, the change in physiological environment can alter the interactions between excipients and proteins, which may be more or less critical depending on the RoA. Since this can have a big impact on bioavailability after SC injection, an *in vitro* system called subcutaneous injection site simulator (Scissor) system was developed to model some of the dynamic physiological properties of the hypodermis and to study the stability of therapeutics in response to external environments upon release [393, 394]. It is a dialysis-based system incorporating extracellular matrix (ECM) components such as hyaluronic acid [393, 394]. Kinnunen et al. demonstrated that introduction of a long acting insulin formulation in the injection chamber resulted in the formation of precipitates, primarily due to the release of excipients into the sink prior to the recovery of insulin solubility [394]. Similarly, due to the large molecular weight of therapeutic proteins, for a diffusion-controlled release system, the release rate of proteins could differ from that of other small molecule excipients. According to the Stokes–Einstein equation, since the diffusion coefficient is inversely proportional to the R_H of a molecule, a larger molecular weight typically will result in a smaller diffusion coefficient [395]. Thus, the extent to which an excipient can provide protective effect within a DDS and the difference in release rates between therapeutic and excipients should be taken into consideration within the context of RoA and site of action. In short, evaluating release from a DDS involves choosing appropriate buffer media and release testing methodologies to ensure product quality. The stability of the API in response to the external environment *in vitro* versus the biological environment at the intended site of injection, as well as the relationship between *in vitro* release kinetics and *in vivo* performance are essential for optimal, analytical development of a therapeutic in combination with a DDS.

Characterization of Drug Release and Integrity

While typical characterization of a DDS is highly focused on the amount of drug released over time, this represents only one quality attribute that should be taken into consideration when holistically evaluating the system during development. When a protein or other API is encapsulated within a DDS via a non-covalent strategy, the API remains unaltered. For example, if a mAb therapeutic is encapsulated within a hydrophilic hydrogel matrix, the mAb is identical to the parent drug despite being formulated within that matrix, and thus is expected to be released in the same state. In many ways, this strategy simplifies analytical characterization of the therapeutic since it is not being changed by the DDS or encapsulation process. Nevertheless, the possibility for unintended modification or degradation of the therapeutic during either encapsulation, residence within the matrix, or release must be taken into consideration and carefully evaluated when designing an analytical characterization strategy. Some examples of potential changes to encapsulated or released therapeutic integrity are demonstrated graphically in Figure 20.14.

Before encapsulation, the therapeutic is expected to be fully characterized for all previously identified quality attributes and in-line with compendial requirements, if they exist. As one representative example, USP<1503> "Quality Attributes of Synthetic Peptide Drug Substances" outlines the tests required for characterization and quality control of a synthetic peptide drug substance. In the case of an encapsulated peptide, this document would serve as the reference point for what characterization would be

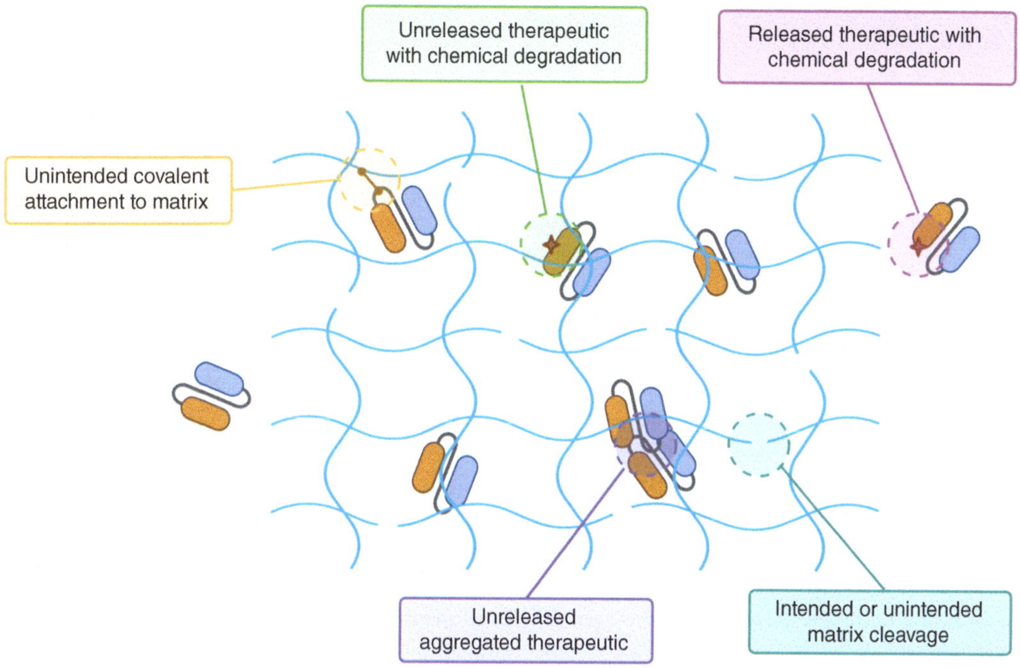

Figure 20.14 Considerations for characterization of a controlled-release drug delivery system with exemplary events that a characterization strategy should capture, including unreleased drug, released drug, and matrix components. Source: Adapted from BioRender.

expected of the peptide that is used in the preparation with the DDS. The exact analytical characterization requirements depend on the nature of the starting API. A small molecule, peptide, protein, or NABM therapeutic each have unique requirements that should align with existing compendial guidelines or best practices.

After encapsulation, it is inappropriate to assume that the drug remains unaltered, no matter how carefully the encapsulation and processing conditions are controlled. Formation of the encapsulating matrix or encapsulation of the drug within the matrix may require the use of particular conditions that have the potential to cause degradation of the therapeutic. For example, many hydrogel chemistries are cross-linked using highly reactive functional groups such as epoxides, to enable rapid and complete cross-linking during processing. Given the high reactivity of these chemistries, including to groups that may be present in a therapeutic molecule's chemical structure, the possibility for unintended modification must be considered. This applies not only to situations when the therapeutic may be directly exposed to the reactive chemicals but also in cases where traces of the reactive chemicals could still be present when the therapeutic is introduced later in the process. To avoid this issue, the use of bioorthogonal chemistries has gained significant interest in recent years, as these groups are highly selective toward their reaction partner, but not reactive toward chemical groups commonly present in therapeutic drug molecules.

Other examples of processing conditions that could directly or indirectly impact the therapeutic molecule include exposure to heat, extreme pH conditions, organic solvents, high shear stresses, and high pressure among others. Thus, after completion of the encapsulation process, it is beneficial to reassess the integrity of the encapsulated therapeutic against many of the same quality attributes that were assessed in the starting drug substance. However, the complicating factor is that the API is now no longer present in its native state but is encapsulated within the DDS. Ideally, a method may be devised to extract the encapsulated therapeutic so that it can be analyzed in the same way as the starting material. Although the exact method to achieve this will depend on the nature of the matrix, it is imperative that it is done in a way that preserves the state of the therapeutic akin to within the DDS. For example, although a small molecule API could be extracted from a PLGA matrix using a solvent that solubilizes the PLGA matrix, such as dimethyl sulfoxide (DMSO), this strategy will not work if the encapsulated API is a protein, which will denature in the presence of organic solvents. In such cases, more elaborate drug extraction procedures are required as part of the analytical strategy.

Similarly, for a DDS that enables extended residence within a matrix, the therapeutic may undergo unintended and unwanted degradation processes that impact its integrity. A common way of monitoring this is by extracting the therapeutic from the DDS at different time points to evaluate it using the same or similar assays used to characterize the starting material. One complexity to note is the therapeutic unintentionally reacting with and covalently or non-covalently binding to components of the DDS matrix. For example, if a bis-functional cross-linking agent is used to form a hydrogel matrix that is later loaded with a protein API, there is a possibility of cross-linker reacting on one end with the other matrix components, but the other end remains unreacted in the final DDS. Over time, encapsulated protein could react with this unreacted cross-linker and be unintentionally covalently attached to the matrix. Moreover, the presence of protein in this state may be very technically challenging to detect as it can resist extraction by standard methods that only extract soluble species. Nevertheless, the modified protein can have particular safety concerns upon administration. Thus, a method to detect and quantify such species is paramount to the complete analytical strategy. This phase of characterization parallels the assessment of stability/shelf-life of traditional dosage forms and may take a similar strategy with accelerated or real-time stability evaluations and employ similar analytical testing requirements.

Finally, the API released from the DDS must also be thoroughly characterized to fully capture both the amount and integrity of the therapeutic molecule. During typical drug release experiments, the rate of release is given considerable weight, and in some cases, it may be the only data presented in support of the DDS' performance. In reality, quantitation of API release at different time points is only one of many critical parameters that should be evaluated. The quantity of drug released as a function of time can be determined by a variety of analytical techniques including direct UV/visible spectroscopy, HPLC, spectroscopic techniques, and labeling-based approaches. However, most of these techniques quantitate only the total amount of API released from the matrix, independent of its integrity. Thus, to gain a better understanding of the system's performance, the amount of API released at each time point should be corrected for the amount of intact, unaltered API, since this is what is to be considered the efficacious component [396]. In order to achieve this, an additional suite of analytical techniques should be implemented as part of a carefully designed drug release study.

The therapeutic that is released should be subjected to a similar suite of analytical tests that were conducted before incorporation into the DDS, as these would be expected to capture the relevant quality attributes for the given therapeutic. For example, a biological drug should be characterized for total amount released and relevant product quality attributes such as size- and charge-based variants (e.g. by SEC and IEC), structural characterization (i.e. secondary structure, ternary structure, etc.), residue-specific characterization (i.e. oxidation, deamidation, glycation, etc.), and antigen binding capacity [396]. Crucially, characterization should also capture the safety and efficacy of the released therapeutic as this is what will be bioavailable in patients. Ideally, these collective data are used to correct the released amount for "intact" or "active" therapeutic such that only released API that is safe and efficacious is considered.

After completion of the release experiments, it is beneficial to extract any remaining therapeutic from the delivery matrix for both quantitation of residual API and an assessment of residual API integrity. The former can be used for formal mass balance calculations (i.e. determining whether cumulative released plus residual drug equals starting drug) as this can help understand blind spots in the release experiment. Mass balance results that significantly deviate from 100% should be investigated thoroughly as they may indicate problems with the experimental setup or analytical methods that may lead to misleading release results. The latter (residual API integrity) parallels previously discussed guidance that not only the released API be subjected to analytical testing but also whatever is remaining in the DDS. This approach allows a more complete evaluation of potential degradation mechanisms within the matrix including potential deleterious interactions between the encapsulated therapeutic and the matrix, as shown graphically in Figure 20.14.

Strategies to Support the Advancement of Targeted LNPs

The approval of mRNA-based therapeutics such as the mRNA vaccines against COVID-19 has paved the way for the successful advancement of LNPs for commercial use. As with other DDSs, appropriate characterization strategies are expected to be implemented on the therapeutic agent before, during, and after encapsulation or delivery of the therapeutic cells. For mRNA LNP vaccines, the appropriate characterization strategy can be derived from the control system developed to support manufacturing of the LNPs and has typically included analytical techniques to test identity (RT-PCR), total and encapsulated RNA content (Ribogreen colorimetric assay), RNA purity/integrity (capillary gel electrophoresis), lipid content and product-related impurities (HPLC-CAD), size (DLS), and potency (cell-based flow cytometry) among others [397]. However, in addition to leveraging a release mechanism of action for mRNA therapeutic action, targeted LNPs also bring about the additional complexities of using a ligand and conjugation strategy, resulting in the need for an even more complex control strategy and rigorous testing requirements. Since none have yet to be approved, the path forward remains largely undefined. However, thinking ahead, it can be envisioned that there would need to be three parallel workflows: (i) the LNP and released mRNA therapeutic, (ii) the linker–ligand conjugation and (iii) the fully assembled mRNA LNP with targeted ligand. With regard to point 2, it can be envisioned that additional techniques that could be implemented might mirror the work that has been done with ADCs [398], while cryo-electron microscopy and small angle X-ray scattering are techniques that have already been leveraged for the fully formed DDS [166].

Conclusions/Future Perspectives

As exemplified by the wide range of topics covered in this chapter, drug delivery is a diverse field, with opportunities for innovation spanning from preclinical activities all the way through to commercially approved products. Although the inclusion of a DDS typically increases the level of complexity of therapeutic development, it also has the potential to greatly improve the patient experience as well as patient compliance. Moreover, including a DDS has the ability to reduce the carbon footprint related to a therapeutic product by decreasing product manufacturing, packaging, distribution, in addition to patient travel to and from an administration site. While there are a plethora of preclinical studies demonstrating their potential, many development factors such as safety, manufacturing process, and critical product attributes need to be taken into consideration in order to realize a DDS for human use.

CMC considerations for developing a DDS are crucial to ensuring the safety, quality, efficacy, and consistency of the drug products. Since DDSs often rely on the use of polydispersed materials, scale-up production from a lab bench scale to a pilot plant scale often can lead to challenges in reproducibility, and a well-controlled manufacturing process is essential to minimize batch to batch variability and ensure product consistency. Product quality attributes such as drug loading, amount of residual solvent after evaporation, as well as stability of proteins during the manufacturing process must be consistently characterized to ensure that the products can maintain minimum required standards within the set specifications [399]. For example, production of polymeric microspheres typically involves an emulsification process with the use of organic solvents [400]. Slight changes in production steps can result in significant differences in the physicochemical properties (e.g. size, shape, and particle stability) of the polymeric carrier, which can have a huge impact on release kinetics and safety profile.

Because a DDS is intended to act for extended periods of time, the potential for toxicity and tolerability issues is higher than with traditional drug development. The stability of drug substance (e.g. protein, peptide, NABM, and cells) and the final drug product (e.g. encapsulated protein in polymer implant, encapsulated NABM, and cells that have been cryogenically frozen) must be characterized concomitantly to determine storage condition specifications and shelf-life of the product. While surfactants, polymers, and lipids are commonly employed in drug delivery technology development to ensure extended *in vivo* stability, in some circumstances these have been linked to injection site reaction and anaphylaxis in patients [401]. Thus, it is crucial to perform in-depth *in vivo* screening by the RoA intended for clinical use to understand the safety of novel excipients in the proposed formulation, including clinical pathology, histopathology, and toxicokinetic analysis [402]. Moreover, as mentioned earlier, with protein therapeutics, aggregation is a key challenge. Not only do protein aggregates lead to a reduction in activity and altered PK, but the formation of soluble and insoluble aggregates can elicit undesirable harmful immune responses [403]. It is therefore important to test the stability of the API (protein or other) within the DDS as well as after release. For further reference, the International Council for Harmonization of Technical Requirements for Pharmaceuticals for Human Use (ICH) has provided guidance for long-term drug stability testing under different storage conditions [404].

With recent advances in drug delivery technologies, there have been numerous developments in sustained release of proteins over the past decades. Notably, the approval of PDS implants with ranibizumab offers an alternative treatment approach for the treatment of back of the eye diseases as opposed to monthly IVT injections, thereby enhancing primary adherence to the prescribed medication. However, special considerations must be taken with the introduction of new delivery systems. For example, although PLGA implants have been approved for IVT administration for sustained release applications in humans, recent toxicology assessment showed that IVT administration of PLGA microspheres in both rabbits and NHPs resulted in a severe immune reaction, characterized by a foreign body response within the vitreous [405]. By contrast, the use of PLGA rods was well tolerated in rabbits, suggesting that different shapes and sizes of PLGA-based intraocular carriers have different clinical safety profiles [405]. On the other hand, during the development of the PDS with ranibizumab, the FDA recommended a chronic 6-month toxicology and biocompatibility study of the PDS in an appropriate animal model in order to support clinical development [406]. Due to the lack of regulatory guidance for a permanent, refillable intraocular implant, proactive interaction with the health authority to identify an acceptable nonclinical toxicology testing strategy was needed. The size of the minipig eye was deemed large enough for a functional implant and was therefore selected as the only relevant species for toxicology evaluation [406]. Bantseev et al. demonstrated that surgical implantation of the PDS device in minipig did not show any toxicology-related safety concerns over a 6-month implantation period, thereby enabling further clinical development [406]. As per the United States Pharmacopeia (USP), specific guidelines need to be followed pertaining to the development of ophthalmic products. LAD products for ophthalmic applications must meet all the specifications such as sterility, potency, purity, identity, and particulate matter [407]. Additional safety assessments are required to confirm biocompatibility for clinical use.

Finally, an example of a DDS earlier in its developmental life cycle is that of targeted LNPs, where there are still many challenges in design, synthesis, and manufacturing that need to be addressed for this strategy to be realized in the clinic. With regard to the targeting ligand itself, there are several aspects that must be carefully considered. First, choosing the appropriate cell-surface target such that it is biologically neutral in the indication is essential; this is especially important for therapeutics that are harnessing the immune system as part of their mechanism of action. Second, choosing the appropriate targeting ligand is another main consideration. For many cell-surface receptors, mAbs may already be readily available, making them an obvious choice, especially for preclinical proof of concept studies. However, careful evaluation must be done to ensure activation of the receptor target is avoided, and in some cases, re-engineering to ablate this may be necessary. While antibody campaigns can support the design of new targeting mAbs, these are typically expensive and time consuming. Thirdly, the antibody-to-particle ratio must also be carefully evaluated and tuned not only to be appropriate for the receptor density, but also to ensure LNP stability. Moreover, the conjugation process must be optimized, and the resulting particles must be fully characterized to ensure antibody receptors are accessible and binding is maintained, as well as unconjugated antibodies removed; both which can be difficult and low-yielding. With so many considerations, it is clear that streamlining CMC is a main consideration for the commercialization of this approach. For example, it can be envisioned that moving away from antibodies toward smaller targeting ligands such as scFvs, nanobodies, or peptides may enable a less complicated clinical path forward. Overall, it remains to be seen whether lipid composition optimization or ligand decorating of LNPs will yield clinical success. With exciting preclinical data already disclosed, the next decade will certainly bring exciting developments in this area of drug delivery.

Abbreviations

AAV	adeno-associated virus
AD	Alzheimer's disease
ADA	anti-drug antibodies
ADA-SCID	adenosine deaminase–deficient severe combined immune deficiency
ADC	antibody–drug conjugate
ADCC	antibody-mediated cell cytotoxicity
ADCS-ADL	Alzheimer's disease cooperative study-activities of daily living
ADME	absorption, distribution, metabolism, and excretion
ASO	antisense oligonucleotide
AS-OCT	anterior-segment optical coherence tomography
ALS	amyotrophic lateral sclerosis
API	active pharmaceutical ingredient
AUC	area under the curve
BACE	β-site amyloid precursor protein cleaving enzyme
BBB	blood–brain barrier
bFGF	basic fibroblast growth factor
BSA	bovine serum albumin
CAD	charged aerosol detector
CAR-T	chimeric antigen receptor T cell
CGM	continuous glucose monitoring
CGRP	human calcitonin gene-related peptide
CLN	ceroid lipofuscinosis
CMC	chemistry, manufacturing, and controls
COVID-19	coronavirus disease 2019
CNS	central nervous system
CPP	cell-penetrating peptide
CSF	cerebrospinal fluid
CQA	critical quality attribute
DAR	drug–antibody ratio
DC	dendritic cell
DDS	drug-delivery system
DOE	design of experiment
DR	dorsal raphe nucleus
DGL	dendrigraft poly-L-lysine
DLS	dynamic light scattering
DMSO	dimethyl sulfoxide
EAE	experimental autoimmune encephalomyelitis
ECM	extracellular matrix
ELSD	evaporative light scattering
ENCP	enveloped nanocomplexes
EVA	ethylene-vinyl acetate
FDA	federal drug agency
FUS	focused ultrasound
GBM	glioblastoma
GDNF	glial cell-line-derived neurotrophic factor
GFP	green fluorescent protein
HD	Huntington's disease
HHC	health hazard category
HIC	hydrophobic interaction chromatography
HPLC	high-performance liquid chromatography
HSA	human serum albumin
HTT	Hereditary hemorrhagic telangiectasia
HuR	human antigen R
ICV	intracerebral ventricular
IGF	insulin-like growth factor
IgG	immunoglobulin G
IM	intramuscular
IN	intranasal
IND	indatraline
IT	intrathecal
IV	intravenous
IVIVC	*in vitro–in vivo* correlation
IVT	intravitreal
kDa	kilodaltons
KD	knockdown
LAD	long-acting delivery
LBL	layer-by-layer
LC	locus coeruleus
LDL	low-density lipoprotein
LNP	lipid nanoparticle
LPR	hybrid lipid-shell polymer core mRNA nanoparticle
LPX	lipoplex
mAb	monoclonal antibody
MALS	multi-angle light scattering
MaS	mass spectrometry
MHC	major histocompatibility complex
miRNA	microRNA
MLNP	mannosylated LNP
MoA	mechanism of action
mRNA	messenger RNA
MS	multiple sclerosis
N/A	not applicable
NABM	nucleic acid-based medicine
NHP	non-human primate
NP	nanoparticle
NLC	nanostructured lipid carrier

OEL	occupational exposure limit
PBS	phosphate-buffered saline
PD	Parkinson's disease
PDS	Port Delivery System
PEG	polyethylene glycol
PEI	polyethylenimine
PFS	prefilled syringe
PK/PD	pharmacokinetics/pharmacodynamics
PLGA	poly(lactic-co-glycolic acid)
PLK1	polo-like kinase 1
PTD	protein transduction domain
PTM	post-translational modification
QELS	quasi-elastic light scattering
RCE	release control element
RoA	route of administration
RP	reverse phase
SC	subcutaneous
SEC	size-exclusion chromatography
SDS-PAGE	sodium dodecyl-sulfate polyacrylamide gel electrophoresis
SERT	serotonin transporter
siRNA	silencing RNA
SNc	substantia nigra compacta
T2D	type 2 diabetes
TA	tannic acid
TAT	trans-activator of transcription
TCR	T-cell receptor
TDM	tetradecyl-β-D-maltoside
T_g	glass transition temperature
TJ	tight junction
TfR	transferrin receptor
TNF	tumor necrosis factor
TRAIL	TNF-related apoptosis inducing ligand
USP	US pharmacopeia
VEGF	vascular endothelial growth factor
VTA	ventral tegmental area

References

1 Wang, L., Wang, N., Zhang, W. et al. (2022). Therapeutic peptides: current applications and future directions. *Signal Transduct. Target. Ther.* 7 (1): 48.

2 Brinkmann, U. and Kontermann, R.E. (2017). The making of bispecific antibodies. *MAbs* 9 (2): 182–212.

3 Jin, S., Sun, Y., Liang, X. et al. (2022). Emerging new therapeutic antibody derivatives for cancer treatment. *Signal Transduct. Target. Ther.* 7 (1): 39.

4 Shaer, D.A., Musaimi, O.A., Albericio, F., and de la Torre, B.G. (2024). 2023 FDA TIDES (peptides and oligonucleotides) harvest. *J. Pharm.* 17 (2): 243.

5 Hilgenfeld, R., Seipke, G., Berchtold, H., and Owens, D.R. (2014). The evolution of insulin glargine and its continuing contribution to diabetes care. *Drugs* 74 (8): 911–927.

6 Global Data. (2024) GLP1 agonists set to become the best-selling drugs in 2024, says Global Data [Internet]. (cited 2024 Jul 2). Available from: //http://www.globaldata.com/media/pharma/glp 1-agonists-set-to-become-the-best-selling-drugs-in-2024-says-globaldata/

7 Ebrahimi, S.B. and Samanta, D. (2023). Engineering protein-based therapeutics through structural and chemical design. *Nat. Commun.* 14 (1): 2411.

8 Brown, T.D., Whitehead, K.A., and Mitragotri, S. (2020). Materials for oral delivery of proteins and peptides. *Nat. Rev. Mater.* 5 (2): 127–148.

9 Baker, M., Reynolds, H.M., Lumicisi, B., and Bryson, C.J. (2010). Immunogenicity of protein therapeutics: the key causes, consequences and challenges. *Self Nonself.* 1 (4): 314–322.

10 Wing, M. (2008). Monoclonal antibody first dose cytokine release syndromes–mechanisms and prediction. *J. Immunotoxicol.* 5 (1): 11–15.

11 Ramakrishnan, M.S., Yu, Y., and Vander Beek, B.L. (2020). Association of visit adherence and visual acuity in patients with neovascular age-related macular degeneration. *JAMA Ophthalmol.* 138 (3): 237–242.

12 Shahzad, H., Mahmood, S., McGee, S. et al. (2023). Non-adherence and non-persistence to intravitreal anti-vascular endothelial growth factor (anti-VEGF) therapy: a systematic review and meta-analysis. *Syst. Rev.* 12 (1): 92.

13 Müller, S., Junker, S., Wilke, T. et al. (2021). Questionnaire for the assessment of adherence barriers of intravitreal therapy: the ABQ-IVT. *Int. J. Retin. Vitr.* 7 (1): 43.

14 Vugmeyster, Y., Xu, X., Theil, F.P. et al. (2012). Pharmacokinetics and toxicology of therapeutic proteins: Advances and challenges. *World J. Biol. Chem.* 3 (4): 73–92.

15 Tibbitts, J., Canter, D., Graff, R. et al. (2016). Key factors influencing ADME properties of therapeutic proteins: a need for ADME characterization in drug discovery and development. *MAbs* 8 (2): 229–245.

16 Richter, W.F., Bhansali, S.G., and Morris, M.E. (2012). Mechanistic determinants of biotherapeutics absorption following SC administration. *AAPS J.* 14 (3): 559–570.

17 Hamuro, L.L. and Kishnani, N.S. (2012). Metabolism of biologics: biotherapeutic proteins. *Bioanalysis* 4 (2): 189–195.

18 Ryman, J.T. and Meibohm, B. (2017). Pharmacokinetics of monoclonal antibodies. *CPT: Pharmacomet. Syst. Pharmacol.* 6 (9): 576–588.

19 Liu, L. (2017). Pharmacokinetics of monoclonal antibodies and fc-fusion proteins. *Protein Cell* 9 (1): 15–32.

20 Bittner, B., Richter, W., and Schmidt, J. (2018). Subcutaneous administration of biotherapeutics: an overview of current challenges and opportunities. *BioDrugs.* 32 (5): 425–440.

21 Doogue, M.P. and Polasek, T.M. (2013). The ABCD of clinical pharmacokinetics. *Ther Adv Drug Saf.* 4 (1): 5–7.

22 Ezan, E., Becher, F., and Fenaille, F. (2014). Assessment of the metabolism of therapeutic proteins and antibodies. *Expert Opin. Drug Metab. Toxicol.* 10 (8): 1079–1091.

23 Kulkarni, J.A., Witzigmann, D., Thomson, S.B. et al. (2021). The current landscape of nucleic acid therapeutics. *Nat. Nanotechnol.* 16 (6): 630–643.

24 El-Kadiry, A.E.H., Rafei, M., and Shammaa, R. (2021). Cell therapy: types, regulation, and clinical benefits. *Front. Med.* 8: 756029.

25 Martinsen, E., Jinnurine, T., Subramani, S., and Rogne, M. (2024). Advances in RNA therapeutics for modulation of 'undruggable' targets. *Prog. Mol. Biol. Transl. Sci.* 204: 249–294.

26 Bashor, C.J., Hilton, I.B., Bandukwala, H. et al. (2022). Engineering the next generation of cell-based therapeutics. *Nat. Rev. Drug Discov.* 21 (9): 655–675.

27 Gauthier, M.A. and Klok, H.A. (2008). Peptide/protein – polymer conjugates: synthetic strategies and design concepts. *Chem. Commun.* (23): 2591–2611.

28 Gao, Y., Joshi, M., Zhao, Z., and Mitragotri, S. (2024). PEGylated therapeutics in the clinic. *Bioeng Transl Med.* 9 (1): e10600.

29 Abuchowski, A., van Es, T., Palczuk, N.C., and Davis, F.F. (1977). Alteration of immunological properties of bovine serum albumin by covalent attachment of polyethylene glycol. *J. Biol. Chem.* 252 (11): 3578–3581.

30 Abuchowski, A., McCoy, J.R., Palczuk, N.C. et al. (1977). Effect of covalent attachment of polyethylene glycol on immunogenicity and circulating life of bovine liver catalase. *J. Biol. Chem.* 252 (11): 3582–3586.

31 Armengol, E.S., Unterweger, A., and Laffleur, F. (2022). PEGylated drug delivery systems in the pharmaceutical field: past, present and future perspective. *Drug Dev. Ind. Pharm.* 48 (4): 129–139.

32 Gupta, V., Bhavanasi, S., Quadir, M. et al. (2019). Protein PEGylation for cancer therapy: bench to bedside. *J Cell Commun Signal* 13 (3): 319–330.

33 D'souza, A.A. and Shegokar, R. (2016). Polyethylene glycol (PEG): a versatile polymer for pharmaceutical applications. *Expert Opin. Drug Deliv.* 13 (9): 1257–1275.

34 Holz, E., Darwish, M., Tesar, D.B., and Shatz-Binder, W. (2023). A review of protein- and peptide-based chemical conjugates: past, present, and future. *Pharm.* 15 (2): 600.

35 Pyzik, M., Kozicky, L.K., Gandhi, A.K., and Blumberg, R.S. (2023). The therapeutic age of the neonatal Fc receptor. *Nat. Rev. Immunol.* 23 (7): 415–432.

36 Oganesyan, V., Damschroder, M.M., Cook, K.E. et al. (2014). Structural insights into neonatal fc receptor-based recycling mechanisms. *J. Biol. Chem.* 289 (11): 7812–7824.

37 Pyzik, M., Rath, T., Lencer, W.I. et al. (2015). FcRn: the architect behind the immune and nonimmune functions of IgG and albumin. *J. Immunol.* 194 (10): 4595–4603.

38 Rogers, B., Dong, D., Li, Z., and Li, Z. (2015). Recombinant human serum albumin fusion proteins and novel applications in drug delivery and therapy. *Curr. Pharm. Des.* 21 (14): 1899–1907.

39 Czajkowsky, D.M., Hu, J., Shao, Z., and Pleass, R.J. (2012). Fc-fusion proteins: new developments and future perspectives. *EMBO Mol. Med.* 4 (10): 1015–1028.

40 Booth, B.J., Ramakrishnan, B., Narayan, K. et al. (2018). Extending human IgG half-life using structure-guided design. *MAbs* 10 (7): 1098–1110.

41 Krohne, T.U., Eter, N., Holz, F.G., and Meyer, C.H. (2008). Intraocular pharmacokinetics of bevacizumab after a single intravitreal injection in humans. *Am. J. Ophthalmol.* 146 (4): 508–512.

42 Krohne, T.U., Liu, Z., Holz, F.G., and Meyer, C.H. (2012). Intraocular pharmacokinetics of ranibizumab following a single intravitreal injection in humans. *Am. J. Ophthalmol.* 154 (4): 682–686.

43 Shatz, W., Aaronson, J., Yohe, S. et al. (2018). Strategies for modifying drug residence time and ocular bioavailability to decrease treatment frequency for back of the eye diseases. *Expert Opin. Drug Deliv.* 16 (1): 43–57.

44 Lanyon-Hogg, T., Faronato, M., Serwa, R.A., and Tate, E.W. (2017). Dynamic protein acylation: new substrates, mechanisms, and drug targets. *Trends Biochem. Sci.* 42 (7): 566–581.

45 Guan, X. and Fierke, C.A. (2011). Understanding protein palmitoylation: biological significance and enzymology. *Sci. China Chem.* 54 (12): 1888–1897.
46 Kurtzhals, P., Østergaard, S., Nishimura, E., and Kjeldsen, T. (2023). Derivatization with fatty acids in peptide and protein drug discovery. *Nat. Rev. Drug Discov.* 22 (1): 59–80.
47 Curry, S., Mandelkow, H., Brick, P., and Franks, N. (1998). Crystal structure of human serum albumin complexed with fatty acid reveals an asymmetric distribution of binding sites. *Nat. Struct. Biol.* 5 (9): 827–835.
48 Jindal, A.B., Bhide, A.R., Salave, S. et al. (2023). Long-acting parenteral drug delivery systems for the treatment of chronic diseases. *Adv. Drug Deliv. Rev.* 198: 114862.
49 Vaishya, R., Khurana, V., Patel, S., and Mitra, A.K. (2015). Long-term delivery of protein therapeutics. *Expert Opin. Drug Deliv.* 12 (3): 415–440.
50 Wan, B., Bao, Q., and Burgess, D. (2023). Long-acting PLGA microspheres: advances in excipient and product analysis toward improved product understanding. *Adv. Drug Deliv. Rev.* 198: 114857.
51 Alsaab, H.O., Alharbi, F.D., Alhibs, A.S. et al. (2022). PLGA-based Nanomedicine: history of advancement and development in clinical applications of multiple diseases. *Pharm* 14 (12): 2728.
52 Schneider, C., Langer, R., Loveday, D., and Hair, D. (2017). Applications of ethylene vinyl acetate copolymers (EVA) in drug delivery systems. *J. Control. Release* 262: 284–295.
53 Diana, J.N., Tao, Y., Du, Q. et al. (2020). PLGA microspheres of hGH of preserved native state prepared using a self-regulated process. *Pharm* 12 (7): 683.
54 Park, K., Skidmore, S., Hadar, J. et al. (2019). Injectable, long-acting PLGA formulations: analyzing PLGA and understanding microparticle formation. *J. Control. Release* 304: 125–134.
55 Butreddy, A., Gaddam, R.P., Kommineni, N. et al. (2021). PLGA/PLA-based long-acting injectable depot microspheres in clinical use: production and characterization overview for protein/peptide delivery. *Int. J. Mol. Sci.* 22 (16): 8884.
56 Pawar, R., Pathan, A., Nagaraj, S. et al. (2023). Polycaprolactone and its derivatives for drug delivery. *Polym. Adv. Technol.* 34 (10): 3296–3316.
57 Li, L., Gatto, G.J., Brand, R.M. et al. (2021). Long-acting biodegradable implant for sustained delivery of antiretrovirals (ARVs) and hormones. *J. Control. Release* 340: 188–199.
58 Tyler, B., Gullotti, D., Mangraviti, A. et al. (2016). Polylactic acid (PLA) controlled delivery carriers for biomedical applications. *Adv. Drug Deliv. Rev.* 107: 163–175.
59 Lu, Y., Cheng, D., Niu, B. et al. (2023). Properties of poly (lactic-co-glycolic acid) and Progress of poly (lactic-co-glycolic acid)-based biodegradable materials in biomedical Research. *J. Pharm.* 16 (3): 454.
60 Makadia, H.K. and Siegel, S.J. (2011). Poly lactic-co-glycolic acid (PLGA) as biodegradable controlled drug delivery carrier. *Polymers* 3 (3): 1377–1397.
61 Fredenberg, S., Wahlgren, M., Reslow, M., and Axelsson, A. (2011). The mechanisms of drug release in poly (lactic-co-glycolic acid)-based drug delivery systems—A review. *Int. J. Pharm.* 415 (1–2): 34–52.
62 Yoo, J. and Won, Y.Y. (2020). Phenomenology of the initial burst release of Drugs from PLGA microparticles. *ACS Biomater Sci. Eng.* 6 (11): 6053–6062.
63 Hines, D.J. and Kaplan, D.L. (2013). Poly (lactic-co-glycolic) acid – controlled-release systems: experimental and modeling insights. *Crit Rev Ther Drug Carrier Syst.* 30 (3): 257–276.
64 Tamani, F., Bassand, C., Hamoudi, M.C. et al. (2019). Mechanistic explanation of the (up to) 3 release phases of PLGA microparticles: diprophylline dispersions. *Int. J. Pharm.* 572: 118819.
65 Washington, M.A., Swiner, D.J., Bell, K.R. et al. (2017). The impact of monomer sequence and stereochemistry on the swelling and erosion of biodegradable poly (lactic-co-glycolic acid) matrices. *Biomaterials* 117: 66–76.
66 Stiepel, R.T., Pena, E.S., Ehrenzeller, S.A. et al. (2022). A predictive mechanistic model of drug release from surface eroding polymeric nanoparticles. *J. Control. Release* 351: 883–895.
67 Ding, A.G. and Schwendeman, S.P. (2008). Acidic microclimate pH distribution in PLGA microspheres monitored by confocal laser scanning microscopy. *Pharm. Res.* 25 (9): 2041–2052.
68 Kang, J., Lambert, O., Ausborn, M., and Schwendeman, S.P. (2008). Stability of proteins encapsulated in injectable and biodegradable poly (lactide-co-glycolide)-glucose millicylinders. *Int. J. Pharm.* 357 (1–2): 235–243.
69 Chang, D.P., Garripelli, V.K., Rea, J. et al. (2015). Investigation of fragment antibody stability and its release mechanism from poly (Lactide-co-Glycolide)–Triacetin depots for sustained-release applications. *J. Pharm. Sci.* 104 (10): 3404–3417.
70 Zhang, S., Nagapudi, K., Shen, M. et al. (2022). Release mechanisms and practical percolation threshold for long-acting biodegradable implants: an image to simulation study. *J. Pharm. Sci.* 111 (7): 1896–1910.

71 Shastri, V. (2003). Non-degradable biocompatible polymers in medicine: past, present and future. *Curr. Pharm. Biotechnol.* 4 (5): 331–337.

72 Kamaly, N., Yameen, B., Wu, J., and Farokhzad, O.C. (2016). Degradable controlled-release polymers and polymeric nanoparticles: mechanisms of controlling drug release. *Chem. Rev.* 116 (4): 2602–2663.

73 Ferrari, G., da Silva, L.S., Cerruti, R. et al. (2024). Development and characterization of 3D printed ethylene vinyl acetate (EVA) as drug delivery device for the treatment of overactive bladder. *Drug Dev. Ind. Pharm.* (ahead-of-print): 1–12.

74 Langer, R. and Folkman, J. (1976). Polymers for the sustained release of proteins and other macromolecules. *Nature* 263 (5580): 797–800.

75 Saltzman, W.M., Mak, M.W., Mahoney, M.J. et al. (1999). Intracranial delivery of recombinant nerve growth factor: release kinetics and protein distribution for three delivery systems. *Pharm. Res.* 16 (2): 232–240.

76 Choi, S.B. (2003). CYPHER coronary stents and risk of thrombosis. *CMAJ* 169 (3): 218.

77 Allyn, M.M., Luo, R.H., Hellwarth, E.B., and Swindle-Reilly, K.E. (2022). Considerations for polymers used in ocular drug delivery. *Front. Med.* 8: 787644.

78 Basle, Y.L., Chennell, P., Tokhadze, N. et al. (2020). Physicochemical stability of monoclonal antibodies: a review. *J. Pharm. Sci.* 109 (1): 169–190.

79 Mueller, M., Loh, M.Q.T., Tee, D.H.Y. et al. (2013). Liquid formulations for long-term storage of monoclonal IgGs. *Appl. Biochem. Biotechnol.* 169 (4): 1431–1448.

80 Chen, Y., Mutukuri, T.T., Wilson, N.E., and Zhou, Q.T. (2021). Pharmaceutical protein solids: drying technology, solid-state characterization and stability. *Adv. Drug Deliv. Rev.* 172: 211–233.

81 Patel, S.M., Nail, S.L., Pikal, M.J. et al. (2017). Lyophilized drug product cake appearance: what is acceptable? *J. Pharm. Sci.* 106 (7): 1706–1721.

82 Wang, W. (2000). Lyophilization and development of solid protein pharmaceuticals. *Int. J. Pharm.* 203 (1–2): 1–60.

83 Carpenter, J.F. and Manning, M.C. (2002). Rational design of stable protein formulations: theory and practice. *Pharm. Biotechnol.*

84 Thakral, S., Sonje, J., Munjal, B., and Suryanarayanan, R. (2021). Stabilizers and their interaction with formulation components in frozen and freeze-dried protein formulations. *Adv. Drug Deliv. Rev.* 173: 1–19.

85 Bjelošević, M., Pobirk, A.Z., Planinšek, O., and Grabnar, P.A. (2020). Excipients in freeze-dried biopharmaceuticals: contributions toward formulation stability and lyophilisation cycle optimisation. *Int. J. Pharm.* 576: 119029.

86 Pinto, J.T., Faulhammer, E., Dieplinger, J. et al. (2021). Progress in spray-drying of protein pharmaceuticals: literature analysis of trends in formulation and process attributes. *Dry. Technol.* 39 (11): 1415–1446.

87 Ameri, M. and Maa, Y.F. (2006). Spray drying of biopharmaceuticals: stability and process considerations. *Dry. Technol.* 24 (6): 763–768.

88 Brom, J.A., Petrikis, R.G., and Pielak, G.J. (2023). How sugars protect dry protein structure. *Biochemistry* 62 (5): 1044–1052.

89 Mensink, M.A., Frijlink, H.W., van Der Voort Maarschalk, K., and WLJ, H. (2017). How sugars protect proteins in the solid state and during drying (review): mechanisms of stabilization in relation to stress conditions. *Eur. J. Pharm. Biopharm.* 114: 288–295.

90 Ahlgren, K., Olsson, C., Ermilova, I., and Swenson, J. (2023). New insights into the protein stabilizing effects of trehalose by comparing with sucrose. *Phys. Chem. Chem. Phys.* 25 (32): 21215–21226.

91 Chang, L.L., Shepherd, D., Sun, J. et al. (2005). Mechanism of protein stabilization by sugars during freeze-drying and storage: native structure preservation, specific interaction, and/or immobilization in a glassy matrix? *J. Pharm. Sci.* 94 (7): 1427–1444.

92 Crowe, J.H., Crowe, L.M., and Chapman, D. (1984). Infrared spectroscopic studies on interactions of water and carbohydrates with a biological membrane. *Arch. Biochem. Biophys.* 232 (1): 400–407.

93 Li, J., Wang, H., Wang, L. et al. (2024). Stabilization effects of saccharides in protein formulations: a review of sucrose, trehalose, cyclodextrins and dextrans. *Eur. J. Pharm. Sci.* 192: 106625.

94 Jovanović, N., Bouchard, A., Hofland, G.W. et al. (2006). Distinct effects of sucrose and trehalose on protein stability during supercritical fluid drying and freeze-drying. *Eur. J. Pharm. Sci.* 27 (4): 336–345.

95 Starciuc, T., Malfait, B., Danede, F. et al. (2020). Trehalose or sucrose: which of the two should be used for stabilizing proteins in the solid state? A dilemma investigated by in situ micro-Raman and dielectric relaxation spectroscopies during and after freeze-drying. *J. Pharm. Sci.* 109 (1): 496–504.

96 Rajagopal, K., Wood, J., Tran, B. et al. (2013). Trehalose limits BSA aggregation in spray-dried formulations at high temperatures: implications in preparing polymer implants for long-term protein delivery. *J. Pharm. Sci.* 102 (8): 2655–2666.

97 Gikanga, B., Hui, A., and Maa, Y.F. (2017). Mechanistic investigation on grinding-induced subvisible particle

formation during mixing and filling of monoclonal antibody formulations. *PDA J. Pharm. Sci. Technol.* 72 (2): 117–133.

98 Alsaleh, F.M., Smith, F.J., Keady, S., and Taylor, K.M.G. (2010). Insulin pumps: from inception to the present and toward the future. *J. Clin. Pharm. Ther.* 35 (2): 127–138.

99 Vora, L.K., Sabri, A.H., Naser, Y. et al. (2023). Long-acting microneedle formulations. *Adv. Drug Deliv. Rev.* 201: 115055.

100 Makwana, S., Basu, B., Makasana, Y., and Dharamsi, A. (2011). Prefilled syringes: an innovation in parenteral packaging. *Int. J. Pharm. Investig.* 1 (4): 200–206.

101 Jaman, R.U. and MstF, S. (2024). Design and analysis of a reservoir-based controllable microneedle for transdermal drug delivery applications. *Drug Deliv. Transl. Res.* 14 (3): 812–825.

102 Mayor, A., Thibert, B., Huille, S. et al. (2021). Inhaled antibodies: formulations require specific development to overcome instability due to nebulization. *Drug Deliv. Transl. Res.* 11 (4): 1625–1633.

103 Chang, M.A., Kapre, A., Kaufman, D. et al. (2022). Patient preference and treatment satisfaction with a port delivery system for Ranibizumab vs Intravitreal injections in patients with Neovascular age-related macular degeneration. *JAMA Ophthalmol.* 140 (8): 771–778.

104 Burckhardt, M., Smith, G.J., Cooper, M.N. et al. (2018). Real-world outcomes of insulin pump compared to injection therapy in a population-based sample of children with type 1 diabetes. *Pediatr. Diabetes* 19 (8): 1459–1466.

105 Sassalos, T.M. and Paulus, Y.M. (2019). Prefilled syringes for intravitreal drug delivery. *Clin. Ophthalmol.* 13: 701–706.

106 Kasi, S.G., Prabhu, S.V., Sanjay, S. et al. (2013). Prefilled syringes versus vials: impact on vaccination efficiency and patient safety in Indian private market. *Pediatr. Infect. Dis.* 5 (4): 181–186.

107 Berget, C., Messer, L.H., and Forlenza, G.P. (2019). A clinical overview of insulin pump therapy for the management of diabetes: past, present, and future of intensive therapy. *Diabetes Spectr.* 32 (3): 194–204.

108 Miller, E.M. (2020). Using continuous glucose monitoring in clinical practice. *Clin Diabetes: A Publ Am. Diabetes Assoc.* 38 (5): 429–438.

109 Lee, T.T.M., Collett, C., Bergford, S. et al. (2023). Automated insulin delivery in women with pregnancy complicated by type 1 diabetes. *N. Engl. J. Med.* 389 (17): 1566–1578.

110 Haidar, A., Legault, L., Raffray, M. et al. (2021). Comparison between closed-loop insulin delivery system (the artificial pancreas) and sensor-augmented pump therapy: a randomized-controlled crossover trial. *Diabetes Technol. Ther.* 23 (3): 168–174.

111 Mosquera-Lopez, C., Wilson, L.M., Youssef, J.E. et al. (2023). Enabling fully automated insulin delivery through meal detection and size estimation using artificial intelligence. *NPJ Digit. Med.* 6 (1): 39.

112 Yohe, S., Maass, K.F., Horvath, J. et al. (2022). In-vitro characterization of ranibizumab release from the port delivery system. *J. Control. Release* 345: 101–107.

113 Ranade, S.V., Wieland, M.R., Tam, T. et al. (2022). The port delivery system with ranibizumab: a new paradigm for long-acting retinal drug delivery. *Drug Deliv.* 29 (1): 1326–1334.

114 Khanani, A.M., Graff, J.M., Marcus, D.M. et al. (2022). Refill-exchange procedure of the port delivery system with Ranibizumab: overview and clinical trial experience. *OSLI RETINA* 53 (5): 257–265.

115 Genentech: Press Releases., (2024). Genentech to Reintroduce Susvimo for People With Wet Age-Related Macular Degeneration (AMD). (cited 2024 Jul 20). Available from: //http://www.gene.com/media/press-releases/15031/2024-07-08/genentech-to-reintroduce-susvimo-for-peo

116 Currie, G.M. (2018). Pharmacology, part 2: Introduction to pharmacokinetics. *J. Nucl. Med. Technol.* 46 (3): 221–230.

117 Hartwig-Otto, H. (1983). Pharmacokinetic considerations of common analgesics and antipyretics. *Am. J. Med.* 75 (5): 30–37.

118 Bruschi, M.L. (2015). *Strategies to Modify the Drug Release from Pharmaceutical Systems*, 63–86. Woodhead Publishing.

119 Zhao, Y.N., Xu, X., Wen, N. et al. (2017). A drug carrier for sustained zero-order release of peptide therapeutics. *Sci. Rep.* 7 (1): 5524.

120 Wilinska, M.E., Chassin, L.J., Schaller, H.C. et al. (2005). Insulin kinetics in type-1 diabetes: continuous and bolus delivery of rapid acting insulin. *IEEE Trans. Biomed. Eng.* 52 (1): 3–12.

121 Wykoff, C.C., Campochiaro, P.A., Pieramici, D.J. et al. (2022). Pharmacokinetics of the port delivery system with Ranibizumab in the ladder phase 2 trial for Neovascular age-related macular degeneration. *Ophthalmol Therapy* 11 (5): 1705–1717.

122 Savjani, K.T., Gajjar, A.K., and Savjani, J.K. (2012). Drug solubility: importance and enhancement techniques. *ISRN Pharm.* 2012: 195727.

123 Han, X., Ning, W., Ma, X. et al. (2020). Improving protein solubility and activity by introducing small

peptide tags designed with machine learning models. *Metab. Eng. Commun.* 11: e00138.
124 Tomar, D.S., Kumar, S., Singh, S.K. et al. (2016). Molecular basis of high viscosity in concentrated antibody solutions: strategies for high concentration drug product development. *MAbs* 8 (2): 216–228.
125 Shinoda, K., Yagura, K., Matsumoto, S. et al. (2021). Intraocular temperature at different sites in eye measured at the beginning of vitreous surgery. *J. Clin. Med.* 10 (15): 3412.
126 Hieb, A.R., Horvath, J., Rea, J. et al. (2024). Stability of ranibizumab during continuous delivery from the port delivery platform. *J. Control. Release* 366: 170–181.
127 FDA [Internet]. (2024). Information Regarding Insulin Storage and Switching Between Products in an Emergencyl. (cited 2024 May 13). Available from: //http://www.fda.gov/drugs/emergency-preparedness-drugs/information-regarding-insulin-storage-and-switching-between-products-emergency#:~:text=Insulin%20contained%20in%20the%20infusion,%C2%B0F%20should%20be%20discarded.
128 Vlieland, N.D., Gardarsdottir, H., Bouvy, M.L. et al. (2016). The majority of patients do not store their biologic disease-modifying antirheumatic drugs within the recommended temperature range. *Rheumatology* 55 (4): 704–709.
129 Yuan, H., Xu, J., van Dam, E.P. et al. (2017). Strategies to increase the thermal stability of truly biomimetic hydrogels: combining hydrophobicity and directed hydrogen bonding. *Macromolecules* 50 (22): 9058–9065.
130 Marco-Dufort, B., Janczy, J.R., Hu, T. et al. (2022). Thermal stabilization of diverse biologics using reversible hydrogels. *Sci. Adv.* 8 (31): eabo 0502.
131 Sridhar, B.V., Janczy, J.R., Hatlevik, Ø. et al. (2018). Thermal stabilization of biologics with Photoresponsive hydrogels. *Biomacromolecules* 19 (3): 740–747.
132 Jenke, D. and Carlson, T. (2014). A compilation of safety impact information for Extractables associated with materials used in pharmaceutical packaging, delivery, administration, and manufacturing systems. *PDA J. Pharm. Sci. Technol.* 68 (5): 407–455.
133 Dorival-García, N., Larsson, I., and Bones, J. (2017). Non-volatile extractable analysis of prefilled syringes for parenteral administration of drug products. *J. Pharm. Biomed. Anal.* 142: 337–342.
134 FDA. Research C for DE and, Research C for BE and, Administration USD of H and HSF and D. Guidance for Industry: Container Closure Systems for Packaging Human Drugs and Biologics. 1999; Available from: //http://www.fda.gov/regulatory-information/search-fda-guidance-documents/container-closure-systems-packaging-human-drugs-and-biologics
135 Saux, S.L., Aubert-Pouëssel, A., Ouchait, L. et al. (2021). Nanotechnologies for intracellular protein delivery: recent Progress in inorganic and organic Nanocarriers. *Adv. Ther.* 4 (6).
136 Janjua, T.I., Cao, Y., Kleitz, F. et al. (2023). Silica nanoparticles: a review of their safety and current strategies to overcome biological barriers. *Adv. Drug Deliv. Rev.* 203: 115115.
137 Croissant, J.G. and Brinker, C.J. (2018). Chapter eight biodegradable silica-based nanoparticles: dissolution kinetics and selective bond cleavage. *Enzyme* 43: 181–214.
138 Lee, J.H., Shin, S.K., Jiang, Y. et al. (2015). Facilitated tau degradation by USP14 Aptamers via enhanced proteasome activity. *Sci. Rep.* 5 (1): 10757.
139 Xu, C., Yu, M., Noonan, O. et al. (2015). Core-cone structured monodispersed mesoporous silica nanoparticles with ultra-large cavity for protein delivery. *Small* 11 (44): 5949–5955.
140 Tyagi, P., Koskinen, M., Mikkola, J. et al. (2018). Silica microparticles for sustained zero-order release of an anti-CD40L antibody. *Drug Deliv. Transl. Res.* 8 (2): 368–374.
141 Janjua, T.I., Cao, Y., Yu, C., and Popat, A. (2021). Clinical translation of silica nanoparticles. *Nat. Rev. Mater.* 6 (12): 1072–1074.
142 Martins, S., Sarmento, B., Ferreira, D.C., and Souto, E.B. (2007). Lipid-based colloidal carriers for peptide and protein delivery – liposomes versus lipid nanoparticles. *Int. J. Nanomedicine.*
143 Viegas, C., Seck, F., and Fonte, P. (2022). An insight on lipid nanoparticles for therapeutic proteins delivery. *J Drug Deliv Sci Technol.* 77: 103839.
144 Almeida, A.J. and Souto, E. (2007). Solid lipid nanoparticles as a drug delivery system for peptides and proteins. *Adv. Drug Deliv. Rev.* 59 (6): 478–490.
145 Chauhan, I., Yasir, M., Verma, M., and Singh, A.P. (2020). Nanostructured lipid carriers: a Groundbreaking approach for transdermal drug delivery. *Adv. Pharm. Bull.* 10 (2): 150–165.
146 Vollrath, M., Engert, J., and Winter, G. (2017). Long-term release and stability of pharmaceutical proteins delivered from solid lipid implants. *Eur. J. Pharm. Biopharm.* 117: 244–255.
147 Vollrath, M., Engert, J., and Winter, G. (2018). New insights into process understanding of solid lipid extrusion (SLE) of extruded lipid implants for sustained protein delivery. *Eur. J. Pharm. Biopharm.* 130: 11–21.
148 Wang, B. and Friess, W. (2018). Lipid-coated mannitol core microparticles for sustained release of protein. *Eur. J. Pharm. Biopharm.* 128: 91–97.

149 Zhai, J., Fong, C., Tran, N., and Drummond, C.J. (2019). Non-lamellar Lyotropic liquid crystalline lipid nanoparticles for the next generation of Nanomedicine. *ACS Nano* 13 (6): 6178–6206.

150 Chavda, V.P., Dawre, S., Pandya, A. et al. (2022). Lyotropic liquid crystals for parenteral drug delivery. *J. Control. Release* 349: 533–549.

151 Rizwan, S.B., Boyd, B.J., Rades, T., and Hook, S. (2010). Bicontinuous cubic liquid crystals as sustained delivery systems for peptides and proteins. *Expert Opin. Drug Deliv.* 7 (10): 1133–1144.

152 Angelova, A., Ollivon, M., Campitelli, A., and Bourgaux, C. (2003). Lipid cubic phases as stable Nanochannel network structures for protein biochip development: X-ray diffraction study. *Langmuir* 19 (17): 6928–6935.

153 Rizwan, S.B., Hanley, T., Boyd, B.J. et al. (2009). Liquid crystalline systems of phytantriol and glyceryl monooleate containing a hydrophilic protein: characterisation, swelling and release kinetics. *J. Pharm. Sci.* 98 (11): 4191–4204.

154 Weiser, P. "FDA Approves New Treatments for Opioid Overdose, Opioid Use Disorder." Formulary Watch, 24 May 2023, p. NA. Gale Academic OneFile, link.gale.com/apps/doc/A776741546/AONE?u=tel_oweb&sid=googleScholar&xid=7285c3c0. Accessed 28 Mar. 2024.

155 Albayaty, M., Linden, M., Olsson, H. et al. (2017). Pharmacokinetic evaluation of once-weekly and once-monthly buprenorphine subcutaneous injection depots (CAM 2038) versus intravenous and sublingual buprenorphine in healthy volunteers under naltrexone blockade: an open-label phase 1 study. *Adv. Ther.* 34 (2): 560–575.

156 Vogel, A.B., Kanevsky, I., Che, Y. et al. (2021). BNT162b vaccines protect rhesus macaques from SARS-CoV-2. *Nature* 592 (7853): 283–289.

157 Jackson, L.A., Anderson, E.J., Rouphael, N.G. et al. (2020). An mRNA vaccine against SARS-CoV-2 — Preliminary report. *N. Engl. J. Med.* 383 (20): 1920–1931.

158 Fiolet, T., Kherabi, Y., Mac Donald, C.J. et al. (2022). Comparing COVID-19 vaccines for their characteristics, efficacy and effectiveness against SARS-CoV-2 and variants of concern: a narrative review. *Clin. Microbiol. Infect.* 28 (2): 202–221.

159 Löfström, E., Eringfält, A., Kötz, A. et al. (2021). Dynamics of IgG-avidity and antibody levels after Covid-19. *J. Clin. Virol.* 144: 104986.

160 Kranz, L.M., Diken, M., Haas, H. et al. (2016). Systemic RNA delivery to dendritic cells exploits antiviral defence for cancer immunotherapy. *Nature* 534 (7607): 396–401.

161 Braiteh, F., LoRusso, P., Balmanoukian, A. et al. (2020). Abstract CT169: a phase Ia study to evaluate RO7198457, an individualized Neoantigen specific immuno therapy (iNeST), in patients with locally advanced or metastatic solid tumors. *Cancer Res.* 80 (16_Supplement): CT169–CT169.

162 Ye, Z., Chen, J., Zhao, X. et al. (2022). In vitro engineering chimeric antigen receptor macrophages and T cells by lipid nanoparticle-mediated mRNA delivery. *ACS Biomater Sci. Eng.* 8 (2): 722–733.

163 Billingsley, M.M., Hamilton, A.G., Mai, D. et al. (2022). Orthogonal Design of Experiments for optimization of lipid nanoparticles for mRNA engineering of CAR T cells. *Nano Lett.* 22 (1): 533–542.

164 Schmid, D., Park, C.G., Hartl, C.A. et al. (2017). T cell-targeting nanoparticles focus delivery of immunotherapy to improve antitumor immunity. *Nat. Commun.* 8 (1): 1747.

165 Billingsley, M.M., Gong, N., Mukalel, A.J. et al. (2024). In Vivo mRNA CAR T cell engineering via targeted Ionizable lipid nanoparticles with extrahepatic tropism. *Small* 20 (11): e2304378.

166 Katakowski, J.A., Mukherjee, G., Wilner, S.E. et al. (2016). Delivery of siRNAs to dendritic cells using DEC205-targeted lipid nanoparticles to inhibit immune responses. *Mol. Ther.* 24 (1): 146–155.

167 Rurik, J.G., Tombácz, I., Yadegari, A. et al. (2022). CAR T cells produced in vivo to treat cardiac injury. *Science* 375 (6576): 91–96.

168 Tarab-Ravski, D., Hazan-Halevy, I., Goldsmith, M. et al. (2023). Delivery of therapeutic RNA to the bone marrow in multiple myeloma using CD38-targeted lipid nanoparticles. *Adv. Sci.* 10 (21): 2301377.

169 Tombácz, I., Laczkó, D., Shahnawaz, H. et al. (2021). Highly efficient CD4+ T cell targeting and genetic recombination using engineered CD4+ cell-homing mRNA-LNPs. *Mol. Ther.* 29 (11): 3293–3304.

170 Tilsed, C.M., Sadiq, B.A., Papp, T.E. et al. (2024). IL7 increases targeted lipid nanoparticle–mediated mRNA expression in T cells in vitro and in vivo by enhancing T cell protein translation. *Proc. Natl. Acad. Sci.* 121 (13): e2319856121.

171 Ramishetti, S., Hazan-Halevy, I., Palakuri, R. et al. (2020). A combinatorial library of lipid nanoparticles for RNA delivery to leukocytes. *Adv. Mater.* 32 (12): e1906128.

172 Veiga, N., Goldsmith, M., Granot, Y. et al. (2018). Cell specific delivery of modified mRNA expressing therapeutic proteins to leukocytes. *Nat. Commun.* 9 (1): 4493.

173 Rosenblum, D., Gutkin, A., Kedmi, R. et al. (2020). CRISPR-Cas 9 genome editing using targeted lipid nanoparticles for cancer therapy. *Sci. Adv.* 6 (47): eabc 9450.

174 Kedmi, R., Veiga, N., Ramishetti, S. et al. (2018). A modular platform for targeted RNAi therapeutics. *Nat. Nanotechnol.* 13 (3): 214–219.

175 Ramishetti, S., Kedmi, R., Goldsmith, M. et al. (2015). Systemic gene silencing in primary T lymphocytes using targeted lipid nanoparticles. *ACS Nano* 9 (7): 6706–6716.

176 Agard, N.J., Prescher, J.A., and Bertozzi, C.R. (2004). A strain-promoted [3 + 2] Azide–alkyne cycloaddition for covalent modification of biomolecules in living systems. *J. Am. Chem. Soc.* 126 (46): 15046–15047.

177 Stranford, D.M., Simons, L.M., Berman, K.E. et al. (2023). Genetically encoding multiple functionalities into extracellular vesicles for the targeted delivery of biologics to T cells. *Nat. Biomed. Eng.* 1–18.

178 Hamilton, J.R., Chen, E., Perez, B.S. et al. (2024). In vivo human T cell engineering with enveloped delivery vehicles. *Nat. Biotechnol.* 1–9.

179 Parayath, N.N. and Stephan, M.T. (2021). In situ programming of CAR T cells. *Annu. Rev. Biomed. Eng.* 23 (1): 1–21.

180 Perez-Villar, J.J., Whitney, G.S., Bowen, M.A. et al. (1999). CD5 negatively regulates the T-cell antigen receptor signal transduction pathway: involvement of SH2-containing Phosphotyrosine phosphatase SHP-1. *Mol. Cell. Biol.* 19 (4): 2903–2912.

181 Hilder, R., Tsai, K., Quandt, Z. et al. (2023). Safety and efficacy of immune checkpoint inhibitor cancer therapy in patients with preexisting type 1 diabetes mellitus. *Front. Endocrinol.* 14: 1242830.

182 Cheadle, E.J., Gornall, H., Baldan, V. et al. (2014). CAR T cells: driving the road from the laboratory to the clinic. *Immunol. Rev.* 257 (1): 91–106.

183 Wang, V., Gauthier, M., Decot, V. et al. (2023). Systematic review on CAR-T cell clinical trials up to 2022: academic Center input. *Cancers* 15 (4): 1003.

184 Zhao, J., Lin, Q., Song, Y., and Liu, D. (2018). Universal CARs, universal T cells, and universal CAR T cells. *J. Hematol. Oncol.* 11 (1): 132.

185 Parayath, N.N., Stephan, S.B., Koehne, A.L. et al. (2020). In vitro-transcribed antigen receptor mRNA nanocarriers for transient expression in circulating T cells in vivo. *Nat. Commun.* 11 (1): 6080.

186 Jung, J.H., Rha, M.S., Sa, M. et al. (2021). SARS-CoV-2-specific T cell memory is sustained in COVID-19 convalescent patients for 10 months with successful development of stem cell-like memory T cells. *Nat. Commun.* 12 (1): 4043.

187 Chen, D.S. and Mellman, I. (2013). Oncology meets immunology: the cancer-immunity cycle. *Immunity* 39 (1): 1–10.

188 Eisenbarth, S.C. (2019). Dendritic cell subsets in T cell programming: location dictates function. *Nat. Rev. Immunol.* 19 (2): 89–103.

189 Sallusto, F., Cella, M., Danieli, C., and Lanzavecchia, A. (1995). Dendritic cells use macropinocytosis and the mannose receptor to concentrate macromolecules in the major histocompatibility complex class II compartment: downregulation by cytokines and bacterial products. *J. Exp. Med.* 182 (2): 389–400.

190 Sasaki, K., Sato, Y., Okuda, K. et al. (2022). mRNA-loaded lipid nanoparticles targeting dendritic cells for cancer immunotherapy. *Pharm* 14 (8): 1572.

191 der Jeught, K.V., Koker, S.D., Bialkowski, L. et al. (2018). Dendritic cell targeting mRNA Lipopolyplexes combine strong antitumor T-cell immunity with improved inflammatory safety. *ACS Nano* 12 (10): 9815–9829.

192 Goswami, R., Chatzikleanthous, D., Lou, G. et al. (2019). Mannosylation of LNP results in improved potency for self-amplifying RNA (SAM) vaccines. *ACS Infect. Dis.* 5 (9): 1546–1558.

193 Kenison, J.E., Stevens, N.A., and Quintana, F.J. (2024). Therapeutic induction of antigen-specific immune tolerance. *Nat. Rev. Immunol.* 24 (5): 338–357.

194 Tahtinen, S., Tong, A.J., Himmels, P. et al. (2022). IL-1 and IL-1ra are key regulators of the inflammatory response to RNA vaccines. *Nat. Immunol.* 23 (4): 532–542.

195 Gomi, M., Nakayama, Y., Sakurai, Y. et al. (2023). Tolerogenic lipid nanoparticles for delivering self-antigen mRNA for the treatment of experimental autoimmune encephalomyelitis. *J. Pharm.* 16 (9): 1270.

196 Krienke, C., Kolb, L., Diken, E. et al. (2021). A noninflammatory mRNA vaccine for treatment of experimental autoimmune encephalomyelitis. *Science* 371 (6525): 145–153.

197 Benne, N., ter Braake, D., Stoppelenburg, A.J., and Broere, F. (2022). Nanoparticles for inducing antigen-specific T cell tolerance in autoimmune diseases. *Front. Immunol.* 13: 864403.

198 Podojil, J.R., Genardi, S., Chiang, M.Y. et al. (2022). Tolerogenic immune-modifying nanoparticles encapsulating multiple recombinant pancreatic β cell proteins prevent onset and progression of type 1 diabetes in nonobese diabetic mice. *J. Immunol.* 209 (3): 465–475.

199 Kelly, C.P., Murray, J.A., Leffler, D.A. et al. (2021). TAK-101 nanoparticles induce gluten-specific tolerance in celiac disease: a randomized, double-blind,

placebo-controlled study. *Gastroenterology* 161 (1): 66–80.e8.

200 Ommaya, A.K. (1963). Subcutaneous reservoir and pump for sterile access to ventricular cerebrospinal fluid. *Lancet* 282 (7315): 983–984.

201 Shapiro, W.R., Young, D.F., and Mehta, B.M. (1975). Methotrexate: distribution in cerebrospinal fluid after intravenous, Ventricular and lumbar injections. *N. Engl. J. Med.* 293 (4): 161–166.

202 Yan, Q., Matheson, C., Sun, J. et al. (1994). Distribution of intracerebral Ventricularly administered Neurotrophins in rat brain and its correlation with Trk receptor expression. *Exp. Neurol.* 127 (1): 23–36.

203 Pardridge, W.M. (2022). A historical review of brain drug delivery. *Pharm* 14 (6): 1283.

204 Blasberg, R.G., Patlak, C., and Fenstermacher, J.D. (1975). Intrathecal chemotherapy: brain tissue profiles after ventriculocisternal perfusion. *J. Pharmacol. Exp. Ther.* 195 (1): 73–83.

205 Fishman, K.A. and Christy, N.P. (1965). Fate of adrenal cortical steroids following intrathecal injection. *Neurology* 15 (1): 1–1.

206 Aird, R.B. (1984). A study of intrathecal, cerebrospinal fluid-to-brain exchange. *Exp. Neurol.* 86 (2): 342–358.

207 Hall, C.D., Dafni, U., Simpson, D. et al. (1998). Failure of Cytarabine in progressive multifocal leukoencephalopathy associated with human immunodeficiency virus infection. *N. Engl. J. Med.* 338 (19): 1345–1351.

208 Nutt, J.G., Burchiel, K.J., Comella, C.L. et al. (2003). Randomized, double-blind trial of glial cell line-derived neurotrophic factor (GDNF) in PD. *Neurology* 60 (1): 69–73.

209 Schulz, A., Ajayi, T., Specchio, N. et al. (2018). Study of intraventricular Cerliponase Alfa for CLN2 disease. *N. Engl. J. Med.* 378 (20): 1898–1907.

210 Meng, Y., Sohar, I., Wang, L. et al. (2012). Systemic administration of tripeptidyl peptidase I in a mouse model of late infantile neuronal ceroid Lipofuscinosis: effect of glycan modification. *PLoS One* 7 (7): e40509.

211 Vuillemenot, B.R., Kennedy, D., Reed, R.P. et al. (2014). Recombinant human tripeptidyl peptidase-1 infusion to the monkey CNS: safety, pharmacokinetics, and distribution. *Toxicol. Appl. Pharmacol.* 277 (1): 49–57.

212 Rapoport, S., Hori, M., and Klatzo, I. (1972). Testing of a hypothesis for osmotic opening of the blood-brain barrier. *Am J Physiol-Legacy Content.* 223 (2): 323–331.

213 Rapoport, S.I., Hori, M., and Klatzo, I. (1971). Reversible osmotic opening of the blood-brain barrier. *Science* 173 (4001): 1026–1028.

214 Brightman, M.W., Hori, M., Rapoport, S.I. et al. (1973). Osmotic opening of tight junctions in cerebral endothelium. *J. Comp. Neurol.* 152 (4): 317–325.

215 Wang, D., Wang, C., Wang, L., and Chen, Y. (2019). A comprehensive review in improving delivery of small-molecule chemotherapeutic agents overcoming the blood-brain/brain tumor barriers for glioblastoma treatment. *Drug Deliv.* 26 (1): 551–565.

216 Rodriguez, A., Tatter, S.B., and Debinski, W. (2015). Neurosurgical techniques for disruption of the blood–brain barrier for glioblastoma treatment. *Pharm* 7 (3): 175–187.

217 Dobrogowska, D. and Vorbrodt, A. (2004). Immunogold localization of tight junctional proteins in normal and osmotically-affected rat blood-brain barrier. *J. Mol. Histol.* 35 (5): 529–539.

218 Wang, M., Etu, J., and Joshi, S. (2007). Enhanced disruption of the blood brain barrier by Intracarotid mannitol injection during transient cerebral Hypoperfusion in rabbits. *J. Neurosurg. Anesthesiol.* 19 (4): 249–256.

219 Neuwelt, E.A., Frenkel, E.P., Rapoport, S., and Barnett, P. (1980). Effect of osmotic blood-brain barrier disruption on methotrexate pharmacokinetics in the dog. *Neurosurgery* 7 (1): 36–43.

220 Kessler, R.M., Goble, J.C., Bird, J.H. et al. (1984). Measurement of blood—brain barrier permeability with positron emission tomography and [68Ga]EDTA. *J. Cereb. Blood Flow Metab.* 4 (3): 323–328.

221 Rapoport, S.I. (1996). Modulation of blood-brain barrier permeability. *J. Drug Target.* 3 (6): 417–425.

222 Zünkeler, B., Carson, R.E., Olson, J. et al. (1996). Hyperosmolar blood-brain barrier disruption in baboons: an in vivo study using positron emission tomography and rubidium-82. *J. Neurosurg.* 84 (3): 494–502.

223 Neuwelt, E.A., Hill, S.A., and Frenkel, E.P. (1984). Osmotic blood-brain barrier modification and combination chemotherapy: concurrent tumor regression in areas of barrier opening and progression in brain regions distant to barrier opening. *Neurosurgery* 15 (3): 362–366.

224 Neuwelt, E.A., Glasberg, M., Frenkel, E., and Barnett, P. (1983). Neurotoxicity of chemotherapeutic agents after blood-brain barrier modification: neuropathological studies. *Ann. Neurol.* 14 (3): 316–324.

225 Tomiwa, K., Hazama, F., and Mikawa, H. (1983). Neurotoxicity of vincristine after the osmotic opening of the blood-brain barrier. *Neuropathol. Appl. Neurobiol.* 9 (5): 345–354.

226 Jeon, J.Y., Kovanlikaya, I., Boockvar, J.A. et al. (2012). Metabolic response of glioblastoma to Superselective intra-arterial cerebral infusion of bevacizumab: a proton MR spectroscopic imaging study. *Am. J. Neuroradiol.* 33 (11): 2095–2102.

227 Kulason, K.O., Schneider, J.R., Chakraborty, S. et al. (2018). Superselective intraarterial cerebral infusion of cetuximab with blood brain barrier disruption combined with Stupp protocol for newly diagnosed glioblastoma. *J. Exp. Ther. Oncol.* 12 (3): 223–229.

228 Mondou, P., Mériaux, S., Nageotte, F. et al. (2023). State of the art on microbubble cavitation monitoring and feedback control for blood-brain-barrier opening using focused ultrasound. *Phys. Med. Biol.* 68 (18): 18TR03.

229 Bathini, P., Sun, T., Schenk, M. et al. (2022). Acute effects of focused ultrasound-induced blood-brain barrier opening on anti-Pyroglu 3 Abeta antibody delivery and immune responses. *Biomol. Ther.* 12 (7): 951.

230 Wasielewska, J.M., Chaves, J.C.S., Johnston, R.L. et al. (2022). A sporadic Alzheimer's blood-brain barrier model for developing ultrasound-mediated delivery of Aducanumab and anti-tau antibodies. *Theranostics* 12 (16): 6826–6847.

231 Ye, D., Yuan, J., Yue, Y. et al. (2021). Focused ultrasound-enhanced delivery of Intranasally administered anti-programmed cell death-ligand 1 antibody to an intracranial murine glioma model. *Pharm* 13 (2): 190.

232 Blesa, J., Pineda-Pardo, J.A., Inoue, K. et al. (2023). BBB opening with focused ultrasound in nonhuman primates and Parkinson's disease patients: targeted AAV vector delivery and PET imaging. *Sci. Adv.* 9 (16): eadf 4888.

233 Kline-Schoder, A.R., Noel, R.L., Phatnani, H. et al. (2023). Focused ultrasound-mediated blood–brain barrier opening best promotes neuroimmunomodulation through brain macrophage redistribution. *Neuroglia.* 4 (2): 141–157.

234 Hughes, A., Khan, D.S., and Alkins, R. (2023). Current and emerging systems for focused ultrasound-mediated blood–brain barrier opening. *Ultrasound Med. Biol.* 49 (7): 1479–1490.

235 Lindgren, M., Hällbrink, M., Prochiantz, A. et al. (2000). Cell-penetrating peptides. *Trends Pharmacol. Sci.* 21 (3): 99–103.

236 Baoum, A., Ovcharenko, D., and Berkland, C. (2012). Calcium condensed cell penetrating peptide complexes offer highly efficient, low toxicity gene silencing. *Int. J. Pharm.* 427 (1): 134–142.

237 Langel, Ü. (2023). *CPP, Cell-Penetrating Peptides*, 133–150. Cham: Springer International Publishing.

238 Lehto, T., Kurrikoff, K., and Langel, Ü. (2012). Cell-penetrating peptides for the delivery of nucleic acids. *Expert Opin. Drug Deliv.* 9 (7): 823–836.

239 Järver, P. and Langel, Ü. (2006). Cell-penetrating peptides—A brief introduction. *Biochim. Biophys. Acta, Biomembr.* 1758 (3): 260–263.

240 Ghorai, S.M., Deep, A., Magoo, D. et al. (2023). Cell-penetrating and targeted peptides delivery systems as potential pharmaceutical carriers for enhanced delivery across the Blood–Brain Barrier (BBB). *Pharm* 15 (7): 1999.

241 Davoodi, Z. and Shafiee, F. (2022). Internalizing RGD, a great motif for targeted peptide and protein delivery: a review article. *Drug Deliv. Transl. Res.* 12 (10): 2261–2274.

242 Morris, M.C., Deshayes, S., Heitz, F., and Divita, G. (2008). Cell-penetrating peptides: from molecular mechanisms to therapeutics. *Biol. Cell.* 100 (4): 201–217.

243 Ruseska, I. and Zimmer, A. (2020). Internalization mechanisms of cell-penetrating peptides. *Beilstein J Nanotechnol.* 11 (1): 101–123.

244 Kaksonen, M. and Roux, A. (2018). Mechanisms of clathrin-mediated endocytosis. *Nat. Rev. Mol. Cell Biol.* 19 (5): 313–326.

245 Futaki, S. and Nakase, I. (2017). Cell-surface interactions on arginine-rich cell-penetrating peptides allow for multiplex modes of internalization. *Acc. Chem. Res.* 50 (10): 2449–2456.

246 Varkouhi, A.K., Scholte, M., Storm, G., and Haisma, H.J. (2011). Endosomal escape pathways for delivery of biologicals. *J. Control. Release* 151 (3): 220–228.

247 Reissmann, S. (2014). Cell penetration: scope and limitations by the application of cell-penetrating peptides. *J. Pept. Sci.* 20 (10): 760–784.

248 Reissmann, S. and Filatova, M.P. (2021). New generation of cell-penetrating peptides: functionality and potential clinical application. *J. Pept. Sci.* 27 (5): e 3300.

249 Stalmans, S., Bracke, N., Wynendaele, E. et al. (2015). Cell-penetrating peptides selectively cross the blood-brain barrier In Vivo. *PLoS One* 10 (10): e0139652.

250 Zou, L.L., Ma, J.L., Wang, T. et al. (2013). Cell-penetrating peptide-mediated therapeutic molecule delivery into the central nervous system. *Curr. Neuropharmacol.* 11 (2): 197–208.

251 Schwarze, S.R., Ho, A., Vocero-Akbani, A., and Dowdy, S.F. (1999). In Vivo protein transduction: delivery of a

biologically active protein into the mouse. *Science* 285 (5433): 1569–1572.

252 Simon, M.J., Kang, W.H., Gao, S. et al. (2011). TAT is not capable of transcellular delivery across an intact endothelial monolayer in vitro. *Ann. Biomed. Eng.* 39 (1): 394–401.

253 Liu, T., Liu, Y., Kao, H.Y., and Pei, D. (2010). Membrane permeable cyclic peptidyl inhibitors against human Peptidylprolyl isomerase pin 1. *J. Med. Chem.* 53 (6): 2494–2501.

254 Mandal, D., Shirazi, A.N., and Parang, K. (2011). Cell-penetrating Homochiral cyclic peptides as nuclear-targeting molecular transporters. *Angew. Chem. Int. Ed.* 50 (41): 9633–9637.

255 Qian, Z., Liu, T., Liu, Y.Y. et al. (2013). Efficient delivery of cyclic peptides into mammalian cells with short sequence motifs. *ACS Chem. Biol.* 8 (2): 423–431.

256 Li, X., Kheirabadi, M., Dougherty, P.G. et al. (2023). The endosomal escape vehicle platform enhances delivery of oligonucleotides in preclinical models of neuromuscular disorders. *Mol. Ther. Nucleic Acids* 33: 273–285.

257 Zhang, Y., Guo, P., Ma, Z. et al. (2021). Combination of cell-penetrating peptides with nanomaterials for the potential therapeutics of central nervous system disorders: a review. *J. Nanobiotechnology.* 19 (1): 255.

258 Xie, J., Bi, Y., Zhang, H. et al. (2020). Cell-penetrating peptides in diagnosis and treatment of human diseases: from preclinical Research to clinical application. *Front. Pharmacol.* 11: 697.

259 Ulapane, K.R., Kopec, B.M., and Siahaan, T.J. (2019). In Vivo brain delivery and brain deposition of proteins with various sizes. *Mol. Pharm.* 16 (12): 4878–4889.

260 Whelan, R., Hargaden, G.C., and Knox, A.J.S. (2021). Modulating the blood–brain barrier: a comprehensive review. *Pharm* 13 (11): 1980.

261 Ulapane, K.R., On, N., Kiptoo, P. et al. (2017). Improving brain delivery of biomolecules via BBB modulation in mouse and rat: detection using MRI, NIRF, and mass spectrometry. *Nano* 1 (2): 217–231.

262 Kiptoo, P., Sinaga, E., Calcagno, A.M. et al. (2011). Enhancement of drug absorption through the blood–brain barrier and inhibition of intercellular tight junction resealing by E-cadherin peptides. *Mol. Pharm.* 8 (1): 239–249.

263 Tabanor, K., Lee, P., Kiptoo, P. et al. (2016). Brain delivery of drug and MRI contrast agent: detection and quantitative determination of brain deposition of CPT-Glu using LC–MS/MS and Gd-DTPA using magnetic resonance imaging. *Mol. Pharm.* 13 (2): 379–390.

264 On, N.H., Kiptoo, P., Siahaan, T.J., and Miller, D.W. (2014). Modulation of blood–brain barrier permeability in mice using synthetic E-cadherin peptide. *Mol. Pharm.* 11 (3): 974–981.

265 de Beco, S., Amblard, F., and Coscoy, S. (2012). Chapter two new insights into the regulation of E-cadherin distribution by endocytosis. *Int. Rev. Cell Mol. Biol.* 295: 63–108.

266 Dithmer, S., Staat, C., Müller, C. et al. (2017). Claudin peptidomimetics modulate tissue barriers for enhanced drug delivery. *Ann. N. York Acad. Sci.* 1397 (1): 169–184.

267 Wong, V. and Gumbiner, B.M. (1997). A synthetic peptide corresponding to the extracellular domain of Occludin perturbs the tight junction permeability barrier. *J. Cell Biol.* 136 (2): 399–409.

268 Sánchez-Navarro, M. and Giralt, E. (2022). Peptide shuttles for blood–brain barrier drug delivery. *Pharm* 14 (9): 1874.

269 Zuchero, Y.J.Y., Chen, X., Bien-Ly, N. et al. (2016). Discovery of novel blood-brain barrier targets to enhance brain uptake of therapeutic antibodies. *Neuron* 89 (1): 70–82.

270 Dehouck, M.P., Tachikawa, M., Hoshi, Y. et al. (2022). Quantitative targeted absolute proteomics for better characterization of an in vitro human blood–brain barrier model derived from hematopoietic stem cells. *Cells* 11 (24): 3963.

271 Kan, S.H., Aoyagi-Scharber, M., Le SQ, V.J. et al. (2014). Delivery of an enzyme-IGFII fusion protein to the mouse brain is therapeutic for mucopolysaccharidosis type IIIB. *Proc. Natl. Acad. Sci.* 111 (41): 14870–14875.

272 Clemmons, D.R. (2016). Role of IGF binding proteins in regulating metabolism. *Trends Endocrinol. Metab.* 27 (6): 375–391.

273 Kurata, T., Miyazaki, K., Morimoto, N. et al. (2013). Atorvastatin and pitavastatin reduce oxidative stress and improve IR/LDL-R signals in Alzheimer's disease. *Neurol. Res.* 35 (2): 193–205.

274 Turley, S.D., Burns, D.K., Rosenfeld, C.R., and Dietschy, J.M. (1996). Brain does not utilize low density lipoprotein-cholesterol during fetal and neonatal development in the sheep. *J. Lipid Res.* 37 (9): 1953–1961.

275 Björkhem, I. and Meaney, S. (2004). Brain cholesterol: long secret life behind a barrier. *ATVB* 24 (5): 806–815.

276 Pardridge, W.M., Kumagai, A.K., and Eisenberg, J.B. (1987). Chimeric peptides as a vehicle for peptide pharmaceutical delivery through the blood-brain barrier. *Biochem. Biophys. Res. Commun.* 146 (1): 307–313.

277 Boado, R.J., Lu, J.Z., Hui, E.K.W., and Pardridge, W.M. (2014). Insulin receptor antibody–sulfamidase fusion protein penetrates the primate blood–brain barrier and reduces glycosoaminoglycans in sanfilippo type A cells. *Mol. Pharm.* 11 (8): 2928–2934.

278 Chew, K.S., Wells, R.C., Moshkforoush, A. et al. (2023). CD98hc is a target for brain delivery of biotherapeutics. *Nat. Commun.* 14 (1): 5053.

279 Puris, E., Gynther, M., Auriola, S., and Huttunen, K.M. (2020). L-type amino acid transporter 1 as a target for drug delivery. *Pharm. Res.* 37 (5): 88.

280 Edavettal, S., Cejudo-Martin, P., Dasgupta, B. et al. (2022). Enhanced delivery of antibodies across the blood-brain barrier via TEMs with inherent receptor-mediated phagocytosis. *Med.* 3 (12): 860–882.

281 Kawabata, H., Yang, R., Hirama, T. et al. (1999). Molecular cloning of transferrin receptor 2. *J. Biol. Chem.* 274 (30): 20826–20832.

282 Chu, C., Li, J.Y., Boado, R.J., and Pardridge, W.M. (2007). Blood—brain barrier genomics and cloning of a novel organic anion transporter. *J. Cereb. Blood Flow Metab.* 28 (2): 291–301.

283 Mayle, K.M., Le, A.M., and Kamei, D.T. (2012). The intracellular trafficking pathway of transferrin. *Biochim. Biophys. Acta Gen. Subj.* 1820 (3): 264–281.

284 Eckenroth, B.E., Steere, A.N., Chasteen, N.D. et al. (2011). How the binding of human transferrin primes the transferrin receptor potentiating iron release at endosomal pH. *Proc. Natl. Acad. Sci.* 108 (32): 13089–13094.

285 Skarlatos, S., Yoshikawa, T., and Pardridge, W.M. (1995). Transport of [125I]transferrin through the rat blood-brain barrier. *Brain Res.* 683 (2): 164–171.

286 Pardridge, W.M. and Chou, T. (2021). Mathematical models of blood-brain barrier transport of monoclonal antibodies targeting the transferrin receptor and the insulin receptor. *J. Pharm.* 14 (6): 535.

287 Hultqvist, G., Syvänen, S., Fang, X.T. et al. (2017). Bivalent brain shuttle increases antibody uptake by monovalent binding to the transferrin receptor. *Theranostics* 7 (2): 308–318.

288 Boado, R.J., Zhang, Y., Wang, Y., and Pardridge, W.M. (2009). Engineering and expression of a chimeric transferrin receptor monoclonal antibody for blood–brain barrier delivery in the mouse. *Biotechnol. Bioeng.* 102 (4): 1251–1258.

289 Pardridge, W.M. (2015). Blood–brain barrier drug delivery of IgG fusion proteins with a transferrin receptor monoclonal antibody. *Expert Opin. Drug Deliv.* 12 (2): 207–222.

290 Ullman, J.C., Arguello, A., Getz, J.A. et al. (2020). Brain delivery and activity of a lysosomal enzyme using a blood-brain barrier transport vehicle in mice. *Sci. Transl. Med.* 12 (545).

291 Niewoehner, J., Bohrmann, B., Collin, L. et al. (2014). Increased brain penetration and potency of a therapeutic antibody using a monovalent molecular shuttle. *Neuron* 81 (1): 49–60.

292 Couch, J.A., Yu, Y.J., Zhang, Y. et al. (2013). Addressing safety liabilities of TfR bispecific antibodies that cross the blood-brain barrier. *Sci. Transl. Med.* 5 (183): 183ra57.

293 Wurster, C.D., Winter, B., Wollinsky, K. et al. (2019). Intrathecal administration of nusinersen in adolescent and adult SMA type 2 and 3 patients. *J. Neurol.* 266 (1): 183–194.

294 Mazur, C., Powers, B., Zasadny, K. et al. (2019). Brain pharmacology of intrathecal antisense oligonucleotides revealed through multimodal imaging. *JCI Insight* 4 (20): e129240.

295 Shadid, M., Badawi, M., and Abulrob, A. (2021). Antisense oligonucleotides: absorption, distribution, metabolism, and excretion. *Expert Opin. Drug Metab. Toxicol.* 17 (11): 1281–1292.

296 Leavitt, B.R. and Tabrizi, S.J. (2020). Antisense oligonucleotides for neurodegeneration. *Science* 367 (6485): 1428–1429.

297 Stephenson, M.L. and Zamecnik, P.C. (1978). Inhibition of Rous sarcoma viral RNA translation by a specific oligodeoxyribonucleotide. *Proc. Natl. Acad. Sci.* 75 (1): 285–288.

298 Chan, J.H., Lim, S., and Wong, W.F. (2006). Antisense oligonucleotides: from design to therapeutic application. *Clin. Exp. Pharmacol. Physiol.* 33 (5–6): 533–540.

299 Southwell, A.L., Skotte, N.H., Bennett, C.F., and Hayden, M.R. (2012). Antisense oligonucleotide therapeutics for inherited neurodegenerative diseases. *Trends Mol. Med.* 18 (11): 634–643.

300 Finkel, R.S., Chiriboga, C.A., Vajsar, J. et al. (2021). Treatment of infantile-onset spinal muscular atrophy with nusinersen: final report of a phase 2, open-label, multicentre, dose-escalation study. *Lancet Child Adolesc. Health* 5 (7): 491–500.

301 DeVos, S.L. and Miller, T.M. (2013). Antisense oligonucleotides: treating neurodegeneration at the level of RNA. *Neurotherapeutics* 10 (3): 486–497.

302 Andres, J.D., Hayek, S., Perruchoud, C. et al. (2022). Intrathecal drug delivery: advances and applications in the Management of Chronic Pain Patient. *Front Pain Res.* 3: 900566.

303 Yaksh, T.L., de Kater, A., Dean, R. et al. (2012). Pharmacokinetic analysis of Ziconotide (SNX-111), an intrathecal N-type Calcium Channel blocking analgesic, delivered by bolus and infusion in the dog. *Neuromodulation Technol. Neural Interface* 15 (6): 508–519.

304 Wallace, M. and Yaksh, T.L. (2012). Characteristics of distribution of morphine and metabolites in cerebrospinal fluid and plasma with chronic intrathecal morphine infusion in humans. *Anesth. Analg.* 115 (4): 797–804.

305 Hocking, G. and Wildsmith, J.A.W. (2004). Intrathecal drug spread. BJA. *Br. J. Anaesth.* 93 (4): 568–578.

306 Kouhi, A., Pachipulusu, V., Kapenstein, T. et al. (2021). Brain disposition of antibody-based therapeutics: dogma, approaches and perspectives. *Int. J. Mol. Sci.* 22 (12): 6442.

307 Hardcastle, N., Boulis, N.M., and Federici, T. (2018). AAV gene delivery to the spinal cord: serotypes, methods, candidate diseases, and clinical trials. *Expert. Opin. Biol. Ther.* 18 (3): 293–307.

308 Zhao, J., Zu, Y., Han, L. et al. (2019). Treatment of Epstein–Barr virus associated central nervous system diseases after allogeneic hematopoietic stem cell transplantation with intrathecal donor lymphocyte infusion. *Bone Marrow Transplant.* 54 (6): 821–827.

309 Logan, M.R., McClure, J.H., and Wildsmith, J.A.W. (1986). Plain bupivacaine: an unpredictable spinal anaesthetic agent. *Br. J. Anaesth.* 58 (3): 292–296.

310 Sallerin-Caute, B., Lazorthes, Y., Monsarrat, B. et al. (1991). CSF baclofen levels after intrathecal administration in severe spasticity. *Eur. J. Clin. Pharmacol.* 40 (4): 363–365.

311 Muenzer, J., Hendriksz, C.J., Fan, Z. et al. (2016). A phase I/II study of intrathecal idursulfase-IT in children with severe mucopolysaccharidosis II. *Genet. Med.* 18 (1): 73–81.

312 Wijburg, F.A., Whitley, C.B., Muenzer, J. et al. (2021). A multicenter open-label extension study of intrathecal heparan-N-sulfatase in patients with Sanfilippo syndrome type A. *Mol. Genet. Metab.* 134 (1–2): 175–181.

313 Alabsi, W., Eedara, B.B., Encinas-Basurto, D. et al. (2022). Nose-to-brain delivery of therapeutic peptides as nasal aerosols. *Pharm* 14 (9): 1870.

314 Chen, Y., Zhang, C., Huang, Y. et al. (2024). Intranasal drug delivery: the interaction between nanoparticles and the nose-to-brain pathway. *Adv. Drug Deliv. Rev.* 115196.

315 Koo, J., Lim, C., and Oh, K.T. (2024). Recent advances in intranasal Administration for Brain-Targeting Delivery: a comprehensive review of lipid-based nanoparticles and stimuli-responsive gel formulations. *Int. J. Nanomedicine* 19: 1767–1807.

316 Jeong, S.H., Jang, J.H., and Lee, Y.B. (2023). Drug delivery to the brain via the nasal route of administration: exploration of key targets and major consideration factors. *J. Pharm. Investig.* 53 (1): 119–152.

317 Maher, R., Moreno-Borrallo, A., Jindal, D. et al. (2023). Intranasal polymeric and lipid-based Nanocarriers for CNS drug delivery. *Pharm.* 15 (3): 746.

318 Patharapankal, E.J., Ajiboye, A.L., Mattern, C., and Trivedi, V. (2023). Nose-to-brain (N2B) delivery: an alternative route for the delivery of biologics in the management and treatment of central nervous system disorders. *Pharm* 16 (1): 66.

319 Du, L., Chen, L., Liu, F. et al. (2023). Nose-to-brain drug delivery for the treatment of CNS disease: new development and strategies. *Int. Rev. Neurobiol.* 171: 255–297.

320 Lochhead, J.J. and Thorne, R.G. (2012). Intranasal delivery of biologics to the central nervous system. *Adv. Drug Deliv. Rev.* 64 (7): 614–628.

321 Salameh, T.S. and Banks, W.A. (2014). Chapter nine delivery of therapeutic peptides and proteins to the CNS. *Adv. Pharmacol.* 71: 277–299.

322 Craft, S., Raman, R., Chow, T.W. et al. (2020). Safety, efficacy, and feasibility of intranasal insulin for the treatment of mild cognitive impairment and Alzheimer disease dementia. *JAMA Neurol.* 77 (9): 1099–1109.

323 Rosenbloom, M., Barclay, T.R., Kashyap, B. et al. (2021). A phase II, single-Center, randomized, double-blind, placebo-controlled study of the safety and therapeutic efficacy of intranasal Glulisine in amnestic mild cognitive impairment and probable mild Alzheimer's disease. *Drugs Aging* 38 (5): 407–415.

324 Kellar, D., Lockhart, S.N., Aisen, P. et al. (2021). Intranasal insulin reduces white matter hyperintensity progression in association with improvements in cognition and CSF biomarker profiles in mild cognitive impairment and Alzheimer's disease. *J. Prev Alzheimer's Dis.* 8 (3): 240–248.

325 Novak, P., Maldonado, D.A.P., and Novak, V. (2019). Safety and preliminary efficacy of intranasal insulin for cognitive impairment in Parkinson disease and multiple system atrophy: a double-blinded placebo-controlled pilot study. *PLoS One* 14 (4): e0214364.

326 Zhang, Y., Zhang, H., Ghosh, D., and Williams, R.O. (2020). Just how prevalent are peptide therapeutic products? A critical review. *Int. J. Pharm.* 587: 119491.

327 Di, L. (2015). Strategic approaches to optimizing peptide ADME properties. *AAPS J.* 17 (1): 134–143.

328 Shaw, P. and Zhang, X. (2013). Intranasal delivery of a cystatin C-peptide as therapy for Alzheimer's disease. *FASEB J.* 27 (S1): 533.1–533.1.

329 von Mentzer, B., Russo, A.F., Zhang, Z. et al. (2020). A CGRP receptor antagonist peptide formulated for nasal administration to treat migraine. *J. Pharm. Pharmacol.* 72 (10): 1352–1360.

330 Li, S., Guo, C., Zhang, X. et al. (2021). Self-assembling modified neuropeptide S enhances nose-to-brain penetration and exerts a prolonged anxiolytic-like effect. *Biomater. Sci.* 9 (13): 4765–4777.

331 Migliore, M.M., Ortiz, R., Dye, S. et al. (2014). Neurotrophic and neuroprotective efficacy of intranasal GDNF in a rat model of Parkinson's disease. *Neuroscience* 274: 11–23.

332 Zhao, Y.Z., Li, X., Lu, C.T. et al. (2014). Gelatin nanostructured lipid carriers-mediated intranasal delivery of basic fibroblast growth factor enhances functional recovery in hemiparkinsonian rats. *Nanomed: Nanotechnol, Biol Med.* 10 (4): 755–764.

333 Buchthal, B., Weiss, U., and Bading, H. (2018). Post-injury nose-to-brain delivery of Activin a and serpin B2 reduces brain damage in a mouse stroke model. *Mol. Ther.* 26 (10): 2357–2365.

334 Sousa, F., Dhaliwal, H.K., Gattacceca, F. et al. (2019). Enhanced anti-angiogenic effects of bevacizumab in glioblastoma treatment upon intranasal administration in polymeric nanoparticles. *J. Control. Release* 309: 37–47.

335 Musumeci, T., Benedetto, G.D., Carbone, C. et al. (2022). Intranasal administration of a TRAIL neutralizing monoclonal antibody adsorbed in PLGA nanoparticles and NLC Nanosystems: an in Vivo study on a mouse model of Alzheimer's disease. *Biomedicine* 10 (5): 985.

336 Correa, D., Scheuber, M.I., Shan, H. et al. (2023). Intranasal delivery of full-length anti-Nogo-a antibody: a potential alternative route for therapeutic antibodies to central nervous system targets. *Proc. Natl. Acad. Sci.* 120 (4): e2200057120.

337 Lopes, J.R., Zhang, X., Mayrink, J. et al. (2023). Nasal administration of anti-CD3 monoclonal antibody ameliorates disease in a mouse model of Alzheimer's disease. *Proc. Natl. Acad. Sci.* 120 (37): e2309221120.

338 Gaikwad, S., Puangmalai, N., Sonawane, M. et al. (2024). Nasal tau immunotherapy clears intracellular tau pathology and improves cognitive functions in aged tauopathy mice. *Sci. Transl. Med.* 16 (754): eadj 5958.

339 Alarcón-Arís, D., Recasens, A., Galofré, M. et al. (2018). Selective α-Synuclein knockdown in monoamine neurons by intranasal oligonucleotide delivery: potential therapy for Parkinson's disease. *Mol. Ther.* 26 (2): 550–567.

340 Alarcón-Arís, D., Pavia-Collado, R., Miquel-Rio, L. et al. (2020). Anti-α-synuclein ASO delivered to monoamine neurons prevents α-synuclein accumulation in a Parkinson's disease-like mouse model and in monkeys. *EBioMedicine* 59: 102944.

341 Borgonetti, V. and Galeotti, N. (2020). Intranasal delivery of an antisense oligonucleotide to the RNA-binding protein HuR relieves nerve injury-induced neuropathic pain. *Pain* 162 (5): 1500–1510.

342 Borgonetti, V. and Galeotti, N. (2023). Posttranscriptional regulation of gene expression participates in the myelin restoration in mouse models of multiple sclerosis: antisense modulation of HuR and HuD ELAV RNA binding protein. *Mol. Neurobiol.* 60 (5): 2661–2677.

343 Samaridou, E., Walgrave, H., Salta, E. et al. (2020). Nose-to-brain delivery of enveloped RNA - cell permeating peptide nanocomplexes for the treatment of neurodegenerative diseases. *Biomaterials* 230: 119657.

344 Ferrés-Coy, A., Galofré, M., Pilar-Cuéllar, F. et al. (2016). Therapeutic antidepressant potential of a conjugated siRNA silencing the serotonin transporter after intranasal administration. *Mol. Psychiatry* 21 (3): 328–338.

345 Yadav, S., Gandham, S.K., Panicucci, R., and Amiji, M.M. (2016). Intranasal brain delivery of cationic nanoemulsion-encapsulated TNFα siRNA in prevention of experimental neuroinflammation. *Nanomed: Nanotechnol, Biol Med.* 12 (4): 987–1002.

346 Sava, V., Fihurka, O., Khvorova, A., and Sanchez-Ramos, J. (2020). Enriched chitosan nanoparticles loaded with siRNA are effective in lowering Huntington's disease gene expression following intranasal administration. *Nanomed: Nanotechnol, Biol Med.* 24: 102119.

347 Yang, Y., Zhang, X., Wu, S. et al. (2022). Enhanced nose-to-brain delivery of siRNA using hyaluronan-enveloped nanomicelles for glioma therapy. *J. Control. Release* 342: 66–80.

348 Yang, X., Yang, W., Xia, X. et al. (2022). Intranasal delivery of BACE1 siRNA and rapamycin by dual targets modified nanoparticles for Alzheimer's disease therapy. *Small* 18 (30): 2203182.

349 Dhaliwal, H.K., Fan, Y., Kim, J., and Amiji, M.M. (2020). Intranasal delivery and transfection of mRNA therapeutics in the brain using cationic liposomes. *Mol. Pharm.* 17 (6): 1996–2005.

350 Belén, L.H., Rangel-Yagui, C.d.O., Lissabet, J.F. et al. (2019). From synthesis to characterization of site-selective PEGylated proteins. *Front. Pharmacol.* 10: 1450.

351 Lee, H., Jang, I.H., Ryu, S.H., and Park, T.G. (2003). N-terminal site-specific mono-PEGylation of epidermal growth factor. *Pharm. Res.* 20 (5): 818–825.

352 Cheng, T.L., Chuang, K.H., Chen, B.M., and Roffler, S.R. (2012). Analytical measurement of PEGylated molecules. *Bioconjug. Chem.* 23 (5): 881–899.

353 Yin, L., Su, C., Ren, T. et al. (2017). MS all strategy for comprehensive quantitative analysis of PEGylated-doxorubicin, PEG and doxorubicin by LC-high resolution q-q-TOF mass spectrometry coupled with all window acquisition of all fragment ion spectra. *Analyst* 142 (22): 4279–4288.

354 Lu, X., Gough, P.C., DeFelippis, M.R., and Huang, L. (2010). Elucidation of PEGylation site with a combined approach of in-source fragmentation and CID MS/MS. *J. Am. Soc. Mass Spectrom.* 21 (5): 810–818.

355 Muneeruddin, K., Bobst, C.E., Frenkel, R. et al. (2016). Characterization of a PEGylated protein therapeutic by ion exchange chromatography with on-line detection by native ESI MS and MS/MS. *Analyst* 142 (2): 336–344.

356 Kou, D., Manius, G., Zhan, S., and Chokshi, H.P. (2009). Size exclusion chromatography with Corona charged aerosol detector for the analysis of polyethylene glycol polymer. *J. Chromatogr. A* 1216 (28): 5424–5428.

357 Zabaleta, V., Campanero, M.A., and Irache, J.M. (2007). An HPLC with evaporative light scattering detection method for the quantification of PEGs and Gantrez in PEGylated nanoparticles. *J. Pharm. Biomed. Anal.* 44 (5): 1072–1078.

358 Shatz, W., Hass, P.E., Peer, N. et al. (2019). Identification and characterization of an octameric PEG-protein conjugate system for intravitreal long-acting delivery to the back of the eye. *PLoS One* 14 (6): e0218613.

359 Gokarn, Y.R., McLean, M., and Laue, T.M. (2012). Effect of PEGylation on protein hydrodynamics. *Mol. Pharm.* 9 (4): 762–773.

360 Klepach, A., Tran, H., Mohammed, F.A., and ElSayed, M.E.H. (2022). Characterization and impact of peptide physicochemical properties on oral and subcutaneous delivery. *Adv. Drug Deliv. Rev.* 186: 114322.

361 Bellmann-Sickert, K., Elling, C.E., Madsen, A.N. et al. (2011). Long-acting Lipidated analogue of human pancreatic polypeptide is slowly released into circulation. *J. Med. Chem.* 54 (8): 2658–2667.

362 Fu, Z., Li, S., Han, S. et al. (2022). Antibody drug conjugate: the "biological missile" for targeted cancer therapy. *Signal Transduct. Target. Ther.* 7 (1): 93.

363 Dugal-Tessier, J., Thirumalairajan, S., and Jain, N. (2021). Antibody-oligonucleotide conjugates: a twist to antibody-drug conjugates. *J. Clin. Forensic Med.* 10 (4): 838.

364 Conilh, L., Sadilkova, L., Viricel, W., and Dumontet, C. (2023). Payload diversification: a key step in the development of antibody–drug conjugates. *J. Hematol. Oncol.* 16 (1): 3.

365 Lehar, S.M., Pillow, T., Xu, M. et al. (2015). Novel antibody–antibiotic conjugate eliminates intracellular *S. aureus*. *Nature* 527 (7578): 323–328.

366 (2023). New guise ADCs couple antibodies to degraders. *Nat. Biotechnol.* 41 (10): 1360–1360.

367 Tsuchikama, K., Anami, Y., Ha, S.Y.Y., and Yamazaki, C.M. (2024). Exploring the next generation of antibody–drug conjugates. *Nat. Rev. Clin. Oncol.* 21 (3): 203–223.

368 Chen, L., Wang, L., Shion, H. et al. (2016). In-depth structural characterization of Kadcyla® (ado-trastuzumab emtansine) and its biosimilar candidate. *MAbs* 8 (7): 1210–1223.

369 Tilly, H., Morschhauser, F., Sehn, L.H. et al. (2021). Polatuzumab Vedotin in previously untreated diffuse large B-cell lymphoma. *N. Engl. J. Med.* 386 (4): 351–363.

370 Herrera, A.F., Patel, M.R., Burke, J.M. et al. (2022). Anti-CD79B antibody-drug conjugate DCDS0780A in patients with B-cell non-Hodgkin lymphoma: phase 1 dose-escalation study. *Clin. Cancer Res.* 28 (7): clincanres. 3261.2021.

371 Ouyang, J. (2013). Antibody-drug conjugates. *Methods Mol. Biol.* 1045: 275–283.

372 Wagh, A., Song, H., Zeng, M. et al. (2018). Challenges and new frontiers in analytical characterization of antibody-drug conjugates. *MAbs* 10 (2): 222–243.

373 Su, Z., Xiao, D., Xie, F. et al. (2021). Antibody–drug conjugates: recent advances in linker chemistry. *Acta Pharm. Sin. B* 11 (12): 3889–3907.

374 Qin, Q. and Gong, L. (2022). Current analytical strategies for antibody–drug conjugates in biomatrices. *Molecules* 27 (19): 6299.

375 Mak, S.Y., Chen, S., Fong, W.J. et al. (2024). A simple and highly sensitive LC–MS workflow for characterization and quantification of ADC cleavable payloads. *Sci. Rep.* 14 (1): 11018.

376 Erickson, H.K., Widdison, W.C., Mayo, M.F. et al. (2010). Tumor delivery and in Vivo processing of Disulfide-linked and Thioether-linked antibody–Maytansinoid conjugates. *Bioconjug. Chem.* 21 (1): 84–92.

377 Sorkin, M.R., Walker, J.A., Kabaria, S.R. et al. (2019). Responsive antibody conjugates enable quantitative determination of intracellular bond degradation rate. *Cell Chem. Biol.* 26 (12): 1643–1651.e4.

378 Leong, S.R., Liang, W.C., Wu, Y. et al. (2015). An anti-B7-H4 antibody–drug conjugate for the treatment of breast cancer. *Mol. Pharm.* 12 (6): 1717–1729.

379 Abbasnezhad, N., Zirak, N., Champmartin, S. et al. (2022). An overview of in vitro drug release methods for drug-eluting stents. *Polymers* 14 (13): 2751.

380 Patel, S., Stracke, J.O., Altenburger, U. et al. (2017). Prediction of intraocular antibody drug stability using ex-vivo ocular model. *Eur. J. Pharm. Biopharm.* 112: 177–186.

381 Sobolewska, B., Heiduschka, P., Bartz-Schmidt, K.U., and Ziemssen, F. (2017). pH of anti-VEGF agents in the human vitreous: low impact of very different formulations. *Int. J. Retin. Vitr.* 3 (1): 22.

382 Kinnunen, H.M. and Mrsny, R.J. (2014). Improving the outcomes of biopharmaceutical delivery via the subcutaneous route by understanding the chemical, physical and physiological properties of the subcutaneous injection site. *J. Control. Release* 182: 22–32.

383 Puntis, M., Reddy, U., and Hirsch, N. (2016). Cerebrospinal fluid and its physiology. *Anaesth Intensiv Care* 17 (12): 611–612.

384 Langer, T., Zanella, A., and Caironi, P. (2016). Understanding the role of the cerebrospinal fluid in acid–base disorders. *Intensive Care Med.* 42 (3): 436–439.

385 Thorat, A.A. and Suryanarayanan, R. (2019). Characterization of phosphate buffered saline (PBS) in frozen state and after freeze-drying. *Pharm. Res.* 36 (7): 98.

386 Sreenivasan, S., Patil, S.S., and Rathore, A.S. (2024). Does aggregation of therapeutic IgGs in PBS offer a true picture of what happens in models derived from human body fluids? *J. Pharm. Sci.* 113 (3): 596–603.

387 Loyet, K.M., Hass, P.E., Sandoval, W. et al. (2018). In vivo stability profiles of anti-factor D molecules support long-acting delivery approaches. *Mol. Pharm.* 16 (1): 86–95.

388 Shen, J. and Burgess, D.J. (2015). In vitro–in vivo correlation for complex non-oral drug products: where do we stand? *J. Control. Release* 219: 644–651.

389 Wan, B., Bao, Q., and Burgess, D.J. (2022). In vitro-in vivo correlation of PLGA microspheres: effect of polymer source variation and temperature. *J. Control. Release* 347: 347–355.

390 Goel, M., Leung, D., Famili, A. et al. (2021). Accelerated in vitro release testing method for a long-acting peptide-PLGA formulation. *Eur. J. Pharm. Biopharm.* 165: 185–192.

391 Jain, K.M.H., Ho, T., Hoe, S. et al. (2024). Accelerated in vitro release testing using USP apparatus 2 to predict rat in vivo performance of single agent and combination long-acting injectable suspension products. *J. Pharm. Sci.*

392 Ohtake, S., Kita, Y., and Arakawa, T. (2011). Interactions of formulation excipients with proteins in solution and in the dried state. *Adv. Drug Deliv. Rev.* 63 (13): 1053–1073.

393 Bown, H.K., Bonn, C., Yohe, S. et al. (2018). In vitro model for predicting bioavailability of subcutaneously injected monoclonal antibodies. *J. Control. Release* 273: 13–20.

394 Kinnunen, H.M., Sharma, V., Contreras-Rojas, L.R. et al. (2015). A novel in vitro method to model the fate of subcutaneously administered biopharmaceuticals and associated formulation components. *J. Control. Release* 214: 94–102.

395 Miyamoto, S. and Shimono, K. (2020). Molecular Modeling to estimate the diffusion coefficients of drugs and other small molecules. *Molecules* 25 (22): 5340.

396 Tesar, D., Luoma, J., Wyatt, E.A. et al. (2017). Protein engineering to increase the potential of a therapeutic antibody fab for long-acting delivery to the eye. *MAbs* 5 (4).

397 Naik R. (2021). Summary Basis for Regulatory Action-COMIRNATY [Internet]. (cited 2024 May 28). Available from: //http://www.fda.gov/vaccines-blood-biologics/comirnaty

398 Bechtold-Peters, K., Ruggiero, A., Vriezen, N. et al. (2023). CMC regulatory considerations for antibody-drug conjugates. *J. Pharm. Sci.* 112 (12): 2965–2980.

399 Herdiana, Y., Wathoni, N., Shamsuddin, S., and Muchtaridi, M. (2022). Scale-up polymeric-based nanoparticles drug delivery systems: development and challenges. *OpenNano* 7: 100048.

400 van der Kooij, R.S., Steendam, R., Frijlink, H.W., and Hinrichs, W.L.J. (2022). An overview of the production methods for core–shell microspheres for parenteral controlled drug delivery. *Eur. J. Pharm. Biopharm.* 170: 24–42.

401 Thacker, S.G., Her, C., Kelley-Baker, L. et al. (2022). Detection of innate immune response modulating impurities (IIRMI) in therapeutic peptides and proteins: impact of excipients. *Front. Immunol.* 13: 970499.

402 Takai, Y., Powlin, S., Awasaki, Y. et al. (2022). In vivo screening of subcutaneous tolerability for the development of novel excipients. *J. Toxicol. Pathol.* 35 (4): 355–360.

403 Pham, N.B. and Meng, W.S. (2020). Protein aggregation and immunogenicity of biotherapeutics. *Int. J. Pharm.* 585: 119523.

404 González-González, O., Ramirez, I.O., Ramirez, B.I. et al. (2022). Drug stability: ICH versus accelerated predictive stability studies. *Pharm.* 14 (11): 2324.

405 Thackaberry, E.A., Farman, C., Zhong, F. et al. (2017). Evaluation of the toxicity of intravitreally injected PLGA microspheres and rods in monkeys and rabbits: effects of depot size on inflammatory response. *IOVS* 58 (10): 4274.

406 Bantseev, V., Horvath, J., Barteselli, G. et al. (2021). Nonclinical toxicology and biocompatibility program supporting clinical development and registration of the port delivery system with Ranibizumab for Neovascular age-related macular degeneration. *Toxicol. Pathol.* 49 (3): 663–672.

407 Chang, D., Park, K., and Famili, A. (2019). Hydrogels for sustained delivery of biologics to the back of the eye. *Drug Discov. Today* 24 (8): 1470–1482.

Index

Characterizing Biotherapeutics: Analytical Methods for Diverse Modalities, First Edition. Edited by Jennie R. Lill and Wendy Sandoval.

l

m